T0258257

Fundamentals of Modern VLSI Devices

Third Edition

A thoroughly updated third edition of a classic and widely adopted text, perfect for practical transistor design and in the classroom. Covering a variety of recent developments, the internationally renowned authors discuss in detail the basic properties and designs of modern VLSI devices, as well as factors affecting performance. Containing around 25% new material, coverage has been expanded to include high-k gate dielectrics, metal gate technology, strained silicon mobility, non-GCA (Gradual Channel Approximation) modeling of MOSFETs, short-channel FinFETS, and symmetric lateral bipolar transistors on SOI. Chapters have been reorganized to integrate the appendices into the main text to enable a smoother learning experience, and numerous additional end-of-chapter homework exercises (+30%) are included to engage students with real-world problems and test their understanding. A perfect text for senior undergraduate and graduate students taking advanced semiconductor devices courses, and for practicing silicon device professionals in the semiconductor industry.

Yuan Taur is Distinguished Professor of Electrical and Computer Engineering at the University of California, San Diego, having previously worked at IBM's T. J. Watson Research Center, New York. He is an IEEE fellow.

Tak H. Ning is IBM Fellow (retired) at the T. J. Watson Research Center, New York. He is a fellow of the IEEE and the American Physical Society, and a member of the US National Academy of Engineering.

Fundamentals of Modern VLSI Devices

Third Edition

YUAN TAUR
University of California, San Diego

TAK H. NING
IBM T. J. Watson Research Center (retired)

CAMBRIDGE
UNIVERSITY PRESS

University Printing House, Cambridge CB2 8BS, United Kingdom

One Liberty Plaza, 20th Floor, New York, NY 10006, USA

477 Williamstown Road, Port Melbourne, VIC 3207, Australia

314–321, 3rd Floor, Plot 3, Splendor Forum, Jasola District Centre, New Delhi – 110025, India

103 Penang Road, #05–06/07, Visioncrest Commercial, Singapore 238467

Cambridge University Press is part of the University of Cambridge.

It furthers the University's mission by disseminating knowledge in the pursuit of education, learning, and research at the highest international levels of excellence.

www.cambridge.org
Information on this title: www.cambridge.org/highereducation/isbn/9781108480024
DOI: 10.1017/9781108847087

© Cambridge University Press 1998, 2009, 2022

First published 1998
Second edition 2009
First paperback edition 2013
Eighth printing 2020
Third edition 2022

Printed in the United Kingdom by TJ Books Limited, Padstow, Cornwall

A catalogue record for this publication is available from the British Library.

Library of Congress Cataloging-in-Publication Data
Names: Taur, Yuan, 1946–, author. | Ning, Tak H., 1943– author.
Title: Fundamentals of modern VLSI devices / Yuan Taur, University of California, San Diego, Tak H. Ning, IBM Fellow (retired).
Description: Third edition. | United Kingdom ; New York : Cambridge University Press, 2022. | Includes bibliographical references and index.
Identifiers: LCCN 2021029073 (print) | LCCN 2021029074 (ebook) | ISBN 9781108480024 (hardback) | ISBN 9781108847087 (epub)
Subjects: LCSH: Metal oxide semiconductors, Complementary. | Bipolar transistors. | Integrated circuits– Very large scale integration. | BISAC: TECHNOLOGY & ENGINEERING / Electronics / Optoelectronics
Classification: LCC TK7871.99.M44 T38 2022 (print) | LCC TK7871.99.M44 (ebook) | DDC 621.39/5–dc23
LC record available at https://lccn.loc.gov/2021029073
LC ebook record available at https://lccn.loc.gov/2021029074

ISBN 978-1-108-48002-4 Hardback

Additional resources for this publication at www.cambridge.org/taur3ed

Contents

Preface to the Third Edition

It has been twenty-four years since the first edition of this book was published, thirteen years since the second edition. Both editions have been translated into Japanese. The second edition has also been translated into Chinese. During this time period, the VLSI integrated-circuit industry marches on. The minimum feature size has shrunk by more than $10\times$ to now reaching <10 nm. The transistor count has well exceeded ten billion in the densest populated IC chips. New technology thrust areas include double-gate MOSFETs (known as FinFETs), thin-film silicon-on-insulator devices (ET-SOI), and 3-D integration in nonvolatile memory chips. Meanwhile, SOI wafers suitable for partially depleted SOI CMOS have found new applications in the integration of n—p—n and p—n—p lateral bipolar, offering the interesting possibility of bipolar as a complement to CMOS for VLSI.

The purpose of writing the third edition is to update the book with additional material developed after the completion of the second edition. Key topics added include high-κ gate dielectrics, metal gate technology, strained silicon mobility, non-GCA (Gradual Channel Approximation) modeling of MOSFETs, and lateral bipolar transistors on SOI. Furthermore, the chapters of the book have been reorganized to consolidate the discussions of the various subjects to the main chapters with no appendices.

We would like to take this opportunity to thank all the friends and colleagues who gave us encouragement and valuable suggestions for improvement of the book; in particular, Professor Sorin Cristoloveanu of the University of Grenoble Alpes and Dr. Kangguo Cheng of IBM on SOI device technology, and Professor Scott Thompson of the University of Florida on strained silicon mobility.

Tak Ning would like to thank many of his colleagues at IBM, particularly Dr. Jin Cai (now at TSMC), Dr. Jeng-Bang Yau, and Dr. Ghavam Shahidi for their contributions to the SOI lateral bipolar project. Yuan Taur would like to thank many of his students at the University of California, San Diego, in particular Chuyang Hong, Qian Xie, and Bo Yu, for their help with the completion of the third edition. He would also like to thank Katie and Ko Taur for their love and support during the course of the work.

Preface to the Second Edition

Since the publication of the first edition of *Fundamentals of Modern VLSI Devices* by Cambridge University Press in 1998, we received much praise and many encouraging reviews on the book. It has been adopted as a textbook for first-year graduate courses on microelectronics in many major universities in the United States and worldwide. The first edition was translated into Japanese by a team led by Professor Shibahara of Hiroshima University in 2002.

During the past ten years, the evolution and scaling of VLSI (very-large-scale-integration) technology has continued. Now, sixty years after the first invention of the transistor, the number of transistors per chip for both microprocessors and DRAM (dynamic random access memory) has increased to over one billion, and the highest clock frequency of microprocessors has reached 5 GHz. In 2007, the worldwide IC (integrated circuits) sales grew to $250 billion. In 2008, the IC industry reached the 45-nm generation, meaning that the leading-edge IC products employ a minimum lithography feature size of 45 nm. As bulk CMOS (complementary metal–oxide–semiconductor field-effect transistor) technologies are scaled to dimensions below 100 nm, the very factor that makes CMOS technology the technology of choice for digital VLSI circuits, namely, its low standby power, can no longer be taken for granted. Not only has the off-state current gone up with the power supply voltage down scaled to the 1 V level, the gate leakage has also increased exponentially from quantum mechanical tunneling through gate oxides only a few atomic layers thick. Power management, both active and standby, has become a key challenge to continued increase of clock frequency and transistor count in microprocessors. New materials and device structures are being explored to replace conventional bulk CMOS in order to extend scaling to 10 nm.

The purpose of writing the second edition is to update the book with additional material developed after the completion of the first edition. Key new material added includes MOSFET scale length theory and high-field transport model, and the section on SiGe-base bipolar devices has been greatly expanded. We have also expanded the discussions on basic device physics and circuits to include metal–silicon contacts, noise margin of CMOS circuits, and figures of merit for RF applications. Furthermore, two new chapters are added to the second edition. Chapter 9 is on memory devices and covers the fundamentals of read and write operations of commonly used SRAM, DRAM, and nonvolatile memory arrays. Chapter 10 is on silicon-on-insulator (SOI) devices, including advanced devices of future potential.

We would like to take this opportunity to thank all the friends and colleagues who gave us encouragement and valuable suggestions for improvement of the book. In particular, Professor Mark Lundstrom of Purdue University, who adopted the first edition early on, and Dr. Constantin Bulucea of the National Semiconductor Corporation, who suggested the treatment on diffusion capacitance. Thanks also go to Professor James Meindl of the Georgia Institute of Technology, Professor Peter Asbeck of the University of California, San Diego, and Professor Jerry Fossum of the University of Florida for their support of the book.

We would like to thank many of our colleagues at IBM, particularly in the areas of advanced silicon-device research and development, for their direct or indirect contributions. Yuan Taur would like to thank many of his students at the University of California, San Diego, in particular Jooyoung Song and Bo Yu, for their help with the completion of the second edition. He would also like to thank Katie Kahng for her love, support, and patience during the course of the work.

We would like to give special thanks to our families for their support and understanding during this seemingly endless task.

Preface to the First Edition

It has been fifty years since the invention of the bipolar transistor, more than forty years since the invention of integrated-circuit (IC) technology, and more than thirty-five years since the invention of the MOSFET. During this time, there has been tremendous and steady progress in the development of IC technology with a rapid expansion of the IC industry. One distinct characteristic in the evolution of IC technology is that the physical feature sizes of the transistors are reduced continually over time as the lithography technologies used to define these features become available. For almost thirty years now, the minimum lithography feature size used in IC manufacturing has been reduced at a rate of $0.7\times$ every three years. In 1997, the leading-edge IC products have a minimum feature size of $0.25\ \mu m$.

The basic operating principles of large and small transistors are the same. However, the relative importance of the various device parameters and performance factors for transistors of the $1\ \mu m$ and smaller generations is quite different from those for transistors of larger-dimension generations. For example, in the case of CMOS, the power-supply voltage was lowered from the standard 5 V, starting with the 0.6 to 0.8 μm generation. Since then CMOS power supply voltage has been lowered in steps once every few years as the device physical dimensions are reduced. At the same time, many physical phenomena, such as short-channel effect and velocity saturation, which are negligible in large-dimension MOSFETs, are becoming more and more important in determining the behavior of MOSFETs of deep-submicron dimensions. In the case of bipolar devices, breakdown voltage and base-widening effects are limiting their performance, and power dissipation is limiting their level of integration on a chip. Also, the advent of SiGe-base bipolar technology has extended the frequency capability of small-dimension bipolar transistors into the range previously reserved for GaAs and other compound-semiconductor devices.

The purpose of this book is to bring together the device fundamentals that govern the behavior of CMOS and bipolar transistors into a single text, with emphasis on those parameters and performance factors that are particularly important for VLSI (very-large-scale-integration) devices of deep-submicron dimensions. The book starts with a comprehensive review of the properties of the silicon material, and the basic physics of p–n junctions and MOS capacitors, as they relate to the fundamental principles of MOSFET and bipolar transistors. From there, the basic operation of MOSFET and bipolar devices, and their design and optimization for VLSI applications are developed. A great deal of the volume is devoted to in-depth discussions of the intricate interdependence and subtle tradeoffs of the various device parameters pertaining to circuit performance and manufacturability. The effects which are particularly important in small-dimension devices, e.g., quantization of the two-dimensional surface inversion layer in a MOSFET device and the heavy-doping effect in the intrinsic base of a bipolar transistor, are covered in detail. Also included in this book are extensive discussions on scaling and limitations to scaling of MOSFET and bipolar devices.

This book is suitable for use as a textbook by senior undergraduate or graduate students in electrical engineering and microelectronics. The necessary background assumed is an introductory understanding of solid-state physics and semiconductor physics. For practicing engineers and scientists actively involved in research and development in the IC industry, this book serves as a reference in providing a body of knowledge in modern VLSI devices for them to stay up to date in this field.

VLSI devices are too huge a subject area to cover thoroughly in one book. We have chosen to cover only the fundamentals necessary for discussing the design and optimization of the state-of-the-art CMOS and bipolar devices in the sub-0.5-µm regime. Even then, the specific topics covered in this book are based on our own experience of what the most important device parameters and performance factors are in modern VLSI devices.

Many people have contributed directly and indirectly to the topics covered in this book. We have benefited enormously from the years of collaboration and interaction we had with our colleagues at IBM, particularly in the areas of advanced silicon-device research and development. These include Douglas Buchanan, Hu Chao, T. C. Chen, Wei Chen, Kent Chuang, Peter Cook, Emmanuel Crabbé, John Cressler, Bijan Davari, Robert Dennard, Max Fischetti, David Frank, Charles Hsu, Genda Hu, Randall Isaac, Khalid Ismail, G. P. Li, Shih-Hsien Lo, Yuh-Jier Mii, Edward Nowak, George Sai-Halasz, Stanley Schuster, Paul Solomon, Hans Stork, Jack Sun, Denny Tang, Lewis Terman, Clement Wann, James Warnock, Siegfried Wiedmann, Philip Wong, Matthew Wordeman, Ben Wu, and Hwa Yu.

We would like to acknowledge the secretarial support of Barbara Grady and the support of our management at IBM Thomas J. Watson Research Center where this book was written. Finally, we would like to give special thanks to our families – Teresa, Adrienne, and Brenda Ning, and Betty, Ying, and Hsuan Taur – for their support and understanding during this seemingly endless task.

Physical Constants and Unit Conversions

Description	Symbol	Value and unit
Electronic charge	q	1.6×10^{-19} C
Boltzmann's constant	k	1.38×10^{-23} J/K
Vacuum permittivity	ε_0	8.85×10^{-14} F/cm
Silicon permittivity	ε_{si}	1.04×10^{-12} F/cm
Oxide permittivity	ε_{ox}	3.45×10^{-13} F/cm
Velocity of light in vacuum	c	3×10^{10} cm/s
Planck's constant	h	6.63×10^{-34} J-s
Free-electron mass	m_0	9.1×10^{-31} kg
Thermal voltage ($T = 300$ K)	kT/q	0.0259 V
Angstrom	Å	$1\text{Å} = 10^{-8}$ cm
Nanometer	nm	1 nm $= 10^{-7}$ cm
Micrometer (micron)	μm	1 μm $= 10^{-4}$ cm
Millimeter	mm	1 mm $= 0.1$ cm
Meter	m	1 m $= 10^2$ cm
Electron-volt	eV	$1\text{eV} = 1.6 \times 10^{-19}$ J
Energy = charge × voltage	$E = qV$	Joule = Coulomb × Volt
Charge = capacitance × voltage	$Q = CV$	Coulomb = Farad × Volt
Power = current × voltage	$P = IV$	Watt = Ampere × Volt
Time = resistance × capacitance	$t = RC$	second = Ω (ohm) × Farad
Current = charge/time	$I = Q/t$	Ampere = Coulomb/second
Resistance = voltage/current	$R = V/I$	Ω (ohm) = Volt/Ampere

A word of caution about the length units: Strictly speaking, MKS units should be used for all the equations in the book. As a matter of convention, electronics engineers often work with centimeters as the unit of length. While some equations work with lengths in either meters or centimeters, not all of them do. It is prudent always to check for unit consistency when doing calculations. It may be necessary to convert the length unit to meters before plugging into the equations.

Symbols

Symbol	Description	Unit
A	Area	cm^2
A_E	Emitter area	cm^2
α	Common-base current gain	None
α_0	Static common-base current gain	None
α_F	Forward common-base current gain in the Ebers–Moll model	None
α_R	Reverse common-base current gain in the Ebers–Moll model	None
α_T	Base transport factor	None
α_n	Electron-initiated rate of electron–hole pair generation per unit distance	cm^{-1}
α_p	Hole-initiated rate of electron–hole pair generation per unit distance	cm^{-1}
BV	Breakdown voltage	V
BV_{CBO}	Collector–base junction breakdown voltage with emitter open circuit	V
BV_{CEO}	Collector–emitter breakdown voltage with base open circuit	V
BV_{EBO}	Emitter–base junction breakdown voltage with collector open circuit	V
β	Current gain	None
β_0	Static common-emitter current gain	None
β_F	Forward common-emitter current gain in the Ebers–Moll model	None
β_R	Reverse common-emitter current gain in the Ebers–Moll model	None
c	Velocity of light in vacuum ($= 3 \times 10^{10}$ cm/s)	cm/s
C	Capacitance	F
C_d	Depletion-layer capacitance per unit area	F/cm^2
C_d	MOS gate depletion-layer capacitance per unit area	F/cm^2
C_{dBC}	Base–collector diode depletion-layer capacitance per unit area	F/cm^2
$C_{dBC,tot}$	Total base–collector diode depletion-layer capacitance	F
C_{dBE}	Base–emitter diode depletion-layer capacitance per unit area	F/cm^2
$C_{dBE,tot}$	Total base–emitter diode depletion-layer capacitance	F
C_{dm}	Maximum MOS gate depletion capacitance per unit area	F/cm^2
C_D	Diffusion capacitance	F
C_D	MOS gate depletion-layer capacitance	F
C_{Dn}	Diffusion capacitance due to excess electrons	F
C_{Dp}	Diffusion capacitance due to excess holes	F
C_{DE}	Emitter diffusion capacitance	F
C_{fb}	MOS capacitance at flat band per unit area	F/cm^2
C_{FC}	Capacitance between the floating gate and the control gate of a MOSFET nonvolatile memory device	F
C_g	Intrinsic gate capacitance per unit area	F/cm^2
C_G	Total gate capacitance of MOSFET	F
C_{inv}	MOS capacitance per unit area in the inversion region	F/cm^2
C_i	Inversion-layer capacitance per unit area	F/cm^2

(cont.)

Symbol	Description	Unit
C_{it}	Interface trap capacitance per unit area	F/cm^2
C_J	Junction capacitance	F
C_L	Load capacitance	F
C_{in}	Equivalent input capacitance of a logic gate	F
C_{inv}	MOSFET capacitance in inversion per unit area	F/cm^2
C_{min}	Minimum MOS capacitance per unit area	F/cm^2
C_{out}	Equivalent output capacitance of a logic gate	F
C_{ov}	Gate-to-source (-drain) overlap capacitance (per edge)	F
C_{ox}	Oxide capacitance per unit area	F/cm^2
C_p	Polysilicon-gate depletion-layer capacitance per unit area	F/cm^2
C_{si}	Silicon capacitance per unit area	F/cm^2
C_w	Wire capacitance per unit length	F/cm
C_π	Base–emitter capacitance in the small-signal hybrid-π equivalent-circuit model	F
C_μ	Base–collector capacitance in the small-signal hybrid-π equivalent-circuit model	F
d	Width of diffusion region in a MOSFET	cm
d_{si}	Depth in silicon in non-GCA model	cm
D_{it}	Interface-state density per unit area per unit energy	1/cm^2-eV
D_n	Electron diffusion coefficient	cm^2/s
D_{nB}	Electron diffusion coefficient in the base of an n–p–n transistor	cm^2/s
D_p	Hole diffusion coefficient	cm^2/s
D_{pE}	Hole diffusion coefficient in the emitter of an n–p–n transistor	cm^2/s
ΔV_t	Threshold voltage rolloff due to short-channel effect	V
ΔE_g	Apparent bandgap narrowing	J
ΔE_{gB}	Bandgap-narrowing parameter in the base region	J
$\Delta E_{g,max}$	Maximum bandgap narrowing due to the presence of Ge	J
$\Delta E_{g,SiGe}$	Local bandgap narrowing due to the presence of Ge	J
ΔL	Channel length modulation in MOSFET	cm
ΔQ_{total}	Total charge stored in a nonvolatile memory device	C
E	Energy	J
E_c	Conduction-band edge	J
E_v	Valence-band edge	J
E_a	Ionized-acceptor energy level	J
E_d	Ionized-donor energy level	J
E_f	Fermi energy level	J
E_g	Energy gap of silicon	J
E_{gB}	Energy gap of base region of a bipolar transistor	J
E_{gE}	Energy gap of emitter region of a bipolar transistor	J
E_i	Intrinsic Fermi level	J
E_{fn}	Electron quasi-Fermi level	J
E_{fp}	Hole quasi-Fermi level	J
$\mathscr{E}$	Electric field	V/cm
$\mathscr{E}_c$	Critical field for velocity saturation	V/cm
$\mathscr{E}_{eff}$	Effective vertical field in MOSFET	V/cm
$\mathscr{E}_{ox}$	Oxide electric field	V/cm
$\mathscr{E}_s$	Electric field at silicon surface	V/cm

(*cont.*)

Symbol	Description	Unit
$\mathscr{E}_x$	Vertical field in silicon	V/cm
$\mathscr{E}_y$	Lateral field in silicon	V/cm
ε_0	Vacuum permittivity ($= 8.85 \times 10^{-14}$ F/cm)	F/cm
ε_i	Permittivity of gate insulator	F/cm
ε_{BOX}	Permittivity of buried oxide layer in SOI	F/cm
ε_{si}	Silicon permittivity ($= 1.04 \times 10^{-12}$ F/cm)	F/cm
ε_{ox}	Oxide permittivity ($= 3.45 \times 10^{-13}$ F/cm)	F/cm
f_D	Probability that an electronic state is filled	None
f	Frequency, clock frequency	Hz
f_{max}	Unity power gain frequency (also called maximum oscillation frequency)	Hz
f_T	Unity current gain frequency (also called cutoff frequency)	Hz
FI	Fan-in	None
FO	Fan-out	None
ϕ	Barrier height	V
ϕ_{bg}	Work function of the back gate in SOI	V
ϕ_{ox}	Silicon–silicon dioxide interface potential barrier for electrons	V
ϕ_{ms}	Work-function difference between metal and silicon	V
ϕ_n	Electron quasi-Fermi potential	V
ϕ_p	Hole quasi-Fermi potential	V
ϕ_{Bn}	Schottky barrier height for electrons	V
ϕ_{Bp}	Schottky barrier height for holes	V
g	Number of degeneracy	None
g_{ds}	Small-signal output conductance per unit width	A/V-cm
g_m	Small-signal transconductance per unit width	A/V-cm
G_E	Emitter Gummel number	s/cm^4
G_B	Base Gummel number	s/cm^4
G_d	Equivalent conductance per unit area in MOS inversion	1/Ω-cm^2
G_n	Electron emission rate (also called electron generation rate)	1/cm^3-s
G_p	Hole emission rate (also called hole generation rate)	1/cm^3-s
γ	Emitter injection efficiency	None
h	Planck's constant ($= 6.63 \times 10^{-34}$ J-s)	J-s
i	Time-dependent current	A
i_B	Time-dependent base current in a bipolar transistor	A
i_b	Time-dependent small-signal base current	A
i_C	Time-dependent collector current in a bipolar transistor	A
i_c	Time-dependent small-signal collector current	A
i_E	Time-dependent emitter current in a bipolar transistor	A
I	Current	A
I_B	Static base current in a bipolar transistor	A
I_C	Static collector current in a bipolar transistor	A
I_E	Static emitter current in a bipolar transistor	A
I_S	Switch current in an ECL circuit	A
I_g	Gate current in a MOSFET	A
I_0	MOSFET current per unit width to length ratio for threshold definition	A
I_{dsat}	MOSFET saturation current	A
I_{on}	MOSFET on current	A

(*cont.*)

Symbol	Description	Unit
I_{off}	MOSFET off current	A
$I_{N/w}$	nMOSFET current per unit width	A/cm
$I_{P/w}$	pMOSFET current per unit width	A/cm
I_N	nMOSFET current	A
I_P	pMOSFET current	A
I_{ds}	Drain-to-source current in a MOSFET	A
I_{sx}	Substrate current in a MOSFET	A
$I_{ds,Vt}$	MOSFET current at threshold	A
$I_{onN/w}$	nMOSFET on current per device width	A/cm
I_{onN}	nMOSFET on current	A
$I_{onP/w}$	pMOSFET on current per device width	A/cm
I_{onP}	pMOSFET on current	A
λ	MOSFET scale length	cm
J	Current density	A/cm^2
J_B	Base current density	A/cm^2
J_C	Collector current density	A/cm^2
J_{CF}	Collector current density injected from emitter	A/cm^2
J_{CR}	Collector current density injected from collector	A/cm^2
J_n	Electron current density	A/cm^2
J_p	Hole current density	A/cm^2
k	Boltzmann's constant ($= 1.38 \times 10^{-23}$ J/K)	J/K
κ	Scaling factor (>1)	None
l	Mean free path	cm
L	Length, MOSFET channel length	cm
L_D	Debye length	cm
L_E	Emitter-stripe length of a bipolar transistor	cm
L_n	Electron diffusion length	cm
L_p	Hole diffusion length	cm
L_{min}	Minimum channel length of MOSFET	cm
L_w	Wire length	cm
m	MOSFET body-effect coefficient	None
m	Ideality factor of current in a Gummel plot	None
m_0	Free-electron mass ($= 9.1 \times 10^{-31}$ kg)	kg
m^*	Electron effective mass	kg
M	Avalanche multiplication factor	None
m_l	Electron effective mass in the longitudinal direction	kg
m_t	Electron effective mass in the transverse direction	kg
μ	Carrier mobility	cm^2/V-s
μ_{eff}	Effective mobility	cm^2/V-s
μ_n	Electron mobility	cm^2/V-s
μ_p	Hole mobility	ccm^2/V-s
n	Density of free electrons	cm^{-3}
n_0	Density of free electrons at thermal equilibrium	cm^{-3}
n_i	Intrinsic carrier density	cm^{-3}
n_{ie}	Effective intrinsic carrier density	cm^{-3}
n_{ieB}	Effective intrinsic carrier density in base of bipolar transistor	cm^{-3}
n_{ieE}	Effective intrinsic carrier density in emitter of bipolar transistor	cm^{-3}
n_n	Density of electrons in n-region	cm^{-3}

(*cont.*)

Symbol	Description	Unit
n_p	Density of electrons in p-region	cm^{-3}
N_a	Acceptor impurity density	cm^{-3}
N_d	Donor impurity density	cm^{-3}
N_b	Impurity concentration in bulk silicon	cm^{-3}
N_{bt}	Density of oxide traps per volume per energy	$cm^{-3}\,eV^{-1}$
N_c	Effective density of states of conduction band	cm^{-3}
N_p	Doping density of polysilicon gate	cm^{-3}
N_v	Effective density of states of valence band	cm^{-3}
N_B	Base doping concentration	cm^{-3}
N_C	Collector doping concentration	cm^{-3}
N_E	Emitter doping concentration	cm^{-3}
$N(E)$	Density of electronic states per unit energy per volume	$1/J\text{-}m^3$
p	Density of free holes	cm^{-3}
p_0	Density of free holes at thermal equilibrium	cm^{-3}
p_n	Density of holes in n-region	cm^{-3}
p_p	Density of holes in p-region	cm^{-3}
P	Power dissipation	W
P_{ac}	Active power dissipation	W
P_{off}	Standby power dissipation	W
q	Electronic charge ($= 1.6 \times 10^{-19}$ C)	C
Q	Charge	C
Q_B	Excess minority charge per unit area in the base	C/cm^2
$Q_{B,tot}$	Total excess minority charge in the base	C
Q_{BE}	Excess minority charge per unit area in the base–emitter space-charge region	C/cm^2
$Q_{BE,tot}$	Total excess minority charge in the base–emitter space-charge region	C
Q_{BC}	Excess minority charge per unit area in the base–collector space-charge region	C/cm^2
$Q_{BC,tot}$	Total excess minority charge in the base–collector space-charge region	C
Q_{DE}	Total stored minority-carrier charge in a bipolar transistor biased in the forward-active mode	C
Q_E	Excess minority charge per unit area in the emitter	C/cm^2
$Q_{E,tot}$	Total excess minority charge in the emitter	C
Q_{pB}	Hole charge per unit area in base of n–p–n transistor	C/cm^2
Q_s	Total charge per unit area in silicon	C/cm^2
Q_d	Depletion charge per unit area	C/cm^2
Q_i	Inversion charge per unit area	C/cm^2
Q_g	Charge on MOS gate per unit area	C/cm^2
Q_t	Interface-state trapped charge per unit area	C/cm^2
Q_{ox}	Equivalent oxide charge density per unit area	C/cm^2
r, R	Resistance	Ω
r_b	Base resistance	Ω
r_{bi}	Intrinsic base resistance	Ω
r_{bx}	Extrinsic base resistance	Ω
r_c	Collector series resistance	Ω
r_e	Emitter series resistance	Ω

(cont.)

Symbol	Description	Unit
r_0	Output resistance in small-signal hybrid-π equivalent-circuit model	Ω
r_π	Input resistance in small-signal hybrid-π equivalent-circuit model	Ω
R_L	Load resistance in a circuit	Ω
R_s	Source series resistance	Ω
R_d	Drain series resistance	Ω
R_n	Electron capture rate (also called electron recombination rate)	$1/\text{cm}^3$-s
R_p	Hole capture rate (also called hole recombination rate)	$1/\text{cm}^3$-s
R_{sd}	Source–drain series resistance	Ω
R_{ch}	MOSFET channel resistance	Ω
R_w	Wire resistance per unit length	Ω/cm
R_{Sbi}	Sheet resistance of intrinsic-base layer	$\Omega/\square$
R_{sw}	Equivalent switching resistance of a CMOS gate	Ω
R_{swn}	Equivalent switching resistance of nMOSFET pulldown	Ω
R_{swp}	Equivalent switching resistance of pMOSFET pullup	Ω
ρ	Resistivity	Ω-cm
ρ_{sh}	Sheet resistivity	$\Omega/\square$
ρ_{ch}	Sheet resistivity of MOSFET channel	$\Omega/\square$
ρ_{sd}	Sheet resistivity of source or drain region	$\Omega/\square$
ρ_c	Specific contact resistivity	Ω-cm^2
ρ_{net}	Volume density of net charge	C/cm^3
S	MOSFET inverse subthreshold current slope	V/decade
S_p	Surface recombination velocity for holes	cm/s
σ_n	Capture cross-section of electron traps	cm^2
σ_p	Capture cross-section of hole traps	cm^2
t	Time	s
t_B	Base transit time	s
t_{BE}	Base–emitter depletion-layer transit time	s
t_{BC}	Base–collector depletion-layer transit time	s
t_{BOX}	Thickness of buried oxide in SOI	cm
t_i	Thickness of gate insulator	cm
t_{inv}	Equivalent oxide thickness for inversion charge calculations	cm
t_{ox}	Oxide thickness	cm
t_r	Transit time	s
t_w	Thickness of wire	cm
t_{si}	Thickness of silicon film	cm
T	Absolute temperature	K
τ	Lifetime	s
τ	Circuit delay	s
τ_b	Buffered delay	s
τ_{int}	Intrinsic, unloaded delay	s
τ_F	Forward transit time of bipolar transistor	s
τ_n	Electron lifetime	s
τ_n	nMOSFET pulldown delay	s
τ_{nB}	Electron lifetime in base of n–p–n transistor	s
τ_p	Hole lifetime	s
τ_p	pMOSFET pullup delay	s
τ_{pE}	Hole lifetime in emitter of n–p–n transistor	s

(*cont.*)

Symbol	Description	Unit
τ_R	Reverse transit time of bipolar transistor	s
τ_w	Wire RC delay	s
τ_E	Emitter delay time	s
τ_B	Base delay time	s
τ_{BE}	Base–emitter depletion-region delay time ($= t_{BE}$)	s
τ_{BC}	Base–collector depletion-region delay time ($= t_{BC}$)	s
U	Net recombination rate	$1/\text{cm}^3\text{-s}$
v	Velocity	cm/s
v	Small-signal voltage	V
v_{th}	Thermal velocity	cm/s
v_d	Carrier drift velocity	cm/s
v_{sat}	Saturation velocity of carriers	cm/s
v_T	Thermal injection velocity at MOSFET source	cm/s
V	Voltage	V
V	Quasi-Fermi potential along MOSFET channel	V
V_A	Early voltage	V
V_{app}	Applied voltage across p–n diode	V
V'_{app}	Applied voltage appearing immediately across p–n junction (smaller than V_{app} by IR drops in series resistances)	V
V_{BE}	Base–emitter bias voltage	V
V_{BC}	Base–collector bias voltage	V
V_{CE}	Collector-to-emitter voltage	V
V_{CG}	Control gate voltage in a nonvolatile memory device	V
V_{FG}	Floating gate voltage in a nonvolatile memory device	V
V_{bg}	Back gate bias voltage in SOI	V
V_{dd}	Power-supply voltage	V
V_{ds}	Source-to-drain voltage	V
V_{dsat}	MOSFET drain saturation voltage	V
V_{fb}	Flat-band voltage	V
V_{ox}	Potential drop across oxide	V
V_g	Gate voltage in MOS	V
V_{gs}	Gate-to-source voltage in a MOSFET	V
V_{bs}	MOSFET body bias voltage	V
V_t	Threshold voltage ($2\psi_B$ definition)	V
V_{on}	Linearly extrapolated threshold voltage	V
V_{in}	Input node voltage of a logic gate	V
V_{out}	Output node voltage of a logic gate	V
V_x	Node voltage between stacked nMOSFETs of a NAND gate	V
$V_{t,high}$	The higher threshold voltage of a nonvolatile memory device	V
$V_{t,low}$	The lower threshold voltage of a nonvolatile memory device	V
W	Width, MOSFET width	cm
W_n	nMOSFET width	cm
W_p	pMOSFET width	cm
W_B	Intrinsic-base width	cm
W_d	Depletion-layer width	cm
W_{dBE}	Base–emitter junction depletion-layer width	cm
W_{dBC}	Base–collector junction depletion-layer width	cm

(*cont.*)

Symbol	Description	Unit
W_{dm}	Maximum depletion-layer width in MOS	cm
W_E	Emitter-layer width (thickness)	cm
W_{E-C}	Emitter-to-collector spacing of a lateral bipolar transistor	cm
W_S	Source junction depletion-layer width	cm
W_D	Drain junction depletion-layer width	cm
ω	Angular frequency	rad/s
x_j	Junction depth	cm
x_c, x_i	Depth of inversion channel	cm
ψ	Potential	V
ψ_B	Difference between Fermi potential and intrinsic potential	V
ψ_{bi}	Built-in potential	V
ψ_f	Fermi potential	V
ψ_i	Intrinsic potential	V
ψ_p	Band bending in polysilicon gate	V
ψ_s	Surface potential in MOS	V
$\psi_{s,\min}$	Minimum surface potential in short-channel MOSFET	V

1 Introduction

Since the invention of the bipolar transistor in 1947, there has been an unprecedented growth of the semiconductor industry, with an enormous impact on the way people work and live. In the last forty years or so, by far the strongest growth area of the semiconductor industry has been in silicon very-large-scale-integration (VLSI) technology. The sustained growth in VLSI technology is fueled by the continued shrinking of transistors to ever smaller dimensions. The benefits of miniaturization – higher packing densities, higher circuit speeds, and lower power dissipation – have been key in the evolutionary progress leading to today's computers, wireless units, and communication systems that offer superior performance, dramatically reduced cost per function, and much reduced physical size, in comparison with their predecessors. On the economic side, the integrated-circuit (IC) business has grown worldwide in sales from $1 billion in 1970 to $20 billion in 1984 and has reached $439 billion in 2020. The electronics industry is now among the largest industries in terms of output as well as employment in many nations. The importance of microelectronics in economic, social, and even political development throughout the world will no doubt continue to ascend. The large worldwide investment in VLSI technology constitutes a formidable driving force that will all but guarantee the continued progress in IC integration density and speed, for as long as physical principles will allow.

1.1 Evolution of VLSI Device Technology

1.1.1 Historical Perspective

An excellent account of the evolution of the metal–oxide–semiconductor field-effect transistor (MOSFET), from its initial conception to VLSI applications in the mid-1980s, can be found in a paper by Sah (1988). Figure 1.1 gives a chronology of the major milestone events in the development of VLSI technology. The vertical bipolar transistor technology was developed early on and was applied to the first integrated-circuit memory in mainframe computers in the 1960s. Vertical bipolar transistors have been used all along where raw circuit speed is most important, for bipolar circuits remain the fastest at the individual-circuit level. However, the large power dissipation

1

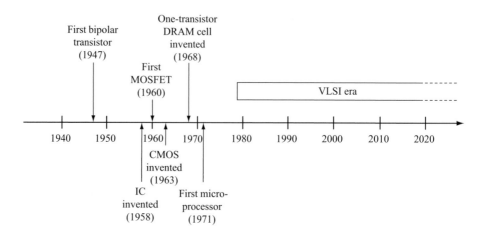

Figure 1.1 A brief chronology of the major milestones in the development of VLSI

of vertical bipolar circuits has severely limited their integration level, to about 10^4 circuits[1] per chip. This integration level is very low by today's VLSI standard.

The idea of modulating the surface conductance of a semiconductor by the application of an electric field was first envisioned in 1930. However, early attempts to fabricate a surface-field-controlled device were not successful because of the presence of large densities of surface states which effectively shielded the surface potential from the influence of an external field. The first MOSFET on a silicon substrate using SiO_2 as the gate insulator was fabricated in 1960 (Kahng and Atalla, 1960). During the 1960s and 1970s, n-channel and p-channel MOSFETs were widely used, along with bipolar transistors, for implementing circuit functions on a silicon chip. Although the MOSFET devices were slow compared to the bipolar devices, they had a higher layout density and were relatively simple to fabricate; the simplest MOSFET chip could be made using only four masks and a single doping step. However, just like vertical bipolar circuits, single-polarity MOSFET circuits suffered from large standby power dissipation, hence were limited in the level of integration on a chip.

The major breakthrough in the level of integration came in 1963 with the invention of CMOS (complementary MOS) (Wanlass and Sah, 1963), in which n-channel and p-channel MOSFETs are constructed side by side on the same substrate. A CMOS circuit typically consists of an n-channel MOSFET and a p-channel MOSFET connected in series between the power-supply terminals, so that there is negligible standby power dissipation. Significant power is dissipated only during switching of the circuit (i.e., only when the circuits are active). By cleverly designing the "switch activities" of the circuits on a chip to minimize active power dissipation, engineers have been able to integrate billions of CMOS transistors on a single chip and still have the chip readily air-coolable. Until the minimum feature size of lithography reached 180 nm, the integration level of CMOS was not limited by chip-level power

---

[1] ECL circuits, discussed in Section 11.2.

dissipation, but by chip fabrication technology. Another advantage of CMOS circuits comes from the ratioless, full rail-to-rail logic swing, which improves the noise margin and makes a CMOS chip easier to design.

As CMOS scaling reached the 0.5-μm level in the early 1990s, the performance of high-end computers built using CMOS started to approach those built using bipolar, due to the much higher integration level of CMOS chips. Designers of high-end computer systems were able to meet their performance targets using CMOS instead of bipolar (Rao *et al.*, 1997). Since then, CMOS has become the technology for digital circuits, and vertical bipolar is used primarily in radio-frequency (RF) and analog circuits only.

Advances in lithography and etching technologies have enabled the industry to scale down transistors in physical dimensions, and to pack more transistors in the same chip area. Such progress, combined with a steady growth in chip size, resulted in an exponential growth in the number of transistors and memory bits per chip. The technology trends up to 2020 in these areas are illustrated in Figure 1.2. Traditionally, dynamic random-access memories (DRAMs) have contained the highest component count of any IC chips. This has been so because of the small size of the one-transistor memory cell (Dennard, 1968) and because of the large and often insatiable demand for more memory in computing systems. It is interesting to note that the entire content of this book can be stored in one 64-Mb DRAM chip, which was in volume production in 1997 and has an area equivalent to a square of about 1.2×1.2 cm^2.

One remarkable feature of silicon devices that fueled the rapid growth of the information technology industry is that their speed increases and their cost decreases as their size is reduced. The transistors manufactured in 2020 were 10-times faster and occupy less than 1% of the area of those built 20 years earlier. This is illustrated in the trend of microprocessor units (MPUs) in Figure 1.2. The increase in the clock frequency of microprocessors is the result of a combination of improvements in microprocessor architecture and improvements in transistor speed.

1.1.2 Recent Developments

Since the publication of the second edition of this book in 2009, there have been major developments in the VLSI industry. Several fabrication technologies emerging at the time have taken hold, enabling the continued chip-level density improvements, resulting in continued reduction of cost per transistor and cost per memory bit. These in turn have driven the continued growth of the semiconductor industry. These recent developments include the following.

- Immersion lithography has been adopted for volume IC manufacturing (Lin, 2004). Immersion lithography is a photolithography resolution enhancement technique where the usual gap between the final lens and the wafer surface is replaced with a liquid medium having a refractive index greater than one. The resolution enhancement is equal to the refractive index of the liquid used. With immersion, deep ultraviolet (DUV) lithography systems remain the work horse for semiconductor manufacturing today.

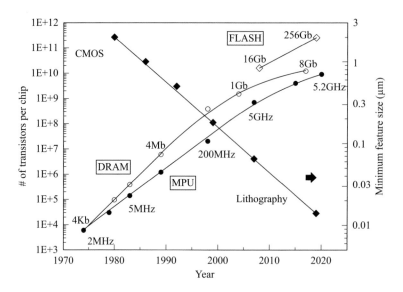

Figure 1.2 Trends in lithographic feature size, number of transistors per chip for DRAM and MPU, and number of memory bits per chip for Flash. The transistor count for DRAM is computed as 1.5-times the number of bits on the chip to account for the peripheral circuits. Data points represent announced leading-edge products

- Driven by the need for low-power and light-weight data storage in battery-operated personal systems, NAND flash (the highest density version of the electrically programmable and erasable nonvolatile memory) development has been on an exceptionally steep trajectory since the mid-1990s. In the past decade or so, NAND flash has overtaken DRAM as the IC chip with the highest component count, as shown in Figure 1.2 (Kim, 2008). Since then, the combination of 3D NAND process technology, where upwards of 100+ layers of NAND flash memory cells are stacked on top of another to form a three-dimensional IC chip, and multi-bit-per-cell design has dramatically increased the chip-level bit density of NAND flash.
- For a long time, a common practice in designing a scaled CMOS device was to allow its off current to increase, by reducing the device threshold voltage, in order to increase the on current to achieve the targeted performance of the scaled device. As a result, the off current of scaled CMOS devices had been increasing from one generation to the next. By the time scaling reached the 65 nm node, the off current of scaled CMOS device had reached 100 nA/μm, the maximum level acceptable to designers of high-end microprocessors. Since then, high-performance CMOS devices have been designed with a nominal off current of 100 nA/μm (Kuhn *et al.*, 2012). Capping the tradeoff between on current and off current in designing scaled CMOS devices severely limits the speed (clock frequency) of microprocessors. Today, the highest speed microprocessors run at 5.2 GHz (Berry *et al.*, 2020), practically the same as those in 2009 (see Figure 1.2).

- Without the ability to increase the device off current, CMOS designers turned to device structures having fully depleted device body, which enables the subthreshold swing of the device to reach the ideal 60 mV/decade at room temperature. Today, FinFET CMOS, where the depleted device body is shaped like a fin, planar ETSOI (extremely thin silicon-on-insulator) CMOS, which is basically fully depleted SOI CMOS, as well as traditional planar bulk CMOS are in volume manufacturing.
- Partially depleted (PD) planar SOI CMOS ran its course over a span of about fifteen years, with the first PD SOI CMOS microprocessor fabricated in the 220 nm node in 1999 (Shahidi *et al.*, 1999), and the last in the 22 nm node in 2015 (Freeman *et al.*, 2015). PD SOI CMOS is judged not scalable to smaller dimensions. However, SOI wafers suitable for PD SOI CMOS have found new applications in complementary (integration of n−p−n and p−n−p) lateral bipolar (Cai *et al.*, 2011), offering interesting possibilities in bipolar for VLSI.

1.2   Scope and Brief Description of the Book

In writing this book, it is our goal to address the factors governing the performance of modern VLSI devices in depth. This is carried out by first discussing the device physics that goes into the design of individual device parameters, and then discussing the effects of these parameters on the performance of small-dimension modern transistors at the basic circuit level. A substantial part of the book is devoted to in-depth discussions on the interdependency among the device parameters and the subtle tradeoffs in the design of modern CMOS and bipolar transistors.

This book contains sufficient background tutorials to be used as a textbook for students taking a graduate or advanced undergraduate course in microelectronics. The prerequisite is one semester of either solid-state physics or semiconductor physics. For the practicing engineer, this book provides an extensive source of reference material that covers the fundamentals of CMOS and bipolar technologies, devices, and circuits. It should be useful to VLSI process engineers and circuit designers interested in learning basic device principles, and to device design or characterization engineers who desire more in-depth knowledge in their specialized areas.

New topics and materials in the third edition include an expanded chapter on ETSOI and FinFETs, non-GCA (Gradual Channel Approximation) model for MOSFETs, and the relatively recent development of symmetric lateral bipolar transistors on SOI. Also added are sections on high-κ gate dielectrics, metal gates, strain effect on mobility, and interface–state models. Much of the materials in the second edition have been restructured by consolidating all the appendices into the main chapters for a more focused coverage of the various subjects. Here is a brief description of each chapter.

Chapter 2: Basic Device Physics

Chapter 2 covers the appropriate level of basic device physics to make the book self-contained, and to prepare the reader with the necessary background on device operation and material physics to follow the discussion in the rest of the book.

Starting with the energy bands in silicon, Chapter 2 introduces the basic concepts of Fermi level, carrier concentration, drift and diffusion current transport, and Poisson's equation. Also addressed in this chapter are generation and recombination, minority carrier lifetime, and current continuity equation.

Chapter 3: p–n Junctions and Metal–Silicon Contacts

Chapter 3 covers the basic physics and operation of p–n junctions and Schottky diodes as well as metal–silicon contacts in general. p–n junctions are basic building blocks of bipolar transistors and key components of MOSFETs. Basic knowledge of their characteristics is a prerequisite to further understand the operation of bipolar devices and for designing MOSFETs. And basic knowledge of Schottky diodes is prerequisite to understanding metal–silicon contacts in general and for designing ohmic contacts with low contact resistance. The chapter ends with a discussion of high-field effects in reverse-biased diodes.

Chapter 4: MOS Capacitors

Chapter 4 covers the fundamentals of MOS capacitors – a prerequisite to MOSFET transistors. Starting with the basic concepts of free electron level and work function, the chapter proceeds to the solution of charge and potential in silicon, followed by a full description of the C–V characteristics. Quantum mechanical effects, important for MOS capacitors of thin oxides, are then discussed. Added in the third edition is a new section on interface states and oxide traps. Lastly, the high field section covers tunneling currents, high-κ gate dielectrics, and gate oxide reliability.

Chapter 5: MOSFETs: Long Channel

Chapter 5 describes the basic characteristics of MOSFET devices, using n-channel MOSFET as an example for most of the discussions. It deals with the more elementary long-channel MOSFETs, with sections on the charge sheet model, regional I–V models, and subthreshold current characteristics. A recently developed non-GCA model gives insights to the saturation region behavior while clarifying the misleading term of "pinch-off" in most standard textbooks. In the section on channel mobility, the strain effects, both biaxial and uniaxial, on electron and hole mobilities are discussed. The last section addresses the body effect, temperature effect, and quantum effect on the long-channel threshold voltage.

Chapter 6: MOSFETs: Short Channel

This chapter deals with the more complex short-channel MOSFETs. Most circuits are built with short-channel devices because of their higher current and lower capacitance. Among the main topics are short-channel effects, scale length model, velocity saturation, and non-local transport. A ballistic MOSFET model is described on the current limit of a MOSFET. Next considered are the major device design issues in a CMOS technology: choice of threshold voltage based on the off-current requirement and on-current performance, power supply voltage, design of nonuniform channel doping, and discrete dopant effects on threshold voltage. The last section discusses high-field effects in a short-channel MOSFET.

Chapter 7: Silicon-on-Insulator and Double-Gate MOSFETs

Chapter 7 deals with fully-depleted SOI and double-gate MOSFETs. A general, asymmetric double-gate model is applied to long channel SOI MOSFETs. For symmetric double-gate MOSFETs – the generic form of FinFETs, an analytic potential model is described that covers all regions of operation continuously. The scale length model first introduced in Chapter 6 for bulk MOSFETs is modified for short-channel DG MOSFETs. Nanowire MOSFET models, both long and short channel, are also discussed. The last section examines the scaling limits of DG and nanowire MOSFETs based on quantum mechanical considerations.

Chapter 8: CMOS Performance Factors

This chapter begins by reviewing MOSFET scaling – the guiding principle for achieving density, speed, and power improvements in VLSI evolution. The implications of the non-scaling factors, specifically, thermal voltage and silicon bandgap, on the path of CMOS evolution are discussed. The rest of the chapter deals with the key factors that govern the switching performance and power dissipation of basic digital CMOS circuits. After a brief description of static CMOS logic gates, their layout and noise margin, Section 8.3 considers the parasitic resistances and capacitances that may adversely affect the delay of a CMOS circuit. These include source and drain series resistance, junction capacitance, overlap capacitance, gate resistance, and interconnect capacitance and resistance. In Section 8.4, a delay equation is formulated and applied to study the sensitivity of CMOS delay to a variety of device and circuit parameters such as wire loading, device width and length, gate oxide thickness, power-supply voltage, threshold voltage, parasitic components, and substrate sensitivity in stacked circuits. The last section addresses the performance factors of MOSFETs in RF circuits, in particular, the unity-current-gain frequency and unity-power-gain frequency.

Chapter 9: Bipolar Devices

The basic components of a bipolar transistor are described in Chapter 9. Both vertical bipolar transistors, including SiGe-base transistors, and symmetric lateral bipolar transistors on SOI are covered. The discussion focuses on the vertical n–p–n transistors, since they are the most commonly used. The difference between n–p–n vertical transistors and symmetric lateral n–p–n transistors are pointed out where appropriate.

The basic operation of a bipolar transistor is described in terms of two p–n diodes connected back to back. The basic theory of a p–n diode is modified and applied to derive the current equations for a bipolar transistor. From these current equations, other important device parameters and phenomena, such as current gain, early voltage, base widening, and diffusion capacitance, are examined. The basic equivalent-circuit models relating the device parameters to circuit parameters are developed. These equivalent-circuit models form the starting point for discussing the performance of a bipolar transistor in circuit applications.

Chapter 10: Bipolar Device Design

Chapter 10 covers the basic design of a bipolar transistor. The design of the individual device regions, namely the emitter, the base, and the collector, are discussed separately. Since the detailed characteristics of a bipolar transistor depend on its operating point, the focus of this chapter is on optimizing the device design according to its intended operating condition and environment, and on the tradeoffs that must be made in the optimization process. The physics and characteristics of vertical SiGe-base bipolar transistors are discussed in depth. The design of symmetric lateral bipolar transistors on SOI is also covered, including the development of analytical models for the device parameters, base and collector currents, and the transit times.

Chapter 11: Bipolar Performance Factors

The major factors governing the performance of bipolar transistors in circuit applications are discussed in Chapter 11. Several of the commonly used figures of merit, namely, cutoff frequency, maximum oscillation frequency, and logic gate delay, are examined, and how a bipolar transistor can be optimized for a given figure of merit is discussed. Sections are devoted to examining the important delay components of a logic gate, and how these components can be minimized. The scaling properties of vertical bipolar transistors for high-speed digital logic circuits are discussed. A discussion of the optimization of bipolar transistors for RF and analog circuit applications is given. The chapter concludes with a discussion of the design tradeoff and optimization of symmetric lateral bipolar transistors for RF and analog circuit applications. Finally, several unique opportunities offered by symmetric lateral bipolar transistors, some of them beyond the capability of CMOS, are discussed.

Chapter 12: Memory Devices

In Chapter 12, the basic operational and device design principles of commonly used memory devices are discussed. The memory devices covered include CMOS SRAM, DRAM, bipolar SRAM, and several commonly used in nonvolatile memories. Typical read, write, and erase operations of the various memory arrays are explained. The issue of noise margin in scaled CMOS SRAM cells is discussed. A brief discussion of more recent developments of NAND flash technologies, including multi-bit per cell, 3D NAND, and wear leveling is given.

2   Basic Device Physics

This chapter reviews the basic concepts of semiconductor device physics. It covers energy bands in silicon, Fermi level, n-type and p-type silicon, electrostatic potential, drift and diffusion current transport, and basic equations governing VLSI device operation. These will serve as the basis for understanding more advanced device concepts discussed in the rest of the book.

2.1 Energy Bands in Silicon

The starting material used in the fabrication of VLSI devices is silicon in the crystalline form. The silicon wafers are cut parallel to either the $<111>$ or $<100>$ planes (Sze, 1981), with $<100>$ material being the most commonly used. This is largely due to the fact that $<100>$ wafers, during processing, produce the lowest charges at the oxide–silicon interface as well as higher mobility (Balk *et al.*, 1965). In a silicon crystal each atom has four valence electrons to share with its four nearest neighboring atoms. The valence electrons are shared in a paired configuration called a *covalent bond*. *The most important result of the application of quantum mechanics to the description of electrons in a solid is that the allowed energy levels of electrons are grouped into bands* (Kittel, 1976). *The bands are separated by regions of energy that the electrons in the solid cannot possess: forbidden gaps*. The highest energy band that is completely filled by electrons at 0 K is called the *valence band*. The next highest energy band, separated by a forbidden gap from the valence band, is called the *conduction band*, as shown in Figure 2.1.

2.1.1 Bandgap of Silicon

What sets a semiconductor such as silicon apart from a metal or an insulator is that, at absolute zero temperature, the valence band is completely filled with electrons, while the conduction band is completely empty, and that the separation between the conduction band and valence band, or the *bandgap*, is on the order of 1 eV. On the one hand, no electrical conduction is possible at 0 K, since there are no electrons in the conduction band, whereas the electrons in the completely filled valence band cannot be accelerated by an electric field and gain energy. On the other hand, the bandgap is small enough that at room temperature a small fraction of the electrons are excited into the conduction band, leaving behind vacancies, or *holes*, in the valence

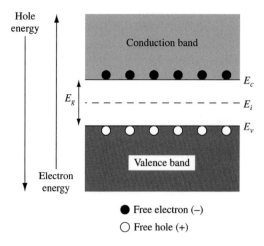

Figure 2.1 Energy-band diagram of silicon

band. This allows limited conduction to take place from the motion of both the electrons in the conduction band and the holes in the valence band. In contrast, an insulator has a much larger forbidden gap of at least several electron volts, making room-temperature conduction virtually nonexistent. Metals, on the contrary, have partially filled conduction bands even at absolute zero temperature, so that the electrons can easily move into states of higher energy in response to an applied electric field. This makes them good conductors at any temperature.

As shown in Figure 2.1, the energy of the electrons in the conduction band increases upward, while the energy of the holes in the valence band increases downward. The bottom of the conduction band is designated E_c, and the top of the valence band E_v. Their separation, or the bandgap, is $E_g = E_c - E_v$. For silicon, E_g is 1.12 eV at room temperature or 300 K. The bandgap decreases slightly as the temperature increases, with a temperature coefficient of $dE_g/dT \approx -2.73 \times 10^{-4}\mathrm{eV/K}$ for silicon near 300 K. Other important physical parameters of silicon and silicon dioxide are listed in Table 2.1 (Green, 1990).

2.1.2   Density of States

The density of available electronic states within a certain energy range in the conduction band is determined by the number of different momentum values that can be acquired by electrons in this energy range. Based on quantum mechanics, there is one allowed state in a phase space of volume $(\Delta x\, \Delta p_x)(\Delta y\, \Delta p_y)(\Delta z\, \Delta p_z) = h^3$, where p_x, p_y, p_z are the x-, y-, z-components of the electron momentum, respectively, and h is Planck's constant. If we let $N(E)dE$ be the number of electronic states per unit volume with an energy between E and $E + dE$ in the conduction band, then

$$N(E)\,dE = 2g\frac{dp_x\,dp_y\,dp_z}{h^3},\tag{2.1}$$

Table 2.1. Physical properties of Si and SiO_2 at room temperature (300 K)

Property	Si	SiO_2
Atomic/molecular weight	28.09	60.08
Atoms or molecules/cm^3	5.0×10^{22}	2.3×10^{22}
Density (g/cm^3)	2.33	2.27
Crystal structure	Diamond	Amorphous
Lattice constant (Å)	5.43	–
Energy gap (eV)	1.12	8–9
Dielectric constant	11.7	3.9
Intrinsic carrier concentration (cm^{-3})	1.0×10^{10}	–
Carrier mobility (cm^2/V-s)	Electron: 1,430	–
	Hole: 480	–
Effective density of states (cm^{-3})	Conduction band, N_c: 2.9×10^{19}	–
	Valence band, N_v: 3.1×10^{19}	–
Breakdown field (V/cm)	3×10^5	$>10^7$
Melting point (°C)	1,415	1600–1700
Thermal conductivity (W/cm-°C)	1.5	0.014
Specific heat (J/g-°C)	0.7	1.0
Thermal diffusivity (cm^2/s)	0.9	0.006
Thermal expansion coefficient (°C^{-1})	2.5×10^{-6}	0.5×10^{-6}

where $dp_x dp_y dp_z$ is the volume in the momentum space within which the electron energy lies between E and $E + dE$, g is the number of equivalent minima in the conduction band, and the factor of two arises from the two possible directions of electron spin. The conduction band of silicon has a sixfold degeneracy, so $g = 6$. Note that MKS units are used here (e.g., length must be in meters, not centimeters).

If the electron kinetic energy is not too high, one can consider the energy–momentum relationship near the conduction-band minima as being parabolic and write

$$E - E_c = \frac{p_x^2}{2m_x} + \frac{p_y^2}{2m_y} + \frac{p_z^2}{2m_y}, \tag{2.2}$$

where $E - E_c$ is the electron kinetic energy, and m_x, m_y, m_z are the effective masses. The constant energy surface in momentum space is an ellipsoid with the lengths of the symmetry axes proportional to the square roots of m_x, m_y, and m_z. For the silicon conduction band in the <100> direction, two of the effective masses are the transverse mass $m_t = 0.19m_0$, and the third is the longitudinal mass $m_l = 0.92m_0$, where m_0 is the free electron mass. The volume of the ellipsoid given by Eq. (2.2) in momentum space is $(4\pi/3)(8m_x m_y m_z)^{1/2}(E - E_c)^{3/2}$. Therefore, the volume $dp_x dp_y dp_z$ within which the electron energy lies between E and $E + dE$ is $4\pi(2m_x m_y m_z)^{1/2}(E - E_c)^{1/2}dE$. Thus, Eq. (2.1) becomes

$$N(E)dE = \frac{8\pi g \sqrt{2m_x m_y m_z}}{h^3} \sqrt{E - E_c}\, dE = \frac{8\pi g \sqrt{2m_t^2 m_l}}{h^3} \sqrt{E - E_c}\, dE. \tag{2.3}$$

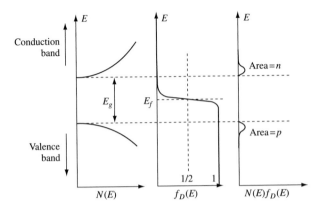

Figure 2.2 Schematic plots of density of states, Fermi–Dirac distribution function, and their products versus electron energy in a band diagram (after Sze, 1981)

The 3-D electron density of states in an energy diagram is then a parabolic function with its downward apex at the conduction-band edge, and vice versa for the hole density of states in the valence band. These are shown schematically in Figure 2.2 (Sze, 1981).

2.1.3   Distribution Function: Fermi Level

The energy distribution of electrons in a solid is governed by the laws of Fermi–Dirac statistics. For a system in thermal equilibrium, the principal result of these statistics is the *Fermi–Dirac distribution function*, which gives the probability that an electronic state at energy E is occupied by an electron,

$$f_D(E) = \frac{1}{1 + e^{(E-E_f)/kT}}.\tag{2.4}$$

Here $k = 1.38 \times 10^{-23}$ J/K is Boltzmann's constant, and T is the absolute temperature. This function contains a parameter, E_f, called the *Fermi level*. The Fermi level is the energy at which the probability of occupation of an energy state by an electron is exactly one-half. At absolute zero temperature, $T = 0$ K, all the states below the Fermi level are filled ($f_D = 1$ for $E < E_f$), and all the states above the Fermi level are empty ($f_D = 0$ for $E > E_f$). At finite temperatures, some states above the Fermi level are filled as some states below become empty, in which case the probability distribution $f_D(E)$ makes a smooth transition from unity to zero as the energy increases across the Fermi level. The width of the transition is governed by the thermal energy, kT. This is plotted schematically in Figure 2.2, with a Fermi level in the middle of the forbidden gap (for reasons that will soon be clear). It is important to keep in mind that the thermal energy at room temperature is 0.026 eV, or roughly $\frac{1}{40}$ of the silicon bandgap. In most cases when the energy is at least several kT above or below the Fermi level, Eq. (2.4) can be approximated by the following:

$$f_D(E) \approx e^{-(E-E_f)/kT} \quad \text{for} \quad E > E_f \tag{2.5}$$

and

$$f_D(E) \approx 1 - e^{-(E_f-E)/kT} \quad \text{for} \quad E < E_f. \tag{2.6}$$

Equation (2.6) should be interpreted as stating that the probability of finding a hole (i.e., an empty state *not* occupied by an electron) at an energy $E < E_f$ is $e^{-(E_f-E)/kT}$. The last two equations follow directly from Maxwell–Boltzmann statistics for classical particles, which are good approximations to Fermi–Dirac statistics when the energy is at least several kT away from E_f.

Fermi level plays an essential role in characterizing the equilibrium state of a system. Consider two electronic systems brought into contact with Fermi levels E_{f1} and E_{f2}, and corresponding distribution functions $f_{D1}(E)$ and $f_{D2}(E)$. If $E_{f1} > E_{f2}$, then $f_{D1}(E) > f_{D2}(E)$, which means that at every energy E where electronic states are available in both systems, a larger fraction of the states in system 1 are occupied by electrons than those in system 2. Equivalently, a larger fraction of the states in system 2 are empty than those in system 1 at energies where electronic states exist. Since the two systems in contact are free to exchange electrons, there is a higher probability for the electrons in system 1 to redistribute to system 2 than vice versa. This leads to a net electron transport from system 1 to system 2, i.e., current flows (defined in terms of positive charges) from system 2 to system 1. If there are no power sources connected to the systems to sustain the Fermi level imbalance, eventually the two systems will come to an equilibrium and $E_{f1} = E_{f2}$. No further net electron flow takes place once the same fractions of the electronic states in the two systems are occupied at every energy E. Note that this conclusion is reached regardless of the specific density of states in each of the two systems. For example, the two systems can be two metals, a metal and a semiconductor, or two semiconductors of different doping or different composition. **When two systems are in thermal equilibrium with no current flow between them, their Fermi levels must be equal**. A direct extension is that, *for a continuous region of metals and/or semiconductors in contact, the Fermi level at thermal equilibrium is flat, i.e., spatially constant, throughout the region*. The role of Fermi level at the contacts when there is an applied voltage driving a steady-state current is further discussed in Section 2.4.2.

2.1.4 Carrier Concentration

Since $f_D(E)$ is the probability that an electronic state at energy E is occupied by an electron, the total number of electrons per unit volume in the conduction band is given by

$$n = \int_{E_c}^{\infty} N(E) f_D(E) dE. \tag{2.7}$$

Here the upper limit of integration is taken as infinity because the top of the conduction band is far above E_c. Both the product $N(E)f_D(E)$ and n, p are shown schematically in Figure 2.2. Equation (2.7) with the full Fermi–Dirac distribution function, Eq. (2.4), is discussed in Section 2.2.3. For nondegenerate silicon with a Fermi level at least $3kT/q$ below E_c, the Fermi–Dirac distribution function can be approximated by the Maxwell–Boltzmann distribution, Eq. (2.5). Equation (2.7) then becomes

$$n = \frac{8\pi g \sqrt{2m_t^2 m_l}}{h^3} \int_{E_c}^{\infty} \sqrt{E - E_c} e^{-(E-E_f)/kT} dE. \qquad (2.8)$$

With a change of variable, the integral can be expressed in the form of a gamma function, $\Gamma(3/2)$, which equals $\pi^{1/2}/2$. The electron concentration in the conduction band is then

$$n = N_c e^{-(E_c - E_f)/kT}, \qquad (2.9)$$

where the pre-exponential factor is defined as the *effective density of states*,

$$N_c = 2g \sqrt{m_t^2 m_l} \left(\frac{2\pi kT}{h^2} \right)^{3/2}. \qquad (2.10)$$

A similar expression can be derived for the hole density in the valence band,

$$p = N_v e^{-(E_f - E_v)/kT}, \qquad (2.11)$$

where N_v is the effective density of states of the valence band, which depends on the hole effective mass and the valence band degeneracy. Both N_c and N_v are proportional to $T^{3/2}$. Their values at room temperature are listed in Table 2.1 (Green, 1990).

For an intrinsic silicon, $n = p$, since for every electron excited into the conduction band, a vacancy or hole is left behind in the valence band. The Fermi level for intrinsic silicon, or the *intrinsic Fermi level*, E_i, is then obtained by equating Eqs. (2.9) and (2.11) and solving for E_f:

$$E_i = E_f = \frac{E_c + E_v}{2} - \frac{kT}{2} \ln \left(\frac{N_c}{N_v} \right). \qquad (2.12)$$

By substituting Eq. (2.12) for E_f in Eq. (2.9) or Eq. (2.11), the intrinsic carrier concentration, $n_i = n = p$ is obtained:

$$n_i = \sqrt{N_c N_v} e^{-(E_c - E_v)/2kT} = \sqrt{N_c N_v} e^{-E_g/2kT}. \qquad (2.13)$$

Since the thermal energy, kT, is much smaller than the silicon bandgap, E_g, and $\ln [N_c/N_v]$ is not a large number, **the intrinsic Fermi level is very close to the midpoint between the conduction band and the valence band**. In fact, E_i is sometimes referred to as the midgap energy level, since the error in assuming E_i to be $(E_c + E_v)/2$ is only about $0.3\,kT$. The intrinsic carrier concentration, n_i, at room temperature is 1.0×10^{10} cm^{-3}, as given in Table 2.1. This is very small compared with the atomic density of silicon.

Equations (2.9) and (2.11) can be rewritten in terms of n_i and E_i:

$$n = n_i e^{(E_f - E_i)/kT}, \tag{2.14}$$

$$p = n_i e^{(E_i - E_f)/kT}. \tag{2.15}$$

These equations give the equilibrium electron and hole densities for any Fermi level position (not too close to the band edges) relative to the intrinsic Fermi level at the midgap. In Section 2.2, we will show how the Fermi level varies with the type and concentration of impurity atoms in silicon. Since any change in E_f causes reciprocal changes in n and p, *a useful, general relationship is that the product*

$$pn = n_i^2 \tag{2.16}$$

in equilibrium is a constant, independent of the Fermi level position.

2.2   n-Type and p-Type Silicon

Intrinsic silicon at room temperature has an extremely low free-carrier concentration; therefore, its resistivity is very high. In practice, intrinsic silicon hardly exists at room temperature, since it would require materials with an unobtainably high purity. Most impurities in silicon introduce additional energy levels in the forbidden gap and can be easily ionized to add either electrons to the conduction band or holes to the valence band, depending on where the impurity level is (Kittel, 1976). *The electrical conductivity of silicon is then dominated by the type and concentration of the impurity atoms, or dopants*, and the silicon is called *extrinsic*.

2.2.1   Donors and Acceptors

Silicon is a column-IV element with four valence electrons per atom. There are two types of impurity in silicon that are electrically active: those from column V, such as arsenic or phosphorus, and those from column III, such as boron. As is shown in Figure 2.3, a column-V atom in a silicon lattice tends to have one extra electron loosely bonded after forming covalent bonds with other silicon atoms. In most cases, the thermal energy at room temperature is sufficient to ionize the impurity atom and free the extra electron to the conduction band. Such types of impurity are called *donors*; they become positively charged when ionized. Silicon material doped with column-V impurities or donors is called *n-type* silicon, and its electrical conductivity is dominated by electrons in the conduction band. On the other hand, a column-III impurity atom in a silicon lattice tends to be deficient by one electron when forming covalent bonds with other silicon atoms (Figure 2.3). Such an impurity atom can also be ionized by accepting an electron from the valence band, which leaves a

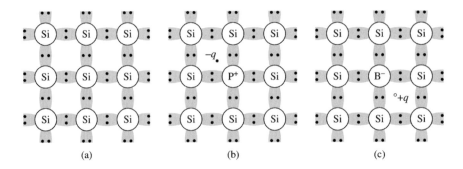

Figure 2.3 Three basic bond pictures of silicon: (a) intrinsic Si with no impurities, (b) n-type silicon with donor (phosphorus), (c) p-type silicon with acceptor (boron) (after Sze, 1981)

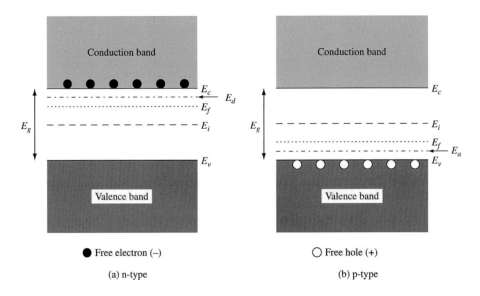

Figure 2.4 Energy-band diagram representation of (a) donor level E_d and Fermi level E_f in n-type silicon, (b) acceptor level E_a and Fermi level E_f in p-type silicon

free-moving hole that contributes to electrical conduction. These impurities are called *acceptors*; they become negatively charged when ionized. Silicon material doped with column-III impurities or acceptors is called *p-type* silicon, and its electrical conductivity is dominated by holes in the valence band. It should be noted that impurity atoms must be in a *substitutional* site (as opposed to *interstitial*) in silicon in order to be electrically active.

In terms of the energy-band diagrams in Figure 2.4, donors add allowed electron states in the bandgap close to the conduction-band edge, while acceptors add allowed states just above the valence-band edge. Donor levels contain positive charge when ionized (emptied). Acceptor levels contain negative charge when ionized (filled). The

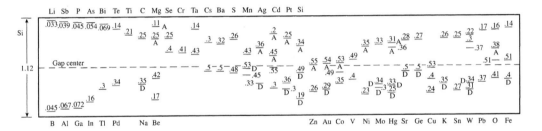

Figure 2.5 Donor and acceptor levels of various impurities in silicon. Numbers next to the level indicate ionization energies $E_c - E_d$ (donors) or $E_a - E_v$ (acceptors) in electron volts (after Sze, 1981)

ionization energies are denoted by $E_c - E_d$ for donors and $E_a - E_v$ for acceptors. Figure 2.5 shows the donor and acceptor levels of common impurities in silicon and their ionization energies (Sze, 1981). Phosphorus and arsenic are commonly used donors, or n-type dopants, with low ionization energies on the order of $2kT$, while boron is a commonly used acceptor or p-type dopant with a comparable ionization energy. Figure 2.6 shows the *solid solubility* of important impurities in silicon as a function of annealing temperature (Trumbore, 1960). Arsenic, boron, and phosphorus have the highest solid solubility among all the impurities, which makes them the most important doping species in VLSI technology.

2.2.2 Fermi Level in Extrinsic Silicon

In contrast to intrinsic silicon, the Fermi level in an extrinsic silicon is not located at the midgap. The Fermi level in n-type silicon moves up toward the conduction band, i.e., $E_c - E_f$ decreases, consistent with the increase of electron density in Eq. (2.9). On the other hand, the Fermi level in p-type silicon moves down toward the valence band, i.e., $E_f - E_v$ decreases, consistent with the increase of hole density in Eq. (2.11). These cases are depicted in Figure 2.4. The exact position of the Fermi level depends on both the ionization energy and the concentration of dopants. For example, for an n-type material with a donor impurity concentration, N_d, the charge neutrality condition in silicon requires that

$$n = N_d^+ + p, \tag{2.17}$$

where N_d^+ is the density of ionized donors given by

$$N_d^+ = N_d[1 - f_D(E_d)] = N_d\left(1 - \frac{1}{1 + \frac{1}{2}e^{(E_d - E_f)/kT}}\right), \tag{2.18}$$

since the probability that a donor state is occupied by an electron (i.e., in the neutral state) is $f_D(E_d)$. The factor $\frac{1}{2}$ in the denominator of $f_D(E_d)$ arises from the spin degeneracy (up or down) of the available electronic states associated with an ionized

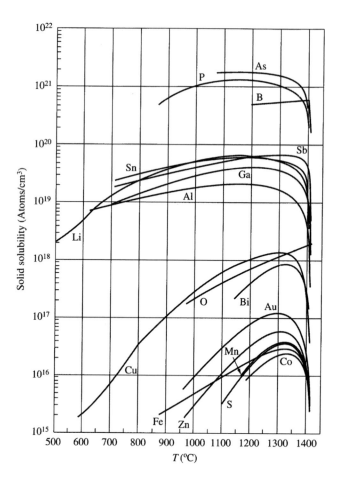

Figure 2.6 Solid solubility of various elements in silicon as a function of temperature (after Trumbore, 1960)

donor level[1] (Ghandhi, 1968). In other words, while a neutral donor atom has only one electron to lose, a positively charged donor atom can recapture an electron in one of its two possible states. Substituting Eqs. (2.9) and (2.11) for n and p in Eq. (2.17) obtains

$$N_c e^{-(E_c-E_f)/kT} = \frac{N_d}{1 + 2e^{-(E_d-E_f)/kT}} + N_v e^{-(E_f-E_v)/kT}, \tag{2.19}$$

which is an algebraic equation that can be solved for E_f. In n-type silicon, electrons are the majority current carriers, while holes are the minority current carriers, which means that the second term on the right-hand side (RHS) of Eq. (2.19) can be neglected. For shallow donor impurities with low-to-moderate concentration at room temperature, $(N_d/N_c) \exp[(E_c - E_d)/kT] \ll 1$, a good approximate solution for E_f is

---

[1] Detailed study showed that there is no other degeneracy with the electronic ground state in a donor except for the spin (Ning and Sah, 1971).

$$E_c - E_f = kT \ln\left(\frac{N_c}{N_d}\right). \tag{2.20}$$

In this case, the Fermi level is at least a few kT below E_d and essentially all the donor levels are empty (ionized), i.e., $n = N_d^+ = N_d$.

It was shown in Eq. (2.16) that, in equilibrium, the product of majority and minority carrier densities equals n_i^2, independent of the dopant type and Fermi level position. The minority hole density in n-type silicon is then given by

$$p = n_i^2/N_d. \tag{2.21}$$

Likewise, for p-type silicon with a shallow acceptor concentration, N_a, the Fermi level is given by

$$E_f - E_v = kT \ln\left(\frac{N_v}{N_a}\right), \tag{2.22}$$

the hole density is $p = N_a^- = N_a$, and the minority electron density is

$$n = n_i^2/N_a. \tag{2.23}$$

Figure 2.7 plots the Fermi level position in the energy gap versus temperature for a wide range of impurity concentration (Grove, 1967). The slight variation of the silicon bandgap with temperature is also incorporated in the figure. It is seen that as the temperature increases, the Fermi level approaches the intrinsic value near midgap. When the intrinsic carrier concentration becomes larger than the doping concentration,

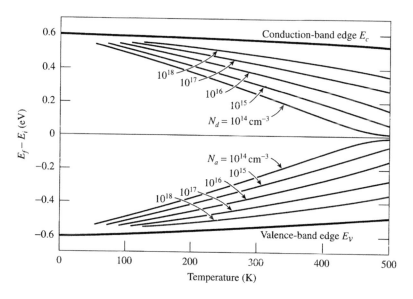

Figure 2.7 The Fermi level in silicon as a function of temperature for various levels of impurity concentration (after Grove, 1967)

the silicon is intrinsic. In an intermediate range of temperature including room temperature, all the donors or acceptors are ionized. The majority carrier concentration is then given by the doping concentration, independent of temperature. For temperatures below this range, *freeze-out* occurs, i.e., the thermal energy is no longer sufficient to ionize all the impurity atoms even with their shallow levels (Sze, 1981). In this case, the majority-carrier concentration is less than the doping concentration, and one would have to solve Eq. (2.19) numerically to find E_f, n, and p (Shockley, 1950).

Instead of using N_c, N_v and referring to E_c and E_v, Eqs. (2.20) and (2.22) can be written in a more useful form in terms of n_i and E_i defined by Eqs. (2.12) and (2.13):

$$E_f - E_i = kT \ln\left(\frac{N_d}{n_i}\right) \tag{2.24}$$

for n-type silicon, and

$$E_i - E_f = kT \ln\left(\frac{N_a}{n_i}\right) \tag{2.25}$$

for p-type silicon. In other words, *the distance between the Fermi level and the intrinsic Fermi level near the midgap is a logarithmic function of doping concentration*. These expressions will be used extensively throughout the book.

2.2.3   Degenerately Doped Silicon

For heavily doped silicon, the impurity concentration N_d or N_a can exceed the effective density of states N_c or N_v, so that $E_f > E_c$ or $E_f < E_v$ according to Eqs. (2.20) and (2.22). In other words, the Fermi level moves into the conduction band for n^+ silicon, and into the valence band for p^+ silicon. Under these circumstances, the silicon is said to be *degenerate*. For degenerately doped silicon, Boltzmann approximation [Eqs. (2.5) and (2.6)] is no longer valid. The full Fermi–Dirac distribution function must be used in Eq. (2.7) for the electron density:

$$n = \frac{8\pi g\sqrt{2m_t^2 m_l}}{h^3} \int_{E_c}^{\infty} \frac{\sqrt{E - E_c}}{1 + e^{(E-E_f)/kT}}\,dE. \tag{2.26}$$

This integral cannot be carried out analytically. In terms of the Fermi–Dirac integral defined by

$$F_{1/2}(u) \equiv \int_0^{\infty} \frac{\sqrt{y}}{1 + e^{y-u}}\,dy, \tag{2.27}$$

Eq. (2.26) takes the form of

$$n = \frac{2}{\sqrt{\pi}} N_c F_{1/2}\left(\frac{E_f - E_c}{kT}\right), \tag{2.28}$$

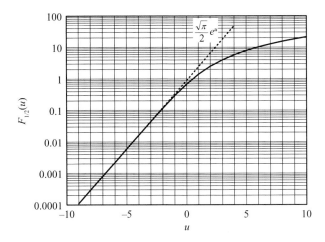

Figure 2.8 Fermi–Dirac integral of the order ½. The dotted line depicts Boltzmann approximation valid for $u \ll -1$

where the effective density of states N_c is given by Eq. (2.10). A numerical plot of the Fermi–Dirac integral is shown in Figure 2.8. Its asymptotic approximations are (Blakemore, 1982)

$$F_{1/2}(u) \begin{cases} \approx \dfrac{\sqrt{\pi}}{2} e^u & \text{for } u \ll -1 \\[2ex] \approx \dfrac{2}{3} u^{3/2} & \text{for } u \gg 1 \end{cases} \tag{2.29}$$

Note that in the non-degenerate limit where $E_c - E_f \gg kT$, Eq. (2.28) is reduced to Eq. (2.9).

Another effect with heavily doped silicon is *bandgap narrowing*. When the impurity concentration is higher than $10^{17}\,\text{cm}^{-3}$, the donor (or acceptor) levels start to broaden into bands. This results in an effective decrease in the ionization energy until finally the impurity band merges with the conduction (or valence) band and the ionization energy becomes zero. Bandgap narrowing has a nonnegligible effect in bipolar devices and is discussed in more detail in Section 9.1.1.2.

2.3 Carrier Transport in Silicon

Carrier transport which gives rise to current flow in silicon is driven by two different mechanisms: (a) the *drift* of carriers, caused by the presence of an electric field, and (b) the *diffusion* of carriers, caused by an electron or hole concentration gradient in silicon. They are discussed in this section.

2.3.1   Drift Current: Mobility

Under thermal equilibrium, electrons possess an average kinetic energy proportional to kT. They move in random directions through the silicon crystal with an average thermal velocity v_{th}. At room temperature, v_{th} is of the order of 10^7 cm/s. In the absence of an electric field, the net velocity of electrons in any particular direction is zero, since the thermal motion is completely random. When an electric field $\mathscr{E}$ is applied, the carriers are accelerated and acquire a drift velocity superimposed upon their random thermal motion. The magnitude of the acceleration is $q\mathscr{E}/m^*$, where m^* is the effective mass of carriers. The drift velocity of carriers, however, does not increase indefinitely under field acceleration because carriers scatter frequently with the lattice (phonons) and ionized impurity atoms. After each collision event, carrier velocities become randomized again and the drift velocity is reset to zero. Consequently, the drift velocity only builds up during the *mean free time* τ between collisions to a value

$$v_d = q\mathscr{E}\tau/m^*. \tag{2.30}$$

τ is related to the *mean free path* l, the average distance carriers travel between collisions, through $\tau = l/v_{th}$. Typically, $l \approx 10$ nm, $\tau \approx 0.1$ ps. This establishes that, at low electric fields, the drift velocity v_d is proportional to the electric field strength $\mathscr{E}$ with a proportionality constant μ, defined as the *mobility*, i.e.,

$$v_d = \mu\mathscr{E}, \tag{2.31}$$

with

$$\mu = \frac{q\tau}{m^*} = \frac{ql}{m^* v_{th}}. \tag{2.32}$$

Electron and hole mobilities in silicon at low impurity concentrations are listed in Table 2.1. The electron mobility is approximately three times the hole mobility, since the effective mass of electrons in the conduction band is much lighter than that of holes in the valence band.

Figure 2.9 plots the electron and hole mobilities at room temperature versus n-type or p-type doping concentration. At low impurity levels, the mobilities are mainly limited by carrier collisions with the silicon lattice or acoustic phonons (Kittel, 1976). As the doping concentration increases beyond $10^{15} - 10^{16}$/cm^3, collisions with the charged (ionized) impurity atoms through Coulomb interaction become more and more important and the mobilities decrease. In general, one can use *Matthiessen's rule* to include different contributions to the mobility:

$$\frac{1}{\mu} = \frac{1}{\mu_L} + \frac{1}{\mu_I} + \cdots, \tag{2.33}$$

where μ_L and μ_I correspond to the lattice- and impurity-scattering-limited components of mobility, respectively. At high temperatures, the mobility tends to be limited by lattice scattering and is proportional to $T^{-3/2}$, relatively insensitive to the doping

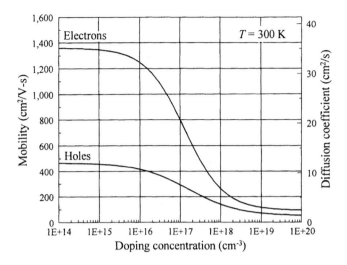

Figure 2.9 Electron and hole mobilities in bulk silicon at 300 K as a function of doping concentration. The right scale is in diffusion coefficients discussed in Section 2.3.4

concentration (Sze, 1981). At low temperatures, the mobility is higher, but is a strong function of doping concentration as it becomes more limited by impurity scattering. What is shown in Figure 2.9 is the *bulk mobility* applicable to conduction in silicon substrates far from the surface. ***In the inversion layer of a MOSFET device, the current flow is governed by the surface mobility, which is much lower than the bulk mobility***. This is mainly due to additional scattering mechanisms between the carriers and the Si − SiO$_2$ interface in the presence of high electric fields normal to the surface. Surface scattering adds another term to Eq. (2.33). Carrier mobility in the surface inversion channel of a MOSFET is discussed in more detail in Section 5.2.

2.3.1.1 Resistivity

For a homogeneous n-type silicon with a mobile electron density n, the drift current density under an electric field $\mathscr{E}$ is

$$J_{n,drift} = qnv_d = qn\mu_n\mathscr{E}, \tag{2.34}$$

where $q = 1.6 \times 10^{-19}$ C is the electronic charge and μ_n is the electron mobility. The resistivity, ρ_n, of n-type silicon defined by $J_{n,drift} = \mathscr{E}/\rho_n$ (a form of Ohm's law) is then

$$\rho_n = \frac{1}{qn\mu_n}. \tag{2.35}$$

Similarly, for p-type silicon,

$$J_{p,drift} = qp\mu_p\mathscr{E} \tag{2.36}$$

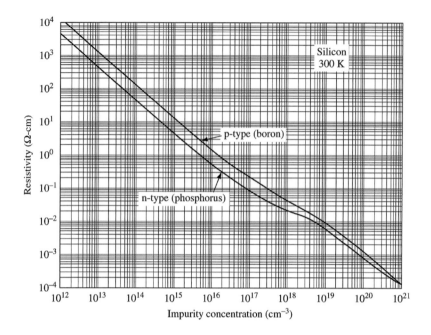

Figure 2.10 Resistivity versus impurity concentration for n-type and p-type silicon at 300 K (after Sze, 1981)

and

$$\rho_p = \frac{1}{qp\mu_p},$$ (2.37)

where μ_p is the hole mobility. In general, the total resistivity includes both the majority and the minority carrier components:

$$\rho = \frac{1}{qn\mu_n + qp\mu_p},$$ (2.38)

since both electrons and holes contribute to electrical conduction. Figure 2.10 shows the measured resistivity of n-type (phosphorus-doped) and p-type (boron-doped) silicon versus impurity concentration at room temperature.

2.3.1.2   Sheet Resistivity

The resistance of a uniform conductor of length L, width W, and thickness t is given by

$$R = \rho \frac{L}{Wt},$$ (2.39)

where ρ is the resistivity in ohm-centimeters. In a planar IC technology, the thickness of conducting regions is often uniform and normally much less than both the length and the width of the region of concern. It is then useful to define a quantity, called the *sheet resistivity*, as

$$\rho_{sh} = \frac{\rho}{t} \tag{2.40}$$

in units of $\Omega/\square$ (ohms per square). Then

$$R = \frac{L}{W}\rho_{sh}, \tag{2.41}$$

i.e., the total resistance is equal to the number of squares ($L/W = 1$ is one square) of the line times the sheet resistivity. Note that sheet resistivity does not depend on the size of the square. The most common technique of measuring the sheet resistivity of a thin film is the four-point method, in which a small current is passed through the two outer probes and the voltage is measured between the two inner probes (Sze, 1981). If the spacing between the probes is much greater than the film thickness but much smaller than the overall size of the conducting film, the resistance measured can be approximated by $V/I = \rho_{sh}(\ln 2)/\pi \approx 0.22\rho_{sh}$, from which ρ_{sh} can be easily determined.

2.3.2   Velocity Saturation

The linear velocity–field relationship discussed in Section 2.3.1 is valid only when the electric field is not too high and the carriers are not too far from thermal equilibrium with the lattice. At high fields, the energy of carriers increases and carriers lose their energy by optical-phonon emission nearly as fast as they gain it from the field. This results in a decrease of the slope of the velocity–field relationship as the field increases until finally the drift velocity reaches a limiting value, $v_{sat} \approx 10^7$ cm/s. This phenomenon is called *velocity saturation*.

Figure 2.11 shows the measured velocity–field relationship of electrons and holes in high-purity bulk silicon at room temperature. At low fields, the drift velocity is proportional to the field (45° slope on a log–log scale) with a proportionality constant given by the electron or the hole mobility. When the field becomes higher than 3×10^3 V/cm for electrons, velocity saturation starts to occur. The saturation velocity of holes is similar to or slightly lower than that of electrons, but saturation for holes takes place at a much higher field because of their lower mobility. For more highly doped material, low-field mobilities are lower because of impurity scattering (Figure 2.9). However, the saturation velocity remains essentially the same, independent of impurity concentration. There is a weak dependence of v_{sat} on temperature. It decreases slightly as the temperature increases (Arora, 1993).

2.3.3   Diffusion Current

Diffusion current in silicon arises when there is a spatial variation of carrier concentration in the material, that is, the carriers tend to move from a region of high

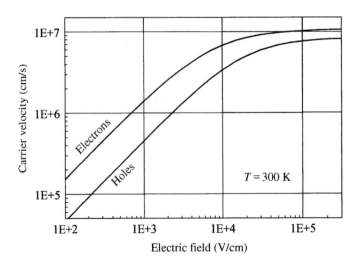

Figure 2.11 Velocity–field relationship of electrons and holes in silicon at 300 K

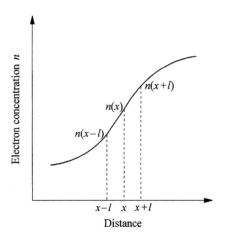

Figure 2.12 Schematic diagram of electron concentration versus distance in a one-dimensional case (after Muller and Kamins, 1977)

concentration to a region of low concentration. To illustrate the diffusion process, let us consider a one-dimensional case shown in Figure 2.12, in which the electron density n varies in the x-direction (Muller and Kamins, 1977). We consider the number of electrons crossing the plane at x per unit area per unit time. The electrons are moving at thermal velocity v_{th} either to the left or to the right and are scattered each time after they travel, on average, a distance equal to the mean free path l. Therefore, the electrons crossing the plane at x from the left start at approximately $x - l$, i.e., one mean free path away on the left side of x. Since electrons have equal chances of moving left or right, half of the electrons at $x - l$ will move across the plane at x before the next collision takes place. The current density per unit area resulting from the motion of those carriers is then

$$J_- = \frac{1}{2} qn(x - l)v_{th}. \tag{2.42}$$

This current is in the negative direction, since it consists of negative charge moving in the positive direction. Similarly, half of the electrons at $x + l$, one mean free path away on the right side of x, will move across the plane x from the right, resulting in a current density in the positive direction:

$$J_+ = \frac{1}{2} qn(x + l)v_{th}. \tag{2.43}$$

The net diffusion current density at x is then

$$J_{diff} = J_+ - J_- = \frac{1}{2} qv_{th}[n(x + l) - n(x - l)]. \tag{2.44}$$

Using a Taylor series expansion of $n(x + l)$ and $n(x - l)$, and keeping only the first-order terms, one obtains

$$J_{diff} = \frac{1}{2} qv_{th}\left[\left(n(x) + l\frac{dn}{dx}\right) - \left(n(x) - l\frac{dn}{dx}\right)\right] = qv_{th}l\frac{dn}{dx}. \tag{2.45}$$

We thus see that the diffusion current density is proportional to the spatial derivative of the electron concentration. In other words, diffusion current results from the random thermal motion of charge carriers in the presence of a concentration gradient. Conventionally, the diffusion currents are expressed as:

$$J_{n,diff} = qD_n\frac{dn}{dx} \tag{2.46}$$

for electrons, and

$$J_{p,diff} = -qD_p\frac{dp}{dx} \tag{2.47}$$

for holes. The proportionality constants D_n and D_p are called the electron and hole *diffusion coefficients* and have units of cm^2/s. There is a negative sign in Eq. (2.47), since diffusion current flows in the direction of decreasing hole (positive charge) concentration.

2.3.4 Einstein Relations

Physically, both drift and diffusion are closely associated with the random thermal motion of carriers and their collisions with silicon lattice or impurity atoms. The implicit assumption here is that they are not too far from the thermal equilibrium. There are one-to-one relationships between the mobility and diffusion coefficient known as *Einstein relations*.

According to Eqs. (2.45) and (2.46), the diffusion coefficient is defined as

$$D \equiv v_{th}l. \tag{2.48}$$

This applies to either electrons or holes with their respective parameters. The ratio of Eq. (2.48) to Eq. (2.32) gives

$$\frac{D}{\mu} = \frac{m^* v_{th}^2}{q}. \tag{2.49}$$

From the theorem for equipartition of energy in a one-dimensional case [see Exercise 2.3(a) for the proof of a three-dimensional case],

$$\frac{1}{2}m^* v_{th}^2 = \frac{1}{2}kT. \tag{2.50}$$

Therefore,

$$\frac{D}{\mu} = \frac{kT}{q} \tag{2.51}$$

for either type of carriers.

Explicitly, Einstein relations for electrons and holes are:

$$D_n = \frac{kT}{q}\mu_n \tag{2.52}$$

and

$$D_p = \frac{kT}{q}\mu_p. \tag{2.53}$$

The values of diffusion coefficient at room temperature can be read from Figure 2.9 using the vertical scale on the right-hand side.

2.4   Basic Equations for Device Operation

2.4.1   Poisson's Equation: Electrostatic Potential

One of the key equations governing the operation of VLSI devices is *Poisson's equation*. It comes from Maxwell's first equation, which in turn is based on Coulomb's law for the electrostatic force associated with a charge distribution. Poisson's equation is expressed in terms of the electrostatic potential, which is defined as the potential energy of a mobile carrier divided by the electronic charge q. The potential energies of carriers are either at the conduction-band edge or at the valence-band edge, as discussed in connection with the energy-band diagram in Figure 2.1. Since only the spatial variation of the electrostatic potential matters, it can be defined with an arbitrary additive constant. It makes no difference whether E_c, E_v, or any other quantity displaced from the band edges by a fixed amount is used to represent the potential. Conventionally, the electrostatic potential is defined in terms of the intrinsic Fermi level,

$$\psi_i = -\frac{E_i}{q}. \tag{2.54}$$

There is a negative sign because E_i is defined as electron energy while ψ_i is defined for a positive charge. The band diagram can thus also be considered as a potential diagram with the potential increasing downward, opposite to the electron energy.

The electric field $\mathscr{E}$, defined as the electrostatic force per unit charge, is equal to the negative gradient of ψ_i,

$$\mathscr{E} = -\frac{d\psi_i}{dx}. \tag{2.55}$$

Now we can write Poisson's equation as

$$\frac{d^2\psi_i}{dx^2} = -\frac{d\mathscr{E}}{dx} = -\frac{\rho_{net}(x)}{\varepsilon_{si}}, \tag{2.56}$$

where $\rho_{net}(x)$ is the net charge density per unit volume at x, and ε_{si} is the permittivity of silicon equal to $11.7\varepsilon_0$. Here $\varepsilon_0 = 8.85 \times 10^{-14}$ F/cm is the vacuum permittivity.

Another form of Poisson's equation is *Gauss's law*, obtained by integrating Eq. (2.56):

$$\mathscr{E}(x_2) - \mathscr{E}(x_1) = \frac{1}{\varepsilon_{si}} \int_{x_1}^{x_2} \rho_{net}(x)\, dx = \frac{Q_s(x_1, x_2)}{\varepsilon_{si}} \tag{2.57}$$

where $Q_s(x_1, x_2)$ is the integrated charge density per unit area within (x_1, x_2).

There are two kinds of charge in silicon: *mobile charge* and *fixed charge*. Mobile charges are electrons and holes, whose densities are represented by n and p. Fixed charges are ionized donor (positively charged) and acceptor (negatively charged) atoms whose densities are represented by N_d^+ and N_a^-, respectively. Equation (2.56) can then be written as

$$\frac{d^2\psi_i}{dx^2} = -\frac{d\mathscr{E}}{dx} = -\frac{q}{\varepsilon_{si}} \left[p(x) - n(x) + N_d^+(x) - N_a^-(x) \right]. \tag{2.58}$$

With no applied field, a uniformly doped n-type or p-type silicon is charge neutral. The RHS of Eq. (2.58) is zero and the potential is constant throughout the sample.

2.4.1.1   Dielectric Boundary Conditions

Equation (2.56) is one dimensional and is for a homogeneous material, silicon. It is adequate for describing most of the basic device operations. In some cases, e.g., in short-channel MOSFETs, two-dimensional Poisson's equation is needed. In addition, in some device regions, there may be two different materials with different dielectric constants, as depicted in Figure 2.13. There are two basic boundary conditions for the components of the electric field at a dielectric interface.

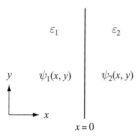

Figure 2.13 Diagram for discussing the boundary conditions of electric field at the interface between two dielectric media

Two-dimensional Poisson's equation takes the following general form:

$$\frac{\partial(\varepsilon\mathscr{E}_x)}{\partial x} + \frac{\partial(\varepsilon\mathscr{E}_y)}{\partial y} = \rho_{net}, \tag{2.59}$$

where $\mathscr{E}_x = -\partial\psi/\partial x$ and $\mathscr{E}_y = -\partial\psi/\partial y$ are the components of the electric field perpendicular and parallel to the dielectric boundary, respectively. Note that in the geometry defined in Figure 2.13, ε is a step function of x only, independent of y. If the potential functions in the two dielectric regions are represented by $\psi_1(x,y)$ and $\psi_2(x,y)$, they must be continuous at the interface, i.e., $\psi_1(0,y) = \psi_2(0,y)$. It then follows that their y-derivatives or **the tangential fields are also continuous**,

$$\mathscr{E}_{1y}(0,y) = \mathscr{E}_{2y}(0,y). \tag{2.60}$$

For a finite volume charge density ρ_{net}, the x-derivative in Eq. (2.59) must be finite at $x = 0$. This means that $\varepsilon\mathscr{E}_x$ must be continuous in the x-direction across the dielectric boundary. Therefore,

$$\varepsilon_1\mathscr{E}_{1x}(0,y) = \varepsilon_2\mathscr{E}_{2x}(0,y), \tag{2.61}$$

i.e., **the perpendicular component of the displacement, $D = \varepsilon\mathscr{E}$, is continuous**. In the presence of a sheet charge at the dielectric interface, ρ_{net} can be considered as a delta function with a multiplier equal to the interface charge density per unit area, Q_{it}. Then the last condition changes to $\varepsilon_2\mathscr{E}_{2x} - \varepsilon_1\mathscr{E}_{1x} = Q_{it}$. Interface trapped charge in an MOS capacitor is discussed in Section 4.5.

2.4.1.2   Carrier Concentration as a Function of Electrostatic Potential

Many of the parameters discussed previously in this chapter can be expressed in terms of the electrostatic potential ψ_i. For example, Eqs. (2.24) and (2.25) can be combined into one equation for both n-type and p-type silicon:

$$\psi_B \equiv |\psi_f - \psi_i| = \frac{kT}{q}\ln\left(\frac{N_b}{n_i}\right), \tag{2.62}$$

where $\psi_f = -E_f/q$ is the Fermi potential and N_b is either the donor or the acceptor concentration. Equation (2.62) is a very useful expression relating the distance of the

Fermi potential from the midgap, ψ_B, to the doping concentration (or the ionized dopant concentration in the case of incomplete ionization). It is based on charge neutrality and is valid only if the local net charge density, mobile plus ionized dopant, or the right-hand side of Eq. (2.58) is zero.

In general, for any Fermi level position in the bandgap, carrier densities are given by Eqs. (2.14) and (2.15), which can be expressed in terms of the electrostatic potential as:

$$n = n_i e^{q(\psi_i - \psi_f)/kT} \tag{2.63}$$

and

$$p = n_i e^{q(\psi_f - \psi_i)/kT}. \tag{2.64}$$

The last two equations are often referred to as *Boltzmann's relations* and are valid for either n-type or p-type silicon in thermal equilibrium. Note that Eqs. (2.63) and (2.64) are derived from the density of states and Fermi level considerations, regardless of charge neutrality. They are generally applicable in the presence of net charge (due to an imbalance between the mobile and the fixed charge densities) and band bending (spatial variation of ψ_i), where $|\psi_f - \psi_i|$ is no longer given by Eq. (2.62).

2.4.1.3 Debye Length

In a silicon region where the doping concentration varies spatially, the bands (E_c, E_v, E_i) are not flat as in the uniformly doped case. Both the intrinsic Fermi level and the bands generally follow the doping variation according to Eq. (2.62). However, if the doping concentration changes abruptly on a very short length scale, the bands do not follow point by point because both ψ_i and its first-order spatial derivative are required to be continuous in Poisson's equation. At a finite temperature, mobile carriers diffuse to smooth out the transition of the bands over the region of abrupt doping change.

Consider an n-type silicon of uniform doping N_d over the region $x < 0$. At $x = 0$, there is a step change of doping to $N_d + \Delta N_d$, extending to the region $x > 0$. For the region immediately to the right of $x = 0$, substituting Eq. (2.63) into Eq. (2.58) yields

$$\frac{d^2 \psi_i}{dx^2} = -\frac{q}{\varepsilon_{si}} \left[N_d + \Delta N_d - n_i e^{q(\psi_i - \psi_f)/kT} \right]. \tag{2.65}$$

To the far left of $x = 0$, $\psi_i(x = -\infty) - \psi_f = (kT/q) \ln(N_d/n_i)$ from Eq. (2.62). In the transition region near $x = 0$, we let $\psi_i(x) = \psi_i(x = -\infty) + \Delta\psi_i(x)$. Without any applied bias or current, the Fermi level is spatially flat, i.e., ψ_f is independent of x, as discussed in Section 2.1.3. By expanding the $\exp(q\Delta\psi_i/kT)$ term on the right-hand side of Eq. (2.65) and keeping only the first-order term (zeroth-order term cancels out with N_d), we obtain

$$\frac{d^2(\Delta\psi_i)}{dx^2} - \frac{q^2 N_d}{\varepsilon_{si} kT} \Delta\psi_i = -\frac{q}{\varepsilon_{si}} \Delta N_d. \tag{2.66}$$

This is a second-order differential equation with the solution $\Delta\psi_i$ in the form of $\exp(-x/L_D)$, where

$$L_D \equiv \sqrt{\frac{\varepsilon_{si}kT}{q^2 N_d}} \qquad (2.67)$$

is called the *Debye length*. Physically, this means that *it takes a distance on the order of L_D for the silicon bands to follow an abrupt spatial change in doping concentration*. A small electric field is set up in this region due to the charge imbalance. The Debye length is usually much smaller than the lateral device dimension. For example, $L_D = 0.04$ μm for $N_d = 10^{16}$ cm^{-3}.

2.4.2   Current–Density Equations

The next set of equations are current–density equations. The total current density is the sum of the drift current density given by Eqs. (2.34) and (2.36) and the diffusion current density given by Eqs. (2.46) and (2.47). In other words,

$$J_n = qn\mu_n \mathscr{E} + qD_n \frac{dn}{dx} \qquad (2.68)$$

for the electron current density, and

$$J_p = qp\mu_p \mathscr{E} - qD_p \frac{dp}{dx} \qquad (2.69)$$

for the hole current density. The total conduction current density is $J = J_n + J_p$.

Using Eq. (2.55) and the Einstein relations, Eqs. (2.52) and (2.53), we can express the current densities as

$$J_n = -qn\mu_n \left(\frac{d\psi_i}{dx} - \frac{kT}{qn} \frac{dn}{dx} \right) \qquad (2.70)$$

and

$$J_p = -qp\mu_p \left(\frac{d\psi_i}{dx} + \frac{kT}{qp} \frac{dp}{dx} \right). \qquad (2.71)$$

If both n and p take on their equilibrium values, Eqs. (2.63) and (2.64) can be substituted into the above to yield:

$$J_n = -qn\mu_n \frac{d\psi_f}{dx} \qquad (2.72)$$

and

$$J_p = -qp\mu_p \frac{d\psi_f}{dx}. \qquad (2.73)$$

The total current is then

$$J = J_n + J_p = -\frac{1}{\rho}\frac{d\psi_f}{dx}, \qquad (2.74)$$

where ρ is the resistivity of silicon given by Eq. (2.38). Equation (2.74) resembles Ohm's law, $J_{drift} = \mathcal{E}/\rho$, discussed in Section 2.3.1.1. With the addition of the diffusion current, **the total current is proportional to the gradient of the Fermi potential** instead of proportional to the electric field, $\mathcal{E} = -d\psi_i/dx$. This reinforces and goes farther than the concept on Fermi level and equilibrium discussed in Section 2.1.3: for a connected system of metals and/or semiconductors in thermal equilibrium with no current flow, the Fermi level is flat, i.e., spatially constant, throughout the system. The key point to keep in mind is that Fermi level difference is the driving force for current flow, much like voltage difference drives currents in a circuit.

Strictly speaking, when current flows, the system is not in equilibrium and the Fermi level is not well defined. The electron distribution function is no longer a function of energy only. It becomes asymmetric in the current flow direction to favor population of the electronic states with a forward momentum. However, if the current is not too large and the net velocity of electron transport is small compared with the thermal velocity, there is only a slight departure from equilibrium. It is then useful to consider a *local* Fermi level, $E_f(x)$ or $\psi_f(x)$, based on the *local* equilibrium state at any given point. In this way, we generalize the Fermi level concept so that the current density equations, Eqs. (2.72) and (2.73), are valid as long as the *local* electron and hole densities are equal to their equilibrium values, Eqs. (2.63) and (2.64).

When an external battery or voltage source is connected to a device, it pumps all the electrons and states at one contact to a higher energy with respect to another contact. By the definition of contacts, both the electron and hole densities equal their equilibrium values at the contacts; therefore, one can define Fermi levels, e.g., E_{f1} and E_{f2}, respectively, for the two contacts being considered. Without any externally applied voltage, $E_{f1} = E_{f2}$ in the steady state. **When a voltage source is connected, $E_{f1} - E_{f2} = qV_{app}$, where V_{app} is the applied voltage** with the lower voltage (higher electron energy) side connected to contact 1. It is this Fermi level differential at the contacts, sustained by an external power source, that drives a steady state current in the device. Fermi level will be used to reference terminal voltage and to establish the alignment relationship of electronic energy bands throughout the book.

2.4.2.1   Quasi-Fermi Potentials

The discussion in this section thus far applies only when both the electron and hole densities take on their local equilibrium values and a local Fermi level can be defined. It is often in VLSI device operation to encounter nonequilibrium situations where the densities of one or both types of carriers depart from their equilibrium values given by Eqs. (2.63) and (2.64). In particular, the minority carrier concentration can be easily overwhelmed by injection from neighboring regions. Because of the slow generation–recombination processes (discussed in Sections 2.4.3 and 2.4.4), it takes a distance scale much larger than VLSI device dimensions to re-establish equilibrium between electrons and holes. Before that happens, while the electrons are in local equilibrium with themselves and so are the holes with themselves, electrons and holes are not in equilibrium with each other. In order to extend the kind of relationship between Fermi level and current densities shown in Eqs. (2.72) and (2.73), one can introduce *separate* Fermi levels for electrons and holes, respectively. They are called *quasi-Fermi levels*, E_{fn} and E_{fp}, defined so as to replace E_f in Eqs. (2.14) and (2.15):

$$n = n_i e^{(E_{fn}-E_i)/kT},\tag{2.75}$$

$$p = n_i e^{(E_i-E_{fp})/kT}.\tag{2.76}$$

In this regard, quasi-Fermi levels have a similar physical interpretation in terms of the state occupancy as the Fermi level. *That is, the electron density in the conduction band can be calculated as if the Fermi level is at E_{fn}, and the hole density in the valence band can be calculated as if the Fermi level is at E_{fp}.* With these definitions, the current densities, Eqs. (2.72) and (2.73), become

$$J_n = -qn\mu_n\frac{d\phi_n}{dx}\tag{2.77}$$

and

$$J_p = -qp\mu_p\frac{d\phi_p}{dx},\tag{2.78}$$

where the *quasi-Fermi potentials* ϕ_n and ϕ_p are defined by

$$\phi_n \equiv -\frac{E_{fn}}{q} = \psi_i - \frac{kT}{q}\ln\left(\frac{n}{n_i}\right)\tag{2.79}$$

and

$$\phi_p \equiv -\frac{E_{fp}}{q} = \psi_i + \frac{kT}{q}\ln\left(\frac{p}{n_i}\right).\tag{2.80}$$

Equations (2.77) and (2.78) are more generally applicable than Eqs. (2.72) and (2.73). They show that electron and hole currents are driven by separate Fermi potentials: *the gradient of electron quasi-Fermi potential drives the electron current, and the gradient of hole quasi-Fermi potential drives the hole current*. When current flows, $E_{fn}(x)$ and $E_{fp}(x)$ [or $\phi_n(x)$ and $\phi_p(x)$] should be interpreted in the same sense of *local* equilibrium at point x as with the case of local $E_f(x)$ discussed with Eqs. (2.72) and (2.73).

When electrons and holes are not in equilibrium with each other, the pn product is given by

$$pn = n_i^2 e^{q(\phi_p-\phi_n)/kT} = n_i^2 e^{(E_{fn}-E_{fp})/kT}.\tag{2.81}$$

It equals n_i^2 when $\phi_p = \phi_n = \psi_f$. Quasi-Fermi potentials are used extensively in the rest of the book for current calculations.

2.4.3   Generation and Recombination

Electron and hole generation and recombination processes play an important role in the operation of many silicon devices and in determining their current–voltage characteristics. Electron and hole generation and recombination can take place directly between the valence band and the conduction band, or indirectly via trap centers in the energy gap. In silicon, the probability of direct band-to-band recombination by a *radiative* (transfer of energy to a photon) or *Auger* (transfer of energy to another carrier) process is very low due to its indirect bandgap. Most of the recombination processes take place indirectly via a trap or a deep impurity level near the middle of the forbidden gap. A model on generation and recombination via trap centers in the bandgap was pioneered by Shockley, Read, and Hall early on (Hall, 1952; Shockley and Read, 1952; Sah *et al.*, 1957).

2.4.3.1   Shockley–Read–Hall (SRH) Theory

Consider a piece of silicon having in it a concentration of N_t trap centers per unit volume. For simplicity, we assume all the trap centers to be identical and located at energy E_t in the bandgap. Also, we assume that each trap center can exist in one of two charge states, namely neutral when it is not occupied by an electron and negatively charged when it is occupied by an electron. Each unoccupied center can capture an electron from the conduction band (electron capture). An electron in an occupied center can be emitted into the conduction band (electron emission). Similarly, each occupied center can capture a hole from the valence band (hole capture), and an unoccupied center can emit a hole into the valence band (hole emission). These four capture and emission processes are illustrated in Figure 2.14. Note that, in hole capture, the center turns from a negatively charged (occupied) state into a neutral (unoccupied) state. Hole capture from the valence band is equivalent to electron emission into the valence band. In hole emission, the center turns from a

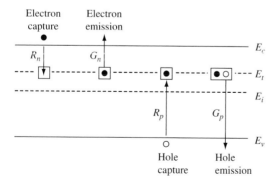

Figure 2.14 Schematic illustrating electron and hole capture and emission processes at a trap center located at an energy level E_t. E_i is the intrinsic Fermi level. The Fermi level E_f, which lies in the upper half of the bandgap for n-type silicon and in the lower half of the bandgap for p-type silicon, is not shown. The physical mechanisms discussed here apply, regardless of the dopant type

neutral (unoccupied) state into a negatively charged (occupied) state. Hole emission is equivalent to electron capture from the valence band.

Assume that out of the total trap density N_t, a density of N_t^- trap centers are occupied by electrons. The density of unoccupied trap centers is then $N_t - N_t^-$. With reference to Figure 2.14,

$$\frac{dN_t^-}{dt} = R_n - G_n - R_p + G_p, \tag{2.82}$$

where R_n is the electron capture rate, G_n is the electron emission rate, R_p is the hole capture rate, and G_p is the hole emission rate. They have the unit of $\text{cm}^{-3}\text{s}^{-1}$. In the SRH theory, R_n is proportional to the electron density n in the conduction band and the density of unoccupied trap centers, $N_t - N_t^-$. The proportional constant, in the unit of cm^3s^{-1}, is conventionally expressed as the product of an electron capture cross-section σ_n and electron thermal velocity v_{th}. In other words,

$$R_n = \sigma_n v_{th} n \left(N_t - N_t^-\right). \tag{2.83}$$

The electron emission rate, G_n, is the density of trapped electrons being emitted to the conduction band per second. Since there are plenty of available states in the conduction band for those electrons to go in, G_n is simply proportional to the density of trap centers occupied by electrons, N_t^-,

$$G_n = \text{constant} \times N_t^-. \tag{2.84}$$

In thermal equilibrium, $N_t^- = \left[N_t^-\right]_0$ is given by the Fermi–Dirac distribution function, i.e.,

$$\left[N_t^-\right]_0 = f_D \times N_t = \frac{N_t}{1 + \exp\left[\left(E_t - E_f\right)/kT\right]}. \tag{2.85}$$

The density of unoccupied trap centers in thermal equilibrium is then $N_t - \left[N_t^-\right]_0 = (1 - f_D)N_t$.

The *principle of detailed balance* requires that, at thermal equilibrium, the capture rate of electrons from the conduction band equals the emission rate of electrons to the conduction band, i.e., $\left[R_n\right]_0 = \left[G_n\right]_0$. This allows the constant in Eq. (2.84) to be expressed in terms of the proportional constant in Eq. (2.83), so that

$$G_n = \frac{\sigma_n v_{th} n_0 \left(N_t - \left[N_t^-\right]_0\right)}{\left[N_t^-\right]_0} \times N_t^- = \frac{\sigma_n v_{th} n_0 (1 - f_D)}{f_D} \times N_t^-. \tag{2.86}$$

Here

$$n_0 = n_i e^{\left(E_f - E_i\right)/kT} \tag{2.87}$$

is the electron density in the conduction band at thermal equilibrium. A parallel treatment for the hole capture and emission rates yields

$$R_p = \sigma_p v_{th} p N_t^-, \tag{2.88}$$

and

$$G_p = \frac{\sigma_p v_{th} p_0 f_D}{1 - f_D} \times \left(N_t - N_t^-\right),$$ (2.89)

with the thermal equilibrium hole density

$$p_0 = n_i e^{\left(E_i - E_f\right)/kT}.$$ (2.90)

In general, when not in thermal equilibrium, $n \neq n_0$ or $p \neq p_0$. There is a net capture or emission of electrons between the traps and the conduction band, as well as a net capture or emission of holes between the traps and the valence band. In the steady state, the density of trap centers occupied by electrons, N_t^-, does not change with time. In view of Eq. (2.82), this means

$$R_n - G_n = R_p - G_p.$$ (2.91)

Substituting Eqs. (2.83), (2.86), (2.88), and (2.89) into the above condition allows N_t^-/N_t to be solved in terms of n and p:

$$\frac{N_t^-}{N_t} = \frac{\sigma_n n + \sigma_p p_0 \dfrac{f_D}{1 - f_D}}{\sigma_n\left(n + n_0 \dfrac{1 - f_D}{f_D}\right) + \sigma_p\left(p + p_0 \dfrac{f_D}{1 - f_D}\right)}$$

$$= \frac{\sigma_n n + \sigma_p n_i \exp\left[(E_i - E_t)/kT\right]}{\sigma_n\{n + n_i \exp\left[(E_t - E_i)/kT\right]\} + \sigma_p\{p + n_i \exp\left[(E_i - E_t)/kT\right]\}}.$$ (2.92)

The last expression made use of $(1 - f_D)/f_D = \exp\left[(E_t - E_f)/kT\right]$ and Eqs. (2.87) and (2.90) for n_0 and p_0. Equation (2.91) means that the net capture rate of electrons from the conduction band equals the net capture rate of holes from the valence band. This rate represents the net recombination rate U of electrons in the conduction band with holes in the valence band via trap centers of density N_t at energy E_t. Using Eq. (2.92) for N_t^-/N_t, we have

$$U \equiv R_n - G_n = R_p - G_p = \frac{\sigma_n \sigma_p v_{th} N_t \left(np - n_i^2\right)}{\sigma_n\{n + n_i \exp[(E_t - E_i)/kT]\} + \sigma_p\{p + n_i \exp[(E_i - E_t)/kT]\}}.$$ (2.93)

It is no surprise to find that the sign of U is the same as that of $np - n_i^2$. At thermal equilibrium, the generation rate is equal to the recombination rate and $np = n_i^2$. When not in thermal equilibrium, there is net recombination if $np > n_i^2$, and net generation if $np < n_i^2$.

2.4.3.2 Midgap Trap Centers

Equation (2.93) shows that the net recombination rate is a function of the electron and hole densities, the electron and hole capture cross-sections, and the energy of the trap center, i.e., $U = U(n, p, \sigma_n, \sigma_p, E_t)$. If we assume the electron and hole capture cross-sections to be independent of E_t, then we find from $\partial U/\partial E_t = 0$ that the recombination rate is maximum when $E_t = E_{t, \max}$ where

$$E_{t,\max} = E_i + \frac{kT}{2} \ln\left(\frac{\sigma_p}{\sigma_n}\right). \tag{2.94}$$

Even for a $|\sigma_n/\sigma_p|$ ratio as large as 1,000, $|E_{t,\max} - E_i|$ is less than 3.5 kT. That is, **effective net recombination centers are midgap states, or states with energy levels close to E_i.** States with energy levels close to either the conduction band or the valence band do not have comparable electron and hole capture rates to function efficiently as net recombination centers.

Since only traps with energy close to mid gap are efficient net recombination centers, **it is physically reasonable and convenient to simply assume $E_t = E_i$ and characterize generation and recombination centers by only their capture cross-sections.** Thus Eq. (2.93) becomes

$$U(n, p, \sigma_n, \sigma_p, E_t = E_i) = \frac{\sigma_n \sigma_p v_{th}(np - n_i^2)N_t}{\sigma_n(n + n_i) + \sigma_p(p + n_i)}. \tag{2.95}$$

2.4.3.3 Minority Carrier Lifetime

When excess minority carriers are generated by light or other means, the recombination rate exceeds the generation rate, with a tendency to return to equilibrium. Consider an n-type silicon region where the majority carrier density $n \gg p$. The denominator in Eq. (2.95) is dominated by the $\sigma_n n$ term. For small perturbations, $n - n_0 \ll n_0$ and $n_i^2/n \approx n_i^2/n_0 = p_0$, Eq. (2.95) is reduced to

$$U(\text{n-region}) \approx \sigma_p v_{th} N_t (p - n_i^2/n) \approx \sigma_p v_{th} N_t (p - p_0). \tag{2.96}$$

This can be expressed in the form of

$$U(\text{n-region}) \equiv \frac{p - p_0}{\tau_p}, \tag{2.97}$$

with a minority hole lifetime defined as

$$\tau_p = \frac{1}{\sigma_p v_{th} N_t}. \tag{2.98}$$

Similarly, for p-type silicon, the minority electron lifetime τ_n is given by

$$\tau_n = \frac{1}{\sigma_n v_{th} N_t}. \tag{2.99}$$

2.4.4 Current Continuity Equations

The next set of equations are *continuity equations* based on the conservation of mobile charge:

$$\frac{\partial n}{\partial t} = \frac{1}{q}\frac{\partial J_n}{\partial x} - R_n + G_n \tag{2.100}$$

and

$$\frac{\partial p}{\partial t} = -\frac{1}{q}\frac{\partial J_p}{\partial x} - R_p + G_p, \tag{2.101}$$

where G_n and G_p are the electron and hole generation rates, R_n and R_p are the electron and hole recombination rates. $\partial J_n/\partial x$ and $\partial J_p/\partial x$ represent the net flux of mobile charges in or out of x. For example, $\partial J_p/\partial x > 0$ means that there are more holes flowing out of x at $x+$ than there are holes flowing into x at $x-$.

In the steady state, $\partial n/\partial t = \partial p/\partial t = 0$. Also, the net electron reduction rate must equal the net hole reduction rate, i.e., $R_n - G_n = R_p - G_p$, so that there is no buildup of net trapped charge with time at any point [see Eq. (2.82)]. Subtracting Eq. (2.101) from Eq. (2.100) then yields $\partial(J_n + J_p)/\partial x = 0$, or continuity of the total current, $J_n + J_p$. In a device region where generation and recombination are negligible, the continuity equations in the steady state are reduced to $dJ_n/dx = dJ_p/dx = 0$, which simply states the continuity of electron current and continuity of hole current, respectively.

2.4.4.1 Dielectric Relaxation Time

The majority-carrier response time, or the *dielectric relaxation time*, is very short in a semiconductor. It can be estimated for a one-dimensional (1-D) homogeneous n-type silicon as follows. Suppose there is a local perturbation in the majority carrier density, Δn. From Poisson's equation, the resulting charge imbalance sets up a field divergence, $\partial \mathscr{E}/\partial x = -q\Delta n/\varepsilon_{si}$, around the point of perturbation. This, in turn, leads to a divergent current according to Ohm's law, $J_n = \mathscr{E}/\rho_n$, which tends to restore the majority carrier concentration back to its equilibrium, charge neutral value. Neglecting R_n and G_n in the continuity equation, Eq. (2.100), we then obtain

$$\frac{\partial \Delta n}{\partial t} = \frac{1}{q}\frac{\partial J_n}{\partial x} = \frac{1}{q\rho_n}\frac{\partial \mathscr{E}}{\partial x} = -\frac{\Delta n}{\rho_n \varepsilon_{si}}. \tag{2.102}$$

The solution to this equation takes the form of $\Delta n(t) \propto \exp(-t/\rho_n \varepsilon_{si})$, where $\rho_n \varepsilon_{si}$ is the *dielectric relaxation time*. **The majority-carrier response time in silicon is typically of the order of 10^{-12} s, which is shorter than most device switching times**.

Note that $\rho_n \varepsilon_{si}$ is the minimum response time for an ideal 1-D case without any parasitic capacitances. In practice, the majority-carrier response time may be limited by the RC delay of the specific silicon device structure and contacts.

2.4.4.2 Minority Carrier Diffusion Length

In contrast to the majority-carrier response time discussed in Section 2.4.4.1, the minority-carrier lifetime is very long. This is because minority carrier density can deviate by several orders of magnitude above or below its thermal equilibrium value without upsetting charge neutrality. The only mechanism to restore minority carrier density to its thermal equilibrium value is through the generation–recombination process which is rather slow.

Consider a p-type silicon in which charge neutrality is largely established by the equality between the majority carrier density and the acceptor doping concentration, i.e., $p_0 = N_a$. The equilibrium minority carrier density, $n_0 = n_i^2/p_0$, is much smaller than p_0 by many orders of magnitude. Suppose at the left end of the sample, an excess minority carrier density is generated and maintained by light or other means. While the out-of-equilibrium minority carrier density n can be much greater than n_0, it is still much smaller than p_0 under low injection conditions. As the electrons diffuse to the right, there is net recombination to restore n to its equilibrium value n_0. In Eq. (2.100) in the steady state, $J_n = qD_n dn/dx$ consists of only the diffusion current since there is no field in the silicon. The electron version of Eq. (2.97) gives $R_n - G_n = (n - n_0)/\tau_n$. We then obtain

$$\frac{d^2 n}{dx^2} - \frac{n - n_0}{D_n \tau_n} = 0. \tag{2.103}$$

The solution $n(x) - n_0$ is of the form $\sim \exp(-x/L_n)$ where

$$L_n \equiv \sqrt{D_n \tau_n} = \sqrt{(kT/q)\mu_n \tau_n} \tag{2.104}$$

is the *minority carrier diffusion length* for electrons in p-type silicon. The last step made use of the Einstein relation, Eq. (2.52). L_n can be interpreted as the average distance excess minority carriers travel before they recombine with majority carriers.

Experimentally, τ_n or τ_p is in the range of 10^{-4} to 10^{-9} s, depending on the quality of the silicon crystal (see Figure 3.14). The diffusion length is typically a few microns to a few millimeters in silicon. ***Since L is much longer than the active dimensions of a VLSI device, generation–recombination in general plays very little role in device operation***. Only in a few special circumstances, such as CMOS latch-up, the floating-body effect in SOI, junction leakage current, and radiation-induced soft error, must the generation–recombination mechanism be taken into account.

Exercises

2.1   Show that the values of the Fermi–Dirac distribution function, Eq. (2.4), at a pair of energies symmetric about the Fermi energy E_f, are complementary, i.e., show that $f_D(E_f - \Delta E) + f_D(E_f + \Delta E) = 1$, independent of temperature.

2.2   For a given donor level E_d and concentration N_d of an n-type silicon, solve the Fermi energy E_f from the charge neutrality condition, Eq. (2.19) (neglecting the hole term). Show that $E_c - E_f$ approaches the complete ionization value, Eq. (2.20), under the condition of shallow donor level with low-to-moderate concentration. What happens if the condition is not satisfied?

2.3   Use the density of states $N(E)$ derived in Section 2.1.2 to evaluate the average kinetic energy of electrons in the conduction band:

$$\langle \text{K.E.} \rangle = \frac{\int_{E_c}^{\infty} (E - E_c) N(E) f_D(E) dE}{\int_{E_c}^{\infty} N(E) f_D(E) dE}.$$

(a) For a nondegenerate semiconductor in which $f_D(E)$ can be approximated by the Maxwell–Boltzmann distribution, Eq. (2.5), show that $\langle \text{K.E.} \rangle = \frac{3}{2} kT$.

(b) For a degenerate semiconductor at 0 K, show that $\langle \text{K.E.} \rangle = \frac{3}{5} (E_f - E_c)$.

2.4 The 3-D Gauss's law is obtained after a volume integration of the 3-D Poisson's equation and takes the form

$$\oiint_{S} \mathcal{E} \cdot dS = \frac{Q}{\varepsilon_{si}},$$

where the left-hand side is an integral of the normal electric field over a closed surface S, and Q is the net charge enclosed within S. Use it to derive the electric field at a distance r from a point charge Q (Coulomb's law). What is the electric potential in this case?

2.5 (a) Use Gauss's law to show that the electric field at a point above a uniformly charged sheet of charge density Q_s per unit area is $Q_s/2\varepsilon$, where ε is the permittivity of the medium.

(b) For two oppositely charged parallel plates with surface charge densities Q_s and $-Q_s$, show that the electric field is uniform and equals Q_s/ε in the region between the two plates and is zero in the regions outside the two plates.

2.6 Work out the details of the Debye length example in Section 2.4.1.3. Solve $\psi_i(x)$ for the entire region from two differential equations, one for $x < 0$ and one for $x > 0$. [Hint: match ψ_i and its derivatives at $x = 0$.]

2.7 In Figure 2.2, the plot to the right shows that the electron density per energy in the conduction band peaks at a small energy δE above E_c. Find δE.

2.8 Figure Ex. 2.8 shows the band diagram of a uniformly doped silicon region.
(a) What are the doping type and doping concentration of this region?
(b) What are the electron and hole densities at point A and point B?
(c) What are the net charge densities (fixed + mobile) at point A and point B?

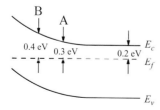

Figure Ex. 2.8

2.9   Consider an n-type silicon with a donor concentration of $10^{15} \, \text{cm}^{-3}$.
(a)   Find the Fermi level position with respect to the conduction band energy E_c.
(b)   Find the Debye length.
(c)   Find the dielectric relaxation time.
(d)   If temperature rises to $100°C$, how does the Fermi level change with respect to E_c? (Ignore the temperature effect on bandgap.)

2.10   At one point of a 1-D silicon device under bias, the doping is $10^{16} \, \text{cm}^{-3}$ n-type, and the Fermi level is 0.2 eV above the intrinsic Fermi level.
(a)   Is it charge neutral at this point?
(b)   What are the electron and hole densities at this point?
(c)   Which one of the diagrams in Figure Ex. 2.10 likely depicts the band diagram in the vicinity of this point?

Figure Ex. 2.10

3   p–n Junctions and Metal–Silicon Contacts

For a semiconductor region to carry electric current, it must be contacted by or connected to a conducting material. The contact between a semiconductor region and a conducting material may be *ohmic* with linear *I–V* behavior for the current passing through the contact, or *rectifying* with current flowing in one direction (forward direction) more readily than in the other direction (reverse direction). Rectifying devices are often referred to as *diodes*. A diode formed with a p-type semiconductor region in contact with an n-type semiconductor region is called a *p–n junction* or a *p–n diode*. A diode formed with a metal contacting a semiconductor region is called a *Schottky diode* or *Schottky barrier diode*. p–n junctions are important devices as well as important components of all MOSFET and bipolar devices. As an active device, p–n diodes are relatively slow. Schottky diodes have much faster response time than p–n diodes. They are often used in microwave circuits. In this chapter, we discuss the basic physics and operation of p–n junctions and Schottky diodes as well as metal–silicon contacts in general.

3.1   p–n Junctions

A p–n diode is formed when one region of a semiconductor substrate is doped n-type and an immediately adjacent region is doped p-type. In practice, a silicon p–n diode is usually formed by counterdoping a local region of a larger region of doped silicon. For instance, a region of a p-type silicon substrate or "well" can be counterdoped with n-type impurities to form the n-type region of a p–n diode. The n-type region thus formed has a donor concentration higher than its acceptor concentration.

A doped semiconductor region is called *compensated* if it contains both donor and acceptor impurities such that neither impurity concentration is negligible compared to the other. For a compensated semiconductor region, it is the *net* doping concentration, i.e., $N_d - N_a$ if it is n-type and $N_a - N_d$ if it is p-type, that determines its Fermi level and its mobile carrier concentration. However, for simplicity, we shall derive the characteristics and behavior of p–n diodes assuming none of the doped regions are compensated, i.e., the n-sides of the diodes have a net donor concentration of N_d and the p-sides have a net acceptor concentration of N_a. The resultant equations can be extended to diodes with compensated doped regions simply by replacing N_d by $N_d - N_a$ for the n-regions and replacing N_a by $N_a - N_d$ for the p-regions.

3.1.1   Energy-Band Diagrams and Built-in Potential for a p—n Diode

It was shown in Section 2.1.3 that Fermi level is spatially constant at thermal equilibrium. Furthermore, as shown in Section 2.4.2, when an external voltage V_{app} is connected to two contacts of a piece of silicon, the Fermi level at the lower voltage contact is shifted relative to the Fermi level at the higher voltage contact by qV_{app}. In this section, we apply these results to establish the energy-band diagrams for a p—n diode.

Consider a p-silicon region and an n-silicon region physically separate from each other. As discussed in Section 2.2.2, the Fermi level for a p-type silicon lies close to its valence band, and that for an n-type silicon lies close to its conduction band. The energy-band diagrams for the two silicon pieces are illustrated schematically in Figure 3.1(a).

If the p-silicon and the n-silicon are brought together to form a p—n diode, the resulting energy-band diagram is as shown in Figure 3.1(b). At thermal equilibrium, the Fermi level must remain flat across the entire p—n diode structure, causing the energy bands of the p-region to lie higher than those of the n-region. Near the physical junction, the energy-bands are bent in order to maintain energy-band continuity between the p-region and the n-region. The band bending implies an electric field, $\mathscr{E} = -d\psi_i/dx$, in this transition region. This electric field causes a drift component of electron and hole currents to flow. At thermal equilibrium, this drift-current component is exactly balanced by a diffusion component of electron and hole currents flowing in the opposite direction

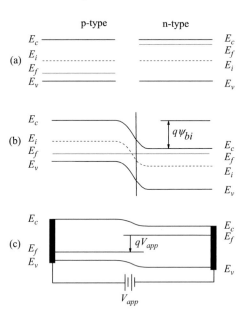

Figure 3.1 Energy-band diagrams for a p—n diode. (a) A uniformly doped p-silicon region and a uniformly doped n-silicon region physically separate from each other. (b) A p—n junction at thermal equilibrium. The vertical solid line indicates the metallurgical junction. (c) A p—n diode connected to a battery, with the n-side connected to the negative end and the p-side connected to the positive end of the battery. The solid vertical bars represent the ohmic contacts of the p- and n-regions. For simplicity and clarity of the figure, E_i is not shown.

caused by the large electron and hole concentration gradients across the junction. The net result is zero electron and hole currents in the p–n junction at thermal equilibrium.

If a battery of voltage V_{app} is connected to the diode, with the p-side connected to the positive end of the battery and the n-side connected to the negative end of the battery, the Fermi level at the n-side contact becomes shifted by qV_{app} relative to the Fermi level at the p-side contact. This is illustrated in Figure 3.1(c).

To facilitate description of both the n-side and the p-side of a diode simultaneously, when necessary for clarity, we shall distinguish the parameters on the n-side from the corresponding ones on the p-side by adding a subscript n to the symbols associated with the parameters on the n-side, and a subscript p to the symbols associated with the parameters on the p-side (Shockley, 1950). For example, n_n and p_n denote the electron concentration and hole concentration, respectively, on the n-side, and n_p and p_p denote the electron concentration and hole concentration, respectively, on the p-side. Thus, n_n and p_p signify majority-carrier concentrations, while n_p and p_n signify minority-carrier concentrations.

Consider the n-side of a p–n diode at thermal equilibrium. If the n-side is non-degenerately doped to a concentration of N_d, Eq. (2.24) gives

$$E_f - E_i = kT \ln\left(\frac{N_d}{n_i}\right) = kT \ln\left(\frac{n_{n0}}{n_i}\right), \tag{3.1}$$

where n_{n0} denotes the n-side electron concentration at thermal equilibrium. Similarly, for the nondegenerately doped p-side of a p–n diode at thermal equilibrium, with a doping concentration of N_a, Eq. (2.25) gives

$$E_i - E_f = kT \ln\left(\frac{N_a}{n_i}\right) = kT \ln\left(\frac{p_{p0}}{n_i}\right), \tag{3.2}$$

where p_{p0} is the p-side hole concentration at thermal equilibrium.

The *built-in potential* across the p–n diode is

$$q\psi_{bi} = E_{i,p\text{-}side} - E_{i,n\text{-}side} = kT \ln\left(\frac{n_{n0}p_{p0}}{n_i^2}\right), \tag{3.3}$$

where $E_{i,p\text{-}side}$ and $E_{i,n\text{-}side}$ represent the intrinsic Fermi level of the p-side and the n-side, respectively. Since $n_{n0}p_{n0} = n_{p0}p_{p0} = n_i^2$, Eq. (3.3) can also be written as

$$q\psi_{bi} = kT \ln\left(\frac{p_{p0}}{p_{n0}}\right) = kT \ln\left(\frac{n_{n0}}{n_{p0}}\right), \tag{3.4}$$

which relates the built-in potential to the electron and hole densities on the two sides of the p–n diode.

3.1.2   Depletion Approximation

The commonly used device model for a p–n diode is to divide the device into three distinct regions: a band-bending region sandwiched between a charge-neutral p-region and a charge-neutral n-region (Shockley, 1950). Since the p- and n-regions are assumed to be charge neutral, their energy bands are flat if they are uniformly doped, as indicated in Figure 3.1.

The spatial dependence of the electrostatic potential $\psi_i(x)$ is governed by Poisson's equation, i.e., Eq. (2.58). As suggested in Figure 3.1(b), $\psi_i(x)$ is independent of x in the uniformly doped quasineutral regions. Within the band-bending region, $\psi_i(x)$ changes from being $-E_{i,p\text{-side}}/q$ at the p-region end of the band-bending region to being $-E_{i,n\text{-side}}/q$ at the n-region end of the band-bending region. Within the band-bending region, Eq. (2.63) suggests that the electron density drops very rapidly as $\psi_i(x)$ changes, being equal to the ionized donor density at the n-region end and dropping $10\times$ at room temperature for every 60-mV change in $\psi_i(x)$. Thus, the density of electrons within the band-bending region is negligible compared to the density of ionized donors except for a very narrow region adjacent to the quasineutral n-region where $q(\psi_i - \psi_f)$ is less than about $3\,kT$. Similarly, Eq. (2.64) suggests that, within the band-bending region, the density of holes is negligible compared to the density of ionized acceptors, except for a very narrow region adjacent to the quasineutral p-region.

A closed-form solution to Poisson's equation can be obtained if the electron and hole densities are assumed to be negligible in the entire band-bending region. This is called the *depletion approximation*. In this case, the abrupt junction is approximated by three well-defined regions, as illustrated in Figure 3.2(a), with the region depleted of mobile carriers extending from $x = -x_p$ to $x = x_n$. Both the p-region, i.e., the region with $x < -x_p$, and the n-region, i.e., the region with $x > x_n$, are assumed to be charge-neutral. The transition region is often referred to as the *depletion region* or *depletion layer*. Since the transition region is not charge-neutral, it is also referred to as the *space-charge region* or *space-charge layer*. The p- and n-regions adjacent to the space-charge region are referred to as *quasineutral regions*.

With no mobile carriers, Poisson's equation, i.e., Eq. (2.58), for the depletion region is

$$-\frac{d^2\psi_i}{dx^2} = \frac{d\mathscr{E}}{dx} = \frac{q}{\varepsilon_{si}}\left[N_d^+(x) - N_a^-(x)\right]. \tag{3.5}$$

For simplicity, we shall assume that all the donors and acceptors within the depletion region are ionized, and that the junction is abrupt and not compensated, i.e., there are no donor impurities on the p-side and no acceptor impurities on the n-side. With these assumptions, Eq. (3.5) becomes

$$-\frac{d^2\psi_i}{dx^2} = \frac{qN_d}{\varepsilon_{si}} \quad \text{for} \quad 0 \le x \le x_n \tag{3.6}$$

and

$$-\frac{d^2\psi_i}{dx^2} = -\frac{qN_a}{\varepsilon_{si}} \quad \text{for} \quad -x_p \le x \le 0. \tag{3.7}$$

Integrating Eq. (3.6) once from $x = 0$ to $x = x_n$, and Eq. (3.7) once from $x = -x_p$ to $x = 0$, subject to the boundary conditions of $d\psi_i/dx = 0$ at $x = -x_p$ and at $x = x_n$, we obtain the maximum electric field, $\mathscr{E}_m$, which is located at $x = 0$. That is,

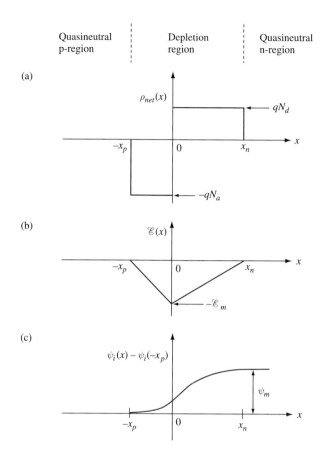

Figure 3.2 Depletion approximation of a p–n junction: (a) charge distribution, (b) electric field, and (c) electrostatic potential

$$\mathscr{E}_m \equiv \left|\frac{-d\psi_i}{dx}\right|_{x=0} = \frac{qN_d x_n}{\varepsilon_{si}} = \frac{qN_a x_p}{\varepsilon_{si}}. \tag{3.8}$$

The total space-charge inside the n-side of the depletion region is equal (but opposite in sign) to the total space-charge inside the p-side of the depletion region. Thus, in Figure 3.2(a), the two charge distribution plots have the same area. Equation (3.8) could have been obtained directly from Gauss's law, i.e., Eq. (2.57).

Let ψ_m be the total potential drop across the p–n junction, i.e., $\psi_m = [\psi_i(x_n) - \psi_i(-x_p)]$. ψ_m can be obtained by integrating Eqs. (3.6) and (3.7) twice. That is,

$$\psi_m = \int_{-x_p}^{x_n} d\psi_i(x) = -\int_{-x_p}^{x_n} \mathscr{E}(x)dx$$

$$= \frac{\mathscr{E}_m(x_n + x_p)}{2} = \frac{\mathscr{E}_m W_d}{2}, \tag{3.9}$$

where $W_d = x_n + x_p$ is the total width of the depletion layer. Equation (3.9) shows that ψ_m is equal to the area in the $\mathscr{E}(x) - x$ plot, i.e., Figure 3.2(b). Eliminating $\mathscr{E}_m$ from Eqs. (3.8) and (3.9) gives

$$W_d = \sqrt{\frac{2\varepsilon_{si}(N_a + N_d)\psi_m}{qN_aN_d}}. \tag{3.10}$$

3.1.2.1 Externally Biased Junctions

At thermal equilibrium, ψ_m is equal to the built-in potential ψ_{bi}, as indicated in Figure 3.1(b). An externally applied voltage across a p–n diode shifts the Fermi level at the n-region contact relative to the Fermi level at the p-region contact. If the applied voltage causes ψ_m to be reduced, the diode is said to be *forward biased*. If the applied voltage causes ψ_m to be increased, the diode is said to be *reverse biased*. In considering a p–n diode in the context of VLSI devices, the forward-bias characteristics are more interesting than the reverse-bias characteristics. Therefore, we shall adopt the convention where a positive applied voltage also means a forward-bias voltage. Physically, this means the external voltage is connected such that the p-side is biased positively relative to the n-side, as in the case illustrated in Figure 3.1(c). The total potential drop ψ_m and the externally applied voltage V_{app} are related by

$$\psi_m = \psi_{bi} - V_{app}, \tag{3.11}$$

where $V_{app} > 0$ means the diode is forward biased and $V_{app} < 0$ means the diode is reverse biased.

The depletion-approximation model can be adapted to situations where the p–n junction is externally biased. If the density of mobile carriers within the space-charge region is small compared to the ionized impurity concentrations, the mobile carriers can simply be ignored and the depletion-approximation results, i.e., Eqs. (3.6) to (3.10), apply. When the density of mobile carriers within the space-charge region is not negligible compared to the ionized impurity concentrations, the mobile carriers should be included in Poisson's equation. This is indeed often done in modeling bipolar transistors (see Chapters 9, 10, and 11).

A quasineutral region has a finite resistivity determined by its dopant impurity concentration (see Figure 2.10). When current flows in a region of finite resistivity, there is a corresponding voltage drop, or *IR* drop, along the current path. In writing Eq. (3.11), the *IR* drops in the quasineutral regions are assumed to be negligible so that V_{app} is the same as the *junction voltage* or *voltage across the space-charge region*, V'_{app}. **If IR drops in the quasineutral regions are not negligible, then V_{app} should be replaced by V'_{app} in Eq. (3.11).**

- p–n *diode as a rectifier.* When a diode is forward biased, the energy barrier limiting current flow is lowered, causing electrons to be injected from the n-side into the p-side and holes injected from the p-side into the n-side, resulting in a current flow

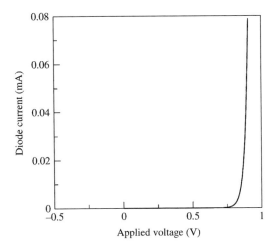

Figure 3.3 A schematic linear plot of the current of a typical silicon diode as a function of its applied voltage. On a linear plot, the reverse current is too low to be observable.

through the diode. This forward-bias current increases exponentially with V'_{app} and hence can be very large (see Section 3.1.5). When a diode is reverse biased, the energy barrier limiting current flow is increased. There is no current flow due to electron and hole injection, only a relatively low background or leakage current. Thus, a diode has rectifying current–voltage characteristics, being conducting when it is forward biased, and nonconducting when it is reverse biased. This is illustrated in Figure 3.3. The equations governing the current–voltage characteristics will be derived in Section 3.1.5.

- *Depletion-layer capacitance.* Consider a small change dV_{app} in the applied voltage. dV_{app} causes a charge per unit area dQ to flow into the p-side, which is equal to the change in the charge in the p-side depletion region. Since all mobile carriers are ignored in our depletion approximation, we can write the charge per unit area in the p-side depletion region as

$$Q_d(\text{p-side}) = -qN_ax_p\left(V_{app}\right),\tag{3.12}$$

- where we have indicated that the p-side depletion-layer width, x_p, is a function of V_{app}. Notice that Q_d for the p-side is negative because ionized acceptors have a charge $-q$. The depletion-layer capacitance per unit area is

$$C_d \equiv \frac{dQ}{dV_{app}} = \frac{dQ_d(\text{p-side})}{dV_{app}} = \frac{\varepsilon_{si}}{W_d}.\tag{3.13}$$

- That is, the depletion-layer capacitance of a diode is equivalent to a parallel-plate capacitor of separation W_d and dielectric constant ε_{si}. Physically, this is due to the fact that only the majority carriers at the edges of the depletion layer, not the space-charge within the depletion region, respond to changes in the applied voltage.

3.1.2.2   One-Sided Junctions

In many applications, such as the source or drain junction of a MOSFET or the emitter–base diode of a bipolar transistor, one side of the p–n diode is degenerately doped while the other side is lightly-to-moderately doped. In this case, practically all the voltage drop and depletion layer occur across the lightly doped side of the diode. That this is the case can be inferred readily from Eq. (3.8), which implies that $x_n = N_a W_d / (N_a + N_d)$ and $x_p = N_d W_d / (N_a + N_d)$. The characteristics of a one-sided p–n diode are therefore determined primarily by the properties of the lightly doped side alone. In this sub-subsection, we shall derive the equations for an n^+–p diode where the characteristics are determined by the p-side. The results can be extended straightforwardly to a p^+–n diode.

It is a good approximation to assume the Fermi level in a degenerately n-doped silicon to be at its conduction-band edge (see Section 2.2.3). That is, $E_f = E_c$ on the degenerately dope n-side at thermal equilibrium. Therefore, the built-in potential for an n^+–p diode, from Eqs. (3.2) and (3.3), is given by

$$q\psi_{bi} = E_f - E_{i,n\text{-side}} + kT \ln\left(\frac{N_a}{n_i}\right)$$

$$= (E_c - E_i)_{n\text{-side}} + kT \ln\left(\frac{N_a}{n_i}\right) \tag{3.14}$$

$$\approx \frac{E_g}{2} + kT \ln\left(\frac{N_a}{n_i}\right),$$

where we have made a further approximation that the intrinsic Fermi level is located half way between the conduction- and valence-band edges (see Section 2.1.4). Figure 3.4 is a plot of ψ_{bi}, as approximated by Eq. (3.14), as a function of the doping concentration of the lightly doped side.

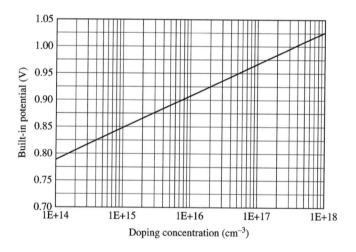

Figure 3.4 Built-in potential for a one-sided p–n junction versus the doping concentration of the lightly doped side

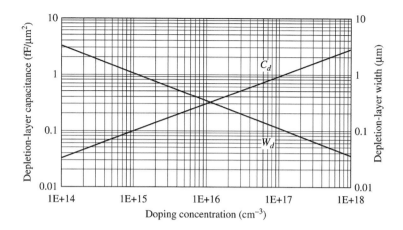

Figure 3.5 Depletion-layer width and depletion-layer capacitance, at zero bias, as a function of doping concentration of the lightly doped side of a one-sided p–n junction

The depletion-layer width, from Eqs. (3.10) and (3.11), is

$$W_d = \sqrt{\frac{2\varepsilon_{si}\left(\psi_{bi} - V_{app}\right)}{qN_a}},\qquad(3.15)$$

where $V_{app} > 0$ if the diode is forward biased and $V_{app} < 0$ if the diode is reverse biased. The depletion-layer capacitance per unit area is given by Eq. (3.13). Figure 3.5 is a plot of the depletion-layer width and capacitance as a function of doping concentration for $V_{app} = 0$. Again, V_{app} in Eq. (3.15) should be replaced by V'_{app} whenever IR drops in the quasineutral regions are not negligible.

3.1.2.3   Thin-i-Layer p–i–n Diodes

Many modern VLSI devices operate at very high electric fields within the depletion regions of some of their p–n diodes. In fact, the junction fields are often so high that detrimental high-field effects, such as avalanche multiplication and hot-carrier effects, limit the attainable device and circuit performance. To overcome the constraints imposed by high fields in a diode, device designers often introduce a thin but lightly doped region between the n- and the p-sides. In practice, this can be accomplished by sandwiching a lightly doped layer during epitaxial growth of the doped layers, or by grading the doping concentrations at or near the junction by ion implantation and/or diffusion. Analyses of such a diode structure become very simple if the lightly doped region is assumed to be intrinsic or undoped, i.e., if the lightly doped region is assumed to be an *i-layer*. This is actually not a bad approximation as long as the net charge concentration in the i-layer is at least several times smaller than the space-charge concentration on either side of the p–n junction, so that the contribution by the i-layer charge to the junction electric field is negligible. Figure 3.6 shows the charge distribution in such a p–i–n diode. The corresponding Poisson equation is

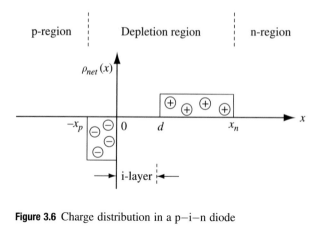

Figure 3.6 Charge distribution in a p–i–n diode

$$-\frac{d^2\psi_i}{dx^2} = \frac{qN_d}{\varepsilon_{si}} \quad \text{for} \quad d < x < x_n, \tag{3.16}$$

$$-\frac{d^2\psi_i}{dx^2} = 0 \quad \text{for} \quad 0 < x < d, \tag{3.17}$$

$$-\frac{d^2\psi_i}{dx^2} = -\frac{qN_a}{\varepsilon_{si}} \quad \text{for} \quad -x_p < x < 0. \tag{3.18}$$

These equations can be solved in the same way as Eqs. (3.6) and (3.7). Thus, integrating the equations once, subject to the boundary conditions that the electric field is zero at $x = -x_p$ and at $x = x_n$, gives

$$\mathscr{E}_m = \frac{qN_a x_p}{\varepsilon_{si}} = \frac{qN_d(x_n - d)}{\varepsilon_{si}}, \tag{3.19}$$

where $\mathscr{E}_m$ is the maximum electric field which exists in the region $0 \le x \le d$. Integrating the equations twice gives the total potential drop ψ_m across the junction as

$$\psi_m = \frac{\mathscr{E}_m(W_d + d)}{2}, \tag{3.20}$$

where $W_d = x_n + x_p$ is the total depletion-layer width. Eliminating $\mathscr{E}_m$ from Eqs. (3.19) and (3.20) gives

$$W_d = \sqrt{\frac{2\varepsilon_{si}(N_a + N_d)\psi_m}{qN_aN_d} + d^2}. \tag{3.21}$$

It is interesting to compare two diodes with the same externally applied voltage and the same p-side and n-side doping concentrations, one with an i-layer and one without. These two diodes have the same ψ_m. From Eq. (3.21), we can write

$$W_d = \sqrt{W_{d0}^2 + d^2} \tag{3.22}$$

for the diode with an i-layer, where W_{d0} is the depletion-layer width, given by Eq. (3.10), for the diode without an i-layer. Therefore,

$$\frac{W_d}{W_{d0}} = \sqrt{1 + \frac{d^2}{W_{d0}^2}}. \tag{3.23}$$

If we denote by $\mathscr{E}_{m0}$ the maximum electric field for the diode without an i-layer, then Eqs. (3.9) and (3.20) give the ratio of the electric fields as

$$\frac{\mathscr{E}_m}{\mathscr{E}_{m0}} = \frac{W_{d0}}{W_d + d} = \sqrt{1 + \frac{d^2}{W_{d0}^2}} - \frac{d}{W_{d0}}. \tag{3.24}$$

Thus, introduction of a lightly doped layer between the n- and p-regions of a diode reduces the maximum electric field in the junction. The depletion-layer charge ratio for the two diodes is, by Gauss's law,

$$\frac{Q_d}{Q_{d0}} = \frac{\mathscr{E}_m}{\mathscr{E}_{m0}} = \sqrt{1 + \frac{d^2}{W_{d0}^2}} - \frac{d}{W_{d0}}, \tag{3.25}$$

where Q_{d0} is the depletion-layer charge for the diode without an i-layer.

The depletion-layer capacitance per unit area can be calculated from Eq. (3.13), i.e., from $dQ_d(\text{p-side})/dV_{app}$, and the result is

$$C_d = \frac{\varepsilon_{si}}{W_d}. \tag{3.26}$$

The junction depletion-layer capacitance is related to the depletion-layer width in exactly the same way with or without an i-layer. This is expected from the physical picture of a parallel-plate capacitor where the capacitance is determined by the separation of the plates and not by any fixed charge distribution between the plates. The ratio of the capacitance with an i-layer to that without an i-layer is

$$\frac{C_d}{C_{d0}} = \frac{W_{d0}}{W_d} = \frac{1}{\sqrt{1 + d^2/W_{d0}^2}}, \tag{3.27}$$

where $C_{d0} = \varepsilon_{si}/W_{d0}$ is the depletion-layer capacitance for the diode without an i-layer.

3.1.3 Spatial Variation of Quasi-Fermi Potentials

In considering the current–voltage characteristics of a p–n diode, it is much more convenient to work with the quasi-Fermi potentials, instead of the intrinsic Fermi potential. The current densities and the quasi-Fermi potentials are given by Eqs. (2.77) to (2.81). These are repeated here for convenience:

$$J_n(x) = -qn(x)\mu_n(x)\frac{d\phi_n(x)}{dx}, \tag{3.28}$$

$$J_p(x) = -qp(x)\mu_p(x)\frac{d\phi_p(x)}{dx}, \tag{3.29}$$

where

$$\phi_n(x) \equiv \psi_i(x) - \frac{kT}{q}\ln\left[\frac{n(x)}{n_i}\right] \tag{3.30}$$

is the quasi-Fermi potential for electrons,

$$\phi_p(x) \equiv \psi_i(x) + \frac{kT}{q}\ln\left[\frac{p(x)}{n_i}\right] \tag{3.31}$$

is the quasi-Fermi potential for holes, and ψ_i is the electrostatic potential. In terms of the quasi-Fermi potentials, the pn product is

$$p(x)n(x) = n_i^2 \exp\left\{\frac{q[\phi_p(x) - \phi_n(x)]}{kT}\right\}. \tag{3.32}$$

In writing Eqs. (3.28) to (3.32), we have indicated explicitly the x dependence of the variables.

A note about the coordinate systems we shall use in deriving the minority-carrier densities and current flow equations for a p−n diode. As the name suggests, the p-region is usually placed on the left side of the n-region in the schematic representation of a p−n diode, such as those in Figures 3.1, 3.2, and 3.6. In the literature, in the derivation of the current equation for minority-carrier current flow in a p−n diode, the usual practice is to examine carriers injected from the left side into the right side so that the minority carriers being considered flow from left to right, i.e., in the x-direction. That means, the minority-carrier current equation is usually derived for holes flowing in the n-side (see e.g., Shockley, 1950; Sah, 1966). The resulting minority-hole current equation is applicable directly for the description of the hole current flowing in the base of a p−n−p bipolar transistor. *However, modern vertical bipolar transistors are of the n−p−n type*, and their emitter−base diodes are one-sided n^+−p diodes. Therefore, we shall refer to the n^+−p diode schematic in Figure 3.7, instead of a generic p−n diode schematic, for derivation of minority-carrier density and current. The insights into the physics of diodes are the same, independent of the diode representation schematic used, but the resultant equations are applicable directly to the description and modeling of modern n−p−n bipolar transistors, which are covered in Chapters 9, 10, and 11.

In theory, the current−voltage characteristics of a diode can be obtained from the coupled Eqs. (3.28) to (3.32). However, simple and close-form equations relating the electron and hole densities and currents to the applied voltage can be obtained if some approximations and assumptions are made. Here we discuss the physical bases for these approximations and assumptions, which will be used later to obtain equations describing the behavior of a diode in response to an applied voltage.

- *Quasineutrality.* As discussed in Section 2.4.4.1, the majority-carrier response time is on the order of 10^{-12} s. As we shall show later [see Figure 3.14(b)], this time is extremely short compared to typical minority-carrier lifetimes. Therefore, as minority carriers are injected into a doped silicon region, the majority carriers respond practically instantaneously to maintain quasineutrality. For instance, let us consider the p-region of a forward-biased diode. As electrons are injected from the n-side, the change in electron concentration in the p-region instantaneously induces a change in the hole concentration in the region such that $\Delta p_p(x) = \Delta n_p(x)$ to maintain quasineutrality. Similarly, for the n-region of a forward-biased diode, we have $\Delta n_n(x) = \Delta p_n(x)$.

- *Low-level injection.* A forward-bias voltage is said to cause low-level injection of minority carriers in a p–n diode if the injected minority-carrier density is small compared to the majority-carrier density. At low-level injection, closed-form solutions can often be obtained which provide deep insight into the operation of p–n diodes. *Unless otherwise stated, low-level injection is assumed in all cases discussed.*

- *High-level injection.* A forward-bias voltage is said to cause a high-level injection of minority carriers in a p–n diode if it results in the injected minority-carrier density being larger than the majority-carrier density. When high-level injection occurs, the space-charge region boundaries are no longer well defined. Usually, Eqs. (3.28) to (3.32) are used in numerical simulation to determine the electron and hole densities at any point in the diode.

3.1.3.1   Spatial Variation of Majority-Carrier Quasi-Fermi Potential and *IR* Drop in a Quasineutral Region

Consider the quasineutral p-region of the diode in Figure 3.7. A bias causes a current to flow in the diode, and hence through the quasineutral p- and n-regions. As the hole current flows in the p-region, it causes the ϕ_p to change according to Eq. (3.29), i.e.,

$$d\phi_p(x) = -\frac{J_p(x)}{qp_p(x)\mu_p(x)}\,dx \qquad \text{(p-region)}. \tag{3.33}$$

Integrating Eq. (3.33) from $x = 0$ (depletion-layer edge) to $x = W_p$ (p-region contact), we have

$$\phi_p(W_p) - \phi_p(0) = -\int_0^{W_p} \frac{J_p(x)}{qp_p(x)\mu_p(x)}\,dx$$
$$= \int_{W_p}^{0} J_p(x)\rho_p(x)\,dx, \tag{3.34}$$

where $\rho_p = 1/qp_p\mu_p$ is the p-region resistivity [see Eq. (2.37)]. Equation (3.34) simply states that the difference in majority-carrier quasi-Fermi potential between two points is equal to the *IR* drop between the two points.

The slope $d\phi_p/dx$ at any point is governed by the hole current density, as indicated in Eq. (3.29). In a forward-biased diode, the holes leaving the p-region end up in the

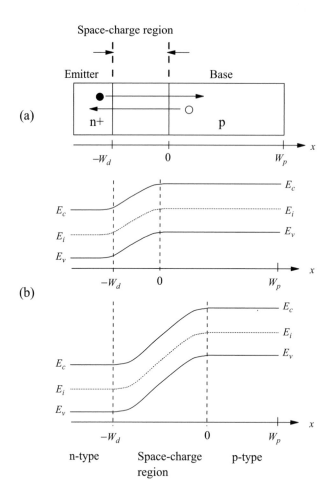

Figure 3.7 Schematic of a n^+–p diode, with a space-charge layer width W_d. The quasineutral p-region extends from $x = 0$ to $x = W_p$. The space-charge region boundary on the n-side is located at $x = -W_d$. (a) Schematic diagram of the physical structure. (b) Schematic of the energy-band diagram when the diode is forward-biased (top) and when the diode is reverse-biased (bottom)

n-region where they continue to flow via diffusion. We shall show in Section 3.1.3.2 that $d\phi_p/dx$ at the space-charge region boundary on the p-side ($x = 0$) is related to $d\phi_p/dx$ at the space-charge region boundary on the n-side ($x = -W_d$), and that

$$\left.\frac{d\phi_p(x)}{dx}\right|_{x=0} \ll \frac{kT}{qL_p} \quad \text{(p-region)}, \tag{3.35}$$

where L_p is the hole diffusion length in the n-region. L_p varies from about 100 μm for $N_d = 1 \times 10^{17}\,\text{cm}^{-3}$, where $kT/qL_p = 0.26\,\text{mV/μm}$, to about 3 μm for $N_d = 1 \times 10^{19}\,\text{cm}^{-3}$, where $kT/qL_n = 9\,\text{mV/μm}$. That is, $\left[d\phi_p(x)/dx\right]_{x=0} \ll 10\,\text{mV/μm}$. This slope is small compared to the change of potential across the space-charge layer, which is about $10\,\mathrm{V}/\mu\mathrm{m}$ (built-in potential of about $1\,\mathrm{V}$ and depletion-layer width of about $0.1\,\mu\mathrm{m}$) (see Figures 3.4 and 3.5). Similar conclusions apply to ϕ_n in the n-region. As a result, in schematic energy-band drawings for a forward-biased diode, ϕ_p and ϕ_n *appear* relatively flat in the p-region and n-region, respectively. For a reverse-biased p–n diode, electron and hole currents are negligible, ϕ_p and ϕ_n are flat in the p-region and n-region, respectively.

3.1.3.2   Spatial Variation of Minority-Carrier Quasi-Fermi Potential in a Quasineutral Region

Let us examine the spatial variation of $\phi_n(x)$ in the p-region of the diode in Figure 3.7. Within the quasineutral p-region, the transport of electrons is governed by diffusion, namely Eq. (2.103),

$$\frac{d^2 n_p(x)}{dx^2} - \frac{n_p(x) - n_{p0}}{L_n^2} = 0, \tag{3.36}$$

where

$$L_n \equiv \sqrt{\tau_n D_n} = \sqrt{\frac{kT \mu_n \tau_n}{q}} \tag{3.37}$$

is the electron diffusion length in the p-region. Solving Eq. (3.36) gives the minority-electron distribution in the p-region as

$$n_p(x) - n_p(W_p) = [n_p(0) - n_p(W_p)] \frac{\sinh[(W_p - x)/L_n]}{\sinh(W_p/L_n)}, \tag{3.38}$$

where $n_p(0)$ is the minority-electron density at the space-charge region boundary located at $x = 0$. At the p-contact at $x = W_p$, the electron density is equal to the equilibrium minority-electron density n_{p0}, i.e., $n_p(W_p) = n_{p0}$. Equation (3.38) applies to both forward bias, where $n_p(0)/n_{p0} > 1$, and reverse bias, where $n_p(0)/n_{p0} < 1$.

Within the uniformly doped p-region, ψ_i is independent of x. Therefore, Eq. (3.30) gives

$$\begin{aligned}
\phi_n(x) - \phi_n(W_p) &= \psi_i(x) - \psi_i(W_p) - \frac{kT}{q} \ln\left[\frac{n_p(x)}{n_i}\right] + \frac{kT}{q} \ln\left[\frac{n_{p0}}{n_i}\right] \\
&= -\frac{kT}{q} \ln\left[\frac{n_p(x)}{n_{p0}}\right] \\
&= -\frac{kT}{q} \ln\left\{ \left[\frac{n_p(0)}{n_{p0}} - 1\right] \frac{\sinh[(W_p - x)/L_n]}{\sinh(W_p/L_n)} + 1 \right\},
\end{aligned} \tag{3.39}$$

where we have used Eq. (3.38) for the ratio $n_p(x)/n_{p0}$. Equation (3.39) governs the variation of ϕ_n within the quasineutral p-region. At the boundary of the space-charge region, the gradient of ϕ_n is

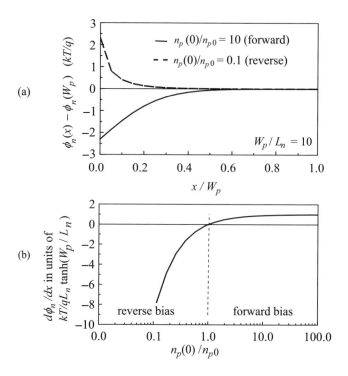

Figure 3.8 (a) The difference $\phi_n(x) - \phi_n(W_p)$ in units kT/q as a function of x/W_p for a wide p-region. (b) The slope $d\phi_n(x)/dx$ at $x = 0$ in units of $kT/qL_n \tanh(W_p/L_n)$ as a function of the ratio $n_p(0)/n_{p0}$. [For a diode with $W_p > 2L_n$, $\tanh(W_p/L_n) \approx 1$, the unit for y-axis is simply kT/qL_n.] In both (a) and (b), $n_p(0)/n_{p0} > 1$ is for forward bias, and $n_p(0)/n_{p0} < 1$ is for reverse bias

$$\left.\frac{d\phi_n(x)}{dx}\right|_{x=0} = \frac{kT}{qL_n \tanh(W_p/L_n)}\left[1 - \frac{n_{p0}}{n_p(0)}\right]. \tag{3.40}$$

Figure 3.8(a) is a plot of $\phi_n(x) - \phi_n(W_p)$ in units of kT/q as a function of x/W_p for a wide $(W_p/L_n = 10)$ p-region. When examined on an energy-band diagram, Figure 3.8(a) suggests the quantity $\{[-q\phi_n(x)] - [-q\phi_n(W_p)]\}$ is positive at forward bias, being largest at the space-charge region boundary at $x = 0$, and decreases toward zero within a distance of several L_n. At reverse bias, Figure 3.8(a) suggests the quantity $\{[-q\phi_n(x)] - [-q\phi_n(W_p)]\}$ is negative, with the difference being largest at the space-charge region boundary at $x = 0$, and decreasing toward zero within a distance of a couple of L_n.

Figure 3.8(b) is a plot of the slope $[d\phi_n(x)/dx]_{x=0}$ in units of $kT/qL_n \tanh(W_p/L_n)$ as a function of the ratio $n_p(0)/n_{p0}$. At forward bias, the slope is positive, increasing with forward bias but quickly reaching a saturated value of $kT/qL_n \tanh(W_p/L_n)$, or kT/qL_n for a wide p-region $(W_p \gg L_n)$, when $n_p(0)/n_{p0}$ reaches about 8, corresponding to a forward bias voltage of about $2 kT/q$.

It is instructive to examine the slope $[d\phi_n(x)/dx]_{x=-W_d}$, i.e., the slope of majority-electron quasi-Fermi potential in the n-region at the space-charge region boundary. From Eq. (3.28), the majority-electron current density leaving the n^+-region at forward bias is

$$J_n(x = -W_d) = -qn_n\mu_n \frac{d\phi_n(x)}{dx}\bigg|_{x=-W_d}. \tag{3.41}$$

If we neglect generation–recombination in the space-charge region, then Eq. (3.41) should be the same as the minority-electron current density $J_n(x = 0)$ entering the p-region. That is

$$n_n \frac{d\phi_n(x)}{dx}\bigg|_{x=-W_d} = n_p(0)\frac{d\phi_n(x)}{dx}\bigg|_{x=0}. \tag{3.42}$$

Since $n_p(0) \ll n_n$ at low injection, $[d\phi_n(x)/dx]_{x=-W_d}$ is much smaller than $[d\phi_n(x)/dx]_{x=0}$, or much smaller than kT/qL_n (Shockley, 1950). The same conclusions can be drawn for the slope $d\phi_p(x)/dx$. The results were used to justify Eq. (3.35) earlier.

Figure 3.8(b) also shows the slope $[d\phi_n(x)/dx]_{x=0}$ in the p-region is negative at reverse bias. The magnitude of the slope increases with increase in reverse bias. We shall return to discuss this point further in Section 3.1.3.3.

3.1.3.3   Spatial Variation of Quasi-Fermi Potentials across the Space-Charge Region

Next, we want to evaluate the difference $\phi_n(-W_d) - \phi_n(0)$ across the space-charge region as a function of V_{app}. From Eqs. (3.3) and (3.11), we have

$$\Delta\psi_i \equiv \psi_i(-W_d) - \psi_i(0) = \psi_{bi} - V_{app}$$
$$= \frac{kT}{q}\ln\left(\frac{N_d N_a}{n_i^2}\right) - V_{app}. \tag{3.43}$$

If we ignore generation–recombination current, then the electron current density in the space-charge region is constant, given by

$$J_n = -q\mu_n n_p(0)\frac{d\phi_n(x)}{dx}\bigg|_{x=0} = \frac{-kT\mu_n[n_p(0) - n_{p0}]}{L_n \tanh(W_p/L_n)}, \tag{3.44}$$

where we have used Eq. (3.40).

From Eq. (2.75), the electron density can be written as

$$n(x) = n_i e^{q[\psi_i(x)-\phi_n(x)]/kT}, \tag{3.45}$$

which can be substituted into Eq. (3.28) to yield

$$e^{-q\phi_n(x)/kT}d\phi_n = \frac{-J_n(x)}{q\mu_n n_i}e^{-q\psi_i(x)/kT}dx. \tag{3.46}$$

Integrating Eq. (3.46) from $x = -W_d$ to $x = 0$, where J_n is constant, we have

$$e^{-q\phi_n(0)/kT} - e^{-q\phi_n(-W_d)/kT} = \frac{J_n}{kT\mu_n n_i}\int_{-W_d}^{0} e^{-q\psi_i(x)/kT}\,dx$$

$$= -\frac{\delta\left[n_p(0) - n_{p0}\right]e^{-q\psi_i(0)/kT}}{n_i},$$

(3.47)

where

$$\delta \equiv \frac{1}{L_n \tanh\left(W_p/L_n\right)}\int_{-W_d}^{0} e^{q[\psi_i(0)-\psi_i(x)]/kT}\,dx.$$

(3.48)

It is left as an exercise (Exercise 3.9) for the reader to show that

$$\frac{n_p(0)}{n_{p0}} = \frac{e^{qV_{app}/kT} + \delta}{1 + \delta},$$

(3.49)

and

$$\phi_n(-W_d) - \phi_n(0) = \Delta\psi_i - \frac{kT}{q}\ln\left[\frac{N_d}{n_p(0)}\right]$$

$$= \frac{kT}{q}\ln\left[\frac{1 + \delta e^{-qV_{app}/kT}}{1 + \delta}\right].$$

(3.50)

Equation (3.50) is applicable to forward bias $(V_{app} > 0)$ and reverse bias $(V_{app} < 0)$. The parameter δ can be obtained by integrating Eq. (3.48), numerically if necessary, if we know the relationship among $\psi_i(x)$, W_d, and V_{app}. It can be inferred readily from Figure 3.7(b) that $\psi_i(0) - \psi_i(x)$ is negative within the space-charge region so that the integral in Eq. (3.48) is less than W_d. That is $\delta < W_d/L_n \tanh(W_p/L_n)$. For a typical silicon diode, $\delta \ll 1$.

As a specific example, let us consider the n^+–p diode in Figure 3.7. The electrostatic potential in the space-charge region is governed by Poisson's equation:

$$\frac{d\psi_i^2}{dx^2} = \frac{q}{\varepsilon_{si}}N_a,$$

(3.51)

where we have neglected the mobile-electron density compared to the ionized acceptor density, which is valid in the low-injection approximation. Equation (3.51) can be integrated twice to give

$$\psi_i(x) = \frac{qN_a}{2\varepsilon_{si}}x^2 + \psi_i(0).$$

(3.52)

From Eq. (3.15), we have

$$W_d = \sqrt{\frac{2\varepsilon_{si}\Delta\psi_i}{qN_a}},$$

(3.53)

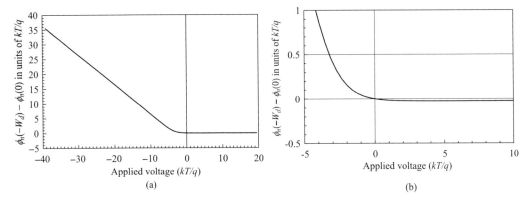

Figure 3.9 (a) Plot of ϕ_n across the space-charge region, i.e., $[\phi_n(-W_d) - \phi_n(0)]$ given by Eq. (3.50) for the diode in Figure 3.7, as a function V_{app}. $V_{app} > 0$ is for forward bias and $V_{app} < 0$ is for reverse bias. The assumptions are: $N_d = 1 \times 10^{20}$ cm^{-3} for the n$^+$-region; $N_a = 1 \times 10^{18}$ cm^{-3}, $L_n = 20$ µm, and $W_p = 1$ µm for the p-region. (b) Same plot as (a) but on greatly expanded scale around $V_{app} = 0$

where $\Delta \psi_i$ is given by Eq. (3.43). Using Eqs. (3.43), (3.52), and (3.53), Eq. (3.48) can be integrated numerically to obtain the parameter δ as a function of V_{app}, which in turn can be used to obtain the ratio $n_p(0)/n_{p0}$ and the difference $\phi_n(-W_d) - \phi_n(0)$ as a function of V_{app}.

Figures 3.9(a) and 3.9(b) are plots of the difference $\phi_n(-W_d) - \phi_n(0)$ as a function of V_{app}. They show that the difference has two distinct regions. For forward bias, the difference is essentially zero (slightly negative on magnified scale). That is, *$\phi_n(x)$ and $\phi_p(x)$ are essentially flat across the space-charge layer at forward bias*.

For reverse bias, $\phi_n(-W_d) - \phi_n(0)$ increases with $|V_{app}|$, first relatively slowly for $|V_{app}|$ up to a couple of kT/q, as indicated in Figure 3.10(b). For further increase in reverse bias, the difference $\phi_n(-W_d) - \phi_n(0)$ increases rapidly with reverse bias. Figure 3.10(a) shows the difference tracks the increase in $|V_{app}|$ linearly one-for-one for $|V_{app}| > 4\,kT/q$. The difference $\phi_n(-W_d) - \phi_n(0)$ is $36\,kT/q$ at $V_{app} = -40\,kT/q$. This result is consistent with Figure 3.8(b), which shows the magnitude of the slope $[d\phi_n(x)/dx]_{x=0}$ increases with increase in junction reverse bias.

When represented in an energy-band diagram, the results in Figure 3.9 suggest the following in a reverse-biased diode. If we assume a reverse-bias voltage of $|V_{app}| = 40\,kT/q$ and follow $\phi_n(x)$ from the p-region contact at $x = W_p$ to the n-side space-charge region boundary at $x = -W_d$, we see the difference $\phi_n(0) - \phi_n(W_p)$ across the p-region is about $4\,kT/q$, while the difference $\phi_n(-W_d) - \phi_n(0)$ is about $36\,kT/q$ across the space-charge layer. That is, *practically all the change in ϕ_n and ϕ_p occurs across the space-charge region in a reverse-biased diode*.

It was shown by Yang and Schroder (2012) that for a reverse-biased p–n junction, ϕ_n and ϕ_p within the space-charge region lie between the band edges E_c and E_v. This, together with the discussion of the plots in Figures 3.8 and 3.9, suggest the spatial

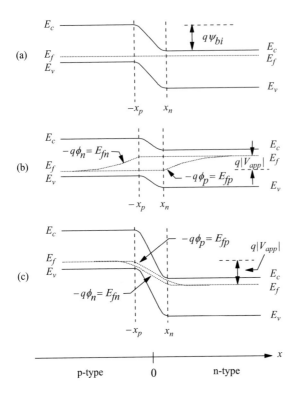

Figure 3.10 Schematics showing the variations of the quasi-Fermi potentials, ϕ_p for holes and ϕ_n for electrons, as a function of distance in a generic p–n diode. (a) A diode at thermal equilibrium with $V_{app} = 0$. (b) A forward-biased diode $(V_{app} > 0)$ with negligible IR drops in the quasineutral regions. (c) A reverse-biased diode $(V_{app} < 0)$

variation of $\phi_n(x)$ and $\phi_p(x)$ for a generic p–n diode, is as shown schematically in Figure 3.10.

3.1.4   The Diode Equation

Consider a generic p–n diode with energy bands and quasi-Fermi levels as depicted in Figure 3.10. When the diode is forward-biased, the voltage V_{app} across the diode terminals, the voltage V'_{app} across the space-charge region, and the quasi-Fermi potentials are related by

$$
\begin{aligned}
V'_{app} &\equiv V_{app} - IR(\text{p-side}) - IR(\text{n-side}) \\
&= V_{app} - [\phi_p(\text{contact}) - \phi_p(-x_p)] - [\phi_n(x_n) - \phi_n(\text{n-contact})] \\
&= \phi_p(-x_p) - \phi_n(x_n).
\end{aligned} \tag{3.54}
$$

V'_{app} is also referred to as the junction voltage or the intrinsic voltage. In Eq. (3.54), we have used the results in Sections 3.1.3.1 and 3.1.3.3. Therefore, for electrons on the p-side at the space-charge region boundary, Eq. (3.32) gives

$$n_p(-x_p) = \frac{n_i^2}{p_p(-x_p)} \exp\{q[\phi_p(-x_p) - \phi_n(-x_p)]/kT\}$$

$$= \frac{n_i^2}{p_p(-x_p)} \exp\{q[\phi_p(-x_p) - \phi_n(x_n)]/kT\} \quad \text{(forward bias)}. \quad (3.55)$$

$$= \frac{n_i^2}{p_p(-x_p)} \exp(qV'_{app}/kT)$$

Similarly, we have for holes on the n-side at the space-charge region boundary

$$p_n(x_n) = \frac{n_i^2}{n_n(x_n)} \exp(qV'_{app}/kT) \quad \text{(forward bias)}. \quad (3.56)$$

Equations (3.55) *and* (3.56) *are the most important boundary conditions governing the operation of a p–n diode*. They relate the minority-carrier densities at the space-charge region boundaries of the quasineutral regions to their majority-carrier densities and to the voltage across the space-charge region. Equations (3.55) and (3.56) are often referred to as the *Shockley diode equations* (Shockley, 1950).

The distinction between V'_{app} and V_{app} is important whenever parasitic resistance is not negligible and there is significant current flow, for instance, in the forward-biased emitter–base diode of a bipolar transistor. In most cases, the parasitic resistance can be modeled as a lump resistor in series with an intrinsic diode having zero resistance, allowing us to quantify the difference between V'_{app} and V_{app} readily. *For simplicity in writing the equations, we will not make the distinction between V'_{app} and V_{app} when we use* Eqs. (3.55) *and* (3.56) *to derive the equations governing current–voltage characteristics*. The distinction between V'_{app} and V_{app} will be pointed out wherever it is important to do so.

3.1.4.1 Diode Equation for Electron Transport in a p-Region

Here we want to set up the diode equation that is readily applicable to electron transport in the base of an n–p–n bipolar transistor. Referring to the n^+–p diode in Figure 3.7, Eq. (3.55) gives

$$p_p(x = 0)n_p(x = 0) = n_i^2 \exp(qV_{app}/kT), \quad (3.57)$$

which can be rewritten as

$$[p_{p0}(0) + \Delta p_p(0)][n_{p0}(0) + \Delta n_p(0)] = n_i^2 \exp(qV_{app}/kT). \quad (3.58)$$

For V_{app} greater than a few kT/q, $n_{p0}(0)$ is negligible compared to $\Delta n_p(0)$, Eq. (3.58) gives

$$[p_{p0}(0) + \Delta p_p(0)]\Delta n_p(0) \approx n_i^2 \exp(qV_{app}/kT). \quad (3.59)$$

With no assumption made of injection level, other than V_{app} is greater than several kT/q, Eq. (3.59) *is mathematically valid for all injection levels*. [However, the applicability of Eq. (3.59) is limited to injection levels where the space-charge region boundaries remain well defined, as discussed at the end of this subsection.] Equation (3.59) is a quadratic equation in $\Delta n_p(0)$, since quasineutrality implies $\Delta n = \Delta p$, with a solution (Cai *et al.*, 2014)

$$\Delta n_p(0) = \frac{p_{p0}(0)}{2}\left[\sqrt{1 + \frac{4n_i^2}{p_{p0}^2(0)}\exp(qV_{app}/kT)} - 1\right] \quad \text{(all injection)}. \quad (3.60)$$

At low injection, the second term within the square root is small compared to unity, and Eq. (3.60) reduces to

$$\Delta n_p(0) \approx n_{p0}(0)\exp(qV_{app}/kT) \quad \text{(low injection)}, \quad (3.61)$$

which is the familiar Shockley equation for a forward-biased diode (Shockley, 1950). [Equation (3.61) can also be obtained directly from Eq. (3.58) by first making the low-injection approximation, namely, by assuming $\Delta n = \Delta p << p_{p0}$. Such an assumption is valid for diodes operating at low-to-intermediate V_{app} values where the condition $\Delta n = \Delta p << p_{p0}$ is satisfied. It is shown in Section 9.2.2.2 that, to avoid operation with deleterious effects associated with high current densities, *the emitter−base diode of a vertical bipolar transistor is usually operated in the low-injection region*.]

At the high-injection limit, $\Delta n_p = \Delta p_p >> p_{p0}$, the second term within the square root becomes large compared to unity, Eq. (3.60) reduces to

$$\Delta n_p(0) \approx n_i\exp(qV_{app}/2kT) \quad \text{(high injection limit)}, \quad (3.62)$$

as expected from Eq. (3.58). Again, V_{app} in all these equations should be replaced by V'_{app} whenever parasitic IR drops are not negligible.

Figure 3.11 illustrates the potential for error when using the conventional diode equation, Eq. (3.61), in device modeling. In Figure 3.11(a), the curves are calculated using Eq. (3.60). The curves deviate from a straight line, i.e., from an $\exp(qV_{app}/kT)$ dependence at large V_{app}, suggesting potential error when using Eq. (3.61) at these large bias voltages. Figure 3.11(b) plots the ratio of $\Delta n_p(0)$ according to Eq. (3.60) to

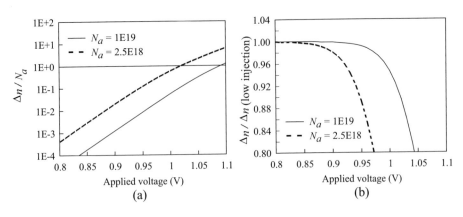

Figure 3.11 (a) A plot of the ratio $\Delta n_p/p_{p0} = \Delta n_p/N_a$ from Eq. (3.60). (b) A plot of the ratio Δn_p given by Eq. (3.60) to Δn_p given by Eq. (3.61). The two values of p-side doping concentration N_a are as indicated

$\Delta n_p(0)$ according to Eq. (3.61). It shows that Eq. (3.61) overestimates the minority-carrier density by 20% at $V_{app} = 0.97$ V for $N_a = 2.5 \times 10^{18}$ cm^{-3}, and at $V_{app} = 1.04$ V for $N_a = 1 \times 10^{19}$ cm^{-3}. That is, the potential for error occurs at lower voltage as N_a is reduced.

While Eq. (3.60) is mathematically valid for all levels of minority-carrier injection, care should be exercised in applying it to real devices when $\Delta n_p = \Delta p_p$ approaches p_{p0}. At such high levels of minority-carrier injection, the concept of a well-defined space-charge region may no longer be valid, and the quasi-Fermi potentials do not have simple behavior in any region of the diode (Gummel, 1967). An example of how the "boundary" of a p–n junction can be "relocated" at high minority-carrier injection can be found in Section 9.3.4 in connection with the discussion of base-widening effects in a vertical bipolar transistor.

3.1.5   Current–Voltage Characteristics Governed by the Diode Equation

In this subsection, we consider the current–voltage characteristics governed by the Shockley diode equations, Eqs. (3.55) and (3.56), which include only diffusion currents. The currents due to generation–recombination within the space-charge region will be covered in Section 3.1.6. The combined diffusion current and space-charge-region current will be examined in Section 3.1.7.

Two types of bipolar transistors are covered in this book: vertical bipolar transistors and symmetric lateral bipolar transistors on SOI (see Chapter 9). *The maximum collector current density for operating a bipolar transistor without speed degradation is limited by the smaller of the base doping concentration or the collector doping concentration.* In Chapter 9, it is shown that low-injection approximation, namely Eq. (3.61), is adequate for describing vertical bipolar transistors. Equation (3.60)*, which is valid for all injection levels, should be used to describe symmetric lateral bipolar transistors on SOI.* Since vertical bipolar transistors are much more widely used than SOI lateral bipolar transistors at this writing, we shall limit to low-injection approximation, i.e. Eq. (3.61), in deriving the minority-carrier transport equations in this chapter. The need to switch to Eq. (3.60) will be pointed out when necessary in the chapters covering bipolar transistors.

Let us refer to the diode in Figure 3.7 and assume the p- and n-regions are uniformly doped. The minority-electron distribution in the p-region is given by Eq. (3.38). When the diode is forward biased, the boundary condition for n_p at $x = 0$ is given by Eq. (3.61), namely

$$n_p(0) \approx \Delta n_p(0) = n_{p0} \exp(qV_{app}/kT) \quad \text{(forward bias)}, \qquad (3.63)$$

where we have used the fact that $n_p = n_{p0} + \Delta n_p \approx \Delta n_p$. The boundary condition for n_p at $x = W_p$ is

$$n_p(W_p) = n_{p0}. \qquad (3.64)$$

Substituting Eqs. (3.63) and (3.64) into Eq. (3.38) gives

$$n_p(x) - n_{p0} = n_{p0}\left[\exp(qV_{app}/kT) - 1\right]\frac{\sinh\left[(W_p - x)/L_n\right]}{\sinh(W_p/L_n)}. \qquad (3.65)$$

The electron diffusion current density entering the p-region is

$$J_n(0) = qD_n\left(\frac{dn_p}{dx}\right)_{x=0}$$
$$= -\frac{qD_n n_{p0}\left[\exp(qV_{app}/kT) - 1\right]}{L_n \tanh(W_p/L_n)}. \qquad (3.66)$$

Note that J_n is negative in sign because electrons have a charge $-q$ and are flowing in the x-direction.

The hole diffusion current density entering the n-region can be derived in an analogous manner. **The total current flowing through a p–n diode is the sum of the minority-electron current on the p-side and the minority-hole current on the n-side.** That is, the forward-bias diode diffusion current density is

$$J_{diode} = -\frac{qD_n n_i^2\left[\exp(qV_{app}/kT) - 1\right]}{N_a L_n \tanh(W_p/L_n)} - \frac{qD_p n_i^2\left[\exp(qV_{app}/kT) - 1\right]}{N_d L_p \tanh(W_n/L_p)}, \qquad (3.67)$$

where we have assumed that all the dopants are ionized so that $p_{p0} = N_a$ and $n_{n0} = N_d$. W_n is the width of the quasineutral n-region. The negative sign in Eq. (3.67) is due to the fact that we placed the p-region to the right of the n-region, causing electrons to flow in the $+x$ direction and holes to flow in the $-x$ direction. The negative sign will not be there if we place the p-region to the left of the n-region.

Ignoring the sign, Eq. (3.67) is often referred to as the *Shockley diode current equation*, or simply the Shockley diode equation. This current is also referred as the *ideal diode current*, increasing with V_{app} at 60 mV/decade at room temperature.

When the diode is reverse-biased, instead of electrons being injected from the n-region into the p-region, electrons in the p-region near the space-charge region boundary diffuse toward the n-region. That is, Δn_p is negative and $n_p < n_{p0}$ near the space-charge region boundary. For reverse-bias voltage of more than several kT/q, the simplest boundary condition for use in Eq. (3.38) is to assume depletion approximation and set

$$n_p(x = 0) = 0 \quad \text{(reverse bias)}. \qquad (3.68)$$

The boundary condition for n_p at $x = W_p$ remains as given by Eq. (3.64). Substituting Eqs. (3.64) and (3.68) into Eq. (3.38) we have

$$n_p(x) - n_{p0} = -n_{p0}\frac{\sinh\left[(W_p - x)/L_n\right]}{\sinh(W_p/L_n)} \quad \text{(reverse bias)}. \qquad (3.69)$$

The current due to electrons back diffusing from the p-region toward the n-region is

$$J_n(0) = qD_n\left(\frac{dn_p}{dx}\right)_{x=0}$$
$$= \frac{qD_n n_i^2}{p_{p0}L_n \tanh(W_p/L_n)} \qquad \text{(reverse bias)}. \qquad (3.70)$$

Notice that $\left[n_p(x) - n_{p0}\right]$ is negative, and J_n is positive in reverse bias because the back-diffused electrons flow in the $-x$ direction. Equation (3.70) is the electron diffusion component of the leakage current in a reverse-biased diode. It is also referred to as the electron *saturation current* of a diode. The hole saturation current can be derived in the same manner. The total diffusion leakage current is the sum of the electron and hole saturation currents. Equation (3.70) does not include electron current due to electron-hole pair generation within the space-charge region. Leakage current due to generation will be discussed in Section 3.1.6.

Equation (3.66) is for forward bias where the quasi-Fermi levels are flat across the space-charge region. If we were to assume Eq. (3.66) to be valid for $V_{app} < 0$, then Eq. (3.66) would lead to Eq. (3.70) for large negative bias. That is, while the total diffusion current in Eq. (3.67) was derived for a forward-biased diode, it gives to the correct reverse-bias electron and hole saturation current for large negative V_{app}. As a result, Eq. (3.67) *is often taken as the equation for current density of an ideal (obeying low-injection approximation and ignoring space-charge-region current) p–n diode for both forward ($V_{app} > 0$) and reverse ($V_{app} < 0$) bias.*

3.1.6   Space-Charge-Region Current

The equations governing the kinetics of generation and recombination are given in Section 2.4.3. The net recombination rate is given by Eq. (2.95), namely

$$U = \frac{\sigma_n \sigma_p v_{th}\left(np - n_i^2\right)N_t}{\sigma_n(n + n_i) + \sigma_p(p + n_i)}. \tag{3.71}$$

In practical silicon diodes, the space-charge-region current could be comparable to or larger than the Shockley diode current given by Eq. (3.67). Here we want to consider the voltage dependence of the space-charge-region current.

3.1.6.1   Recombination Current in Forward Bias

In forward bias, ϕ_n and ϕ_p are constant inside the space-charge region, as indicated in Figure 3.10(b). However, the electrostatic potential $\psi_i(x) = -E_i(x)/q$ is not spatially constant. This is illustrated in Figure 3.12. The $n(x)p(x)$ product inside the space-charge region is spatially constant, and is given by

$$n(x)p(x) = n_i^2 e^{q(\phi_p - \phi_n)/kT} = n_i^2 e^{qV_{app}/kT}. \tag{3.72}$$

From Eqs. (2.75) and (2.76), we have

$$n(x) = n_i e^{[-q\phi_n + q\psi_i(x)]/kT} \tag{3.73}$$

and

$$p(x) = n_i e^{[q\phi_p - q\psi_i(x)]/kT}, \tag{3.74}$$

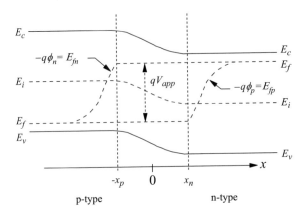

Figure 3.12 Schematic showing the spatial dependence of $-q\phi_p$, $-q\phi_n$, and $E_i = -q\psi_i$ in the space-charge region of a forward-biased p–n diode

which implies that the electron and hole densities have an implicit spatial dependence through $\psi_i(x)$. The net recombination rate, Eq. (3.71), in turn has an implicit spatial dependence through $n(x)$ and $p(x)$. Notice that the net recombination rate is also a function of the applied voltage because $\psi_i(x)$ depends on applied voltage. For a given applied voltage, we want to determine the maximum net recombination rate inside the space-charge region.

Using Eq. (3.72), we can rewrite Eq. (3.71) as a function of V_{app} and n, i.e.,

$$U = U\left(V_{app}, x\right) = U\left[V_{app}, n(x)\right] = \frac{\sigma_n \sigma_p v_{th} n_i^2 \left(e^{qV_{app}/kT} - 1\right)N_t}{\sigma_n(n + n_i) + \sigma_p\left(\dfrac{n_i^2 e^{qV_{app}/kT}}{n} + n_i\right)}. \tag{3.75}$$

Let us denote by x_m the location where the net recombination rate is maximum. x_m is determined from the condition

$$\left.\frac{\partial U\left[V_{app}, n(x)\right]}{\partial n(x)}\right|_{x=x_m} = 0, \tag{3.76}$$

which gives

$$n(x_m) = \sqrt{\frac{\sigma_p}{\sigma_n}} n_i e^{qV_{app}/2kT}, \tag{3.77}$$

and

$$p(x_m) = \sqrt{\frac{\sigma_n}{\sigma_p}} n_i e^{qV_{app}/2kT}. \tag{3.78}$$

Notice that $\sigma_n n(x_m) = \sigma_p p(x_m)$. That is, the net recombination rate is maximum at locations where the probability of electron capture is equal to that of hole capture.

Elsewhere, the net recombination rate is less because the electron and hole capture events are not balanced. Substituting Eq. (3.77) into Eq. (3.75), we have

$$U_{max} \equiv U(V_{app}, x_m) = \frac{\sigma_n \sigma_p v_{th} n_i^2 (e^{qV_{app}/kT} - 1) N_t}{2\sqrt{\sigma_n \sigma_p} n_i e^{qV_{app}/2kT} + n_i(\sigma_n + \sigma_p)}. \tag{3.79}$$

For V_{app} larger than a few kT/q, Eq. (3.79) reduces to

$$U_{max} \approx \frac{\sigma_n \sigma_p v_{th} n_i^2 e^{qV_{app}/kT} N_t}{2\sqrt{\sigma_n \sigma_p} n_i e^{qV_{app}/2kT}}$$

$$= \frac{1}{2}\sqrt{\sigma_n \sigma_p} v_{th} n_i e^{qV_{app}/2kT} N_t. \tag{3.80}$$

The diode current density due to generation and recombination from the space-charge region is a function of the applied voltage, and is given by

$$J_{SC}(V_{app}) = q \int_{-x_p}^{x_n} U(V_{app}, x) dx. \tag{3.81}$$

It is not practical to determine the recombination current from Eq. (3.81). However, we expect the current to be some fraction of the product of U_{max} and the space-charge layer thickness, W_d, i.e., we expect

$$J_{SC}(V_{app} > 0) \sim q U_{max} W_d. \tag{3.82}$$

Using Eq. (3.80) for U_{max}, we have

$$J_{SC}(qV_{app}/kT \gg 1) \sim q\sqrt{\sigma_n \sigma_p} v_{th} n_i e^{qV_{app}/2kT} N_t W_d. \tag{3.83}$$

That is, **the recombination current in the space-charge region has an** $exp(qV_{app}/2kT)$ **dependence**.

3.1.6.2 Generation Current in Reverse Bias

In reverse bias, depletion approximation means $n = p = 0$ in the space-charge region, and Eq. (3.71) becomes

$$U(n = p = 0) = \frac{-\sigma_n \sigma_p v_{th} n_i N_t}{\sigma_n + \sigma_p}, \tag{3.84}$$

and the space-charge-region current due to generation is

$$J_{SC}(V_{app} < 0) = q \int_{-x_p}^{x_n} U dx = \frac{-q n_i W_d}{\tau_n + \tau_p}, \tag{3.85}$$

where we have used the relationship between capture cross-section and lifetime in Eqs. (2.98) and (2.99). Equation (3.85) is negative because the generated electrons flow toward the n-side and the generated holes flow toward the p-side.

3.1.6.3   Voltage Dependence of Space-Charge-Region Current

At zero bias, $np = n_0 p_0$ and $U(x) = 0$. There is zero space-charge-region current, as expected. Combining this with the results in Eqs. (3.83) and (3.85), Sah (Sah *et al.*, 1957; Sah, 1991) proposed writing the space-charge-region current in the form

$$J_{SC}(V_{app}) = J_{SC0}\left(e^{qV_{app}/2kT} - 1\right), \tag{3.86}$$

with

$$J_{SC0} = \frac{qn_i W_d}{\tau_n + \tau_p}. \tag{3.87}$$

Equation (3.86) is referred to as the *Sah–Noyce–Shockley diode equation* (Sah *et al.*, 1957; Sah, 1991). It gives a good description of the observed space-charge-region currents in practical silicon diodes.

3.1.7   Measured Diode Current and Ideality Factor

The measured current I_{total} of a p–n diode of junction area A_{diode} is the sum of its diffusion current, given by Eq. (3.67), and its space-charge-region current, given by Eq. (3.86). That is,

$$I_{total} = I_{diode} + I_{SC}, \tag{3.88}$$

where

$$I_{diode} = I_0\left[\exp(qV_{app}/kT) - 1\right], \tag{3.89}$$

with

$$I_0 = A_{diode}qn_i^2\left[\frac{D_n}{N_a L_n \tanh(W_p/L_n)} + \frac{D_p}{N_d L_p \tanh(W_n/L_p)}\right], \tag{3.90}$$

and

$$I_{SC} = I_{SC0}\left[\exp(qV_{app}/2kT) - 1\right], \tag{3.91}$$

with

$$I_{SC0} = \frac{A_{diode}qn_i W_d}{\tau_n + \tau_p}. \tag{3.92}$$

Figure 3.13 is a schematic semi-log plot, or *Gummel plot*, of the measured current of a diode as a function of its forward-bias terminal voltage, with series resistances neglected. The slope in a Gummel plot is often used to infer the ideality of a diode. That is, the measured forward diode current is often expressed in the form

$$I_{total}(\text{forward}) \sim \exp(qV_{app}/mkT), \tag{3.93}$$

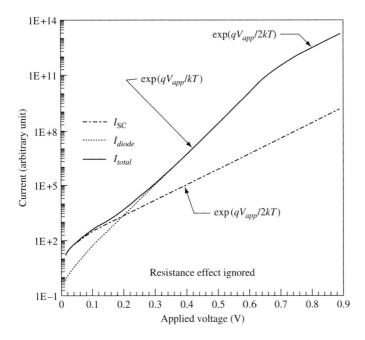

Figure 3.13 A schematic Gummel plot of the forward-bias current of a p–n diode. Series resistance effects are ignored. I_{diode} is the Shockley diode current. I_{SC} is the space-charge-region current

where m is called the *ideality factor*. When m is unity, the current is considered "ideal." Figure 3.13 suggests that a forward diode current is ideal except at very small and very large forward biases. The nonideality at small forward bias is caused by the space-charge-region current. Space-charge-region current leads to $m \sim 2$. The nonideality with $m \sim 2$ at very large forward bias is due to high-injection effect in the Shockley diode current [see Eq. (3.62)].

3.1.8   Temperature Dependence and Magnitude of Diode Leakage Currents

For a reverse-biased diode, the total leakage current is the sum of the diffusion saturation current I_0 and the space-charge region saturation current I_{SC0} in Eqs. (3.90) and (3.92), respectively. The temperature dependence of I_0 is dominated by the temperature dependence of the n_i^2 factor, which is proportional to $\exp(-E_g/kT)$, where E_g is the bandgap energy. The space-charge-region leakage current I_{SC0}, being proportional to n_i, has a temperature dependence of $\exp(-E_g/2kT)$. In other words, the diffusion leakage current has an activation energy of about 1.1 eV while the generation leakage current has an activation energy of about 0.5 eV. This difference in activation energy can be used to distinguish the sources of the observed leakage current (Grove and Fitzgerald, 1966).

Measurements from properly processed n^+–p diodes show that the diffusion leakage current is comparable to the space-charge-region leakage current at room temperature, both being on the order of 10^{-13} A/cm^2 (Kircher, 1975). However, the diffusion leakage current, due to its larger activation energy, is usually larger than the space-charge-region leakage current at elevated temperatures.

3.1.9 Minority-Carrier Mobility, Lifetime, and Diffusion Length

Among the three minority-carrier transport parameters, namely, lifetime, mobility, and diffusion length, only lifetime and mobility are independent. Diffusion length is given by $L = (kT\mu\tau/q)^{1/2}$ [see Eq. (2.104)]. There have been many attempts to measure the minority-carrier lifetimes, mobilities, or diffusion lengths. For doping concentrations greater than about 1×10^{19} cm^{-3}, the experiments are quite difficult because the minority-carrier concentrations are very small. As a result, there is quite a bit of spread in the reported data (Dziewior and Schmid, 1977; Dziewior and Silber, 1979; del Alamo *et al.*, 1985a, 1985b). For the purposes of device modeling, the following empirical equations have been proposed for minority-carrier electrons (Swirhun *et al.*, 1986) and minority-carrier holes (del Alamo *et al.*, 1985a, 1985b):

$$\mu_n = 232 + \frac{1,180}{1 + \left(N_a/8 \times 10^{16}\right)^{0.9}} \text{ cm}^2\text{-V-s}^{-1} \tag{3.94}$$

$$\mu_p = 130 + \frac{370}{1 + \left(N_d/8 \times 10^{17}\right)^{1.25}} \text{ cm}^2\text{-V-s}^{-1} \tag{3.95}$$

$$\frac{1}{\tau_n} = 3.45 \times 10^{-12}N_a + 0.95 \times 10^{-31}N_a^2 \text{ s}^{-1} \tag{3.96}$$

$$\frac{1}{\tau_p} = 7.8 \times 10^{-13}N_d + 1.8 \times 10^{-31}N_d^2 \text{ s}^{-1}. \tag{3.97}$$

The minority-carrier mobilities, lifetimes, and diffusion lengths from these equations are plotted as a function of doping concentration in Figures 3.14(a), (b), and (c), respectively.

There are more recent models of minority-carrier mobilities. In particular, there is a physics-based model that describes the mobilities of both majority and minority carriers in a consistent manner (Klaassen, 1990). For minority electrons, the Klaassen model is about the same as that in Figure 3.14(a). For minority holes, Klaassen's model gives about the same mobilities as in Figure 3.14(a) for high $\left(>2 \times 10^{18}\text{ cm}^{-3}\right)$ doping concentrations and about 30% lower mobilities at low $\left(<1 \times 10^{18}\text{ cm}^{-3}\right)$ doping concentrations (Klaassen *et al.*, 1992).

In device application, e.g., when used as a photodetector, the response of a diode is via minority-carrier diffusion and recombination processes. Unless the quasineutral region involved is very narrow compared to the diffusion length, the minority carriers respond primarily via recombination, with a response time determined by the minority-carrier

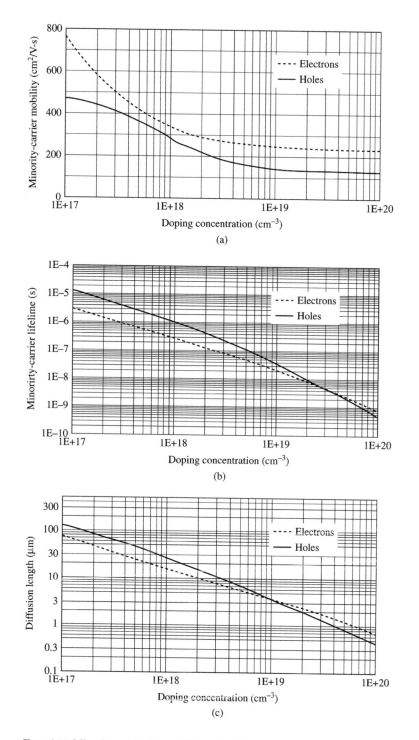

Figure 3.14 Minority-carrier (a) mobilities, (b) lifetimes, and (c) diffusion lengths as a function of doping concentration, calculated using the empirical equations, Eqs. (3.94) to (3.97)

lifetime. In the literature, there are many reports of using special process techniques to reduce the lifetime of a silicon diode, to values much shorter than those in Figure 3.14(b), in order to improve its response time. Often, gold, which forms trap centers with an energy level 0.55 eV below E_c, is added to the silicon as a "lifetime killer" (see e.g., Forbes, 1977). In modern bipolar transistors, the base region of the emitter–base diode is much narrower than its minority-carrier diffusion length. In this case, the diode response time is determined not by lifetime but by diffusion process, or by a combination of diffusion and drift in vertical SiGe-base bipolar transistors (see Chapters 9 and 10).

3.2   Metal–Silicon Contacts

Metal–semiconductor contacts are critically important elements in all semiconductor devices. As a contact to a silicon device terminal, a metal–silicon contact should be non-rectifying and have a small contact resistance in order to minimize voltage drop across the contact. In general, a metal–semiconductor contact has rectifying current–voltage characteristics similar to those of a p–n diode. In this section, we discuss the basic physics and operation of a metal–silicon contact, focusing on its current–voltage characteristics as a Schottky diode and as an ohmic contact.

3.2.1   Static Characteristics of a Schottky Diode

The surface potential of the semiconductor in a Schottky diode is affected by the electron occupation of the surface states on the semiconductor surface. Therefore, the characteristics of a Schottky diode depend on the properties of the metal and the properties of the semiconductor and its surface states. Here we first discuss the static characteristics of a Schottky diode ignoring all surface states, and then discuss how surface states can modify the diode characteristics.

3.2.1.1   Schottky Diodes without Surface States

The energy-band diagrams for a metal–n-silicon contact at thermal equilibrium is illustrated in Figure 3.15. From consideration of free electron level at the interface, the Schottky barrier height for electrons, $q\phi_{Bn}$, is

$$q\phi_{Bn} = q(\phi_m - \chi) \quad \text{(no surface states)}, \tag{3.98}$$

where $q\phi_m$ is the metal work function and $q\chi$ is the electron affinity of silicon. The built-in potential ψ_{bi} is

$$q\psi_{bi} = q\phi_{Bn} - \left(E_c - E_f\right)_{bulk}$$

$$= q\left(\phi_{Bn} - \frac{E_g}{2q} + \psi_B\right), \tag{3.99}$$

where $\psi_B = |\psi_f - \psi_i|$ [see Eq. (2.62)]. The built-in potential implies an amount of depletion-layer charge in the silicon. To maintain overall charge neutrality of the

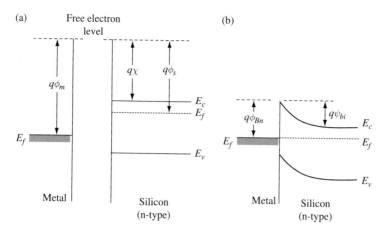

Figure 3.15 Energy-band diagrams of a metal–silicon system where *the silicon surface is assumed to be absent of any surface states*. (a) When the metal and the silicon are far apart. (b) When the metal is in contact with the silicon, with no externally applied voltage

metal–silicon system, this depletion charge induces a sheet of opposite-polarity charge of the same density per unit area in the metal at the metal–silicon interface.

The physical picture for the Schottky diode illustrated in Figure 3.15 and Eq. (3.98) are based on energy-band diagrams for *bulk* silicon and metal. It does not take into consideration any surface properties of the semiconductor. *Experimentally measured barrier heights are not consistent with* Eq. (3.98). For some semiconductors, the measured barrier heights show little dependence on metal work function. For others, the dependence on work function is weaker than suggested by Eq. (3.98). That the measured barrier height has a weaker dependence on metal work function than suggested by Eq. (3.98) is attributable to the presence of surface states. This is discussed in Section 3.2.1.2.

3.2.1.2 Schottky Barrier Diodes with Surface States

Many ideas and models for improving the understanding of real Schottky diodes have been proposed. In terms of explaining the relationship between measured barrier heights and metal work functions, the models all include the effects of surface states. Here we discuss qualitatively how surface states can influence the barrier height. A good reference where many ideas and models are discussed at length can be found in Henisch (1984).

The inclusion of surface states makes the physical picture of a metal–semiconductor contact more complex, as illustrated in Figure 3.16. Figure 3.16(a) illustrates the situation where the metal and the silicon are physically and electrically separate. As electrically separate systems, there can be no charge exchange between the metal and the silicon. Occupation of some of the surface states by electrons induces a positive depletion-layer charge of Q_d per unit area in the silicon near the surface, causing the energy bands to bend upward near the surface. The amount of band bending is represented by the surface potential ψ_s. The relationship between the depletion charge density and surface potential can be inferred from Eq. (3.15) for a one-sided diode, namely

$$-q\psi_s(\text{free surface}) = \frac{[Q_d(\text{free surface})]^2}{2\varepsilon_{si}N_d}, \qquad (3.100)$$

where N_d is the doping concentration of the n-type silicon.

Figure 3.16(b) illustrates the situation where the metal and the silicon are electrically connected but physically separated, with a physical gap between the metal and

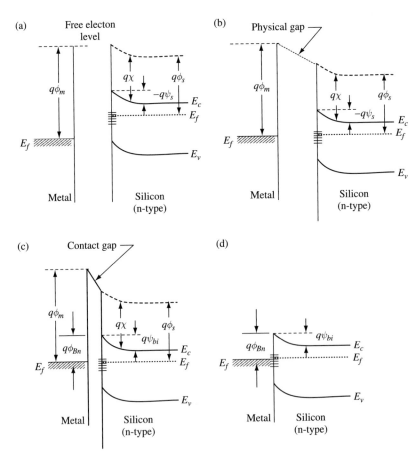

Figure 3.16 Energy-band diagram of a metal–silicon system where the silicon is assumed to have a large density of surface states. (a) When the metal and the silicon are not electrically connected as one system. Some of the interface states are filled with electrons, causing the bands to bend upward. (b) When the metal and the silicon are electrically connected to form one system, but the metal is physically separate from the silicon surface by a gap space. (c) When the metal is in contact with the silicon to form a Schottky diode. A contact gap of atomic dimension is shown. The contact-gap region is discussed in the text. (d) A simplified diagram where the contact gap is omitted.

silicon surfaces. (In the literature, for the purposes of establishing a model for a Schottky diode, this physical gap is often replaced by an oxide layer, similar to an MOS capacitor.) A charge exchange between the metal and the silicon occurs until the Fermi level is spatially constant across the entire system. The free electron levels across the interfaces are continuous, suggesting the presence of an electric field in the gap. This electric field is supported by a *net* charge Q_M on the metal surface, as required by Gauss's law. Q_M is a negative charge for the electric field direction indicated in Figure 3.16(b). If the amount of net charge per unit area in the surface states is Q_{it}, then overall charge neutrality of the metal–silicon system requires that $Q_M = -(Q_{it} + Q_d)$. As the gap between the metal surface and the silicon surface is reduced, the electric field in the gap increases, and hence the magnitude of Q_M increases, requiring the magnitude of $(Q_{it} + Q_d)$ to increase.

The Bardeen model (Bardeen, 1947) for the roles played by surface states provides a physical picture for explaining how the charges Q_M, Q_{it}, and Q_d may change together. According to the Bardeen model, the charges involved in the charge-transfer process in the formation of a metal–semiconductor contact come from four double layers. (A double layer consists of a layer of positive charge and a layer of negative charge of the same magnitude.) These layers are: (i) a double layer at the metal surface, (ii) a double layer at the semiconductor surface, (iii) a double layer formed from the surface charges on the metal and the semiconductor, and (iv) a double layer formed from a surface charge on the semiconductor surface and a depletion charge layer in the semiconductor. According to this model, Q_M has contributions from double layers (i) and (iii); Q_{it} has contributions from double layers (ii), (iii), and (iv); and Q_d has contribution from double layer (iv).

Bardeen showed that the strength of double layer (iii) is small for semiconductors where the density of surface states is small (less than about $10^{13}\,\mathrm{cm}^{-2}$), and a change in Q_M is balanced by charge transfer primarily among double layers (ii) and (iv). In this case, as the physical gap in Figure 3.16(b) is reduced, the change in Q_M is balanced primarily by a change in Q_d, with little change in Q_{it}. For semiconductors with sufficiently high (greater than $10^{13}\,\mathrm{cm}^{-2}$) density of surface states, double layer (iii) can be the primary source for any change in Q_M. In this case, as the physical gap in Figure 3.16(b) is reduced, the change in Q_M is balanced primarily by a change in Q_{it}, with little change in Q_d. When there is little change in Q_d, there is little change in the surface potential, which in turn implies little change in the energy-band edges at the semiconductor surface relative to the Fermi level. The Fermi level at the semiconductor surface is said to be more or less pinned by the high density of surface states.

When the metal makes contact with the silicon to form a Schottky diode, the energy-band diagram is as illustrated in Figure 3.16(c). Note that a contact gap is shown to represent the region containing the double layers (i), (ii), and (iii) in the Bardeen model. The width of the contact gap is of atomic dimensions, at least for good contacts where there is no unintended interfacial material. The contact gap is assumed to be sufficiently thin to play no role in the transport of electrons between the metal and the silicon. In Figure 3.16(b), before the Schottky diode is formed, the surface potential ψ_s depends on the physical gap space. If there is weak Fermi level pinning,

ψ_s changes with the physical gap space. If there is strong Fermi level pinning, ψ_s is relatively insensitive to the physical gap space. In Figure 3.16(c), we have $\psi_s = -\psi_{bi}$, where ψ_{bi} is the built-in potential of the Schottky diode. The electron energy barrier is $q\phi_{Bn}$. (The contact gap is transparent to electron transport between the metal and the silicon.) Since the contact gap in a good metal–semiconductor contact is assumed to be transparent to electron transport, we can omit it from the energy-band diagram completely for purposes of modeling device characteristics of a Schottky diode. The *simplified* energy-band diagram for a Schottky diode is as illustrated in Figure 3.16(d).

It should be noted that Figure 3.16(c) suggests that $\phi_{Bn} \neq \phi_m - \chi$ in cases of high surface-state density. This should be contrasted with the case of no surface states shown in Figure 3.15(b), where we have $\phi_{Bn} = \phi_m - \chi$. If we had drawn the contact gap as a layer of zero thickness, the free electron level would have appeared discontinuous at the metal–silicon interface. Therefore, while in common practice Figure 3.16(d) is shown and used by itself for description of a Schottky diode, the correct and complete physical picture must include a contact gap of finite thickness in order to maintain continuity of free electron level across the metal–semiconductor system. A more detailed discussion on Schottky barriers and surface states, with a mathematical model for the barrier height taking into consideration the work function difference, the density of surface states, and the contact-gap thickness, can be found in Cowley and Sze (1965).

The energy-band diagrams in Figure 3.16 are sketched based on the assumption of well-defined metal and silicon surfaces and a well-defined contact-gap region. For an intimate contact, the metal will tend to broaden the surface levels. Furthermore, Heine (1965) showed that the wavefunction of an electron in a surface state does spread over some finite distance and that the concepts of band bending in a metal–semiconductor contact must not be taken too seriously over distances of the spread of localized electron wavefunctions. Therefore, for an intimate metal–semiconductor contact, the interface boundaries that define the contact-gap region are not as well defined as implied in Figure 3.16(c). Fortunately, for the purposes of describing and establishing the electrical characteristics of a Schottky diode, the details of the contact-gap region are not important because the region is transparent to electron transport.

For a metal–silicon contact represented by Figures 3.16(c) or 3.16(d), the built-in potential is given by

$$q\psi_{bi} = -q\psi_s(\text{contacted surface}) = \frac{[Q_d(\text{contacted surface})]^2}{2\varepsilon_{si}N_d}. \tag{3.101}$$

Once the built-in potential is known, the electron energy barrier can be determined from Eq. (3.99).

3.2.1.3   Measured Barrier Heights

The degree of Fermi level pinning by surface states varies with the semiconductor and often depends on the details of the process used for forming the metal–semiconductor contact. For metal contacts to many "clean" group IV and III–V semiconductor surfaces, the pinning appears to be total, with the measured electron barrier heights practically independent of the metal used (Mead and Spitzer, 1964). In the case of silicon, annealing a metal–silicon system to form a metal–silicide–silicon contact can lead to barrier heights that are different from but more reproducible than the corresponding metal–silicon contact (Andrews, 1974; Andrews and Phillips, 1975). Tung (1992) suggested that there can be lateral inhomogeneity in the distribution of surface states or surface charge as well. Thus, a metal–semiconductor interface can be modeled as consisting of nanometer-sized local patches, with each patch having its own local electron energy barrier (Im *et al.*, 2001). The measured barrier height represents the averaged barrier height of the entire contact.

As can be inferred readily from Figures 3.15(b) and 3.16(d), the hole energy barrier $q\phi_{Bp}$ is related to the electron energy barrier $q\phi_{Bn}$ through

$$q\phi_{Bn} + q\phi_{Bp} = E_g, \qquad (3.102)$$

where E_g is the energy gap of the semiconductor. We focus our discussion on metal contacts to n-type silicon where the barrier height is $q\phi_{Bn}$. Metal contacts to p-type silicon where the barrier height is $q\phi_{Bp}$ will not be discussed explicitly.

3.2.1.4 Image-Force-Induced Barrier Lowering

Consider an electron at a distance x from a metal surface in free space, as illustrated in Figure 3.17. This electron induces an image charge $+q$ located at $-x$ in the metal. The attractive force exerted by the image charge on the electron is

$$F_{image}(x) = \frac{-q^2}{4\pi\varepsilon_0(2x)^2} = \frac{-q^2}{16\pi\varepsilon_0 x^2}, \qquad (3.103)$$

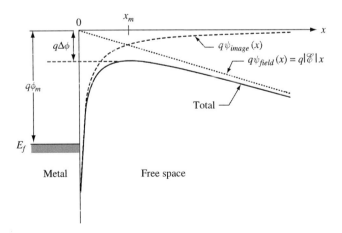

Figure 3.17 Schematic showing the energy-band diagram between a metal and a free space appropriate for electron emission from the metal into the free space. The metal work function is $q\phi_m$. However, the energy barrier for electron emission is $q(\phi_m - \Delta\phi)$

where ε_0 is the vacuum permittivity. The electric field (electrostatic force per unit charge) due to this image force is

$$\mathscr{E}_{image}(x) = \frac{F_{image}(x)}{-q} = \frac{q}{16\pi\varepsilon_0 x^2} . \tag{3.104}$$

The corresponding electrostatic potential $\psi_{image}(x)$ is given by

$$\mathscr{E}_{image}(x) = -\frac{d\psi_{image}(x)}{dx}, \tag{3.105}$$

which can be integrated, subject to the boundary condition of $\psi_{image}(\infty) = 0$, to give

$$\psi_{image}(x) = \int_x^{\infty} \mathscr{E}_{image}(x)\,dx = \frac{q}{16\pi\varepsilon_0 x} . \tag{3.106}$$

When a voltage is applied to pull an electron away from the metal surface, it results in a constant electric field $\mathscr{E}$ in the free space, as indicated in Figure 3.17. The associated electrostatic potential $\psi_{field}(x)$ is given by

$$\mathscr{E} = -\frac{d\psi_{field}(x)}{dx}, \tag{3.107}$$

which can be integrated, subject to the boundary condition of $\psi_{field}(0) = 0$, to give

$$\psi_{field}(x) = |\mathscr{E}|x. \tag{3.108}$$

The total electrostatic potential as seen by an electron being pulled away from a metal surface in free space by an electric field $\mathscr{E}$ is $\psi_{image}(x) + \psi_{field}(x)$. The electron energy associated with this potential is

$$PE(x) = -q\left[\psi_{image}(x) + \psi_{field}(x)\right] = -\left(\frac{q^2}{16\pi\varepsilon_0 x} + q|\mathscr{E}|x\right). \tag{3.109}$$

Note that $PE(x)$ has a peak that lies below the free electron level $(E_f + q\phi_m)$. That is, the combination of an electric field and the image force lowers the energy barrier for electron emission from a metal surface into the free space. This is known as *Schottky effect* or *Schottky barrier lowering*. The amount of energy barrier lowering $q\Delta\phi$ and the location x_m of the peak of $PE(x)$, both indicated in Figure 3.17, can be obtained from $dPE(x)/dx = 0$. The results are:

$$x_m = \sqrt{\frac{q}{16\pi\varepsilon_0|\mathscr{E}|}}, \tag{3.110}$$

and

$$q\Delta\phi = \sqrt{\frac{q^3|\mathscr{E}|}{4\pi\varepsilon_0}}. \tag{3.111}$$

Schottky barrier lowering applies also to a metal–semiconductor system (Schottky diode). In this case, ε_0 in Eq. (3.103) is replaced by the permittivity of the

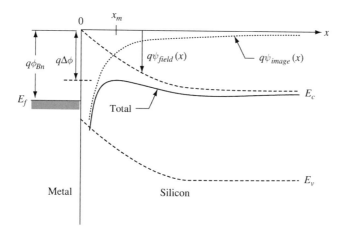

Figure 3.18 Schematic showing the energy-band diagram of a metal–silicon Schottky diode. Also shown are the electrostatic potentials involved in determining the energy barrier for electron emission

semiconductor (Sze *et al.*, 1964). The energy-band diagram of a metal–silicon Schottky barrier diode for illustrating Schottky effect is shown schematically in Figure 3.18. There is an electric field associated with the band bending near the surface. The electric field is not constant because of the charge in the depletion region. The electrostatic potential associated with this field is $q\psi_{field}(x)$. The maximum electric field is located at the metal–silicon interface, and has an absolute value of

$$\mathscr{E}_m = \frac{qN_dW_d}{\varepsilon_{si}}, \tag{3.112}$$

where N_d is the dopant concentration, ε_{si} is the permittivity of silicon, and W_d is the width of the space-charge layer [see Eq. (3.8)]. It is left as an exercise (Exercise 3.10) for the reader to show that, for a typical Schottky diode, the location of the potential barrier peak, x_m, is small compared with W_d. With $x_m \ll W_d$, we have

$$x_m \approx \sqrt{\frac{q}{16\pi\varepsilon_{si}\mathscr{E}_m}}, \tag{3.113}$$

and

$$q\Delta\phi \approx \sqrt{\frac{q^3\mathscr{E}_m}{4\pi\varepsilon_{si}}}. \tag{3.114}$$

That is, *for a metal–semiconductor system, it is the maximum electric field at the surface that determines the amount of Schottky barrier lowering*. For a maximum electric field of $\mathscr{E}_m = 1 \times 10^5$ V/cm, the barrier lowering is $q\Delta\phi = 35$ meV for a silicon Schottky diode.

In the literature, the Schottky effect has also been invoked in the studies of electron injection from a metal or semiconductor into an insulator (Berglund and Powell, 1971) by replacing ε_0 in Eqs. (3.110) and (3.111) by the permittivity of the insulator. However, there are also publications questioning the validity of the concept of image force at the interface between a semiconductor and an insulator. Interested readers are referred to the publications for more detailed discussions (Fischetti *et al.*, 1995, and the references therein).

As a result of Schottky effect, the actual energy barrier in a Schottky diode, as seen by an electron entering the semiconductor from the metal, is $(q\phi_{Bn} - q\Delta\phi)$, with $q\Delta\phi$ given by Eq. (3.114). The total band bending in the silicon is $q(\psi_{bi} - V_{app})$, where V_{app} is the applied voltage across the Schottky diode. The space-charge layer thickness W_d corresponding to the total band banding can be obtained from Eq. (3.15). The maximum electric field $\mathcal{E}_m$ in turn can be obtained using W_d in Eq. (3.8). A forward bias $(V_{app} > 0)$ across a diode reduces the electric field and hence increases the effective energy barrier, while a reverse bias $(V_{app} < 0)$ increases the electric field and hence reduces the effective energy barrier.

3.2.2   Current–Voltage Characteristics of a Schottky Diode

In this subsection, we first examine the mobile carriers involved in the current flow in a Schottky diode. We then discuss the physical mechanisms by which the mobile carriers are transported across the metal–semiconductor interface, and derive the current–voltage characteristics of a Schottky diode.

3.2.2.1   Schottky Diode As a Majority-Carrier Device

The energy-band diagrams illustrating the flow of electrons across an n-silicon Schottky diode are shown schematically in Figure 3.19. For an electron in the metal having an energy $E = E_f$, it sees an energy barrier of $q\phi_{Bn}$ (barrier-lowering effect is ignored for simplicity of discussion). For an electron having an energy $E = E_f + \Delta E$ it sees an energy barrier of $q\phi_{Bn} - \Delta E$. Similarly, for an electron in the quasineutral silicon having an energy of $E = E_c$, it sees an energy barrier of $q(\psi_{bi} - V_{app})$. For an

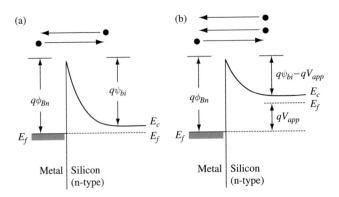

Figure 3.19 Schematic energy-band diagrams illustrating the flow of electrons in an n-type Schottky diode. (a) At thermal equilibrium, there is an equal and opposite flow of electrons. (b) At forward bias, there is a net flow of electrons from the silicon into the metal. For simplicity of illustration, the barrier-lowering effect is not shown

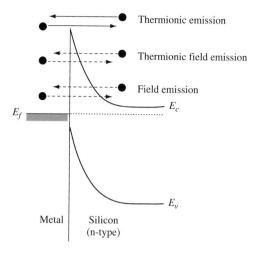

Figure 3.20 Schematic energy-band diagram of a Schottky diode illustrating the principal transport processes

electron having an energy ΔE above E_c, it sees an energy barrier of $q(\psi_{bi} - V_{app}) - \Delta E$. These energy barriers for current flow should not be confused with the energy barrier of the Schottky diode itself, which is $q\phi_{Bn}$.

At thermal equilibrium, there is no net electron flow in either direction in the diode, as indicated in Figure 3.19(a). If a forward voltage V_{app} is applied to the diode, there will be a net electron flow from the n-silicon to the metal, but there are no holes (minority carriers) flowing into the n-silicon, as indicated in Figure 3.19(b). Similarly, in a p-silicon Schottky diode, there are no electrons (minority carriers) flowing into the p-silicon. That is, *the current transport in a Schottky barrier diode is due to majority carriers in the semiconductor*. This should be contrasted with a p–n diode where current transport is mainly due to minority carriers (electrons injected into the conduction band of the p-side and holes injected into the valence band of the n-side).

Being a majority-carrier device, the *response time of a Schottky diode is much shorter than that of a p–n diode*, which is a minority-carrier device. Schottky diodes are often used as microwave diodes and as gates of microwave transistors where speed is important (see e.g., Irvin and Vanderwal, 1969).

The processes by which electrons are transported across the interface in an n-silicon Schottky diode are illustrated in Figure 3.20. Thermionic emission refers to the electrons having sufficient energy to surmount the effective (image-force effect included) energy barrier. Field emission refers to the tunneling of electrons from around the conduction-band edge. Thermionic-field emission describes the tunneling of electrons having energy above the conduction band, but not enough energy to surmount the barrier. For a diode designed to function as an active device or a circuit component, the doping concentration is usually sufficiently light, and therefore the depletion layer thickness sufficiently large, so that thermionic emission is the dominant process for electron transport. Field emission is the dominant transport process in metal–semiconductor contacts where the doping levels are very high (e.g., in ohmic contacts, which will be discussed in Section 3.2.3).

3.2.2.2   Thermionic Emission

In thermionic emission, the simplest model is to treat the electrons as an ideal gas that follows Boltzmann statistics in energy distribution. Electron collision within the semiconductor depletion region is ignored, and only those electrons traveling in the direction of emission and having sufficient energy to surmount the barrier are emitted. In the case of a multi-valley semiconductor like silicon, we should consider the emission current from each valley and then sum the currents to obtain the total current.

The conduction band of silicon has six identical valleys located on the k_x-, k_y-, and k_z-axes. Each valley is an ellipsoid, with a longitudinal mass of $m_l = 0.92\, m_0$ and a transverse mass of $m_t = 0.19\, m_0$, where m_0 is the free-electron mass. Derivation of the thermionic electron emission current is simplest for <100> silicon. Since <100> is the most commonly used silicon orientation, we shall only consider this orientation here. The reader interested in other orientations is referred to the papers by Crowell (1965, 1969).

With electron collision within the depletion region ignored, we need to consider only electrons in the silicon having sufficient starting energy to surmount the energy barrier. From Eq. (2.1), the number of electronic states per unit volume having momenta between p_x and $p_x + dp_x$, between p_y and $p_y + dp_y$, and between p_z and $p_z + dp_z$ in one of the conduction-band valleys is

$$N\left(p_x, p_y, p_z\right) dp_x dp_y dp_z = \frac{2}{h^3} dp_x dp_y dp_z. \tag{3.115}$$

Note that the factor g denoting the number of equivalent minima in the conduction band is unity in Eq. (3.115) because only one valley is being considered. The kinetic energy of an electron in the quasineutral region is $E - E_c$, and the relationship between kinetic energy and momenta is given by Eq. (2.2). In terms of Boltzmann statistics, the probability that an electronic state at energy E is occupied by an electron is $\exp\left[-(E - E_f)/kT\right]$ [see Eq. (2.5)]. Therefore, Eq. (3.115) gives the number of electrons per unit volume having momenta between p_x and $p_x + dp_x$, between p_y and $p_y + dp_y$, and between p_z and $p_z + dp_z$ in one of the conduction-band valleys as

$$
\begin{aligned}
dn\left(p_x, p_y, p_z\right) &= \frac{2}{h^3} e^{-\left(E - E_f\right)/kT} dp_x dp_y dp_z \\
&= \frac{2}{h^3} e^{-\left(E_c - E_f\right)/kT} e^{-\left(p_x^2/2m_x kT + p_y^2/2m_y kT + p_z^2/2m_z kT\right)} dp_x dp_y dp_z.
\end{aligned}
\tag{3.116}
$$

The number of electrons per unit volume having momenta between p_x and $p_x + dp_x$ is given by integrating Eq. (3.116) over all values of p_y and p_z. That is,

$$
\begin{aligned}
dn(p_x) &= \frac{2}{h^3} e^{-\left(E_c - E_f\right)/kT} e^{-p_x^2/2m_x kT} dp_x \int_{-\infty}^{\infty} e^{-p_y^2/2m_y kT} dp_y \int_{-\infty}^{\infty} e^{-p_z^2/2m_z kT} dp_z \\
&= \frac{4\sqrt{m_y m_z}}{h^3} \pi kT e^{-\left(E_c - E_f\right)/kT} e^{-p_x^2/2m_x kT} dp_x.
\end{aligned}
\tag{3.117}
$$

- *Ignoring Barrier Lowering Effect.* In this case, when a voltage V_{app} is applied to a Schottky diode, the minimum energy an electron traveling perpendicularly to the emission surface must have in order to surmount the emission barrier is $q(\psi_{bi} - V_{app})$. For <100> silicon, the emission surface is perpendicular to the k_x-axis. The current density due to thermionic emission of electrons from a single conduction-band valley into the metal is

$$J_{1,s \to m}(V_{app}) = q \int_{p_x=p_{x0}}^{\infty} v_x dn(p_x)$$

$$= \frac{4\sqrt{m_y m_z}}{h^3} q\pi k Te^{-(E_c-E_f)/kT} \int_{p_{x0}}^{\infty} \frac{p_x}{m_x} e^{-p_x^2/2m_x kT} dp_x \qquad (3.118)$$

$$= \frac{4\pi q \sqrt{m_y m_z}}{h^3} k^2 T^2 e^{-(E_c-E_f)/kT} e^{-q(\psi_{bi}-V_{app})/kT},$$

where $v_x = p_x/m_x$ is the velocity of an electron traveling in the x-direction, and the lower integration limit p_{x0} is given by $p_{x0}^2/2m_x = q(\psi_{bi} - V_{app})$. The subscript 1 indicates that the current density is from only one conduction-band valley. Since $q\phi_{Bn} = q\psi_{bi} + E_c - E_f$, Eq. (3.118) can be rewritten as

$$J_{1,s \to m}(V_{app}) = \frac{4\pi q \sqrt{m_y m_z}}{h^3} k^2 T^2 e^{-q\phi_{Bn}/kT} e^{qV_{app}/kT}$$

$$= A \frac{\sqrt{m_y m_z}}{m_0} T^2 e^{-q\phi_{Bn}/kT} e^{qV_{app}/kT}, \qquad (3.119)$$

where the quantity

$$A \equiv \frac{4\pi q m_0 k^2}{h^3} = 120 \text{ A/cm}^2/\text{K}^2 \qquad (3.120)$$

is the Richardson constant for free electrons.

To obtain the total thermionic electron emission current from the silicon, we need to sum the currents from all six valleys. For <100> silicon, the two valleys on the k_x-axis have $m_y = m_z = m_t$, and the four valleys on the k_y- and k_z-axes have $m_y = m_l$ and $m_z = m_t$. The total thermionic electron emission current density from silicon into metal is

$$J_{n-Si<100>,s \to m}(V_{app}) = \sum_1^6 J_{i,n-Si<100>,s \to m}(V_{app})$$

$$= A \left(\frac{2m_t}{m_0} + \frac{4\sqrt{m_t m_l}}{m_0} \right) T^2 e^{-q\phi_{Bn}/kT} e^{qV_{app}/kT} \qquad (3.121)$$

$$= A^*_{n-Si<100>} T^2 e^{-q\phi_{Bn}/kT} e^{qV_{app}/kT},$$

where

$$A^*_{n-Si<100>} = A\left(\frac{2m_t}{m_0} + \frac{4\sqrt{m_t m_l}}{m_0}\right) = 2.05A \qquad (3.122)$$

is the Richardson constant for n-type <100> silicon.

For silicon, the orientation dependence is relatively weak. For n-type <111> silicon, the Richardson's constant is $2.15A$ (Crowell, 1965). The simple thermionic emission model gives only a qualitative description of experimentally measured currents in a typical Schottky diode. The measured Richardson's constant should not be used to infer information about the effective mass tensor (Crowell, 1969). In practice, the Richardson constant is often treated as an adjustable parameter for fitting experimental data (Henisch, 1984).

At zero applied bias, the electron emission current from the metal into the silicon is equal in magnitude but opposite in direction to the electron emission current from the silicon into the metal. That is,

$$J_{n-Si<100>,m\to s}\left(V_{app} = 0\right) = -J_{n-Si<100>,s\to m}\left(V_{app} = 0\right),$$
$$= -A^*_{n-Si<100>}T^2 e^{-q\phi_{Bn}/kT}. \qquad (3.123)$$

When barrier-lowering effect is ignored, the energy barrier for electron emission from metal into silicon is independent of V_{app}. Therefore, we expect the electron emission current from metal into silicon to be independent of V_{app} when barrier-lowering effect is ignored. The total thermionic emission current density for an n-type <100> silicon Schottky barrier diode, when barrier-lowering effect is ignored, is therefore

$$J_{thermionic,n-Si<100>}\left(V_{app}\right) = J_{n-Si<100>,s\to m}\left(V_{app}\right) + J_{n-Si<100>,m\to s}\left(V_{app}\right)$$
$$= A^*_{n-Si<100>}T^2 e^{-q\phi_{Bn}/kT}\left(e^{qV_{app}/kT} - 1\right). \qquad (3.124)$$

For a forward-biased diode $\left(V_{app} > 0\right)$, the current is dominated by the emission from the semiconductor into the metal. For a reverse-biased diode $\left(V_{app} < 0\right)$, the current is dominated by the emission from the metal into the semiconductor. Equation (3.124) shows that, when barrier-lowering effect is ignored, a Schottky diode has $I - V$ characteristics similar to those of a p–n diode [cf. Eq. (3.67)], with an $\exp(qV_{app}/kT)$ dependence on V_{app} in forward bias, and a saturation current that is independent of V_{app} in reverse bias.

- Including Barrier Lowering Effect. There is a subtle difference between a Schottky diode and a p–n diode when the barrier-lowering effect is included. When the barrier-lowering effect is included, the Schottky barrier $q\phi_{Bn}$ in Figure 3.19 and in Eqs. (3.119) to (3.124) should be replaced by an effective Schottky barrier $q(\phi_{Bn} - \Delta\phi)$. The barrier-lowering term $q\Delta\phi$ depends on the applied voltage through the electric field $\mathscr{E}_m$ [see Eq. (3.114)]. A forward bias $\left(V_{app} > 0\right)$ reduces $q\Delta\phi$ and hence increases the effective Schottky barrier, while a reverse bias $\left(V_{app} < 0\right)$ increases $q\Delta\phi$ and hence reduces the effective Schottky barrier. Thus, replacing $q\phi_{Bn}$ by $q(\phi_{Bn} - \Delta\phi)$ in Eq. (3.124) suggests that the forward-bias current of a Schottky diode increases with V_{app} at a rate somewhat slower than $\exp(qV_{app}/kT)$.

In the literature, more complex theories have been proposed for describing the transport process in Schottky diodes. There is a diffusion emission theory which includes the effect of electron collisions within the semiconductor depletion region. There is also a theory which combines the physics involved in the simple thermionic emission process and the diffusion emission process (Crowell and Sze, 1966a). All theories result in an equation similar to Eq. (3.124), with the difference only in the pre-exponential factor. The interested reader is referred to the literature for the details (Sze, 1981; Henisch, 1984). From a device point of view, the important $I–V$ characteristics of a Schottky diode are contained in the exponential factors in Eq. (3.124), namely in the $\exp(-q\phi_{Bn}/kT)$ dependence on $q\phi_{Bn}$ and the $\left[\exp(qV_{app}/kT) - 1\right]$ dependence on qV_{app}.

3.2.2.3   Field Emission and Thermionic-Field Emission

If the semiconductor is heavily doped, the depletion region thickness will be thin, and the electron transport can become dominated by a combination of field emission and/or thermionic-field emission. In this case, large currents can flow even at low applied biases. In general, when field emission and thermionic-field emission dominate the electron transport, a metal–semiconductor contact is no longer useful as a rectifying diode. As a result, we will not consider the theory of thermionic-field emission any further. The interested reader is referred to the literature for details (Padovani and Stratton, 1966; Crowell and Rideout, 1969). Field emission is considered in Section 3.2.3 in the context of ohmic contact.

3.2.3   Ohmic Contacts

Ohmic contacts are usually made with metal or metal silicide in contact with a heavily doped semiconductor. The electron transport process in this case is dominated by field emission. Let us first consider the tunneling of a conduction-band electron from the quasineutral semiconductor region into the metal. The band bending near the metal–semiconductor contact is illustrated in Figure 3.21. The total band bending is $q\psi_m = q(\psi_{bi} - V_{app})$ when a forward bias of V_{app} is applied. For a given V_{app}, let us assume the conduction-band starts to bend upward at $x = 0$, and the interface is located at $x = W_d$, where W_d is the depletion-layer thickness. Since we are considering an electron in the conduction band, it is convenient to use the conduction-band edge of the quasineutral silicon region $E_c(x < 0)$ as the energy reference, as indicated in Figure 3.21. $\psi(x)$ is the electrostatic potential at location x relative to $E_c(x < 0)$, i.e., $-q\psi(0) = E_c(x < 0)$, and $-q\psi(x)$ is the potential energy of an electron at location x. The Poisson's equation [Eq. (2.58)] can be integrated twice to give

$$\psi(x) = -\frac{qN_d x^2}{2\varepsilon_{si}} - \frac{E_c(x < 0)}{q},$$

(3.125)

where N_d is the semiconductor doping concentration. From Eq. (3.15), we have

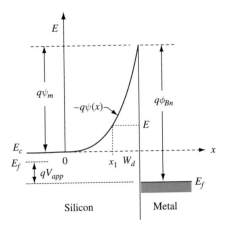

Figure 3.21 Schematic showing the energy bands appropriate for considering field emission in a metal–silicon contact. As illustrated, the Schottky diode is forward biased, as indicated by the Fermi level in the silicon being higher than that in the metal

$$W_d = \sqrt{\frac{2\varepsilon_{si}\left(\psi_{bi} - V_{app}\right)}{qN_d}}. \tag{3.126}$$

In the WKB (Wentzel–Kramers–Brillouin) approximation for tunneling through an energy barrier, the transmission coefficient through the energy barrier represented by $-q\psi(x)$ for an electron with energy E is (Liboff, 2003)

$$
\begin{aligned}
T(E) &= \exp\left[\frac{-4\pi}{h}\int_{x_1}^{W_d}\sqrt{2m^*}\sqrt{-q\psi(x) - E}dx\right]\\
&= \exp\left[\frac{-4\pi}{h}\int_{x_1}^{W_d}\sqrt{2m^*}\sqrt{\frac{q^2N_dx^2}{2\varepsilon_{si}} + E_c(x < 0) - E}dx\right],
\end{aligned}
\tag{3.127}
$$

where the lower integration limit x_1 is given by $-q\psi(x_1) = E$.

In considering an ohmic contact, we are interested in the current due to electrons tunneling from the quasineutral region of the silicon through the potential barrier into the metal at small applied voltages. These electrons have only thermal energy ($kT \approx 26$ meV at room temperature) which is small compared to the maximum tunneling barrier height $q\left(\psi_{bi} - V_{app}\right)$, which is approximately equal to $q\psi_{bi}$ at small V_{app}. Therefore, we can assume the tunneling electrons to have an energy $E \approx E_c(x < 0)$. For these electrons, the tunneling process starts at $x_1 = 0$ and the corresponding transmission coefficient is

$$T[E = E_c(x < 0)] = \exp\left(\frac{-q\left(\psi_{bi} - V_{app}\right)}{E_{00}}\right), \tag{3.128}$$

where

$$E_{00} \equiv \frac{qh}{4\pi} \sqrt{\frac{N_d}{m^* \varepsilon_{si}}}. \tag{3.129}$$

For a heavily doped silicon region, its Fermi level and conduction-band edge are about equal. That is, referring to Figure 3.19(a), we have ψ_{bi}(heavily doped) $\approx \phi_{Bn}$. Therefore, for a heavily doped Schottky diode, we have

$$T[(E = E_c(x < 0)] = \exp\left(\frac{-q(\phi_{Bn} - V_{app})}{E_{00}}\right). \tag{3.130}$$

That is, for an ohmic contact, the current density varies as

$$J_{ohmic} \propto \exp\left(\frac{-q(\phi_{Bn} - V_{app})}{E_{00}}\right). \tag{3.131}$$

The specific contact resistance, or contact resistivity, ρ_c, is an important figure of merit for ohmic contacts:

$$\rho_c \equiv \left(\frac{\partial J}{\partial V_{app}}\right)^{-1}_{V_{app}=0}. \tag{3.132}$$

Using Eq. (3.131) for ohmic contacts, we have

$$\rho_c \propto \frac{E_{00}}{q} \exp(q\phi_{Bn}/E_{00}). \tag{3.133}$$

The behavior of ρ_c is dominated by the exponential factor. That is, to ensure a low contact resistance, a low-barrier metal should be used and the silicon should be as heavily doped as possible (to maximize E_{00}). For $N_d = 10^{20}$ cm^{-3} and $T = 300\ K$, ρ_c is about 8×10^{-6} Ω-cm^2 for $q\phi_{Bn} = 0.6$ eV, and about 8×10^{-8} Ω-cm^2 for $q\phi_{Bn} = 0.4$ eV (Yu, 1970; Chang et al., 1971). Recent experiments suggest that using a combination of contact-barrier and contact-metallurgy engineering, and maximizing doping concentration, contact resistivity approaching 10^{-9} Ω-cm^2 may be obtained (Zhang et al., 2013). Experimental determination of the resistance of a contact can be very involved for low-resistance contacts. The reader is referred to the literature for a discussion and comparison of the many contact-resistance measurement methods (Schroder, 1990).

3.3 High-Field Effects in Reverse-Biased Diodes

The electric field in the space-charge region of a diode increases with its reverse-bias voltage. In the presence of an electric field, carriers gain energy from the field as they drift along. These carriers in turn lose energy by emitting phonons. As the field increases, the average energy of the carriers increases. At sufficiently high fields, a number of physical phenomena which have important implications on the design and operation of VLSI devices can occur. In the case of high fields in bulk silicon, away from the silicon/oxide interface, these phenomena include impact ionization, or generation of electron–hole pairs, junction breakdown, and band-to-band tunneling. The basic physics of these phenomena is discussed in this section. High-field effects related to silicon/oxide interface are covered in Section 4.6.

3.3.1   Impact Ionization and Avalanche Breakdown

Consider the space-charge region of a p–n diode. At sufficiently high fields, an electron in the conduction band can gain enough energy to "lift" an electron from the valence band into the conduction band, thus generating one free electron in the conduction band and one free hole in the valence band. This process is known as *impact ionization*. Similarly, a hole in the valence band can gain enough energy to cause impact ionization. If the field is high enough, these secondary electrons and holes can themselves cause further impact ionization, thus beginning a process of carrier multiplication in the high-field region. The p–n diode breaks down when the multiplication process runs away or becomes an avalanche.

Consider a reverse-biased p–n diode with an electric field in its space-charge region high enough to cause impact ionization. Let $x = 0$ and $x = W$ be the locations of the two boundaries of the space-charge region. Suppose there is a hole current I_{p0} entering the space-charge region at $x = 0$, as illustrated in Figure 3.22. This hole current generates electron–hole pairs. The secondary electrons and holes in turn cause further impact ionization as they traverse the space-charge region. Thus, the hole current will increase with distance, reaching a value of $M_p I_{p0}$ at $x = W$, where M_p is the multiplication factor for holes. At steady state, the total current I is constant and independent of distance, i.e., $I = M_p I_{p0}$. Within the space-charge region, the total current is the sum of the hole and electron currents (Moll, 1964), i.e.,

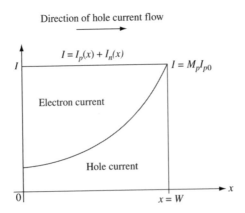

Figure 3.22 Schematic illustrating the steady-state current caused by hole-initiated impact ionization within the space-charge region of a p–n diode

$$I = I_p(x) + I_n(x). \tag{3.134}$$

These current components are illustrated in Figure 3.22. The field is such that holes move toward the right $(x = W)$, and electrons move toward the left $(x = 0)$.

Consider a differential distance between x and $x + dx$. There are $I_p(x)/q$ holes and $I_n(x)/q$ electrons crossing this differential distance per unit time. In crossing this differential distance, the holes cause $\alpha_p(x)I_p(x)dx/q$ electron–hole pairs to be generated, where α_p is the *hole-initiated rate of electron–hole pair generation per unit distance*. Similarly, the number of electron–hole pairs generated by the electrons is $\alpha_n(x)I_n(x)dx/q$, where α_n is the *electron-initiated rate of electron–hole pair generation per unit distance*. Thus, the increase in hole current as the electrons and holes cross the differential distance dx is

$$dI_p = \alpha_p I_p dx + \alpha_n I_n dx. \tag{3.135}$$

Equations (3.134) and (3.135) give

$$\frac{dI_p}{dx} = (\alpha_p - \alpha_n)I_p + \alpha_n I, \tag{3.136}$$

which, subject to the boundary condition $I_p(0) = I/M_p$, has a solution (Sze, 1981)

$$I_p(x) = I\left[\frac{1}{M_p} + \int_0^x \alpha_n \exp\left(-\int_0^{x'} (\alpha_p - \alpha_n)dx''\right)dx'\right]\exp\left(\int_0^x (\alpha_p - \alpha_n)dx'\right). \tag{3.137}$$

Since we are considering hole-initiated impact ionization, and there is no electron current entering the depletion region at $x = W$, the hole current at $x = W$ is simply equal to I. Therefore, Eq. (3.137) gives

$$\frac{1}{M_p} = \exp\left(-\int_0^W (\alpha_p - \alpha_n)dx\right) - \int_0^W \alpha_n \exp\left(-\int_0^x (\alpha_p - \alpha_n)dx'\right)dx. \tag{3.138}$$

Similarly, for impact ionization initiated by electrons, the electron multiplication factor M_n is given by

$$\frac{1}{M_n} = \exp\left(-\int_0^W (\alpha_n - \alpha_p)dx\right) - \int_0^W \alpha_p \exp\left(-\int_x^W (\alpha_n - \alpha_p)dx'\right)dx. \tag{3.139}$$

Avalanche breakdown occurs when carrier multiplication by impact ionization runs away, i.e., when the multiplication factors become infinite. It can be shown (see Exercise 3.11) that *the condition for avalanche breakdown is the same whether the breakdown process is initiated by electrons or by holes*.

It should be noted that avalanche multiplication of electrons and holes is a positive feedback process where both the electrons and holes generated by impact ionization take part in generating additional electrons and holes. As a result, it is possible to have avalanche breakdown, i.e., M_n and M_p being infinite, for finite values of W, α_n and α_p at some large reverse bias voltage. If the feedback process by either the secondary electrons or the secondary holes were absent, avalanche breakdown would not occur for finite values of α_n and α_p. To demonstrate this point, let us consider the case of impact ionization initiated by holes, where the multiplication factor M_p is given by Eq. (3.138). If there were no positive feedback by the secondary electrons, we would have $\alpha_n = 0$. In this case, Eq. (3.138) shows that M_p would be finite (no breakdown) for any finite values of α_p.

If the high-field region where impact ionization occurs is sufficiently wide, the positive feedback process will eventually lead to avalanche breakdown. It is left to the reader to show that, for the special case of both α_n and α_p being constant, independent of distance or electric field, avalanche breakdown occurs when the width of the high-field region approaches a value of $\left[\ln(\alpha_p/\alpha_n) \right] / \left(\alpha_p - \alpha_n \right)$ (see Exercise 3.1).

In theory, Eqs. (3.138) and (3.139) can be used to calculate the multiplication factors, and hence the breakdown voltage. In practice, however, the ionization rates, as well as the junction doping profiles, are simply not known accurately enough for calculation of breakdown voltages to be made with sufficient accuracy for VLSI device design purposes. Breakdown voltages are usually determined experimentally.

3.3.1.1   Empirical Impact Ionization Rates

The measured ionization rates are often expressed in the ***empirical*** form of

$$\alpha = A \exp(-b/\mathscr{E}) \tag{3.140}$$

where A and b are constants, and $\mathscr{E}$ is the electric field (Chynoweth, 1957). There is quite a bit of spread in the measured impact ionization rates reported in the literature. However, the most recent measurements give similar results (van Overstraeten and de Man, 1970; Grant, 1973). These results are shown in Table 3.1 and plotted in Figure 3.23.

Two points are clear from Figure 3.23. First, α_n is much larger than α_p, particularly at low electric fields. This is due to the effective mass of holes being much larger than that of electrons. Second, the impact ionization rates increase very rapidly with electric field. For the space-charge region of a p–n diode where the electric field is not constant, it is the small region surrounding the maximum-field point that contributes

Table 3.1. Impact-ionization rates in silicon

Data	Field Range (V/cm)	$\alpha_n\,(\mathrm{cm}^{-1})$		$\alpha_p\,(\mathrm{cm}^{-1})$	
		$A_n\,(\mathrm{cm}^{-1})$	$b_n\,(\mathrm{V/cm})$	$A_p\,(\mathrm{cm}^{-1})$	$b_p\,(\mathrm{V/cm})$
van Overstraeten and de Man (1970)	$1.75 \times 10^5 < \mathscr{E} < 4.0 \times 10^5$	7.03×10^5	1.231×10^6	1.582×10^6	2.036×10^6
	$4.0 \times 10^5 < \mathscr{E} < 6.0 \times 10^5$	7.03×10^5	1.231×10^6	6.71×10^5	1.693×10^6
Grant (1973)	$2.0 \times 10^5 < \mathscr{E} < 2.4 \times 10^5$	2.6×10^6	1.43×10^6	2.0×10^6	1.97×10^6
	$2.4 \times 10^5 < \mathscr{E} < 5.3 \times 10^5$	6.2×10^5	1.08×10^6	2.0×10^6	1.97×10^6
	$5.3 \times 10^5 < \mathscr{E}$	5.0×10^5	0.99×10^6	5.6×10^5	1.32×10^6

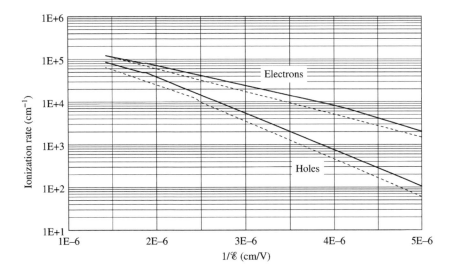

Figure 3.23 Impact-ionization rates in silicon. The solid curves are data of Grant (1973), and the dash curves are data of van Overstraeten and de Man (1970)

the most to the impact-ionization currents. Thus, to minimize impact ionization in a p–n diode, the maximum electric field should be minimized. As mentioned in Section 3.1.2.3, doping-profile grading, or using lightly doped regions or i-layers can effectively reduce the peak electric field in a p–n junction.

Impact ionization rates decrease as temperature increases (Grant, 1973). This is due to the increased lattice scattering at higher temperatures. The data in Table 3.1 and Figure 3.23 are for room temperature.

3.3.2 Band-to-Band Tunneling

When the electric field across a reverse-biased p–n junction approaches 10^6 V/cm, significant current flow can occur due to tunneling of electrons from the valence band of the p-region into the conduction band of the n-region. This phenomenon is illustrated schematically in Figure 3.24. In silicon this tunneling process usually involves the emission or absorption of phonons (Chynoweth *et al.*, 1960; Kane, 1961), and the tunneling current density is given by (Fair and Wivell, 1976)

$$J_{b-b} = \frac{\sqrt{2m^*}\, q^3 \mathscr{E} V_R}{4\pi^3 \hbar^2 E_g^{1/2}} \exp\left(-\frac{4\sqrt{2m^*}\, E_g^{3/2}}{3q\mathscr{E}\hbar}\right) \qquad (3.141)$$

where $\mathscr{E}$ is the electric field, E_g is the bandgap energy, and V_R is the reverse bias voltage across the junction. An upper-bound estimate of the peak electric field can be made by assuming a one-sided junction. In this case, the analyses in Section 3.1.2.2 give the upper-bound for the electric field as

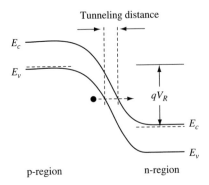

Figure 3.24 Schematic illustrating band-to-band tunneling in a p−n junction

$$\mathscr{E}_{max} = \sqrt{\frac{2qN_a(V_R + \psi_{bi})}{\varepsilon_{si}}}, \tag{3.142}$$

where N_a is the doping concentration of the lightly doped side (assumed p-type) of the diode and ψ_{bi} is the built-in potential of the diode. With these approximations, the band-to-band tunneling current density is about 1 A/cm^2 for $N_a = 5 \times 10^{18}$ cm^{-3} and $V_R = 1$ V (Taur *et al.*, 1995).

More recently, Solomon *et al.* (2004) showed that band-to-band tunneling current can be modeled using the concept of an effective tunneling distance. In this model, the tunneling current is assumed to be proportional to $\exp(-w_T/\lambda_T)$, where w_T is the tunneling distance, illustrated schematically in Figure 3.24, and λ_T is an effective tunneling decay length. The reader is referred to Solomon's paper for the details.

In scaling down the dimensions of a transistor, the doping concentrations increase and the junction doping profiles become more abrupt, and hence band-to-band tunneling effect increases. Once the leakage current due to band-to-band tunneling is appreciable, it increases very rapidly with electric field or reduction of the tunneling distance. For modern VLSI devices, band-to-band tunneling is becoming one of the most important leakage−current components, particularly for applications such as DRAM and battery-operated systems where leakage currents must be kept extremely low.

Exercises

3.1   The multiplication factors for holes and for electrons are given by Eqs. (3.138) and (3.139), respectively. For the special case of constant α_p and α_n, show that $M_p \to \infty$ occurs when the depletion-layer width approaches the value of $W = \ln(\alpha_n/\alpha_p)/(\alpha_n - \alpha_p)$. Also show that the condition for $M_n \to \infty$ gives the same result for W.

3.2 Prove the following mathematical identities:

$$\int_0^W f(x) \exp\left(-\int_0^x f(x')\, dx'\right) dx = 1 - \exp\left(-\int_0^W f(x)\, dx\right)$$

and

$$\int_0^W f(x) \exp\left(-\int_x^W f(x')\, dx'\right) dx = 1 - \exp\left(-\int_0^W f(x)\, dx\right).$$

These identities are used in Exercise 3.11 to show that the condition for hole-initiated avalanche breakdown, namely $1/M_p \to 0$, is the same as that for electron-initiated avalanche breakdown, namely $1/M_n \to 0$.

3.3 The depletion-layer capacitance per unit area C_d of a uniformly doped abrupt p–n diode and its dependence on doping concentration and applied voltage are given in Eqs. (3.10), (3.11), and (3.13). Sketch $1/C_d^2$ as a function of the applied reverse-bias voltage V_{app}. Show how this plot can be used to determine N_a and N_d.

3.4 The depletion-layer capacitance of a one-sided p–n diode is often used to determine the doping profile of the lightly doped side. Consider an n^+–p diode, with a nonuniform p-side doping concentration of $N_a(x)$. If $Q_d(V)$ is the depletion-layer charge per unit area at bias voltage V, the capacitance per unit area at bias voltage V is $C = dQ_d/dV$. In terms of the depletion-layer width W, we have $C(V) = \varepsilon_{si}/W$, where W is a function of V. (For simplicity, we have dropped the subscripts in C, W, and V here.) Show that the doping concentration at the depletion-layer edge is given by

$$N_a(W) = \frac{2}{q\varepsilon_{si} d\left(1/C^2\right)/dV}.$$

3.5 The charge distribution of a p–i–n diode is shown schematically in Figure 3.6. The i-layer thickness is d. The depletion-layer capacitance is given by Eq. (3.26), namely $C_d = \varepsilon_{si}/W_d$, where $W_d = x_n + x_p$ is the total depletion-layer width. Derive this result from $C_d = dQ_d/dV$.

3.6 Show that the band-to-band tunneling exponent in Eq. (3.141) can be derived from the WKB approximation, i.e. Eq. (3.127), for tunneling through a triangular barrier of height E_g, slope $q\mathscr{E}$, and tunneling distance $E_g/q\mathscr{E}$.

3.7 Assume silicon, room temperature, complete ionization. An abrupt p–n junction with $N_a = N_d = 10^{17}\,\text{cm}^{-3}$ is reverse-biased at 2.0 V.
(a) Draw the band diagram. Label the Fermi levels and indicate where the voltage appears.
(b) What is the total depletion layer width?
(c) What is the maximum field in the junction?

3.8 For an abrupt n^+–p diode in Si, the n^+ doping is $10^{20}\,\text{cm}^{-3}$, the p-type doping is $3 \times 10^{16}\,\text{cm}^{-3}$. Assume room temperature and complete ionization.

(a) Draw the band diagram at zero bias. Indicate $x = 0$ as the boundary where the doping changes from n^+ to p. Also indicate where the Fermi level is with respect to the midgap.

(b) Write the equation and calculate the built-in potential.

(c) Write the equation and calculate the depletion-layer width.

(d) Will the built-in potential increase or decrease if the temperature goes up and why?

3.9 Following the discussion in Section 3.1.3.3, show that

$$\frac{n_p(0)}{n_{p0}} = \frac{e^{qV_{app}/kT} + \delta}{1 + \delta}$$

and

$$\phi_n(-W_d) - \phi_n(0) = \Delta\psi_i - \frac{kT}{q} \ln\left[\frac{N_d}{n_p(0)}\right]$$

$$= \frac{kT}{q} \ln\left[\frac{1 + \delta e^{-qV_{app}/kT}}{1 + \delta}\right],$$

where

$$\delta \equiv \frac{1}{L_n \tanh(W_p/L_n)} \int_{-W_d}^{0} e^{q[\psi_i(0)-\psi_i(x)]/kT} dx.$$

3.10 The energy-band diagram of a metal–silicon Schottky barrier diode and the electrostatic potentials involved in determining the energy barrier for electron emission is shown schematically in Figure 3.18. Derive the equation for the total potential energy $PE(x) = -q[\psi_{image}(x) + \psi_{field}(x)]$ in terms of the dopant concentration N_d and the depletion layer thickness W_d. Show that the location x_m of the peak of $PE(x)$ is small compared with the depletion layer thickness W_d in the silicon. Show that x_m is given by

$$x_m \approx \sqrt{\frac{q}{16\pi\varepsilon_{si}\mathscr{E}_m}}$$

and the image-force-induced barrier lowering $q\Delta\phi$ is given by

$$q\Delta\phi \approx \sqrt{\frac{q^3\mathscr{E}_m}{4\pi\varepsilon_{si}}}.$$

where $\mathscr{E}_m$ is absolute value of the electric field in the silicon at the metal–silicon interface.

3.11 In Section 3.3.1, we showed that the multiplication factor M_p for hole-initiated impact ionization is given by

$$\frac{1}{M_p} = \exp\left(-\int_0^W (\alpha_p - \alpha_n)dx\right) - \int_0^W \alpha_n \exp\left(-\int_0^x (\alpha_p - \alpha_n)dx'\right)dx,$$

and the multiplication factor M_n for electron-initiated impact ionization is given by

$$\frac{1}{M_n} = \exp\left(-\int_0^W (\alpha_n - \alpha_p)dx\right) - \int_0^W \alpha_p \exp\left(-\int_x^W (\alpha_n - \alpha_p)dx'\right)dx.$$

Show that when $1/M_p = 0$, $1/M_n$ is also zero. That is, *the condition for avalanche breakdown is the same whether the breakdown process is initiated by electrons or by holes* [Hint: Use the results from Exercise 3.2].

3.12 Consider an abrupt p–n junction with $N_a = 5 \times 10^{16}$ cm^{-3} and $N_d = 10^{17}$ cm^{-3} at zero bias. Assume $x = 0$ is the metallurgical junction, i.e., p-type on $x < 0$ and n-type on $x > 0$. Apply the depletion approximation to answer the following.
(a) What is $E_i - E_f$ at $x = 0$? Include the sign with the value.
(b) At what x does E_i cross E_f, i.e., $E_i = E_f$?

3.13 An abrupt p–n junction with $N_a = 10^{17}$ cm^{-3} and $N_d = 10^{18}$ cm^{-3} is forward biased at 0.7 V. The diffusivities are $D_n = 10$ cm^2/s and $D_p = 5$ cm^2/s. The minority carrier diffusion lengths are $L_n = 20$ μm and $L_p = 5$ μm.

Consider the quasineutral p-side (positioned to the right of the n-side), with the edge of the quasineutral p-region located at $x = 0$. Assume the contacts are far away (∞) from the junction and there is no generation–recombination in the depletion region.
(a) What are the electron and hole current densities at $x = 0$?
(b) What are the electron and hole current densities at $x = 10$ μm?

3.14 An abrupt p-n junction with $N_d = 10^{20}$ cm^{-3} and $N_a = 10^{17}$ cm^{-3} is forward biased at 0.6 V. The diffusivities are $D_n = 10$ cm^2/s and $D_p = 5$ cm^2/s. The minority carrier diffusion lengths are $L_n = 20$ μm and $L_p = 5$ μm.

Assume the p-side is positioned to the right of the n-side, with the edge of the quasineutral p-region located at $x = 0$. Assume the contacts are far away (∞) from the junction and there is no generation–recombination in the depletion region.
(a) What is the depletion region capacitance at this bias?
(b) What is the total current density at this bias?
(c) What is the electron current density at $x = 10$ μm on the p-side?

3.15 An abrupt p-n junction with $N_a = 3 \times 10^{16}$ cm^{-3} and $N_d = 10^{17}$ cm^{-3} is forward biased at 0.5 V.
(a) What is the depletion charge per unit area on the p-side of the junction? Is it positive or negative?
(b) What is the depletion layer capacitance per unit area at this bias?

3.16 An abrupt p–n junction with $N_d = 10^{20}$ cm^{-3} and $N_a = 10^{17}$ cm^{-3} is forward-biased at 0.6 V.
(a) What are the total depletion layer width and the maximum field in the junction?
(b) What is the electron density at the edge of the depletion region on the p-side?

3.17   An abrupt p–n junction with $N_d = 10^{18}$ cm^{-3} and $N_a = 10^{17}$ cm^{-3} is forward-biased at 0.5 V.

(a)   Draw the band diagram, with the p-region positioned to the left of the n-region. Label the Fermi levels and indicate where the voltage appears.

(b)   What are the total depletion layer width and the maximum field in the junction?

(c)   What is the electron density at the edge of the depletion region on the p-side?

3.18   An abrupt p–n junction with $N_a = 5 \times 10^{15}$ cm^{-3} and $N_d = 10^{16}$ cm^{-3} is biased at 0V.

(a)   What is the built-in potential ψ_{bi}?

(b)   What is the depletion region width W_d, and how is it divided between n- and p-side?

(c)   What is the maximum electric field in the depletion region?

4 MOS Capacitors

The basis of a MOSFET device lies in the metal–oxide–semiconductor (MOS) capacitor. The Si – SiO$_2$ MOS system has been studied extensively (Nicollian and Brews, 1982) because it is directly related to most planar devices and integrated circuits. In this chapter, we review the fundamental properties of MOS capacitors and the basic equations that govern their operation. The effects of charge traps in the oxide layer and at the oxide–silicon interface are discussed in Section 4.5.

For CMOS logic technologies beyond the 22-nm node (meaning lithographic line-width resolution below 22 nm), the VLSI industry moved from SiO$_2$ to high-κ dielectric and metal gate technology to deal with the gate oxide tunneling and polysilicon gate depletion problems. These subjects are discussed in Sections 4.3.5 and 4.6.1.

4.1   Energy Band Diagram of an MOS System

4.1.1   Free Electron Level, Work Functsion, and Flatband Voltage

The cross-section of an MOS capacitor is shown in Figure 4.1. It consists of a conducting gate electrode (metal or heavily doped polysilicon) on top of a thin layer of silicon dioxide grown on a silicon substrate. The energy band diagrams of the three components when separate are shown in Figure 4.2. Before we discuss the energy band diagram of an MOS device, it is necessary to first introduce the concept of *free electron level* and *work function* which play key roles in the relative energy band placement when two different materials are brought into contact. Figure 4.2(c) shows the band diagram of a p-type silicon with the addition of the free electron level at some energy above the conduction band. The free electron level is defined as the energy level at which the electron is free, i.e., no longer bonded to the lattice.[1] In silicon, the free electron level is 4.05 eV above the conduction band edge, as shown in Figure. 4.2(c). In other words, an electron at the conduction band edge must gain an additional energy of 4.05 eV (called the *electron affinity*, $q\chi$) in order to break free from the crystal field of silicon. Figure 4.2(b) shows the band diagram of silicon dioxide – an

---

[1]   In other texts, the free electron level is often referred to as the *vacuum level*. Here we use a different term to avoid the implication that the vacuum level is universal.

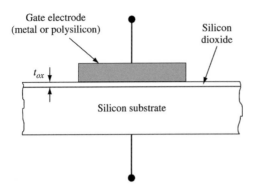

Figure 4.1 Schematic cross-section of an MOS capacitor

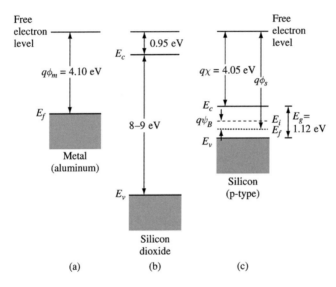

Figure 4.2 Energy-band diagram of the three components of an MOS capacitor: (a) metal (aluminum), (b) silicon dioxide, and (c) p-type silicon

insulator with a large energy gap in the range of 8–9 eV. The free electron level in silicon dioxide is 0.95 eV above its conduction band.

Work function is defined as the energy difference between the free electron level and the Fermi level. For the p-type silicon example in Figure 4.2(c), the work function, $q\phi_s$, can be expressed as:

$$q\phi_s = q\chi + \frac{E_g}{2} + q\psi_B.$$

(4.1)

Here ψ_B, given by Eq. (2.62), is the difference between the Fermi potential and the intrinsic potential. The same definition of work function, $q\phi_m$, applies to metals (remember that the conduction band is half filled in metals), as shown in

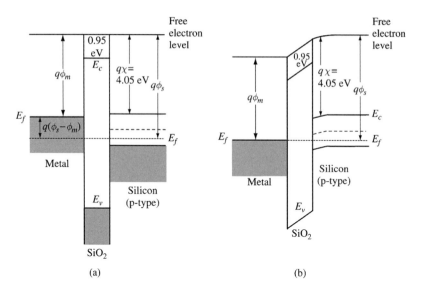

Figure 4.3 Band diagrams of an MOS system under (a) the flatband condition, and (b) zero gate-voltage condition

Figure 4.2(a). It means that an electron at the Fermi level needs to receive an energy equal to $q\phi_m$ to be free from the metal. Different metals have different work functions.

When two different materials are brought into contact, they must share the same free electron level at the interface, i.e., the free electron level is continuous from one material to the next. This is because, at the interface of two materials, an electron that is free from the crystal field of one material is also free from the crystal field of the other material. Figure 4.3(a) shows the band diagram of an MOS system under the *flatband condition* in which there is no field in all three materials. Since for this example the metal work function is less than the silicon work function, the flatband condition is reached by applying a negative gate voltage, $-(\phi_s - \phi_m) \equiv \phi_{ms}$, called the *flatband voltage*, with respect to the silicon substrate. This is seen in Figure 4.3(a) as the displacement between the two Fermi levels. In general, the flatband voltage of an MOS device is given by

$$V_{fb} = (\phi_m - \phi_s) - \frac{Q_{ox}}{C_{ox}} = \phi_{ms} - \frac{Q_{ox}}{C_{ox}}, \tag{4.2}$$

where Q_{ox} is the equivalent oxide charge per unit area at the oxide–silicon interface (discussed in Section 4.5), and C_{ox} is the oxide capacitance per unit area,

$$C_{ox} = \frac{\varepsilon_{ox}}{t_{ox}} \tag{4.3}$$

for an oxide film of thickness t_{ox} and permittivity ε_{ox}. In modern VLSI technologies, Q_{ox}/q at the $Si - SiO_2$ interface can be controlled to below 10^{10} cm^{-2} for <100>-oriented surfaces. Its contribution to the flatband voltage is negligible for gate oxides thinner than ~10 nm. Therefore, the flatband voltage is mainly determined by the work function difference ϕ_{ms}.

At zero gate voltage when the Fermi levels line up, electric fields are developed in both the oxide and silicon, as shown in Figure 4.3(b). One should note that the electron affinity, $q\chi$, is a material property which depends only on the type of the semiconductor and does not change with either the doping type or the depth. When there is band bending, as in Figure 4.3(b) or, e.g., as in the depletion region of a p–n junction, the free electron level would bend in parallel with the conduction band such that the distance between the free electron level and the conduction band remains constant, much like the energy gap. While in this case the free electron level is not flat within one material, it must still be continuous between adjacent materials. Because of the common free electron level at the interface, *the electron energy barrier is* $q\phi_{ox} = 4.05\,eV - 0.95\,eV = 3.1\,eV$ *between the conduction bands of silicon and silicon dioxide*. This figure has important significance when discussing the reliability of $Si - SiO_2$ systems (Section 4.6). For simplicity, the free electron level is mostly omitted in subsequent MOS band diagrams. One should keep in mind its essential role in setting the relative band placement between different materials.

The same principle of continuity of free-electron level at the interface applies to metal–semiconductor junctions (Schottky diodes discussed in Section 3.2) and hetero-junction devices (different bandgaps in a device) for establishing their band alignment relationships.

4.1.2   Gate Voltage, Surface Potential, and Charge in Silicon

Figure 4.4 shows the band diagram when a nonzero gate voltage is applied to an MOS device. The free electron level is only shown at the top of the oxide region, not over

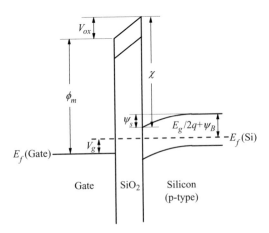

Figure 4.4 Energy band and potential diagrams of an MOS capacitor when a positive gate voltage is applied. Note that in the diagram the electron energy increases upward while the potential or voltage increases downward

the gate or silicon. With the applied gate voltage, > 0 in this case, the Fermi level of the gate is displaced downward by qV_g with respect to that of the p-type substrate. Because of the fixed band relationship between the gate and oxide, and between the oxide and silicon, a field is developed in the oxide as well as in silicon. There is a downward bending of the bands toward the surface of the p-type silicon. The amount of band bending in silicon is defined as the *surface potential*, ψ_s, i.e., the potential at the silicon surface relative to that in the bulk substrate. The potential drop across the oxide is denoted as V_{ox} in Figure 4.4. By adding up the potential components on the gate side and on the silicon side, and equating their difference to V_g, we obtain

$$(V_{ox} + \phi_m) - (\chi - \psi_s + E_g/2q + \psi_B) = V_g \tag{4.4}$$

The above terms can be regrouped in terms of ϕ_s and V_{fb}, using Eqs. (4.1) and (4.2), to yield

$$V_g - (\phi_m - \phi_s) = V_g - V_{fb} = V_{ox} + \psi_s, \tag{4.5}$$

where zero oxide charge (Q_{ox}) has been assumed. How $V_g - V_{fb}$ is partitioned into ψ_s and V_{ox} depends on both the oxide thickness and the doping concentration of the p-type silicon.

If the trapped charge at the oxide–silicon interface is negligible, the field in the oxide $\mathscr{E}_{ox}$ and the surface field in silicon $\mathscr{E}_s$ are related by the dielectric boundary condition discussed in Section 2.4.1.1:

$$\varepsilon_{ox}\mathscr{E}_{ox} = \varepsilon_{si}\mathscr{E}_s. \tag{4.6}$$

Note that this equation applies to both the magnitude and the direction of the fields. The field in the oxide is uniform because there is no charge in the oxide and Poisson's equation becomes $d\mathscr{E}/dx = 0$. Therefore, $\mathscr{E}_{ox} = V_{ox}/t_{ox}$ and the left-hand side of the above equals $\varepsilon_{ox}V_{ox}/t_{ox} = C_{ox}V_{ox}$, with C_{ox} given by Eq. (4.3). From Gauss's law, the right-hand side of the above equation is $-Q_s = \varepsilon_{si}\mathscr{E}_s$, where Q_s is the total charge per unit area induced in silicon. By substituting the relation between V_{ox} and Q_s in Eq. (4.5), *the gate bias equation is expressed as*

$$V_g - V_{fb} = V_{ox} + \psi_s = \frac{-Q_s}{C_{ox}} + \psi_s. \tag{4.7}$$

There is a negative sign in front of Q_s in Eq. (4.7) because the charge on the metal gate is always equal but opposite to the charge in silicon, i.e., Q_s is negative when $V_g - V_{fb}$ is positive and vice versa. Equation (4.7) represents the central relationship between the applied gate voltage and the resulting potential and charge in silicon. To obtain the potential and charge density separately as a function of V_g, Poisson's equation in silicon needs to be solved. This is done in Section 4.2.

4.1.3   Accumulation, Depletion, and Inversion

Figure 4.5 shows the band diagrams of p-type ((a)–(d)) and n-type ((e)–(h)) MOS capacitors under different gate bias voltages with respect to the flatband voltage. For

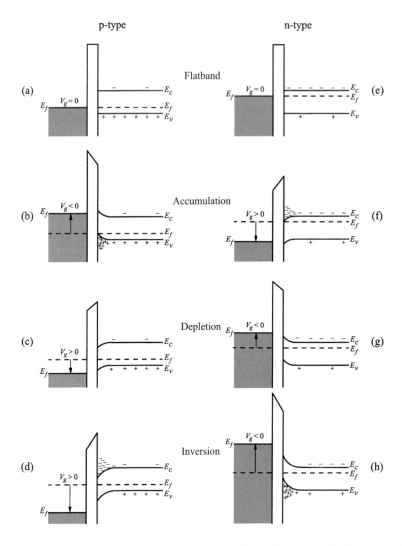

Figure 4.5 Energy-band diagrams for ideal (zero flatband voltage) (a)–(d) p-type and (e)–(h) n-type MOS capacitors under different bias conditions: (a), (e), flat band; (b), (f), accumulation; (c), (g), depletion; and (d), (h), inversion (after Sze, 1981)

simplicity, the flatband voltage is taken to be zero for all cases. The flatband condition for p-type MOS discussed in Figure 4.3 is shown in Figure 4.5(a). There is no charge, no field, and the carrier concentration equals the ionized acceptor concentration throughout the silicon. Now consider the case when a negative voltage is applied to the gate of a p-type MOS capacitor, as shown in Figure 4.5(b). This raises the metal Fermi level (i.e., electron energy) with respect to the silicon Fermi level and creates an electric field in the oxide that would accelerate a negative charge toward the silicon substrate. A field is also induced at the silicon surface in the same direction as the oxide field. Because of the low carrier concentration in silicon (compared with metal), the bands bend upward toward the oxide interface. *The Fermi level stays flat within the silicon, since there is no net flow of conduction current*, as was discussed in Section 2.4.2. Due to the band bending, the valence band at the surface is much closer to the Fermi level than is the valence band in the bulk silicon. This results in a hole concentration much higher at the surface than the equilibrium hole concentration in the bulk. Since excess holes are accumulated at the surface, this is referred to as the *accumulation* condition. One can think of the excess holes as being attracted toward the surface by the negative gate voltage. An equal amount of negative charge appears on the metal side of the MOS capacitor, as required for charge neutrality.

On the other hand, if a positive voltage is applied to the gate of a p-type MOS capacitor, the metal Fermi level moves downward, which creates an oxide field in the direction of accelerating a negative charge toward the metal electrode. A similar field is induced in the silicon, which causes the bands to bend downward toward the surface, as shown in Figure 4.5(c). Since the valence band at the surface is now farther away from the Fermi level than is the valence band in the bulk, the hole concentration at the surface is lower than the concentration in the bulk. This is referred to as the *depletion* condition. One can think of the holes as being repelled away from the surface by the positive gate voltage. The situation is similar to the depletion layer in a p–n junction discussed in Section 3.1.2. The depletion of holes at the surface leaves the region with a net negative charge arising from the unbalanced acceptor ions. An equal amount of positive charge appears on the metal side of the capacitor.

As the positive gate voltage increases, the band bending also increases, resulting in a wider depletion region and more (negative) depletion charge. This goes on until the bands bend downward so much that, at the surface, the conduction band is closer to the Fermi level than the valence band is, as shown in Figure 4.5(d). When this happens, not only are the holes depleted from the surface, but the surface potential is such that it is energetically favorable for electrons to populate the conduction band. In other words, the surface behaves like n-type material with an electron concentration given by Eq. (2.63). *Note that this n-type surface is formed not by doping, but instead by inverting the original p-type substrate with an applied electric field*. This condition is called *inversion*. The negative charge in the silicon consists of both the ionized acceptors and the thermally generated electrons in the conduction band. Again, it is balanced by an equal amount of positive charge on the metal gate. The surface is inverted as soon as $E_i = (E_c + E_v)/2$ crosses E_f. This is called *weak inversion* because the electron concentration remains small until E_i is considerably below E_f. If the gate voltage is increased further, the concentration of electrons at the surface will be equal to, and then exceed, the hole concentration in the substrate. This condition is called *strong inversion*.

So far we have discussed the band bending for accumulation, depletion, and inversion of silicon surface in a p-type MOS capacitor. Similar conditions hold true in an n-type MOS capacitor, except that the polarities of voltage, charge, and band bending are reversed, and the roles of electrons and holes are interchanged. The band diagrams for flatband, accumulation, depletion, and inversion conditions of an n-type

MOS capacitor are shown in Figures 4.5(e)–(h), where the metal work function per electron charge ϕ_m is assumed to be equal to that of the n-type silicon, given by

$$\phi_s = \chi + \frac{E_g}{2q} - \psi_B, \tag{4.8}$$

instead of Eq. (4.1). Accumulation occurs when a positive voltage is applied to the metal gate and the silicon bands bend downward at the surface. Depletion and inversion occur when the gate voltage is negative and the bands bend upward toward the surface.

4.2   Electrostatic Potential and Charge Distribution in Silicon

4.2.1   Solving Poisson's Equation

In this section, the relations among the surface potential, charge, and electric field are derived by solving Poisson's equation in the surface region of silicon. A more detailed band diagram at the surface of a p-type silicon is shown in Figure 4.6. The potential $\psi(x) = \psi_i(x) - \psi_i(x = \infty)$ is defined as the amount of band bending at position x, where $x = 0$ is at the silicon surface and $\psi_i(x = \infty)$ is the intrinsic potential in the bulk silicon. Remember that $\psi(x)$ is positive when the bands bend downward. The boundary conditions are $\psi = 0$ in the bulk silicon, and $\psi = \psi(0) = \psi_s$ at the surface. The surface potential ψ_s depends on the applied gate voltage, as discussed in Section 4.1.2. Poisson's equation, Eq. (2.58), is

$$\frac{d^2\psi}{dx^2} = -\frac{d\mathscr{E}}{dx} = -\frac{q}{\varepsilon_{si}}\left[p(x) - n(x) + N_d^+(x) - N_a^-(x)\right]. \tag{4.9}$$

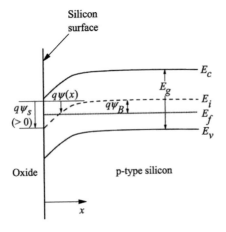

Figure 4.6 Energy-band diagram near the silicon surface of a p-type MOS device. The band bending ψ is defined as positive when the bands bend downward with respect to the bulk. Accumulation occurs when $\psi_s < 0$. Depletion and inversion occur when $\psi_s > 0$

For a uniformly doped p-type substrate of acceptor concentration N_a with complete ionization, $N_d^+(x) - N_a^-(x) = -N_a$, independent of x. Charge neutrality condition deep in the bulk substrate requires

$$N_d^+(x) - N_a^-(x) = -N_a = -p_0 + \frac{n_i^2}{p_0}, \tag{4.10}$$

where $p_0 = p(x = \infty)$ and $n_i^2/p_0 = n(x = \infty)$ are the majority (holes) and minority (electrons) carrier densities in the bulk substrate, respectively. In general, $p(x)$ and $n(x)$ are given by Eqs. (2.63) and (2.64), which can be expressed in terms of $\psi(x)$ using $\psi(x) = \psi_i(x) - \psi_i(x = \infty)$ and $\psi_B = \psi_f - \psi_i(x = \infty)$, as defined in Figure 4.6:

$$p(x) = n_i e^{q[\psi_f - \psi_i(x)]/kT} = n_i e^{q[\psi_B - \psi(x)]/kT} = p_0 e^{-q\psi(x)/kT} \tag{4.11}$$

and

$$n(x) = n_i e^{q[\psi_i(x) - \psi_f]/kT} = n_i e^{q[\psi(x) - \psi_B]/kT} = \frac{n_i^2}{p_0} e^{q\psi(x)/kT}. \tag{4.12}$$

Note that ψ_f is independent of x because there is no net current flow perpendicular to the surface and the Fermi level stays flat. Also note that $p_0 = p(x = \infty) = n_i \exp(q\psi_B/kT)$ and $n_i^2/p_0 = n(x = \infty) = n_i \exp(-q\psi_B/kT)$ in the last steps of Eqs. (4.11) and (4.12).

In practice, $N_a \gg n_i$, and $p_0 \approx N_a$ from Eq. (4.10). Substituting the last three equations into Eq. (4.9) and replacing p_0 by N_a yields

$$\frac{d^2\psi}{dx^2} = -\frac{q}{\varepsilon_{si}} \left[N_a \left(e^{-q\psi/kT} - 1 \right) - \frac{n_i^2}{N_a} \left(e^{q\psi/kT} - 1 \right) \right]. \tag{4.13}$$

Multiplying $(d\psi/dx)dx$ on both sides of Eq. (4.13) and integrating from the bulk $(\psi = 0, d\psi/dx = 0)$ toward the surface, one obtains

$$\int_0^{d\psi/dx} \frac{d\psi}{dx} d\left(\frac{d\psi}{dx} \right)$$

$$= -\frac{q}{\varepsilon_{si}} \int_0^\psi \left[N_a \left(e^{-q\psi/kT} - 1 \right) - \frac{n_i^2}{N_a} \left(e^{q\psi/kT} - 1 \right) \right] d\psi, \tag{4.14}$$

which gives the electric field at x, $\mathscr{E} = -d\psi/dx$, in terms of ψ:

$$\mathscr{E}^2(x) = \left(\frac{d\psi}{dx} \right)^2 = \frac{2kTN_a}{\varepsilon_{si}} \left[\left(e^{-q\psi/kT} + \frac{q\psi}{kT} - 1 \right) \right.$$

$$\left. + \frac{n_i^2}{N_a^2} \left(e^{q\psi/kT} - \frac{q\psi}{kT} - 1 \right) \right]. \tag{4.15}$$

At $x = 0$, we let $\psi = \psi_s$ and $\mathscr{E} = \mathscr{E}_s$. From Gauss's law, Eq. (2.57), the total charge per unit area induced in the silicon (equal and opposite to the charge on the metal gate) is

$$Q_s = -\varepsilon_{si}\mathscr{E}_s = \pm\sqrt{2\varepsilon_{si}kTN_a}\left[\left(e^{-q\psi_s/kT} + \frac{q\psi_s}{kT} - 1\right)\right.$$

$$\left. + \frac{n_i^2}{N_a^2}\left(e^{q\psi_s/kT} - \frac{q\psi_s}{kT} - 1\right)\right]^{1/2}$$

(4.16)

This function is plotted in Figure 4.7. At the flat-band condition, $\psi_s = 0$ and $Q_s = 0$. In accumulation, $\psi_s < 0$ (bands bending upward) and the first term in the square brackets dominates once $-q\psi_s/kT > 1$. However, Eq. (4.16) is based on the Boltzmann approximation, not accurate when the Fermi level is into the valence band or the conduction band. The charge density is overestimated in Figure 4.7 when $\psi_s < -0.2$ V or > 0.9 V. A proper treatment with Fermi–Dirac distribution is given in Section 4.2.1.3. In depletion, $\psi_s > 0$ and $q\psi_s/kT > 1$, but $\exp(q\psi_s/kT)$ is not large enough to make the n_i^2/N_a^2 term appreciable. Therefore, the $q\psi_s/kT$ term in the square brackets dominates and the negative depletion charge density (from ionized acceptor atoms) is proportional to $\psi_s^{1/2}$. When ψ_s increases further, the $(n_i^2/N_a^2)\exp(q\psi_s/kT)$ term eventually becomes larger than the $q\psi_s/kT$ term and dominates the square bracket. This is when inversion occurs.

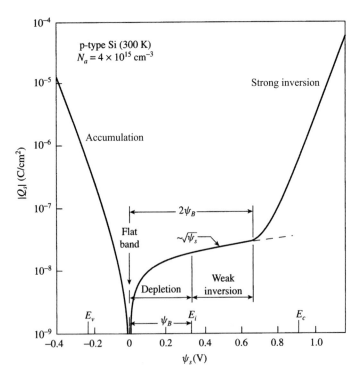

Figure 4.7 Variation of total charge density (fixed plus mobile) in silicon as a function of surface potential ψ_s for a p-type MOS device. The labels E_v, E_i, and E_c indicate the surface potential values where the valence band, the intrinsic level, and the conduction band cross the Fermi level. $\psi_B = 0.33$ V in this example (after Sze, 1981)

A popular criterion for the onset of strong inversion is for the surface potential to reach a value such that $(n_i^2/N_a^2) \, exp(q\psi_s/kT) = 1$, i.e.,

$$\psi_s(\text{inv}) = 2\psi_B = 2\frac{kT}{q}\ln\left(\frac{N_a}{n_i}\right). \tag{4.17}$$

Under this condition, the electron concentration given by Eq. (4.12) at the surface becomes equal to the depletion charge density N_a. *After inversion takes place, even a slight increase in the surface potential results in a large buildup of electron density at the surface.* The inversion layer effectively shields the silicon from further penetration of the gate field. Since almost all of the incremental charge is taken up by electrons, there is no further increase of either the depletion charge or the depletion-layer width. The expression in Eq. (4.17) is a rather weak function of the substrate doping concentration. For typical values of $N_a = 10^{16} - 10^{18}$ cm^{-3}, $2\psi_B$ varies only slightly, from 0.70 to 0.94 V.

4.2.1.1 Depletion Approximation

In general, Eq. (4.15) must be solved numerically to obtain $\psi(x)$. In the particular region of depletion where $2\psi_B > \psi > kT/q$, only the $q\psi/kT$ term in the square bracket needs to be kept, which allows the differential equation to be solved analytically:

$$\frac{d\psi}{dx} = -\sqrt{\frac{2qN_a\psi}{\varepsilon_{si}}}. \tag{4.18}$$

It can be integrated after rearranging the factors:

$$\int_{\psi_s}^{\psi} \frac{d\psi}{\sqrt{\psi}} = -\int_0^x \sqrt{\frac{2qN_a}{\varepsilon_{si}}}\, dx, \tag{4.19}$$

where ψ_s is the surface potential at $x = 0$ as assumed before. Therefore,

$$\psi = \psi_s\left(1 - \sqrt{\frac{qN_a}{2\varepsilon_{si}\psi_s}}x\right)^2, \tag{4.20}$$

which can be written as

$$\psi = \psi_s\left(1 - \frac{x}{W_d}\right)^2. \tag{4.21}$$

This is a parabolic equation with the minimum point at $\psi = 0$, $x = W_d$, where

$$W_d = \sqrt{\frac{2\varepsilon_{si}\psi_s}{qN_a}} \tag{4.22}$$

is the depletion-layer width defined as the distance to which the band bending extends. The total depletion charge density in silicon, Q_d, is equal to the charge per unit area of ionized acceptors in the depletion region:

$$Q_d = -qN_aW_d = -\sqrt{2\varepsilon_{si}qN_a\psi_s}. \qquad (4.23)$$

These results are very similar to those of the one-sided abrupt p − n junction under the depletion approximation, discussed in Section 3.1.2.2. *In the MOS case, however, W_d reaches a maximum value W_{dm} at the onset of strong inversion when $\psi_s = 2\psi_B$.* Substituting Eq. (4.17) into Eq. (4.22) gives the maximum depletion width:

$$W_{dm} = \sqrt{\frac{4\varepsilon_{si}kT\ln(N_a/n_i)}{q^2N_a}}. \qquad (4.24)$$

4.2.1.2   Strong Inversion

Beyond strong inversion, the $\left(n_i^2/N_a^2\right)\exp(q\psi/kT)$ term representing the inversion charge in Eq. (4.15) becomes appreciable and must be kept, along with the depletion charge term:

$$\frac{d\psi}{dx} = -\sqrt{\frac{2kTN_a}{\varepsilon_{si}}\left(\frac{q\psi}{kT} + \frac{n_i^2}{N_a^2}e^{q\psi/kT}\right)}. \qquad (4.25)$$

This equation can only be integrated numerically. The boundary condition is $\psi = \psi_s$ at $x = 0$. After $\psi(x)$ is solved, the electron distribution $n(x)$ in the inversion layer can be calculated from Eq. (4.12). Examples of the numerically calculated $n(x)$ are plotted in Figure 4.8 for two values of ψ_s with $N_a = 10^{16}$ cm^{-3}. *The electrons are distributed extremely close to the surface with an inversion-layer width less than 50 Å.* A higher surface potential or field tends to confine the electrons even closer to the surface. In

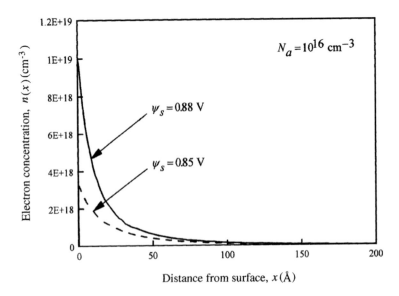

Figure 4.8 Electron concentration versus distance in the inversion layer of a p-type MOS device

general, electrons in the inversion layer must be treated quantum-mechanically as a 2-D gas (Stern and Howard, 1967). This is discussed in more detail in Section 4.4.

4.2.1.3   Fermi–Dirac Distribution

In solving Poisson's equation thus far in this section, Boltzmann approximation has been used to simplify the math. But the carrier density in strong inversion is overestimated when the Fermi level is close to or above E_c. In Section 2.2.3, we derived the electron density with full Fermi–Dirac distribution. Consider the MOS capacitor in Figure 4.6 biased in inversion. By applying Eqs. (2.27) and (2.28) to Poisson's equation, we obtain

$$\frac{d^2\psi}{dx^2} = \frac{q}{\varepsilon_{si}}[n(x) + N_a] = \frac{q}{\varepsilon_{si}}\left[\frac{2}{\sqrt{\pi}}N_c \int_0^\infty \frac{\sqrt{y}}{1 + e^{y-u}}dy + N_a\right] \qquad (4.26)$$

where

$$u(x) = \frac{-E_g/2 - q\psi_B + q\psi(x)}{kT} \qquad (4.27)$$

is linearly related to $\psi(x)$. Here, only the depletion charge and inversion charge are included. As with Eq. (4.13), Eq. (4.26) can be integrated once by multiplying $d\psi/dx$ or du/dx on both sides:

$$\left(\frac{du}{dx}\right)^2 = \frac{2q^2}{\varepsilon_{si}kT}\int_{u(\infty)}^{u(x)}\left[\frac{2}{\sqrt{\pi}}N_c\int_0^\infty \frac{\sqrt{y}}{1 + e^{y-u}}dy + N_a\right]du$$

$$= \frac{2q^2}{\varepsilon_{si}kT}\left[\frac{2}{\sqrt{\pi}}N_c\int_0^\infty \sqrt{y}\ln(1 + e^{u-y})dy + N_a u\right]\Bigg|_{u(\infty)}^{u(x)}. \qquad (4.28)$$

For $u(\infty) = -(E_g/2 + q\psi_B)/kT$, the first term in the square bracket can be shown to be n_i^2/N_a and neglected. Then

$$\left(\frac{d\psi}{dx}\right)^2 = \frac{2kTN_a}{\varepsilon_{si}}\left\{\left[\frac{2}{\sqrt{\pi}}\frac{N_c}{N_a}\int_0^\infty \sqrt{y}\ln(1 + e^{u-y})dy\right] + \frac{q\psi}{kT}\right\}, \qquad (4.29)$$

and the total charge in silicon is

$$-Q_s = \varepsilon_{si}\left|\frac{d\psi}{dx}\right|_{x=0} = \sqrt{2\varepsilon_{si}kTN_a}\left\{\left[\frac{2}{\sqrt{\pi}}\frac{N_c}{N_a}\int_0^\infty \sqrt{y}\ln(1 + e^{u_s-y})dy\right] + \frac{q\psi_s}{kT}\right\}^{1/2}$$

$$(4.30)$$

where $u_s = (-E_g/2 - q\psi_B + q\psi_s)/kT$. With numerical integration, $Q_s(\psi_s)$ for the Fermi–Dirac distribution is plotted in Figure 4.9, assuming the same N_a as in Figure 4.7. It is clear that Eq. (4.16) with Boltzmann approximation greatly overestimates the inversion charge density when $\psi_s > 0.9$ V where the Fermi level moves into the conduction band. Asymptotically, from Eq. (2.29), $F_{1/2}(u) \propto u^{3/2}$ for $u \gg 1$, Fermi–Dirac distribution gives $Q_s \propto u_s^{5/4} \propto (\psi_s - E_g/2q - \psi_B)^{5/4}$ instead of the $\sim \exp(q\psi_s/2kT)$ dependence from the Boltzmann approximation.

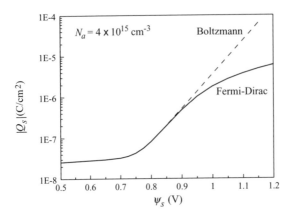

Figure 4.9 Total charge density in silicon versus surface potential given by Eq. (4.30) with Fermi–Dirac distribution compared to that by Eq. (4.16) with Boltzmann approximation

4.2.2   Surface Potential and Charge Density as a Function of Gate Voltage

Figure 4.10 shows schematically the band diagram and the charge distribution of an MOS capacitor with p-type substrate biased in inversion. The total charge Q_s in silicon includes depletion (Q_d) and inversion (Q_i) components. Both are negative, with their sum equal and opposite to the positive charge on the gate. For simplicity of discussion, oxide and interface trapped charges are ignored here. They will be discussed in detail in Section 4.5. As discussed in Sections 4.2.1.1 and 4.2.1.2, the depletion charge is distributed uniformly over the depth of the depletion region, i.e., the region where band bending takes place. The inversion charge, on the other hand, is distributed in a very thin layer at the silicon surface.

In Section 4.2.1, charge and potential distributions in silicon were solved in terms of the surface potential ψ_s as a boundary condition. ψ_s is not directly measurable, but is controlled by and can be determined from the applied gate voltage. The gate bias equation, Eq. (4.7), relates the charge density induced in silicon Q_s and the band bending ψ_s in silicon to the departure of the gate voltage V_g from the flatband voltage. Combining Eqs. (4.7) and (4.16) yields

$$V_g - V_{fb} = \psi_s + \frac{-Q_s}{C_{ox}} = \psi_s + \frac{\sqrt{2\varepsilon_{si}kTN_a}}{C_{ox}}\left[\left(e^{-q\psi_s/kT} + \frac{q\psi_s}{kT} - 1\right) + \frac{n_i^2}{N_a^2}\left(e^{q\psi_s/kT} - \frac{q\psi_s}{kT} - 1\right)\right]^{1/2}.$$

$$(4.31)$$

For a given V_g, the above implicit equation can be solved for ψ_s.

An explicit solution of ψ_s can be found in depletion where only the term $q\psi_s/kT$ under the square root is significant:

$$V_g - V_{fb} = \frac{qN_aW_d}{C_{ox}} + \psi_s = \frac{\sqrt{2\varepsilon_{si}qN_a\psi_s}}{C_{ox}} + \psi_s.$$

$$(4.32)$$

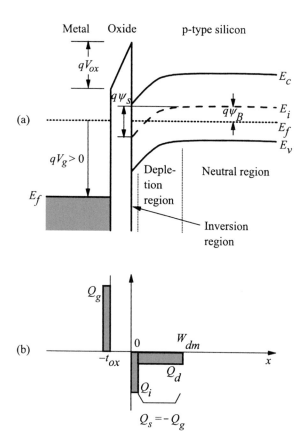

Figure 4.10 (a) Band diagram of a p-type MOS capacitor with a positive voltage applied to the gate $(V_{fb} = 0)$ to reach strong inversion. (b) The corresponding charge distribution

This is a quadratic equation in $\sqrt{\psi_s}$ with the solution

$$\sqrt{\psi_s} = \sqrt{V_g - V_{fb} + \frac{\varepsilon_{si} q N_a}{2 C_{ox}^2}} - \frac{\sqrt{\varepsilon_{si} q N_a / 2}}{C_{ox}}. \tag{4.33}$$

In general, Eq. (4.31) needs to be solved numerically. An example with the focus on the depletion and inversion regions is shown in Figure 4.11. Below the condition for strong inversion, $\psi_s = 2\psi_B$, ψ_s increases more or less linearly with V_g, as expected from Eq. (4.33). Beyond $\psi_s = 2\psi_B$, ψ_s nearly saturates – increasing by less than 0.2 V while V_g increases by 2 V. After ψ_s is solved, Q_s is calculated and plotted as a function of V_g in Figure 4.11. By numerically evaluating the integrals in Exercise 4.1, Q_s is separated into its two components, the depletion charge density Q_d and the inversion charge density Q_i, which are also plotted in Figure 4.11. **It is clear that before the $\psi_s = 2\psi_B$ condition, the charge in the silicon is predominantly of the depletion type.** Under such depletion conditions, $Q_s(\psi_s) = Q_d(\psi_s) = -\sqrt{2\varepsilon_{si} q N_a \psi_s}$ from Eq. (4.23). After $\psi_s = 2\psi_B$, the depletion charge no longer increases with V_g

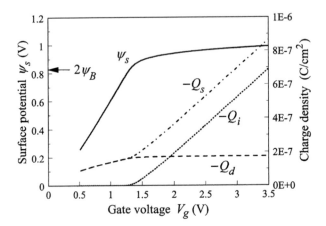

Figure 4.11 Numerical solutions of surface potential, total silicon charge density, inversion charge density, and depletion charge density from the gate bias equation, Eq. (4.31). The MOS device parameters are $N_a = 10^{17}$ cm^{-3}, $t_{ox} = 10$ nm, and $V_{fb} = 0$

because of screening by the inversion layer discussed before. ***Almost all of the increase of $-Q_s$ beyond $\psi_s = 2\psi_B$ is taken up by $-Q_i$ with a slope $-dQ_i/dV_g \approx C_{ox}$.*** While on linear scale it appears that $-Q_i$ is zero below the $\psi_s = 2\psi_B$ threshold, on log scale it is clear that $-Q_i$ actually remains finite and decreases exponentially with V_g. This is the source of the subthreshold leakage current in MOSFETs – an important design consideration further addressed in detail in Section 5.3. Under extreme accumulation and inversion conditions, $-Q_s \approx C_{ox}(V_g - V_{fb})$, since both V_g and V_{ox} can be much larger than the silicon bandgap, $E_g/q = 1.12$ V, for CMOS technologies with $V_{dd} \gg 1$ V, while ψ_s does not significantly exceed E_g/q (surface potential pinned to either the valence band or the conduction band edge).

4.3   Capacitance–Voltage Characteristics of MOS Capacitors

4.3.1   Measurement Setup

Since no conduction current flows in an MOS device, the only means of characterization is by a capacitance–voltage (C–V) measurement. Figure 4.12 shows the measurement scheme. For every dc bias voltage V_g, the small-signal ac admittance of the MOS is measured by superimposing a voltage source of frequency ω and reading the current at the same frequency. The in-phase and out-of-phase current components give the conductance and capacitance of MOS at this dc gate voltage.

4.3.2   Capacitance Components in MOS

Figure 4.13 shows the equivalent capacitance circuit of an MOS device. MOS capacitances are defined as the small-signal differential of charge per an incremental voltage or potential. The total MOS capacitance per unit area is

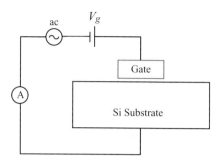

Figure 4.12 Setup for MOS *C-V* measurement. A small-signal ac voltage of frequency typically in the range of 1 kHz to 1 MHz is applied. The capacitance is proportional to the small-signal current at the same frequency 90° out of phase with the ac voltage

$$C_g = \frac{d(-Q_s)}{dV_g}.$$ (4.34)

If we differentiate Eq. (4.7) with respect to $-Q_s$ and define the silicon part of the capacitance as

$$C_{si} = \frac{d(-Q_s)}{d\psi_s},$$ (4.35)

we obtain

$$\frac{1}{C_g} = \frac{1}{C_{ox}} + \frac{d\psi_s}{d(-Q_s)} = \frac{1}{C_{ox}} + \frac{1}{C_{si}}.$$ (4.36)

In other words, the total capacitance equals the oxide capacitance and the silicon capacitance connected in series, as shown in Figure 4.13(a). The capacitances are defined in such a way that they are all positive quantities. Here, we assume an ideal MOS capacitor with no interface states or charge traps in the oxide. Capacitances arising from charging and discharging of Si − SiO$_2$ interface traps will be discussed in Section 4.5.

Along the same line of $Q_s = Q_d + Q_i$ in Figure 4.10(b), Figure 4.13(b) further breaks C_{si} into parallel depletion and inversion charge capacitances C_d and C_i. While charging and discharging of C_d involve majority carriers hence are essentially instantaneous (Section 2.4.4.1, Dielectric Relaxation Time), charging and discharging of C_i require minority carrier generation and recombination which are rather slow (Section 2.4.4.2, Minority Carrier Diffusion Length). This is taken into account by an equivalent conductance G_d in series with C_i in Figure 4.13(b). As a result, $C-V$ characteristics become frequency dependent when biased in inversion. The expression for G_d in terms of physical parameters is given in Section 4.3.3.5.

4.3.3   *C–V* Characteristics in Different Bias Regions

Figure 4.14 shows a typical $C–V$ curve of an MOS capacitor on p-type silicon substrate, assuming zero flatband voltage. Various expressions of capacitance in different bias regions are covered here.

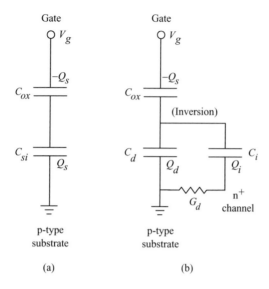

Figure 4.13 Equivalent circuits of an MOS capacitor. (a) All the silicon capacitances are lumped into C_{si}. (b) C_{si} is broken up into a depletion charge capacitance C_d and an inversion-layer capacitance C_i. C_d arises from the majority carriers, which can respond to high-frequency as well as low-frequency signals. C_i arises from the minority carriers, which can only respond to low-frequency signals

4.3.3.1 Capacitance–Voltage Characteristics: Accumulation

When the gate voltage is negative (by more than a few kT/q) with respect to the flatband voltage, the p-type MOS capacitor is in accumulation. In this case, $Q_s > 0$, with a magnitude going up rapidly with negative ψ_s, as shown in Figure 4.7. Initially, $Q_s \propto \exp(-q\psi_s/2kT)$ as the first term in the square root of Eq. (4.16) becomes dominant. However, owing to Fermi–Dirac distribution, Q_s no longer increases exponentially when $\psi_s < -(E_g/2q - \psi_B)$ and the Fermi level is in the valence band. A parallel treatment as that in Section 4.2.1.3 gives $Q_s \propto (-\psi_s - E_g/2q + \psi_B)^{5/4}$ in the degenerate limit. Therefore, $C_{si} = -dQ_s/d\psi_s \propto (-\psi_s - E_g/2q + \psi_B)^{1/4}$. The MOS capacitance rapidly approaches C_{ox} when the gate voltage is on the negative side of the flat-band voltage.

4.3.3.2 Capacitance–Voltage Characteristics: Flatband

When the gate bias is zero in Figure 4.14, the MOS is near the flat-band condition; therefore, $q\psi_s/kT \ll 1$. The inversion charge term in Eq. (4.16) can be neglected and the first exponential term can be expanded into a power series. By keeping the first three terms of the series, we obtain $Q_s = -(\varepsilon_{si}q^2 N_a/kT)^{1/2}\psi_s$. From Eq. (4.36), the flatband capacitance per unit area is given by

$$\frac{1}{C_{fb}} = \frac{1}{C_{ox}} + \sqrt{\frac{kT}{\varepsilon_{si}q^2 N_a}} = \frac{1}{C_{ox}} + \frac{L_D}{\varepsilon_{si}}, \qquad (4.37)$$

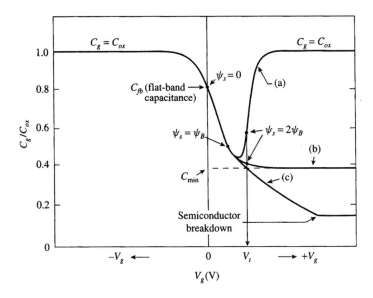

Figure 4.14 MOS capacitance–voltage curves: (a) low frequency, (b) high frequency, (c) deep depletion. $V_{fb} = 0$ is assumed (after Sze, 1981)

where L_D is the Debye length defined in Eq. (2.67). In most cases, C_{fb} is somewhat less than C_{ox}. For very thin oxides and low substrate doping, C_{fb} can be much smaller than C_{ox}.

4.3.3.3 Capacitance–Voltage Characteristics: Depletion

When the gate voltage is more positive than the flatband voltage in a p-type MOS capacitor, the surface starts to be depleted of holes; $1/C_{si}$ becomes appreciable and the capacitance decreases. Using the depletion approximation, we can find an analytical expression for C_g in this case. From Eqs. (4.22) and (4.23),

$$C_d = \frac{d(-Q_d)}{d\psi_s} = \sqrt{\frac{\varepsilon_{si}qN_a}{2\psi_s}} = \frac{\varepsilon_{si}}{W_d}. \tag{4.38}$$

This expression is identical to the depletion-layer capacitance per unit area in the p–n junction case discussed in Section 3.1.2. Applying Eq. (4.36) with $C_{si} = C_d$ and expressing C_d of Eq. (4.38) in terms of $V_g - V_{fb}$ using Eq. (4.33) for $\sqrt{\psi_s}$, we obtain

$$C_g = \frac{C_{ox}}{\sqrt{1 + [2C_{ox}^2(V_g - V_{fb})/(\varepsilon_{si}qN_a)]}}. \tag{4.39}$$

This equation shows how the MOS capacitance decreases with increasing V_g under the depletion condition. It serves as a good approximation to the middle portions of the C–V curve in Figure 4.14, provided that the MOS capacitor is not biased near the flat-band or the inversion condition.

4.3.3.4   Capacitance–Voltage Characteristics: Inversion (Low Frequency)

As the gate voltage increases further, the capacitance stops decreasing when $\psi_s = 2\psi_B$ [Eq. (4.17)] is reached and inversion occurs. *Once the inversion layer forms, the capacitance starts to increase since, due to the proximity of inversion channel to the surface, the inversion charge capacitance is much larger than the depletion capacitance*. The MOS capacitance rapidly increases back to C_{ox} when the gate voltage goes beyond the $\psi_s = 2\psi_B$ condition, as shown in the low-frequency C–V curve (a) in Figure 4.14.

The above consideration is based on $Q_s(\psi_s)$ of Eq. (4.16) [or Eq. (4.30) for Fermi–Dirac] which is quasi-static in nature. When ψ_s is modulated at ac frequency ω in Figure 4.12, Q_s made of inversion charge can only respond if the frequency is low enough for the slow process of minority carrier generation and recombination to follow. The conductance G_d in series with the inversion charge capacitance C_i in Figure 4.13(b) gives the criterion for low-frequency C–V: $\omega < G_d/C_i$.

4.3.3.5   Capacitance–Voltage Characteristics: Inversion (High Frequency)

For an estimate on G_d, consider an MOS on p-type silicon of doping N_a. For minority carrier diffusion, we derived in Section 2.4.4.2 that

$$\frac{d^2 n}{dx^2} - \frac{n - n_0}{L_n^2} = 0, \tag{4.40}$$

where L_n is the minority carrier diffusion length and $n_0 = n_i^2/N_a$ is the thermal equilibrium minority carrier concentration. If the excess minority carrier density at the end of depletion region, defined as $x = 0$, is Δn, the solution $n(x) - n_0$ to the above equation is $\Delta n \exp(-x/L_n)$. The diffusion current density at $x = 0$ is then

$$\Delta J_n = q D_n \frac{dn}{dx}\bigg|_{x=0} = -\mu_n kT \frac{\Delta n}{L_n}. \tag{4.41}$$

Δn also gives rise to a change in the electron quasi-Fermi potential ϕ_n of Eq. (2.79). Taking the differential of the diffusion term obtains

$$\Delta \phi_n = -\frac{kT}{q} \Delta \left[\ln\left(\frac{n}{n_i}\right) \right] = -\frac{kT}{q} \frac{\Delta n}{n_0}. \tag{4.42}$$

From the above,

$$G_d = \frac{\Delta J_n}{\Delta \phi_n} = \frac{q \mu_n n_0}{L_n} = \frac{q \mu_n n_i^2}{L_n N_a} \tag{4.43}$$

(Nicollian and Brews, 1982). Experimentally, L_n is at least 1 μm (Figure 3.14). G_d is then a very small number, $\sim 10^{-8}\ \Omega^{-1}/\text{cm}^2$. Since $C_i \sim 10^{-6}\ \text{F}/\text{cm}^2$ (Figure 4.9), the measurement frequency would have to be $< 1\,\text{Hz}$ to obtain low-frequency C–V results.

From a separate point of view, the minority-carrier response time can be estimated from the generation–recombination current density in the space-charge region,

$J_R = q n_i W_d / \tau$, where τ is the minority-carrier lifetime discussed in Section 3.1.6. The time it takes to generate enough minority carriers to replace something comparable to the depletion charge, $Q_d = q N_a W_d$, is on the order of $Q_d / J_R = (N_a / n_i) \tau$ (Jund and Poirier, 1966). This is typically $0.1 - 10$ s.

It is clear from this discussion that *for frequencies higher than 100 Hz or so, the inversion charge cannot respond to the applied ac signal*. Only the depletion charge (majority carriers) can respond to the signal, which means that the silicon capacitance is given by C_d of Eq. (4.38) with W_d equal to its maximum value, W_{dm}, in Eq. (4.24). The high-frequency capacitance per unit area thus approaches a constant minimum value, C_{min}, in inversion given by

$$\frac{1}{C_{min}} = \frac{1}{C_{ox}} + \sqrt{\frac{4kT \ln(N_a/n_i)}{\varepsilon_{si} q^2 N_a}}. \tag{4.44}$$

This is shown in the high-frequency $C-V$ curve (b) in Figure 4.14.

Typically, $C-V$ curves are traced by applying a slow-varying ramp voltage to the gate with a small ac signal superimposed on it. However, if the ramp rate is fast enough that the ramping time is shorter than the minority-carrier response time, then there is insufficient time for the inversion layer to form, and the MOS capacitor is biased into deep depletion as shown by curve (c) in Figure 4.14. In this case, the depletion width can exceed the maximum value given by Eq. (4.24), and the MOS capacitance decreases further below C_{min} until impact ionization takes place (Sze, 1981). Note that deep depletion is not a steady-state condition. If an MOS capacitor is held under such bias conditions, its capacitance will gradually increase toward C_{min} as the thermally generated minority charge builds up in the inversion layer until an equilibrium state is established. The time it takes for an MOS capacitor to recover from deep depletion and return to equilibrium is referred to as the *retention time*. It is a good indicator of the defect density in the silicon wafer and is often used to qualify processing tools in a facility.

4.3.4   Split *C–V* Measurement

It is possible to obtain *low-frequency-like* $C-V$ curves at high measurement frequencies. One way is to expose the MOS capacitor to intense illumination, which generates a large number of minority carriers in the silicon. Another commonly used technique is to form an n^+ region adjacent to the MOS device and connect it electrically to the p-type substrate (Grove, 1967). The n^+ region then acts like a reservoir of electrons which can exchange minority carriers freely with the inversion layer. In other words, the n^+ region is connected to the *surface channel* of the inverted MOS device. This structure is similar to that of a gated diode, to be discussed in Section 4.6.3. Based on the equivalent circuit in Figure 4.13(b), the total MOS capacitance per unit area is given by

$$C_g = \frac{C_{ox}(C_d + C_i)}{C_{ox} + C_d + C_i}. \tag{4.45}$$

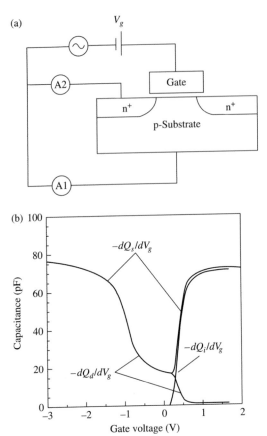

Figure 4.15 (a) Setup of the split $C-V$ measurement. Both the dc bias and the small-signal ac voltage are applied to the gate. Small signal ac currents are measured by two ammeters, A1 and A2, connected separately as shown. (b) Measured $C-V$ curves where the $-dQ_d/dV_g$ component is obtained from A1, and the $-dQ_i/dV_g$ component is obtained from A2. The sum is the total capacitance per unit area, $-dQ_s/dV_g$

The majority and minority carrier contributions to the total capacitance can be separately measured in a *split C–V* setup shown in Figure 4.15(a) (Sodini *et al.*, 1982). With a small signal ac voltage applied to the gate, the out-of-phase ac currents are sensed by two ammeters: one (A1) connected to the p-type substrate for the hole current, and another (A2) connected to the n^+ region for the electron current. Typical measured results are shown in Figure 4.15(b). The hole contribution to the capacitance measured by A1 is

$$-\frac{dQ_d}{dV_g} = \frac{C_{ox}C_d}{C_{ox} + C_d + C_i}. \tag{4.46}$$

And the electron contribution to the capacitance measured by A2 is

$$-\frac{dQ_i}{dV_g} = \frac{C_{ox}C_i}{C_{ox} + C_d + C_i}. \tag{4.47}$$

They add up to the total capacitance per unit area, $C_g = -dQ_s/dV_g$. Note that the $-dQ_d/dV_g$ curve decreases to zero soon after strong inversion when C_i becomes dominant ($\gg C_{ox}$). To put it another way, the highly conductive inversion channel shields the majority carriers in the bulk silicon from responding to the modulation of gate voltage. The $-dQ_i/dV_g$ curve can be integrated to yield the inversion charge density as a function of the gate voltage. It is used, for example, in channel mobility measurements where the inversion charge density must be determined accurately.

4.3.5 Polysilicon Gate: Work Function and Depletion Effects

4.3.5.1 Work Function and Flatband Voltage of Polysilicon Gates

The use of polysilicon gates is a key advance in modern CMOS technology, since it allows the source and drain regions to be self-aligned to the gate, thus eliminating parasitics from overlay errors (Kerwin et al., 1969). Furthermore, their work function can be tailored by doping: a n^+-polysilicon gate has been used for nMOSFET and a p^+-polysilicon gate used for pMOSFET to obtain threshold voltages of low magnitude in both devices (Wong et al., 1988).

The Fermi level of heavily-doped n^+ polysilicon is near the conduction band edge, so its work function is given by the electron affinity, $q\chi$. From Eq. (4.1), the work function difference for an n^+ polysilicon gate on a p-type substrate of doping concentration N_a is

$$\phi_{ms} = -\frac{E_g}{2q} - \psi_B = -0.56 - \frac{kT}{q}\ln\left(\frac{N_a}{n_i}\right) \tag{4.48}$$

in volts. Similarly, the work function difference for a p^+ polysilicon gate on an n-type substrate of doping concentration N_d is

$$\phi_{ms} = \frac{E_g}{2q} + \psi_B = 0.56 + \frac{kT}{q}\ln\left(\frac{N_d}{n_i}\right), \tag{4.49}$$

which is symmetric to Eq. (4.48). These relations give rise to flatband voltages with key implications on the scalability of MOSFET devices, as will be discussed in Section 6.3.

The band diagram of a n^+-polysilicon-gated p-type MOS capacitor at zero gate voltage is shown in Figure 4.16(a), where the Fermi levels line up and the free electron level of the bulk p-type silicon is higher in electron energy than the free electron level of the n^+ polysilicon gate. This sets up an oxide field in the direction of accelerating electrons toward the gate, and at the same time a downward bending of the silicon bands (depletion) toward the surface to produce a field in the same direction. The flatband condition is reached by applying a negative voltage equal to the work function difference to the gate, as shown in Figure 4.16(b).

4.3.5.2 Polysilicon-Gate Depletion Effects

The drawback of polysilicon gates is that they can only be doped to a level of about 10^{20} cm^{-3}, at which there can be a non-negligible depletion width in the gate itself.

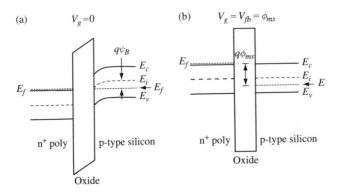

Figure 4.16 Band diagram of an n$^+$-polysilicon-gated p-type MOS capacitor biased at (a) zero gate voltage and (b) flatband condition

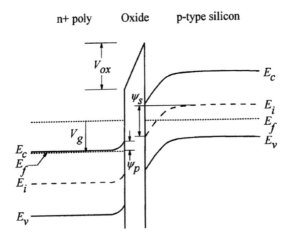

Figure 4.17 Band and potential diagram showing polysilicon-gate depletion effects when a positive voltage is applied to the n$^+$ polysilicon gate of a p-type MOS capacitor

This is especially a concern with the dual n$^+$–p$^+$ polysilicon-gate process in which the gates are doped by ion implantation (Wong *et al.*, 1988). Gate depletion results in an additional capacitance in series with the oxide capacitance, which in turn leads to a reduced inversion-layer charge density and degradation of the MOSFET transconductance.

Consider the band diagram of a n$^+$-polysilicon-gated p-type MOS capacitor biased into inversion shown in Figure 4.17. Since the oxide field points in the direction of accelerating a negative charge toward the gate, the bands in the n$^+$ polysilicon bend slightly upward toward the oxide interface. This depletes the surface of electrons and forms a thin space-charge region in the polysilicon layer, thereby lowering the total capacitance.

4.3.5.3 Effect of Polysilicon Doping Concentration on C–V Characteristics

Typical low-frequency C–V curves with the presence of gate depletion effects are shown in Figure 4.18 (Rios and Arora, 1994). A distinct feature is that the capacitance at inversion does not return to the full oxide capacitance, as in Figure 4.14. Instead, the inversion capacitance exhibits a maximum value somewhat less than C_{ox}, depending on the effective doping concentration of the polysilicon gate. The higher the doping concentration is, the less the gate depletion effect and the closer the maximum capacitance is to the oxide capacitance.

The existence of a local maximum in the low-frequency C–V curve can be understood by expanding the analysis of MOS capacitance in Section 4.3.2 to include the polysilicon depletion effect. In Figure 4.17, we assume ψ_s to be the amount of band bending in the bulk silicon and ψ_p to be that in the n^+ polysilicon. From charge neutrality, the total charge density Q_g of the ionized donors in the depletion region of the n^+ polysilicon gate is equal and opposite to the combined inversion and depletion charge density Q_s in the silicon substrate, i.e., $Q_g = -Q_s$ [Figure 4.10(b)]. The gate bias equation for an applied voltage V_g is obtained by adding an additional term, ψ_p, for the band bending in the polysilicon gate to Eq. (4.7):

$$V_g = V_{fb} + \psi_s + \psi_p - \frac{Q_s}{C_{ox}}. \tag{4.50}$$

Differentiating this equation with respect to $-Q_s$ and using the capacitance definitions, Eqs. (4.34) and (4.35), we obtain

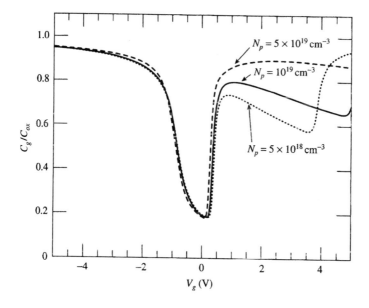

Figure 4.18 Low-frequency C–V curves of a p-type MOS capacitor with n^+ polysilicon gate doped at several different concentrations (after Rios and Arora, 1994)

$$\frac{1}{C_g} = \frac{1}{C_{ox}} + \frac{1}{C_{si}} + \frac{1}{C_p}, \tag{4.51}$$

where $C_p = -dQ_s/d\psi_p = dQ_g/d\psi_p$ is the capacitance of the polysilicon depletion region.

The same depletion approximation can be applied to polysilicon gate depletion as well. Similar to Eq. (4.23), the total gate depletion charge density is $Q_g(\psi_p) = \sqrt{2\varepsilon_{si}qN_p\psi_p}$, where N_p is the doping concentration of the polysilicon gate. The polysilicon depletion capacitance can be expressed in terms of Q_g as

$$C_p = \sqrt{2\varepsilon_{si}qN_p} \frac{1}{2\sqrt{\psi_p}} = \frac{\varepsilon_{si}qN_p}{Q_g}. \tag{4.52}$$

As V_g becomes more positive, Q_g and C_{si} increase but C_p decreases in Eq. (4.51). This results in a local maximum of the low-frequency capacitance as observed in Figure 4.18. For $N_p < 10^{19}$ cm^{-3}, another abrupt rise in the MOS capacitance may be observed at a much higher gate voltage. This is due to the onset of inversion (to p$^+$) at the n$^+$ polysilicon surface.

In practice, polysilicon gates cannot be doped much higher than 1×10^{20} cm^{-3}. Modern CMOS devices often operate at a maximum oxide field of 5 MV/cm, corresponding to a sheet electron density of $Q_g/q = 10^{13}$ cm^{-2}. For these values, the polysilicon depletion capacitance has the equivalent effect as adding $3 - 4$ Å to the gate oxide thickness. The discussion here applies to p$^+$ polysilicon gates on n-type silicon as well as to n$^+$ polysilicon gates on p-type silicon. Note that for n$^+$ polysilicon gates on n-type silicon and p$^+$ polysilicon gates on p-type silicon, gate depletion occurs when the substrate is accumulated.

4.3.5.4   Metal Gate Technology

MOSFET technology started in its early days with metal gates, namely, aluminum. Due to the low melting point of aluminum, it must be deposited and patterned after the source-drain implantation and annealing above 800 °C. The misalignments between the separate lithographic patterning of the gate and of the source-drain resulted in excessive overlap capacitance which significantly degraded the MOSFET circuit performance (Section 8.4). As mentioned in Section 4.3.5.1, the introduction of polysilicon gate technology represented a major advance that allowed the source-drain regions to be self-aligned to the already patterned gates. Also, polysilicon gates can be doped n$^+$ for nMOS and p$^+$ for pMOS in a CMOS circuit.

When the gate oxide is scaled to 1.5 nm or so, however, the performance loss due to polysilicon depletion effects became too much to tolerate. Hence, along with high-κ gate dielectrics, the VLSI industry moves back to the metal gate technology starting with the 45-nm node. To preserve the benefits of polysilicon gates, new metal gates need to deliver at least two different work functions for n- and pMOS, respectively, while maintaining self-aligned source-drain regions for control of the overlap capacitance. This puts too much burden on the choice of metal gate material for a gate-first process in which the metal must endure the source-drain anneal at high-temperatures (Robertson and Wallace,

2015). At the expense of process complexity, the industry adopted a gate-last or replacement gate process in which dummy polysilicon gates are patterned prior to source-drain implantation as before (Chang *et al.*, 2000). After source-drain anneal, oxide deposition, and etch back, the polysilicon gates are etched out, leaving the open trenches to be refilled with the final metal gates in one or several Chemical–Mechanical–Polish (CMP) steps. Further variations include the introduction of a dipole layer between the high-κ dielectric and metal gate to deliver the effective work function desired (Bao *et al.*, 2018).

4.3.6   MOS under Nonequilibrium

Consider an MOS capacitor where there is an n^+ region adjacent to the gated p-type region like that shown in Figure 4.15(a) for the split C–V measurement. The n^+ region and the p-type region form an n^+–p diode. Both the ungated and gated structures are shown schematically in Figure 4.19, along with their corresponding band diagrams.

4.3.6.1   Inversion Condition of an MOS under Nonequilibrium

As discussed in Section 3.1.1, when both the n^+ region and the p-type substrate are connected to the same potential (grounded), the p–n junction is in equilibrium and the Fermi level is constant across the p–n junction. If the gate voltage is large enough to invert the p-type surface, which occurs for a surface potential bending of $\psi_s(\text{inv}) = 2\psi_B$, the inverted channel is connected to the n^+ region and has the same potential as the n^+ region. In other words, *the electron quasi-Fermi level in the channel region is the same as the Fermi level in the n^+ region* as well as the Fermi level in the p-type substrate. The depletion region now extends from the p–n junction to the region under the gate between the inverted channel and the substrate, as shown in Figure 4.19(b).

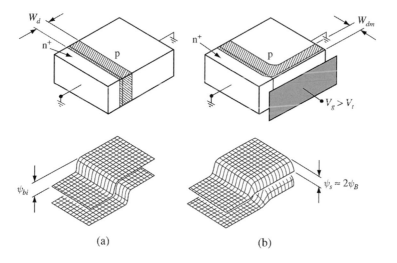

(a)   (b)

Figure 4.19 Gated diode or p-type MOS with adjacent n^+ region in equilibrium (zero voltage across the p–n junction). The gate is biased at (a) flatband and (b) inversion conditions (after Grove, 1967)

If, on the other hand, the p–n junction is reverse-biased at a voltage V_R, as shown in Figure 4.20, the MOS is in a nonequilibrium condition in which $np \ll n_i^2$. From Eq. (3.49), the electron concentration on the p-type side of the junction is

$$n = \frac{e^{-qV_R/kT} + \delta}{1 + \delta}\left(\frac{n_i^2}{N_a}\right) \approx \left[e^{-qV_R/kT} + \delta\right]\left(\frac{n_i^2}{N_a}\right), \tag{4.53}$$

where $\delta \ll 1$. For V_R more than a few kT/q, $n \ll n_i^2/N_a$. Now consider the case when a positive voltage large enough to bend the bands by $2\psi_B$ is applied to the gate, as shown in Figure 4.20(b). This brings the conduction band at the surface $2\psi_B$ closer to the electron quasi-Fermi level in the n^+ region. As far as the electron density at the surface is concerned, the reverse bias is effectively reduced by $2\psi_B$, which amounts to multiplying the $\exp(-qV_R/kT)$ term in Eq. (4.53) by a factor of $\exp(2q\psi_B/kT) = (N_a/n_i)^2$:

$$n = \left[e^{-qV_R/kT}e^{2q\psi_B/kT} + \delta\right]\left(\frac{n_i^2}{N_a}\right) = e^{-qV_R/kT}N_a + \delta\left(\frac{n_i^2}{N_a}\right). \tag{4.54}$$

For V_R more than a few kT/q, this is still $\ll N_a$, i.e., much lower than the hole density in the p-type substrate. Even though the positive voltage is sufficient to invert the surface in the equilibrium case, it is not enough to cause inversion in the reverse-biased case. This is because *the reverse bias lowers the quasi-Fermi level of electrons so that even if the bands at the surface are bent as much as in the equilibrium case in Figure 4.19(b), the conduction band is still not close enough to the quasi-Fermi level of electrons for inversion to occur.*

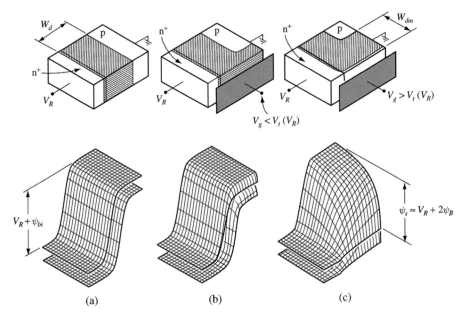

(a)   (b)   (c)

Figure 4.20 Gated diode or p-type MOS with adjacent n^+ region under nonequilibrium (reverse bias across the p–n junction). The gate is biased at (a) flat-band, (b) depletion, and (c) inversion conditions (after Grove, 1967)

To reach inversion in the nonequilibrium case, a much larger gate voltage, sufficient to bend the bands by $2\psi_B + V_R$, must be applied so that the conduction band of p-type region is as close to the Fermi level of n-type region as that in Figure 4.19(b). This is the case shown in Figure 4.20(c), where the electron concentration at the surface is now $n = N_a$, the same as the condition for inversion introduced in Section 4.2.1. Notice that the surface depletion layer is much wider than in the equilibrium case, just like in a reverse-biased p $-$ n junction.

4.3.6.2 Band Bending and Charge Distribution of an MOS under Nonequilibrium

The situations in Figure 4.20 are further illustrated in Figure 4.21, where the charge distribution and band bending in a cross-section perpendicular to the gate through the neutral p-type region are shown for both the equilibrium and the nonequilibrium cases. The equilibrium case is the same as that discussed in Section 4.2.1. In the

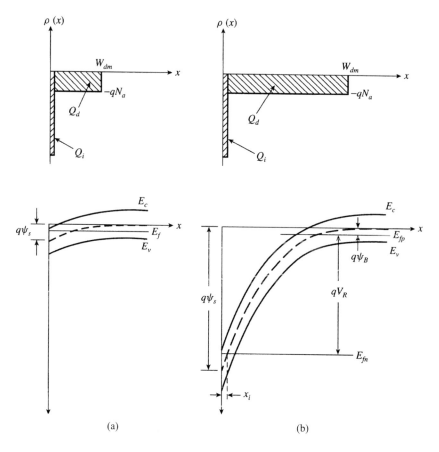

(a)

(b)

Figure 4.21 Comparison of charge distribution and energy-band variation of an inverted p-type region for (a) the equilibrium case and (b) the nonequilibrium case (after Grove, 1967)

nonequilibrium case, the hole quasi-Fermi level is the same as the Fermi level in the bulk p-type silicon, but the electron quasi-Fermi level is dictated by the Fermi level in the n$^+$ region (not shown in Figure 4.21), which is qV_R lower than the p-type Fermi level. As a result, surface inversion occurs at a band bending

$$\psi_s(\text{inv}) = V_R + 2\psi_B, \tag{4.55}$$

and the maximum depletion width is a function of the reverse bias V_R,

$$W_{dm} = \sqrt{\frac{2\varepsilon_{si}(V_R + 2\psi_B)}{qN_a}}, \tag{4.56}$$

from Eq. (4.22).

For a p-type MOS capacitor, the effect of an adjacent reverse biased n$^+$ region on the $C–V$ characteristics is shown in Figure 4.22. With increasing V_g, the surface potential ψ_s also increases, as labeled under the curve. At $V_R = 0$, the $C–V$ curve resembles a regular low-frequency $C–V$ curve [curve (a) in Figure 4.14] with inversion (sharp rise of the capacitance to C_{ox}) taking place at $V_g \approx 13$ V where $\psi_s \approx 0.7$ V $(= 2\psi_B)$. Note that as V_R increases, the onset of inversion shifts to increasingly more positive gate voltages as the MOS goes into deeper and deeper depletion (lower C_{min}). It can be seen that the value of surface potential at inversion increases by approximately V_R, consistent with Eq. (4.55). The decrease of C_{min}, the serial combination of C_{ox} and $C_d = \varepsilon_{si}/W_{dm}$, with V_R follows from Eq. (4.56).

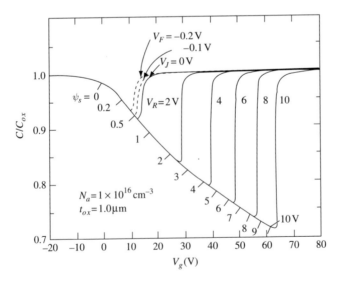

Figure 4.22 Normalized $C–V$ characteristics of a p-type MOS capacitor with an adjacent n$^+$ region as that shown in Figure 4.20(c). V_g is the voltage applied to the gate with respect to the p-type substrate. A series of $C–V$ curves are shown for a range of V_R: the reverse bias voltage applied to the n$^+$/p junction

4.4   Quantum Mechanical Effects in MOS

4.4.1   Coupled Poisson–Schrodinger's Equations

In Section 4.2.1.2 we discussed that in the inversion layer of a MOSFET, carriers are confined in a potential well very close to the silicon surface. The well is formed by the oxide barrier (essentially infinite except for tunneling calculations) and the silicon conduction band, which bends down severely toward the surface due to the applied gate field. Because of the confinement of motion in the direction normal to the surface, inversion-layer electrons should be treated quantum-mechanically as a 2-D gas (Stern and Howard, 1967), especially at high normal fields. Thus the energy levels of the electrons are grouped in discrete *subbands*, each of which corresponds to a quantized level for motion in the normal direction, with a continuum for motion in the plane parallel to the surface.

To solve for the charge and potential, Poisson's equation must be coupled to Schrodinger's equation. The electron density in Poisson's equation is no longer given by the classical 3-D expression, $n = N_c e^{-(E_c - E_f)/kT}$ of Eq. (2.9). Instead, it is given by the square of the magnitude of wave functions summed over all the discrete subbands. The potential energy in Schrodinger's equation, in turn, comes from the electric potential in Poisson's equation. In general, a numerical routine is employed to solve the coupled equations self-consistently. A special case in which the two equations are decoupled is discussed in Section 4.4.3.

An example of the numerically solved quantum-mechanical energy levels and band bending is shown in Figure 4.23. The electron concentration peaks below the silicon–oxide interface and goes to nearly zero at the interface, as dictated by the boundary condition of the electron wave function. This is in contrast to the classical model in which the electron concentration peaks at the surface, as depicted in Figure 4.24. Quantum-mechanical behavior of inversion-layer electrons affects MOSFET operation in two ways. *First, at high fields, threshold voltage becomes higher, since more band bending is required to populate the lowest subband at some energy above the bottom of the conduction band. Second, once the inversion layer forms below the surface, it takes a higher gate-voltage overdrive to produce a given level of inversion charge density*. In other words, the effective gate oxide thickness is slightly larger than the physical thickness. This reduces the transconductance and the current drive of a MOSFET. A more recent work on quantum mechanical modeling of MOS took into account polysilicon gate depletion and direct tunneling through thin gate oxides (Lo *et al.*, 1999).

4.4.2   Quantum Effect on Inversion-Layer Depth

After strong inversion, the inversion charge density builds up rapidly which in turn further enhances the field at surface. If the separation between the minimum energies of the lowest and the next higher subbands is large enough that only the lowest subband is populated, a variational approach leads to an approximate expression for the average distance of electrons from the surface (Stern, 1972):

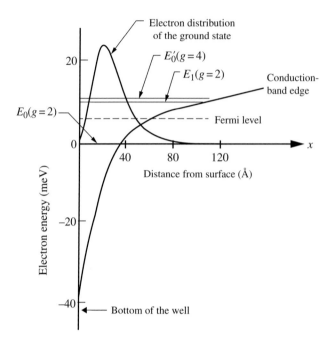

Figure 4.23 An example of quantum-mechanically calculated band bending and energy levels of inversion-layer electrons near the surface of an MOS device. The ground state is about 40 meV above the bottom of the conduction band at the surface. The dashed line indicates the Fermi level for 10^{12} electrons/cm^2 in the inversion layer (after Stern and Howard, 1967)

$$x_{av}^{QM} = \left(\frac{9\varepsilon_{si}h^2}{16\pi^2 m_x qQ^*}\right)^{1/3}, \tag{4.57}$$

where $Q^* = Q_d + \frac{11}{32}Q_i$ is a combination of the depletion and inversion charge per unit area in the channel. In general, the solution must be obtained numerically. Figure 4.25 shows a comparison of the classical and QM inversion-layer depths versus the effective normal field in silicon (Ohkura, 1990). The QM value is consistently larger than the classical value by about $10 - 12\,\text{Å}$ for a wide range of channel doping (uniform) and effective fields. This degrades the inversion layer capacitance, $-dQ_i/d\psi_s = \varepsilon_{si}/x_{av}$ (Section 4.3.4),[2] and therefore the inversion charge component of the gate capacitance, $-dQ_i/dV_{gs} = C_{inv} \equiv \varepsilon_{ox}/t_{inv}$. **Effectively, the quantum-mechanical effect adds** $\Delta t_{ox} = (\varepsilon_{ox}/\varepsilon_{si})\Delta x_{av} = (x_{av}^{QM} - x_{av}^{CL})/3$ **or about** $3 - 4\text{Å}$ **to** t_{inv}. The C_{inv} measured in a split C–V setup lumps the polysilicon gate depletion effect into t_{inv} as well. Overall, t_{inv} is $5 - 10\,\text{Å}$ thicker than the physical t_{ox} – a degrading factor on the current drive and transconductance of thin-oxide MOSFETs.

---

[2] Strictly speaking, the x_{av} factor in the capacitance is the center of mass of the differential inversion charge responding to a differential change of ψ_s. Here we neglect the subtle difference.

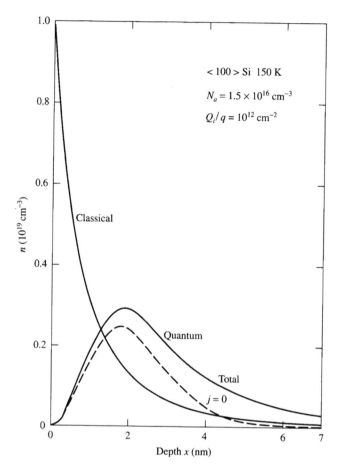

Figure 4.24 Classical and quantum-mechanical electron density versus depth for a $\langle 100 \rangle$ silicon inversion layer. The dashed curve shows the electron density distribution for the lowest subband (after Stern, 1974)

4.4.3   Quantum-Mechanical Solution in Weak Inversion

In weak inversion below the $2\psi_B$ threshold condition, the inversion charge density is not high enough to affect the local field. In other words, the electron density can be neglected in Poisson's equation. The potential or band bending is solely determined by the depletion charge. Poisson's and Schrodinger's equations are then decoupled, making it possible to derive the QM inversion charge density analytically. Note that MKS units are used throughout this section (e.g., length must be in meters, not centimeters).

4.4.3.1   Triangular Potential Approximation for Subthreshold

Since the inversion electrons are located in a narrow region close to the surface where the electric field is nearly constant ($\mathscr{E}_s$), it is a good approximation to consider the

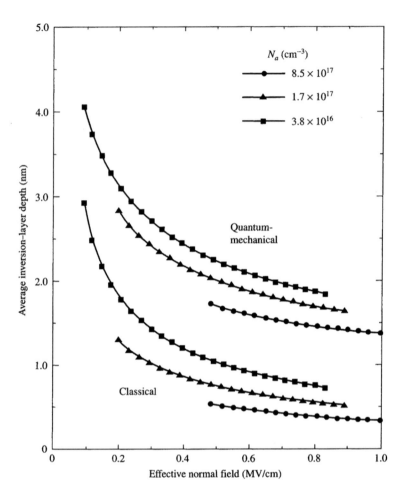

Figure 4.25 Calculated QM and classical inversion-layer depth versus effective normal field for several uniform doping concentrations (after Ohkura, 1990)

potential well as composed of an infinite oxide barrier for $x < 0$, and a triangular potential $V(x) = q\mathscr{E}_s x$ due to the depletion charge for $x > 0$, as depicted in Figure 4.26. The Schrödinger equation is solved with the boundary conditions that the electron wave function goes to zero at $x = 0$ and at infinity. The solutions are Airy functions with eigenvalues E_j given by (Stern, 1972)

$$E_j = \left[\frac{3hq\mathscr{E}_s}{4\sqrt{2m_x}} \left(j + \frac{3}{4} \right) \right]^{2/3}, \quad j = 0, 1, 2, \ldots, \tag{4.58}$$

where $h = 6.63 \times 10^{-34}$ J-s is Planck's constant, and m_x is the effective mass of electrons in the direction of confinement. The average distance from the surface for electrons in the j th subband is given by

$$x_j = \frac{2E_j}{3q\mathscr{E}_s}. \tag{4.59}$$

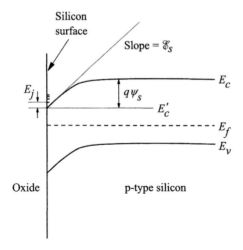

Figure 4.26 Schematic band diagram showing band bending in the subthreshold region and quantized electron energy levels in the inversion layer at the silicon surface

For silicon in the $\langle 100 \rangle$ direction, there are two groups of subbands, or *valleys*. The lower valley has a twofold degeneracy $(g = 2)$ with $m_x = m_l = 0.92\,m_0$, where $m_0 = 9.1 \times 10^{-31}$ kg is the free-electron mass. These energy levels are designated as $E_0, E_1, E_2 \ldots$. The higher valley has a fourfold degeneracy $(g' = 4)$ with $m'_x = m_t = 0.19\,m_0$. The energy levels are designated as $E'_0, E'_1, E'_2, \ldots$, given by

$$E'_j = \left[\frac{3hq\mathscr{E}_s}{4\sqrt{2m'_x}} \left(j + \frac{3}{4}\right) \right]^{2/3}, \quad j = 0, 1, 2, \ldots . \tag{4.60}$$

At room temperature, several subbands in both valleys are occupied near threshold, with a majority of the electrons in the lowest subband of energy E_0 above the bottom of the conduction band.

4.4.3.2 2-D Density of States

2-D density of states is needed for calculating the density of electrons in an inversion layer as they are free to move in a continuum of energy in the plane parallel to the surface. The density of states of a 2-D gas can be derived along the same line as in Section 2.1.2 for 3-D density of states. Based on quantum mechanics, there is one allowed state in a phase space of volume $(\Delta y \Delta p_y) \times (\Delta z\, \Delta p_z) = h^2$, where p_y and p_z are the y- and z-components of the electron momentum and h is Planck's constant. If we let $N(E)dE$ be the number of electronic states per unit area with an energy between E and $E + dE$, then

$$N(E)dE = 2g\frac{dp_y dp_z}{h^2}, \tag{4.61}$$

where $dp_y\, dp_z$ is the area in the momentum space within which the electron energy lies between E and $E + dE$, g is the degeneracy of the subband, and the factor of two arises from the two possible directions of electron spin.

If E_{min} is the ground-state energy of a particular subband, the energy–momentum relationship near the bottom of that subband is

$$E - E_{min} = \frac{p_y^2}{2m_y} + \frac{p_z^2}{2m_z}, \tag{4.62}$$

where $E - E_{min}$ is the electron kinetic energy, and m_y and m_z are the effective masses. The area of the ellipse given by Eq. (4.62) in momentum space is $2\pi \left(m_y m_z\right)^{1/2} (E - E_{min})$. Therefore, the area $dp_y\, dp_z$ within which the electron energy lies between E and $E + dE$ is $2\pi \left(m_y m_z\right)^{1/2} dE$ and Eq. (4.61) becomes

$$N(E)dE = \frac{4\pi g \sqrt{m_y m_z}}{h^2} dE. \tag{4.63}$$

The number of electrons per unit area in this subband is then given by

$$n = \int_{E_{min}}^{\infty} N(E) f_D(E) dE, \tag{4.64}$$

where $f_D(E)$ is the Fermi–Dirac distribution function, Eq. (2.4). Since $N(E)$ is a constant and can be taken out of the integral, Eq. (4.64) can be easily integrated to yield

$$n = \frac{4\pi g k T \sqrt{m_y m_z}}{h^2} \ln\left[1 + e^{\left(E_f - E_{min}\right)/kT}\right]. \tag{4.65}$$

4.4.3.3   QM Inversion Charge Density

In Figure 4.26, the bottom of the conduction band at surface E_c' is lower than the conduction band E_c in the bulk by an amount $q\psi_s$ due to the applied field from the gate. Here ψ_s is the surface potential or band bending described in Section 4.2.1. For the jth sub-band, the minimum energy is

$$E_{min} = E_c' + E_j = E_c - q\psi_s + E_j. \tag{4.66}$$

Likewise for the higher, primed valley. Summing over all the subbands in both valleys based on Eq. (4.65) gives the total inversion charge density per unit area (Stern and Howard, 1967):

$$-Q_i^{QM} = \frac{4\pi q k T}{h^2} \left\{ g m_t \sum_j \ln\left[1 + e^{\left(E_f - E_c - E_j + q\psi_s\right)/kT}\right] \right.$$
$$\left. + g' \sqrt{m_t m_l} \sum_j \ln\left[1 + e^{\left(E_f - E_c - E_j' + q\psi_s\right)/kT}\right] \right\}. \tag{4.67}$$

Note that for the first group of subbands in Eq. (4.67), the confinement (x) is in the longitudinal direction and the density-of-state effective mass $\left(m_y m_z\right)^{1/2}$ is $\left(m_t m_t\right)^{1/2}$

or $m_t = 0.19\, m_0$. For the second group of subbands in Eq. (4.67), the confinement (x) is perpendicular to the longitudinal direction and the density-of-state effective mass $(m_y m_z)^{1/2}$ is $(m_l m_t)^{1/2} = 0.42\, m_0$.

In the subthreshold region, the Fermi level is at least a few kT below the lowest subband of energy $E_c - q\psi_s + E_0$, and the factors $\ln(1 + e^x)$ can be approximated by e^x in both terms of Eq. (4.67). Furthermore, from Eq. (2.9), $E_f - E_c$ in the bulk silicon equals $kT \ln(n/N_c)$ or $kT \ln[n_i^2/(N_a N_c)]$, where N_c is the conduction band effective density of states, and $n = n_i^2/N_a$ is the equilibrium electron concentration in the bulk silicon. For a nonuniformly doped substrate, N_a refers to the p-type concentration at the edge of the depletion layer. Equation (4.67) is then simplified to

$$-Q_i^{QM} = \frac{4\pi q k T n_i^2}{h^2 N_c N_a} \left[2m_t \sum_j e^{-E_j/kT} + 4\sqrt{m_t m_l} \sum_j e^{-E_j'/kT} \right] e^{q\psi_s/kT}, \qquad (4.68)$$

where $g = 2$ and $g' = 4$ have been substituted.

4.4.3.4   Convergence of the QM solution to 3-D Continuum Results

It is instructive to show that the QM inversion charge density, Eq. (4.68), converges to the classical 3-D results in the limit of $E_j << kT$, where E_j is the energy of lower valley given by Eq. (4.58). When the surface field is low ($\mathscr{E}_s < 10^4\,\mathrm{V/cm}$ at room temperature), the spacings between the quantized energy levels are small compared with kT and the 2-D quantum effect is weak. In this case, the serial summations in Eq. (4.68) can be replaced by integrals using the identity

$$\sum_n e^{-(n\Delta y)^{2/3}} \Delta y = \int_0^\infty e^{-y^{2/3}} dy \qquad (4.69)$$

in the limit of $\Delta y \to 0$. By a simple change of variable $(u = y^{1/3})$, the integral on the right-hand side of Eq. (4.69) can be converted to a gamma function, whose value is $3\pi^{1/2}/4$. Therefore,

$$\sum_j e^{-E_j/kT} \to \frac{3\sqrt{\pi}}{4} \left(\frac{4\sqrt{2m_l}(kT)^{3/2}}{3hq\mathscr{E}_s} \right) \qquad (4.70)$$

in the low-field limit. Likewise for the primed valley. Equation (4.68) can then be evaluated as

$$-Q_i^{QM} = \frac{4\pi q k T n_i^2}{h^2 N_c N_a} \left[2m_t \frac{\sqrt{2\pi m_l}(kT)^{3/2}}{hq\mathscr{E}_s} + 4\sqrt{m_t m_l} \frac{\sqrt{2\pi m_t}(kT)^{3/2}}{hq\mathscr{E}_s} \right] e^{q\psi_s/kT}. \qquad (4.71)$$

Substituting N_c from Eq. (2.10) with the 3-D degeneracy factor $g = 6$ into Eq. (4.71) yields

$$-Q_i^{QM} = \frac{kTn_i^2}{\mathscr{E}_s N_a} e^{q\psi_s/kT}. \qquad (4.72)$$

The 3-D classical inversion charge density in weak inversion can be derived from Eq. (4.16) for the total charge density in silicon per unit area:

$$-Q_s = \varepsilon_{si}\mathscr{E}_s = \sqrt{2\varepsilon_{si}kTN_a}\left[\left(e^{-q\psi_s/kT} + \frac{q\psi_s}{kT} - 1\right) + \frac{n_i^2}{N_a^2}\left(e^{q\psi_s/kT} - \frac{q\psi_s}{kT} - 1\right)\right]^{1/2}.$$

(4.73)

Since $(n_i^2/N_a^2)e^{q\psi_s/kT} \ll q\psi_s/kT$, the inversion charge term can be separated out from the total charge by power series expansion of the square root expression. The zeroth order term gives the depletion charge density. The first order term gives the inversion charge density, which is identical to Eq. (4.72), noting that $\mathscr{E}_s = \sqrt{2qN_a\psi_s/\varepsilon_{si}}$. This shows that when the surface field is low and/or the temperature is high, the 2-D quantum solution converges to the 3-D continuum case. At high fields, however, Q_i^{QM} is lower than the 3-D inversion charge density for the same band bending, which results in a higher threshold voltage than that of the classical model.

4.5   Interface States and Charge Traps in Oxide

Thus far we have treated gate oxide as an ideal insulator, with no space-charge in or associated with it, and no charge exchange between it and the silicon it covers. The gate oxide and the oxide–silicon interface in real devices are never completely electrically neutral. There can be mobile ionic charges, electrons, or holes trapped in the oxide layer. There can also be fabrication-process-induced fixed oxide charges near the oxide–silicon interface, and charges trapped at the so-called *surface states* at the oxide–silicon interface. Electrons and holes can make transitions from the crystalline states near the oxide–silicon interface to the surface states, and vice versa. Since every device has some regions that are covered by oxide, the electrical characteristics of a device are very sensitive to the density and properties of the charges inside its oxide regions and at its silicon–oxide interface.

For SiO_2, the density of surface states, and hence the density of interface traps, is a function of silicon substrate orientation and a strong function of the device fabrication process (EMIS, 1988; Razouk and Deal, 1979). In general, for a given device fabrication process, the dependence of the interface trap density on substrate orientation is $\langle 100 \rangle < \langle 110 \rangle < \langle 111 \rangle$. Also, a postmetallization or "final" anneal in hydrogen, or in a hydrogen-containing ambient, at temperatures around $400\,°C$ is quite effective in minimizing the density of interface traps. Consequently, <100> silicon and postmetallization anneal in hydrogen are commonly used in modern VLSI device fabrication. For modern MOS devices, by the time a device fabrication process is ready for manufacturing, the amount of oxide charge and surface states is usually quite low, with Q_{ox}/q typically about 10^{11} cm^{-2} or less. For an MOS device having an oxide thickness of 10 nm, the corresponding flatband voltage shift is only 46 mV. At such low surface-state densities, the MOS $C–V$ characteristics are quite ideal in that the measured $C–V$ curves match well with the calculated ones.

The Silicon/SiO$_2$ system has been intensely studied for many decades since the 1960s. The vast knowledge generated enabled the VLSI industry to scale SiO$_2$ in MOSFETs down to an amazing thickness of ~1 nm– only a couple of atomic layers thick. But at that dimension, fundamental physics of quantum mechanical tunneling sets in to limit further scaling of SiO$_2$. In the past decade or two, the industry turned to high-κ gate insulators. A high-κ layer of the same thickness as a SiO$_2$ layer has much larger capacitance. Thus, for the same gate capacitance, a high-κ gate insulator is thicker than SiO$_2$ and has a much lower tunneling current. This opens up a new set of issues on gate insulator–semiconductor interface states and charge traps.

The presence of oxide charges and interface traps has three major effects on the characteristics of devices. First, the charge in the oxide, or in the interface traps, interacts with the charge in the silicon near the surface and thus changes the silicon charge distribution and the surface potential. Second, as the density of interface trapped charge changes with changes in the surface potential, it gives rise to an additional capacitance component in parallel with the silicon capacitance C_{si} in Figure 4.13. Third, the interface traps can act as generation–recombination centers, or assist in the band-to-band tunneling process, and thus contribute to the leakage current in a gated-diode structure. These effects are discussed more quantitatively in Sections 4.5.1–4.5.3.

4.5.1   Effect of Oxide Charge on Flatband Voltage

Equation (4.2) in Section 4.1.1 stated that, in addition to the gate-silicon work function difference ϕ_{ms}, fixed charge in the oxide Q_{ox} may also contribute to the flatband voltage. In Figure 4.27, we consider a sheet charge per unit area, δQ, in the oxide at a distance x from the gate. This inflicts a change of slope at x given by

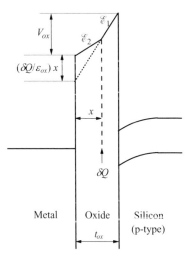

Figure 4.27 Schematic illustrating the effect of a sheet charge of areal density δQ within the oxide layer of an MOS capacitor at a distance x from the gate

$$\mathscr{E}_1 - \mathscr{E}_2 = \frac{\delta Q}{\varepsilon_{ox}} \tag{4.74}$$

from Gauss's law. Gauss's law also gives $Q_g = \varepsilon_{ox}\mathscr{E}_2$ for the charge on the gate and $Q_s = -\varepsilon_{ox}\mathscr{E}_1$ for the charge in silicon. Combining them yields the charge neutrality condition,

$$Q_g + \delta Q + Q_s = 0, \tag{4.75}$$

as expected. With the change of slope, the potential across the oxide becomes

$$V_{ox} = \mathscr{E}_1(t_{ox} - x) + \mathscr{E}_2 x = \mathscr{E}_1 t_{ox} - \frac{\delta Q}{\varepsilon_{ox}} x = \frac{-Q_s}{C_{ox}} - \frac{\delta Q}{\varepsilon_{ox}} x, \tag{4.76}$$

as indicated in Figure 4.27. Substituting the above V_{ox} in the gate voltage equation, Eq. (4.5), obtains

$$V_g - (\phi_m - \phi_s) + \frac{\delta Q}{\varepsilon_{ox}} x = \psi_s + \frac{-Q_s}{C_{ox}}. \tag{4.77}$$

The left-hand side is defined as $V_g - V_{fb}$ in accordance with Eq. (4.7). Therefore,

$$V_{fb} = (\phi_m - \phi_s) - \frac{\delta Q}{\varepsilon_{ox}} x = \phi_{ms} - \frac{\delta Q}{\varepsilon_{ox}} x. \tag{4.78}$$

The effect of a sheet charge δQ at x is then to shift the flatband voltage negatively by $(\delta Q / \varepsilon_{ox})x$. The closer the charge is to the oxide–silicon interface, the larger is its effect on the flatband voltage.

If the charge in oxide is somewhat mobile, its position may be shifted by the applied gate field. For example, a positive V_g tends to push mobile positive oxide charge toward the silicon interface resulting in a more negative V_{fb}. This can give rise to hysteresis in C–V characteristics, namely, V_{fb} differs between a positive going ramp and a negative going ramp of dc bias.

If there is a distribution of oxide charge over the thickness with density $\rho_{ox}(x)$, the combined effect on flatband voltage is expressed by an integral,

$$V_{fb} = \phi_{ms} - \frac{1}{\varepsilon_{ox}} \int_0^{t_{ox}} x \rho_{ox}(x) dx \equiv \phi_{ms} - \frac{Q_{ox}}{C_{ox}}. \tag{4.79}$$

The last step defines Q_{ox} as the equivalent charge at the oxide–silicon interface that produces the same effect as the distribution $\rho_{ox}(x)$. It is clear that for a thinner oxide, the same magnitude of oxide charge has less of an effect on V_{fb}.

4.5.2   Interface–State Capacitance and Conductance

At the semiconductor–oxide interface, the lattice of bulk semiconductor and all the properties associated with its periodicity terminate. As a result, localized states with energy in the forbidden energy gap of the semiconductor are introduced at or very near the oxide interface (Many *et al.*, 1965). Just like the impurity energy levels in bulk silicon discussed in Section 2.1.3, the probability of occupation of a surface state by an electron or by a hole is determined by the surface-state energy relative to the Fermi level. Thus, as the surface potential is changed, the energy level of a surface state, which is fixed relative to the energy-band edges at the surface, moves with it. This is shown schematically in Figure 4.28. The change relative to the Fermi level causes a change in the probability of occupation of the surface state by an electron. Such change of the interface trapped charge with surface potential gives rise to an additional capacitance component. The interface–state capacitance is connected through conductance components since its charging and discharging is not instantaneous.

Electrons in silicon but near an oxide–silicon interface can make transitions between the conduction-band states and the surface states. An electron in the conduction band can contribute readily to electrical conduction current, while an electron in a surface state, i.e., an interface trapped electron, does not contribute readily to electrical conduction current. Similarly, holes in silicon but near an oxide–silicon interface can make transitions between the valence-band states and the surface states, and trapped interface holes do not contribute readily to electrical conduction. By trapping electrons and holes, surface states can reduce the conduction current in MOSFETs. Furthermore, trapped electrons and holes can act like charged scattering centers, located at the interface, for the mobile carriers in a surface channel, thus lowering their mobility (Sah *et al.*, 1972).

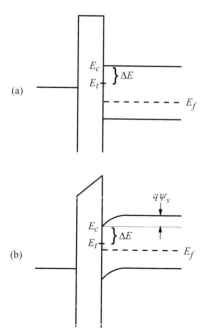

Figure 4.28 Schematic energy-band diagram of an MOS capacitor with surface states at energy E_t in the bandgap. (a) Flatband condition, (b) $V_g > V_{fb}$ is applied to cause a band bending of $q\psi_s$. The position of E_t with respect to E_c is fixed, i.e., $\Delta E = E_c - E_t$ is constant, regardless of band bending

4.5.2.1 Traps of a Single Energy Level

Consider in Figure 4.28 a density of N_t interface states (traps) per area at energy E_t in the bandgap (Nicollian and Brews, 1982). The probability of occupancy of electrons is given by

$$f_D = \frac{1}{1 + \exp\left[(E_t - E_f)/kT\right]}. \tag{4.80}$$

The total charge in those states is

$$Q_t = -qN_t f_D. \tag{4.81}$$

For an incremental change in the surface potential, $\delta\psi_s = -\delta E_t/q$,

$$\delta Q_t = -qN_t \delta f_D = qN_t \frac{\exp\left[(E_t - E_f)/kT\right]}{\left\{1 + \exp\left[(E_t - E_f)/kT\right]\right\}^2}\left(-\frac{q\delta\psi_s}{kT}\right) = -\frac{q^2}{kT}N_t f_D(1 - f_D)\delta\psi_s. \tag{4.82}$$

The equivalent capacitance (per unit area) due to those interface traps is therefore

$$C_{it} = -\frac{\delta Q_t}{\delta\psi_s} = \frac{q^2}{kT}N_t f_D(1 - f_D). \tag{4.83}$$

Note that $f_D(1 - f_D)$ ($\propto df_D/dE_t$) is maximum when $f_D = \frac{1}{2}$. It is not surprising to find *that C_{it} sharply peaks at $E_t = E_f$, i.e., when the trap energy level crosses the Fermi level.*

The interface state capacitance is in parallel with the depletion capacitance C_d since δQ_t represents additional charging and discharging that must take place when ψ_s is modulated. However, unlike C_d, C_{it} cannot be charged and discharged instantaneously. It takes a capture and emission process much like that described in Section 2.4.3.1 for bulk traps. In thermal equilibrium, Eq. (2.83) gives the rate of electron capture from the conduction band:

$$R_n = \sigma_n v_{th} n_s N_t (1 - f_D), \tag{4.84}$$

where n_s is the electron concentration at the surface. R_n is balanced by the rate of electron emission back to the conduction band. Now consider a sudden incremental change of the surface potential $\delta\psi_s$. Instantaneously, the trap occupancy has not changed. Only the electron density n_s in the conduction band changes instantaneously. Since $n_s \propto \exp(q\psi_s/kT)$,

$$\delta R_n = \sigma_n v_{th} \delta n_s N_t (1 - f_D) = \sigma_n v_{th} n_s N_t (1 - f_D) q\delta\psi_s/kT. \tag{4.85}$$

There is no instantaneous change of the emission rate G_n because it only depends on the density of occupied traps [Eq. (2.84)]. From Eq. (4.85), the instantaneous charging current is

$$\delta J = q\delta R_n = \sigma_n v_{th} n_s N_t (1 - f_D) q^2 \delta\psi_s/kT, \tag{4.86}$$

which corresponds to a conductance (per unit area) of

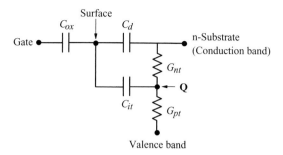

Figure 4.29 Equivalent circuit of surface state traps at a single energy level. The substrate is assumed to be n-type

$$G_{nt} = \delta J/\delta \psi_s = \sigma_n v_{th} n_s N_t (1 - f_D) q^2 / kT. \tag{4.87}$$

Likewise, there is also a conductance due to the capture of holes from the valence band:

$$G_{pt} = \sigma_p v_{th} p_s N_t f_D q^2 / kT, \tag{4.88}$$

where p_s is the hole density at the surface. The equivalent circuit is shown in Figure 4.29.

The connection by G_{nt} and G_{pt} between the conduction band and valence band in Figure 4.29 means that surface states can also act like localized generation–recombination centers. Depending on the surface potential, a surface state can first capture an electron from the conduction band, or a hole from the valence band. This captured electron can subsequently recombine with a hole from the valence band, or the captured hole can recombine with an electron from the conduction band. In this way, the surface state acts like a recombination center. Similarly, a surface state can act like a generation center by first emitting an electron followed by emitting a hole, or by first emitting a hole followed by emitting an electron. Thus, the presence of surface states can lead to surface generation–recombination leakage currents.

4.5.2.2 **Traps of a Continuum Energy Distribution**

In practice, interface states are found to have a continuum energy distribution in the bandgap, characterized by D_{it}, the areal trap density per energy. Here we adopt the industry standard unit of $cm^{-2} eV^{-1}$ for D_{it} so the trap density in $cm^{-2} joule^{-1}$ is D_{it}/q. The total trap capacitance and conductance are then obtained by replacing N_t with $(D_{it}/q)dE_t$ and integrating each component over E_t (in joule). Circuit elements C_{it}, G_{nt}, and G_{pt} in Figure 4.29, however, cannot be directly integrated because the center node Q is not common for all the different energies. They must first be transformed from a Y configuration to a Δ configuration using the general rules outlined in Figure 4.30. Nodes A, B, and C are the surface, conduction band, and valence band, as depicted in Figure 4.29. Figure 4.31 shows the circuit after the transformation, with the interface state components C_{Tn}, C_{Tp}, and G_{gr} enclosed in a

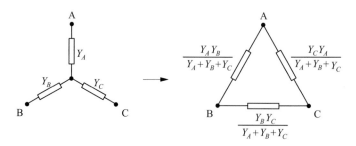

Figure 4.30 Y to Δ admittance transformation

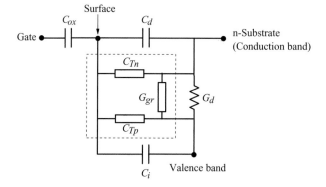

Figure 4.31 Equivalent circuit of surface state traps distributed over a continuum of energy. The dashed-line box contains the components of admittance from interface states: C_{Tn}, C_{Tp}, and G_{gr}. The substrate is assumed to be n-type

rectangular box of dashed lines. It is more convenient to change the integration variable from E_t to f_D using $df_D/dE_t = -f_D(1 - f_D)/kT$. Then we have

$$C_{Tn} = q \int_0^1 \frac{\sigma_n v_{th} n_s (1 - f_D) D_{it} df_D}{j\omega f_D(1 - f_D) + \sigma_n v_{th} n_s (1 - f_D) + \sigma_p v_{th} p_s f_D}, \qquad (4.89)$$

$$C_{Tp} = q \int_0^1 \frac{\sigma_p v_{th} p_s f_D D_{it} df_D}{j\omega f_D(1 - f_D) + \sigma_n v_{th} n_s (1 - f_D) + \sigma_p v_{th} p_s f_D}, \qquad (4.90)$$

and

$$G_{gr} = q \int_0^1 \frac{\sigma_n v_{th} n_s \sigma_p v_{th} p_s D_{it} df_D}{j\omega f_D(1 - f_D) + \sigma_n v_{th} n_s (1 - f_D) + \sigma_p v_{th} p_s f_D}. \qquad (4.91)$$

The transformation has mixed up the capacitance and conductance such that C_{Tn}, C_{Tp}, and G_{gr} represent admittances $j\omega C_{Tn}$, $j\omega C_{Tp}$, and G_{gr} in general. Only at low frequencies are C_{Tn} and C_{Tp} pure capacitance and G_{gr} pure conductance. The rest of the components in Figure 4.31 are the intrinsic MOS elements in Figure 4.13(b).

It is a good approximation to take D_{it} outside of the integral with its value at $E_t = E_f$. This is justified by that over a wide range of $f_D \in (0,1)$, E_t is within $(1-2)kT$ of E_f (see Figure 2.2), and that $D_{it}(E_t)$ does not vary significantly on the scale of kT. It is inferred from Figure 4.28 that D_{it} is a function of the surface potential ψ_s, hence a function of the gate voltage. By introducing

$$\tau_n = \frac{1}{\sigma_n v_{th} n_s} \tag{4.92}$$

and

$$\tau_p = \frac{1}{\sigma_p v_{th} p_s}, \tag{4.93}$$

the above expressions become (Nicollian and Brews, 1982)

$$C_{Tn} = \frac{qD_{it}}{\tau_n} \int_0^1 \frac{(1-f_D)df_D}{j\omega f_D(1-f_D) + (1-f_D)/\tau_n + f_D/\tau_p} \tag{4.94}$$

$$C_{Tp} = \frac{qD_{it}}{\tau_p} \int_0^1 \frac{f_D df_D}{j\omega f_D(1-f_D) + (1-f_D)/\tau_n + f_D/\tau_p} \tag{4.95}$$

$$G_{gr} = \frac{qD_{it}}{\tau_n \tau_p} \int_0^1 \frac{df_D}{j\omega f_D(1-f_D) + (1-f_D)/\tau_n + f_D/\tau_p}. \tag{4.96}$$

In general, as functions of frequency, admittances $j\omega C_{Tn}$, $j\omega C_{Tp}$, and G_{gr} are completely specified by three parameters: D_{it}, τ_n, and τ_p. They all vary with the surface potential, hence with the dc bias of V_g.

4.5.2.3   **The "Conductance Method"**

Multiple-frequency admittance (C–V and G–V) measurement is an indispensable tool for characterization of interface states. For an intrinsic MOS with no surface states biased in accumulation or depletion, there is no conductance nor any dispersion of the capacitance. Time constants [Eqs. (4.92) and (4.93)] associated with the interface states give rise to C–V dispersion – a sure sign of the presence of surface states.

For an MOS capacitor on n-type substrate biased in depletion, the hole density p_s is very low, hence $1/\tau_p$ is negligible. The only significant interface state component is C_{Tn}, which can be simplified to

$$C_{Tn} = \frac{qD_{it}}{\tau_n} \int_0^1 \frac{df_D}{j\omega f_D + 1/\tau_n} = \frac{qD_{it}}{j\omega \tau_n} \ln(1 + j\omega\tau_n). \tag{4.97}$$

By separating it into real and imaginary parts and combining with C_d, the equivalent circuit of Figure 4.32 is obtained where

$$C_p = C_d + \frac{qD_{it}}{\omega\tau_n} \tan^{-1}(\omega\tau_n) \tag{4.98}$$

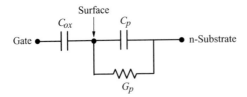

Figure 4.32 Equivalent circuit of MOS biased in depletion with surface states

and

$$G_p = \frac{qD_{it}}{2\tau_n} \ln\left[1 + (\omega\tau_n)^2\right]. \tag{4.99}$$

Examples of C_p and G_p/ω versus ω plots are shown in Figure 4.33. In Figure 4.33(a), C_p varies from C_d at high frequencies $(\omega\tau_n \gg 1)$ where no surface states respond to the ac modulation to $C_d + qD_{it}$ at low frequencies $(\omega\tau_n \ll 1)$ where all surface states respond to the ac signal. This is known as the "high–low capacitance method" for D_{it} determination. In most cases, however, the C–V measurement system does not reach low enough frequencies to cover the high C_p plateau. D_{it} **can also be read from the peak of G_p/ω versus ω curve**, as shown in Figure 4.33(b). The peak position, at $\omega\tau_n \approx 2$, gives the value of τ_n at this bias point. This is known as the "conductance method." As the V_g bias moves further into depletion, the surface electron density n_s decreases precipitously so τ_n becomes longer. The G_p/ω peak then moves below the low end of the measured frequency range at, e.g., 1 kHz. C_p at 1 kHz would decrease with bias not because D_{it} is decreasing but because of the effect of longer τ_n (see Figure 4.35).

Note that C_p and G_p are not the raw admittance data measured between the gate and the substrate. Rather, the raw data have to be processed to first take out the serial contribution from C_{ox}. It is not a simple task to determine C_{ox} accurately, especially for very thin oxides. The MOS device cannot be biased too far into accumulation due to breakdown concerns. The measured maximum capacitance is significantly below C_{ox} because of the inversion layer capacitance, polysilicon depletion, and quantum mechanical effects. Experimental data must be compared to a modeled ideal C–V curve taking the above into account. Errors in C_{ox} could lead to errors in C_p, G_p, and therefore in D_{it} (Chen *et al.*, 2013).

4.5.2.4   C–V "Stretch Out" by Interface States

The C–V ramp is usually done statically, in which case all surface states are charged or discharged when V_g or ψ_s is varied. The MOS capacitance circuit is that shown in Figure 4.34. For incremental static changes,

$$\Delta V_g = \Delta\psi_s + \Delta V_{ox} = \Delta\psi_s\left[1 + \frac{C_d + qD_{it}}{C_{ox}}\right]. \tag{4.100}$$

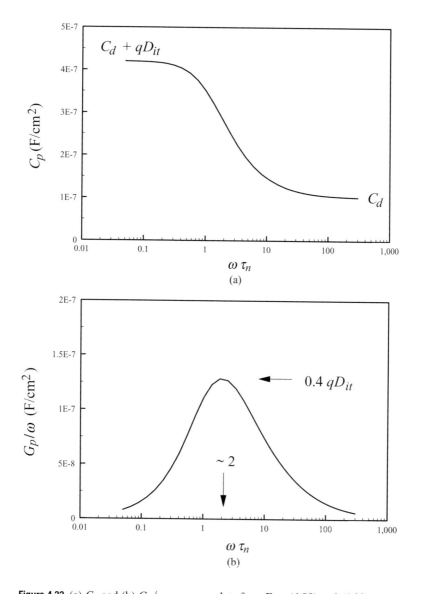

Figure 4.33 (a) C_p and (b) G_p/ω versus ω plots from Eqs. (4.98) and (4.99)

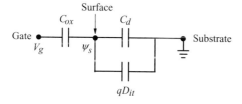

Figure 4.34 dc capacitance circuit with interface states

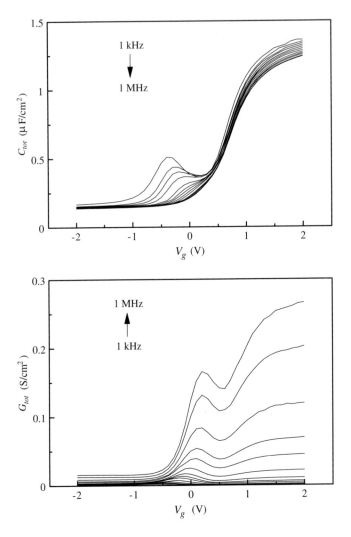

Figure 4.35 Measured capacitance–voltage and conductance–voltage data from an Al$_2$O$_3$ MOS capacitor on n-type InGaAs (after Taur *et al.*, 2015)

This means that the presence of D_{it} makes it more difficult for the gate voltage to control the surface potential. If there happens to be a high density of local D_{it} at a particular energy in the bandgap, it is said that ψ_s is "pinned" by the surface states to that energy because it takes a very large ΔV_g to move ψ_s beyond that point. More typically, the high-to-low transition of the *C*–*V* curve is "stretched out" because it takes a larger V_g swing to deplete the substrate to a deeper depth.

Consider a high-frequency *C*–*V* at, e.g., 1 MHz, so that no surface states can respond to the ac signal. In other words, the MOS capacitance is simply given by C_d in series with C_{ox}. The only effect of D_{it} is the dc stretch out of the *C*–*V* curve. A technique to extract D_{it} from such stretch out has been proposed in "Terman method" (Terman, 1962). The experimental C–V is compared to an ideal, simulated C–V with no D_{it}. The capacitance is an indication of ψ_s. By comparing the slopes dC/dV_g of the two curves at the same C,

$$D_{it} = \left[\frac{(\Delta V_g)_{data}}{(\Delta V_g)_{ideal}} - 1 \right] \frac{C_{ox} + C_d}{q} = \left[\frac{(dC/dV_g)_{ideal}}{(dC/dV_g)_{data}} - 1 \right] \frac{C_{ox} + C_d}{q}, \qquad (4.101)$$

based on Eq. (4.100). In practice, however, the highest frequency of a standard C–V system, typically 1 MHz, is not high enough to rid all the D_{it} contributions to C. This is because of the slow, $1/\omega$ decay of the D_{it} term in Eq. (4.98). Figure 4.33(a) indicates that it is necessary to go to $\omega\tau_n > 100$ for the D_{it} contribution to C_p to be negligible. Further complicating the matter is the fact that τ_n is bias dependent, increasing with the degree of depletion by V_g. Such varying contributions of D_{it} to the measured 1 MHz capacitance causes errors in dC/dV_g, hence the extracted D_{it}.

A key implication of Eq. (4.100) is that **the impact of interface states on device performance is measured by the factor** qD_{it}/C_{ox}. Ideally, if $qD_{it} \ll C_{ox}$, the D_{it} effect is negligible. A scaled MOSFET with thinner gate oxides can thus tolerate a higher D_{it} level.

4.5.3 Distributed Circuit Model for Oxide Traps

Silicon technology has been blessed by the fact that the charge trap density in SiO_2 is very low. In general, traps in the gate insulator can be charged and discharged by conduction band electrons via tunneling. An example is shown in Figure 4.35 for an n-type InGaAs MOS capacitor with Al_2O_3 as the gate dielectric. Here, the C–V dispersion in the depletion–inversion region over negative V_g is due to interface states discussed in Section 4.5.2 (Chen *et al.*, 2012). The dispersion in the accumulation–depletion region is caused by oxide traps. It has a signature $\sim \ln \omega$ dependence on frequency.

Figure 4.36 depicts a distribution of oxide traps at a distance x from the surface. The trap density at an energy E_t is N_{bt} in units of $cm^{-3}\,eV^{-1}$. The capture and emission of electrons by oxide traps is the same as those of interface states, except with an extra factor of the tunneling exponential, $\exp(-2\kappa x)$, coming from decaying of the electron wavefunction into the oxide. Here, $\kappa = \hbar^{-1}\sqrt{2m^*\phi_B}$ depends on the barrier height ϕ_B and the effective mass m^* [Eq. (3.127)]. The incremental capacitance and conductance components due to N_{bt} are then given by Eqs. (4.83) and (4.87) with N_t replaced by $N_{bt}\Delta x\Delta E_t$ and n_s replaced by $n_s \exp(-2\kappa x)$:

$$\Delta C_{bt} = \frac{q}{kT} f_D(1 - f_D)N_{bt}\Delta x\Delta E_t, \qquad (4.102)$$

$$\Delta G_{bt} = \frac{q}{kT} \sigma_n v_{th} n_s e^{-2\kappa x}(1 - f_D)N_{bt}\Delta x\Delta E_t, \qquad (4.103)$$

where f_D is the same as that of Eq. (4.80). The incremental admittance from oxide traps between x and $x + \Delta x$ is therefore

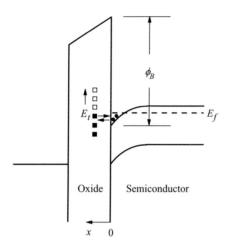

Figure 4.36 Schematic diagram showing tunneling of electrons between the conduction band and oxide traps a small distance from the interface

$$\Delta Y_{bt}(x) = \int_{\Delta E_t} \frac{j\omega \Delta C_{bt} \Delta G_{bt}}{j\omega \Delta C_{bt} + \Delta G_{bt}} = \frac{q}{kT} \int \frac{j\omega f_D (1 - f_D) N_{bt} \Delta x}{1 + j\omega f_D e^{2\kappa x}/(\sigma_n v_{th} n_s)} dE_t. \quad (4.104)$$

As done with the D_{it} integrals in Section 4.5.2.2, by applying $df_D/dE_t = -f_D(1 - f_D)/kT$ to convert dE_t into df_D, N_{bt} can be taken outside the integral with its value at $E_t = E_f$: Thus

$$\Delta Y_{bt}(x) = j\omega q N_{bt} \Delta x \int_0^1 \frac{df_D}{1 + j\omega f_D e^{2\kappa x}/(\sigma_n v_{th} n_s)} = \frac{q N_{bt} \ln[1 + j\omega \tau_n e^{2\kappa x}]}{\tau_n e^{2\kappa x}} \Delta x, \quad (4.105)$$

where τ_n is the same as that defined in Eq. (4.92).

The distributed circuit model for oxide traps is shown in Figure 4.37 (Yuan et al., 2012). The oxide is partitioned into capacitors of incremental thickness Δx connected in series. $Y(x)$ is defined as the admittance between a point x in the oxide and the substrate. The recursive relation for $Y(x)$ is

$$Y(x + \Delta x) = \Delta Y_{bt}(x) + \frac{(\varepsilon_{ox}/\Delta x) Y(x)}{(\varepsilon_{ox}/\Delta x) + Y(x)}. \quad (4.106)$$

By expanding the last term into a power series and keeping only the first-order terms, a differential equation is obtained:

$$\frac{dY}{dx} = -\frac{Y^2}{j\omega \varepsilon_{ox}} + \frac{q N_{bt} \ln[1 + j\omega \tau_n e^{2\kappa x}]}{\tau_n e^{2\kappa x}}. \quad (4.107)$$

It is solved numerically with the boundary condition $Y(x = 0) = j\omega C_s$ where C_s is the semiconductor capacitance. If the MOS is biased in accumulation, C_s is the accumulation capacitance discussed in Section 4.3.3.1. After $Y(x)$ is solved, $Y(x = t_{ox}) = G_{tot} + j\omega C_{tot}$ gives the total MOS admittance.

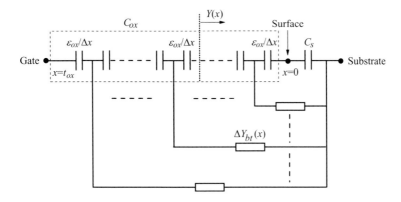

Figure 4.37 A distributed circuit model for oxide traps (after Yuan *et al.*, 2012)

Figure 4.38 shows an example of the results versus frequency for the parameter set given. At high enough frequencies where no oxide traps respond to the ac signal, the MOS capacitance equals the serial combination of C_{ox} and C_s. C_{tot} **then goes up more or less linearly with $ln(1/\omega)$** extending far below the measured frequency range. The conductance G_{tot} is essentially a linear function of ω over the frequency range. Most experimental data can be explained by a uniform N_{bt} with no depth dependence. The log-frequency dependence of C_{tot} can be understood by noting that the capacitive part of the N_{bt} term in Eq. (4.107) decays rapidly with x once $\omega \tau_n e^{2\kappa x} > 1$. In other words, **the depth of oxide traps contributing to the capacitance at ω is $x \approx (1/2\kappa) \ln(1/\omega \tau_n)$**. For the example in Figure 4.38, this depth is 1.3 nm for a signal frequency of 1 kHz.

4.6 High-Field Effects in Oxide and Oxide Degradation

The effects of high field in a reverse biased p–n diode is covered in Section 3.3. In this section, we discuss the effects of high field across an oxide layer in an MOS capacitor and in a gated p–n diode. In the case of high field in gate oxide layer, the important phenomena include tunneling through the oxide layer, degradation of the oxide layer due to the tunneling current, and dielectric breakdown. In the case of high field in a gated diode structure, the important phenomena include injection of hot electrons or hot holes from silicon into the gate oxide, and degradation of the gate oxide due to trapping of the injected electrons and holes in the gate oxide layer. The basic physics of these phenomena as they relate to VLSI devices are discussed.

4.6.1 Tunneling into and through Silicon Dioxide

Consider an MOS capacitor with heavily doped n-type polysilicon as its gate electrode. When biased at the flatband condition, the energy-band diagram is as

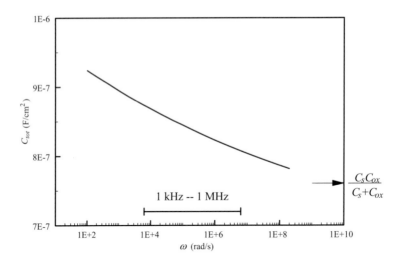

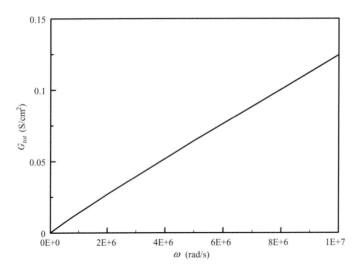

Figure 4.38 C_{tot} and G_{tot} versus ω solved from Eq. (4.107). The parameters are $C_{ox} = 1.06\ \mu F/cm^2$, $C_s = 2.7\ \mu F/cm^2$, $\varepsilon_{ox}/\varepsilon_0 = 6$, $\kappa = 5.1\ nm^{-1}$, $N_{bt} = 4.2 \times 10^{19}\ cm^{-3}\ eV^{-1}$, and $\tau_n = 2.3 \times 10^{-10}$ s. The arrow indicates the $\omega\tau_n \gg 1$ limit of $0.76\ \mu F/cm^2$

shown in Figure 4.39(a), where $q\phi_{ox}$ denotes the $Si - SiO_2$ interface energy barrier for electrons which, as indicated in Figure 4.3, is about 3.1 eV. When a large positive bias is applied to the gate electrode, electrons in the strongly inverted surface can tunnel through the oxide layer and hence give rise to a gate current. Similarly, if a large negative voltage is applied to the gate electrode, electrons from the n^+ polysilicon can tunnel through the oxide layer, and again give rise to a gate current.

(a)

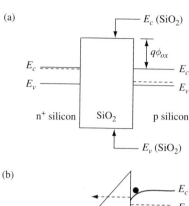

(b)

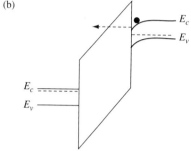

(c)

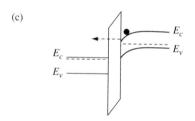

Figure 4.39 Tunneling effects in an MOS capacitor structure: (a) energy-band diagram of an n-type polysilicon-gate MOS structure at flat band; (b) Fowler–Nordheim tunneling; (c) direct tunneling

4.6.1.1 Fowler–Nordheim Tunneling

Fowler–Nordheim tunneling occurs when electrons tunnel into the conduction band of the oxide layer and then drift through the oxide layer. Figure 4.39(b) illustrates Fowler–Nordheim tunneling of electrons from the silicon surface inversion layer. The complete theory of Fowler–Nordheim tunneling is rather complicated (Good and Müller, 1956). For the simple case, where the effects of finite temperature and image–force barrier lowering (which is discussed in Section 3.2.1.4) are ignored, the tunneling current density is given by (Lenzlinger and Snow, 1969)

$$J_{FN} = \frac{q^2 \mathscr{E}_{ox}^2}{16\pi^2 \square \phi_{ox}} \exp\left(-\frac{4(2qm^*)^{1/2}\phi_{ox}^{3/2}}{3\square \mathscr{E}_{ox}}\right), \qquad (4.108)$$

where $\mathscr{E}_{ox}$ is the electric field in the oxide. Equation (4.108) shows that Fowler–Nordheim tunneling current is characterized by a straight line in a plot of $\log\left(J/\mathscr{E}_{ox}^2\right)$ versus $1/\mathscr{E}_{ox}$.

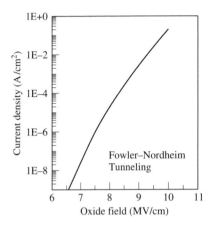

Figure 4.40 Fowler–Nordheim tunneling current density as a function of electric field in oxide (Pavan *et al.*, 1997)

As discussed in Section 4.6.1.3, electrons tunneling into an oxide layer can be trapped in an oxide layer. If the tunneling current is measured at a constant voltage, then the trapped electrons in turn can cause the observed tunneling current to decrease with time. Depending on the thickness of the oxide layer and its formation process, this decrease in tunneling current can go on for some time before it reaches a more-or-less steady state. The tunneling currents reported in the classic paper by Lenzlinger and Snow were taken after the samples were first subjected to a current density of about $10^{-10}\,\mathrm{A/cm^2}$ for two hours, during which time the tunneling currents decreased by about one order of magnitude from their initial values (Lenzlinger and Snow, 1969). Figure 4.40 shows a typical plot of Fowler–Nordheim tunneling current density as a function of electric field, measured from oxides used in flash memory devices (see Section 12.3).

The characteristics of the tunneling currents represented by Eqs. (3.141) and (4.108) are determined primarily by their exponential factors. It should be noted that the exponents of the two equations are basically the same. The Fowler–Nordheim tunneling is through a triangular barrier of height $q\phi_{ox}$, slope $q\mathscr{E}_{ox}$, and tunneling distance $\phi_{ox}/\mathscr{E}_{ox}$. See Exercise 3.6 for derivation of the tunneling exponent from the WKB approximation.

4.6.1.2　Direct Tunneling

If the oxide layer is very thin, say 4 nm or less, then, instead of tunneling into the conduction band of the SiO_2 layer, electrons from the inverted silicon surface can tunnel directly through the forbidden energy gap of the SiO_2 layer. This is illustrated in Figure 4.39(c). The theory of *direct tunneling* is even more complicated than that of Fowler–Nordheim tunneling, and there is no simple dependence of the tunneling current density on voltage or electric field (Chang *et al.*, 1967; Schuegraf *et al.*, 1992). Direct-tunneling current can be very large for thin oxide layers. Figure 4.41

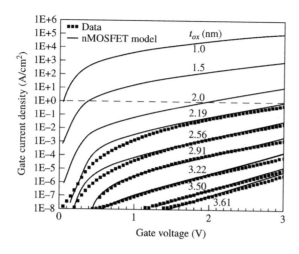

Figure 4.41 Measured (dots) and calculated (solid lines) tunneling currents in thin-oxide polysilicon-gate MOS devices. The dashed line indicates a tunneling-current level of 1 A/cm^2 (after Lo *et al.*, 1997)

is a plot of the measured and simulated thin-oxide tunneling current versus voltage in polysilicon-gate MOSFETs (Lo *et al.*, 1997). For the gate-voltage range shown in Figure 4.41, the current is primarily a direct-tunneling current. Direct-tunneling current is important in MOSFETs of very small dimensions, where the gate oxide layers can be scaled below 2 nm in thickness.

4.6.1.3   Defect Generation by Tunneling Current

The tunneling of electrons into and through a silicon dioxide layer can cause "defects" to be generated within the oxide layer and/or at the oxide–silicon interface. These defects can take the form of electron traps, hole traps, trapped electrons, trapped holes, or interface states (DiStefano and Shatzkes, 1974; Harari, 1978; Chen *et al.*, 1986; DiMaria *et al.*, 1993). These defects govern the time-dependent behavior of the tunneling current and play an important role in the wear-out and eventual breakdown of the oxide layer. Here, we briefly discuss how these defects can influence the tunneling process. The reader is referred to the vast literature on the subject for more details (DiMaria and Cartier, 1995, and Suehle, 2002, and the references therein).

- *Tunneling into an electron trap.* As electrons tunnel into an oxide layer, some of the electrons can get trapped. The trapped electrons modify the oxide field such that the field near the cathode (the electrode that acts as an electron source) is decreased, while the field near the anode (the electrode that acts as an electron sink) is increased. This is illustrated in Figure 4.42. The reduced field near the cathode, in turn, causes the tunneling current to decrease. In a constant-voltage tunneling current measurement, electron trapping is what causes the current to decrease with time. In a ramped-voltage (voltage increasing with time at a constant rate) tunneling current measurement, electron trapping often leads to a hysteresis in the current–voltage plot.

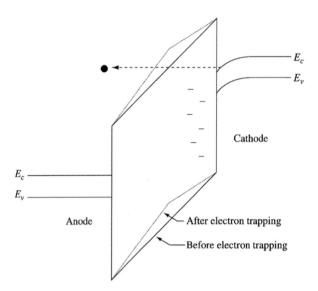

Figure 4.42 Schematic illustrating the trapping of tunneling electrons. As electrons are trapped, the oxide field near the cathode (electron source) is decreased, while the oxide field near the anode (electron sink) is increased

- *Hole generation, injection, and trapping.* As an electron travels in the conduction band of an oxide layer, it gains energy from the oxide field. If the voltage drop across the oxide layer is larger than the bandgap energy of silicon dioxide, which, as indicated in Figure 4.2, is about 9 eV, the electron can gain sufficient energy to cause impact ionization in the oxide. The holes generated by impact ionization can be trapped in the oxide. Holes can be injected indirectly into the oxide layer during electron tunneling as well. A tunneling electron arriving at the anode can cause impact ionization in the anode near the oxide–anode interface. Depending on the energy of the tunneling electron, the hole thus generated can be from deep down in the valence band of the anode, and thus can be "hot." A hot hole in the anode near the anode–oxide interface can be injected into the oxide layer. This process is illustrated in Figure 4.43. The injected hole can be trapped in the oxide layer as it travels toward the cathode. The trapped holes in the oxide layer cause the oxide field near the cathode to increase, which in turn causes the tunneling current to increase. This is illustrated in Figure 4.44. Thus the trapping of holes provides a positive feedback to the electron tunneling process. In a constant-voltage tunneling current measurement, hole trapping is the primary reason the current increases with time.
- *Trap and interface-state generation.* Traps can also be generated in the silicon dioxide layer and at the oxide–silicon interface by the electron current (Harari, 1978; DiMaria, 1987; Hsu and Ning, 1991). In addition to increasing electron and hole trapping, these traps can enhance the tunneling current by assisting in the tunneling process, as discussed in Sections 4.6.1.4 and 4.6.1.5.

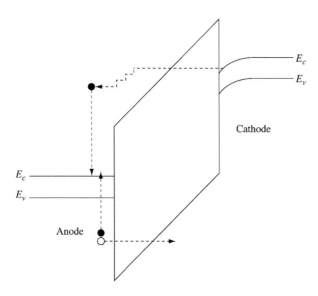

Figure 4.43 Schematic illustrating the generation of an electron–hole pair in the anode by a tunneling electron. The hole thus generated can then be injected (by tunneling in this example) into the oxide layer

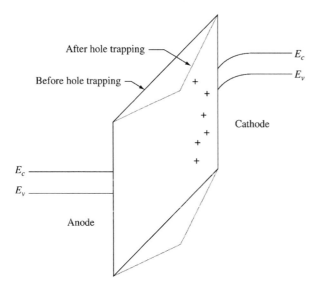

Figure 4.44 Schematic showing the trapping of holes in the oxide layer. The trapped holes enhance the electric field near the cathode, and decrease the electric field near the anode

4.6.1.4   Bulk-Trap-Assisted Tunneling

Instead of tunneling directly through an oxide layer, an electron can first tunnel from the cathode into an electron trap in the oxide and then tunnel from the trap to the anode. Thus, traps in the oxide layer can act as stepping stones for the tunneling electrons. This is illustrated in Figure 4.45. The enhanced tunneling current in turn can increase the generation of traps in the oxide. Thus, trap-assisted tunneling plays an important role in the degradation of an oxide under voltage stress because of the positive feedback between trap-generation and trap-assisted tunneling (DiMaria and Cartier, 1995).

4.6.1.5   Interface-Trap-Assisted Direct Tunneling at Low Voltages

Interface traps can also assist in the direct tunneling process. This is illustrated in Figure 4.46. Interface states exist at both the substrate silicon–oxide interface and the

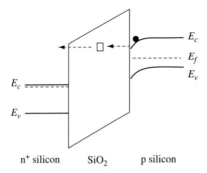

n⁺ silicon   SiO₂   p silicon

Figure 4.45 Schematic illustrating bulk-trap-assisted tunneling in an MOS capacitor structure

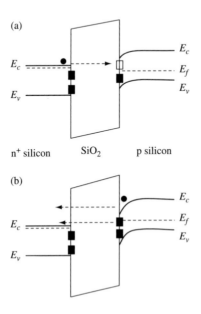

Figure 4.46 Schematics illustrating interface-trap-assisted direct tunneling of electrons in a silicon-gate MOS structure. (a) The gate electrode has a small negative bias. The Fermi level in the p-silicon lies somewhat below that in the n⁺ silicon gate. (b) The gate electrode has a small positive bias. The Fermi level in the p-silicon lies somewhat above that of the n⁺ silicon gate

gate silicon–oxide interface. In general, states above the Fermi level are empty of electrons and states below the Fermi level are filled with electrons. In Figure 4.46(a), the gate electrode is biased slightly negatively (toward flatband condition and surface accumulation of the p-silicon). As the surface of the p-silicon is driven toward flat band and surface accumulation, more and more states at the interface become empty of electrons. An electron in the conduction band of the gate electrode can tunnel into an empty state at the p-silicon surface. Also, not shown in the figure, an electron in an occupied surface state of the gate electrode can also tunnel into an empty surface state of the p-silicon. These interface-trap-assisted direct tunneling processes are in addition to the normal direct tunneling of electrons from the conduction band of the gate electrode into the conduction band of the p-substrate.

In Figure 4.46(b), the gate is biased somewhat positively (toward surface inversion of the p-silicon). As the surface of the p-silicon is driven toward inversion, more and more states at the surface become filled with electrons. Until the p-silicon surface is inverted, there are very few electrons available to tunnel from the conduction band of the p-silicon to the conduction band of the gate electrode. However, electrons from the filled surface states of the p-silicon can tunnel into the conduction band of the gate electrode. Since interface-trap-assisted tunneling is a direct tunneling process, it is important only for thin oxides. Also, as can be inferred from Figure 4.46, interface-trap-assisted tunneling is effective only when the p-silicon surface potential lies between inversion and weak accumulation. That is, interface-trap-assisted tunneling is important only at low voltages. *Interface-trap-assisted tunneling can enhance the low-voltage tunneling current in thin oxides whether the gate electrode is biased positively or negatively.* Interface-trap-assisted tunneling in modern CMOS devices is a widely studied subject. The reader is referred to the literature for more details (see e.g. Crupi *et al.*, 2002, and the references therein).

4.6.1.6 High-κ Gate Insulators

As seen in Figure 4.41, the direct gate tunneling current rises sharply when SiO_2 is thinner than 4 nm or so. *Below a thickness of $t_{ox} \sim 1.5$ nm, the standby power dissipation of a VLSI chip due to gate tunneling leakage becomes unacceptable* (Taur *et al.*, 1997). To alleviate this problem, the VLSI industry has developed high-κ gate insulators in the past two decades. The principle is to achieve a higher gate capacitance, ε_{ox}/t_{ox}, for MOSFET operation not with a thinner t_{ox} but with an insulator of higher permittivity, $\varepsilon_i \equiv \kappa\varepsilon_0$, where ε_0 is the vacuum permittivity. The physical thickness of high-κ insulator, t_i, can thus remain thick enough with acceptable levels of tunneling current. A frequently cited parameter, called the Equivalent Oxide Thickness (EOT), defined by

$$\text{EOT} \equiv \frac{\varepsilon_{ox}}{\varepsilon_i} t_i \qquad (4.109)$$

is used to indicate the effectiveness of the high-κ insulator.

Other than physical thickness, another factor that affects gate tunneling current is the barrier height between silicon and the gate insulator (also known as the band offset). From Eq. (3.127) based on the WKB approximation, the negative tunneling exponent is proportional to the physical thickness and the square root of effective mass and the barrier height. For the same capacitance or EOT, the physical thickness is proportional to the dielectric constant of the high-κ insulator. A figure of merit for high-κ insulators can be defined in terms of $\varepsilon_i \sqrt{m^* \phi_{ox}}$ with respect to that of SiO_2. The barrier height usually scales with the bandgap of the high-κ insulator. A plot of the insulator bandgap versus dielectric constant shows that larger dielectric constants tend to correlate with smaller bandgaps (Robertson and Wallace, 2015). Among the high-κ materials, a good choice is HfO_2, with a κ value of 20–25 and a bandgap of ~6 eV. However, an interfacial SiO_2 layer is still needed to maintain interface integrity and reliability. This adds to the EOT of high-κ, making it difficult for the composite EOT to go significantly below 1.0 nm.

While the EOT relates to the vertical field in the gate insulator perpendicular to the silicon interface, the lateral field (sometimes called the fringing field) also plays a role from the 2-D scaling point of view. A higher κ of the insulator has no effect on the lateral field, as is evident from the dielectric boundary conditions in Section 2.4.1.1. It will be discussed in Section 6.1 that *the physical thickness of the gate insulator, t_i, still needs to scale with the MOSFET channel length no matter how high κ is.*

4.6.2   Injection of Hot Carriers from Silicon into Silicon Dioxide

If a silicon region of sufficiently high electric field is located near the $Si - SiO_2$ interface, some electrons or holes in the silicon region can gain enough energy from the electric field to surmount the $Si - SiO_2$ interface barrier and enter the SiO_2 layer. In general, injection from Si into SiO_2 is much more likely for hot electrons than for hot holes because (a) electrons can gain energy from the electric field much more readily than holes due to their smaller effective mass, and (b) the $Si - SiO_2$ interface energy barrier is larger for holes (≈ 4.6 eV) than for electrons (≈ 3.1 eV), as indicated in Figure 4.2. The process of hot-electron and hot-hole injection from silicon into silicon dioxide is much too complex to model quantitatively. Thus far, quantitative agreement has been shown only for the special case of hot electrons traveling from the silicon substrate perpendicularly toward the $Si - SiO_2$ interface, and only with Monte-Carlo models that take into account the correct band structures, all the relevant scattering processes, and nonlocal transport properties (Fischetti *et al.*, 1995). Here we discuss a simple model for the injection of hot electrons from Si into SiO_2. The same model can be modified readily to describe the injection of hot holes from Si into SiO_2.

4.6.2.1   Energy Barrier for Hot Electron Injection

The energy barrier $q\phi_{ox}$ shown in Figure 4.39(a) is the difference in energy between the conduction band of SiO_2 and the conduction band of Si. In Section 3.2.1.4, it is shown that the image-force effect causes the barrier for injection of a hot electron from Si into SiO_2 to be lowered by an amount equal to

$$q\Delta\phi = \sqrt{\frac{q^3 \mathcal{E}_{ox}}{4\pi\varepsilon_{ox}}}. \tag{4.110}$$

The actual energy barrier for hot electron emission is therefore $(q\phi_{ox} - q\Delta\phi)$.

For $\mathcal{E}_{ox} = 1 \times 10^6$ V/cm, $q\Delta\phi = 0.19$ eV. Thus, for practical oxide fields of 10^6 V/cm or larger, image–force barrier lowering is not negligible compared to the interface energy barrier of 3.1 eV. Image–force barrier lowering is included in the more accurate theories of Fowler–Nordheim tunneling (Lenzlinger and Snow, 1969) and direct tunneling (Chang *et al.*, 1967). However, in the literature there are also publications questioning the validity of the concept of image potential at the interface between a semiconductor and an insulator (see e.g. Fischetti *et al.*, 1995, and the references therein).

4.6.2.2 The Lucky Electron Model

The simple one-dimensional injection process is illustrated in Figure 4.47. The simplest model for describing the injection process is the *lucky electron* model proposed by Shockley (1961). It is an empirical model, but it describes the measured data surprisingly well (Ning *et al.*, 1977a). In this model, the probability that a hot electron at a distance d from the Si $-$ SiO$_2$ interface will be emitted into the SiO$_2$ layer is expressed as

$$P(d) = A\exp(-d/\lambda), \tag{4.111}$$

where λ is an effective mean free path for energy loss by hot electrons in silicon, and A is a fitting constant to the experimental data. The relation between the parameter d, the effective hot-electron emission energy barrier, and the electron potential energy, is illustrated in Figure 4.47. The parameter d can be obtained as follows. Referring to Figure 4.47, $qV(x)$ is the potential energy of an electron at x. An electron at $x = d$ has just enough potential energy to overcome the effective energy barrier for emission if it can travel from

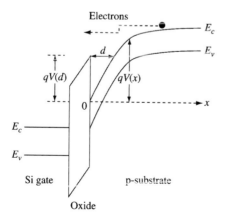

Figure 4.47 Schematic illustrating hot electrons traveling perpendicularly toward the Si $-$ SiO$_2$ interface and being injected into the SiO$_2$ layer

$x = d$ to the interface at $x = 0$ without encountering any energy-losing collision. That is, $qV(d)$ is equal to the effective energy barrier for emission. The injection of hot holes from Si into SiO_2 can be described by a similar *lucky hole* model (Selmi *et al.*, 1993).

It was determined empirically that the effective energy barrier for electron emission can be written as

$$qV(d) = q\phi_{ox} - q\Delta\phi - \alpha\mathscr{E}_{ox}^{2/3}, \qquad (4.112)$$

where the first two terms are the $Si - SiO_2$ interface energy barrier and the image–force barrier lowering discussed in Section 4.6.2.1, and the third term is introduced to account for the fact that hot electrons without enough energy to surmount the image–force-lowered energy barrier can still tunnel into the oxide layer. It was found that setting $\alpha = 1 \times 10^{-5}\,\text{eV}(\text{cm/V})^{2/3}$ and $A = 2.9$ fits a wide range of measured emission probabilities (Ning *et al.*, 1977a).

The temperature dependence of the hot-electron injection process is contained in the temperature dependence of the effective mean free path (Crowell and Sze, 1966b)

$$\lambda(T) = \lambda_0 \tanh(E_R/2kT), \qquad (4.113)$$

where $E_R = 63$ meV is the optical-phonon energy, and λ_0 is the low-temperature limit of λ. It was found empirically that $\lambda_0 = 10.8$ nm (Ning *et al.*, 1977a). The mean free path associated with hot-hole injection has a comparable value (Selmi *et al.*, 1993).

4.6.3   High-Field Effects in Gated Diodes

Thus far, the effects of high fields have been considered for p–n diodes, in Section 3.3, and MOS capacitors, in Sections 4.6.1 and 4.6.2, separately. In a gated diode structure, both the location of the peak-field region and the magnitude of the peak field vary with gate voltage. Let us consider a gated n^+–p diode. As discussed in Section 4.3.6, when the gate is biased to invert the silicon surface, the inverted surface region has about the same potential as the n^+ region, and the gated diode behaves like a large-area n^+–p diode. If the p-region is uniformly doped, the depletion-layer width is about the same below the n^+ silicon region as below the surface inversion region, hence the electric field is rather uniformly distributed and is about the same as in a simple p–n diode. This is illustrated schematically in Figure 4.48(a).

When the gate is biased somewhat negatively to accumulate the silicon surface, the silicon surface under the gate has about the same potential as the p-type substrate. Owing to the presence of the accumulated holes at the surface, the surface behaves like a p-region more heavily doped than the substrate, causing the depletion layer at the surface to become narrower than elsewhere. This is illustrated schematically in Figure 4.48(b). The narrowing of the depletion layer at or near the intersection of the p–n junction and the Si–SiO_2 interface causes *field crowding*, or an increase in the local electric field.

When the negative gate bias is large enough, the n^+ region under the gate can become depleted, and even inverted. This is illustrated in Figure 4.48(c). In this case,

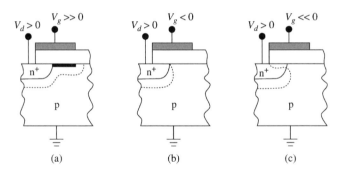

Figure 4.48 Schematics illustrating a gated n^+–p diode when the surface is (a) inverted and (b) accumulated, and (c) when the surface of the n^+ region is depleted or inverted. The dashed lines indicate the boundary of the depletion region

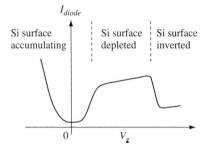

Figure 4.49 Schematic illustrating an n^+–p gated-diode leakage current as a function of gate voltage

the gate and the n^+ region behave like an MOS capacitor with a heavily doped n-type "substrate." There is more field crowding, and the peak field increases.

As the local electric field in and around the gated p–n junction is increased by the gate voltage, all the high-field effects, such as avalanche multiplication and band-to-band tunneling, can increase very dramatically. Thus the leakage current of a reverse-biased gated diode can increase dramatically when the gate voltage begins to cause field crowding in and around the junction region. This is illustrated in Figure 4.49 which shows the expected gated-diode leakage current as a function of gate voltage. In Figure 4.49, aside from the increase in current due to field crowding at negative gate voltage, the diode current is simply the sum of the leakage current from the depletion region of a diode and the leakage current from the exposed surface states (Grove and Fitzgerald, 1966). When the Si surface is in accumulation (at $V_g \approx 0$), the leakage current is from the depletion region of the bulk diode alone. When the Si surface is inverted (at large V_g), the leakage current is higher because of the additional leakage current coming from the depletion region of the diode formed by the surface inversion layer and the substrate. When the Si surface is depleted (at intermediate values of V_g), the leakage current is the largest because of the additional leakage current coming from the exposed surface states of the depleted silicon surface.

When field crowding occurs in the drain junction of a MOSFET, caused by the gate voltage modulating the local electric field in the diode formed between the drain region and the device body, the increased junction leakage current is called *gate-induced drain leakage*, or GIDL (Chan *et al.*, 1987a; Noble *et al.*, 1989). **GIDL is an important leakage–current component that must be minimized in modern CMOS devices.**

It should be noted that for the gated $n^+ - p$ diode considered here, when the gate is biased to accumulate the silicon surface, the oxide field favors the injection of hot holes from the silicon substrate into the silicon dioxide layer (Verwey, 1972). Similarly, for a gated $p^+ - n$ diode, the oxide field favors the injection of hot electrons when the gate is biased to accumulate the silicon surface. Thus injection of majority carriers, instead of minority carriers, from the silicon substrate into the silicon dioxide layer takes place when significant gate-voltage-induced avalanche multiplication occurs in a gated diode. These hot-carrier injection mechanisms are employed in the programming and/or erasure of some EPROM and EEPROM devices (see Section 12.3).

4.6.4   Dielectric Breakdown

As discussed in Section 4.6.1, significant electron tunneling can take place when a large electric field is applied across an oxide layer. Figure 4.50 illustrates schematically the typical time dependence of the tunneling current when a constant voltage is applied across an oxide layer. A sudden jump in the tunneling current indicates that the oxide sample has suffered a *dielectric breakdown* event. For oxides thicker than about 10 nm, the tunneling current typically decreases gradually with time until the oxide breaks down. For these thick oxides, the voltages used to measure tunneling current are usually so large that, unless special care is taken to limit the current, a breakdown event usually leads to the oxide being physically damaged (Shatzkes *et al.*, 1974). There is a distribution in the measured oxide breakdown time (Harari, 1978). This is illustrated in Figure 4.50(a).

For oxides thinner than about 5 nm, the voltages used to measure tunneling current are usually sufficiently small so that not every breakdown event leads to catastrophic breakdown. For most samples, *successive breakdown* events are observed before final or catastrophic breakdown (Suñé *et al.*, 2004). This is illustrated in Figure 4.50(b). An oxide layer ceases to be a good electrical insulator after it suffers final or catastrophic breakdown.

In thin oxides, the increase in tunneling current from the first breakdown event to final breakdown can occur quite gradually. When the thin gate oxide of a modern MOS transistor in a circuit starts showing signs of breaking down, often the gate tunneling current can grow to a sufficiently large value to cause the circuit to fail long before the gate oxide layer suffers final breakdown (Kaczer *et al.*, 2000; Linder *et al.*, 2001). Figure 4.51 is a schematic illustrating the time dependence of the tunneling current in a typical thin-oxide MOS device at constant voltage stress. There are roughly three stages in the evolution of the tunneling current. In the initial stage

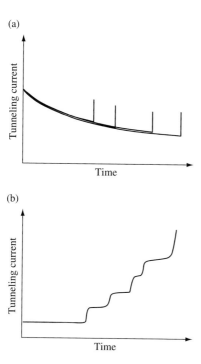

Figure 4.50 Schematics illustrating the typical time dependence of the tunneling current at constant voltage in (a) a thick oxide layer and (b) a thin oxide layer. A sudden jump in tunneling current signals a dielectric breakdown event

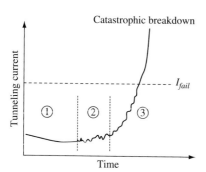

Figure 4.51 Schematic illustrating the evolution of the tunneling current in a thin oxide MOS device at constant applied voltage. I_{fail} indicates the tunneling current at which the device fails to function properly in a circuit. The three stages marked 1, 2, and 3 are discussed in the text

(stage 1 in Figure 4.51), the current is relatively featureless, typically first decreasing, due to electron trapping, and then rising, due to hole trapping and trap-assisted tunneling, as a function of time. As the current continues to rise, it becomes noisy (stage 2). Finally, the current rises much more rapidly (stage 3) for some time before final breakdown. In Figure 4.51, I_{fail} denotes the tunneling-current level at which a circuit using the MOS transistor fails to function properly. In the literature, stage 1 is often referred to as the *defect generation* stage or *stress-induced leakage current* stage (DiMaria, 1987; Stathis and DiMaria, 1998); stage 2 as the *soft breakdown* stage (Depas *et al.*, 1996), and stage 3 as the *successive breakdown* or *progressive breakdown* stage (Okada, 1997; Linder *et al.*, 2002; Suñé and Wu, 2002), which will be discussed further in Section 4.6.4.3.

4.6.4.1   Breakdown Field

In the literature, the quality of an oxide film is often measured in terms of the electric field at which dielectric breakdown, usually the first breakdown event, occurs. "Good-quality" thick $(>100 \text{ nm})\text{SiO}_2$ films typically break down at fields greater than 10 MV/cm, while "good-quality" thin $(<10 \text{ nm})\text{SiO}_2$ films usually show larger breakdown fields, often in excess of 15 MV/cm. In bipolar transistors, because there are normally no thin oxide components, the electric fields across the oxide layers are usually so small that dielectric breakdown is not a concern. In CMOS devices, the maximum oxide field varies widely, depending on the application. For devices used in logic and memory circuits, the maximum oxide field is typically in the $3-6 \text{ MV/cm}$ range in normal operation, and can reach as high as $5-9 \text{ MV/cm}$ in special operations (such as during a device burn-in process). For devices used in electrically programmable nonvolatile memory applications, where normal operation involves tunneling through a thin dielectric layer, the maximum electric field across the thin dielectric layer is in excess of 10 MV/cm. Dielectric breakdown is a real concern in many CMOS devices.

4.6.4.2   Time to Breakdown and Charge to Breakdown

The breakdown characteristics of an oxide film are often described in terms of its *time to breakdown*, which measures the time needed for the film to reach breakdown, or its *charge to breakdown*, which measures the integrated total tunneling charge leading up to breakdown. In product design, we want to ensure that the gate current of a CMOS device will not grow to the point of causing circuit failure before the product end of life. Therefore, in product design, we want to know the time to breakdown. However, it appears easier to develop physical models relating charge to breakdown to the physical mechanisms involved in the dielectric breakdown process, such as hole current, trapping, trap generation, and interface state generation (Schuegraf and Hu, 1994; DiMaria and Stathis, 1997; Stathis and DiMaria, 1998), than to develop physical models relating time to breakdown to these physical mechanisms. Most publications on the physics of dielectric breakdown discuss the breakdown process in terms of charge to breakdown instead of time to breakdown. Therefore, we will not discuss time to breakdown any further here. The reader is referred to the literature for discussions on time to breakdown and the breakdown statistics in the time domain (see e.g. Suñé *et al.*, 2004, and the references therein).

As discussed in Section 4.6.1.3, a tunneling electron current can generate a hole current. Thus, the charge to breakdown, Q_{BD}, is the sum of the charges due to electrons and holes. If an MOS capacitor structure is used to measure Q_{BD}, then,

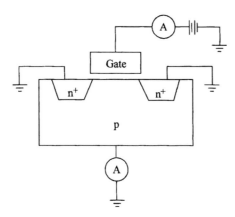

Figure 4.52 Schematic illustrating the bias configuration of an n-channel MOSFET for measuring the charge to breakdown and its hole-charge component

owing to the two-terminal nature of the device, only the total charge can be measured. However, if an n-channel MOSFET or an n^+–p gated-diode structure is used to measure Q_{BD}, then both the total charge and the hole-charge component can be determined. For the case of an n-channel MOSFET, the bias configuration for such measurements is illustrated in Figure 4.52. The basic concept of this *charge separation* method is that electron current is measured at the n-type terminal and hole current is measured at the p-type terminal, much like in the split *C*–*V* setup in Section 4.3.4. Integration of the gate current gives the total charge, and integration of the substrate current gives the charge due to the holes. It is shown that, charge for charge, hot holes are much more effective than tunnel electrons in generating defects that lead to oxide breakdown (Li *et al.*, 1999).

4.6.4.3 Progressive Breakdown and Successive Breakdown

Referring to Figure 4.51, the evolution of the tunneling current can be described as a process of positive feedback between defect generation and trap-assisted tunneling. When a stress voltage is applied across an oxide layer, at first there are few defects in the oxide and the tunneling current is relatively low, decreasing with time as electrons are trapped. As trapped holes start to accumulate in the oxide, the current will start increasing. The tunneling current generates defects which in turn assist in the tunneling process (DiMaria, 1987; Stathis and DiMaria, 1998). At some point, soft breakdown starts when the defects in the oxide become dense enough such that an electron can tunnel relatively easily from one defect center to another across the oxide layer. This trap-assisted tunneling current tends to be noisy (Depas *et al.*, 1996). As the defect density continues to grow, hard breakdown (a breakdown event or a series of breakdown events) starts when there is a connected path of overlapping defects all the way across the oxide layer (Degraeve *et al.*, 1995; Stathis, 1999). This connected path of defects acts as a low-resistance conduction path for the electrons. Once hard breakdown starts, the electron current is completely dominated by the flow along this

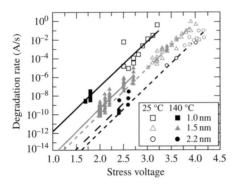

Figure 4.53 Measured rate of increase of breakdown current for typical thin oxides (after Linder *et al.*, 2002)

low-resistance path and the magnitude of the current is more-or-less independent of the device area. The electron current causes the diameter of the path of connected defects to grow, which in turn causes the current to grow. The current does not grow smoothly, but in a staircase manner. Each time the tunneling current jumps, it represents a breakdown event. Eventually, catastrophic breakdown of the oxide layer occurs. Thus, a thin oxide can go through many successive breakdown events before it breaks down catastrophically (Suñé and Wu, 2002). The oxide degradation rate (the rate at which the average breakdown current increases with time) is a strong function of the stress voltage (Linder *et al.*, 2002; Lombardo *et al.*, 2003). Figure 4.53 is a plot of typical measured degradation rates for thin oxides. It suggests that *even after hard breakdown has commenced, the tunneling current in a thin-oxide and low-voltage MOSFET may take a long time to grow to a value sufficiently large to cause circuit failure.*

In the literature, most charge to breakdown measurements are made by integrating the tunneling current until the current shows a sudden jump in magnitude, or until the first breakdown event. For a given oxide film, the charge to breakdown Q_{BD} is often plotted as a function of oxide voltage. Figure 4.54 is a typical plot for oxide thickness in the 2.5–10 nm range (Schuegraf and Hu, 1994). It shows that, for these relatively thick oxides, Q_{BD} decreases with increasing oxide voltage. It has also been shown that Q_{BD} is about the same for n-channel and p-channel MOSFETs (DiMaria and Stathis, 1997). Since these published Q_{BD} values do not take into account the progressive nature of the breakdown process, they project a lower allowed voltage for an oxide than is justified from a device reliability point of view. The progressive breakdown and the successive breakdown models, which take into account the oxide degradation rate after hard breakdown has commenced (stage 3 in Figure 4.51), project a larger but more accurate allowed voltage (Linder *et al.*, 2002). The reader is referred to the literature on thin oxide reliability for more details (Degraeve *et al.*, 1998; Suehle, 2002; Suñé *et al.*, 2004).

As illustrated in Figure 4.51, the gate leakage current of a MOSFET in a circuit has to reach a certain critical level, indicated by I_{fail}, before the circuit ceases to function

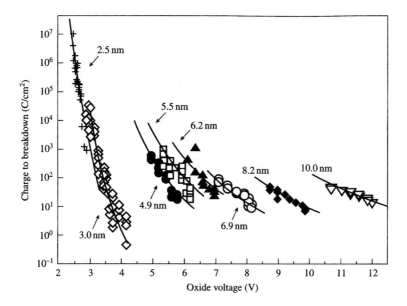

Figure 4.54 Typical plot of charge to breakdown versus oxide voltage for several oxide thickness values (after Schuegraf and Hu, 1994)

properly. Thus, *in circuit applications, what designers really need to know is the time to critical current instead of time to first breakdown or charge to breakdown.*

Exercises

4.1 The total depletion charge and inversion charge densities of a p-type MOS capacitor can be expressed as

$$Q_d = -qN_a \int_0^{W_d} \left(1 - e^{-q\psi/kT}\right) dx = -qN_a \int_0^{\psi_s} \frac{1 - e^{-q\psi/kT}}{\mathscr{E}} d\psi,$$

and

$$Q_i = -q\frac{n_i^2}{N_a} \int_0^{W_d} \left(e^{q\psi/kT} - 1\right) dx = -q\frac{n_i^2}{N_a} \int_0^{\psi_s} \frac{e^{q\psi/kT} - 1}{\mathscr{E}} d\psi,$$

using Eqs. (4.11) and (4.12). Here $\mathscr{E} = -d\psi/dx$ is given by Eq. (4.15).

(a) Write down the expressions for the small-signal depletion capacitance, $C_d = -dQ_d/d\psi_s$, and the small-signal inversion capacitance (low frequency), $C_i = -dQ_i/d\psi_s$, in silicon as represented in the equivalent circuit in Figure 4.13.

(b) Show that $C_d + C_i = C_{si}$, where $C_{si} = -dQ_s/d\psi_s$ is evaluated using Eq. (4.16).

(c) Show that $C_d \approx C_i$ at the condition of strong inversion, $\psi_s = 2\psi_B$. (This allows one to use a split C–V measurement to determine the gate voltage where $\psi_s = 2\psi_B$).

(d) From the behavior of C_d beyond strong inversion, explain the "screening" of depletion charge (incremental) by the inversion layer.

4.2 Shown in Figure Ex. 4.2 is the high-frequency C–V of a uniformly doped MOS capacitor. The oxide is SiO_2 and the interface state charge Q_{ox} is zero.
(a) Determine the oxide thickness.
(b) Determine the substrate doping concentration. Is it p-type or n-type?
(c) Determine the flatband voltage.
(d) What is the work function of the metal gate?

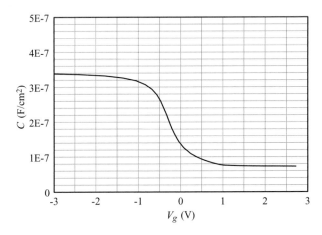

Figure Ex. 4.2

4.3 $\psi_s(V_g)$ under the depletion condition is given by Eq. (4.33). Show that for incremental changes, $\Delta\psi_s = \Delta V_g/(1 + C_d/C_{ox})$, where C_d is the depletion charge capacitance given by Eq. (4.38).

4.4 When the gate voltage greatly exceeds the threshold for strong inversion, a first-order solution of $\psi_s(V_g)$ can be obtained from the coupled equations, Eqs. (4.7) and (4.16), by keeping only the inversion charge term. Show that

$$\psi_s \approx 2\psi_B + \frac{2kT}{q} \ln\left(\frac{C_{ox}\left(V_g - V_{fb} - 2\psi_B\right)}{\sqrt{2\varepsilon_{si}kT N_a}}\right)$$

under these circumstances. Estimate how much higher ψ_s can be over $2\psi_B$ by substituting some typical values in the logarithmic expression.

4.5 In the split C–V measurement in Figure 4.15, show that the n^+ channel part of the small-signal gate capacitance is

$$-\frac{dQ_i}{dV_g} = \frac{C_{ox}C_i}{C_{ox} + C_i + C_d},$$

where C_i and C_d are as defined in Figure 4.13(b). Sketch the functional behavior of $-dQ_i/dV_g$ versus V_g, and from it describe the behavior of $-Q_i$ versus V_g.

4.6 Sketch the $C-V$ curve (high frequency) of an MOS capacitor consisting of n$^+$ poly gate on n-type Si doped to $N_d = 10^{16}$ cm^{-3}. Calculate and show the flatband voltage on the $C-V$. Draw the band diagram for $V_g = 0$. Given $t_{ox} = 10$ nm, what is V_{ox} (potential across oxide) at the onset of inversion ($\psi_s = 2\psi_B$)? Ignore quantum and poly depletion effects.

4.7 Consider an MOS device with 20 nm thick gate oxide and uniform p-type substrate doping of 10^{17} cm^{-3}. The gate work function is that of n$^+$ Si.
(a) What is the flatband voltage? What is the threshold voltage for strong inversion?
(b) Sketch the high frequency $C-V$ curve. Label where the flatband voltage and threshold voltage are.
(c) Calculate the maximum and the minimum capacitance (per area) values.

4.8 If the device in Exercise 4.7 is biased at zero gate voltage, determine the surface potential and the electron and hole densities at the surface.

4.9 Consider an MOS on p-type silicon of $N_a = 10^{16}$ cm^{-3}. $t_{ox} = 10$ nm, and $V_{fb} = 0$. The interface state charge Q_{ox} is zero. Ignore quantum effect.
(a) At what gate voltage does the surface potential ψ_s equal 0.5 V?
(b) For the bias in (a), what is the total capacitance of the MOS?
(c) What is the potential drop over oxide, V_{ox}, approximately at $V_g = 2$ V?
(d) What is the potential drop over oxide, V_{ox}, approximately at $V_g = -2$ V?

4.10 For an MOS with $t_{ox} = 10$ nm on a uniformly doped p-type substrate of 10^{17} cm^{-3}, the gate is n$^+$ polysilicon doped to 10^{20} cm^{-3}. Estimate the depletion layer width in the polysilicon gate at a gate voltage of 3 V.

4.11 Consider an MOS on uniformly doped p-type substrate of $N_a = 10^{18}$ cm^{-3} biased at the threshold condition. Calculate the first three quantum mechanical energy levels for inversion electrons in the lower valley with an effective mass of $0.92\,m_0$ where m_0 is the free electron mass. Express the answers in eV.

4.12 As electrons are injected from silicon into silicon dioxide, some of these electrons become trapped in the oxide. Let N_T be the electron trap density, n_T be the density of trapped electrons, and j_G/q be the injected electron particle current density. The rate equation governing $n_T(t)$ is

$$\frac{dn_T}{dt} = \frac{j_G}{q}\sigma(N_T - n_T),$$

where σ is the capture cross-section of the traps. If the initial condition for n_T is $n_T(t = 0) = 0$, show that the time dependence of the trapped electron density is given by

$$n_T(t) = N_T\{1 - \exp[-\sigma N_{inj}(t)]\},$$

where

$$N_{inj}(t) \equiv \int_0^t \frac{j_G(t')}{q} dt'$$

is the number of injected electrons per unit area. Assume $N_T = 5 \times 10^{12}\,\text{cm}^{-3}$ and $\sigma = 1 \times 10^{-13}\,\text{cm}^2$, sketch a log–log plot of n_T as a function of N_{inj}. (The capture cross-section is often measured by fitting to such a plot.)

5 MOSFETs

Long Channel

The metal–oxide–semiconductor field-effect transistor (MOSFET) is the building block of VLSI circuits in microprocessors and dynamic memories. Because the current in a MOSFET is transported predominantly by carriers of one polarity only (e.g., electrons in an n-channel device), the MOSFET is usually referred to as a unipolar or majority-carrier device. Throughout this chapter, n-channel MOSFETs are used as an example to illustrate device operation and derive drain current equations. The results can easily be extended to p-channel MOSFETs by exchanging the dopant types and reversing the voltage polarities.

The basic structure of a MOSFET is shown in Figure 5.1. It is a four-terminal device with the terminals designated as *gate* (subscript *g*), *source* (subscript *s*), *drain* (subscript *d*), and *substrate* or *body* (subscript *b*). An n-channel MOSFET, or nMOSFET, or nMOS in short, consists of a p-type silicon substrate into which two n^+ regions, the source and the drain, are formed (e.g., by ion implantation). The gate electrode is usually made of metal or heavily doped polysilicon and is separated from the substrate by a thin silicon dioxide film, the *gate oxide*. The gate oxide is usually formed by thermal oxidation of silicon. In VLSI circuits, a MOSFET is surrounded by a thick oxide called the *field oxide* to isolate it from the adjacent devices. The surface region under the gate oxide between the source and drain is called the *channel* region and is critical for current conduction in a MOSFET. The basic operation of a MOSFET device can be easily understood from the MOS capacitor discussed in Chapter 4. When there is no voltage applied to the gate or when the gate voltage is zero, the p-type silicon surface is either in accumulation or in depletion and there is negligible current flow between the source and drain. The MOSFET device acts like two back-to-back p–n junction diodes with only low-level leakage currents present. When a sufficiently large positive voltage is applied to the gate, the silicon surface is inverted to n-type, which forms a conducting channel between the n^+ source and drain. If there is a voltage difference between them, an electron current will flow from the source to the drain. A MOSFET device therefore operates like a switch ideally suited for digital circuits. *Since the gate electrode is electrically insulated from the substrate, there is effectively no dc gate current whether the MOSFET is on or off.*[1] *The*

---

[1] Except for the gate tunneling current when the oxide is very thin, as discussed in Section 4.6.1.

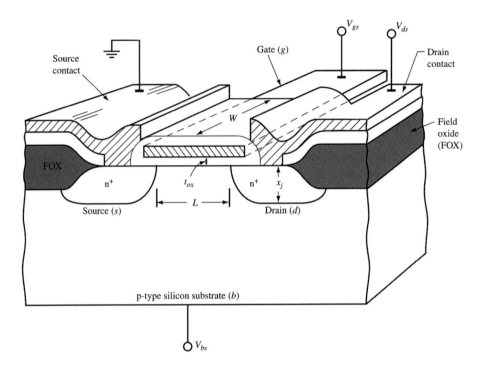

Figure 5.1 Three-dimensional view of the basic MOSFET device structure (after Arora, 1993)

channel is capacitively coupled to the gate via the electric field in the oxide (hence the name *field-effect transistor*).

This chapter describes the basic characteristics of a long-channel MOSFET, which will serve as the foundation for understanding the more important but more complex short-channel MOSFETs in Chapter 6.

5.1   MOSFET *I–V* Characteristics

In this section, a general MOSFET current model based on the *gradual channel approximation* (GCA) is first formulated in Section 5.1.1. The GCA is valid for most regions of MOSFET operation except beyond the saturation point. A *charge-sheet model* is then introduced in Section 5.1.2 to obtain implicit equations for the source–drain current as a function of gate and drain voltages. Regional approximations are applied in Section 5.1.3 to derive explicit but piecewise *I–V* expressions for the linear, parabolic, and subthreshold regions. A non-GCA model is described in Section 5.1.4 to model the *I–V* characteristics beyond the point of saturation.

Figure 5.2 shows the schematic cross-section of an n-channel MOSFET in which the source is the n^+ region on the left, and the drain is the n^+ region on the right. A thin oxide film separates the gate from the channel region between the source and drain. We choose an x–y coordinate system consistent with Section 4.2 on MOS capacitors,

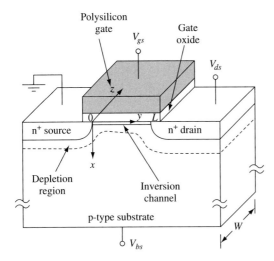

Figure 5.2 A schematic MOSFET cross-section, showing the axes of coordinates and the bias voltages at the four terminals for the drain current model

namely, the x-axis is perpendicular to the gate electrode and is pointing into the p-type substrate with $x = 0$ at the silicon surface. The y-axis is parallel to the channel or the current flow direction, with $y = 0$ at the source and $y = L$ at the drain. L is called the *channel length* and is a key indicator of the MOSFET performance. The MOSFET is assumed to be uniform along the z-axis over a distance called the *channel width*, W, determined by the boundaries of the thick field oxide.

Conventionally, the source voltage is defined as the ground potential. The drain voltage is V_{ds}, the gate voltage is V_{gs}, and the p-type substrate is biased at V_{bs}. Initially, we assume $V_{bs} = 0$, i.e., the substrate contact is grounded to the source potential. Later on (Section 5.3.1), we will discuss the effect of substrate bias on MOSFET characteristics. The p-type substrate is assumed to be uniformly doped with an acceptor concentration N_a.

5.1.1   Gradual Channel Approximation

A major assumption in any 1-D MOSFET model is the *gradual channel approximation (GCA), which assumes that the variation of the electric field in the y-direction (along the channel) is much less than the corresponding variation in the x-direction (perpendicular to the channel)* (Pao and Sah, 1966). This allows us to reduce the 2-D Poisson equation to 1-D slices (x-component only) as in Eq. (4.9). The GCA is valid for most of the channel regions except beyond the saturation point.

As defined in Section 4.2, $\psi(x, y)$ is the band bending, or intrinsic potential, at (x, y) with respect to the intrinsic potential of the bulk substrate. From Eq. (4.12), the electron density at (x, y) is proportional to the factor $e^{q\psi/kT}$. We further assume that $V(y)$ is the electron quasi-Fermi potential at a point y along the channel with respect to the Fermi potential of the n^+ source. In other words, $V(y) \equiv \phi_n(y) - \phi_n(0)$. The assumption that V is independent of x in the direction perpendicular to the surface is justified by the consideration that current is proportional to the gradient of the quasi-Fermi potential and that MOSFET current flows predominantly in the source-to-drain, or y-direction. At the source end of the channel, $V(y = 0) = 0$. At the drain end of the channel, $V(y = L) = V_{ds}$. The electron quasi-Fermi potential at a point in the channel is essentially flat in the vertical direction across the n-type inversion layer. From Eq. (2.75), the effect of V is to multiply the electron density by $e^{-qV/kT} \propto e^{-q\phi_n/kT} = e^{E_{fn}/kT}$ over its $V = 0$ value. The electron concentration at any point (x, y) is therefore

$$n(x, y) = \frac{n_i^2}{N_a} e^{q(\psi - V)/kT}. \tag{5.1}$$

Following the same approach as in Section 4.2.1, one obtains an expression for the electric field similar to that of Eq. (4.15):

$$\mathscr{E}^2(x, y) = \left(\frac{d\psi}{dx}\right)^2 = \frac{2kTN_a}{\varepsilon_{si}} \left[\left(e^{-q\psi/kT} + \frac{q\psi}{kT} - 1\right) + \frac{n_i^2}{N_a^2} \left(e^{-qV/kT}\left(e^{q\psi/kT} - 1\right) - \frac{q\psi}{kT}\right) \right]. \tag{5.2}$$

The condition for surface inversion, Eq. (4.55), becomes

$$\psi(0, y) = V(y) + 2\psi_B, \tag{5.3}$$

since $V(y)$ plays the role of the reverse bias at y. From Eq. (4.56), the maximum depletion layer width is

$$W_{dm}(y) = \sqrt{\frac{2\varepsilon_{si}[V(y) + 2\psi_B]}{qN_a}}. \tag{5.4}$$

Both of the above are functions of y. However, those equations employing $2\psi_B$ are only approximations for the strong inversion region, thus not suitable for the general model formulations of this section and Section 5.1.2. They will be applied in Section 5.1.3 to obtain regional, explicit I–V characteristics.

5.1.1.1   Pao and Sah's Double Integral

Under the assumption that both the hole current and the generation and recombination current are negligible, the current continuity equation can be applied to the electron current in the y-direction. In other words, the total drain-to-source current I_{ds} is the same at any point along the channel. From Eq. (2.77), the electron current density at a point (x, y) is

$$J_n(x, y) = -q\mu_n n(x, y) \frac{dV(y)}{dy}, \tag{5.5}$$

where $n(x, y)$ is the electron density, and μ_n is the electron mobility in the channel. The carrier mobility in the channel is generally much lower than the mobility in the bulk, due to additional surface scattering mechanisms, as will be addressed in Section 5.2. *With V(y) defined as the quasi-Fermi potential, i.e., playing the role of ϕ_n in* Eq. (2.77), Eq. (5.5) *includes both the drift and diffusion currents.* The total current at a point y along the channel is obtained by multiplying Eq. (5.5) with the channel width W and integrating over the depth of the current-carrying layer. The integration is carried out from $x = 0$ to x_i, where x_i is a depth into the p-type substrate much deeper than the inversion layer but not infinity:

$$I_{ds}(y) = qW \int_0^{x_i} \mu_n n(x, y) \frac{dV}{dy} dx. \tag{5.6}$$

There is a sign change, as we define $I_{ds} > 0$ to be the drain-to-source current in the $-y$ direction. Since V is a function of y only, dV/dy can be taken outside the integral. We also assume that μ_n can be taken outside the integral by defining an *effective mobility*, μ_{eff}, at some average gate and drain fields. What remains in the integral is the electron concentration, $n(x, y)$. Its integration over the inversion layer gives the inversion charge per unit gate area, Q_i:

$$Q_i(y) = -q \int_0^{x_i} n(x, y) dx. \tag{5.7}$$

Equation (5.6) then becomes

$$I_{ds}(y) = -\mu_{eff} W \frac{dV}{dy} Q_i(y) = -\mu_{eff} W \frac{dV}{dy} Q_i(V). \tag{5.8}$$

In the last step, Q_i is expressed as a function of V; V is interchangeable with y, since V is a function of y only. Multiplying both sides of Eq. (5.8) by dy and integrating from 0 to L (source to drain) yields

$$\int_0^L I_{ds} dy = \mu_{eff} W \int_0^{V_{ds}} [-Q_i(V)] dV. \tag{5.9}$$

Current continuity requires that I_{ds} be a constant, independent of y. Therefore, the drain-to-source current is

$$I_{ds} = \mu_{eff} \frac{W}{L} \int_0^{V_{ds}} [-Q_i(V)] dV. \tag{5.10}$$

An alternative form of $Q_i(V)$ can be derived if $n(x, y)$ is expressed as a function of (ψ, V) using Eq. (5.1), i.e.,

$$n(x, y) = n(\psi, V) = \frac{n_i^2}{N_a} e^{q(\psi - V)/kT}, \tag{5.11}$$

and substituted into Eq. (5.7):

$$Q_i(V) = -q \int_{\psi_s}^{\delta} n(\psi, V) \frac{dx}{d\psi} d\psi$$

$$= -q \int_{\delta}^{\psi_s} \frac{(n_i^2/N_a) e^{q(\psi-V)/kT}}{\mathscr{E}(\psi, V)} d\psi.$$

(5.12)

Here, ψ_s is the surface potential at $x = 0$ and $\mathscr{E}(\psi, V) = -d\psi/dx$ is given by the square root of Eq. (5.2). The lower integration limit, δ, represents any small potential $\ll kT/q$, but not zero as the integral is unbounded at $\psi = 0$. [An alternative is to replace the factor $e^{q\psi/kT}$ with $(e^{q\psi/kT} - 1)$ to remove the singularity.] Substituting Eq. (5.12) into Eq. (5.10) yields

$$I_{ds} = q\mu_{eff} \frac{W}{L} \int_0^{V_{ds}} \left(\int_{\delta}^{\psi_s} \frac{(n_i^2/N_a) e^{q(\psi-V)/kT}}{\mathscr{E}(\psi, V)} d\psi \right) dV.$$

(5.13)

This is referred to as *Pao and Sah's double integral* (Pao and Sah, 1966). The boundary value ψ_s is determined by two coupled equations: Eq. (4.7) and $Q_s = -\varepsilon_{si}\mathscr{E}_s(\psi_s)$ or Gauss's law, where $\mathscr{E}_s(\psi_s)$ is obtained by letting $\psi = \psi_s$ in Eq. (5.2). In depletion and inversion where $q\psi_s/kT \gg 1$, only two of the terms in Eq. (5.2) are significant and need to be kept. The merged equation is then

$$V_{gs} = V_{fb} + \psi_s - \frac{Q_s}{C_{ox}}$$

$$= V_{fb} + \psi_s + \frac{\sqrt{2\varepsilon_{si}kTN_a}}{C_{ox}} \left[\frac{q\psi_s}{kT} + \frac{n_i^2}{N_a^2} e^{q(\psi_s-V)/kT} \right]^{1/2},$$

(5.14)

which is an implicit equation for $\psi_s(V)$. Equations (5.13) and (5.14) can only be solved numerically.

5.1.2   Charge Sheet Model

Pao and Sah's double integral can be simplified to a single integral if the inversion charge density Q_i can be expressed as a function of ψ_s. This is the approach taken by the *charge-sheet model* (Brews, 1978). It is based on the fact that the inversion layer is located very close to the silicon surface like a thin sheet of charge. There is a sharp increase of the field (spatial integration of the volume charge density) across the thin inversion layer, but very little change of the potential (spatial integration of the field). As shown in the example in Figure 4.11, neither the surface potential nor the depletion charge density changes much after strong inversion. The central assumption of the charge-sheet model is that Eq. (4.23) for the depletion charge density,

$$Q_d = -qN_aW_d = -\sqrt{2\varepsilon_{si}qN_a\psi_s},$$

(5.15)

can be extended to strong inversion and beyond. (Actually, once the inversion charge dominates, Q_d hardly changes with ψ_s (see Exercise 4.1). Since the total silicon charge density Q_s is given by Eq. (5.14) or Eq. (4.7), Eq. (5.15) allows the inversion charge density to be expressed as

$$Q_i = Q_s - Q_d = -C_{ox}(V_{gs} - V_{fb} - \psi_s) + \sqrt{2\varepsilon_{si}qN_a\psi_s}. \tag{5.16}$$

It should be noted that the charge sheet model does not literally assume all the inversion charge is located at the silicon surface with a zero depth. That would mean $d|Q_i|/dV_{gs} = C_{ox}$, which is not the case with Eq. (5.16) since ψ_s also increases with V_{gs} as described by Eq. (5.14).

The variable in the drain current integral, Eq. (5.10), can be transformed from V to ψ_s,

$$I_{ds} = \mu_{eff}\frac{W}{L}\int_{\psi_{s,s}}^{\psi_{s,d}}[-Q_i(\psi_s)]\frac{dV}{d\psi_s}d\psi_s, \tag{5.17}$$

where $\psi_{s,s}$ and $\psi_{s,d}$ are the values of the surface potential at the source end and the drain end of the channel. For given V_{gs} and V_{ds}, they can be solved numerically from the implicit equation, Eq. (5.14), by setting $V = 0$ (for $\psi_{s,s}$) and $V = V_{ds}$ (for $\psi_{s,d}$), respectively. Equation (5.14) can also be used to solve for $V(\psi_s)$,

$$V = \psi_s - \frac{kT}{q}\ln\left\{\frac{N_a^2}{n_i^2}\left[\frac{C_{ox}^2(V_{gs}-V_{fb}-\psi_s)^2}{2\varepsilon_{si}kTN_a} - \frac{q\psi_s}{kT}\right]\right\}, \tag{5.18}$$

and evaluate its derivative:

$$\frac{dV}{d\psi_s} = 1 + \frac{2kT}{q}\frac{C_{ox}^2(V_{gs}-V_{fb}-\psi_s) + \varepsilon_{si}qN_a}{C_{ox}^2(V_{gs}-V_{fb}-\psi_s)^2 - 2\varepsilon_{si}qN_a\psi_s}. \tag{5.19}$$

Substituting Eqs. (5.16) and (5.19) into Eq. (5.17) yields

$$\begin{aligned}I_{ds} = \mu_{eff}\frac{W}{L}\int_{\psi_{s,s}}^{\psi_{s,d}}\Bigg[&C_{ox}(V_{gs}-V_{fb}-\psi_s) - \sqrt{2\varepsilon_{si}qN_a\psi_s} \\ &+ \frac{2kT}{q}\frac{C_{ox}^2(V_{gs}-V_{fb}-\psi_s) + \varepsilon_{si}qN_a}{C_{ox}(V_{gs}-V_{fb}-\psi_s) + \sqrt{2\varepsilon_{si}qN_a\psi_s}}\Bigg]d\psi_s,\end{aligned} \tag{5.20}$$

which expresses the drain current in a single integral.

It is too tedious to carry out the integral in Eq. (5.20) analytically. A second approximation is introduced in the charge sheet model (Brews, 1978) to obtain an algebraic expression for the drain current. Note that the first two terms in the square bracket of Eq. (5.20) are simply $-Q_i$. Because of the kT/q multiplier, the last term in the square bracket is usually much smaller than the first two unless $Q_i \approx 0$, which happens when $C_{ox}(V_{gs}-V_{fb}-\psi_s) \approx \sqrt{2\varepsilon_{si}qN_a\psi_s}$. It is then a good approximation to apply this relation to the last term in the square bracket so that the integral can be carried out analytically:

$$I_{ds} = \mu_{eff}\frac{W}{L}\left\{C_{ox}\left(V_{gs}-V_{fb}+\frac{kT}{q}\right)\psi_s - \frac{1}{2}C_{ox}\psi_s^2 - \frac{2}{3}\sqrt{2\varepsilon_{si}qN_a}\psi_s^{3/2} + \frac{kT}{q}\sqrt{2\varepsilon_{si}qN_a}\psi_s\right\}\Bigg|_{\psi_{s,s}}^{\psi_{s,d}}. \tag{5.21}$$

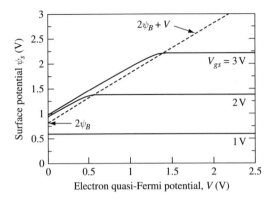

Figure 5.3 Numerical solutions of the implicit Eq. (5.14) for three values of V_{gs}. The dashed line represents the regional approximation used in Section 5.1.3. The MOS device parameters are $N_a = 10^{17} \text{ cm}^{-3}$, $t_{ox} = 10 \text{ nm}$, and $V_{fb} = 0$

Because Eq. (5.21) covers all regions of MOSFET operation: subthreshold, linear, and saturation in a single, continuous function, it has become the basis of all surface potential based compact models (SPICE) for circuit simulations (Gildenblat *et al.*, 2006). Many numerical methods have been developed to solve the implicit Eq. (5.14) for $\psi_{s,s}$ and $\psi_{s,d}$, given V_{gs} and V_{ds}. They employ either explicit approximations or iterative procedures. An example of the general behavior of the solution $\psi_s(V)$ is shown in Figure 5.3. The same set of device parameters as those of Figure 4.11 are used. For $V_{gs} = 1\text{V}$, the device is below threshold where the inversion charge is negligible, i.e., the $\left(n_i^2/N_a^2\right)e^{q(\psi_s - V)/kT}$ term in the square bracket of Eq. (5.14) is negligible. The solution ψ_s depends on V_{gs} [see Eq. (4.33)], but is totally insensitive to V. For $V_{gs} = 2\text{V}$, the MOSFET is turned on. Here, ψ_s increases more or less linearly with V when V is not too large. As V increases (for large enough V_{ds}), ψ_s reaches a saturation value beyond which it becomes independent of V. This is called the *saturation* condition where Q_i given by Eq. (5.16) becomes very small. The argument of the log function in Eq. (5.18) also approaches zero. For $V_{gs} = 3\text{V}$, the saturation value of ψ_s increases while saturation happens at a higher V (or V_{ds}).

Because of the two simplifying approximations used, the current calculated from the charge sheet model, Eq. (5.21), deviates from that of Pao and Sah's double integral, Eq. (5.13). The error is a function of doping concentration, oxide thickness, gate and drain bias voltages. Typically, it can be of the order of 10% under certain conditions when biased above threshold (Kyung, 2005). The error is generally larger in subthreshold where the current levels are low and high accuracy is not a paramount issue.

5.1.3   Regional *I–V* Models

To obtain the explicit $I_{ds}(V_{gs}, V_{ds})$ characteristics of a MOSFET in the linear, parabolic, and subthreshold regions, it is necessary to apply regional approximations to the charge sheet model and break it into piecewise models. Some accuracy is inevitably lost in the regions adjoining them.

5.1.3.1 Linear Region

After the onset of inversion but before saturation, the surface potential can be approximated by $\psi_s = 2\psi_B + V(y)$, or Eq. (5.3). This relation is plotted in Figure 5.3 (dashed line) for comparison with the more exact curves. It then follows that $dV/d\psi_s = 1$ and Eq. (5.17) can be readily integrated. By applying $\psi_{s,s} = 2\psi_B$ and $\psi_{s,d} = 2\psi_B + V_{ds}$, we obtain the drain current as a function of gate and drain voltages:

$$I_{ds} = \mu_{eff} C_{ox} \frac{W}{L} \left\{ \left(V_{gs} - V_{fb} - 2\psi_B - \frac{V_{ds}}{2} \right) V_{ds} \right.$$

$$\left. - \frac{2\sqrt{2\varepsilon_{si}qN_a}}{3C_{ox}} \left[(2\psi_B + V_{ds})^{3/2} - (2\psi_B)^{3/2} \right] \right\}. \tag{5.22}$$

Equation (5.22) represents the basic *I–V* characteristics of a MOSFET device based on the charge-sheet model. It indicates that, for a given V_{gs}, the drain current I_{ds} first increases linearly with the drain voltage V_{ds} (called the *linear* or *triode* region), then gradually levels off to a saturated value (*parabolic* region).

When V_{ds} is small, we can expand Eq. (5.22) into a power series in V_{ds} and keep only the lowest-order (first-order) terms:

$$I_{ds} = \mu_{eff} C_{ox} \frac{W}{L} \left(V_{gs} - V_{fb} - 2\psi_B - \frac{\sqrt{4\varepsilon_{si}qN_a\psi_B}}{C_{ox}} \right) V_{ds}$$

$$= \mu_{eff} C_{ox} \frac{W}{L} \left(V_{gs} - V_t \right) V_{ds}, \tag{5.23}$$

where V_t is the *threshold voltage* given by

$$V_t = V_{fb} + 2\psi_B + \frac{\sqrt{4\varepsilon_{si}qN_a\psi_B}}{C_{ox}}. \tag{5.24}$$

Comparing this equation with Eq. (4.32), we can see that V_t *is simply the gate voltage when the surface potential or band bending reaches $2\psi_B$ and the charge in silicon (the square root) is equal to the bulk depletion charge for that potential*. As a reminder, $2\psi_B = (2kT/q) \ln(N_a/n_i)$, which is typically 0.6–0.9 V. When V_{gs} is below V_t, there is very little current flow and the MOSFET is said to be in the *subthreshold* region, to be discussed in Section 5.1.3.3. Equation (5.23) indicates that, *in the linear region, the MOSFET simply acts like a resistor with a sheet resistivity, $\rho_{sh} = 1/[\mu_{eff} C_{ox} (V_{gs} - V_t)]$, modulated by the gate voltage*.

Figure 5.4 compares the low-drain I_{ds}–V_{gs} characteristics of the piecewise model with a numerical solution of the all-region continuous model without the $2\psi_B$ simplification. The latter is obtained by solving the implicit Eq. (5.14) for $\psi_s(V_{gs})$ with $V = 0$, then calculating $Q_i(\psi_s)$ from the Eq. (5.12) integral also with $V = 0$. The

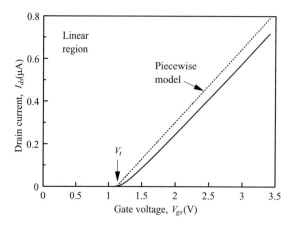

Figure 5.4 MOSFET I_{ds}–V_{gs} characteristics in the linear region. The dotted line is the piecewise model, Eq. (5.23). The solid curve is from the numerical solutions of Eqs. (5.12) and (5.14)

discrepancy can be ascribed as the *inversion layer capacitance* effect. The $2\psi_B$ model, or $|Q_i| = C_{ox}(V_{gs} - V_t)$, basically assumes that the inversion layer has zero thickness so all the inversion charge is at the silicon surface a distance t_{ox} away from the gate. It is equivalent to setting the inversion-layer capacitance, $C_i = |dQ_i/d\psi_s|$, to infinity once above the threshold. This causes significant errors, especially in the region near threshold. Note that there is no "threshold voltage" defined or specified in the continuous charge sheet model. The *linearly extrapolated threshold voltage* (V_{on}), obtained by extrapolating the linear portion of the numerical I_{ds}–V_{gs} curve in Figure 5.4 to its intercept at $I_{ds} = 0$, is slightly (a few kT/q) higher than the "$2\psi_B$" V_t.

An explicit expression for the inversion charge density taking this effect into account can be derived by considering the small-signal capacitances in Figure 4.13(b) (Wordeman, 1986),

$$\frac{d(-Q_i)}{dV_{gs}} = \frac{C_{ox}C_i}{C_{ox} + C_i + C_d} \approx C_{ox}\left(1 - \frac{1}{1 + C_i/C_{ox}}\right). \tag{5.25}$$

Here $C_d \approx 0$ after the onset of strong inversion because of screening by the inversion charge. With the Boltzmann approximation of Eq. (4.16), $|Q_i| \propto \exp(q\psi_s/2kT)$, hence $C_i \approx |Q_i|/(2kT/q)$. Since $|Q_i| \approx C_{ox}(V_{gs} - V_t)$, one can write $C_i/C_{ox} = (V_{gs} - V_t)/(2kT/q)$. Substituting it into Eq. (5.25) and integrating with respect to V_{gs} yields

$$-Q_i = C_{ox}\left[(V_{gs} - V_t) - \frac{2kT}{q}\ln\left(1 + \frac{q(V_{gs} - V_t)}{2kT}\right)\right]. \tag{5.26}$$

This agrees well with the numerically calculated curve [Q_i times $\mu_{eff}(W/L)V_{ds}$] in Figure 5.4 for V_{gs} above V_t. Equation (5.26) indicates that the inversion layer capacitance is related to the kT/q factor. Note that Eq. (5.26) is only a good approximation for V_{gs} above V_t, not valid for V_{gs} at or below V_t.

5.1.3.2 Parabolic Region

For larger values of V_{ds}, the second-order terms in the power series expansion of Eq. (5.22) are also important and must be kept. A good approximation to the drain current is then

$$I_{ds} = \mu_{eff} C_{ox} \frac{W}{L} \left((V_{gs} - V_t) V_{ds} - \frac{m}{2} V_{ds}^2 \right), \tag{5.27}$$

where

$$m = 1 + \frac{\sqrt{\varepsilon_{si} q N_a / 4 \psi_B}}{C_{ox}} \tag{5.28}$$

is a factor greater than one. It will be discussed in Sections 5.1.3.3 and 5.3.1 that the factor m plays a key role in the *subthreshold slope* and the *body effect* of a MOSFET device. Equation (5.28) can be converted to several alternative expressions by using Eq. (4.38) for the bulk depletion capacitance C_{dm} at $\psi_s = 2\psi_B$:

$$m = 1 + \frac{C_{dm}}{C_{ox}} = 1 + \frac{3 t_{ox}}{W_{dm}}. \tag{5.29}$$

The last expression follows from $C_{dm} = \varepsilon_{si}/W_{dm}$, $C_{ox} = \varepsilon_{ox}/t_{ox}$, and $\varepsilon_{si}/\varepsilon_{ox} \approx 3$. A graphical interpretation of m is given in Figure 5.5. Near the threshold condition, $\psi_s = 2\psi_B$, the MOSFET acts like two capacitors, C_{ox} and C_{dm}, in series as the inversion charge capacitance is still negligible. The factor m equals $\Delta V_{gs}/\Delta \psi_s$, where $\Delta \psi_s$ is the incremental change of surface potential due to ΔV_{gs}, an incremental change of gate voltage. ΔV_{gs} induces sheet charge densities $+\Delta Q$ at the gate and $-\Delta Q$ at the far edge of the depletion region. They cause a field change of $\Delta \mathscr{E} = \Delta Q/\varepsilon_{si}$ in the

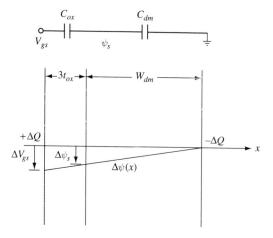

Figure 5.5 Incremental change of potential in a MOSFET due to a gate voltage modulation near or below threshold. Grounding of the body anchored the potential on the bulk side of the depletion region where $\Delta \psi = 0$. The potential drop across the oxide, $(\Delta Q/\varepsilon_{ox}) t_{ox}$, is equivalent to $(\Delta Q/\varepsilon_{si})[(\varepsilon_{si}/\varepsilon_{ox}) t_{ox}]$. The factor m is defined as $\Delta V_{gs}/\Delta \psi_s$, which equals $(W_{dm} + 3 t_{ox})/W_{dm}$

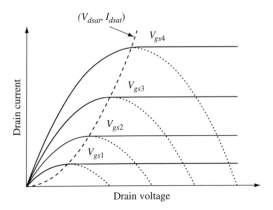

Figure 5.6 Long-channel MOSFET $I_{ds}-V_{ds}$ characteristics (solid curves) for several different values of V_{gs}. The dashed curve shows the trajectory of drain voltage beyond which the current saturates. The dotted curves help to illustrate the parabolic behavior of the characteristics

silicon and $\Delta Q/\varepsilon_{ox}$ in the oxide, which give rise to an incremental change of potential $\Delta \psi(x)$, as shown in Figure 5.5. Here, the oxide width is expanded to $\varepsilon_{si}/\varepsilon_{ox} \approx$ 3-times its physical width so there is no change of slope at the silicon–oxide interface. *While Eq. (5.28) is only valid for uniform bulk doping, Eq. (5.29) is more generally valid for nonuniform doping profiles*, to be discussed in Section 6.3.3. Since $1/m$ measures the efficiency of the gate voltage in modulating the surface potential, m should be kept close to one, e.g., between 1.1 and 1.4, in MOSFET design.

Equation (5.27) indicates that as V_{ds} increases, I_{ds} follows a parabolic curve, as shown in Figure 5.6, until a maximum or saturation value is reached. This occurs when

$$V_{ds} = V_{dsat} = \frac{V_{gs} - V_t}{m}, \tag{5.30}$$

at which

$$I_{ds} = I_{dsat} = \mu_{eff} C_{ox} \frac{W}{L} \frac{\left(V_{gs} - V_t\right)^2}{2m}. \tag{5.31}$$

Equation (5.31) reduces to the more widely known expression for the MOSFET saturation current when the bulk depletion charge is neglected (valid for low substrate doping) so $m = 1$. The dashed curve in Figure 5.6 shows the trajectory of V_{dsat} through the various $I_{ds}-V_{ds}$ curves for different V_{gs}. The decrease of I_{ds} past the apex is clearly unphysical. The dashed (V_{dsat}, I_{dsat}) curve marks the boundary of validity of the regional approximation, $\psi_s = 2\psi_B + V$, used in the derivation of Eq. (5.22) and therefore Eq. (5.27).

To gain more insight into the saturation of I_{ds} with V_{ds} for a given V_{gs}, we examine the inversion charge density Q_i as a function of the quasi-Fermi potential V. Within the piecewise model, we can substitute $\psi_s = 2\psi_B + V$ in Eq. (5.16), expand the square-root term into a power series in V, and keep only the two lowest terms,

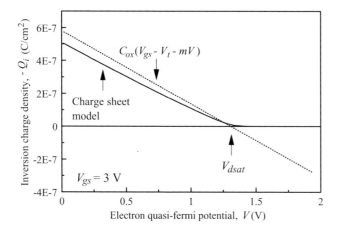

Figure 5.7 Inversion charge density as a function of the quasi-Fermi potential V. The dotted curve is Eq. (5.32). The solid curve is generated from the charge sheet model. The parameters are the same as those of Figure 5.3

$$-Q_i(V) = C_{ox}\left(V_{gs} - V_t - mV\right). \qquad (5.32)$$

The above is plotted in Figure 5.7 for a fixed V_{gs}. Note from Eq. (5.10) that the drain current is proportional to the area under the $|Q_i(V)|$ curve between $V = 0$ and V_{ds}. When V_{ds} is small (linear region), the inversion charge density at the drain end of the channel is only slightly lower than that at the source end. As the drain voltage increases (for a fixed gate voltage), the area or current increases, but the inversion charge density at the drain decreases until finally it goes to zero when $V_{ds} = V_{dsat} = (V_{gs} - V_t)/m$. At this point, I_{ds} reaches its maximum value, I_{dsat} of Eq. (5.31).

Also plotted in Figure 5.7 is the continuous $-Q_i(V)$ curve of the charge sheet model generated by numerically solving the implicit Eq. (5.14) for $\psi_s(V)$, then calculating $Q_i(\psi_s)$ from Eq. (5.16). At $V = 0$, $-Q_i$ is slightly lower than $C_{ox}(V_{gs} - V_t)$ due to the inversion layer capacitance effect discussed in Section 5.1.3.1. Instead of $-Q_i = 0$ at $V = V_{dsat}$ then going negative as in the piecewise model, $-Q_i$ of the charge sheet model approaches 0 continuously as $V \to \infty$. This means that I_{ds}, proportional to the area under the charge sheet $-Q_i(V)$, converges continuously to the saturation value as V_{ds} becomes $\gg V_{dsat}$. Such behavior, however, is a consequence of the GCA made in the beginning of the charge sheet model. To properly model the drain current beyond V_{dsat}, a non-GCA model is needed. This will be discussed in Section 5.1.4.

5.1.3.3   Subthreshold Region

Depending on the gate and source–drain voltages, a MOSFET device can be biased in one of the three regions shown in Figure 5.8. Linear and parabolic region characteristics have been described in Sections 5.1.3.1 and 5.1.3.2. In this subsection, we discuss the characteristics of a MOSFET device in the subthreshold region where $V_{gs} < V_t$.

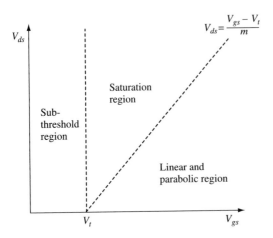

Figure 5.8 Three regions of MOSFET operation in the V_{ds}–V_{gs} plane

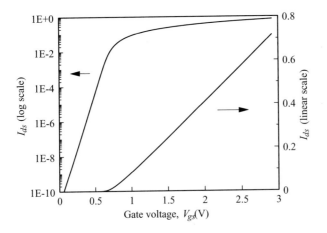

Figure 5.9 Low drain bias I_{ds}–V_{gs} characteristics generated by the charge sheet model. The same current is plotted in both linear and logarithmic scales

Figure 5.9 plots the low-drain I_{ds}–V_{gs} characteristics generated by the continuous model Eqs. (5.14) and (5.12) in both linear and logarithmic current scales. On the linear scale, the drain current appears to drop to zero immediately below the threshold voltage. However, the logarithmic scale reveals that the descending drain current remains at nonnegligible levels for several tenths of a volt below V_t. This is because the inversion charge density does not drop to zero abruptly. Rather, it follows an exponential dependence on ψ_s or V_{gs}, as is evident from Eq. (5.11). The subthreshold region immediately below V_t, in which $\psi_B \leq \psi_s \leq 2\psi_B$, is also called the *weak inversion* region. Subthreshold region behavior describes how a MOSFET device switches off. Although the off current, or I_{ds} at $V_{gs} = 0$, is very low in each individual device, it adds up collectively over a VLSI chip of $> 10^9$ transistors to a significant level of standby power dissipation. This is further considered in Section 6.3.

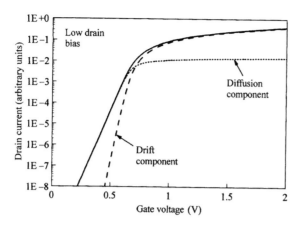

Figure 5.10 Drift and diffusion components of current in an I_{ds}–V_{gs} plot. Their sum is the total current represented by the solid curve

Unlike the strong inversion region, in which the drift current dominates, sub-threshold conduction is dominated by the diffusion current. This can be seen by considering that the total current is proportional to dV/dy while the drift current is proportional to the electric field or $d\psi_s/dy$. In Figure 5.3, from the charge sheet model, $d\psi_s/dV \approx 1$ above threshold, meaning that the current is predominantly drift. Below threshold (the $V_{gs} = 1$ V case), on the other hand, $d\psi_s/dV \approx 0$, the current is all diffusion. In general, current continuity only applies to the total current, not to its individual components. In other words, the fractional ratio between the drift and the diffusion components may vary from one point of the channel to another. At low drain bias, i.e., in the limit of $V \rightarrow 0$, however, Eq. (5.14) becomes a one-to-one relationship between V_{gs} and ψ_s. In that case, $d\psi_s/dV$ has the same value from the source to the drain. By taking a differential of Eq. (5.14) then setting $V = 0$, it can be shown that

$$\frac{d\psi_s}{dV} = \frac{\left(n_i^2/N_a^2\right)e^{q\psi_s/kT}}{1 + \left(n_i^2/N_a^2\right)e^{q\psi_s/kT} + \left(C_{ox}^2/\varepsilon_{si}qN_a\right)\left(|Q_s|/C_{ox}\right)}, \tag{5.33}$$

where $|Q_s|/C_{ox}$ above is the voltage drop across the oxide given by the last term of Eq. (5.14). Both $d\psi_s/dV$ and $1 - d\psi_s/dV$ are plotted in Figure 5.10 versus V_{gs}. It is clear that in subthreshold where $\psi_s < 2\psi_B$, the numerator in Eq. (5.33) is much less than unity and the diffusion component dominates. Conversely, beyond strong inversion, $d\psi_s/dV \approx 1$ and the drift current dominates.

To find an expression for the subthreshold current, we note from Eq. (5.14) that the total charge density in silicon is

$$-Q_s = \varepsilon_{si}\mathscr{E}_s = \sqrt{2\varepsilon_{si}kTN_a}\left[\frac{q\psi_s}{kT} + \frac{n_i^2}{N_a^2}e^{q(\psi_s - V)/kT}\right]^{1/2}. \tag{5.34}$$

In weak inversion, the second term in the square bracket arising from the inversion charge density is much less than the first term from the depletion charge density.

The square root can then be expanded into a power series: the zeroth-order term is identified as the depletion charge density $-Q_d$ by Eq. (5.15), and the first-order term gives the inversion charge density

$$-Q_i = \sqrt{\frac{\varepsilon_{si}qN_a}{2\psi_s}} \left(\frac{kT}{q}\right) \left(\frac{n_i}{N_a}\right)^2 e^{q(\psi_s - V)/kT}. \tag{5.35}$$

The surface potential ψ_s is related to the gate voltage through Eq. (5.14). Since the inversion charge density is small, ψ_s is a function of V_{gs} only, independent of V.

Substituting Q_i into Eq. (5.10) and carrying out the integration, we obtain the drain current in the subthreshold region:

$$I_{ds} = \mu_{eff} \frac{W}{L} \sqrt{\frac{\varepsilon_{si}qN_a}{2\psi_s}} \left(\frac{kT}{q}\right)^2 \left(\frac{n_i}{N_a}\right)^2 e^{q\psi_s/kT} \left(1 - e^{-qV_{ds}/kT}\right). \tag{5.36}$$

ψ_s can be expressed in terms of V_{gs} using Eq. (5.14), where only the depletion charge term needs to be kept:

$$V_{gs} = V_{fb} + \psi_s + \frac{\sqrt{2\varepsilon_{si}qN_a\psi_s}}{C_{ox}}. \tag{5.37}$$

This equation is the same as Eq. (4.32) for an MOS in depletion. Its solution for ψ_s is given by Eq. (4.33). To further simplify the result, we consider ψ_s as only slightly deviated from the threshold value, $2\psi_B$ (Swanson and Meindl, 1972). Using the concept of $m = \Delta V_{gs}/\Delta \psi_s$ in Figure 5.5, we can approximate V_{gs} as $V_{gs} = V_t + m(\psi_s - 2\psi_B)$. Solving for ψ_s and substituting it into Eq. (5.36) yields the subthreshold current as a function of V_{gs}:

$$I_{ds} = \mu_{eff} \frac{W}{L} \sqrt{\frac{\varepsilon_{si}qN_a}{4\psi_B}} \left(\frac{kT}{q}\right)^2 e^{q(V_{gs} - V_t)/mkT} \left(1 - e^{-qV_{ds}/kT}\right), \tag{5.38}$$

or

$$I_{ds} = \mu_{eff} C_{ox} \frac{W}{L} (m - 1) \left(\frac{kT}{q}\right)^2 e^{q(V_{gs} - V_t)/mkT} \left(1 - e^{-qV_{ds}/kT}\right). \tag{5.39}$$

The subthreshold current is independent of the drain voltage once V_{ds} is larger than a few kT/q, as would be expected for a diffusion-dominated current transport. The dependence on gate voltage, on the other hand, is exponential with an *inverse subthreshold slope* or *subthreshold swing* (Figure 5.10),

$$S = \left(\frac{d(\log_{10} I_{ds})}{dV_{gs}}\right)^{-1} = 2.3 \frac{mkT}{q} = 2.3 \frac{kT}{q} \left(1 + \frac{C_{dm}}{C_{ox}}\right), \tag{5.40}$$

of typically 70–100 mV/decade. Here, $m = 1 + (C_{dm}/C_{ox})$ from Eq. (5.29). For VLSI circuits, a steep subthreshold slope is desirable for the ease of switching the transistor current off. In MOSFET design, therefore, the gate oxide thickness and the bulk doping concentration should be chosen such that the factor m is not too much larger than unity, e.g., between 1.1 and 1.4. ***The subthreshold swing has a lower bound of***

2.3 kT/q, or 60 mV/decade at room temperature, that does not change with device dimensions. This has significant implications on threshold voltage design and device scaling, as will be discussed in Chapters 6 and 8.[2]

5.1.4   Non-GCA Model for the Saturation Region

The MOSFET current model covered thus far has been developed under the framework of Gradual Channel Approximation (GCA). It assumes that the field gradient in the *y*-direction or the channel direction is negligible compared to the field gradient in the *x*-direction or the gate direction so 2-D Poisson's equation,

$$\frac{\partial^2 \psi}{\partial x^2} + \frac{\partial^2 \psi}{\partial y^2} = -\frac{q}{\varepsilon_{si}} \left[p(x) - n(x) + N_d^+(x) - N_a^-(x) \right], \tag{5.41}$$

is reduced to the 1-D MOS equation of Eq. (4.9). The GCA model works fine in the linear, parabolic, and subthreshold regions, but fails in the saturation region when $V_{ds} > V_{dsat}$. To see that, note in Figure 5.7 that the mobile charge density $(-Q_i)$ from the charge sheet model approaches zero when $V > V_{dsat}$. However, the current continuity equation, Eq. (5.8):

$$I_{ds} = -\mu_{eff} W \frac{dV}{dy} Q_i(V) \tag{5.42}$$

demands that the product $(-Q_i) \times dV/dy$ be a constant throughout the channel. When $-Q_i \to 0$, $dV/dy \to \infty$, thus invalidating the GCA.

In most standard texts, this is called the *"pinch-off"* condition. Pinch-off is a term originally applied to JFETs (Junction Field-Effect Transistors) in the early days of transistor development (Shockley, 1952). It describes how a p- or n-type conducting path is squeezed to zero by the encroaching depletion regions of reverse-biased p–n junctions on both sides of the path. It is rather misleading to use *"pinch-off"* to describe the point of current saturation in MOSFETs because what goes to zero at $V = V_{dsat}$ is the vertical field, $\mathscr{E}_x = -(\partial \psi / \partial x)|_{x=0}$, or the gate induced charge density, not the entire mobile charge density. As a matter of fact, *both $\partial^2 \psi / \partial x^2$ and $\mathscr{E}_x$ become negative beyond $V = V_{dsat}$,* as seen in Figure 5.11(a) from 2-D numerical simulations (TCAD[2]). This shows that the above V_{dsat} behavior of the charge-sheet curve in Figure 5.7 is a consequence of the GCA model not allowing $\mathscr{E}_x$ to go negative, rather than being physically correct. Also shown in Figure 5.11(a) is that *the electron density is never zero whether $(\partial \psi / \partial x)|_{x=0}$ is positive or negative.* From the perspective of 2-D Eq. (5.41), when $\partial^2 \psi / \partial x^2$ is negative, the $\partial^2 \psi / \partial y^2$ term becomes more positive to overcome the negative $\partial^2 \psi / \partial x^2$, thus making the total sum positive. In this regard, "pinch-off" never happens; $\partial^2 \psi / \partial y^2$ and therefore the lateral field increase sharply while the vertical field takes on negative values when $V_{ds} > V_{dsat}$.

---

[2] TCAD (Technology Computer Aided Design) is a 2D (or 3D) semiconductor device simulation tool that solves a set of coupled partial differential equations (including Poisson's and current continuity equations) numerically over a user specified device geometry by finite-element analysis. There are several commercial versions available.

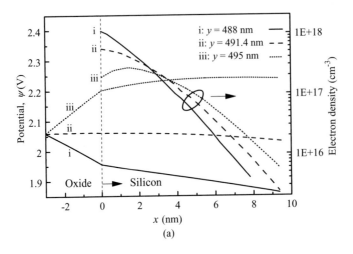

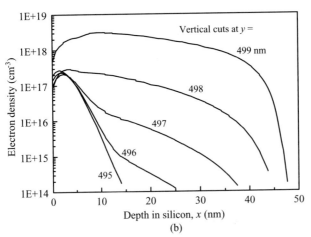

Figure 5.11 Plots from TCAD simulations. (a) Potential $\psi(x)$ and electron density $n(x)$ (right scale) along three vertical cuts: (i) before the saturation point, (ii) at the saturation point, (iii) beyond the saturation point. For this plot, ψ is defined as the intrinsic potential with respect to the Fermi potential of the source (see Fig. 7.13). The MOSFET parameters are $L = 500$ nm, $t_{inv} = 3.3$ nm, $N_a = 10^{18}$ cm^{-3} (uniform), $V_{gs} = 1.5$ V, $V_{ds} = 2.0$ V. The gate work function is that of n$^+$ silicon. (b) Electron density versus depth in silicon along five vertical cuts between the saturation point and the drain ($y = 500$ nm). The junction depth is $x_j = 50$ nm in this case

5.1.4.1   A Continuous Non-GCA Model into the Saturation Region

To construct a non-GCA model, a $\partial^2\psi/\partial y^2$ term is added to $-Q_i$ in the current continuity equation (Taur and Lin, 2018):

$$I_{ds} = \mu_{eff} W \left[-Q_i(V) + \varepsilon_{si} d_{si} \frac{d^2\psi}{dy^2} \right] \frac{dV}{dy}. \qquad (5.43)$$

Here, d_{si} is an effective depth in silicon to convert the per volume charge density, $\varepsilon_{si} d^2\psi/dy^2$, to an areal charge density. For double-gate MOSFETs with thin silicon film, the clear choice for d_{si} is the silicon thickness. For bulk MOSFETs, d_{si} is some fraction of the junction depth x_j. This can be seen in the TCAD plot in Figure 5.11(b) of the depth distribution of the electron density beyond the point of saturation. When the vertical cut moves closer to the drain junction, the electron density spreads deeper toward the junction depth, $x_j = 50$ nm, indicating a similar spread of the current density. In this regard, d_{si} is an effective or average depth rather than a physical depth. For this example, a depth parameter of $d_{si} = 20$ nm serves as a good approximation. Also seen in Figure 5.11 is that the electron density per area, i.e., $n(x)$ integrated over x, which has been decreasing before the saturation point ($y \approx 491.4$ nm), keeps on decreasing through the saturation point until a point of minimum at $y \approx 497$ nm, very close to the drain junction edge.

For the expression of $-Q_i(V)$ in Eq. (5.43), a good choice is Eq. (5.32):

$$-Q_i(V) = C_{inv}\left(V_{gs} - V_t - mV\right), \tag{5.44}$$

which does go negative beyond $V = V_{dsat} = (V_{gs} - V_t)/m$ (see the dotted line in Figure 5.7). Here, C_{inv} is used in place of C_{ox} to take the inversion layer capacitance into account. At the source, $C_{inv}(V_{gs} - V_t)$ is given by Eq. (5.16) of the charge sheet model with $\psi_s = \psi_{s,s}$ for $V = 0$. The linear slope of $Q_i(V)$ is a reasonable approximation for $V_{gs} - V_t$ larger than several kT/q, e.g., $V_{gs} - V_t > 0.2$ V. For near-threshold bias conditions, the decrease of $|Q_i|$ with V is much softer due to inversion layer capacitance effects (see Figure 5.4) (Ren and Taur, 2020).

To make Eq. (5.43) easier to solve, an approximation, $d^2\psi/dy^2 \approx d^2V/dy^2$, is made on the grounds that near the drain where the $\varepsilon_{si} d^2\psi/dy^2$ term is important, the current is mostly drift, i.e., $d\psi/dy \approx dV/dy$. With that substitution, Eq. (5.43) can be integrated once to yield:

$$\frac{I_{ds}}{\mu_{eff} W} y = C_{inv}\left[(V_{gs} - V_t)V - \frac{m}{2}V^2\right] + \frac{\varepsilon_{si} d_{si}}{2}\left[\left(\frac{dV}{dy}\right)^2 - \mathscr{E}_0^2\right], \tag{5.45}$$

where $\mathscr{E}_0$ is dV/dy at $y = 0$. Since the non-GCA term in Eq. (5.43) is negligible at $y = 0$, we have

$$\mathscr{E}_0 = \frac{I_{ds}}{\mu_{eff} W C_{inv}\left(V_{gs} - V_t\right)}. \tag{5.46}$$

Equation (5.45) is a first-order ordinary differential equation valid for all regions above threshold, both before and after saturation. For a given I_{ds}, it solves for $V(y)$ numerically from $y = 0$ to $y = L$, yielding $V_{ds} = V(L)$ as the result. The standard method of evaluating $dV/dy = f(y, V)$ and applying it to get to the next point runs into the trouble of magnifying the numerical imprecision in the region of $V \ll V_{dsat}$ where $y(V)$ is simply given by the first term (GCA) on the right-hand side of Eq. (5.45) with the $(dV/dy)^2$ term (non-GCA) negligible. Instead, to go from a point (y, V) to the next point $(y + \delta y, V + \delta V)$, the following difference equation is used:

$$\frac{I_{ds}}{\mu_{eff}W}\delta y = C_{inv}\left\{(V_{gs}-V_t)\delta V-\frac{m}{2}\left[(2V\delta V)+(\delta V)^2\right]\right\}$$

$$+\frac{\varepsilon_{si}d_{si}}{2}\left\{\left[2\left(\frac{\delta V}{\delta y}\right)-\frac{dV}{dy}\Big|_{y,V}\right]^2-\left[\frac{dV}{dy}\Big|_{y,V}\right]^2\right\},\qquad(5.47)$$

where $(dV/dy)|_{y,V}$ is the value of dV/dy at (y, V). For a given δy, the above can be re-grouped into a quadratic equation for δV with standard solutions. This procedure can be repeated for a large number of steps on a spreadsheet to produce a continuous transition from the GCA dominated region to the non-GCA region.

Examples of the solution $V(y)$ for two different values of I_{ds}, both slightly over I_{dsat}, are plotted in Figure 5.12 as y versus V so that y can be decomposed into its two components: the first term on the right-hand side of Eq. (5.45) stemming from $-Q_i$ (labeled GCA) and the second term from $(dV/dy)^2$ (labeled non-GCA). Consider first the GCA curve. It has a peak value of $y = (I_{dsat}/I_{ds})L$ at $V = V_{dsat} = (V_{gs} - V_t)/m$, then decreases toward zero. This would be unphysical, like the downturn of I_{ds} past V_{dsat} in Figure 5.6, were the $-Q_i$ component solely responsible for the current. In the non-GCA model, the additional component from $(dV/dy)^2$, while negligible for $V < V_{dsat}$, increases sharply beyond V_{dsat} so the sum y (solid curves) continues to increase toward L, as seen in Figure 5.12. The slope dy/dV is, of course, never negative, although it is much reduced in the saturation region than before saturation.

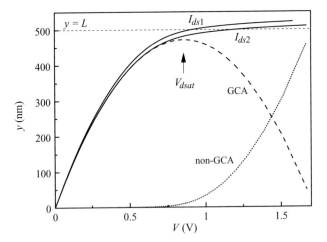

Figure 5.12 $y(V)$ solution to Eq. (5.45) for two values of I_{ds}: I_{ds1} is 3% over I_{dsat}, I_{ds2} is 6% over I_{dsat}. The device is the same as that of Figure 5.11, with $V_t = 0.4$ V, $m = 1.28$, biased at $V_{gs} = 1.5$ V so $V_{dsat} = 0.86$ V and $I_{dsat} = 2.0$ A/cm. d_{si} is chosen to be 20 nm. The crossover with the $y = L$ line gives the V_{ds} solution for the particular I_{ds}. The I_{ds2} result is further partitioned into two curves, according to the two terms on the right-hand side of Eq. (5.45). The dashed curve labeled GCA is the first term divided by $(I_{ds}/\mu_{eff}W)$. The dotted curve labeled non-GCA is the second term divided by the same

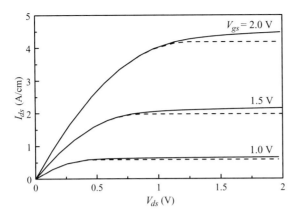

Figure 5.13 $I_{ds}-V_{ds}$ curves (solid) solved from Eq. (5.45) for the device described in the caption to Figure 5.12. The dashed curves are from the GCA model for which currents saturate at I_{dsat} [Eq. (5.31)]

The notion of *Channel Length Modulation* (CLM) is based on the fact that the peak y-value of the GCA curve, $(I_{dsat}/I_{ds})L$, at $V = V_{dsat}$ becomes $< L$ if $I_{ds} > I_{dsat}$. If we let this y value be $L - \Delta L$, we obtain $I_{ds} = I_{dsat}/(1 - \Delta L/L)$. In view of the full non-GCA model, CLM only serves as an approximation as the y value at $V = V_{dsat}$ on the solid curve in Figure 5.12 is slightly higher than the y value at $V = V_{dsat}$ on the dashed (GCA) curve.

Figure 5.13 shows the $I_{ds}-V_{ds}$ curves generated from this model. They are continuous from the linear and parabolic regions into the saturation region.

5.1.4.2 Regional Approximation for the Saturation Region

Equation (5.45) can be greatly simplified in the saturation region where $y \approx L$ as is evident in Figure 5.12. The $\mathscr{E}_0^2$ term can also be dropped. Further rearrangement yields

$$\frac{L}{\mu_{eff}W}(I_{ds}-I_{dsat}) + \frac{m}{2}C_{inv}(V-V_{dsat})^2 = \frac{\varepsilon_{si}d_{si}}{2}\left(\frac{dV}{dy}\right)^2, \tag{5.48}$$

where V_{dsat} and I_{dsat} are given by Eqs. (5.30) and (5.31). If V_{ds} is not too close to V_{dsat}, the first term on the left-hand side is much smaller than the second term. *It then follows that in the saturation region, dV/dy increases linearly with $V - V_{dsat}$. Further integration gives V(y) as an exponential function of y,* $\propto exp\left[y/\sqrt{\varepsilon_{si}d_{si}/(mC_{inv})}\right]$. The correlation between the characteristic lateral length $\sqrt{\varepsilon_{si}d_{si}/(mC_{inv})}$ and the vertical dimensions reflects the 2-D nature of the non-GCA effect (Ko *et al.*, 1981).

Based on the CLM picture, there is a correspondence of Δy with ΔI_{ds}. Specifically, $\Delta y/L = \Delta I_{ds}/I_{dsat}$. Equation (5.48) then gives the output conductance in the saturation region:

$$\frac{dI_{ds}}{dV_{ds}} = \frac{I_{dsat}}{L}\left(\frac{dV}{dy}\right)^{-1} = \sqrt{\frac{\varepsilon_{si}d_{si}}{mC_{inv}L^2}\frac{I_{dsat}}{V_{ds}-V_{dsat}}}. \tag{5.49}$$

For not too short channel devices, the dimensionless square root factor is $\ll 1$, e.g., ~1/40 for the device in Figure 5.13. The slope in the saturation region increases with V_{gs} through the I_{dsat} factor and decreases with V_{ds} for a given V_{gs}.

5.1.5   pMOSFET I–V Characteristics

Thus far we have used an n-channel device as an example to discuss MOSFET operation and I–V characteristics. A p-channel MOSFET operates similarly, except that it is fabricated inside an n-well with implanted p^+ source and drain regions (cf. Figure 8.3), and that the polarities of all the voltages and currents are reversed. For example, I_{ds}–V_{ds} characteristics for a pMOSFET (cf. Figure 8.4) have negative gate and drain voltages with respect to the source terminal for a hole current to flow from the source to the drain.

Since the source of a pMOSFET is at the highest potential compared to the other terminals, it is usually connected to the power supply V_{dd} in a CMOS circuit so that all the voltages are positive (or zero). In that case, the device conducts if the gate voltage is lower than $V_{dd} + V_t$, where V_t (< 0) is the pMOS threshold voltage. The ohmic contact to the n-well is also connected to V_{dd}, in contrast to an nMOSFET, for which the p-type substrate is usually tied to the ground potential. This leaves the n-well-to-p-substrate junction reverse biased. More about nMOS and pMOS bias conditions in a CMOS circuit configuration will be given in Section 8.2.

5.2   MOSFET Channel Mobility

The current of a long-channel MOSFET is directly proportional to the carrier mobility in the channel [Eq. (5.8)]. Channel mobility is thus a figure of merit for MOSFET performance. Since electrons typically have a lighter effective mass, hence a higher mobility than holes, nMOSFETs have a higher current per device width than pMOSFETs. Carrier mobilities in an unstrained silicon are discussed in Section 5.2.1. The strain effect on mobility is discussed in Section 5.2.2.

5.2.1   Empirical Universal Mobility

Carrier mobility in a MOSFET channel is significantly lower than that in bulk silicon, due to additional scattering mechanisms. Lattice or phonon scattering is aggravated by the crystalline discontinuity at the surface boundary, and surface roughness scattering severely degrades mobility at high normal fields.

5.2.1.1   Effective Mobility and Effective Normal Field

In Section 5.1.1.1, the channel mobility was taken out of the integral by defining an effective mobility as

$$\mu_{eff} = \frac{\int_0^{x_i} \mu_n n(x)dx}{\int_0^{x_i} n(x)dx},$$ (5.50)

which is essentially an average value weighted by the carrier concentration in the inversion layer. Empirically, it has been found that *when μ_{eff} is plotted against an effective normal field $\mathscr{E}_{eff}$, there exists a universal relationship for a wide range of substrate bias, doping concentration, and gate oxide thickness* (Sabnis and Clemens, 1979). The effective normal field is defined as the average electric field perpendicular to the Si–SiO$_2$ interface experienced by the carriers in the channel. Using Gauss's law, we can express $\mathscr{E}_{eff}$ in terms of the depletion and inversion charge densities:

$$\mathscr{E}_{eff} = \frac{1}{\varepsilon_{si}}\left(|Q_d| + \frac{1}{2}|Q_i|\right),$$ (5.51)

where $|Q_d| + \frac{1}{2}|Q_i|$ is the total silicon charge inside a Gaussian surface through the middle of the inversion layer. By using Eqs. (4.23) and (5.24), the depletion charge density can be expressed as

$$|Q_d| = \sqrt{4\varepsilon_{si}qN_a\psi_B} = C_{ox}(V_t - V_{fb} - 2\psi_B).$$ (5.52)

Substituting this expression and $|Q_i| \approx C_{ox}(V_{gs} - V_t)$ into Eq. (5.51) yields

$$\mathscr{E}_{eff} = \frac{V_t - V_{fb} - 2\psi_B}{3t_{ox}} + \frac{V_{gs} - V_t}{6t_{ox}},$$ (5.53)

where $C_{ox} = \varepsilon_{ox}/t_{ox}$ and $\varepsilon_{si} \approx 3\varepsilon_{ox}$ were used. Equation (5.53) can further be simplifed if the gate electrode is n$^+$ polysilicon (for nMOSFETs) such that $V_{fb} = -E_g/2q - \psi_B$. For submicron CMOS technologies, $\psi_B = 0.30$–0.42 V. Therefore, the effective normal field can be expressed in terms of explicit device parameters as

$$\mathscr{E}_{eff} = \frac{V_t + 0.2}{3t_{ox}} + \frac{V_{gs} - V_t}{6t_{ox}}.$$ (5.54)

5.2.1.2 Electron Mobility Data

A typical set of data on mobility versus effective normal field for nMOSFETs is shown in Figure 5.14 (Takagi *et al.*, 1988). At room temperature, the mobility follows a $\mathscr{E}_{eff}^{-1/3}$ dependence below 5×10^5 V/cm. A simple, approximate expression for this case is (Baccarani and Wordeman, 1983)

$$\mu_{eff} \approx 32500 \times \mathscr{E}_{eff}^{-1/3}.$$ (5.55)

Beyond $\mathscr{E}_{eff} = 5 \times 10^5$ V/cm, μ_{eff} decreases much more rapidly with increasing $\mathscr{E}_{eff}$ because of increased surface roughness scattering as carriers are distributed closer to the surface under high normal fields. For each doping concentration, there exists an effective field below which the mobility falls off the universal curve. This is presumably due to Coulomb (or impurity) scattering, which becomes more important when

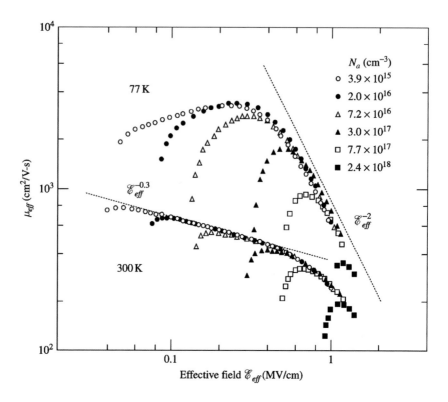

Figure 5.14 Measured electron mobility at 300 and 77 K versus effective normal field for several substrate doping concentrations (after Takagi *et al.*, 1988)

the doping concentration is high and the gate voltage or the normal field is low. There is less effect of Coulomb scattering on mobility when the inversion charge density is high because of charge screening effects. At 77 K, μ_{eff} is an even stronger function of $\mathscr{E}_{eff}$ and N_a. At low temperatures, surface scattering is the dominant mechanism at high fields, while Coulomb scattering dominates at low fields.

Since the mobility decreases with increasing gate voltage, $I_{ds}-V_{gs}$ characteristics measured at low drain bias exhibit a downward curvature, as shown in the example in Figure 5.15. There is a point of maximum slope or linear transconductance about 0.5 V above the threshold voltage. It is conventional to define the linearly extrapolated threshold voltage, V_{on}, by the intercept of a tangent through this point. Typically, V_{on} is $(2–4)kT/q$ higher than the threshold voltage V_t at $\psi_s(\text{inv}) = 2\psi_B$.

5.2.1.3 Hole Mobility Data

Similar mobility–field data for pMOSFETs are shown in Figure 5.16. In this case, however, the effective normal field is defined by

$$\mathscr{E}_{eff} = \frac{1}{\varepsilon_{si}}\left(|Q_d| + \frac{1}{3}|Q_i|\right), \tag{5.56}$$

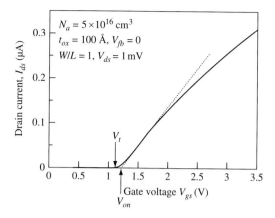

Figure 5.15 Low-drain I_{ds}–V_{gs} curve with inversion-layer capacitance and mobility degradation effects. The dotted line shows the linearly extrapolated threshold voltage V_{on}

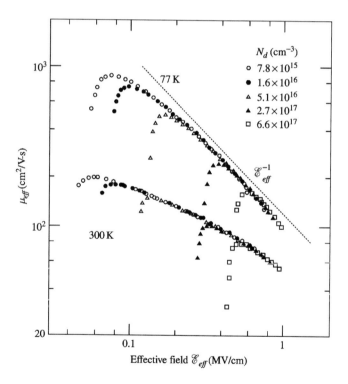

Figure 5.16 Measured hole mobility at 300 K and 77 K versus effective normal field (with a factor 1/3) for several substrate doping concentrations (after Takagi *et al.*, 1988)

which has been found necessary in order for the measured hole mobilities to fall on a universal curve when plotted against $\mathscr{E}_{eff}$ (Arora and Gildenblat, 1987). Note that the factor 1/3 is entirely empirical with no physical basis behind it. Like electron mobility, hole mobility is also influenced by Coulomb scattering at low fields, depending on the doping concentration. The field dependence is also stronger at 77 K, but not quite as strong as in the electron case. It should be noted that the hole mobility data were taken from surface-channel pMOSFETs with p^+ polysilicon gate.

At higher temperatures, MOSFET channel mobility decreases because of increased phonon scattering. The temperature dependence is similar to that of bulk mobility discussed in Section 2.3.1, i.e., $\mu_{eff} \propto T^{-3/2}$.

5.2.2   Strain Effect on Mobility

It has long been established that strain, either tensile or compressive, alters the band structure of silicon, hence affects the electron and hole mobilities. For the past two decades or so, the IC industry has taken advantage of this effect to engineer various strains in n- and pMOSFETs to better their performance. Generally speaking, strain breaks the crystal symmetry of an unstrained silicon thereby lifting the degeneracy of the conduction band and the valence band. It can be engineered such that the band or subband with a lighter effective mass thus a higher mobility becomes more populated. Lifting the degeneracy also reduces the inter-valley scattering. In some cases, the band curvature is also changed by strain to that of a lighter effective mass than the unstrained case. There are two types of strain employed in silicon: biaxial strain and uniaxial strain, as discussed in Sections 5.2.2.1 and 5.2.2.2. The default wafer orientation is in the <100> plane.

5.2.2.1   Biaxial Strain

As shown in Figure 5.17(a), biaxial strain means stretching or compressing in the <100> wafer plane two-dimensionally. This is usually achieved in heterostructures at a wafer level. For example, SiGe has a larger lattice constant than silicon when both are unstrained. If a thin SiGe film is grown epitaxially on a silicon wafer, the SiGe film is compressed biaxially. Conversely, biaxial tensile strain is present in a thin epitaxial silicon film grown on top of a relaxed SiGe buffer layer (Rim *et al.*, 1995).

The constant-energy ellipsoids of the silicon conduction band with six-fold degeneracy in a 3-D crystal are shown in Figure 5.18. Section 4.4 discussed that in an MOS

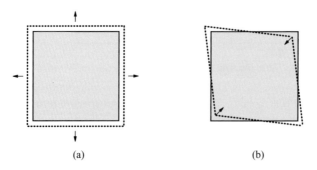

(a)   (b)

Figure 5.17 Schematic drawing of (a) biaxial tensile strain in an *x*–*y* plane perpendicular to [001], and (b) uniaxial compressive strain along a [110] direction (after Sun *et al.*, 2007)

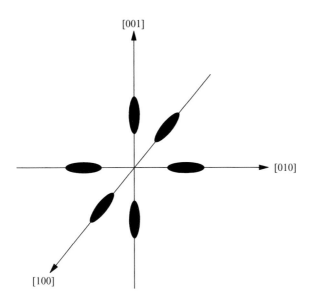

Figure 5.18 Constant energy surfaces of a silicon conduction band in unstrained state

inversion layer, the six-fold degeneracy breaks into a Δ_2 valley and a Δ_4 valley. The lower energy state is the Δ_2 valley which has a confinement effective mass of $m_l = 0.92m_0$, larger than the confinement effective mass of $m_t = 0.19m_0$ of the Δ_4 valley.

Biaxial tensile strain is beneficial to electron mobility since it lifts the six-fold degeneracy of the conduction band and moves the Δ_2 valley below the Δ_4 valley. For the MOS inversion layer, this further separates the energy difference between the two valleys. A more populated Δ_2 valley results in a higher overall mobility because its effective mass in the transport direction is $m_t = 0.19m_0$, lower than that of the Δ_4 valley.

In the unstrained state, the valence band of silicon is also degenerate, consisting of a heavy-hole band (HH) and a light-hole band (LH) at the Γ-point ($k = 0$). Tensile strain moves the LH band up while compressive strain moves the HH band up in energy. In this case, both types of strain are beneficial to the hole mobility because the curvature of the top band is also altered, resulting in a lighter effective mass.

5.2.2.2 Uniaxial Strain

Uniaxial compressive strain in the $<110>$ direction is shown schematically in Figure 5.17(b). It is process induced, usually along the channel direction in $<110>$ (Thompson *et al.*, 2006). In early experimental reports, an etch stop nitride layer over the device area exerts tensile strain which increases the electron mobility in nMOSFETs. For pMOSFETs, selectively grown epitaxial SiGe source-drain regions are shown to produce compressive strain on the channel and raise the hole mobility (Gannavaram *et al.*, 2000). Wafer bending experiments have demonstrated that uniaxial compressive strain along $<110>$ is much more effective than along $<100>$ in improving the hole mobility (Uchida *et al.*, 2004). In a more elaborate scheme, the nitride film is engineered to deliver either tensile or compressive strain on the device. By masking and Ge implantation, the nitride film can be made tensile on nMOSFETs and compressive on pMOSFETs for best effects (Shimizu *et al.*, 2001).

Overall, the VLSI industry has utilized strain to gain about 10–25% on the drive current of nMOSFET and 50% or more on the drive current of pMOSFET. The higher gain on the latter helps to mitigate the mismatch between n- and pMOSFETs in a CMOS circuit.

5.3   MOSFET Threshold Voltage

Threshold voltage is among the most important parameters of a MOSFET device. In this section, we examine the substrate bias, temperature, and quantum effects on threshold voltage.

5.3.1   Substrate Sensitivity (Body Effect)

The drain current equations in Section 5.1.3 were derived assuming zero substrate bias (V_{bs}). If $V_{bs} \neq 0$, the previously discussed MOSFET equations can be modified by considering that applying V_{bs} to the substrate is equivalent to subtracting all other voltages, namely, gate, source, and drain voltages, by V_{bs} while keeping the substrate grounded. This is shown in Figure 5.19.

Using the charge sheet model with the regional approximation $\psi_s = 2\psi_B + V$ as before, Eq. (5.16) becomes

$$Q_i = -C_{ox}\left(V_{gs} - V_{bs} - V_{fb} - 2\psi_B - V\right) + \sqrt{2\varepsilon_{si}qN_a(2\psi_B + V)}, \tag{5.57}$$

where V is the reverse bias voltage between a point in the channel and the substrate.

The current is obtained by integrating Q_i from $V = -V_{bs}$ (source) to $V_{ds} - V_{bs}$ (drain):

$$I_{ds} = \mu_{eff}C_{ox}\frac{W}{L}\left\{\left(V_{gs} - V_{fb} - 2\psi_B - \frac{V_{ds}}{2}\right)V_{ds}\right.$$
$$\left. - \frac{2\sqrt{2\varepsilon_{si}qN_a}}{3C_{ox}}\left[(2\psi_B - V_{bs} + V_{ds})^{3/2} - (2\psi_B - V_{bs})^{3/2}\right]\right\}. \tag{5.58}$$

At low drain voltages (linear region), the current is still given by Eq. (5.23), except that the threshold voltage is now

$$V_t = V_{fb} + 2\psi_B + \frac{\sqrt{2\varepsilon_{si}qN_a(2\psi_B - V_{bs})}}{C_{ox}}. \tag{5.59}$$

Thus **the effect of a reverse substrate bias ($V_{bs} < 0$) is to widen the bulk depletion region and raise the threshold voltage**. Figure 5.20 plots V_t as a function of $-V_{bs}$. The slope of the curve,

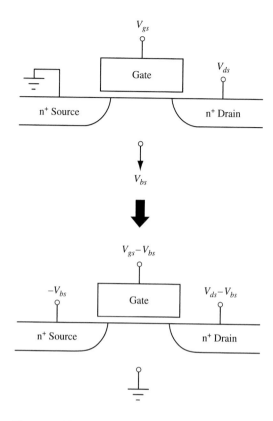

Figure 5.19 Equivalent circuits for considering the effect of substrate bias on MOSFET *I–V* characteristics

$$\frac{dV_t}{d(-V_{bs})} = \frac{\sqrt{\varepsilon_{si}qN_a/[2(2\psi_B - V_{bs})]}}{C_{ox}}, \tag{5.60}$$

is referred to as the *substrate sensitivity*. At $V_{bs} = 0$, the slope equals C_{dm}/C_{ox}, or $m - 1$ [Eq. (5.28)]. The substrate sensitivity is higher for higher bulk doping concentrations. It is clear from Figure 5.20 that the substrate sensitivity decreases as the substrate reverse bias increases. Based on Eq. (5.40), a reverse substrate bias also makes the subthreshold slope slightly steeper, since it widens the depletion region and lowers C_{dm}.

5.3.2   Temperature Dependence of Threshold Voltage

Next, we examine the temperature dependence of threshold voltage. The flat-band voltage of an nMOSFET with n⁺ polysilicon gate is $V_{fb} = -E_g/2q - \psi_B$ [Eq. (4.48)], assuming there is no oxide charge. Substituting it into Eq. (5.24) yields the threshold voltage,

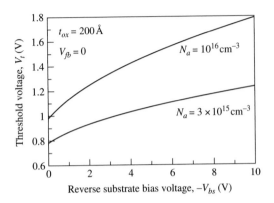

Figure 5.20 Threshold-voltage variation with reverse substrate bias for two uniform substrate doping concentrations

$$V_t = -\frac{E_g}{2q} + \psi_B + \frac{\sqrt{4\varepsilon_{si}qN_a\psi_B}}{C_{ox}}, \tag{5.61}$$

at zero substrate bias. The temperature dependence of V_t is related to the temperature dependence of E_g and ψ_B:

$$\begin{aligned} \frac{dV_t}{dT} &= -\frac{1}{2q}\frac{dE_g}{dT} + \left(1 + \frac{\sqrt{\varepsilon_{si}qN_a/\psi_B}}{C_{ox}}\right)\frac{d\psi_B}{dT} \\ &= -\frac{1}{2q}\frac{dE_g}{dT} + (2m-1)\frac{d\psi_B}{dT}. \end{aligned} \tag{5.62}$$

$d\psi_B/dT$ stems from the temperature dependence of intrinsic carrier concentration, which can be evaluated using Eqs. (2.62) and (2.13):

$$\begin{aligned} \frac{d\psi_B}{dT} &= \frac{d}{dT}\left[\frac{kT}{q}\ln\left(\frac{N_a}{\sqrt{N_cN_v}e^{-E_g/2kT}}\right)\right] \\ &= -\frac{k}{q}\ln\left(\frac{\sqrt{N_cN_v}}{N_a}\right) - \frac{kT}{q\sqrt{N_cN_v}}\frac{d\sqrt{N_cN_v}}{dT} + \frac{1}{2q}\frac{dE_g}{dT}. \end{aligned} \tag{5.63}$$

Since both N_c and N_v are proportional to $T^{3/2}$, we have $d(N_cN_v)^{1/2}/dT = \frac{3}{2}(N_cN_v)^{1/2}/T$. Substituting Eq. (5.63) into Eq. (5.62) yields

$$\frac{dV_t}{dT} = -(2m-1)\frac{k}{q}\left[\ln\left(\frac{\sqrt{N_cN_v}}{N_a}\right) + \frac{3}{2}\right] + \frac{m-1}{q}\frac{dE_g}{dT}. \tag{5.64}$$

From Section 2.1.1 and Table 2.1, $dE_g/dT \approx -2.7 \times 10^{-4}$ eV/K and $(N_cN_v)^{1/2} \approx 3 \times 10^{19}$ cm^{-3}. For $N_a \sim 10^{16}$ cm^{-3} and $m \approx 1.1$, *dV$_t$/dT is typically -1 mV/K.* Note that the temperature coefficient decreases slightly as N_a increases: for $N_a \sim 10^{18}$ cm^{-3} and $m \approx 1.3$, dV_t/dT is about -0.7 mV/K. These numbers imply that, at an elevated temperature of, for example, 100 °C, the threshold voltage is 55–75 mV lower than the room temperature value. Since digital VLSI circuits often operate at elevated temperatures due to heat generation, this effect, plus the degradation of subthreshold slope with temperature, causes the leakage current at $V_{gs} = 0$ to increase considerably over its room-temperature value. Typically, the off-state leakage current of a MOSFET at $100\,°C$ is 20–50-times larger than the leakage current at $25\,°C$. These are important design considerations, to be addressed in Section 6.3.

5.3.3   Quantum Effect on Threshold Voltage

At room temperature, the quantum effect is weak when the surface electric field $\mathscr{E}_s$ is below $10^4 – 10^5$ V/cm. Both the lowest energy level E_0 and the spacings between the subbands are comparable to or less than kT. A large number of subbands are occupied. It was shown in Section 4.4.3.4 that in this case, Q_i^{QM} is essentially the same as the classical inversion charge density per unit area given by Eq. (5.35) for the subthreshold region,[3]

$$Q_i = \frac{kTn_i^2}{\mathscr{E}_s N_a} e^{q\psi_s/kT}. \qquad (5.65)$$

When $\mathscr{E}_s > 10^5$ V/cm, however, the subband spacings become greater than kT and Q_i^{QM} is significantly less than Q_i. *The $Q_i^{QM} - \psi_s$ curve [Eq. (4.68)] exhibits a positive parallel shift with respect to the classical $Q_i - \psi_s$ curve [Eq. (5.65)] on a semilogarithmic scale, which means that additional band bending is needed to achieve the same inversion charge per unit area as the classical value.* The classical threshold condition, $\psi_s = 2\psi_B$, should therefore be modified to $\psi_s = 2\psi_B + \Delta\psi_s^{QM}$, where $Q_i^{QM}\left(\psi_s = 2\psi_B + \Delta\psi_s^{QM}\right) = Q_i(\psi_s = 2\psi_B)$. From this definition,

$$\Delta\psi_s^{QM} = \frac{kT}{q} \ln\left(\frac{Q_i(\psi_s = 0)}{Q_i^{QM}(\psi_s = 0)}\right) \qquad (5.66)$$

can be evaluated from the preexponential factors in Eqs. (5.65) and (4.68). Figure 5.21 shows the calculated $\Delta\psi_s^{QM}$ as a function of $\mathscr{E}_s$. Beyond 10^6 V/cm, most of the electrons are in the lowest subband. Equation (5.66) then becomes

$$\Delta\psi_s^{QM} \approx \frac{E_0}{q} - \frac{kT}{q} \ln\left(\frac{8\pi q m_t \mathscr{E}_s}{h^2 N_c}\right), \qquad (5.67)$$

shown in Figure 5.21 as the dotted curve. Knowing $\Delta\psi_s^{QM}$, we can easily calculate the threshold voltage shift due to quantum effect:

$$\Delta V_t^{QM} = \frac{dV_{gs}}{d\psi_s} \Delta\psi_s^{QM} = m\Delta\psi_s^{QM}, \qquad (5.68)$$

where $m = 1 + (3t_{ox}/W_{dm})$ as before.

---

[3] This expression can be generalized to cover nonuniformly doped cases where $\mathscr{E}_s$ is the electric field at the surface and N_a is the doping concentration at the edge of the depletion layer.

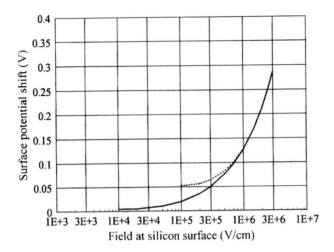

Figure 5.21 Additional band bending $\Delta\psi_s^{QM}$ (over the classical $2\psi_B$ value) required for reaching the threshold condition as a function of the surface electric field. The dotted curve is calculated by keeping only the lowest term (twofold degeneracy) in Eq. (4.68)

As an example, consider a 50 nm MOSFET with a uniform doping of $N_a = 3 \times 10^{18}\,\mathrm{cm}^{-3}$, which gives $W_{dm} = 20$ nm for control of the short-channel effect (see Section 6.1). For this device, $\mathscr{E}_s \approx 10^6$ V/cm, so $\Delta\psi_s^{QM} = 0.13$ V from Figure 5.21. If $m = 1.3$, then $\Delta V_t^{QM} = 0.17\,\mathrm{V}$, resulting in a significantly higher threshold voltage than the classical value. A retrograde doping profile (to be discussed in Section 6.3.2) not only reduces the depletion charge density (for a given W_{dm}) but also lowers the surface field, hence ΔV_t^{QM}.

5.4   MOSFET Capacitance

The capacitance of a MOSFET has a major impact on the switching delay of a logic gate. For a given current, the capacitance affects how fast the gate can be charged (or discharged) to a certain voltage which turns on (or off) the source-to-drain current of the receiving stage. MOSFET capacitances can be divided into two main categories: intrinsic capacitances and parasitic capacitances. This section focuses on the intrinsic MOSFET capacitances arising from the inversion and depletion charges in the channel region. Parasitic capacitances are discussed in Section 8.3. Similar to the earlier discussion on drain currents, gate capacitances are also considered separately in the three regions of MOSFET operation: subthreshold region, linear region, and saturation region (Figure 5.8).

- *Subthreshold region.* In the subthreshold region, the inversion charge is negligible. Only the depletion charge changes with the gate potential. Therefore, the intrinsic gate-to-source–drain capacitance is essentially zero (the extrinsic gate-to-source–drain overlap capacitance is discussed in Section 8.3), while the gate-to-body capacitance is given by the serial combination of C_{ox} and C_d (Figure 4.13), i.e.,

$$C_G = WL\left(\frac{1}{C_{ox}} + \frac{1}{C_d}\right)^{-1} \approx WLC_d, \tag{5.69}$$

where C_d is the depletion capacitance per unit area given by Eq. (4.38). Here, the upper-case subscript is used to distinguish the total gate capacitance C_G from the gate capacitance per unit area, C_g.

- *Linear region.* Once the surface channel forms, there is no more capacitive coupling between the gate and the body due to screening by the inversion charge. All the gate capacitances are to the channel, i.e., to the source and drain terminals. Within the framework of the regional charge-sheet model, Eq. (5.32), the inversion charge density Q_i at low drain biases varies linearly from $-C_{ox}(V_{gs} - V_t)$ at the source end to $-C_{ox}(V_{gs} - V_t - mV_{ds})$ at the drain end where $V_{ds} << (V_{gs} - V_t)/m$. The gate-to-channel capacitance is simply given by the oxide capacitance,

$$C_G = WLC_{ox}. \tag{5.70}$$

- *Saturation region.* When V_{ds} is appreciable, the inversion charge density $Q_i(V)$ is given by $-C_{ox}(V_{gs} - V_t - mV)$ [Eq. (5.31)] and shown in Figure 5.7. At the pinch-off condition, $V_{ds} = V_{dsat} = (V_{gs} - V_t)/m$, and $Q_i = 0$ at the drain. The total inversion charge in the MOSFET is obtained by integrating in the length and width directions, $Q_I = W \int_0^L Q_i(V)dy$. The integration over dy can be converted to integration over dV using the current continuity, Eq. (5.8), $I_{ds} = -\mu_{eff}WQ_idV/dy$. Therefore,

$$Q_I = -\frac{\mu_{eff}W^2}{I_{ds}} \int_0^{V_{dsat}} Q_i^2(V)dV = -\frac{\mu_{eff}W^2C_{ox}^2}{3mI_{ds}}(V_{gs} - V_t)^3 = -\frac{2}{3}WLC_{ox}(V_{gs} - V_t). \tag{5.71}$$

The last step follows from Eq. (5.31). The gate-to-channel capacitance in the saturation region is then

$$C_G = \frac{2}{3}WLC_{ox}. \tag{5.72}$$

Exercises

5.1 Consider an n-channel MOSFET with 20 nm thick gate oxide and uniform p-type substrate doping of 10^{17} cm^{-3}. The gate work function is that of n$^+$ Si.

(a) What is the threshold voltage? Sketch the band diagram at threshold condition, $\psi_s = 2\psi_B$.

(b) What is the threshold voltage if a reverse bias of 1 V is applied to the substrate? Sketch the band diagram at threshold.

5.2   Fill in the steps that lead to Eq. (5.33), the fraction of drift current component in the limit of $V \to 0$.

5.3   The effective field $\mathscr{E}_{\text{eff}}$ plays an important role in MOSFET channel mobility. Show that the definition

$$\mathscr{E}_{\text{eff}} \equiv \frac{\int_0^{x_i} n(x)\mathscr{E}(x)dx}{\int_0^{x_i} n(x)dx}$$

leads to Eq (5.51), i.e., $\mathscr{E}_{\text{eff}} = (|Q_d| + |Q_i|/2)/\varepsilon_{si}$. Note that

$$|Q_i| = q \int_0^{x_i} n(x)dx$$

and

$$\mathscr{E}(x) = \frac{1}{\varepsilon_{si}} \left(|Q_d| + q \int_x^{x_i} n(x')dx' \right)$$

from Gauss's law. The inversion-layer depth x_i is assumed to be much smaller than the bulk depletion width.

5.4   An alternative threshold definition is based on the rate of change of inversion charge density with gate voltage. Equation (5.25) from Figure 4.13(b) states that $d|Q_i|/dV_{gs}$ is given by the serial combination of C_{ox} and $C_i \equiv d|Q_i|/d\psi_s$. Below threshold, $C_i \ll C_{ox}$, so that $d|Q_i|/dV_{gs} \approx C_i$ and $|Q_i|$ increases exponentially with V_{gs}. Above threshold, $C_i \gg C_{ox}$, so that $d|Q_i|/dV_{gs} \approx C_{ox}$ and $|Q_i|$ increases linearly with V_{gs}. The change of behavior occurs at an *inversion charge threshold voltage*, V_t^{inv}, where $C_i = C_{ox}$. Show that at $V_{gs} = V_t^{inv}$ one has $d|Q_i|/dV_{gs} = C_{ox}/2$ and $|Q_i| \approx (2kT/q)C_{ox}$. Note that such an inversion charge threshold is independent of depletion charge and is slightly higher than the conventional $2\psi_B$ threshold.

5.5   In the charge sheet model, the drain current is expressed in terms of $\psi_{s,s}$ and $\psi_{s,d}$, the surface potentials at the source end and the drain end of the channel. Compute $\psi_{s,s}$ and $\psi_{s,d}$ for an nMOSFET with a gate oxide thickness of 7 nm and p-type substrate doping of $2 \times 10^{17}/\text{cm}^3$, biased at $V_{gs} - V_{fb} = 2\,\text{V}$ and V_{ds} in saturation. (Assume Si, room temperature, complete ionization.)

5.6   Consider an nMOSFET on uniformly doped p-type substrate of $N_a = 2 \times 10^{17}/\text{cm}^3$. The gate oxide thickness is 10 nm. (Assume silicon, room temperature, complete ionization.)
(a)   What is the inverse subthreshold current slope of this device?
(b)   If the gate is biased above the threshold such that the total charge density in silicon at the source is $-10^{-6}\,\text{C/cm}^2$, how much of that is the inversion charge density?
(c)   Estimate the effective mobility for the bias condition in (b).

5.7   If instead of integrating from 0 to L and 0 to V_{ds}, the current continuity Eq. (5.8) is integrated from 0 to y and 0 to V with $Q_i(V)$ given by Eq. (5.31), a relationship between y and V is obtained. Solve for $V(y)$ explicitly given Eq. (5.27) for I_{ds}.

5.8 Strictly speaking, if $V_{ds} > 2\psi_B$, the last term of Eq. (5.22) cannot be expanded into a power series in V_{ds}. A more exact form of the saturation voltage is obtained by letting $Q_i = 0$ in Eq. (5.16) and solve for $\psi_s = 2\psi_B + V_{dsat}$. Derive this V_{dsat} as a function of V_{gs} and estimate how much difference it makes on I_{dsat} with respect to that of Eq. (5.30).

5.9 Show that in the subthreshold region when the drain bias is low, Eq. (5.12) leads to Eq. (5.65):

$$-Q_i = \frac{kTn_i^2}{\mathcal{E}_s N_a} e^{q\psi_s/kT},$$

where ψ_s is the surface potential and $\mathcal{E}_s$ is the surface electric field. This equation is more general than Eq. (5.35) since it is valid for nonuniform (vertically) dopings with N_a being the p-type concentration at the edge of the depletion layer. (Note that the factor N_a merely reflects the fact in Figure 4.6 that the band bending ψ_s is defined with respect to the bands of the neutral bulk region of doping N_a.)

5.10 In a short-channel device or in a nonuniformly doped (laterally) MOSFET, ψ_s may vary along the channel length direction from the source to drain. Generalize the expression in Exercise 5.9 and show that

$$\frac{V_{ds}}{I_{ds}} = \frac{1}{\mu_{eff} W} \int_0^L \frac{dy}{Q_i(y)} = \frac{N_a}{\mu_{eff} W k T n_i^2} \int_0^L \mathcal{E}_s(y) e^{-q\psi_s(y)/kT} dy$$

for the subthreshold region at low drain biases. Since $\mathcal{E}_s(y) \approx [V_{gs} - V_{fb} - \psi_s(y)]/3t_{ox}$ is not a strong function of ψ_s, the exponential factor dominates. This implies that the subthreshold current is controlled by the point of highest barrier (lowest ψ_s) in the channel.

5.11 Figure 5.3 shows that for a given V_{gs} above threshold, the solution of $\psi_s(V)$ from Eq. (5.14) saturates as $V \to \infty$. Derive an expression for this maximum ψ_s as a function of V_{gs}.

6 MOSFETs
Short Channel

From the discussion of long-channel MOSFETs in Chapter 5, it is clear that for a given supply voltage, MOSFET current increases with decreasing channel length. The intrinsic capacitance of a short-channel MOSFET is also lower, making it easier to switch. However, for a given process, channel length cannot be arbitrarily shortened even if within the lithography capability. This chapter covers the key aspects of short-channel devices that are important for device design considerations. Discussed first is the short-channel effect, followed by velocity saturation.

6.1 Short-Channel Effect

6.1.1 Threshold Voltage Roll-off

Short-channel effect (SCE) is the decrease of MOSFET threshold voltage as the channel length is reduced below a certain value. An example is shown in Figure 6.1 (Taur *et al.*, 1985). The short-channel effect is especially pronounced when the drain is biased at a voltage equal to that of the power supply (high drain bias). In a CMOS VLSI technology, channel length varies statistically from chip to chip, wafer to wafer, and lot to lot due to process tolerances. The short-channel effect is therefore an important consideration in device design; one must ensure that the threshold voltage does not become too low for the minimum-channel-length device on the chip.

The key difference between a short-channel and a long-channel MOSFET is that the field pattern in the depletion region of a short-channel MOSFET is two-dimensional, as shown in Figure 6.2. The constant-potential contours in a long-channel device in Figure 6.2(a) are largely parallel to the oxide–silicon interface or along the channel length direction (y-axis), so that the electric field is one-dimensional (in the vertical direction or along the x-axis) over most part of the device. The constant-potential contours in a short-channel device in Figure 6.2(b), however, are more curvilinear, and the resulting electric field pattern is of a two-dimensional nature. In other words, both the x- and y-components of the electric field are appreciable in a short-channel MOSFET. It is also important to note that, for a given gate voltage, the surface potential is higher in a short-channel device than in a long-channel device. Specifically, the maximum surface potential is slightly over 0.65 V (the fourth contour

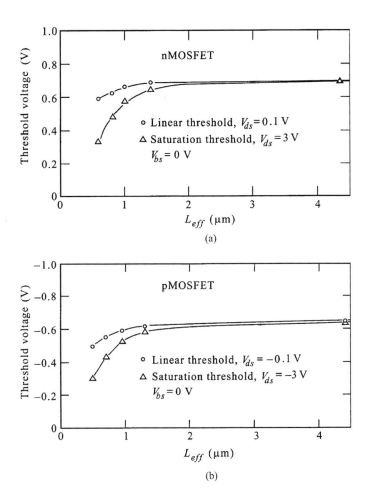

Figure 6.1 Short-channel threshold rolloff: Measured low- and high-drain threshold voltages of (a) n- and (b) p-MOSFETs versus channel length from a 1-μm CMOS technology (after Taur *et al.*, 1985)

from the bottom) in Figure 6.2(b), but below 0.65 V in Figure 6.2(a). This points to a lower threshold voltage in the short-channel MOSFET.

The two-dimensional field pattern in a short-channel device arises from the proximity of source and drain regions. Just like the depletion region under an MOS gate, there are also depletion regions surrounding the source and drain junctions (see Figure 4.20). In a long-channel device, the source and drain are far enough separated that their depletion regions have no effect on the potential or field pattern in most parts of the device. *In a short-channel device, however, the source–drain distance is comparable to the MOS depletion width in the vertical direction, hence the source–drain potential has a strong effect on the band bending over a significant portion of the device.*

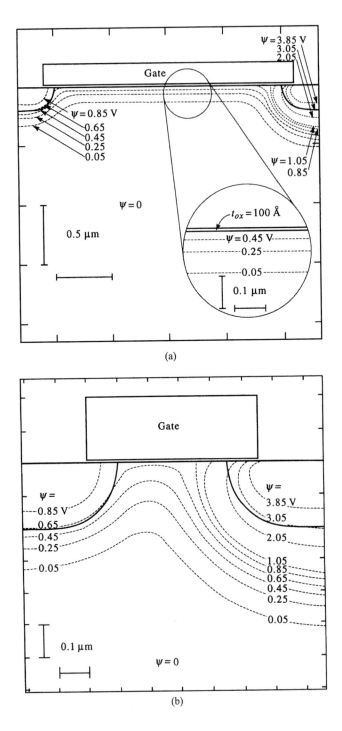

Figure 6.2 Simulated constant potential contours of (a) a long-channel and (b) a short-channel nMOSFET. The contours are labeled by the potential or band bending with respect to the neutral p-type region. The solid lines indicate the location of the source and drain junctions (metallurgical). The drain is biased at 3.0 V. Both devices are biased at the same gate voltage slightly below the threshold

6.1.1.1   Drain-Induced Barrier Lowering

The physics of the short-channel effect can be understood from another angle in Figure 6.3 by considering the potential barrier (to electrons for an n-channel MOSFET) at the surface between the source and drain. Under the off condition, this potential barrier (p-type region) prevents the electrons in the source from flowing to the drain. The height of the surface barrier is mainly controlled by the gate voltage. When the gate voltage is below the threshold voltage, there are only a limited number of electrons injected from the source over the barrier and collected by the drain (subthreshold current). In the long-channel case, the potential barrier is flat over most parts of the device. Source and drain fields only affect the very ends of the channel. As the channel length is shortened, however, the source and drain fields penetrate into the middle of the channel, which lowers the potential barrier between the source and drain. This leads to a substantial increase of the subthreshold current. In other words, the threshold voltage becomes lower than the long-channel value. The region of maximum potential barrier also shrinks to a point toward the source.

For a long-channel MOSFET, the barrier height for the most part is completely controlled by the gate voltage, independent of the drain voltage. *For a short-channel MOSFET, however, the maximum barrier height is also sensitive to the drain voltage, namely, the barrier height decreases with increasing drain voltage.* This effect is referred to as *drain-induced barrier lowering* (DIBL). It explains the experimentally observed increase of subthreshold current with drain voltage in a short-channel MOSFET. Figure 6.4 shows the subthreshold characteristics of long- and short-channel devices with different drain bias voltages. For long-channel devices, the subthreshold current is independent of drain voltage ($\geq 2\ kT/q$), as expected from Eq. (5.39). For short-channel devices, however, there is a negative shift of the curve to a lower threshold voltage. The shift is drain voltage dependent, with a larger shift under high drain bias conditions. At even shorter channel lengths, the subthreshold slope starts to degrade as the surface potential is more controlled by the drain than by the gate.

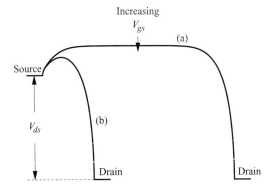

Figure 6.3 Conduction band energy at surface versus lateral distance from the source to the drain for (a) a long-channel MOSFET and (b) a short-channel MOSFET under the same applied gate voltage (in subthreshold)

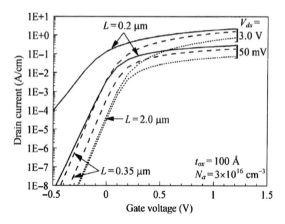

Figure 6.4 Subthreshold characteristics of long- and short-channel devices at low and high drain bias in a 0.35-μm technology

Eventually, the *punch-through* condition is reached in which the gate totally loses control of the channel and high drain current persists independent of gate voltage.

6.1.2   Analytic Solutions to 2-D Poisson's Equation in Subthreshold

In this section, an analytic approach is described for solving the 2-D Poisson's equation of electrostatic potential subject to the source and drain boundary conditions. Subthreshold region is assumed in which the mobile charge density is negligible. This keeps the partial differential equation linear and decouples it from the current continuity equation involving the quasi-Fermi potential in the channel.

6.1.2.1   A Simplified Rectangular Geometry

Figure 6.5 shows a simplified rectangular geometry for solving the 2-D Poisson's equation (Nguyen, 1984). A potential function $\psi(x, y)$ is to be solved for every point (x, y) inside the rectangle *BEGH* given the values of ψ along the four boundaries. The width of the rectangle is given by the channel length or the distance between the n⁺ source and n⁺ drain, L. Both the source and drain junctions are assumed to be abrupt. The height of the rectangle consists of t_{ox}, the thickness of gate oxide, and W_d, the gate depletion depth in silicon. The depth of the source and drain junctions is assumed to be deeper than W_d, usually the case in most CMOS technologies. Since in general, the permittivity of the oxide region *AFGH*, ε_{ox}, differs from that of the silicon region *ABEF*, ε_{si}, it is necessary to define two potential functions: $\psi_1(x, y)$ for the oxide region and $\psi_2(x, y)$ for the silicon region. Along the interface *AF*, the derivatives of ψ_1 and ψ_2 or the fields satisfy the 2-D dielectric boundary conditions discussed in Section 2.4.1.1. Thus for the oxide region, *AFGH*,

$$\frac{\partial^2 \psi_1}{\partial x^2} + \frac{\partial^2 \psi_1}{\partial y^2} = 0, \tag{6.1}$$

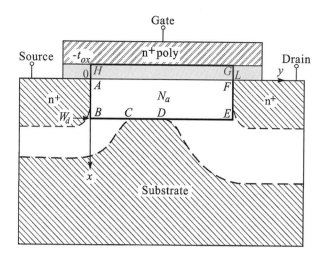

Figure 6.5 Simplified geometry for analytically solving Poisson's equation in a short-channel MOSFET. The x-axis is in the vertical direction into the depth of silicon, y-axis is in the lateral direction along the channel. The origin is point A at the silicon surface and the source boundary. The hashed areas represent conductor-like regions of constant potential (after Nguyen, 1984)

and for the silicon region, $ABEF$,

$$\frac{\partial^2 \psi_2}{\partial x^2} + \frac{\partial^2 \psi_2}{\partial y^2} = \frac{qN_a}{\varepsilon_{si}}, \tag{6.2}$$

where N_a is the uniform p-type doping of the substrate. Note that there is only the depletion charge density on the right-hand side of Eq. (6.2), compared to Eq. (4.9). This is only valid when the gate is biased below threshold.

The four boundary conditions are

$$\text{Gate}: \ \psi_1(-t_{ox}, y) = V_{gs} - V_{fb} \ \text{ along } GH, \tag{6.3}$$

$$\text{Source}: \ \psi_2(x, 0) = \psi_{bi} \qquad \text{ along } AB, \tag{6.4}$$

$$\text{Drain}: \ \psi_2(x, L) = \psi_{bi} + V_{ds} \qquad \text{ along } EF, \tag{6.5}$$

$$\text{Substrate}: \ \psi_2(W_d, y) = 0 \qquad \text{ along } CD. \tag{6.6}$$

As with the 1-D MOS case in Section 4.2.1, $\psi_2(x, y) = \psi_i(x, y) - \psi_i(x = \infty, y)$ is defined as the intrinsic potential at a point (x, y) with respect to the intrinsic potential of the p-type substrate. V_{gs} and V_{ds} are the gate-to-source and drain-to-source voltages, V_{fb} is the flatband voltage dependent on the gate work function, and ψ_{bi} is the built-in potential of the source or drain-to-substrate junction. For an abrupt n^+–p junction, $\psi_{bi} = E_g/2q + \psi_B$, where $\psi_B = (kT/q)\ln(N_a/n_i)$. If there is a substrate bias V_{bs}, then ψ_{bi} is replaced by $\psi_{bi} - V_{bs}$, and V_{gs} by $V_{gs} - V_{bs}$. The bottom boundary is actually a movable one, as W_d is dependent on the gate voltage V_{gs}. For a given V_{gs}, the rectangle is fixed. There are four segments on the boundary where the potentials are not specified in Eqs. (6.3)–(6.6). For BC and DE, we simply assume the same condition as CD, i.e., we stretch Eq. (6.6) to be valid for the entire bottom boundary BE. The boundary conditions for the two small gaps FG and HA are obtained by linear interpolation between the end-point potentials.

6.1.2.2   Solving the Boundary-Value Problem by Superposition

The solution technique makes use of the superposition principle by breaking the electrostatic potential into the following terms:

$$\psi_1(x, y) = v_1(x) + u_{L1}(x, y) + u_{R1}(x, y) \tag{6.7}$$

for the oxide region, and

$$\psi_2(x, y) = v_2(x) + u_{L2}(x, y) + u_{R2}(x, y) \tag{6.8}$$

for the silicon region. Here, $v_1(x)$ and $v_2(x)$ are the solutions to the 1-D MOS as those discussed in Section 4.2.1.1. They satisfy the top and bottom boundary conditions, as well as the dielectric boundary conditions at the interface, $x = 0$. $u_{L1}, u_{R1}, u_{L2}, u_{R2}$ are all solutions to the homogeneous (Laplace) equation, and are zero at the top and bottom boundaries, respectively. In addition, u_{L1} and u_{L2} are zero on the right boundary and u_{R1} and u_{R2} are zero on the left boundary. They are chosen so that $v_1 + u_{L1}$ and $v_2 + u_{L2}$ satisfy the source boundary condition on the left, and $v_1 + u_{R1}$ and $v_2 + u_{R2}$ satisfy the drain boundary condition on the right. They must also satisfy the 2-D dielectric boundary conditions along $x = 0$.

For the 1-D MOS solution, the depletion approximation yields

$$v_1(x) = \psi_s - \frac{V_{gs} - V_{fb} - \psi_s}{t_{ox}} x \tag{6.9}$$

for the oxide region $-t_{ox} \leq x \leq 0$, and

$$v_2(x) = \psi_s \left(1 - \frac{x}{W_d}\right)^2 \tag{6.10}$$

for the silicon region $0 \leq x \leq W_d$. Here ψ_s is the long-channel surface potential dependent on V_{gs}. It is related to the depletion region depth W_d by

$$\psi_s = \frac{qN_aW_d^2}{2\varepsilon_{si}} \tag{6.11}$$

for Eq. (6.10) to satisfy the differential equation, Eq. (6.2). Note that Eq. (6.9) satisfies the top boundary condition Eq. (6.3), and Eq. (6.10) satisfies the bottom boundary condition, Eq. (6.6). v_1 and v_2 are continuous at $x=0$. The dielectric boundary condition requires that the displacement be continuous, i.e., $\varepsilon_{ox}dv_1/dx = \varepsilon_{si}dv_2/dx$ at $x = 0$, which yields a relationship between ψ_s and V_{gs}:

$$\varepsilon_{ox}\frac{V_{gs} - V_{fb} - \psi_s}{t_{ox}} = \varepsilon_{si}\frac{2\psi_s}{W_d} = \sqrt{2\varepsilon_{si}qN_a\psi_s}. \tag{6.12}$$

The last step made use of Eq. (6.11). Note that this is just the gate bias equation in subthreshold, Eq. (5.37), or Eq. (4.32), with the solution $\psi_s(V_{gs})$ given by Eq. (4.33). The rest of the solutions for $\psi_1(x, y)$ are the following series:

$$u_{L1}(x, y) = \sum_{n=1}^{\infty} b_{1,n} \frac{\sinh\left(\frac{\pi(L-y)}{\lambda_n}\right)}{\sinh\left(\frac{\pi L}{\lambda_n}\right)} \sin\left(\frac{\pi(x+t_{ox})}{\lambda_n}\right) \qquad (6.13)$$

$$u_{R1}(x, y) = \sum_{n=1}^{\infty} c_{1,n} \frac{\sinh\left(\frac{\pi y}{\lambda_n}\right)}{\sinh\left(\frac{\pi L}{\lambda_n}\right)} \sin\left(\frac{\pi(x+t_{ox})}{\lambda_n}\right). \qquad (6.14)$$

Note that every term of the u_{L1} and u_{R1} series satisfies Laplace's equation, e.g., $\partial^2 u_{L1}/\partial x^2 + \partial^2 u_{L1}/\partial y^2 = 0$, for any λ_n. Also note that u_{L1} vanishes on the top ($x = -t_{ox}$) and the right boundaries ($y = L$) while u_{R1} vanishes on the top and the left ($y = 0$) boundaries. Likewise, the left (source) and right (drain) components of $\psi_2(x, y)$ are

$$u_{L2}(x, y) = \sum_{n=1}^{\infty} b_{2,n} \frac{\sinh\left(\frac{\pi(L-y)}{\lambda_n}\right)}{\sinh\left(\frac{\pi L}{\lambda_n}\right)} \sin\left(\frac{\pi(W_d-x)}{\lambda_n}\right) \qquad (6.15)$$

$$u_{R2}(x, y) = \sum_{n=1}^{\infty} c_{2,n} \frac{\sinh\left(\frac{\pi y}{\lambda_n}\right)}{\sinh\left(\frac{\pi L}{\lambda_n}\right)} \sin\left(\frac{\pi(W_d-x)}{\lambda_n}\right), \qquad (6.16)$$

so that u_{L2} vanishes on the bottom ($x = W_d$) and the right boundaries ($y = L$) while u_{R2} vanishes on the bottom and the left ($y = 0$) boundaries.

At the common boundary shared by both $\psi_1(x, y)$ and $\psi_2(x, y)$, $x = 0$, the normal displacement, $\varepsilon \partial \psi/\partial x$, as well as the potential ψ (hence the tangential field, $\partial \psi/\partial y$) must be continuous from one dielectric medium to the other at any y. Because for each n, the sinh factor is a different function of y, every term in u_{L1} and $\varepsilon_{ox}\partial u_{L1}/\partial x$ must equal its counterpart in u_{L2} and $\varepsilon_{si}\partial u_{L2}/\partial x$ at $x = 0$. Therefore,

$$b_{1,n} \sin\left(\frac{\pi t_{ox}}{\lambda_n}\right) = b_{2,n} \sin\left(\frac{\pi W_d}{\lambda_n}\right) \qquad (6.17)$$

and

$$b_{1,n}\varepsilon_{ox} \cos\left(\frac{\pi t_{ox}}{\lambda_n}\right) = -b_{2,n}\varepsilon_{si} \cos\left(\frac{\pi W_d}{\lambda_n}\right). \qquad (6.18)$$

Similar relations in terms of $c_{1,n}$ and $c_{2,n}$ are obtained from u_{R1}, u_{R2}, and their derivatives. For nonzero solutions of $b_{1,n}$ and $b_{2,n}$, Eqs. (6.17) and (6.18) can be divided to yield an eigenvalue equation for λ_n:

$$\frac{1}{\varepsilon_{ox}} \tan\left(\frac{\pi t_{ox}}{\lambda_n}\right) + \frac{1}{\varepsilon_{si}} \tan\left(\frac{\pi W_d}{\lambda_n}\right) = 0. \tag{6.19}$$

This is an implicit equation for λ_n which has a series of discrete solutions, $\lambda_1, \lambda_2, \ldots$, etc., in descending order (Frank *et al.*, 1998).

The full 2-D potential solution in silicon is then

$$\psi_2(x,y) = \psi_s \left(1 - \frac{x}{W_d}\right)^2 + \sum_{n=1}^{\infty} \frac{b_{2,n} \sinh\left(\frac{\pi(L-y)}{\lambda_n}\right) + c_{2,n} \sinh\left(\frac{\pi y}{\lambda_n}\right)}{\sinh\left(\frac{\pi L}{\lambda_n}\right)} \sin\left(\frac{\pi(W_d - x)}{\lambda_n}\right) \tag{6.20}$$

The coefficients $b_{2,n}$ and $c_{2,n}$ are determined by the source and drain boundary conditions, Eqs. (6.4) and (6.5), in Section 6.1.2.4. Consider, e.g., the surface potential, $\psi_2(0, y)$. The first term on the right-hand side of Eq. (6.20) is the long channel surface potential ψ_s. The second term, the series, represents SCE which adds to ψ_s, thus making the surface potential of a short-channel device higher, i.e., the barrier height lower, than the long channel device. To gain more insight into the y-dependence of the SCE term, the approximation $\sinh(z) \approx \exp(z)/2$ for $z > 0$ is taken:

$$\frac{b_{2,n} \sinh\left(\frac{\pi(L-y)}{\lambda_n}\right) + c_{2,n} \sinh\left(\frac{\pi y}{\lambda_n}\right)}{\sinh\left(\frac{\pi L}{\lambda_n}\right)} \sin\left(\frac{\pi W_d}{\lambda_n}\right)$$

$$\approx \left[b_{2,n} \exp\left(\frac{-\pi y}{\lambda_n}\right) + c_{2,n} \exp\left(\frac{\pi(y-L)}{\lambda_n}\right)\right] \sin\left(\frac{\pi W_d}{\lambda_n}\right)$$

$$\geq 2\sqrt{b_{2,n} c_{2,n}} \exp\left(-\frac{\pi L}{2\lambda_n}\right) \sin\left(\frac{\pi W_d}{\lambda_n}\right). \tag{6.21}$$

This indicates that ***there is a point of minimum potential or maximum barrier in the channel***, much as that depicted in Figure 6.3(b). Because of the $\exp(-\pi L/2\lambda_n)$ factor, almost the entire short-channel contribution to the minimum potential comes from the leading term with the longest λ_n, i.e., λ_1. The point of minimum potential is at $y_{min} = \frac{L}{2} - \frac{\lambda_1}{2\pi} \ln\left(\frac{c_{2,1}}{b_{2,1}}\right)$, where the two exponential terms are equal.

6.1.2.3   MOSFET Scale Length

The longest eigenvalue is defined as the ***scale length***, $\lambda \equiv \lambda_1$, as the severity of SCE hinges on the ratio L/λ. In Section 6.1.2.4, after the pre-exponential factor is worked out, we will see that ***the minimum channel length is $L_{min} \approx 2\lambda$ for the SCE to be tolerable***. To remove the V_{gs} dependence of λ, W_d is taken as the maximum depletion width at the $2\psi_B$ condition, Eq. (4.24):

$$W_{dm} = \sqrt{\frac{4\varepsilon_{si} kT \ln(N_a/n_i)}{q^2 N_a}}. \tag{6.22}$$

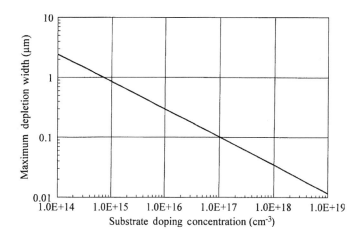

Figure 6.6 Depletion region width at the $2\psi_B$ threshold condition versus doping concentration for uniformly doped substrates

The justification is that the SCE of most practical interest is at the gate voltages not too far below the threshold voltage. Figure 6.6 plots W_{dm} versus N_a for a uniformly doped substrate. Generally, the slope is $-1/2$ on the log–log scale. To reduce the scale length λ, both t_{ox} and W_{dm} must be scaled down. Thus the substrate or body doping concentration has been going up as the CMOS technology moves to shorter and shorter channel lengths.

With the replacement of W_d by W_{dm} and λ_n by λ, ***the scale length equation,*** Eq. (6.19), ***becomes***

$$\frac{1}{\varepsilon_{ox}} tan\left(\frac{\pi t_{ox}}{\lambda}\right) + \frac{1}{\varepsilon_{si}} tan\left(\frac{\pi W_{dm}}{\lambda}\right) = 0. \tag{6.23}$$

This equation cannot be solved in explicit forms. The numerical solutions for the longest λ are shown in normalized units in Figure 6.7 for several representative values of $\varepsilon_{ox}/\varepsilon_{si}$. The following characteristics of the solution to Eq. (6.23) are observed in Figure 6.7.

- $\lambda \geq W_{dm}$ and $\lambda \geq t_{ox}$, i.e., λ is larger than the larger of W_{dm}, t_{ox}.
- In the special case of $\varepsilon_{ox} = \varepsilon_{si}$, $\lambda = W_{dm} + t_{ox}$, the physical height of the rectangular box in Figure 6.5.
- In the special case of $W_{dm} = t_{ox}$, $\lambda = 2W_{dm} = 2t_{ox}$, regardless of $\varepsilon_{ox}, \varepsilon_{si}$.
- If $t_{ox} \ll W_{dm}$ (lower right corner of Figure 6.7), $\lambda \approx W_{dm} + (\varepsilon_{si}/\varepsilon_{ox})t_{ox}$.
- If $W_{dm} \ll t_{ox}$ (upper left corner of Figure 6.7), $\lambda \approx t_{ox} + (\varepsilon_{ox}/\varepsilon_{si})W_{dm}$.

While Eq. (6.23) and thus Figure 6.7 are symmetric with respect to t_{ox} and W_{dm}, only the λ solutions for which $t_{ox} < W_{dm}$, or in the lower right region of Figure 6.7, is technologically feasible. Consideration of $\Delta V_{gs}/\Delta \psi_s$ or the m-factor in Figure 5.5 also requires that $t_{ox}/\varepsilon_{ox} < W_{dm}/\varepsilon_{si}$. In the lower right corner, $\lambda \approx W_{dm} + (\varepsilon_{si}/\varepsilon_{ox})t_{ox}$. High-$\kappa$ gate dielectric helps because for the same t_{ox}/ε_{ox} contribution to λ, higher ε_{ox} allows a thicker t_{ox} hence a much lower tunneling current. However, because of the extreme

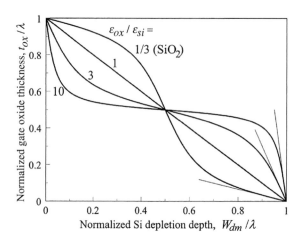

Figure 6.7 Numerical solutions to Eq. (6.23) for different values of $\varepsilon_{ox}/\varepsilon_{si}$. The dotted lines at the lower right corner depict the asymptotic solution behavior, $\lambda \approx W_{dm} + (\varepsilon_{si}/\varepsilon_{ox})t_{ox}$, for $t_{ox} \ll W_{dm}$

nonlinearity of the curves for $\varepsilon_{ox}/\varepsilon_{si} \gg 1$, the λ solution quickly departs from the above one-region, linear approximation (dotted lines in Figure 6.7) as t_{ox}/λ increases. There exists a limit of $t_{ox}/\lambda = \frac{1}{2}$, or $\lambda = 2t_{ox}$, where t_{ox} is the physical thickness of the insulator, no matter how high the dielectric constant is. Physically, this is owing to the lateral fields which, unlike the normal fields, are not related to the dielectric constant of the material (Section 2.4.1.1). **With very high-κ gate insulators, therefore, the physical thickness of the insulator would still have to be scaled down to contain the SCE due to lateral fields** (Xie et al., 2012). Note that such an effect is captured by the 2-D scale length equation, Eq. (6.23), but not by Eq. (4.109) of the 1-D EOT.

6.1.2.4   Short-Channel Potential Solution

The 2-D potential in silicon is given by Eq. (6.20) with coefficients $b_{2,n}$ and $c_{2,n}$ yet to be determined. For the oxide, the 2-D potential is

$$\psi_1(x,y) = \psi_s - \frac{V_{gs} - V_{fb} - \psi_s}{t_{ox}}x + \sum_{n=1}^{\infty} \frac{b_{1,n}\sinh\left(\frac{\pi(L-y)}{\lambda_n}\right) + c_{1,n}\sinh\left(\frac{\pi y}{\lambda_n}\right)}{\sinh\left(\frac{\pi L}{\lambda_n}\right)}\sin\left(\frac{\pi(x+t_{ox})}{\lambda_n}\right)$$

(6.24)

Note that $b_{1,n}$ and $b_{2,n}$ are related by Eq. (6.17). Likewise for $c_{1,n}$ and $c_{2,n}$.

At the source, $y = 0$, $\psi_2(x,0) = \psi_{bi}$ from Eq. (6.4). $\psi_1(x,0)$ is assumed to vary linearly between $\psi_1(-t_{ox},0) = V_{gs} - V_{fb}$ of Eq. (6.3) and $\psi_1(0,0) = \psi_2(0,0) = \psi_{bi}$, which gives

$$\psi_1(x,0) = \psi_{bi} - \frac{V_{gs} - V_{fb} - \psi_{bi}}{t_{ox}}x.$$

(6.25)

Matching of the solutions, Eqs. (6.20) and (6.24), with the boundary conditions, Eqs. (6.4) and (6.25), while setting W_d to W_{dm}, yields

$$\psi_2(x,0) = \psi_{bi} = \psi_s \left(1 - \frac{x}{W_{dm}}\right)^2 + \sum_{n=1}^{\infty} b_{2,n} \sin\left(\frac{\pi(W_{dm} - x)}{\lambda_n}\right) \qquad (6.26)$$

and

$$\psi_1(x,0) = \psi_{bi} - \frac{V_{gs} - V_{fb} - \psi_{bi}}{t_{ox}} x = \psi_s - \frac{V_{gs} - V_{fb} - \psi_s}{t_{ox}} x$$

$$+ \sum_{n=1}^{\infty} b_{2,n} \frac{\sin(\pi W_{dm}/\lambda_n)}{\sin(\pi t_{ox}/\lambda_n)} \sin\left(\frac{\pi(x + t_{ox})}{\lambda_n}\right), \qquad (6.27)$$

where Eq. (6.17) has been used to express $b_{1,n}$ in terms of $b_{2,n}$.

To evaluate the coefficients separately, the following orthogonality relation is utilized:

$$\varepsilon_{ox} \int_{-t_{ox}}^{0} \frac{\sin(\pi W_{dm}/\lambda_n)}{\sin(\pi t_{ox}/\lambda_n)} \sin\left(\frac{\pi(x + t_{ox})}{\lambda_n}\right) \frac{\sin(\pi W_{dm}/\lambda_m)}{\sin(\pi t_{ox}/\lambda_m)} \sin\left(\frac{\pi(x + t_{ox})}{\lambda_m}\right) dx$$

$$+ \varepsilon_{si} \int_{0}^{W_{dm}} \sin\left(\frac{\pi(W_{dm} - x)}{\lambda_n}\right) \sin\left(\frac{\pi(W_{dm} - x)}{\lambda_m}\right) dx = 0 \qquad \text{if } n \neq m. \quad (6.28)$$

$b_{2,n}$ is solved by multiplying Eqs. (6.26) and (6.27) with the corresponding factors, integrating over $(0, W_{dm})$ and $(-t_{ox}, 0)$ respectively, then adding them together:

$$b_{2,n} = \frac{2[(1 + A)\psi_{bi} - (A + B)\psi_s]}{C \times \pi}, \qquad (6.29)$$

where the dimensionless parameters are

$$A = \cos[\pi(1 - W_{dm}/\lambda_n)] \frac{\tan(\pi t_{ox}/\lambda_n)}{\pi t_{ox}/\lambda_n}, \qquad (6.30)$$

$$B = \frac{4\sin(\pi W_{dm}/2\lambda_n)}{\pi W_{dm}/\lambda_n}\left[\cos(\pi W_{dm}/2\lambda_n) - \frac{\sin(\pi W_{dm}/2\lambda_n)}{\pi W_{dm}/\lambda_n}\right], \qquad (6.31)$$

and

$$C = \frac{W_{dm}}{\lambda_n} - \frac{t_{ox}}{\lambda_n} \frac{\sin(2\pi W_{dm}/\lambda_n)}{\sin(2\pi t_{ox}/\lambda_n)}. \qquad (6.32)$$

Similarly, $c_{2,n}$ is evaluated from the drain boundary condition at $y = L$. The result can simply be obtained by replacing ψ_{bi} in Eq. (6.29) with $\psi_{bi} + V_{ds}$. Since $A \sim \lambda_n$, $B \sim \lambda_n$, and $C \sim 1/\lambda_n$, both $b_{2,n}$ and $c_{2,n}$ decrease with λ_n as n goes up.

Consider a typical case of $\varepsilon_{ox}/\varepsilon_{si} = 1/3$ (SiO$_2$), $t_{ox}/W_{dm} = 1/10$, so $m = 1 + (\varepsilon_{si}/\varepsilon_{ox}) t_{ox}/W_{dm} = 1.3$. The leading term λ_1 or λ is solved from Eq. (6.23) to be $1.26 \times W_{dm}$. Then $A = 0.81$, $B = -0.09$, and $C = 0.95$ for this case. These suggest simplifying Eqs. (6.29)–(6.32) by approximating $A \approx 1$, $B \approx 0$, and $C \approx 1$:

$$b_{2,1} \approx \frac{4}{\pi} [\psi_{bi} - 0.5\psi_s], \tag{6.33}$$

$$c_{2,1} \approx \frac{4}{\pi} [(\psi_{bi} + V_{ds}) - 0.5\psi_s]. \tag{6.34}$$

With the above, the minimum surface potential at $y_{min} = \frac{L}{2} - \frac{\lambda}{2\pi} \ln\left(1 + \frac{V_{ds}}{\psi_{bi} - 0.5\psi_s}\right)$, given by Eqs. (6.20) and (6.21), is then

$$\psi_2(0, y_{min}) \approx \psi_s + 2\sqrt{b_{2,1} c_{2,1}} \exp\left(-\frac{\pi L}{2\lambda}\right) \sin\left(\frac{\pi W_{dm}}{\lambda}\right)$$

$$\approx \psi_s + \frac{8}{\pi}\sqrt{(\psi_{bi} - 0.5\psi_s)(\psi_{bi} + V_{ds} - 0.5\psi_s)} \sin\left(\frac{\pi W_{dm}}{\lambda}\right) \exp\left(-\frac{\pi L}{2\lambda}\right). \tag{6.35}$$

Take $\psi_{bi} \sim \psi_s \sim V_{ds} \sim 1$ V, the pre-exponential factor is ~1.5 V. To limit the above SCE to below 0.1 V, the $\exp(-\pi L/2\lambda)$ factor needs to be $<1/20$. **A general guideline on the minimum channel length is therefore $L_{min} \approx 2\lambda$ for the SCE to be tolerable.**

The implication of the $-\psi_s$ dependence of the SCE term in Eq. (6.35) is noteworthy. First, SCE becomes more severe deeper into the subthreshold where ψ_s is lower. Second, the control of the minimum surface potential in a short-channel device by the gate voltage, $d\psi_2(0, y_{min})/dV_{gs}$, is reduced to below $d\psi_s/dV_{gs}$ because of the $-\psi_s$ factor in the SCE term. Since $d\psi_s/dV_{gs}$ gives the subthreshold slope of a long channel device, this means degradation of the short-channel subthreshold slope. Because the $\exp(-\pi L/2\lambda)$ factor is small at $L = L_{min}$, the slope degradation at L_{min} is not severe.

The minimum potential of Eq. (6.35) only gives a rough estimate on the short-channel effect on threshold voltage. More accurately, an I_{ds}–V_{gs} curve (in subthreshold) can be generated from the 2-D potential $\psi_2(x, y)$ of Eq. (6.20) and the current continuity equation,

$$I_{ds} = \mu_{eff} W(-Q_i) \frac{dV}{dy}. \tag{6.36}$$

Here, I_{ds} is independent of y and V is a function y only, independent of x. Q_i is the mobile charge density per area given by

$$-Q_i(y) = q \int_0^{W_{dm}} \frac{n_i^2}{N_a} \exp\left(\frac{q[\psi_2(x,y) - V]}{kT}\right) dx = q \frac{n_i^2}{N_a} \exp\left(-\frac{qV}{kT}\right) \int_0^{W_{dm}} \exp\left(\frac{q\psi_2(x,y)}{kT}\right) dx. \tag{6.37}$$

Equation (6.36) can thus be re-arranged and integrated with respect to y from 0 to L on one side and with respect to V from 0 to V_{ds} on the other side to yield

$$I_{ds} = \mu_{eff} kTW \frac{n_i^2}{N_a} \frac{1 - e^{-qV_{ds}/kT}}{\int_0^L \frac{dy}{\int_0^{W_{dm}} e^{q\psi_2(x,y)/kT} dx}}. \tag{6.38}$$

Without SCE, it can be shown that Eq. (6.38) is reduced to the long channel result, Eq. (5.36) [Exercise 6.3], for which $I_{ds} \propto W/L$. Threshold voltage roll-off is seen as a negative shift of the short-channel I_{ds}–V_{gs} curve (with W/L normalized to 1) on a semi-logarithmic scale with respect to that of the long channel device. The shift is no longer parallel when L is short enough as to degrade the subthreshold slope, as seen in the example in Figure 6.4.

In all of the above discussions, the source and drain junction depth, x_j, is assumed to be deeper than the depletion region width, W_{dm}. It led to the result that the source and drain junction depth does not play a role in SCE. It is the technologically relevant case since in practice it is difficult to scale down the junction depth without degrading the device current from increased series resistance. But if a MOSFET is made with $x_j < W_{dm}$, e.g., in a raised source–drain process, V_t rolloff can be linearly improved in proportion to x_j/W_{dm} (Sleva and Taur, 2005).

Another source–drain effect on SCE not covered in the above analytic model is the depletion in the source–drain region. When the channel length L is scaled to 20 nm or so, the source–drain depletion width, on the order of 2–3 nm for ~10^{20} cm^{-3} doping, becomes an appreciable fraction of the channel length and starts to have an effect on the SCE (Lin and Taur, 2017). Not surprisingly, lower doping softens the source–drain to channel boundary and lessens the SCE. The SCE predicted by the analytic scale length model is at the same level as that of a 10^{20} cm^{-3} source–drain doping. Too low a source–drain doping would result in a higher series resistance and lower device current.

6.2   High-Field Transport

6.2.1   Velocity Saturation

It has been discussed in Section 5.1.3 that in a long-channel MOSFET, the drain current saturates at a drain voltage equal to $V_{dsat} = (V_{gs} - V_t)/m$. ***In a short-channel MOSFET, the saturation of drain current occurs at a much lower voltage due to velocity saturation***. This causes the saturation current I_{dsat} to deviate from the $1/L$ dependence depicted in Eq. (5.31) for long-channel devices. Velocity–field relationships in bulk silicon are plotted in Figure 2.11. Saturation velocities of electrons and holes in a MOSFET channel are slightly lower than their bulk values. $v_{sat} \approx 7$–8×10^6 cm/s for electrons and $v_{sat} \approx 6$–7×10^6 cm/s for holes have been reported in the literature (Coen and Muller, 1980; Taur *et al.*, 1993a). Figure 6.8 shows the experimentally measured I_{ds}–V_{ds} curves of a 0.25-μm nMOSFET. The dashed curve represents the long-channel-like current given by Eq. (5.27) for $V_{gs} = 2.5$ V. Due to velocity saturation, the drain current saturates at a drain voltage much lower than $(V_{gs} - V_t)/m$, thus significantly limiting the saturation current of a short-channel device.

6.2.1.1   Velocity–Field Relationship

Experimental measurements show that the velocity–field relationship for electrons and holes takes the empirical form (Caughey and Thomas, 1967)

$$v = \frac{\mu_{eff}\mathscr{E}}{[1 + (\mathscr{E}/\mathscr{E}_c)^n]^{1/n}},$$ (6.39)

where $n = 2$ for electrons and $n = 1$ for holes. $n\ (\geq 1)$ is a measure of how rapidly the carriers approach saturation, as depicted in Figure 6.9. The parameter $\mathscr{E}_c$ is called the

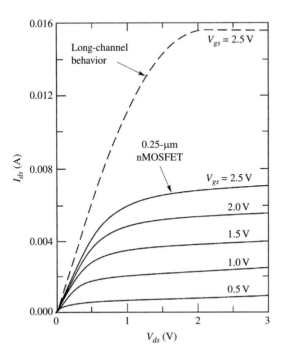

Figure 6.8 Experimental I–V curves of a 0.25-μm nMOSFET (solid lines). The device width is 9.5 μm. The dashed curve shows the long-channel-like drain current expected for this channel length if there were no velocity saturation (after Taur *et al.*, 1993a)

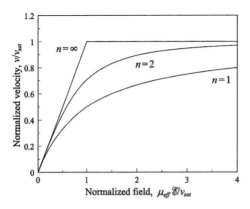

Figure 6.9 Velocity–field relationship of various velocity saturation models plotted in normalized units. The rate of approaching saturation velocity differs in different models

critical field. When the field strength is comparable to or greater than $\mathscr{E}_c$, velocity saturation becomes important. At low fields, $v = \mu_{eff} \mathscr{E}$, which is simply Ohm's law. As $\mathscr{E} \to \infty$, $v = v_{sat} = \mu_{eff} \mathscr{E}_c$. Therefore,

$$\mathscr{E}_c = \frac{v_{sat}}{\mu_{eff}}. \tag{6.40}$$

It was discussed in Section 5.2.1 that the effective mobility μ_{eff} is a function of the vertical (or normal) field $\mathscr{E}_{eff}$. Since v_{sat} is a constant independent of $\mathscr{E}_{eff}$, the critical field $\mathscr{E}_c$ is a function of $\mathscr{E}_{eff}$ as well. More specifically, *for a higher vertical field, the effective mobility is lower, but the critical field for velocity saturation becomes higher* (Sodini *et al.*, 1984). Similarly, holes have a critical field higher than that of electrons, since the hole mobility is lower.

In another version of the velocity saturation relationship, the driving force $\mathscr{E} = |d\psi_i/dy|$ in Eq. (6.39) is replaced by the gradient of quasi-Fermi potential, $|dV/dy|$. This ensures that the total current (drift + diffusion), Eq. (5.8), not just the drift current, does not exceed $WQ_i v_{sat}$, where W is the MOSFET width and Q_i is the mobile charge density per area. In practice, there is only a slight difference between the two models because the current in the velocity saturation region is predominantly a drift current. The dV/dy version of Eq. (6.39) is adopted in the rest of this section.

6.2.1.2 $n = 1$ Velocity Saturation Model

Mathematically, the simplest velocity saturation model to analyze is that of $n = 1$, in which the carrier velocity is given by

$$v = \frac{\mu_{eff}(dV/dy)}{1 + (\mu_{eff}/v_{sat})(dV/dy)}. \tag{6.41}$$

The GCA model approach is discussed first. Following similar steps to those in Section 5.1.1, we replace the low-field velocity, $\mu_{eff} \, dV/dy$, in Eq. (5.8) with Eq. (6.41) to obtain:

$$I_{ds} = -WQ_i(V)\frac{\mu_{eff}dV/dy}{1 + (\mu_{eff}/v_{sat})dV/dy}. \tag{6.42}$$

Here V is the quasi-Fermi potential at a point y in the channel, and $Q_i(V)$ is the integrated (over the depth) inversion charge density at that point. Note that $dV/dy > 0$. Current continuity requires that I_{ds} be a constant, independent of y.

Equation (6.42) can be rearranged to yield

$$I_{ds} = -\left(\mu_{eff}WQ_i(V) + \frac{\mu_{eff}I_{ds}}{v_{sat}}\right)\frac{dV}{dy}. \tag{6.43}$$

After multiplying dy to the left-hand side, the above can be integrated from $y = 0$ to L and from $V = 0$ to V_{ds} to solve I_{ds}:

$$I_{ds} = \frac{-\mu_{eff}(W/L)\int_0^{V_{ds}} Q_i(V)dV}{1 + (\mu_{eff}V_{ds}/v_{sat}L)}. \tag{6.44}$$

The numerator is simply the long-channel current, Eq. (5.10), without velocity saturation. It is clear that if the "average" field along the channel, V_{ds}/L, is much less than the critical field $\mathcal{E}_c = v_{sat}/\mu_{eff}$, the drain current is hardly affected by velocity saturation. When V_{ds}/L becomes comparable to or greater than $\mathcal{E}_c$, however, the drain current is significantly reduced. A convenient, approximate expression for $Q_i(V)$ is Eq. (5.44):

$$-Q_i(V) = C_{inv}(V_{gs} - V_t - mV), \tag{6.45}$$

where $C_{inv}(V_{gs} - V_t)$ is given by Eq. (5.16) of the charge sheet model with $\psi_s = \psi_{s,s}$ for $V = 0$. The integration in Eq. (6.44) can then be carried out to yield

$$I_{ds} = \frac{\mu_{eff}C_{inv}(W/L)\left[(V_{gs} - V_t)V_{ds} - (m/2)V_{ds}^2\right]}{1 + (\mu_{eff}V_{ds}/v_{sat}L)}. \tag{6.46}$$

For a given V_{gs}, I_{ds} increases with V_{ds} until a maximum current is reached. The saturation voltage, V_{dsat}, is found by solving $dI_{ds}/dV_{ds} = 0$. To compact the equations, a dimensionless parameter

$$z \equiv \frac{2\mu_{eff}(V_{gs} - V_t)}{mv_{sat}L} \tag{6.47}$$

is introduced. It is a measure of the severity of velocity saturation. Then,

$$V_{dsat} = \frac{2(V_{gs} - V_t)/m}{1 + \sqrt{1 + 2\mu_{eff}(V_{gs} - V_t)/(mv_{sat}L)}} \equiv \frac{Lv_{sat}}{\mu_{eff}}\left(\sqrt{1+z} - 1\right) \tag{6.48}$$

This expression is always less than the long-channel saturation voltage, $(V_{gs} - V_t)/m$. Substituting Eq. (6.48) into Eq. (6.46), we find the saturation current,

$$I_{dsat} = C_{inv}Wv_{sat}(V_{gs} - V_t)\frac{\sqrt{1+z} - 1}{\sqrt{1+z} + 1}. \tag{6.49}$$

For $z \ll 1$, Eq. (6.49) is reduced to the long-channel saturation current,

$$I_{dsat} = \mu_{eff}C_{inv}\frac{W}{L}\frac{(V_{gs} - V_t)^2}{2m}. \tag{6.50}$$

For $z \gg 1$, Eq. (6.49) becomes the velocity-saturation-limited current,

$$I_{dsat} = C_{inv}Wv_{sat}(V_{gs} - V_t). \tag{6.51}$$

Note that in this limit, I_{dsat} is independent of channel length L and varies linearly with V_{gs}–V_t instead of quadratically as in the long-channel case. This is consistent with observations of the experimental curves in Figure 6.8.

At the saturation point, $V(y = L) = V_{dsat}$. It can be shown that

$$I_{dsat} = C_{inv}Wv_{sat}(V_{gs} - V_t - mV_{dsat}) = -Wv_{sat}Q_i(y = L).$$ (6.52)

In other words, carriers at the drain are traveling at the saturation velocity, which means $dV/dy \to \infty$ in Eq. (6.41). Note that $-Q_i$ of Eq. (6.45) is positive at this point. The commonality between the current saturation in the case of constant mobility and in the case of velocity saturation is therefore not $-Q_i \to 0$, but the divergence of dV/dy under the GCA model.

For $V_{ds} > V_{dsat}$, the GCA model breaks down. It is necessary to invoke the non-GCA model discussed in Section 5.1.4. Like with Eq. (5.43), $\varepsilon_{si}d_{si}d^2\psi/dy^2 \approx \varepsilon_{si}d_{si}d^2V/dy^2$ is added to $-Q_i(V)$ of Eq. (6.42) to obtain

$$I_{ds} = W\left[-Q_i(V) + \varepsilon_{si}d_{si}\frac{d^2V}{dy^2}\right]\frac{\mu_{eff}dV/dy}{1 + (\mu_{eff}/v_{sat})dV/dy}.$$ (6.53)

The depth parameter d_{si} has been discussed in the paragraph following Eq. (5.43). With $-Q_i(V)$ of Eq. (6.45), the above can be integrated once to yield

$$\frac{I_{ds}}{\mu_{eff}W}y + \frac{I_{ds}}{v_{sat}W}V = C_{inv}\left[(V_{gs} - V_t)V - \frac{m}{2}V^2\right] + \frac{\varepsilon_{si}d_{si}}{2}\left[\left(\frac{dV}{dy}\right)^2 - \mathscr{E}_0^2\right].$$ (6.54)

This is similar to Eq. (5.45) except the second term on the left-hand side from velocity saturation. Here, $\mathscr{E}_0$ is dV/dy at $y = 0$. Since the non-GCA term in Eq. (6.53) is negligible at $V = 0$ $(y = 0)$, we have

$$\mathscr{E}_0 = \frac{I_{ds}}{\mu_{eff}WC_{inv}(V_{gs} - V_t) - (\mu_{eff}/v_{sat})I_{ds}}.$$ (6.55)

Equation (6.54) is a first-order ordinary differential equation valid for all regions above threshold, both before and after saturation. For a given I_{ds}, it solves for $V(y)$ numerically from $y = 0$ to $y = L$, yielding $V(L) = V_{ds}$ as the output. The numerical procedure for solving Eq. (6.54) follows the same line as that of Eq. (5.47). *The addition of the d^2V/dy^2 term in* Eq. (6.53) *removes the GCA singularity at V_{dsat} and keeps dV/dy finite and positive for $V_{ds} > V_{dsat}$.* Figure 6.10 shows that the I_{ds}-V_{ds} characteristic produced by the GCA model exhibits an unphysical downturn past the peak point (V_{dsat}, I_{dsat}), whereas with the non-GCA model, a continuous I_{ds}-V_{ds} curve is generated from the triode region throughout the velocity saturation region with finite, positive output conductance (Taur *et al.*, 2019).

6.2.1.3 $n = 2$ Velocity Saturation Model

It has been known that the $n = 1$ velocity saturation model has a discontinuity problem with the second order derivative around $V_{ds} = 0$ because the dV/dy factor in the denominator of Eq. (6.41) should in fact be $|dV/dy|$ to keep it always positive (Joardar *et al.*, 1998). To satisfy the continuity requirement, n needs to be an even integer. The least of which is 2. For the GCA model with $n = 2$ velocity saturation, Eq. (6.42) becomes

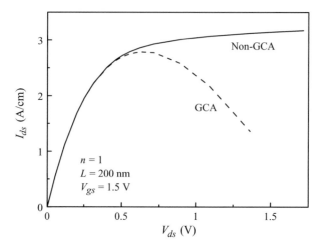

Figure 6.10 I_{ds}–V_{ds} characteristics generated by the GCA and non-GCA models under the $n = 1$ velocity saturation model. The MOSFET parameters are $t_{inv} = 3.3$ nm, $N_a = 10^{18}$ cm^{-3} (uniform), n$^+$ silicon gate work function, so $V_t = 0.4$ V and $m = 1.28$. Other parameters are $\mu_{eff} = 200$ cm^2/V-s, $v_{sat} = 10^7$ cm/s, and $d_{si} = 20$ nm

$$I_{ds} = -WQ_i(V)\frac{\mu_{eff}(dV/dy)}{\sqrt{1 + (\mu_{eff}/v_{sat})^2(dV/dy)^2}}. \tag{6.56}$$

It can be re-arranged to yield an integral equation between I_{ds} and V_{ds} for a given V_{gs},

$$LI_{ds} = \mu_{eff}\int_0^{V_{ds}}\sqrt{[WQ_i(V)]^2 - (I_{ds}/v_{sat})^2}\,dV. \tag{6.57}$$

With $Q_i(V)$ of Eq. (6.45), the above integral can be carried out by transforming V to an intermediary variable u,

$$WC_{inv}(V_{gs} - V_t - mV) = (I_{ds}/v_{sat})\cosh u. \tag{6.58}$$

Then,

$$L = \frac{\mu_{eff}I_{ds}}{2mWC_{inv}v_{sat}^2}[\sinh u \cosh u - u]|_{u_d}^{u_s}, \tag{6.59}$$

where u_s and u_d are given by Eq. (6.58) with $V = 0$ and V_{ds}, respectively.

The I_{ds}–V_{ds} curve generated for a fixed V_{gs} is shown in Figure 6.11. There is a maximum $V_{ds} = V_{dsat}$ where I_{ds} reaches a peak value I_{dsat} beyond which no solution exists. This corresponds to $u_d = 0$ where the factor in the square root of Eq. (6.57) is zero, meaning carriers are traveling at v_{sat} and $dV/dy \to \infty$. The peak current is

$$I_{dsat} = WC_{inv}(V_{gs} - V_t)v_{sat}/\cosh u_s, \tag{6.60}$$

where u_s (for the peak point) is solved by the implicit equation,

$$L = \frac{\mu_{eff}\left(V_{gs} - V_t\right)}{2mv_{sat}}\left[\sinh u_s - \frac{u_s}{\cosh u_s}\right]. \tag{6.61}$$

To go beyond the saturation point, it is necessary to employ the non-GCA model. The $n = 2$ version of Eq. (6.53) is

$$I_{ds} = W\left[-Q_i(V) + \varepsilon_{si}d_{si}\frac{d^2V}{dy^2}\right]\frac{\mu_{eff}dV/dy}{\sqrt{1 + \left(\mu_{eff}/V_{sat}\right)^2\left(dV/dy\right)^2}}. \tag{6.62}$$

The second-order derivative can be converted to

$$\frac{d^2V}{dy^2} = \frac{dV}{dy}\frac{d}{dV}\left(\frac{dV}{dy}\right) = \frac{1}{2}\frac{d}{dV}\left(\frac{dV}{dy}\right)^2. \tag{6.63}$$

By defining

$$g(V) = \left(\frac{dV}{dy}\right)^2, \tag{6.64}$$

Eq. (6.62) becomes a first order differential equation for $g(V)$:

$$I_{ds}^2\left[1 + \left(\frac{\mu_{eff}}{v_{sat}}\right)^2 g\right] = W^2\mu_{eff}^2 g\left[-Q_i(V) + \frac{\varepsilon_{si}d_{si}}{2}\frac{dg}{dV}\right]^2. \tag{6.65}$$

For a given I_{ds}, the above is solved numerically in a manner similar to that described by Eq. (5.47), except that a cubic equation in δg is solved at every small step of δV instead of a quadratic equation. Once $g(V)$ is solved, it is straightforward to numerically integrate $g^{-1/2}$ with respect to V to obtain $y(V)$, hence V_{ds} for which $y(V_{ds}) = L$. Figure 6.11 shows the I_{ds}–V_{ds} curve, continuous from the triode region throughout the

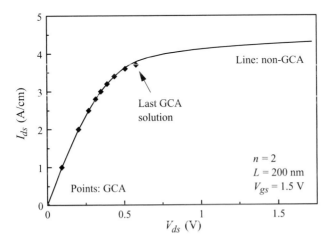

Figure 6.11 I_{ds}–V_{ds} characteristics generated by the GCA and non-GCA models under the $n = 2$ velocity saturation relation. The device parameters are the same as those described in the caption to Figure 6.10

saturation region, generated by the non-GCA $n = 2$ velocity saturation model (Taur *et al.*, 2019).

6.2.1.4   Regional Solution for the Velocity Saturation Region

To derive an analytical solution for the velocity saturation region, a regional approximation is applied: $(\mu_{eff}/v_{sat})(dV/dy) \gg 1$ so that carrier velocity $\approx v_{sat}$, regardless of the order n of the velocity–field relationship. With Q_i given by Eq. (6.45), either Eq. (6.53) or Eq. (6.62) can be simplified to

$$I_{ds} = Wv_{sat}\left[C_{inv}\left(V_{gs} - V_t - mV\right) + \varepsilon_{si}d_{si}\frac{d^2V}{dy^2}\right]. \tag{6.66}$$

By applying Eq. (6.63), Eq. (6.66) can be integrated from V_{dsat} to V to obtain

$$\left(\frac{dV}{dy}\right)^2 = \frac{mC_{inv}}{\varepsilon_{si}d_{si}}\left[(V - V_{dsat} + a)^2 + (b^2 - a^2)\right], \tag{6.67}$$

where

$$a \equiv \frac{I_{ds} - I_{dsat}}{mC_{inv}Wv_{sat}} = \frac{I_{ds} - I_{dsat}}{I_{dsat}}\left(\frac{V_{gs} - V_t}{m} - V_{dsat}\right) \tag{6.68}$$

and

$$b = \sqrt{\frac{\varepsilon_{si}d_{si}}{mC_{inv}}}\left(\frac{dV}{dy}\right)\Bigg|_{V_{dsat}}. \tag{6.69}$$

Both a and b are small voltages. The boundary condition $(dV/dy)|_{V_{dsat}}$ at the beginning of the velocity saturation region cannot be determined analytically. We take an approximation $b \approx a$ to simplify the results. If $V(y = L) = V_{ds}$ is not too close to V_{dsat}, *dV/dy at the drain increases linearly with* V_{ds}–V_{dsat}. *Further integration of Eq. (6.67) gives V as an exponential function of y beyond the point of saturation*:

$$V - V_{dsat} + a \approx (V_{ds} - V_{dsat} - a)\exp\left[\sqrt{\frac{mC_{inv}}{\varepsilon_{si}d_{si}}}(y - L)\right]. \tag{6.70}$$

In terms of CLM, $y = L - \Delta L$ at the point where $V = V_{dsat}$. This point moves toward the source as V_{ds} increases. So we have

$$\Delta L = \sqrt{\frac{\varepsilon_{si}d_{si}}{mC_{inv}}}\ln\left(\frac{V_{ds} - V_{dsat} - a}{a}\right). \tag{6.71}$$

Similar expressions have been derived by applying 2-D Gauss's law to the velocity saturation region (Ko *et al.*, 1981).

The conventional expression of the CLM effect on drain current, $I_{ds} = I_{dsat}/(1 - \Delta L/L)$, is only valid for constant mobility in which $I_{ds} \propto 1/L$. For the $n = 1$ velocity saturation case, Eq. (6.49) gives $\Delta I_{ds}/I_{ds} = (1 + z)^{-1/2}\Delta z/z = (1 + z)^{-1/2}\Delta L/L$, where z is given by Eq. (6.47). Hence

$$I_{ds} = \frac{I_{dsat}}{1 - \left(\frac{1}{\sqrt{1+z}}\right)\frac{\Delta L}{L}}. \tag{6.72}$$

Likewise, it can be derived for $n = 2$ that

$$I_{ds} = \frac{I_{dsat}}{1 - \left(\frac{\sinh u_s \cosh u_s - u_s}{\sinh u_s \cosh u_s + u_s}\right)\frac{\Delta L}{L}}, \tag{6.73}$$

where u_s is solved from Eq. (6.61).

6.2.1.5   Short-Channel Effects above Threshold

The short channel MOSFET model described in Section 6.1.2 is only valid for the subthreshold region where the mobile charge density is negligible. For the region above threshold, the 2-D Poisson's equation takes the following general form:

$$\frac{\partial^2 \psi}{\partial x^2} + \frac{\partial^2 \psi}{\partial y^2} = \frac{q}{\varepsilon_{si}}\left[N_a + n_i e^{q(\psi - V)/kT}\right]. \tag{6.74}$$

This is a nonlinear partial differential equation for which superposition is not applicable. Furthermore, it is coupled to the current continuity equation for $V(y)$ connected to the mobile charge density. In this section, ψ is defined as the intrinsic potential with respect to the Fermi potential of the source (see Fig. 7.13).

While the 2-D effect near the drain is covered by the non-GCA model in Section 5.1.4.1, there is also 2-D effect near the source. Basically, the n$^+$ bands of the source (for nMOS) would encroach into the channel since $\psi(y)$ does not change abruptly to the channel potential which is a function of the gate voltage (GCA model). In contrast to near the drain, where the current is predominantly drift thus permitting the approximation $d^2\psi/dy^2 \approx d^2V/dy^2$, the current near the source is mostly diffusion, hence $d^2\psi/dy^2$ must be kept.

Following a similar approach as that in Section 5.1.4.1, we apply the GCA solution with $-Q_i(V)$ of Eq. (5.44) to the 1-D equation in the x-direction:

$$\frac{d^2\psi}{dx^2} = \frac{q}{\varepsilon_{si}}N_a + \frac{-Q_i}{\varepsilon_{si}d_{si}} = \frac{q}{\varepsilon_{si}}N_a + \frac{C_{inv}\left(V_{gs} - V_t - mV\right)}{\varepsilon_{si}d_{si}}. \tag{6.75}$$

The difference between Eqs. (6.74) and (6.75) is

$$\frac{d^2\psi}{dy^2} = \frac{q}{\varepsilon_{si}}n_i e^{q(\psi-V)/kT} - \frac{C_{inv}\left(V_{gs} - V_t - mV\right)}{\varepsilon_{si}d_{si}}. \tag{6.76}$$

This can be treated as an ordinary differential equation in the lateral or the source–drain direction, coupled to the current continuity equation, e.g., for the $n = 1$ velocity saturation relation,

$$I_{ds} = W d_{si}q n_i e^{q(\psi-V)/kT}\frac{\mu_{eff}dV/dy}{1 + \left(\mu_{eff}/V_{sat}\right)dV/dy}. \tag{6.77}$$

Note that $qn_id_{si}e^{q(\psi-V)/kT}$ is the total mobile charge density per area that includes both the gate induced component $-Q_i$ and the lateral component from $d^2\psi/dy^2$.

In general, for given values of V_{gs}, V_{ds}, there are four boundary conditions: $\psi(0) = E_g/2q$ and $V(0) = 0$ at the source, and $\psi(L) = E_g/2q + V_{ds}$ and $V(L) = V_{ds}$ at the drain. The approximation here is that the Fermi level of n^+ source-drain is at the conduction band edge E_c, with negligible depletion in the source and drain. The coupled Eqs. (6.76) and (6.77) are then solved for the unknown I_{ds}. Numerically, a good way is to define a function $u(y) = \psi(y) - V(y)$ and transform Eq. (6.76) into a second-order differential equation in u. The boundary conditions then become $u(0) = u(L) = E_g/2q$. For a given I_{ds}, the u-equation is solved along with Eq. (6.77) to obtain $V_{ds} = V(L)$. Note that this is a non-GCA model that generates I_{ds}–V_{ds} characteristics continuous from the triode region throughout the velocity saturation region.

For a useful insight into the above model, we consider the case of low-drain bias. With $qV/kT << 1$, Eq. (6.76) is reduced to

$$\frac{d^2\psi}{dy^2} = \frac{q}{\varepsilon_{si}}n_ie^{q\psi/kT} - \frac{C_{inv}(V_{gs} - V_t)}{\varepsilon_{si}d_{si}}, \tag{6.78}$$

and is no longer coupled to Eq. (6.77). Figure 6.12 shows an example of the numerical solutions of $\psi(y)$ for several values of V_{gs}. The source–drain encroachment into the channel can be viewed as an effective reduction of channel length as far as the low-drain channel conductance is concerned. However, *the reduction of channel length is a function of the gate voltage V_{gs}. At a higher V_{gs}, the channel is more conductive hence less the short-channel effect due to source–drain encroachment.*

At high drain biases, the model is more cumbersome to solve. A finite difference method implemented in a standard mathematics tool can be employed (LeVeque, 2007). Figure 6.13 shows the potential profile of the device described in Figure 6.12

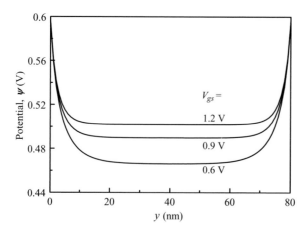

Figure 6.12 Low-drain-bias potential solution from the source to the drain. The device is the same as that described in the caption to Figure 6.10, except that $L = 80$ nm. The boundary conditions are $\psi(0) = \psi(L) \approx 0.60$ V for 10^{20} cm^{-3} source–drain doping. If L is not too short, the potential at mid-channel is largely controlled by the gate: $\psi(L/2) \approx (kT/q)\ln[C_{inv}(V_{gs} - V_t)/(qn_id_{si})]$

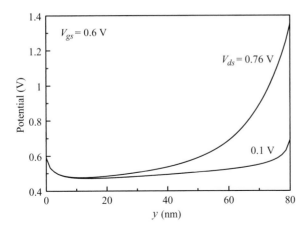

Figure 6.13 Plots of $\psi(y)$ from the source to the drain of a short-channel MOSFET above threshold (Hong and Taur, 2021)

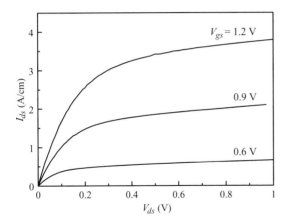

Figure 6.14 I_{ds}-V_{ds} characteristics of the MOSFET described in the caption to Figure 6.12. They are solved numerically from the coupled Eqs. (6.76) and (6.77) (Hong and Taur, 2021)

for two values of V_{ds}. Drain induced barrier lowering at the point of maximum barrier (virtual cathode) is apparent. The corresponding I_{ds}–V_{ds} characteristics generated by the model are shown in Figure 6.14.

6.2.2   Nonlocal Transport

6.2.2.1   Velocity Overshoot

All the MOSFET current formulations discussed thus far, including the mobility definition and velocity saturation, are under the realm of the drift–diffusion approximation. It applies when the device dimension is much longer than the carrier mean free path between collisions, in which case carriers are close to local equilibrium with their environment. ***The drift–diffusion model is not accurate in ultrashort-channel devices where high field or rapid spatial variation of potential is present***. In such cases, there are not enough scattering events to establish local equilibrium, and some fraction of the carriers may acquire much higher energy over a portion of the device, for example, near the drain. These carriers are far from thermal equilibrium with the silicon lattice and are generally referred to as *hot carriers*. Under these circumstances, it is possible for the carrier velocity to exceed the saturation velocity. This phenomenon is called *velocity overshoot*.

A more rigorous treatment of the carrier transport under spatially nonuniform high-field conditions has been carried out by a Monte Carlo solution of the Boltzmann transport equation for the electron distribution function (Laux and Fischetti, 1988). Local velocity overshoot first starts to occur near the drain. At channel lengths of 50 nm or below, velocity overshoot takes place over a substantial portion of the device such that the terminal saturation transconductance may exceed the velocity saturation limited value,

$$g_{msat} \equiv \frac{dI_{dsat}}{dV_g} = C_{ox} W v_{sat},$$ (6.79)

based on Eq. (6.51) (Sai-Halasz *et al.*, 1988).

6.2.2.2   Ballistic MOSFET

A ballistic MOSFET is a hypothetical device in which mobile carriers suffer no collisions in the channel. This may happen, in principle, when the channel length is shorter than the mean free path, the average distance carriers travel between collisions. In an ordinary MOSFET, carriers moving from the source to the drain under the influence of the applied field collide with the silicon lattice, impurity (dopant) atoms, and surfaces. These collisions limit the velocity they can acquire from the field (Section 2.3.1), resulting in a reduced drain current. Under low field conditions, the effect of these collisions is lumped into a mobility factor proportional to the mean free time between successive collisions (Section 2.3.1). For long-channel MOSFETs, the drain current is proportional to mobility (Section 5.1.3). For short-channel MOSFETs under high drain bias conditions, high-field scattering becomes important. This is usually modeled by velocity saturation (Section 6.2.1). In the absence of any scattering, carriers entering the channel from the source are accelerated by the applied field ballistically toward the drain. They can attain very high speeds, especially in the high-field region near the drain. However, such high speeds (velocity overshoot) do not necessarily translate into large currents. Since current must be continuous from source to drain, it is bounded by the rate at which carriers are injected from the source. ***In a ballistic MOSFET then, the bottleneck is near the source where the carriers move into the channel at relatively low velocities (before field acceleration).*** Current continuity down the path is satisfied by a decreased carrier density near the drain such that the product of carrier density and velocity at the drain is the same as that at the source (see Figure 6.17).

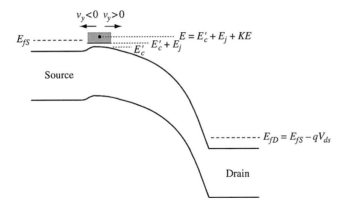

Figure 6.15 A schematic MOSFET band diagram under high-drain bias conditions. Dashed lines indicate the Fermi levels of the degenerately doped source and drain. At the point of highest energy barrier near the source, electrons populate states allowed in discrete subbands

The key to modeling the drain current of a ballistic MOSFET is to consider the average carrier velocity moving into the channel in the low field region near the source (Natori, 1994). In fact, as shown in the MOSFET band diagram in Figure 6.15, there is typically a point of zero field near the source where the electron energy barrier is the highest. The conduction band energy at this point is designated as E_c'. Quasi-equilibrium condition holds at this point since there is no net force acting on the carriers. One expects the electron distribution function to take a Fermi–Dirac-like form here. Without any collisions in the channel, all the electrons moving from left to right ($v_y > 0$) at this point will eventually make it to the drain. This constitutes the positive component of the current. Conversely, the electrons moving from right to left ($v_y < 0$) at this point must have all originated from the drain, which make up the negative component of the current. Therefore, the left to right ($v_y > 0$) electron population is in quasi-equilibrium with the source and their distribution function is controlled by the source Fermi level E_{fS}, while the right to left ($v_y < 0$) electron distribution function is controlled by the drain Fermi level E_{fD} ($= E_{fS} - qV_{ds}$).

It was discussed in Section 4.4 that electrons confined in an inversion layer populate discrete subbands with a minimum energy E_j (jth subband) above the conduction band E_c' (Figure 6.15). Consider electrons in the jth subband moving from left to right with a velocity between (v_y, v_z) and ($v_y + dv_y$, $v_z + dv_z$), where z is in the device width direction. From Eq. (4.61), the electronic states per unit area is $(2g/h^2)m_ym_zdv_ydv_z$. The probability of each state being populated by an electron is given by the Fermi–Dirac distribution function, Eq. (2.4), with $E = E_c' + E_j + KE = E_c' + E_j + (1/2)m_yv_y^2 + (1/2)m_zv_z^2$ and $E_f = E_{fS}$. The charge density per unit area of these electrons is then

$$dQ_{j+} = \frac{(2g/h^2)qm_ym_zdv_ydv_z}{1 + e^{(E_c' + E_j + m_yv_y^2/2 + m_zv_z^2/2 - E_{fS})/kT}}. \qquad (6.80)$$

Since current per width equals the product of charge density and carrier velocity, the left-to-right current component is given by the summation over all subbands of the following double integral:

$$
\begin{aligned}
I_+ &= W \sum_j \int_{allv_z} \int_{v_y > 0} v_y dQ_{j+} \\
&= \sum_j \int_{-\infty}^{\infty} \int_0^{\infty} \frac{(2g/h^2) q W m_y m_z v_y}{1 + e^{(E_c' + E_j + m_y v_y^2/2 + m_z v_z^2/2 - E_{fS})/kT}} dv_y dv_z.
\end{aligned}
\tag{6.81}
$$

The integration with respect to v_y can be done analytically by a simple change of variable, $u = v_y^2$. Using integration by parts, the result of the second integration can be expressed in terms of the Fermi–Dirac integral, defined by Eq. (2.27), as

$$
I_+ = \sum_j (4g/h^2) q W \sqrt{2m_z} (kT)^{3/2} F_{1/2} \left(\frac{E_{fS} - E_c' - E_j}{kT} \right).
\tag{6.82}
$$

Likewise, the right-to-left current component is obtained by replacing E_{fS} with E_{fD},

$$
I_- = \sum_j (4g/h^2) q W \sqrt{2m_z} (kT)^{3/2} F_{1/2} \left(\frac{E_{fD} - E_c' - E_j}{kT} \right).
\tag{6.83}
$$

The full expression for the net drain to source current, $I_{ds} = I_+ - I_-$, is of the following form:

$$
\begin{aligned}
I_{ds} &= \frac{4\sqrt{2} q W (kT)^{3/2}}{h^2} \left\{ g\sqrt{m_z} \sum_j \left[F_{1/2} \left(\frac{E_{fS} - E_c' - E_j}{kT} \right) - F_{1/2} \left(\frac{E_{fS} - qV_{ds} - E_c' - E_j}{kT} \right) \right] \right. \\
&\left. + g'\sqrt{m_z'} \sum_j \left[F_{1/2} \left(\frac{E_{fS} - E_c' - E_j'}{kT} \right) - F_{1/2} \left(\frac{E_{fS} - qV_{ds} - E_c' - E_j'}{kT} \right) \right] \right\}.
\end{aligned}
\tag{6.84}
$$

Here, for silicon in the $\langle 100 \rangle$ direction, $g = 2$, $m_z = m_t$ for the unprimed valley, and $g' = 4$, $m_z' = (\sqrt{m_l} + \sqrt{m_t})^2/4$ for the primed valley. The unknowns in Eq. (6.84) are $E_{fS} - E_c' - E_j$ and $E_{fS} - E_c' - E_j'$, the relative position of the various subbands at the point of highest electron barrier with respect to the source Fermi level. They are controlled by the gate voltage and, in general, must be solved numerically from the coupled Poisson's and Schrödinger's equations (Section 4.4).

With the same partition of charge into the positive and negative directions, the total integrated density of inversion charge per unit area can be expressed as

$$
Q_i = Q_+ + Q_- = \sum_j \int_{allv_z} \int_{v_y > 0} dQ_{j+} + \sum_j \int_{allv_z} \int_{v_y < 0} dQ_{j-}.
\tag{6.85}
$$

After the integrations are carried out,[1] an expression similar to Eq. (4.67) is obtained:

---

[1] The double integration can be carried out by transforming v_y, v_z to elliptical-polar coordinates, as done in Section 4.4.3.2.

$$
\begin{aligned}
Q_i = \frac{2\pi qkT}{h^2} \Bigg\{ & gm_t \sum_j \left[\ln\left(1 + e^{\left(E_{fS} - E_c' - E_j\right)/kT}\right) + \ln\left(1 + e^{\left(E_{fS} - qV_{ds} - E_c' - E_j\right)/kT}\right) \right] \\
& + g'\sqrt{m_l m_t} \sum_j \left[\ln\left(1 + e^{\left(E_{fS} - E_c' - E_j'\right)/kT}\right) + \ln\left(1 + e^{\left(E_{fS} - qV_{ds} - E_c' - E_j'\right)/kT}\right) \right] \Bigg\}.
\end{aligned}
$$

$$ (6.86) $$

On the other hand, the inversion charge density at the source can be solved from electrostatics, e.g., Eq. (5.14) with $V = 0$, as a function of gate voltage. Above the threshold voltage V_t, Q_i varies approximately linearly with V_{gs} and can be expressed as $C_{inv}(V_{gs} - V_t)$, where C_{inv} is the effective gate capacitance per unit area discussed in Section 4.4.2. It lumps the inversion layer depth and polysilicon gate depletion effects together with the oxide capacitance.

In the case of strong quantum effects when either the temperature is low or the field is high, the spacing between E_j of successive subbands becomes larger than kT. It is a good approximation to keep only the lowest subband term, i.e., $j = 0$ of the unprimed valley (Section 4.4.3.1). Equating Eq. (6.86) to $C_{inv}(V_{gs} - V_t)$ then allows $E_{fS} - E_c' - E_0$ to be solved analytically:

$$
e^{\left(E_{fS} - E_c' - E_0\right)/kT} = \frac{1}{2} \left\{ \sqrt{\left(e^{qV_{ds}/kT} - 1\right)^2 + 4\exp\left[\frac{h^2 C_{inv}\left(V_{gs} - V_t\right)}{4\pi qkTm_t} + \frac{qV_{ds}}{kT}\right]} - 1 - e^{qV_{ds}/kT} \right\}.
$$

$$ (6.87) $$

Under the same one subband approximation, Eq. (6.84) is reduced to

$$
I_{ds} = \frac{8\sqrt{2m_t}\, qW(kT)^{3/2}}{h^2} \left[F_{1/2}\left(\frac{E_{fS} - E_c' - E_0}{kT}\right) - F_{1/2}\left(\frac{E_{fS} - E_c' - E_0}{kT} - \frac{qV_{ds}}{kT}\right) \right].
$$

$$ (6.88) $$

For given values of V_{gs} and V_{ds}, the current of a ballistic MOSFET can be calculated from Eqs. (6.87) and (6.88). An example is given in Figure 6.16.[2] Obviously, $I_{ds} = 0$ when $V_{ds} = 0$ because $E_{fS} = E_{fD}$ and the positive moving ($v_y > 0$ in Figure 6.15) electron flux from the source exactly cancels the negative moving ($v_y < 0$) electron flux from the drain. As V_{ds} increases, I_{ds} increases because the negative moving electron population decreases with decreasing $E_{fD} = E_{fS} - qV_{ds}$, while the positive moving electron population increases in order to conserve the total Q_i. The current saturates at a drain voltage V_{dsat} when the negative moving electron flux at the point of maximum barrier in Figure 6.15 becomes negligibly small.

While the drain current of a ballistic MOSFET exhibits saturation just like that of an ordinary MOSFET, the underlying physics is very different. In an ordinary MOSFET, the drain current saturates due to velocity saturation at the drain. *The current of a ballistic MOSFET saturates because, for a given gate-induced electron*

---

[2] An explicit analytic function that approximates Fermi–Dirac integrals numerically is used (Blakemore, 1982).

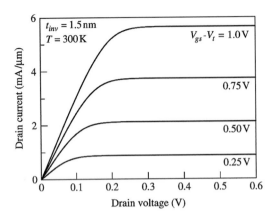

Figure 6.16 I_{ds}–V_{ds} curves of a ballistic MOSFET calculated from Eqs. (6.87) and (6.88). Here $C_{inv} = \varepsilon_{ox}/t_{inv}$

density, there is a maximum electron flux that can be extracted from the source, given by the positive half of the thermal distribution. Also note that the current of a ballistic MOSFET is independent of channel length, as it represents the highest current limit when $L \to 0$.

Further analytical expressions of V_{dsat} and I_{dsat} of a ballistic MOSFET can be derived under the *degenerate* condition in which the source Fermi level is at least a few kT above the minimum energy of the lowest subband, i.e., when $(E_{fS} - E_c' - E_0)/kT > 1$. This happens at relatively high gate voltages when the dimensionless parameter in Eq. (6.87), $h^2 C_{inv}(V_{gs} - V_t)/(4\pi qkTm_t)$, is greater than one, i.e., when the electron sheet density, $C_{inv}(V_{gs} - V_t)/q$, is higher than half of the 2-D effective density of states, $4\pi kTm_t/h^2 \approx 2 \times 10^{12}$ cm^{-2}. For example, for $t_{inv} = 1.5$ nm and $V_{gs} - V_t = 1$ V, $C_{inv}(V_{gs} - V_t)/q$ is approximately 1.3×10^{13} cm^{-2}. Fermi–Dirac integrals can then be approximated by $F_{1/2}(u) \approx (2/3)u^{3/2}$ for $u \gg 1$ (Blakemore, 1982). In saturation, $qV_{ds}/kT \gg 1$, Eq. (6.87) gives $(E_{fS} - E_c' - E_0)/kT \approx h^2 C_{inv}(V_{gs} - V_t)/(4\pi qkTm_t)$. Substituting it in Eq. (6.88) yields the saturation current,

$$I_{dsat} = \frac{2\sqrt{2}hW}{3\pi\sqrt{\pi q}m_t} \left[C_{inv}\left(V_{gs} - V_t\right) \right]^{3/2}. \tag{6.89}$$

Note that this limiting current is independent of temperature. The saturation voltage can be estimated by equating $(E_{fS} - E_c' - E_0)/kT$ to qV_{ds}/kT in Eq. (6.88), yielding

$$V_{dsat} \approx \frac{h^2 C_{inv}\left(V_{gs} - V_t\right)}{4\pi q^2 m_t}. \tag{6.90}$$

This is much less than $V_{gs} - V_t$ since in practice the effective gate oxide (t_{inv}) is much thicker than the constant, $h^2\varepsilon_{ox}/(4\pi q^2 m_t) \approx 0.27$ nm, in Eq. (6.90).

In the linear region, $qV_{ds}/kT \ll 1$, Eq. (6.87) gives $(E_{fS} - E_c' - E_0)/kT \approx h^2 C_{inv}(V_{gs} - V_t)/(8\pi qkTm_t)$. Substituting it in Eq. (6.88) yields the linear region conductance,

$I_{dlin}/V_{ds} = \frac{4q^2W}{\sqrt{\pi qh}}\sqrt{C_{inv}(V_{gs}-V_t)}$. Note that **the ballistic-limited conductance is independent of mobility and channel length.** It can be shown that it equals the quantum of conductance, $2q^2/h$, times the number of modes in a width W (Lundstrom and Jeong, 2013).

6.2.2.3 Scattering Theory

It should be noted that, while carrier velocity can reach rather high values in the high-field region near the drain, it does not lead to proportionately high currents. This is illustrated in Figure 6.17 where the band diagram of a MOSFET biased in saturation is shown. At a point near the drain, the average carrier velocity v_d is high, but the inversion charge sheet density $Q_i = C_{ox}(V_{gs} - V_t - mV_{dsat})$ is low. V_{dsat} will assume a value to maintain current continuity such that

$$I_{ds} = WC_{ox}(V_{gs} - V_t - mV_{dsat})v_d \tag{6.91}$$

at the drain equals

$$I_{ds} = WC_{ox}(V_{gs} - V_t)v_s \tag{6.92}$$

at the source where $Q_i = C_{ox}(V_{gs} - V_t)$ is high but the carrier velocity v_s is low. In this picture, MOSFET current is more directly related to the average carrier velocity v_s at the bottleneck region near the source.

In the ballistic MOSFET model discussed in Section 6.2.2.2, the saturation current is limited by the thermal injection velocity v_T from the source. Equating I_{dsat} of Eq. (6.89) in the one subband degenerate limit to $WC_{inv}(V_{gs} - V_t)v_T$ yields

$$v_T = \frac{2h}{3\pi m_t}\sqrt{\frac{2C_{inv}(V_{gs} - V_t)}{\pi q}}. \tag{6.93}$$

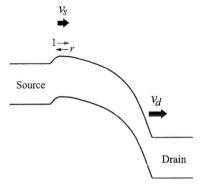

Figure 6.17 Band diagram of a MOSFET biased in saturation; v_s and v_d are the average carrier velocities near the source and the drain, respectively. Carriers near the source can be considered as made up of an incident flux ($1\rightarrow$) and a reflected flux ($\leftarrow r$) in the scattering theory

This limit is independent of the field and scattering parameters, and is not enhanced by velocity overshoot near the drain. Note that the ballistic saturation current takes the same form as the velocity-saturation-limited current, Eq. (6.51), with the parameter v_{sat} replaced by the thermal injection velocity v_T. For an electron sheet density of $C_{inv}(V_{gs} - V_t)/q \approx 10^{13}$ cm^{-2}, $v_T \approx 2 \times 10^7$ cm/s, or about twice v_{sat}.

In the scattering theory (Lundstrom, 1997) for an ordinary MOSFET, carriers are injected into the channel from the source (toward the right) and from the drain (toward the left) at their respective injection velocities. They may be back scattered due to random collisions in the channel. Under high drain bias conditions, $V_{ds} \gg kT/q$, carriers originated from the drain have virtually no possibility of making it uphill all the way to the source. Carriers at the point of highest barrier near the source can then be considered as made up of an incident flux (from the source) of amplitude 1 and a reflected flux of amplitude $r < 1$, as shown in Figure 6.17. Both fluxes are moving at the thermal velocity v_T, but in opposite directions. The average carrier velocity v_s near the source is therefore

$$v_s = \frac{1 - r}{1 + r} v_T. \tag{6.94}$$

The drain current is obtained by substituting v_s into Eq. (6.92). The reflection coefficient r is determined by the field and scattering rates (mobility) in the low-field channel region near the source (Lundstrom, 1997). Once the carriers are a few kT below the highest energy barrier in Figure 6.17, they are unlikely to be scattered back to the source. In a ballistic MOSFET, there is no scattering, so $r = 0$, $v_s = v_T$, and the current reaches the highest limit.

Note that r is a phenomenological parameter that cannot be predicted from the scattering theory. A Monte Carlo type of solution to the Boltzmann transport equation is needed to calculate r from the detailed physical processes. Recent experimental data suggest that $r \approx 1/3$ and $v_s \approx v_T/2$ in state-of-the-art sub-100 nm MOSFETs (Lochtefeld and Antoniadis, 2001). It has been argued that r is rather insensitive to scaling because of the inevitable loss of mobility (Figure 5.14) due to increased fields (both lateral and vertical) in the shorter device. This implies that scaling silicon MOSFETs to shorter channel lengths, e.g., 10 nm, may not result in a drain current closer to the ballistic limit.

6.3   MOSFET Threshold Voltage and Channel Profile Design

A key design parameter in a MOSFET device is the threshold voltage. In this section, threshold voltage requirements with regard to off- and on-currents are discussed, leading to the design of MOSFET channel profile with nonuniform body doping. Section 6.3.4 deals with the effect of dopant number fluctuations on threshold voltage in a minimum geometry device.

6.3.1   Threshold Voltage Requirement

6.3.1.1   Various Definitions of Threshold Voltage

First, we examine the various definitions of threshold voltage and the threshold voltage requirement from a technology point of view. There are quite a number of different ways to define the threshold voltage of a MOSFET device. In Chapter 5, we followed the most commonly used definition, $\psi_s(\text{inv}) = 2\psi_B$, of V_t. The advantage of this definition lies in its popularity and ease of incorporation into analytical solutions. However, it is not directly measurable from experimental I–V characteristics (it can be determined from a split C–V measurement; see Exercise 4.1). In Section 5.1.3, we introduced the linearly extrapolated threshold voltage, V_{on}, determined by the intercept of a tangent through the point of maximum-slope (linear transconductance) on a low-drain I_{ds}–V_{gs} curve. This is easily measured experimentally, but is several kT/q higher than the $2\psi_B$ threshold voltage, due to the inversion-layer capacitance effect illustrated in Figure 5.15.

Another commonly employed definition of threshold voltage is based on the subthreshold I_{ds}–V_{gs} characteristics, Eq. (5.39). For a given constant current level I_0 (say, 50 nA), one can define a threshold voltage V_t^{sub} such that $I_{ds}\left(V_{gs} = V_t^{sub}\right) = I_0(W/L)$. The advantages of such a threshold voltage definition are two-fold. First, it is easy to extract from the hardware data and is therefore suitable for automated measurements of a large number of devices. Second, the device off current, $I_{off} = I_{ds}(V_{gs} = 0)$, can be directly calculated from I_0, V_t^{sub}, and the subthreshold slope. In subsequent discussions, we will adhere to the $2\psi_B$ definition of V_t. In general, V_t depends on temperature (temperature coefficient), substrate bias (body-effect coefficient), channel length, and drain voltage (short-channel effect, or SCE).

6.3.1.2   Off-current and Standby Power

By definition, the off-current of a MOSFET is the drain-to-source subthreshold leakage current when the gate-to-source voltage V_{gs} is zero and the drain-to-source voltage V_{ds} is V_{dd}, the power supply voltage. From Eq. (5.39), the expression for the off-current with $V_{ds} = V_{dd} \gg kT/q$ is

$$I_{off} = I_{ds}|_{V_{gs}=0, V_{ds}=V_{dd}} = I_{ds, Vt}e^{-qV_t/mkT}, \tag{6.95}$$

where

$$I_{ds, Vt} = \mu_{eff} C_{ox} \frac{W}{L}(m-1)\left(\frac{kT}{q}\right)^2 \tag{6.96}$$

is defined as the drain-to-source current at threshold ($V_{gs} = V_t$, $V_{ds} = V_{dd}$). In the worst case, the drain–source voltage of the transistors in the off-state is the power supply voltage V_{dd}. The standby power dissipation due to I_{off} is then $V_{dd} \times I_{off}$. For order-of-magnitude estimates, $V_{dd} \approx 1\,\text{V}$. If it is required that the standby power of a VLSI chip containing 10^9 transistors be no higher than 10 W, the off-current per transistor should be kept less than 10 nA.[3]

Note that $I_{ds,Vt}$ is rather insensitive to the temperature since $\mu_{eff} \propto T^{-3/2}$. However, it does depend on technology. For a 0.1 μm CMOS technology with $t_{ox} \approx 3$ nm, $\mu_{eff} \approx 350$ cm^2/V-s, $m \approx 1.3$, and $W/L = 10$, $I_{ds,Vt}$ is approximately 1 μA. VLSI chips are usually specified for a worst-case temperature of 100°C where the off-current is much higher than the room temperature value because not only does V_t decrease with temperature, but the slope of the log(I_{ds})–V_{gs} curve also degrades in proportion to q/kT. Typically, the inverse slope of subthreshold current is 100 mV/decade at 100 °C. For the factor $\exp(-qV_t/mkT)$ in Eq. (6.95) to deliver a two-orders-of-magnitude reduction from $I_{ds,Vt} = 1$ μA to $I_{off} = 10$ nA, V_t (100 °C) needs to be at least 0.2 V. Because V_t has a negative temperature coefficient of ≈ -0.7 mV/°C (Section 5.3.2), this means V_t (25 °C) ≥ 0.25 V.[4]

The figures just discussed are acceptable for CMOS logic technologies. In a dynamic memory technology (Dennard, 1984), however, the off-current requirement is much more stringent for the access transistor in the cell: on the order of $I_{off} \approx 10^{-13}$–10^{-14} A (see Section 12.2.2). This means V_t (100 °C) ≥ 0.6V for a DRAM access device with $W = L = 0.1$ μm. It should be noted that Eqs. (6.95) and (6.96) are analytical expressions derived under some simplifying approximations, e.g., long channel, uniform doping, etc. They are used here for order-of-magnitude estimates. More exact values of the off-current for a particular design should be obtained by numerical simulations.

Another consideration that may further limit how low the threshold voltage can be is the *burn-in* procedure. Burn-in is required in most VLSI technologies to remove early failures and ensure product reliability. It is usually carried out at elevated temperatures and over voltages to accelerate the degradation process. Both of these conditions further lower the threshold voltage, resulting in increased leakage current of the chip. Ideally, a burn-in procedure should be designed such that it does not require a compromise on the device performance.

6.3.1.3   On-current and MOSFET Performance

While the lower bound of threshold voltage is set by standby power constraints, the upper bound is imposed by considerations of on-current and switching delay. The on-current of a MOSFET is defined in the saturation region as

$$I_{on} = I_{ds}|_{V_{gs}=V_{dd}, V_{ds}=V_{dd}}. \tag{6.97}$$

Consider an nMOSFET initially in the off state with the source grounded and the drain charged to $V_{ds} = V_{dd}$ (e.g., in one of the CMOS inverter states in Figure 8.9). If a gate voltage $V_{gs} = V_{dd}$ is applied to turn it on, the drain node will be discharged by the current I_{on} (initially) and the drain voltage will decrease at a rate given by

---

[3] For a small fraction of transistors on the chip or for a larger standby power budget, higher off-current per transistor can be allowed.

[4] Lower threshold voltages are allowed in the scenario under footnote #3.

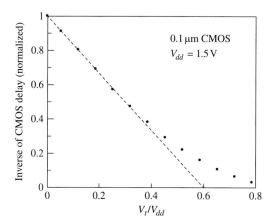

Figure 6.18 The reciprocal of CMOS delay in normalized units versus V_t/V_{dd}. The dots are from SPICE model simulations. The dashed line is a fitting proportional to $0.6 - V_t/V_{dd}$

$$C\frac{dV_{ds}}{dt} = -I_{on},\qquad(6.98)$$

where C is the total effective capacitance of the drain node. The switching delay for an incremental change of V_{ds} is then $-C\Delta V_{ds}/I_{on} \propto 1/I_{on}$. From a CMOS performance perspective, therefore, the lower the threshold voltage is, the higher the current drive I_{on}, hence the faster the switching speed.

It will be discussed in Chapter 8 that because of the finite rise time of V_{gs} at the input, the current that goes into the discharge equation, Eq. (6.98), is somewhat less than I_{on}. A circuit simulation model is used in Chapter 8 to analyze the delay sensitivity to threshold voltage. Figure 6.18 shows a typical example of CMOS performance, defined as the reciprocal of CMOS delay, versus normalized threshold voltage, V_t/V_{dd}. For $V_t/V_{dd} < 0.5$, the result can be fitted to a linear function proportional to $0.6 - V_t/V_{dd}$. This indicates, for example, about 30% of the performance is lost if V_t/V_{dd} is increased from 0.2 to 0.3. Because of such delay sensitivity, *the V_t/V_{dd} ratio is usually kept $\leq 1/4$ for high performance CMOS circuits.*

6.3.1.4 I_{on} versus I_{off} Characteristics

Since the choice of threshold voltage hinges on the tradeoff between I_{off} and I_{on}, we plot in Figure 6.19 a typical I_{ds} versus V_{gs} curve in both linear and logarithmic scales for the ease of reading both I_{on} and I_{off} in one figure. In essence, adjusting the threshold voltage of the device is equivalent to parallel shifting the I_{ds}–V_{gs} curves horizontally along the V_{gs}-axis.

Note that for an incremental shift of $\Delta V_t > 0$, I_{off} decreases by a factor of $\exp(q\Delta V_t/mkT)$ while I_{on} decreases by an amount of $g_m\Delta V_t$, where $g_m = dI_{ds}/dV_{gs}$ is the saturation transconductance or the slope of the I_{ds}–V_{gs} curve at $V_{gs} = V_{dd}$. In this regard, the often cited I_{on}/I_{off} ratio is not a meaningful figure of merit because it changes with ΔV_t, i.e., with the shift of the curves. In fact, to maximize the I_{on}/I_{off} ratio for a given V_{dd}, one

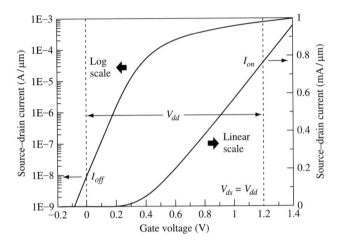

Figure 6.19 I_{ds}–V_{gs} characteristics in both linear and log scales; $V_{dd} = 1.2$ V in this example

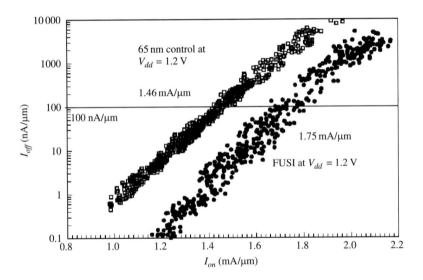

Figure 6.20 An experimental I_{off}–I_{on} plot for 65 nm nMOSFETs (Ranade *et al.*, 2005)

would want to shift to as high a threshold voltage as possible so that the entire $0 \leq V_{gs} \leq V_{dd}$ range is in the subthreshold. That, however, is not a desired mode of operation for high performance CMOS because I_{on} would then be so low that the delay is severely degraded by parasitic capacitances (discussed in Chapter 8).

It is a common practice in the VLSI industry to publish the I_{on}–I_{off} characteristics of their fabricated transistors thus circumventing any need of defining a threshold voltage. An example is shown in Figure 6.20 (Ranade *et al.*, 2005). The off-current criterion, $I_{off} = 100$ nA/μm, indicated by the horizontal line became a standard for high-performance MOSFETs (Kuhn *et al.*, 2012).

6.3.2   Channel Profile Design

In this section, we consider the design of MOSFET doping profile that satisfies the threshold voltage and other device requirements. Parameters that come into play include the gate length, power supply voltage, and gate oxide thickness. Trends of the power supply voltage, the threshold voltage, and the channel profile over the CMOS technology generations are discussed.

6.3.2.1   CMOS Design Considerations

CMOS device design involves choosing a set of parameters that are connected to a variety of circuit characteristics to be optimized. The choice of these device parameters is further subject to technology constraints and system compatibility requirements. Because various circuit characteristics are interrelated through the device parameters, tradeoffs among them are often necessary. For example, reduction of W_{dm} improves the short-channel effect, but degrades subthreshold current slope; thinner t_{ox} increases current drive, but has reliability concerns, etc. There is no unique way of designing CMOS devices for a given technology generation. Nevertheless, we attempt here to give a general guideline of how these device parameters should be chosen.

Since threshold voltage plays a key role in determining I_{off} and I_{on}, it is important to minimize the V_t tolerance, i.e., the spread between the high and low threshold voltages on the chip. The most dominant source of threshold voltage tolerances in a CMOS technology is from the short-channel effect. Channel length variations on a chip due to process imperfections give rise to threshold voltage variations. From Eq. (6.35), the gate-controlled potential barrier in a MOSFET of channel length L is lower than that of the long-channel by

$$\approx \frac{8}{\pi} \sqrt{(\psi_{bi} - 0.5\psi_s)(\psi_{bi} + V_{ds} - 0.5\psi_s)} \sin\left(\frac{\pi W_{dm}}{\lambda}\right) \exp\left(-\frac{\pi L}{2\lambda}\right), \qquad (6.99)$$

where λ is the scale length solved from Eq. (6.23). Since the pre-exponential factor is $\sim 1.5\,\mathrm{V}$, the general requirement is $L/\lambda > 2$ for the SCE of Eq. (6.99) to be less than $0.1\,\mathrm{V}$. For thin SiO$_2$ gate insulators, $\lambda \approx W_{dm} + 3t_{ox}$, where W_{dm} is the maximum depletion width at the threshold condition, $\psi_s = 2\psi_B$.

On the other hand, W_{dm} also affects the body factor $m = \Delta V_{gs}/\Delta\psi_s = 1 + 3t_{ox}/W_{dm}$, illustrated in Figure 5.5. It was discussed in Sections 5.1.3 and 5.3.1 that from the saturation current, subthreshold slope, and substrate sensitivity considerations, m should not be too much greater than unity, e.g., $m \leq 1.4$. These considerations are depicted in a plot of the t_{ox}–W_{dm} design plane in Figure 6.21. The intercept of the two lines, $W_{dm} + 3t_{ox} = L/2$ and $3t_{ox}/W_{dm} = m - 1 = 0.4$, defines an upper bound for the oxide thickness, $t_{ox,max} \approx L/20$. The lower limit of t_{ox} is imposed by technology constraints to $V_{dd}/\mathscr{E}_{ox,max}$, where $\mathscr{E}_{ox,max}$ is the maximum allowable oxide field based on breakdown and reliability considerations. *For a given L and V_{dd}, the allowable parameter space in a t_{ox}–W_{dm} design plane is a triangular area bounded by SCE, oxide field, and subthreshold slope (also substrate sensitivity) requirements*.

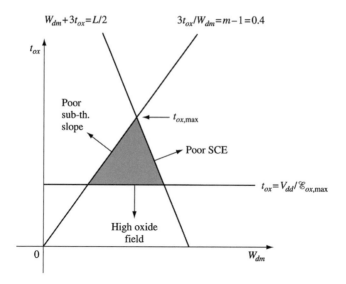

Figure 6.21 A t_{ox}–W_{dm} design plane. Some tradeoff among the various factors can be made within the parameter space bounded by SCE, body effect, and oxide field considerations

In addition to the oxide field limitation, direct quantum mechanical tunneling (Figure 4.40) also sets a lower limit to the thickness of gate oxide. Gate current density increases sharply as t_{ox} decreases below 2 nm. From Figure 4.40, the gate tunneling current density for a 1 nm thick oxide biased at 1 V is 10^3–10^4A/cm^2. Assume $L \approx 30$ nm, the gate current of an individual transistor, ~3 μA/μm, is still small compared with the typical on current, ~1 mA/μm, of the preceding stage so the switching delay of active transistors is hardly affected. But consider 10^9 transistors each with $W/L \approx 10$ and $L \approx 30$ nm, the total gate area per chip is of the order of 0.1 cm^2. The standby power dissipation of all the turned-on transistors[5] in the chip has reached an intolerable level above 100 W.

With high-κ gate insulators (Section 4.6.1.6), the EOT of the composite dielectric including the interfacial oxide layer can be reduced to 1 nm or lower with acceptable tunneling leakage. *Given the $t_{ox,max} \approx L/20$ criterion discussed with* Figure 6.21, *the 1 nm EOT limit translates into a channel length limit of ≈ 20 nm for planar bulk MOSFETs,* about what has been reported experimentally in the literature (Cho *et al.*, 2011; Shang *et al.*, 2012).

6.3.2.2 Trends of Power Supply Voltage and Threshold Voltage

For a design window to exist in Figure 6.21, it is required that $V_{dd}/\mathcal{E}_{ox,max} \leq t_{ox,max} \approx L/20$. This imposes an upper limit on the power supply voltage, namely,

$$V_{dd} \leq L\mathcal{E}_{ox,\,max}/20. \tag{6.100}$$

---

[5] The worst case gate leakage occurs with nMOSFETs biased at $V_{gs} = V_{dd}$ and $V_{ds} = 0$ (electrons tunnel from the inversion channel to the gate).

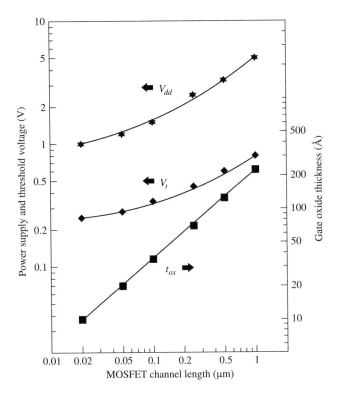

Figure 6.22 Trends of power-supply voltage, threshold voltage, and gate oxide thickness versus channel length for CMOS technologies from 1 μm to 0.02 μm (after Taur *et al.*, 1995)

For $L = 1\,\mu m$ CMOS technology, the gate oxides are relatively thick (~25 nm in Figure 6.22) and $\mathscr{E}_{ox,max} \approx 3\,MV/cm$. Equation (6.100) gives $V_{dd} \le 15$ V. There is plenty of design room to choose the power supply and threshold voltages that satisfy both the off-current and the performance requirements discussed in Sections 6.3.1.2 and 6.3.1.3. For example, $V_{dd} = 5$ V and $V_t = 0.8\text{–}1.0$ V, as shown in Figure 6.22, in which the history and trends of power supply voltage, threshold voltage, and oxide thickness are plotted for CMOS logic technologies from $1.0\,\mu m$ to $0.02\,\mu m$ channel lengths (Taur *et al.*, 1995). At shorter channel lengths, V_{dd} must be reduced. It becomes increasingly more difficult to satisfy both the performance and the off-current requirements. Fortunately, $\mathscr{E}_{ox,max}$ tends to increase for thinner oxides (see Section 4.6.4) as L is scaled down. This allows V_{dd} to be scaled at a slower rate than the channel length. Experimentally, it was found that $\mathscr{E}_{ox,max} \approx 6\,MV/cm$ for oxides thinner than 3 nm. Equation (6.100) then requires, e.g., that $V_{dd} \le 1.5$ V for $L = 50\,nm$ CMOS technology. With such a low supply voltage, one often faces a tradeoff of circuit speed versus leakage current. Scaling down V_t for speed causes I_{off} to increase exponentially. Even for the same V_t, I_{off} would increase over the generations since $I_{ds,Vt}$ of Eq. (6.96) goes up as the devices are scaled down – a manifestation of subthreshold current nonscalability to be discussed in Section 8.1. For this reason and for compatibility with the standardized power supply voltage of earlier generation

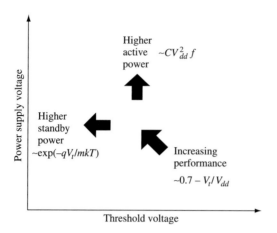

Figure 6.23 CMOS performance, active power, and standby power tradeoff in a V_{dd}–V_t design plane. The performance here is defined as the reciprocal of CMOS delay

systems, *the general trend is that V_{dd} has not been scaled down in proportion to L, and V_t has not been scaled down in proportion to V_{dd}*, as is evident in Figure 6.22. At $L = 20$ nm, $\mathcal{E}_{ox,\max}$ is pushed to 10 MV/cm for operation at $V_{dd} = 1$V.

As a result of the non-scaled V_{dd}, not only does the field increase over the CMOS generations (Section 8.1), the increasing power density also becomes more difficult to manage. It is discussed in Section 8.2 that the active or switching power of a CMOS circuit is given by

$$P_{ac} = CV_{dd}^2 \, f, \tag{6.101}$$

where C is the total equivalent capacitance being charged and discharged in a clock cycle, and f is the clock frequency. The power versus delay tradeoff can be represented conceptually in a V_{dd}–V_t plane shown in Figure 6.23 (Mii *et al.*, 1994). For higher performance, i.e., shorter delay, it is necessary to go for higher V_{dd} or lower V_t, which inevitably results in higher active power or higher standby power, or both. *Depending on the specific requirements of the application, CMOS technologies can be tailored to some extent by choosing an appropriate set of power supply and threshold voltages.* High performance CMOS operates at the upper left-hand corner of the design space, pushing both power limits. Low power CMOS can operate at lower supply voltages and possibly at a higher threshold voltage if the standby power is of primary concern. It is a common practice in the state-of-the-art CMOS technologies to provide multiple threshold voltages on a chip to allow the design flexibility of using different types of devices for different functions, e.g., in memory and logic circuits. This comes, of course, at the expense of additional process complexity and cost.

6.3.2.3   Channel Profile Requirement and Trends

For a uniformly doped p-type substrate of doping concentration N_a, the threshold voltage of an nMOSFET is given by Eq. (5.24):

$$V_t = V_{fb} + 2\psi_B + \frac{\sqrt{4\varepsilon_{si}qN_a\psi_B}}{C_{ox}}, \qquad (6.102)$$

where

$$V_{fb} = \phi_m - \phi_s = \phi_m - \left(\chi + \frac{E_g}{2q} + \psi_B\right). \qquad (6.103)$$

It was discussed in Section 4.3.5.1 that to obtain low threshold voltages in surface channel operation, n$^+$-polysilicon gates have been used for n-channel MOSFETs, and vice versa for p-channel (Taur *et al.*, 1993b). This means $V_{fb} = -E_g/2q - \psi_B$, hence

$$V_t = -\frac{E_g}{2q} + \psi_B + \frac{\sqrt{4\varepsilon_{si}qN_a\psi_B}}{C_{ox}}. \qquad (6.104)$$

Since the sum of the first two terms on the RHS is a small number ~ -0.1–0.2 V, V_t is largely determined by the third term which is proportional to the depletion charge density at the $2\psi_B$ condition.

By using Eq. (4.22) with $\psi_s = 2\psi_B$, or

$$W_{dm} = \sqrt{\frac{4\varepsilon_{si}\psi_B}{qN_a}}, \qquad (6.105)$$

Eq. (6.104) can be expressed in terms of W_{dm} and t_{ox} as

$$V_t = -\frac{E_g}{2q} + \psi_B + \frac{4\varepsilon_{si}\psi_B}{W_{dm}} \frac{t_{ox}}{\varepsilon_{ox}} = -\frac{E_g}{2q} + \psi_B + 2(m-1)(2\psi_B), \qquad (6.106)$$

where Eq. (5.29) has been used for m. Channel profile trends can be inferred by noticing that Eq. (6.106) does not scale much with device generation as neither m nor ψ_B changes significantly with channel length or doping. In fact, both m and ψ_B tend to increase slightly as the CMOS channel length scales down and higher doping is required to reduce W_{dm} thus the scale length λ. This is contrary to the downward trend of the V_t requirement depicted in Figure 6.22. For example, for a typical $m \approx 1.4$ and $\psi_B \approx 0.45$ V, $V_t \approx 0.6$ V with n$^+$-polysilicon gates. While this value happens to meet the V_t requirement for the 0.5 μm CMOS generation, it is too low for 1 μm CMOS and too high for CMOS generations 0.25 μm and below. It is therefore necessary to employ nonuniform doping to adjust the depletion charge density and obtain the desired V_t.

Equation (6.102) can be generalized to the case of a nonuniformly doped body as

$$V_t = V_{fb} + 2\psi_B + \frac{-Q_d}{C_{ox}}, \qquad (6.107)$$

where $Q_d < 0$ is the depletion charge density in silicon at the $2\psi_B$ condition. The next subsection shows that a high–low doping profile increases the depletion charge density for a given W_{dm} and therefore raises V_t over the uniformly doped value, whereas a low–high profile reduces the depletion charge and lowers V_t.

6.3.3   Nonuniform Channel Doping

In this section, we consider the threshold voltage and the maximum depletion width of a nonuniformly doped MOSFET. Specific examples include high–low and low–high doping profiles.

By employing the depletion approximation in subthreshold, the electric field, surface potential, and threshold voltage can be solved for an arbitrary p-type doping profile $N(x)$. The x-coordinate is in the depth direction as that defined in Figure 4.10. The electric field is obtained by integrating Poisson's equation once (neglecting mobile carriers in the depletion region):

$$\mathscr{E}(x) = \frac{q}{\varepsilon_{si}} \int_x^{W_d} N(x)dx, \tag{6.108}$$

where W_d is the depletion-layer width. Integrating again gives the surface potential,

$$\psi_s = \frac{q}{\varepsilon_{si}} \int_0^{W_d} \int_x^{W_d} N(x')dx'dx \tag{6.109}$$

Using integration by parts, one can show that the above is equivalent to (Brews, 1979)

$$\psi_s = \frac{q}{\varepsilon_{si}} \int_0^{W_d} xN(x)dx. \tag{6.110}$$

The integral of $xN(x)$ equals the center of mass of $N(x)$ within $(0, W_d)$ times the integral of $N(x)$.

The maximum depletion-layer width (long-channel) W_{dm} is determined by the condition $\psi_s = 2\psi_B$ when $W_d = W_{dm}$. **The threshold voltage of a nonuniformly doped MOSFET is then determined by both the integral (depletion charge density) and the center of mass of $N(x)$ within $(0, W_{dm})$.**

6.3.3.1   A High–Low Step Profile

Consider the idealized step doping profile shown in Figure 6.24 (Rideout *et al.*, 1975). It can be formed by making one or more low-dose, shallow implants into a uniformly doped substrate of concentration N_a. After drive-in, the implanted profile is approximated by a region of constant doping N_s that extends from the surface to a depth x_s. If the entire depletion region at the threshold condition is contained within x_s, the MOSFET can be considered as uniformly doped with a concentration N_s. The case of particular interest analyzed here is when the depletion width W_d exceeds x_s, so that part of the depletion region has a charge density N_s and part of it N_a. The integration in Eq. (6.110) can be easily carried out for this profile to yield the surface potential, or the band bending at the surface,

$$\psi_s = \frac{qN_s}{2\varepsilon_{si}}x_s^2 + \frac{qN_a}{2\varepsilon_{si}}\left(W_d^2 - x_s^2\right). \tag{6.111}$$

This equation can be solved for W_d as a function of ψ_s:

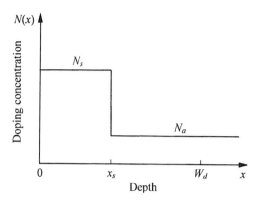

Figure 6.24 A schematic diagram showing the high–low step doping profile. $x = 0$ denotes the silicon–oxide interface

$$W_d = \sqrt{\frac{2\varepsilon_{si}}{qN_a}\left(\psi_s - \frac{q(N_s - N_a)x_s^2}{2\varepsilon_{si}}\right)}. \tag{6.112}$$

This is less than the depletion width in the uniformly doped (N_a) case for the same surface potential.

The electric field at the surface is obtained by evaluating the integral in Eq. (6.108) with $x = 0$:

$$\mathscr{E}_s = \frac{qN_sx_s}{\varepsilon_{si}} + \frac{qN_a(W_d - x_s)}{\varepsilon_{si}}. \tag{6.113}$$

From Gauss's law, the total depletion charge per unit area in silicon is given by

$$Q_d = -\varepsilon_{si}\mathscr{E}_s = -qN_sx_s - qN_a(W_d - x_s), \tag{6.114}$$

as would be expected from Figure 6.24. *The effect of the additional surface doping is then to increase the depletion charge within $0 \le x \le x_s$ by $(N_s - N_a)x_s$ and, at the same time, reduce the depletion layer width as given by* Eq. (6.112).

The maximum depletion width at threshold is given by Eq. (6.112) with $\psi_s = 2\psi_B$:

$$W_{dm} = \sqrt{\frac{2\varepsilon_{si}}{qN_a}\left(2\psi_B - \frac{q(N_s - N_a)x_s^2}{2\varepsilon_{si}}\right)}. \tag{6.115}$$

The threshold voltage is then obtained by substituting Eqs. (6.114) and (6.115) into Eq. (6.107):

$$V_t = V_{fb} + 2\psi_B + \frac{1}{C_{ox}}\sqrt{2\varepsilon_{si}qN_a\left(2\psi_B - \frac{q(N_s - N_a)x_s^2}{2\varepsilon_{si}}\right)} + \frac{q(N_s - N_a)x_s}{C_{ox}}. \tag{6.116}$$

There is an ambiguity as to whether $2\psi_B$ is defined in terms of N_s or N_a. We adopt the convention that $2\psi_B$ is defined in terms of the p-type concentration at the depletion-layer edge, i.e., $2\psi_B = (2kT/q) \ln(N_a/n_i)$. In fact, it makes very little difference which concentration to use, since $2\psi_B$ is a rather weak function of the doping concentration anyway.[6]

In Section 5.1.3.3, we showed that the inverse subthreshold slope is given by 2.3 mkT/q per decade where $m = dV_{gs}/d\psi_s$ at $\psi_s = 2\psi_B$. In the nonuniformly doped case, m is still given by Eq. (5.29), i.e.,

$$m = 1 + \frac{\varepsilon_{si}/W_{dm}}{C_{ox}} = 1 + \frac{C_{dm}}{C_{ox}} = 1 + \frac{3t_{ox}}{W_{dm}}, \tag{6.117}$$

but with W_{dm} given by Eq. (6.115). This is to be expected from the basic concept of m in Figure 5.5, which applies regardless of the doping specifics. Similarly, the substrate sensitivity is

$$\left.\frac{dV_t}{d(-V_{bs})}\right|_{V_{bs}=0} = \frac{\varepsilon_{si}/W_{dm}}{C_{ox}} = \frac{C_{dm}}{C_{ox}} = m - 1 \tag{6.118}$$

as before. In short, *all the previous expressions for the depletion capacitance, subthreshold slope, and body-effect coefficient in terms of W_{dm} for the uniformly doped case remain valid for the nonuniformly doped case*. The only difference is that the maximum depletion layer width W_{dm} in the high–low step doping case is given by Eq. (6.115) instead of Eq. (6.105).

The results of the high–low step profile discussed can be generalized to other profiles as well. As far as the threshold voltage and depletion width are concerned, only the integrated doping density within $(0, W_{dm})$ [Eq. (6.108)] and the center of mass [Eq. (6.110)] matter. For the step profile in Figure 6.24, the added dose (number of dopants per area) over a constant background doping of N_a is

$$D_I = (N_s - N_a)x_s \tag{6.119}$$

centered at $x_c = x_s/2$. Any implanted Gaussian profile (assuming it's entirely in silicon) with the same dose D_I and peak depth x_c would result in the same W_{dm} and V_t as that of Eqs. (6.115) and (6.116). For a given dose D_I, the resulting threshold voltage shift depends on the location of the implant, x_c. *For shallow surface implants, $x_c = 0$, there is no change in the depletion width. The V_t shift is simply given by qD_I/C_{ox}, like that from a sheet of charge qD_I at the silicon–oxide interface*. All other device parameters, e.g., substrate sensitivity and subthreshold slope, remain unchanged.

Although the above analysis on nonuniform doping assumes $N_s > N_a$, the results remain equally valid if $N_s < N_a$. Such a profile is referred to as the *retrograde channel doping*, discussed in Section 6.3.3.2.

---

[6] The "$2\psi_B$" definition of threshold voltage is only a historical convention. Actually, the channel "turns on" when the conduction band at surface is within 0.1 V (a few kT/q) of the conduction band of the n⁺ source, regardless of the p-type body doping. In that respect, the approximation $2\psi_B \sim 1$ V is often used in the discussions.

6.3.3.2 Retrograde (Low–High) Channel Profile

When the channel length is scaled to 0.25 μm and below, a higher doping concentration is needed in the channel to reduce W_{dm} and control the short-channel effect. If a uniform profile were used, the threshold voltage [Eq. (6.102)] would be too high even with dual polysilicon gates. The problem is further aggravated by quantum effects, which, as discussed in Section 5.3.3, can add another 0.1–0.2V to the threshold voltage because of the increasing fields (van Dort *et al.*, 1994).

To reduce the threshold voltage without significantly increasing the gate depletion width, a retrograde channel profile, i.e., a low–high doping profile, as shown schematically in Figure 6.25, is required (Sun *et al.*, 1987; Shahidi *et al.*, 1989). Such a profile is formed using higher-energy implants that peak below the surface. It is assumed that the maximum gate depletion width extends into the higher-doped region. All the equations in Section 6.3.3.1 remain valid for $N_s < N_a$. For simplicity, we assume an ideal retrograde channel profile for which $N_s = 0$. Equation (6.116) then becomes

$$V_t = V_{fb} + 2\psi_B + \frac{qN_a}{C_{ox}}\sqrt{\frac{4\varepsilon_{si}\psi_B}{qN_a} + x_s^2} - \frac{qN_a x_s}{C_{ox}}. \tag{6.120}$$

Similarly, Eq. (6.115) with $N_s = 0$ gives the maximum depletion width,

$$W_{dm} = \sqrt{\frac{4\varepsilon_{si}\psi_B}{qN_a} + x_s^2}. \tag{6.121}$$

The net effect of low–high doping is that the threshold voltage is reduced, but the depletion width has increased, just opposite to that of high–low doping. Note that Eq. (6.121) has the same form as Eq. (3.21) for a p–i–n diode. All other expressions, such as those for the subthreshold swing and the substrate sensitivity, in Section 6.3.3.1 apply with W_{dm} replaced by Eq. (6.121).

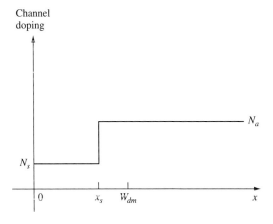

Figure 6.25 A schematic diagram showing the low–high (retrograde) step doping profile. $x = 0$ denotes the silicon–oxide interface

6.3.3.3   Extreme Retrograde Profile and Ground-Plane MOSFET

Two limiting cases are worth discussing. If $x_s \ll (4\varepsilon_{si}\psi_B/qN_a)^{1/2}$, then W_{dm} remains essentially unchanged from the uniformly doped value [Eq. (6.121)], while V_t is lowered by a net amount equal to $qN_a x_s/C_{ox}$ [Eq. (6.120)]. In the other limit, N_a is sufficiently high that $x_s \gg (4\varepsilon_{si}\psi_B/qN_a)^{1/2}$. In that case, $W_{dm} \approx x_s$, and the entire depletion region is undoped. All the depletion charge is located at the edge of the depletion region. The square root term in Eq. (6.120) can be expanded into a power series to yield

$$V_t = V_{fb} + 2\psi_B + \frac{\varepsilon_{si}(2\psi_B/x_s)}{C_{ox}}. \tag{6.122}$$

The last term, due to the depletion charge density in silicon, $\varepsilon_{si}(2\psi_B/x_s)$, can also be derived from Gauss's law by considering that the field in the undoped region is constant and equals $2\psi_B/x_s$ at threshold. Note that the work function difference that goes into V_{fb} is between the gate and the p$^+$ silicon at the edge of the depletion region. Using $m = 1 + 3t_{ox}/W_{dm} = 1 + 3t_{ox}/x_s$, we can rewrite Eq. (6.122) as

$$V_t = V_{fb} + 2\psi_B + (m-1)2\psi_B. \tag{6.123}$$

Comparison with Eq. (6.106) shows that, with the extreme retrograde profile, the depletion charge (the third) term of V_t is reduced to half of the uniformly doped value. If there is a substrate bias V_{bs} present, the $2\psi_B$ factor in the last term of Eq. (6.123) is replaced by $(2\psi_B - V_{bs})$, i.e.,

$$V_t = V_{fb} + 2m\psi_B - (m-1)V_{bs}. \tag{6.124}$$

Since ψ_B is a weak function of N_a, the results of Eqs. (6.122)–(6.124) are independent of the exact value of N_a as long as it is high enough to satisfy $x_s \gg (4\varepsilon_{si}\psi_B/qN_a)^{1/2}$. All the essential device characteristics, such as SCE (W_{dm}), subthreshold slope (m), and threshold voltage, are determined by the depth of the undoped layer, x_s. **The limiting case of the retrograde channel profile therefore degenerates into a ground-plane MOSFET** (Yan et al., 1991). The band diagram and charge distribution of such a device at the threshold condition are shown schematically in Figure 6.26. Note that the field is constant (no curvature in potential) in the undoped region between the surface and x_s. There is an abrupt change of field at $x = x_s$, where a delta function of depletion charge (area $= 2\varepsilon_{si}\psi_B/x_s$) resides. Beyond x_s, the bands are essentially flat. It is desirable not to extend the p$^+$ region under the source and drain junctions, since that will increase the parasitic junction capacitance. The ideal channel doping profile is then that of a low–high–low type, in which the narrow p$^+$ region is used only to confine the gate depletion width. Such a profile is also referred to as *pulse-shaped doping* or *delta doping* in the literature.

When CMOS devices are scaled to 20 nm channel lengths and below, the field is so high and the quantum effect so strong that even the extreme retrograde profile cannot deliver a $V_t \approx 0.2$ V with n$^+$ and p$^+$ polysilicon gates. Besides finding new gate materials with work functions outside of n$^+$ and p$^+$ silicon, further reduction of V_t can be accomplished, by either counterdoping the channel or forward biasing the

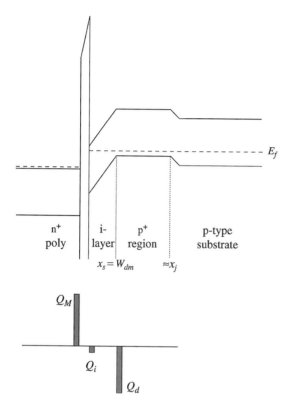

Figure 6.26 Band diagram and charge distribution of an extreme retrograde-doped or ground-plane nMOSFET at the threshold condition

substrate. A forward substrate bias also helps improve short-channel effects as it effectively reduces the built-in potential, ψ_{bi} in Eq. (6.35), between the source–drain and the p-type substrate. The flip side is that it causes source junction leakage, increases drain-to-substrate capacitance, and degrades subthreshold current slope and body effect.

Instead of analyzing the counter-doped channel mathematically by letting $N_s < 0$ for the profile in Figure 6.25, it is more instructive to give a graphical interpretation of the potential, field, and depletion charge by plotting the electric field versus depth, as shown in Figure 6.27. In the uniformly doped case, $\mathscr{E}(x)$ is a straight line with a negative slope whose magnitude is proportional to the substrate doping concentration N_a. The x-intercept gives the depletion layer width where $\mathscr{E} = 0$. The y-intercept gives the surface electric field $\mathscr{E}_s$ which by Gauss's law is proportional to the total depletion charge per unit area. Since $\mathscr{E} = -d\psi/dx$, the triangular area under $\mathscr{E}(x)$ equals the surface potential or the band bending ψ_s. As the gate voltage increases, both W_d and $\mathscr{E}_s$ increase, so does ψ_s until it reaches $2\psi_B$. At this point, surface inversion occurs and the depletion layer width has reached its maximum value. The depletion charge term of the threshold voltage, $V_{ox} = \varepsilon_{si}\mathscr{E}_s/C_{ox}$, is proportional to the y-intercept or the surface field $\mathscr{E}_s$ when $\psi_s = 2\psi_B$.

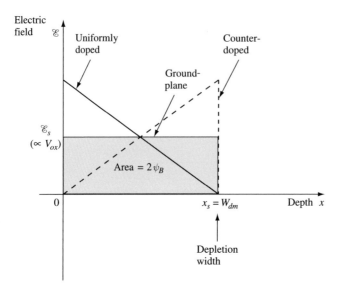

Figure 6.27 Graphical interpretation of uniformly doped, extreme retrograde or ground-plane, and counterdoped profiles. The band bending is given by the area under $\mathcal{E}(x)$ which equals $2\psi_B$ at threshold for all three cases

In the extreme retrograde or the ground-plane case, $\mathcal{E}(x)$ is constant within the undoped region, $0 < x < x_s$, where there is no depletion charge. At the threshold condition, the shaded rectangular area for the ground-plane case is approximately the same as the triangular area under the uniformly doped $\mathcal{E}(x)$ since $2\psi_B$ is a rather weak function of N_a and can be considered as a constant for practical purposes. It is then clear that the depletion charge term of V_t or the y-intercept of the ground-plane case is half of that of the uniformly doped case – exactly as indicated by Eqs. (6.123) and (6.106).

Also shown in Figure 6.27 is a specific example of a counter-doped channel. The slope $d\mathcal{E}/dx$ has the same magnitude as the uniformly doped case, but of the opposite polarity. Both the depletion width and the band bending [area under $\mathcal{E}(x)$] are the same as the previous two cases. But the y-intercept ($\mathcal{E}_s$) is zero which means that the net charge in silicon is zero due to cancellation of the charge in the counter-doped region with the depletion charge at the edge of the depletion region. This yields a very low V_t. Further counter-doping would result in $\mathcal{E}_s < 0$ or a buried channel MOSFET.

The band diagrams of these three doping cases at the threshold condition are further illustrated in Figure 6.28. Both the depletion width and the band bending are kept the same for all three. However, the surface fields (slopes) are very different, leading to dramatically different potential drops across the gate oxide.

6.3.3.4   Laterally Nonuniform Channel Doping

So far we have discussed nonuniform channel doping in the vertical direction. Another type of nonuniform doping used in very short-channel devices is in the

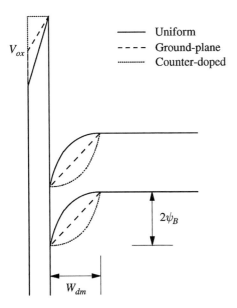

Figure 6.28 Band diagrams of uniformly doped, ground-plane (extreme retrograde), and counter-doped MOSFETs at threshold

lateral direction. For nMOSFETs, this is achieved by a medium-dose p-type implant carried out at the same time as the n^+ source–drain implant after gate patterning. As shown in Figure 6.29, the p-type doping peaks near the source and drain ends of the channel but dips in the middle because of blocking of the implant by the gate. Such a self-aligned, laterally nonuniform channel doping is often referred to as *halo* or *pocket* implants (Ogura *et al.*, 1982). Figure 6.29 shows how halo works to counteract the short-channel effect, i.e., threshold voltage rolloff toward the shorter devices within a spread of the channel length (or gate length) tolerance. At the longer end of the spread shown in Figure 6.29(a), the two p^+ pockets are farther apart than the pockets at the shorter end of the spread in Figure 6.29(b). This creates a higher average p-type doping in the shorter device than in the longer device. Higher average doping means higher threshold voltage. ***So laterally nonuniform halo doping produces a tendency for the threshold voltage to increase toward the shorter devices, which works to counteract the short-channel effect in the opposite direction***. With an optimally designed 2-D nonuniform doping profile called the *superhalo*, it is possible in principle to neutralize the short-channel effect and achieve essentially the same threshold voltage for channel lengths within 20-30 nm (Taur *et al.*, 1998).

6.3.4   Discrete Dopant Effects on Threshold Voltage

As CMOS devices are scaled down, the number of dopant atoms in the depletion region of a minimum geometry device decreases. Due to the discreteness of atoms, there is a random statistical fluctuation of the number of dopants within a given

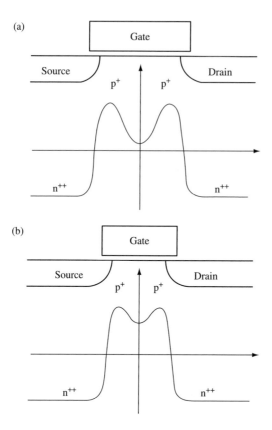

Figure 6.29 Laterally nonuniform halo doping in nMOSFETs. For a given design length on the mask, there is a spread of the actual gate length on the wafer. The longer end of the spread is shown in (a), the shorter in (b). The sketch below each cross-section shows the schematic doping variation along a horizontal cut through the source and drain regions

volume around its average value. For example, in a uniformly doped $W = L = 100$ nm nMOSFET, if $N_a = 10^{18}$ cm^{-3} and $W_{dm} = 35$ nm, the average number of acceptor atoms in the depletion region is $N = N_a LW W_{dm} = 350$. The actual number fluctuates from device to device with a standard deviation $\sigma_N = \langle (\Delta N)^2 \rangle^{1/2} = N^{1/2} = 18.7$, which is a significant fraction of the average number N. Since the threshold voltage of a MOSFET depends on the ionized dopant charge in the depletion region, this translates into a threshold-voltage fluctuation which could affect the operation of VLSI circuits.

6.3.4.1   A Simple First-Order Model

To estimate the effect of depletion charge fluctuation on threshold voltage, we consider a small volume $dx\,dy\,dz$ at a point (x, y, z) in the depletion region of a uniformly doped (N_a) MOSFET. The x-axis is in the depth direction, the y-axis in the length direction, and the z-axis in the width direction. The average number of dopant atoms in this small volume is $N_a \, dx \, dy \, dz$. The actual number fluctuates around this value with a standard deviation of $\sigma_{dN} = (N_a \, dx \, dy \, dz)^{1/2}$. This fluctuation can be thought of as a small delta function of nonuniform doping (either positive or negative) at (x, y, z) superimposed on the uniformly doped background N_a. To the first order, the effect of such a small fluctuation on the threshold voltage is equivalent to a sheet doping density or dose of $\Delta D = \sigma_{dN}/WL = (N_a \, dx \, dy \, dz)^{1/2}/WL$ at depth x but spread out uniformly in the length (L) and width (W) directions. The threshold-voltage shift due to ΔD at x is obtained by replacing $(N_s - N_a)x_s$ with ΔD and $x_s/2$ with x in Eq. (6.116), then extracting the first-order terms in ΔD:

$$\Delta V_t = \frac{q\Delta D}{C_{ox}}\left(1 - x\sqrt{\frac{qN_a}{2\varepsilon_{si}(2\psi_B)}}\right) = \frac{q\Delta D}{C_{ox}}\left(1 - \frac{x}{W_{dm}}\right). \tag{6.125}$$

This expression is quite general and is applicable even if the background doping is nonuniform. It indicates that the effect of an incremental doping on the threshold voltage depends on what the depth is. Near the surface, the incremental doping just adds to the total depletion charge with very little effect on the surface potential or depletion region width. Near the depletion region boundary, however, the incremental doping also adds to ψ_s according to Eq. (6.110). This reduces the depletion region width W_{dm} needed to reach the $2\psi_B$ condition. Because of the two offsetting effects, an incremental doping near the depletion region boundary has very little net effect on threshold voltage.

With the earlier expression of $\Delta D = (N_a \, dx \, dy \, dz)^{1/2}/WL$, the mean square deviation (variance) of threshold voltage due to dopant number fluctuation in a volume $dx \, dy \, dz$ at (x, y, z) is then

$$\left\langle \Delta V_t^2 \right\rangle_{x,y,z} = \frac{q^2 N_a}{C_{ox}^2 L^2 W^2}\left(1 - \frac{x}{W_{dm}}\right)^2 dx \, dy \, dz. \tag{6.126}$$

Since dopant number fluctuations at various points are completely random and uncorrelated, the total mean square fluctuation of the threshold voltage is obtained by integrating Eq. (6.126) over the entire depletion region:

$$\sigma_{V_t}^2 = \frac{q^2 N_a}{C_{ox}^2 L^2 W^2}\int_0^W \int_0^L \int_0^{W_{dm}}\left(1 - \frac{x}{W_{dm}}\right)^2 dx \, dy \, dz. \tag{6.127}$$

It is straightforward to carry out the integration and obtain

$$\sigma_{V_t} = \frac{q}{C_{ox}}\sqrt{\frac{N_a W_{dm}}{3LW}}. \tag{6.128}$$

In the above 100-nm example, $\sigma_{Vt} = 17.5$ mV if $t_{ox} = 3.5$ nm. This is small compared with the short-channel threshold rolloff in Section 6.1.2.4, but can be significant in minimum-geometry devices, for example, in an SRAM cell. Note that Eq. (6.128) is likely to underestimate the effect of dopant number fluctuations because of the simplification of the model.

In the above analysis, it was assumed that the surface potential is uniform in both the length and the width directions of the device. In other words, all the lumpiness due to local fluctuations of the depletion charge is smoothed out and the surface potential depends only on the average (or total) depletion charge of the device. This assumption is not quite valid in the subthreshold region, where current injection is dominated by the highest potential barrier in the channel rather than by the average value (Nguyen, 1984). In general, the problem needs to be solved by 3-D numerical simulations (Wong and Taur, 1993). The results indicate that, in addition to the threshold fluctuations of a similar magnitude to that expected from Eq. (6.128), there is also a negative shift of the average threshold voltage, especially in the subthreshold region. This is due to the inhomogeneity of surface potential resulting from the microscopic random distribution of discrete dopant atoms in the channel.

6.3.4.2 Discrete Dopant Effects in a Retrograde-Doped Channel

Threshold voltage fluctuations due to discrete dopants are greatly reduced in a retrograde-doped channel. Consider the profile in Figure 6.25 with $N_s = 0$, i.e., the channel is undoped within $0 < x < x_s$. The average threshold voltage and the maximum depletion width W_{dm} are given by Eqs. (6.120) and (6.121), respectively. For a small volume of dopants at (x, y, z) where $x_s < x < W_{dm}$, Eq. (6.126) still holds. The x-integral in Eq. (6.127), however, is carried out from x_s to W_{dm}, yielding

$$\sigma_{V_t} = \frac{q}{C_{ox}} \sqrt{\frac{N_a W_{dm}}{3LW}} \left(1 - \frac{x_s}{W_{dm}}\right)^{3/2} \tag{6.129}$$

for a retrograde-doped channel. *In the extreme retrograde or ground-plane limit shown in Figure 6.27, $x_s = W_{dm}$ and the threshold voltage fluctuation goes to zero.* This is also clear from Eq. (6.122), where the threshold voltage is essentially independent of N_a. Of course, the technological challenge is then to control the tolerance of the undoped-layer thickness x_s so that it does not introduce a different kind of threshold voltage variation.

In reality, retrograde channel doping reduces the threshold fluctuations due to discrete dopants, but does not eliminate them. For an optimally designed 25 nm MOSFET with superhalo (Taur *et al.*, 1998), a 3-D Monte-Carlo simulation has shown that the 1σ threshold voltage fluctuation due to discrete, random dopants is $10 \times W^{-1/2}$ mV, where W is the device width in microns (Frank *et al.*, 1999). This is tolerable for logic devices with $W/L > 10$, but could be problematic for SRAM cell transistors which have minimum widths and require a 6σ guard band for large arrays on a chip. For example, for $W = 25$ nm, the above figure gives a 1σ threshold variation of 63 mV and a 6σ threshold variation of 0.38 V. The above numbers also indicate that Eq. (6.128) may be as much as $2\times$ too low.

6.4 MOSFET Degradation and Breakdown at High Fields

Besides causing dielectric breakdown, the high electric fields in a MOSFET can also cause degradation of the device characteristics. *Hot-carrier effects* (HCE) and *negative-bias-temperature instability* (NBTI) are two of the most important degradation phenomena in modern CMOS devices. Here we describe briefly the phenomena and the physical mechanisms involved.

6.4.1 Hot-Carrier Effects

Consider an n-channel MOSFET with $V_{gs} > V_t$ applied to the gate and $V_{ds} > 0$ applied to the drain. A high-field space-charge region is established in the silicon near the drain, as illustrated in Figure 6.30. As the electrons drift toward the drain, they gain energy from the electric field in the channel and become hot. The hot electrons can cause impact ionization near the drain, or they can be injected into the gate insulator (see Section 4.6.2). The secondary holes from impact ionization contribute to a substrate current (Abbas, 1974). Substrate currents at drain voltages less than the silicon bandgap voltage E_g/q have been observed, suggesting that some electrons can gain additional energy from electron–electron and/or electron–phonon collisions (Chung *et al.*, 1990).

Figure 6.31 is a typical plot of the channel current and substrate current as a function of the gate voltage, in this case for an n-channel MOSFET with 0.25 μm channel length having a threshold voltage, V_t, of about 0.4 V (Chang *et al.*, 1992). The V_{ds} is fixed at 2.5 V. Notice that the substrate current increases with the gate voltage in the subthreshold region, peaking at a gate voltage somewhat beyond V_t, and then decreases with further increases in the gate voltage. This complex dependence on gate voltage can be understood as follows. The electrons available for initiating impact ionization, to first order, come from the drain current. At $V_{gs} < V_t$, there is no surface inversion channel and the maximum electric field in the silicon near the drain end is relatively independent of the gate voltage. As a result, the substrate current is roughly

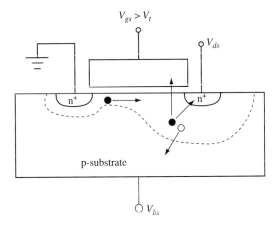

Figure 6.30 Schematic illustrating the physical processes that give rise to the channel hot-electron effect in an n-channel MOSFET. The dotted line indicates the boundary of the space-charge region

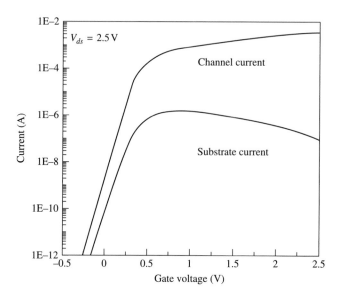

Figure 6.31 Typical plots of the channel current and substrate current of a MOSFET. The example shown here is for an n-channel FET having 0.25 μm channel length and 10 μm channel width (after Chang *et al.*, 1992)

proportional to the drain current, as can be seen in Figure 6.31. At $V_{gs} > V_t$, a surface inversion channel is formed. Since V_{ds} is fixed to a large value beyond the saturation voltage V_{dsat}, e.g., of Eq. (6.48), the MOSFET is biased in the velocity saturation region. The lateral field, $\sim dV/dy$, in this region grows exponentially with y, as indicated by Eq. (6.70). The highest lateral field, at the drain, is proportional to $V_{ds} - V_{dsat}$. *For a given* V_{ds}, *as* V_{gs} *increases, so does* V_{dsat}, *which means* $V_{ds} - V_{dsat}$ *decreases, hence the maximum electric field at the drain decreases.* As shown in Figure 3.23, the rate of impact ionization decreases rapidly with a decrease in the electric field. The net result is that the substrate current decreases with increasing gate voltage for V_{gs} substantially higher than V_t, as seen in Figure 6.31. The reader interested in detailed models for the substrate current in a MOSFET is referred to the literature (e.g., Hu *et al.*, 1985; Kolhatkar and Dutta, 2000).

The hot electrons injected into the gate insulator near the drain region contribute to a gate current. The gate current can cause bulk and interface traps to be generated (see Section 4.6.1.3), and some of the injected electrons can become trapped in the gate insulator near the drain. The trapped electrons and the interface states cause the surface potential near the drain to shift. Since the source of the hot electrons is the channel current, this device degradation is referred to as *channel hot electron (CHE) effect* (Ning *et al.*, 1979). The turn on characteristics of a MOSFET are determined primarily by the surface potential near the source end of the channel (see Section 5.1.2). With the damage localized near the drain junction, a MOSFET that has suffered significant CHE damage shows a larger damage-induced threshold voltage shift when it is operated in the source–drain-reversed mode than in the normal mode

(Abbas and Dockerty, 1975; Ning *et al.*, 1977b). In the case of a p-channel MOSFET, the same device degradation is referred to as *channel hot hole effect*.

In certain circuit configurations, the terminal voltages of a MOSFET are such that V_{ds} is small or zero, while both the source and the drain are reverse biased with respect to the substrate, i.e., $V_{bs} < 0$ for nMOS. (An example of this bias configuration is when a MOSFET is used as a pass-gate.) In this case, the channel electrons do not gain much energy traveling from source to drain. However, the minority electrons from the p-substrate can gain energy as they drift toward the silicon−oxide interface. These hot electrons can be injected into the gate insulator, as depicted in Figure 4.47. The injected electrons can generate bulk and interface traps or become trapped in the oxide layer. Since the hot electrons come from the substrate, the associated device degradation is referred to as *substrate hot electron (SHE) effect*. By symmetry, SHE damage to the oxide spreads more or less uniformly over the entire device channel region. Also, since the minority electron density in the substrate increases with temperature, SHE effect increases with temperature (Ning *et al.*, 1976). In the case of a p-channel MOSFET, instead of hot electrons, we have hot holes causing device degradation. As a device degradation effect, SHE is appreciable only in CMOS devices designed to operate with voltages larger than 2.5 V, and when the reverse substrate bias is large enough to cause high probability of hot electron injection (see Figure 4.47).

Hot-carrier degradation is one of the major effects limiting the voltage that can be applied to CMOS devices. In general, the approach to minimize hot-carrier damage is to (a) reduce the peak electric field in silicon to reduce the energy of hot carriers, and (b) minimize the trap density in the gate insulator. Designers often employ the so-called *lightly doped drain* (LDD) design to suppress CHE effect (Ogura *et al.*, 1982). In a LDD design, the drain region adjacent to the channel has a lower doping concentration than the drain region away from the channel. This laterally graded drain doping profile reduces the peak electric field near the drain. The physics is qualitatively the same as that for a p−i−n diode, discussed in Section 3.1.2.3. However, both the trap density of the gate insulator and its susceptibility to hot-carrier damage depend on the growth process of the gate insulator, as well as on the subsequent processes used to complete the integrated-circuit chip fabrication. As a result, hot-carrier effects cannot be predicted sufficiently accurately in advance. Instead, for each CMOS technology generation, the effects are characterized, modeled, and then included in the design of circuits. The reader is referred to the literature on the subject for more details (see, e.g., Takeda *et al.*, 1995).

6.4.2 Negative-Bias-Temperature Instability

It was reported by Deal *et al.* (1967) that a negative voltage applied to the gate of an MOS capacitor at elevated temperatures can cause both a build up of positive charge in the oxide and an increase in the density of surface states. To turn on a p-channel MOSFET, the gate electrode is biased negatively with respect to the n-type body. Also, in many applications, the device temperature can be rather high, often close to 100 °C. Thus, NBTI can occur naturally in the operation of a p-channel MOSFET, causing its threshold voltage to become more negative with time. Many papers have been published on the various aspects of NBTI. It is found that device degradation due to NBTI is a function of the gate insulator process, and the degradation worsens with both the temperature and oxide field (Jeppson and Svensson, 1977; Blat *et al.*, 1991). For short time periods, the degradation has approximately a $t^{1/4}$ time dependence (Jeppson and Svensson, 1977). It then saturates to a value depending on the oxide field and the temperature (Zafar *et al.*, 2004).

If a p-channel MOSFET is stressed with zero or relatively small voltages between the source and drain, there is only NBTI effect and the channel hot hole effect is negligible. In this stress mode, the device degradation or NBTI by itself is relatively insensitive to channel length. However, a pMOSFET in the off state in a CMOS circuit typically has a relatively large voltage between its source and drain. Therefore, the device degradation experienced by a short-channel pMOSFET is often caused by a combination of channel hot hole effect and NBTI (La Rosa *et al.*, 1997). NBTI must be characterized for each technology so its effect can be included in the design of circuits, particularly for CMOS circuits that depend on good matching of the threshold voltages of p-channel MOSFETs (Rauch III, 2002). An excellent review of the current understanding of the physical mechanisms of NBTI in modern CMOS devices has been given by Schroder and Babcock (2003).

6.4.3   MOSFET Breakdown

Breakdown occurs in a short-channel MOSFET when the drain voltage exceeds a certain value, as shown in Figure 6.32. At high drain voltages, the peak electric field in

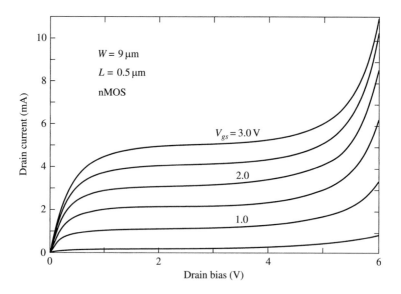

Figure 6.32 Example I_{ds}–V_{ds} curves of a short-channel nMOSFET showing breakdown at high drain voltages (after Sun *et al.*, 1987)

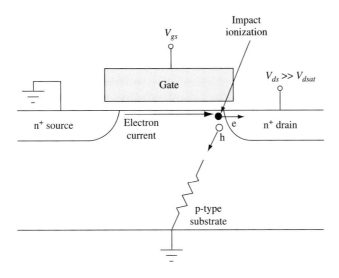

Figure 6.33 Schematic diagram showing impact ionization at the drain

the saturation region can attain large values. When the field exceeds mid-10^5 V/cm, impact ionization (Section 3.3.1) takes place at the drain, leading to an abrupt increase of the drain current. The breakdown voltage of nMOSFETs is usually lower than that of pMOSFETs because electrons have a higher rate of impact ionization (Figure 3.23) and because n^+ source and drain junctions are more abrupt than p^+ junctions. There is also a weak dependence of the breakdown voltage on channel length; shorter devices have a lower breakdown voltage.

The breakdown process in an nMOSFET is shown schematically in Figure 6.33. Electrons gain energy from the field as they move down the channel. Before they lose energy through collisions, they possess high kinetic energy and are capable of generating secondary electrons and holes by impact ionization. The generated electrons are attracted to the drain, adding to the drain current, while the holes are collected by the substrate contact, resulting in a substrate current. The substrate current in turn can produce a voltage (IR) drop from the spreading resistance in the bulk, which tends to forward-bias the source junction. This lowers the threshold voltage of the MOSFET and triggers a positive feedback effect, which further enhances the channel current. Substrate current (Figure 6.31) is usually a good indicator of hot carriers generated by low-level impact ionization before runaway breakdown occurs.

Breakdown often results in permanent damage to the MOSFET as large amounts of hot carriers are injected into the oxide in the gate-to-drain overlap region. MOSFET breakdown is particularly a problem for VLSI technology during the elevated-voltage burn-in process. It can be relieved to some extent by using a LDD structure (Ogura *et al.*, 1982), which introduces additional series resistance and reduces the peak field in a MOSFET. However, drain current and therefore device performance are traded off as a result. Ultimately, the devices should operate at a power-supply voltage far enough below the breakdown condition. This is one of the key CMOS design considerations in Section 6.3.2.

Exercises

6.1 For the MOSFET scale length given by Eq. (6.23), show that
(a) $\lambda \geq W_{dm}$ and $\lambda \geq t_{ox}$, i.e., λ is larger than the larger of W_{dm}, t_{ox}.
(b) In the special case of $\varepsilon_{ox} = \varepsilon_{si}$, $\lambda = W_{dm} + t_{ox}$.
(c) If $t_{ox} \ll W_{dm}$ (lower right corner of Figure 6.7), $\lambda \approx W_{dm} + (\varepsilon_{si}/\varepsilon_{ox})t_{ox}$.

6.2 The MOSFET scale length is given by Eq. (6.23). For $\varepsilon_{si} = 11.7\varepsilon_0$, $\varepsilon_{ox} = 7.8\ \varepsilon_0$, $t_{ox} = 5.0$ nm, and $W_{dm} = 10.0$ nm, find the three longest eigenvalues λ_1, λ_2, and λ_3. Take $L \approx 2\lambda_1$; what's the ratio between $\exp(-\pi L/2\lambda_1)$ and $\exp(-\pi L/2\lambda_2)$?

6.3 Show that Eq. (6.38) for short-channel subthreshold current is reduced to Eq. (5.36) of long-channel subthreshold current if $\psi_2(x, y)$ is taken to be the long channel potential $v_2(x)$ of Eq. (6.10). [Hint: Convert dx to $d\psi/\mathscr{E}$ and approximate $\mathscr{E}$ by $\mathscr{E}_s$, the field at the surface.]

6.4 Equation (6.38) is derived under the assumption of constant mobility. Show that the differential equation for I_{ds} under the $n = 1$ velocity saturation model is

$$\frac{dV}{dy} = \frac{I_{ds}/\mu_{eff}}{qW(n_i^2/N_a)\left[\int_0^{W_d} e^{q\psi_2(x,y)/kT}dx\right]e^{-qV/kT} - I_{ds}/v_{sat}},$$

with the boundary condition $V(0) = 0$. This equation needs to be solved iteratively until an I_{ds} value is found which satisfies $V(L) = V_{ds}$.

6.5 For an nMOSFET with $t_{ox} = 10$ nm, $\mu_{eff} = 500$ cm^2/V-s, $v_{sat} = 10^7$ cm/s, $W = 10\ \mu$m, and $L = 1\ \mu$m, assume $m = 1$.
(a) Use the $n = 1$ velocity saturation model to generate I_{ds} versus V_{ds} (0–5 V) curves for $V_{gs} - V_t = 1, 2, 3, 4,$ and 5 V. (Note: $I_{ds} = I_{dsat}$ beyond V_{dsat}.)
(b) Now let L vary from 0.5 μm to 5 μm. Calculate and plot the saturation current for $V_{gs} - V_t = 3$ V versus L. Compare it with the long-channel saturation current (without velocity saturation) for the same $V_{gs} - V_t$ and range of L.

6.6 The small-signal transconductance in the saturation region is defined as $g_{msat} \equiv dI_{dsat}/dV_{gs}$. Derive an expression for g_{msat} using Eq. (6.49) based on the $n = 1$ velocity saturation model. Show that g_{mast} approaches the saturation–velocity-limited value, Eq. (6.79), when $L \to 0$. What becomes of the expression for g_{mast} in the long-channel limit when $v_{sat} \to \infty$?

6.7 From Eq. (6.49) based on the $n = 1$ velocity saturation model, what is the carrier velocity at the source end of the channel? What are the limiting values when $L \to 0$ and when $v_{sat} \to \infty$?

6.8 Assuming the $n = 1$ velocity saturation model, show that the total integrated inversion charge under the gate is

$$Q_i(\text{total}) = -WLC_{ox}(V_{gs} - V_t) \frac{\sqrt{1 + 2\mu_{eff}(V_{gs} - V_t)/(mv_{sat}L)} + \frac{1}{3}}{\sqrt{1 + 2\mu_{eff}(V_{gs} - V_t)/(mv_{sat}L)} + 1}$$

in the saturation region. Evaluate the intrinsic gate-to-channel capacitance, and show that it approaches Eq. (5.72) in the long-channel limit.

6.9 Consider an n-channel MOSFET with n^+ polysilicon gate (neglect poly depletion effect). The gate oxide is 7 nm thick, and the p-type body (or substrate) has a retrograde doping, as shown in Figure 6.25, with $N_s = 0$. Take $2\psi_B = 1$V.
(a) Choose the values of x_s and N_a such that the maximum depletion width is $W_{dm} = 0.1$ μm and the threshold voltage (at $2\psi_B$) is $V_t = 0.3$ V.
(b) Following (a), what is the body effect coefficient, m, and the inverse slope of log subthreshold current versus gate voltage (long-channel device)?
(c) Following (a), how short a channel length can the device be scaled to before the short-channel effect becomes severe?

6.10 Consider a low–high (retrograde) doping profile of an nMOSFET in Figure 6.25 with $x_s = 50$ nm, $N_s = 0$, and $N_a = 2 \times 10^{17}$ cm^{-3}. At a gate voltage below threshold, the depletion width is $W_d = 70$ nm. What are the surface potential and the surface field in silicon at this gate voltage?

7 Silicon-on-Insulator and Double-Gate MOSFETs

This chapter considers silicon-on-insulator (SOI) MOSFETs and double-gate MOSFETs. Along with bulk MOSFETs, they have been in high-volume production of VLSI circuits and systems.

There are three main types of SOI materials: SIMOX, BESOI, and Smart Cut® (Celler and Cristoloveanu, 2003). SIMOX stands for Separation by IMplantation of OXygen. It is formed by first implanting a high dose of high-energy oxygen ions into a silicon substrate. A high temperature anneal subsequently drives the chemical reaction which forms a stochiometric oxide layer buried in the silicon wafer. The anneal also regenerates the crystalline quality of the silicon layer remaining over the buried oxide. The main advantage of this technique is the thickness uniformity of the thin SOI layer. The main drawback is the high defect densities in the regrown silicon and in the buried oxide. BESOI stands for bond and etch back. It starts with two silicon wafers. After oxidation, the two wafers are bonded together by heating them to high temperatures. One of them is then etched back until only a thin film of silicon remains over the oxide and the other wafer. The crystalline quality of BESOI material is in principle as good as that of bulk silicon wafer. But there are usually significant thickness variations within the wafer. They can be tolerated for thick film SOI devices. For thin film SOI, some kind of etch stop technique needs to be employed in the etch back step to obtain better thickness uniformity. In the Smart Cut® process, both ion implantation and bonding are used. Before bonding, a high dose hydrogen implantation is made to wafer A to weaken the silicon bond strength at the implanted depth. After oxidation and bonding of wafers A and B, they are pulled apart mechanically. They break apart at the weakened cleavage plane, thus leaving a thin silicon film of A over the oxide and wafer B. The rest of A can be reused to save the cost. High temperature anneal and chemical–mechanical polish steps are done to the SOI wafer before device fabrication.

The basic difference between bulk and SOI or DG MOSFETs is that the latter are built in thin film silicon. The thin silicon film has the same single crystalline quality as the bulk substrate. We discussed in Section 6.1 that bulk silicon must be doped at high enough concentration to scale down the depletion width and therefore the channel length. For thin-film silicon, the device depth is limited by the physical thickness without any need for doping. This circumvents the problem of threshold voltage

---

® Trade mark, SOITEC.

uncertainty due to discrete dopants discussed in Section 6.3.4. Other inherent advantages of SOI or DG MOSFETs over bulk CMOS are listed here.

• **Very low junction capacitance**

In SOI MOSFETs, the source and drain junction capacitance through the thick buried oxide layer (BOX) to the substrate is very small.

• **Small or no body effect**

There is no contact to the silicon film other than the source and drain contacts in either SOI or DG MOSFETs. What serves as the body contact in fully-depleted SOI MOSFETs is the back gate or the substrate below the BOX. Since the BOX is usually very thick, the body effect factor m is very small (see Figure 5.5 and Section 7.1.1.2). DG MOSFETs have no body effect.

• **Soft error immunity**

In bulk devices, minority carriers are generated along the track of any high-energy particle or ionizing radiation that strikes through the silicon. If the collected charge of a junction node exceeds a certain threshold, it may cause an upset of the stored logic state. This is commonly referred to as a *soft error*. SOI devices offer a substantial improvement in the soft error rate since the presence of the buried oxide greatly reduces the volume susceptible to ionizing radiation.

Partially depleted SOI MOSFETs built in a thicker silicon film with a neutral (undepleted) body region at the bottom will not be discussed here. They operate similarly as bulk MOSFETs, in the sense that both require high enough body doping to scale down the gate depletion width. While there have been past chip products employing partially depleted SOI CMOS, the technology is no longer being practiced beyond the 22 nm node.

7.1   SOI MOSFETs

SOI CMOS involves building more or less conventional MOSFETs on a thin layer of crystalline silicon, as illustrated in Figure 7.1. The thin layer of silicon is separated from the substrate by a thick layer (typically 25 nm or more) of buried SiO_2 film, thus electrically isolating the devices from the underlying silicon substrate and from each other. An SOI CMOS process can be readily developed due to the compatibility with established bulk processing technology.

7.1.1   Long-Channel SOI MOSFETs

We first discuss the operation of long-channel SOI MOSFETs. Figure 7.2 shows a schematic cross-section in the vertical or the depth direction. The parameters in addition to those of bulk MOSFETs are: silicon thickness (t_{si}), thickness of the buried

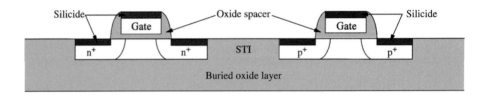

Figure 7.1 A schematic cross-section of SOI CMOS, with shallow trench isolation, dual polysilicon gates, and self-aligned silicide

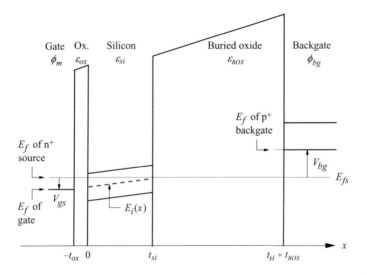

Figure 7.2 A schematic cross-section of SOI MOSFETs. The long dotted line across all thicknesses represents E_{fs}, the Fermi level of the n$^+$ source. The source and drain regions are not shown as they are in the perpendicular direction

oxide (t_{BOX}), permittivity of the buried oxide (ε_{BOX}), work function of the backgate (ϕ_{bg}), and backgate bias (V_{bg}), as labeled in Figure 7.2.

7.1.1.1 Solution to Poisson's Equation

Under the gradual channel approach outlined in Section 5.1.1, the 1-D Poisson's equation along a vertical cut perpendicular to the silicon film is:

$$\frac{d^2\psi}{dx^2} = \frac{q}{\varepsilon_{si}} n_i e^{q(\psi-V)/kT}. \tag{7.1}$$

The right-hand side consists of only the mobile charge (electrons) term under the assumption that the doping density in the silicon film is at a negligible level. We consider an nMOSFET here with $q\psi/kT \gg 1$, hence the hole density is negligible. $\psi(x)$ is the electrostatic potential, defined as the intrinsic potential at x with respect to the Fermi potential at the source, i.e., $\psi(x) \equiv -[E_i(x) - E_{fs}]/q$, referring to Figure 7.2. V is the electron quasi-Fermi potential at a point in the channel. Since the gradient of V is in the direction of current flow along the channel (y direction), V is independent of x. At the source, $V(y = 0) = 0$. At the drain, $V(y = L) = V_{ds}$. $V(y)$ satisfies the current continuity equation, Eq. (5.8). The boundary conditions can be constructed from Figure 7.2:

$$\varepsilon_{ox} \frac{V_{gs} - (\phi_m - \phi_{si}) - \psi(0)}{t_{ox}} = -\varepsilon_{si} \frac{d\psi}{dx}\bigg|_{x=0} \tag{7.2}$$

and

$$\varepsilon_{BOX} \frac{V_{bg} - (\phi_{bg} - \phi_{si}) - \psi(t_{si})}{t_{BOX}} = \varepsilon_{si} \frac{d\psi}{dx}\bigg|_{x=t_{si}}, \tag{7.3}$$

where ϕ_{si} is the work function of intrinsic silicon.

Integration of Eq. (7.1) yields

$$\frac{d\psi}{dx} = -\sqrt{\frac{2kTn_i}{\varepsilon_{si}} e^{q(\psi - V)/kT} + \mathscr{E}_0^2}, \tag{7.4}$$

with a positive constant of integration, $\mathscr{E}_0^2$. This is the appropriate choice for the SOI MOSFET depicted in Figure 7.2, namely, with only the front surface close to inversion. The case with a negative constant of integration is applied later to double-gate MOSFETs in which both silicon surfaces invert. Equation (7.4) can be integrated again to obtain (Taur, 2001)

$$\psi(x) = V + \frac{2kT}{q} \ln\left\{ \frac{\mathscr{E}_0\sqrt{\varepsilon_{si}/(2kTn_i)}}{\sinh[q(\mathscr{E}_0 x + v_0)/2kT]} \right\}, \tag{7.5}$$

where v_0 is the second constant of integration. Substituting the above in Eqs. (7.2) and (7.3), we obtain

$$V_{gs} - \phi_{mi} - V = \frac{2kT}{q} \ln\left\{ \frac{\mathscr{E}_0\sqrt{\varepsilon_{si}/(2kTn_i)}}{\sinh[qv_0/2kT]} \right\} + \frac{\varepsilon_{si}}{\varepsilon_{ox}} t_{ox}\mathscr{E}_0 \coth\left(\frac{qv_0}{2kT}\right) \tag{7.6}$$

and

$$V_{bg} - \phi_{bgi} - V = \frac{2kT}{q} \ln\left\{ \frac{\mathscr{E}_0\sqrt{\varepsilon_{si}/(2kTn_i)}}{\sinh[q(\mathscr{E}_0 t_{si} + v_0)/2kT]} \right\} - \frac{\varepsilon_{si}}{\varepsilon_{BOX}} t_{BOX}\mathscr{E}_0 \coth\left(\frac{q(\mathscr{E}_0 t_{si} + v_0)}{2kT}\right), \tag{7.7}$$

where $\phi_{mi} \equiv \phi_m - \phi_{si}$ and $\phi_{bgi} \equiv \phi_{bg} - \phi_{si}$. These are coupled implicit equations for solving $\mathscr{E}_0$ and v_0 given the structural and the bias parameters. From Gauss's law, the total inversion charge density in silicon is

$$-Q_i = -\varepsilon_{si}\frac{d\psi}{dx}\bigg|_{x=0} + \varepsilon_{si}\frac{d\psi}{dx}\bigg|_{x=t_{si}} = \varepsilon_{ox}\frac{V_{gs} - \phi_{mi} - \psi(0)}{t_{ox}} + \varepsilon_{BOX}\frac{V_{bg} - \phi_{bgi} - \psi(t_{si})}{t_{BOX}}.$$

(7.8)

7.1.1.2 Subthreshold Region

In the subthreshold region, the inversion charge density is negligible, hence the field is constant in the silicon region. By taking the difference of Eqs. (7.2) and (7.3), the field is determined to be

$$-\frac{d\psi}{dx} = \mathscr{E}_0 = \frac{(V_{gs} - \phi_{mi}) - (V_{bg} - \phi_{bgi})}{t_{si} + (\varepsilon_{si}/\varepsilon_{ox})t_{ox} + (\varepsilon_{si}/\varepsilon_{BOX})t_{BOX}},$$

(7.9)

where $\psi(t_{si}) - \psi(0) = t_{si}(d\psi/dx)$ has been applied. Substituting it back in Eq. (7.2) obtains

$$\psi(0) = \frac{t_{si} + (\varepsilon_{si}/\varepsilon_{BOX})t_{BOX}}{t_{si} + (\varepsilon_{si}/\varepsilon_{ox})t_{ox} + (\varepsilon_{si}/\varepsilon_{BOX})t_{BOX}}(V_{gs} - \phi_{mi})$$

$$+ \frac{(\varepsilon_{si}/\varepsilon_{ox})t_{ox}}{t_{si} + (\varepsilon_{si}/\varepsilon_{ox})t_{ox} + (\varepsilon_{si}/\varepsilon_{BOX})t_{BOX}}(V_{bg} - \phi_{bgi})$$

(7.10)

for the front surface potential. A body-effect factor m, similar to that depicted in Figure 5.5 for bulk MOSFETs, can be defined as

$$m \equiv \frac{\Delta V_{gs}}{\Delta \psi(0)} = \frac{t_{si} + (\varepsilon_{si}/\varepsilon_{ox})t_{ox} + (\varepsilon_{si}/\varepsilon_{BOX})t_{BOX}}{t_{si} + (\varepsilon_{si}/\varepsilon_{BOX})t_{BOX}} = 1 + \frac{(\varepsilon_{si}/\varepsilon_{ox})t_{ox}}{t_{si} + (\varepsilon_{si}/\varepsilon_{BOX})t_{BOX}}.$$

(7.11)

It is the same as Eq. (5.29) with W_{dm} replaced by $t_{si} + (\varepsilon_{si}/\varepsilon_{BOX})t_{BOX}$. Equation (7.10) then becomes

$$\psi(0) = \frac{1}{m}(V_{gs} - \phi_{mi}) + \left(1 - \frac{1}{m}\right)(V_{bg} - \phi_{bgi}).$$

(7.12)

For the above assumption of subthreshold condition to hold, the conduction band edge E_c at the front silicon surface must be below the source Fermi level E_{fs}. This means $\psi(0)$ should be less than $E_g/2q$ by some amount. A reasonable definition of the threshold voltage is the V_{gs} value for which Eq. (7.12) gives $\psi(0) = E_g/2q$, i.e.,

$$V_t = \phi_{mi} + m\frac{E_g}{2q} - (m - 1)(V_{bg} - \phi_{bgi}).$$

(7.13)

This V_t choice is examined later in an example in Figure 7.3.

With the introduction of the parameter m, all the previous results discussed for bulk MOSFETs apply. For example, the subthreshold swing is $2.3m(kT/q)$, as that of Eq. (5.40) (Mazhari et al., 1991). The sensitivity of threshold voltage to backgate bias is (Lim and Fossum, 1983)

$$\frac{\Delta V_t}{\Delta\left(-V_{bg}\right)} = m - 1 = \frac{(\varepsilon_{si}/\varepsilon_{ox})t_{ox}}{t_{si} + (\varepsilon_{si}/\varepsilon_{BOX})t_{BOX}}. \tag{7.14}$$

This is the same as the substrate bias sensitivity derived in Eq. (5.60) for bulk MOSFETs with m given by Eq. (5.28). In fact, Eq. (7.13) is very much parallel to Eq. (6.124) for a ground-plane MOSFET. In most cases, t_{BOX} is much thicker than t_{ox}, hence $m \approx 1$ and the subthreshold swing is close to the ideal 60 mV/decade value for SOI MOSFETs.[1]

One particular feature of SOI MOSFETs is that the threshold voltage can be controlled by the backgate bias within certain limits. This allows optimization of switching speed versus standby power. For the low activity block of the chip, the threshold voltage can be adjusted higher to reduce the standby power. For standard SOI substrates with a BOX thickness of 25 nm or more, however, it takes a large V_{bg} to shift V_t because $m - 1 \ll 1$. There is a recent push for thin BOX of ~ 10 nm thick so that a V_{bg} ~ 1 V can shift V_t by ~ 0.1 V for a front gate EOT of $\approx$ 1 nm (Cheng and Khakifirooz, 2015).

There are limits to the modulation of V_t by V_{bg}. Equation (7.13) is valid only if the back surface is neither accumulated nor inverted. In other words, $\psi(t_{si})$ cannot exceed $E_g/2q$ or below $-E_g/2q$. With $\psi(0) = E_g/2q$, the threshold condition at the front surface, the field $\mathscr{E}_0$ is constrained by $0 < \mathscr{E}_0 < (E_g/q)/t_{si}$, except for very thin silicon films (Cristoloveanu et al., 2017). From Eq. (7.2), this means

$$\phi_{mi} + \frac{E_g}{2q} \leq V_t \leq \phi_{mi} + \frac{E_g}{2q} + \left(\frac{\varepsilon_{si}t_{ox}}{\varepsilon_{ox}t_{si}}\right)\frac{E_g}{q}. \tag{7.15}$$

When either bound is reached, $\psi(t_{si})$ is pinned and V_t is no longer sensitive to further changes of V_{bg}.

7.1.1.3   Above Threshold Region

Near and above the threshold voltage, the full model of Eq. (7.5) with constants $\mathscr{E}_0$ and v_0 solved from Eqs. (7.6) and (7.7) must be used. An example is shown in Figure 7.3, with n^+ silicon work function on the front gate and p^+ silicon work function on the back gate. The V_t given by Eq. (7.13) is very close to the linearly extrapolated x-intercept of the Q_i-V_{gs} curve. The linear slope, $d(-Q_i)/dV_{gs}$ at V_{gs} ~ 1 V, is ~ 0.94 × C_{ox}, less than C_{ox} due to the effect of inversion layer capacitance (Figure 5.4).

The dependence of the inversion charge density Q_i on electron Fermi potential V is plotted in Figure 7.4 for several fixed values of V_{gs}. It is essentially the same as the bulk relation, Eq. (5.44),

$$-Q_i(V) = C_{inv}\left(V_{gs} - V_t - mV\right), \tag{7.16}$$

---

[1] This is only for long-channel SOI MOSFETs. See Section 7.1.2 for discussion on subthreshold swing of short-channel SOI MOSFETs.

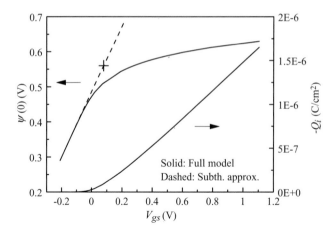

Figure 7.3 Front surface potential $\psi(0)$ and inversion charge density Q_i versus gate voltage for an SOI MOSFET. The film thicknesses are $t_{ox} = 2$ nm, $t_{si} = 10$ nm, $t_{BOX} = 25$ nm with $\varepsilon_{si}/\varepsilon_{ox} = \varepsilon_{si}/\varepsilon_{BOX} = 3$, so $m = 1.07$. The front gate has n$^+$ silicon work function and the back gate p$^+$ silicon work function so $\phi_{mi} = -E_g/2q = -0.56$ V and $\phi_{bgi} = E_g/2q = 0.56$ V. $V_{bg} = 0$, $V = 0$. The dashed line is the subthreshold approximation of Eq. (7.12). The cross indicates the threshold condition of $\psi(0) = E_g/2q = 0.56$ V, where $V_{gs} = V_t = 0.08$ V

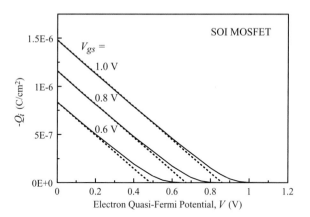

Figure 7.4 Inversion charge density Q_i versus electron Fermi potential for three fixed values of gate voltage. The solid curves are numerical solutions of the full model equations. The dotted lines are the linear equation of Eq. (7.16)

with m given by Eq. (7.11) for SOI MOSFETs. This can be understood in view of Eqs. (7.6) and (7.7). An incremental ΔV has the effect of retarding V_{gs} and V_{bg} by the same amount. The degradation of V_{bg}, in turn, is equivalent to degrading V_{gs} by $(m - 1)\Delta V_{bg}$, or $(m - 1)\Delta V$. The total effect of ΔV on V_{gs} is therefore $m\Delta V$. With the same form of $Q_i(V)$ as bulk MOSFETs, I_{ds}–V_{ds} characteristics of SOI MOSFETs simply follow those of Section 5.1.3 with a redefined m. Likewise, the non-GCA model in Section 5.1.4 can be applied to generate the saturation region characteristics of SOI

MOSFETs. Note that the linear approximation of Eq. (7.16) is less accurate at lower V_{gs}. This is because neither the slope $d(-Q_i)/dV_{gs}$ in Figure 7.3 nor the slope dQ_i/dV in Figure 7.4 is truly a constant. They tend to decrease smoothly to zero as $Q_i \to 0$.

7.1.2   Short-Channel SOI MOSFETs

It has long been reported in the literature that fully-depleted SOI MOSFETs are more susceptible to short-channel effects (SCE) for lack of a conducting plane not too far below the device region (Su *et al.*, 1994; Wong *et al.*, 1994). The 2-D scale length model discussed in Section 6.1.2, however, does not apply to SOI MOSFETs because no closed rectangular region can be defined with known potential values on its boundary. TCAD has become a necessary tool for investigating SCE in SOI MOSFETs (Xie *et al.*, 2013).

7.1.2.1   2-D Fields in the Buried Oxide

Figure 7.5 compares the constant potential contours of a bulk ground-plane MOSFET with an SOI MOSFET side by side. In the bulk case, the 2-D fields are confined to the depletion (undoped) region bounded below by the conducting substrate. In the SOI case, on the other hand, the 2-D fields from the source and drain penetrate into the thick BOX region. Conceptually, since the scale length is given by the effective vertical distance between the gate and the bottom conductor, deeper field penetration

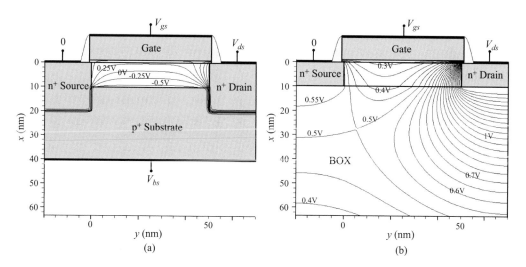

Figure 7.5 2-D constant potential contours of (a) bulk and (b) SOI MOSFETs. For both devices, $t_{ox} = 1$ nm, $L = 50$ nm, $V_{ds} = 1.0$ V, $V_{bs} = V_{bg} = 0$. For the bulk, the depletion region (undoped) depth is 10 nm. For the SOI, the silicon thickness is 10 nm, and the BOX thickness is 200 nm. The labels refer to the potential as that defined in Eq. (7.1), i.e., $\psi(x, y) \equiv -[E_i(x, y) - E_{fs}]/q$. The value of V_{gs} is such that the minimum surface potential between the source and drain, $\psi_{s,min}$, is 0.29 V (after Xie *et al.*, 2013)

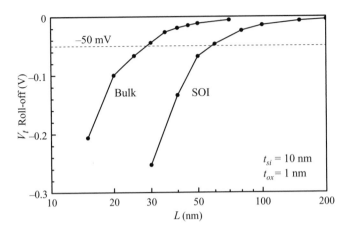

Figure 7.6 Short-channel V_t roll-off of the bulk and SOI MOSFETs in Figure 7.5. Here, V_t is defined as the V_{gs} value where $I_{ds} = 10^{-8}$ A ($W/L = 1$), and V_t roll-off is defined as $\Delta V_t = V_t(L) - V_t$(Long channel). The –50 mV intercepts are $L = 29$ nm for bulk and 58 nm for SOI

would worsen the SCE. The mitigating factor is that the depth of field penetration is channel length dependent. Only for very long channel devices is the vertical distance given by the entire BOX thickness. For short channel devices where it matters, the effective depth of field penetration is much less than the BOX thickness.

Figure 7.6 compares the V_t roll-off curves of the bulk and SOI MOSFETs in Figure 7.5. By defining an L_{min} where the V_t roll-off is $\Delta V_t = -50$ mV, we obtain $L_{min} = 29$ nm for the bulk MOSFET[2] and $L_{min} = 58$ nm for the SOI MOSFET. To gain further insight, $\Delta\psi_{s,min}$, the minimum surface potential between the source and drain of a short channel device with respect to that of the long channel device [see Eq. (6.21)], is plotted as a function of L in Figure 7.7. For the bulk MOSFET, $\Delta\psi_{s,min}$ versus L is largely proportional to $\exp[-\pi L/(2\lambda)]$, as expected from the scale length model with a λ of 12.6 nm given by Eq. (6.23) for $W_{dm} = 10$ nm and $t_{ox} = 1$ nm. For the SOI MOSFET, first, the exponential slope is far less steep compared to that of the bulk device, indicating longer λ and worse SCE. Second, the exponential slope is not constant, but increases toward shorter L, i.e., the effective λ decreases with decreasing L. This is attributed to the decrease of the depth of 2-D field penetration in Figure 7.5(b) as L is shortened.

7.1.2.2   Effect of Back Gate Bias

Note in Figures 7.6 and 7.7 that for the bulk MOSFET, the channel length where $\Delta\psi_{s,min} = 50$ mV is not too different from the L_{min} where $\Delta V_t = 50$ mV. For the SOI MOSFET, on the other hand, the L_{min} where $\Delta V_t = 50$ mV is much longer than the channel length where $\Delta\psi_{s,min} = 50$ mV. A closer look at the SOI potential contours in

---

[2] While the bulk MOSFET depicted in Figure 7.5 is a hypothetical one, bulk MOSFETs of $L \sim 20$ nm have been fabricated (Cho *et al.*, 2011; Shang *et al.*, 2012).

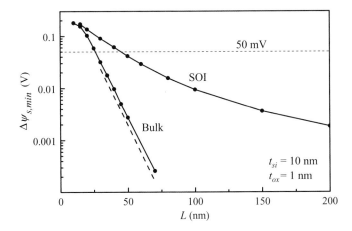

Figure 7.7 $\Delta\psi_{s,min}$, the minimum surface potential between the source and drain of a short channel device with respect to that of the long channel device for the SOI and bulk MOSFETs in Figure 7.5. The dashed line is $\exp[-\pi L/(2\lambda)]$ with $\lambda = 12.6$ nm. The 50 mV intercepts are $L = 26$ nm for bulk and 45 nm for SOI

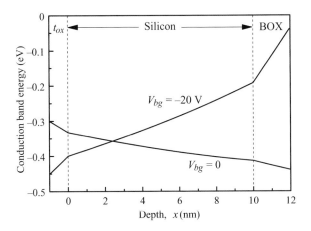

Figure 7.8 Conduction band energy versus depth in silicon along a vertical cut through the point of $\psi_{s,min}$. The x coordinate is the same as that of Figure 7.2

Figure 7.5(b) tells that the potential at the back surface of silicon is actually higher than ψ_s at the front surface. This does not happen in long-channel SOI MOSFETs. Figure 7.8 plots the conduction band energy E_c along a vertical cut in silicon at a gate voltage below the threshold of a 50 nm SOI MOSFET. *Because the source and drain lateral fields have a stronger effect in lowering the backside energy barrier, the inversion charge density is actually highest at the back surface (the case of $V_{bg} = 0$).* This weakens the gate control of the subthreshold current integrated over the silicon thickness and explains why ΔV_t is significantly worse than $\Delta\psi_{s,min}$ for the SOI MOSFET.

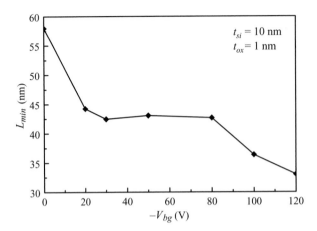

Figure 7.9 L_{min} versus $-V_{bg}$. L_{min} is defined by a V_t roll-off of -50 mV, as that depicted in Figure 7.6. The BOX thickness is 200 nm

When a reverse bias is applied to the substrate of an SOI device, a field is established in the silicon to drive electrons to the front oxide–silicon interface. The magnitude of substrate bias needed for this to happen is proportional to the BOX thickness. Here, we assume a BOX 200 nm thick with substrate bias voltages on the order of -10 to -100 V for nMOSFETs. If the BOX is thinned to 10 nm, a few volts of V_{bg} will be effective. By overcoming the barrier lowering at the back surface due to source and drain fields, the reverse substrate bias restores the maximum potential to the front surface, as shown in Figure 7.8. This makes it bulk-like and enables the gate to have better control of the integrated inversion charge density hence of the source–drain current. In this case, with a -20 V of V_{bg}, the L_{min} of SOI is reduced by ~30% from 58 nm for $V_{bg} = 0$ to 44 nm.

Further increase of the reverse substrate bias brings little SCE improvement, as shown in Figure 7.9, until eventually a p$^+$ accumulation layer is formed at the back silicon surface. Beyond that point the backside potential is pinned and the vertical field in silicon no longer increases with $-V_{bg}$. The backside accumulation layer serves as a 'ground plane' at a distance t_{si} away from the front surface. The long-channel subthreshold swing now becomes bulk-like, with t_{si} effectively the depletion layer width. In other words, the m factor in Eq. (7.11) increases to $1 + (\varepsilon_{si}/\varepsilon_{ox})(t_{ox}/t_{si})$. The SCE, on the other hand, improves substantially because the source and drain fields no longer penetrate into the BOX. As shown in Figure 7.9, this ultimately reduces L_{min} to ~33 nm, nearly the same as that of the bulk MOSFET in Figure 7.6. In practice, the BOX thickness would have to be greatly reduced, e.g., to ~10 nm, for this to happen at a reasonable V_{bg}.

7.1.2.3   Empirical Minimum Channel Length

Based on Eq. (6.21) of the scale length model, the lateral field, $\partial \psi / \partial y$, in silicon is proportional to $\exp(-\pi y/\lambda)$ on the source side and to $\exp[-\pi(L - y)/\lambda]$ on the drain side,

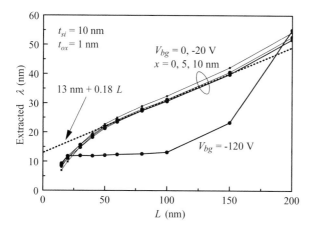

Figure 7.10 $\lambda(L)$ extracted from individual devices, for different depths and V_{bg}. For $V_{bg} = -120$ V, λ becomes a constant ≈ 13 nm where a p$^+$ accumulation layer is formed at the back silicon surface

insensitive to the depth or the gate voltage in subthreshold. It measures the penetration of source and drain fields into the middle of the channel which is central to the short-channel effect. This allows the extraction of $\lambda(L)$ from the exponential slope of lateral fields in a series of TCAD devices of varying channel lengths. For bulk, λ extracted from the lateral fields is independent of L. For SOI, λ decreases with decreasing L, as shown in Figure 7.10, with an empirical fitting of

$$\lambda = t_{si} + (\varepsilon_{si}/\varepsilon_{ox})t_{ox} + 0.18L, \tag{7.17}$$

for the critical region of 40 nm $< L <$ 100 nm (Xie et al., 2013). Note that for the region $L < 2\lambda$ ($L \sim 41$ nm in this case), higher order terms become non-negligible and the extracted λ deviates from the above empirical line. In practice, the SCE in this region is too severe for the device to be useful. An added curve for $V_{bg} = -120$ V in Figure 7.10 shows that bulk-like scale length, $\lambda \approx t_{si} + (\varepsilon_{si}/\varepsilon_{ox})t_{ox}$, is reached when the backside of silicon is accumulated. For this case, $m = 1.3$ and the long channel subthreshold swing is $SS \approx 80$ mV/decade, worse than the ≈ 60 mV/decade SS of long channel SOI MOSFETs without backside accumulation. However, the latter goes up quickly toward shorter channel lengths because of the more severe SCE. If we plot the SS versus channel length for the two devices, there is a cross-over of SS at certain channel length below which the backside accumulated device has better subthreshold swing.

The relationship between L_{min} and λ in SOI MOSFETs is V_{bg} dependent. The worst case is $L_{min} \approx 2.5\lambda$ for zero V_{bg} in Figure 7.6. A -20 V V_{bg} improves it to $L_{min} \approx 2.1\lambda$ by driving the potential maximum to the front surface. If we assume a parameter k for L_{min}/λ, its intersection with Eq. (7.17) yields an empirical expression for L_{min}:

$$L_{min} = \frac{1}{k^{-1} - 0.18}[t_{si} + (\varepsilon_{si}/\varepsilon_{ox})t_{ox}] \qquad \text{where } 2.1 < k < 2.5. \tag{7.18}$$

With no backgate bias, it gives $L_{min} \approx 4.5[t_{si} + (\varepsilon_{si}/\varepsilon_{ox})t_{ox}]$, consistent with an earlier 2-D numerical study (Lu and Taur, 2006). *A moderate backgate bias can achieve $L_{min} \approx 3.4[t_{si} + (\varepsilon_{si}/\varepsilon_{ox})t_{ox}]$* by reaching the flat region in Figure 7.9. Note that while λ of Eq. (7.17) is independent of V_{ds}, the factor k does depend on V_{ds}. The range shown in Eq. (7.18) is for $V_{ds} = 1.0$ V. The recently reported experimental data on ultra-thin SOI MOSFETs: 20 nm L_{min} with $t_{si} = 3.5$ nm and high-κ gate insulator (Khakifirooz et al., 2012), are generally in line with the empirical Eq. (7.18).

7.2　Double-Gate and Nanowire MOSFETs

In Section 7.1.2, we discussed that fully depleted thin film SOI MOSFETs do not scale well to short channel lengths for lack of a conducting plane under the device region. If a conductor is placed under the device, it needs to be only a thin insulator away from the silicon film to be effective. But in such a "ground-plane" configuration, the capacitance between the drain and the bottom conductor would be excessively high, resulting in poor switching delays. Ultimately, a second gate placed under the silicon film, with minimal overlap to the source and drain regions, can serve as the bottom conductor. This forms a double gate (DG) MOSFET (Frank et al., 1992). In the symmetric case, both gates have identical work function. When they switch together, two inversion channels are formed: one at the top of the silicon film next to the top gate insulator, and the other at the bottom next to the bottom insulator. Note that as far as the intrinsic delay is concerned, there is no inherent benefit of a symmetric double-gate MOSFET over a single-gate MOSFET of the same channel length because both the current and the inversion-charge capacitance (and the gate fringe capacitance for that matter) are doubled in the double-gate case.

The advantages of DG MOSFETs are multifold. First, like SOI, there is no contact to the silicon body, thus the m factor, $\Delta V_{gs}/\Delta \psi_s$, depicted in Figure 5.5 is unity, ensuring the ideal 60 mV/decade subthreshold swing. Second, a DG MOSFET with sub-10 nm thick silicon film can scale to shorter channel lengths than a planar MOSFET can. The latter relies on heavy doping of the body to scale down the gate depletion width and is reaching its limit when $L \sim 20$ nm with the body doping exceeding 10^{19} cm^{-3} (see Figure 6.6). Third, since no doping is needed to confine the depletion region, discrete dopant number fluctuation is not a problem in DG MOSFETs. The threshold voltage is completely determined by the gate work function.[3]

In a high-performance DG MOSFET, both gates must be self-aligned to the source–drain regions in the silicon film. Any random misalignment could cause either excessive gate-to-drain overlap capacitance or underlapping in which an ungated gap would add large series resistance to the channel. The silicon film must be very thin (<10 nm) in order for the scale length λ to scale to 10 nm and below (see Section 7.2.2). Atomic dimension tolerances would be required for the control of silicon film

---

[3] To provide multiple threshold voltages on the same chip, however, some light doping of either polarity may be employed.

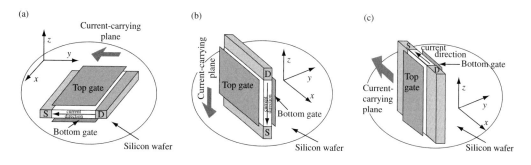

Figure 7.11 Three different orientations of double-gate MOSFETs (Wong *et al.*, 1999)

thickness. Figure 7.11 shows the three possible orientations of a DG MOSFET on a silicon wafer (Wong *et al.*, 1999). In type A configuration, both the gates and the silicon film are parallel to the wafer plane. In terms of device and circuit layout, this geometry is most compatible to the conventional planar bulk CMOS technology. The key challenge is how to place the bottom gate under the active silicon film while self aligned to the top gate and source–drain regions. Attempts have been made to grow an epitaxial silicon film through a tunnel pre-formed between the sacrificial gates (Wong *et al.*, 1997). In type B configuration, both the gates and the silicon film are perpendicular to the wafer plane. This forms a vertical MOSFET with the two gates on both sides of the silicon film. It is easier to deposit the two gates. However, it is difficult to form the source and drain by ion implantation and have them self-aligned to the gates. In type C configuration, the gates and the silicon film are also perpendicular to the wafer plane, but the current direction is in the wafer plane. This geometry is often referred to as FinFET (Hisamoto *et al.*, 2000). The conventional ion implantation process can be employed to form the source and drain regions on the same plane. The MOSFET width is the fin height in the vertical direction. Since the fin height is limited by structural stability, multiple fins connected in parallel are needed to build up the current drive. The main difficulty is to pattern a fin thickness <10 nm, below the lithographic resolution. This can be accomplished by a sidewall imaging technique (Ogura *et al.*, 1982). FinFETs have been in volume production of VLSI chips below the 22 nm node (Auth *et al.*, 2012).

7.2.1 Analytic Potential Model for Symmetric DG MOSFETs

In Section 4.2.1, the 1-D Poisson's equation for a bulk MOS capacitor can be integrated once to yield Eq. (4.15), but it cannot be integrated again to solve the potential function $\psi(x)$ in a closed form. This had necessitated the charge sheet approximation (Section 5.1.1) to formulate the drain current model for bulk MOSFETs. Because there is no depletion charge term, 1-D Poisson's equation for an undoped DG MOS device is analytically integratable to yield a closed-form solution of the potential everywhere in the silicon film (Taur, 2000). By extending this solution, a continuous, analytic *I–V* model for double-gate MOSFETs, called the

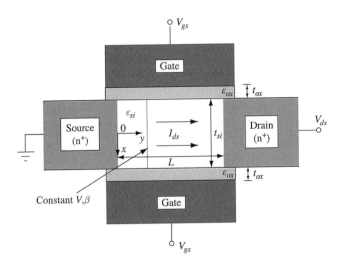

Figure 7.12 Schematic diagram of a double-gate MOSFET. $V(y)$ is the quasi-Fermi potential at a point in the channel. $V(0) = 0$ at the source and $V(L) = V_{ds}$ at the drain. β is a function of V

analytic potential model, has been formulated directly from the current continuity equation without the charge sheet approximation (Taur *et al.*, 2004). This analytic solution covers all regions of MOSFET operation, thus maintaining strong continuity while retaining all the essential device physics.

Consider a generic, symmetric double-gate MOSFET, shown schematically in Figure 7.12. The silicon is undoped or lightly doped. The same voltage is applied to the two gates having the same work function. The band diagrams along a cut perpendicular to the silicon film and the two gates are shown in Figure 7.13 for two gate voltages. At zero gate voltage below threshold, the bands are essentially flat throughout the silicon film as well as in the gate insulators because both the depletion charge and the inversion charge are negligible. *Since there is no contact to the silicon body, the Fermi level of the lightly doped body is determined by the Fermi levels of the source and drain, as depicted in Figures 7.2 and 7.13.* At $V_{ds} = 0$, the body is in thermal equilibrium with the source–drain and has the same Fermi level. While the body is not charge neutral in general, the potential or field due to the depletion charge in a lightly doped body is negligible. For example, the band bending produced by an inadvertent doping of 10^{16} cm^{-3} (n- or p-type) in a 10 nm silicon film is only $\sim qN_a t_{si}^2/2\varepsilon_{si} < 1$ mV. As the gate voltage increases toward the threshold voltage in the right half of Figure 7.13, mobile charge or electron density becomes appreciable when the conduction band of the silicon body moves to near the Fermi level of the source.

Under the GCA, Poisson's equation along a vertical cut perpendicular to the silicon film (Figure 7.12) takes the same form as Eq. (7.1) of SOI with only the mobile charge (electrons) term:

$$\frac{d^2\psi}{dx^2} = \frac{q}{\varepsilon_{si}} n_i e^{q(\psi - V)/kT}, \tag{7.19}$$

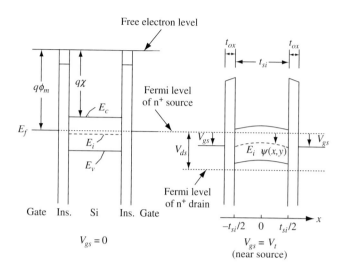

Figure 7.13 Band diagram along a vertical cut in Figure 7.12 for two gate voltages. The gate work function in this example is slightly lower than that of intrinsic silicon. For $V_{gs} = V_t$ on the right, the cut is near the source where $V = 0$ (after Taur *et al.*, 2004)

where $\psi(x)$ is the electrostatic potential, defined as the intrinsic potential at x with respect to the Fermi potential at the source, $\psi(x) \equiv -[E_i(x) - E_{fs}]/q$, as shown in Figure 7.13. $V(y)$ is the electron quasi-Fermi potential at y with respect to that of the source. Here we consider an nMOSFET with $q\psi/kT \gg 1$ so the hole density is negligible.

Since the current flows predominantly from the source to the drain along the y-direction, the gradient of the electron quasi-Fermi potential is also in the y-direction. This justifies the assertion that V is constant in the x-direction. Equation (7.19) can then be integrated twice under the symmetry condition $d\psi/dx|_{x=0} = 0$ to yield the solution

$$\psi(x) = V - \frac{2kT}{q} \ln\left[\frac{t_{si}}{2\beta}\sqrt{\frac{q^2 n_i}{2\varepsilon_{si}kT}}\cos\left(\frac{2\beta x}{t_{si}}\right)\right], \qquad (7.20)$$

where β is a constant (independent of x) to be determined from the boundary condition

$$\varepsilon_{ox}\frac{V_{gs} - \phi_{mi} - \psi(x = \pm t_{si}/2)}{t_{ox}} = \pm\varepsilon_{si}\frac{d\psi}{dx}\bigg|_{x=\pm t_{si}/2}. \qquad (7.21)$$

Here V_{gs} is the voltage applied to both gates, and ϕ_{mi} is the work function of both the top and bottom gate electrodes with respect to that of the intrinsic silicon. In other words, $\phi_{mi} \equiv \phi_m - (\chi + E_g/2q)$ as defined before. Since Eq. (7.20) is symmetric about $x = 0$, substituting it into Eq. (7.21) yields the same condition from either surface:

$$\frac{q(V_{gs} - \phi_{mi} - V)}{2kT} - \ln\left(\frac{2}{t_{si}}\sqrt{\frac{2\varepsilon_{si}kT}{q^2 n_i}}\right) = \ln\beta - \ln(\cos\beta) + \frac{2\varepsilon_{si}t_{ox}}{\varepsilon_{ox}t_{si}}\beta\tan\beta. \quad (7.22)$$

For a given V_{gs}, β can be solved from the implicit equation, Eq. (7.22), as a function of V. Along the channel direction (y), V varies from the source to the drain. So does β. The functional dependence of $V(y)$ and $\beta(y)$ is determined by the current continuity condition which requires that the current, $I_{ds} = \mu W Q_i dV/dy$, be a constant, independent of V or y. Here μ is the effective mobility, W is the device width, and Q_i is the total mobile charge per unit area including both channels. Integrating $I_{ds} dy$ from the source to the drain and expressing dV/dy as $(dV/d\beta)(d\beta/dy)$, we obtain

$$I_{ds} = \mu \frac{W}{L} \int_0^{V_{ds}} [-Q_i(V)] dV = \mu \frac{W}{L} \int_{\beta_s}^{\beta_d} [-Q_i(\beta)] \frac{dV}{d\beta} d\beta, \qquad (7.23)$$

where β_s and β_d are solutions to Eq. (7.22) corresponding to $V = 0$ and $V = V_{ds}$, respectively. From Gauss's law,

$$-Q_i = 2\varepsilon_{si} \frac{d\psi}{dx}\bigg|_{x=t_{si}/2} = 2\varepsilon_{si} \frac{2kT}{q} \frac{2\beta \tan\beta}{t_{si}}, \qquad (7.24)$$

where the last step follows from the differentiation of Eq. (7.20). Note that $dV/d\beta$ can also be expressed as a function of β by differentiating Eq. (7.22). Substituting these factors in Eq. (7.23) and carrying out the integration analytically yields the drain current:

$$I_{ds} = \mu \frac{W}{L} \frac{4\varepsilon_{si}}{t_{si}} \left(\frac{2kT}{q}\right)^2 \int_{\beta_d}^{\beta_s} \left[\tan\beta + \beta \tan^2\beta + \frac{2\varepsilon_{si} t_{ox}}{\varepsilon_{ox} t_{si}} \beta \tan\beta \frac{d}{d\beta}(\beta \tan\beta) \right] d\beta$$

$$= \mu \frac{W}{L} \frac{4\varepsilon_{si}}{t_{si}} \left(\frac{2kT}{q}\right)^2 \left[\beta \tan\beta - \frac{\beta^2}{2} + \frac{\varepsilon_{si} t_{ox}}{\varepsilon_{ox} t_{si}} \beta^2 \tan^2\beta \right]_{\beta_d}^{\beta_s}. \qquad (7.25)$$

The range of β_s, β_d is $(0, \pi/2)$. **MOSFET characteristics for all regions: linear, saturation, and subthreshold, can be generated from this continuous, analytic solution** (Taur et al., 2004). For example, in the linear region above threshold, the left-hand side of Eq. (7.22) $\gg 1$ for both $V = 0$ and V_{ds}, so β_s, $\beta_d \sim \pi/2$. The last terms on the right-hand side of Eqs. (7.22) and (7.25) dominate; therefore,

$$I_{ds} = \mu \frac{\varepsilon_{ox}}{t_{ox}} \frac{W}{L} \left[(V_{gs} - V_t)^2 - (V_{gs} - V_t - V_{ds})^2 \right] = 2\mu \frac{\varepsilon_{ox}}{t_{ox}} \frac{W}{L} (V_{gs} - V_t - V_{ds}/2) V_{ds}, \qquad (7.26)$$

where

$$V_t = \phi_{mi} + \frac{2kT}{q} \ln \left[\frac{2}{t_{si}} \sqrt{\frac{2\varepsilon_{si} kT}{q^2 n_i}} \right] + \frac{2kT}{q} \ln \left[\frac{q\varepsilon_{ox} t_{si} (V_{gs} - V_t)}{4\varepsilon_{si} t_{ox} kT} \right]. \qquad (7.27)$$

The last term is a second-order term coming from the $\ln[\beta/\cos\beta]$ term in Eq. (7.22). It is kept here to show that the t_{si} factor cancels the $1/t_{si}$ factor in the previous term so V_t is independent of t_{si}. Equation (7.27) is not a conventional definition of V_t as it involves a $(V_{gs} - V_t)$ factor. Like the charge sheet model, the analytic potential model is a continuous model that needs no specific introduction of V_t. In the saturation region, where $\beta_s \sim \pi/2$, but $\beta_d \ll 1$, we obtain

$$I_{ds} = \mu \frac{\varepsilon_{ox}}{t_{ox}} \frac{W}{L} \left[\left(V_{gs} - V_t \right)^2 - \frac{8\varepsilon_{si}t_{ox}k^2T^2}{\varepsilon_{ox}t_{si}q^2} e^{q\left(V_{gs}-V_t-V_{ds}\right)/kT} \right]. \qquad (7.28)$$

Note that in this continuous model, the current approaches the saturation value with a difference term that decreases exponentially with V_{ds}.

In the subthreshold region, both β_s, $\beta_d \ll 1$, so the $\ln \beta$ term dominates on the right-hand side of Eq. (7.22), and

$$I_{ds} = \mu \frac{W}{L} kTn_i t_{si} e^{q\left(V_{gs}-\phi_{mi}\right)/kT} \left(1 - e^{-qV_{ds}/kT} \right), \qquad (7.29)$$

as would be expected from the basic diffusion current density, $J_{diff} = qD_n \, dn/dx$. Note that the subthreshold current is proportional to the silicon thickness, but independent of ε_{ox}/t_{ox} – a manifestation of "volume inversion" that the potential in silicon is near constant over the thickness and directly follows the gate modulation when the mobile carrier density is low. In contrast, the currents above threshold, Eqs. (7.26) and (7.28), are proportional to ε_{ox}/t_{ox}, but independent of silicon thickness. This is the same as in a bulk MOSFET in that once a high density of inversion charge appears at the surface, it electrostatically screens the interior of silicon from the gate.

While doping in the silicon film has been neglected in the above analytic potential model, its first-order effect can be incorporated simply as a threshold voltage shift of magnitude $qN_a t_{si}/(2C_{ox})$ from the depletion charge [see Eq. (7.41)]. P-type dopants shift the threshold positively and n-type negatively.

7.2.2 Short-Channel DG MOSFETs

The scale length theory described in Section 6.1.2 for bulk MOSFETs can be applied to DG MOSFETs by modifying the boundary conditions. In fact, the boundary of a DG MOSFET is more of an ideal rectangle for the application of scale length model, compared to that of a bulk MOSFET in Figure 6.5. Although a DG MOSFET consists of three dielectric regions, one silicon region, and two oxide regions, the symmetry condition reduces the potential boundary condition to effectively one dielectric interface.

7.2.2.1 Scale Length of DG MOSFETs

Referring to the 2-D geometry of a symmetric DG MOSFET in Figure 7.12, we divide the potential function into three regions: $\psi(x, y)$ for the silicon region, $-t_{si}/2 < x < t_{si}/2$, and $\psi^\pm(x, y)$ for the bottom and top oxide regions, $t_{si}/2 < x < t_{si}/2 + t_{ox}$, and $-t_{si}/2 - t_{ox} < x < -t_{si}/2$, respectively. In subthreshold, the mobile charge density is negligible. For a uniform p-type doping density of N_a, the boundary value problem is specified as follows:

$$\frac{\partial^2 \psi}{\partial x^2} + \frac{\partial^2 \psi}{\partial y^2} = \frac{q}{\varepsilon_{si}} N_a \qquad \text{in silicon} \qquad (7.30)$$

$$\frac{\partial^2 \psi^{\pm}}{\partial x^2} + \frac{\partial^2 \psi^{\pm}}{\partial y^2} = 0 \qquad \text{in oxides.} \qquad (7.31)$$

For n-type doping, N_a is replaced by $-N_d$. At the two dielectric boundaries, the usual continuity conditions for the potential and normal displacement hold:

$$\psi(\pm t_{si}/2, y) = \psi^{\pm}(\pm t_{si}/2, y), \qquad (7.32)$$

$$\varepsilon_{si} \frac{\partial \psi}{\partial x}(\pm t_{si}/2, y) = \varepsilon_{ox} \frac{\partial \psi^{\pm}}{\partial x}(\pm t_{si}/2, y). \qquad (7.33)$$

The four (effectively three) boundary conditions for the rectangular region are:

$$\psi^{\pm}(\pm t_{si}/2 \pm t_{ox}, y) = V_{gs} - \phi_{mi}. \qquad (7.34)$$

$$\psi(x, 0) = \frac{E_g}{2q}. \qquad (7.35)$$

$$\psi(x, L) = \frac{E_g}{2q} + V_{ds}. \qquad (7.36)$$

Here, $\psi(x, y) \equiv -[E_i(x, y) - E_{fs}]/q$ and $\phi_{mi} \equiv \phi_m - [\chi + E_g/(2q)]$, as defined in Section 7.1.1.1.

There are four gaps in the 2-D region not covered by conductors: two on the source side and two on the drain side. The BCs in those gaps are filled in by linear interpolation between the gate and the source or the drain potential, namely,

$$\psi^{\pm}(x, 0) = \left(V_{gs} - \phi_{mi} - \frac{E_g}{2q} \right) \frac{\pm x - t_{si}/2}{t_{ox}} + \frac{E_g}{2q}, \qquad (7.37)$$

and

$$\psi^{\pm}(x, L) = \left(V_{gs} - \phi_{mi} - \frac{E_g}{2q} - V_{ds} \right) \frac{\pm x - t_{si}/2}{t_{ox}} + \frac{E_g}{2q} + V_{ds}. \qquad (7.38)$$

This is a good approximation when the gap width t_{ox} is small compared with the other dimensions, t_{si} and L. But it eventually fails in SOI MOSFETs in which case one of the t_{ox} becomes the BOX (buried oxide) thickness (see Section 7.1.2).

The 2-D short-channel potential is constructed by superposition (Liang and Taur, 2004). For silicon,

$$\psi(x, y) = v(x) + u_S(x, y) + u_D(x, y) \qquad (7.39)$$

For oxide,

$$\psi^{\pm}(x, y) = v^{\pm}(x) + u_S^{\pm}(x, y) + u_D^{\pm}(x, y). \qquad (7.40)$$

The long channel solutions are:

$$v(x) = V_{gs} - \phi_{mi} - \frac{qN_a t_{si} t_{ox}}{2\varepsilon_{ox}} - \frac{qN_a}{2\varepsilon_{si}} \left[\left(\frac{t_{si}}{2} \right)^2 - x^2 \right], \qquad (7.41)$$

and

$$v^{\pm}(x) = V_{gs} - \phi_{mi} - \frac{qN_a t_{si}}{2\varepsilon_{ox}}(t_{si}/2 + t_{ox} \mp x). \qquad (7.42)$$

They satisfy Poisson's Eqs. (7.30) and (7.31), the dielectric boundary conditions, Eqs. (7.32) and (7.33), and the top and bottom gate bias condition, Eq. (7.34). Functions u_S, u_D, $u_S^{\pm}$, and $u_D^{\pm}$ are to satisfy the homogeneous Eq. (7.31), with $u_S^{\pm}$, $u_D^{\pm}$ to vanish at the top and bottom gates. Furthermore, u_S and u_D are symmetric with respect to $x = 0$, u_S and $u_S^{\pm}$ are zero at $y = L$, and u_D and $u_D^{\pm}$ are zero at $y = 0$. The functions thus constructed are:

$$u_S(x, y) = \sum_{n=1}^{\infty} s_n \frac{\sinh[\pi(L-y)/\lambda_n]}{\sinh(\pi L/\lambda_n)} \cos(\pi x/\lambda_n), \qquad (7.43)$$

$$u_D(x, y) = \sum_{n=1}^{\infty} d_n \frac{\sinh[\pi y/\lambda_n]}{\sinh(\pi L/\lambda_n)} \cos(\pi x/\lambda_n), \qquad (7.44)$$

$$u_S^{\pm}(x, y) = \pm \sum_{n=1}^{\infty} s_n^{\pm} \frac{\sinh[\pi(L-y)/\lambda_n]}{\sinh(\pi L/\lambda_n)} \sin[\pi(\pm x - t_{si}/2 - t_{ox})/\lambda_n], \qquad (7.45)$$

$$u_D^{\pm}(x, y) = \pm \sum_{n=1}^{\infty} d_n^{\pm} \frac{\sinh[\pi y/\lambda_n]}{\sinh(\pi L/\lambda_n)} \sin[\pi(\pm x - t_{si}/2 - t_{ox})/\lambda_n]. \qquad (7.46)$$

The eigenvalues λ_n are solved by applying the dielectric boundary conditions, Eqs. (7.32) and (7.33), to Eqs. (7.43) and (7.45), as well as to Eqs. (7.44) and (7.46) at $x = \pm t_{si}/2$. Since the equalities are satisfied at every point of y, the corresponding terms of the two series must match each other. For non-zero solutions of s_n, d_n, etc., λ_n must satisfy the eigenvalue equation,

$$tan\left(\frac{\pi t_{ox}}{\lambda_n}\right) tan\left(\frac{\pi t_{si}}{2\lambda_n}\right) = \frac{\varepsilon_{ox}}{\varepsilon_{si}}. \qquad (7.47)$$

As was with bulk MOSFETs, the longest λ_n dominates the short-channel contribution to the minimum potential, and is designated as the scale length λ. The numerical solutions for λ are plotted in Figure 7.14 in normalized units with $\varepsilon_{ox}/\varepsilon_{si}$ as a parameter. In the most straightforward case where $\varepsilon_{ox} = \varepsilon_{si}$, the scale length is simply $\lambda = t_{si} + 2t_{ox}$, the physical height of the rectangular region between the two gates in Figure 7.12. For $\varepsilon_{ox}/\varepsilon_{si} < 1$, $\lambda > t_{si} + 2t_{ox}$, and for $\varepsilon_{ox}/\varepsilon_{si} > 1$, $\lambda < t_{si} + 2t_{iox}$. But in no case can λ be smaller than t_{si} or $2t_{ox}$, whichever is larger. *This means that even extremely high-κ insulators would have to be physically thin to be effective in scaling down λ.* In the very high-κ limit, the scale length approaches $2t_{ox}$, insensitive to the silicon thickness as long as $t_{si} < 2t_{ox}$. In the lower right corner of the curves where silicon is thick and the insulator is thin, Eq. (7.47) can be approximated to $\lambda = t_{si} + 2(\varepsilon_{si}/\varepsilon_{ox})t_{ox}$, since $\pi t_{ox}/\lambda \ll 1$ for the first tangent factor and $\pi t_{si}/2\lambda \approx \pi/2$ for the second tangent factor. High-κ gate dielectric is even more beneficial to DG than

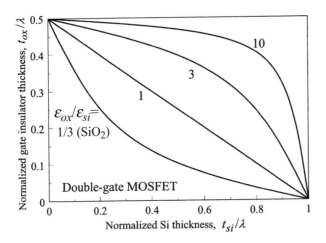

Figure 7.14 Numerical solutions to Eq. (7.47) for different values of $\varepsilon_{ox}/\varepsilon_{si}$

to bulk MOSFETs because of the double thickness. In the opposite, upper left corner, $\pi t_{si}/2\lambda \ll 1$ and $\pi t_{ox}/\lambda \approx \pi/2$, Eq. (7.47) can be approximated to $\lambda = 2t_{ox} + (\varepsilon_{si}/\varepsilon_{ox})t_{si}$.

7.2.2.2   2-D Potential in Subthreshold

To determine the 2-D potential $u_S + u_D$, we make use of the relationship

$$s_n^\pm = \mp s_n \frac{\cos[\pi t_{si}/(2\lambda_n)]}{\sin(\pi t_{ox}/\lambda_n)} \tag{7.48}$$

obtained from the matching conditions at $x = \pm t_s/2$. It enables $u_S^\pm$ to be expressed in terms of s_n:

$$u_S^\pm(x,y) = \sum_{n=1}^{\infty} s_n \frac{\cos[\pi t_{si}/(2\lambda_n)]}{\sin(\pi t_{ox}/\lambda_n)} \frac{\sinh[\pi(L-y)/\lambda_n]}{\sinh(\pi L/\lambda_n)} \sin[\pi(t_{si}/2 + t_{ox} \mp x)/\lambda_n]. \tag{7.49}$$

The same can be done for $d_n^\pm$, d_n, and $u_D^\pm$. Now s_n and d_n can be solved from the source and drain boundary conditions Eqs. (7.35)–(7.38). At the source, both u_D and $u_D^\pm$ vanish. Hence

$$\psi(x,0) - v(x) = u_S(x,0) = \sum_{n=1}^{\infty} s_n \cos(\pi x/\lambda_n) \tag{7.50}$$

and

$$\psi^\pm(x,0) - v^\pm(x) = u_S^\pm(x,0) = \sum_{n=1}^{\infty} s_n \frac{\cos[\pi t_{si}/(2\lambda_n)]}{\sin(\pi t_{ox}/\lambda_n)} \sin[\pi(t_{si}/2 + t_{ox} \mp x)/\lambda_n]. \tag{7.51}$$

where $v(x)$ and $v^\pm(x)$ are given by Eqs. (7.41) and (7.42). To evaluate s_n, an orthogonality relation is needed. It can be expressed in terms of an eigenfunction for the entire range of x:

$$f_{\lambda n}(x,0) = \begin{cases} \dfrac{\cos[\pi t_{si}/(2\lambda_n)]}{\sin(\pi t_{ox}/\lambda_n)}\sin[\pi(t_{si}/2+t_{ox}+x)/\lambda_n] & -\dfrac{t_{si}}{2}-t_{ox} < x < -\dfrac{t_{si}}{2} \\[2ex] \cos(\pi x/\lambda_n) & -\dfrac{t_{si}}{2} < x < \dfrac{t_{si}}{2} \\[2ex] \dfrac{\cos[\pi t_{si}/(2\lambda_n)]}{\sin(\pi t_{ox}/\lambda_n)}\sin[\pi(t_{si}/2+t_{ox}-x)/\lambda_n] & \dfrac{t_{si}}{2} < x < \dfrac{t_{si}}{2}+t_{ox} \end{cases}$$

$$(7.52)$$

which satisfies

$$\int_{-t_{si}/2-t_{ox}}^{t_{si}/2+t_{ox}} \varepsilon(x)f_{\lambda m}(x,0)f_{\lambda n}(x,0)dx = 0 \qquad \text{except } \lambda_m = \lambda_n \qquad (7.53)$$

Here, $\varepsilon(x)$ is a step function that equals ε_{si} in the silicon region and equals ε_{ox} in the two oxide regions. The results for s_n are:

$$s_n = \left\{ \frac{2\varepsilon_{ox}\lambda_n^2\cos[\pi t_{si}/(2\lambda_n)]}{\pi^2\varepsilon_{si}t_{ox}}\left(\frac{E_g}{2q}+\phi_{mi}-V_{gs}\right) + \frac{2qN_a\lambda_n^3}{\varepsilon_{si}\pi^3}\sin[\pi t_{si}/(2\lambda_n)] \right\}$$
$$\Big/ \left[\frac{t_{si}}{2}+t_{ox}\frac{\sin(\pi t_{si}/\lambda_n)}{\sin(2\pi t_{ox}/\lambda_n)}\right].$$

$$(7.54)$$

For d_n, simply replace $E_g/(2q)$ in Eq. (7.54) with $E_g/(2q) + V_{ds}$.

To gain more insights, consider a simplified case of $\varepsilon_{si} = \varepsilon_{ox}$ for which $\lambda = t_{si} + 2t_{ox}$. By setting $N_a = 0$, Eq. (7.54) is reduced to

$$s_n = \frac{4}{\pi}\left(\frac{\sin(\pi t_{ox}/\lambda_n)}{\pi t_{ox}/\lambda_n}\right)\left(\frac{E_g}{2q}+\phi_{mi}-V_{gs}\right). \qquad (7.55)$$

Likewise,

$$d_n = \frac{4}{\pi}\left(\frac{\sin(\pi t_{ox}/\lambda_n)}{\pi t_{ox}/\lambda_n}\right)\left(\frac{E_g}{2q}+\phi_{mi}+V_{ds}-V_{gs}\right). \qquad (7.56)$$

The dominant terms in the short-channel potential, $u_S + u_D$, are those of $n = 1$. Following the same steps as in Eq. (6.21), we obtain the minimum potential between the source and drain:

$$2\sqrt{s_1 d_1}e^{-\pi L/(2\lambda)}\cos\left(\frac{\pi x}{\lambda}\right) = \frac{8}{\pi}\left(\frac{\sin(\pi t_{ox}/\lambda_n)}{\pi t_{ox}/\lambda_n}\right)$$
$$\times \sqrt{\left(\frac{E_g}{2q}+\phi_{mi}-V_{gs}\right)\left(\frac{E_g}{2q}+\phi_{mi}+V_{ds}-V_{gs}\right)}e^{-\pi L/(2\lambda)}\cos\left(\frac{\pi x}{\lambda}\right).$$

$$(7.57)$$

This equation is similar to Eq. (6.35) for bulk MOSFETs. It shows that the potential at the center of silicon, $x = 0$, is most influenced by the source–drain fields. Since the pre-exponential factor is ~1.5 V, *the scaling limit is given by $L_{min} \approx 2\,\lambda$ for the SCE to be below 100 mV, just like in bulk MOSFETs.*

The implications of the $-V_{gs}$ dependence of s_n and d_n are two-fold. First, the SCE becomes more severe at lower V_{gs}, i.e., farther below V_t. Second, it eventually leads to the degradation of $d\psi/dV_{gs}$, hence the subthreshold current slope. A simplified expression for the subthreshold swing can be derived by considering the V_{gs} dependence of the potential at the minimum point,

$$\psi_{min} \approx \left(V_{gs} - \phi_{mi}\right) + \frac{8}{\pi}\left(\frac{E_g}{2q} + \phi_{mi} - V_{gs} + \frac{V_{ds}}{2}\right)e^{-\pi L/(2\lambda)}. \tag{7.58}$$

The first term is the long channel potential from $v(x)$ in Eq. (7.41). The second term is the short channel potential simplified from Eq. (7.57). The subthreshold swing is then given by $(d\psi_{min}/dV_{gs})^{-1}$ or

$$S \approx \frac{1}{1 - (8/\pi)e^{-\pi L/(2\lambda)}} \times 60 \text{ mV/decade} \tag{7.59}$$

At $L_{min} = 2\lambda$, the subthreshold swing is ~10% degraded from the long-channel value.

With the potential function $\psi(x, y)$ solved, subthreshold $I_{ds}-V_{gs}$ characteristics for a given L can be generated numerically from a double integral similar to that of Eq. (6.38). The effect of V_{ds} on $I_{ds}-V_{gs}$ or DIBL is built into the model through the parameter d_n. Figure 7.15 compares the results of the scale length model to TCAD simulations. It includes a case of $L/\lambda \approx 0.85$ with severe SCE. Note that the model is no longer accurate above $I_{ds} \sim 10^{-6}$ A because the mobile charge density is not negligible when too close to the threshold.

The sign of the doping dependent term in Eq. (7.54) needs explanation because it appears that p-type doping aggravates SCE. But, as N_a goes up, V_{gs} should go up also because the long channel threshold voltage goes up with N_a [Eq. (7.41)]. If, instead of keeping V_{gs} constant, we keep the center potential,

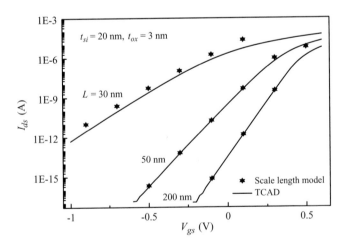

Figure 7.15 Subthreshold $I_{ds}-V_{gs}$ characteristics generated by the scale length model with only the $n = 1$ term compared to TCAD. The device parameters are $t_{si} = 20$ nm, $t_{ox} = 3$ nm, $\varepsilon_{ox}/\varepsilon_{si} = 1/3$ for a scale length λ of 35.4 nm. All I_{ds} are normalized to $W/L = 1$.

$$v(0) = V_{gs} - \phi_{mi} - \frac{qN_a t_{si} t_{ox}}{2\varepsilon_{ox}} - \frac{qN_a t_{si}^2}{8\varepsilon_{si}}, \tag{7.60}$$

constant as N_a varies, s_n and d_n would decrease with p-type doping and increase with n-type doping, as expected. In any case, for $t_{si} = 10$ nm and below, the doping effect on SCE is insignificant until above the 10^{19} cm^{-3} level (Pandey et al., 2018).

7.2.2.3   Short-Channel Model above Threshold

For short channel DG MOSFETs above threshold, the non-GCA model discussed in Section 6.2.1.5 for bulk MOSFETs can be employed. The basic equations are Poisson's equation

$$\frac{\partial^2 \psi}{\partial x^2} + \frac{\partial^2 \psi}{\partial y^2} = \frac{q}{\varepsilon_{si}} n_i e^{q(\psi - V)/kT}, \tag{7.61}$$

and the current continuity equation, e.g., for the $n = 1$ velocity saturation model,[4]

$$I_{ds} = Wqn_i t_{si} e^{q(\psi - V)/kT} \frac{\mu_{eff} dV/dy}{1 + (\mu_{eff}/v_{sat}) dV/dy}. \tag{7.62}$$

Here, t_{si} is used to convert between the volume charge density and the areal charge density. The mobile charge density is divided into two parts: the $\partial^2 \psi/\partial x^2$ part from the gate field, and the $\partial^2 \psi/\partial y^2$ part from the source–drain fields. As before, the gate induced part, $-Q_i$, can be expressed in terms of the quasi-Fermi potential V based on the GCA model,

$$\frac{d^2 \psi}{dx^2} = \frac{-Q_i}{\varepsilon_{si} t_{si}} = \frac{2C_{inv}(V_{gs} - V_t - V)}{\varepsilon_{si} t_{si}}. \tag{7.63}$$

For a given V_{gs}, C_{inv} is evaluated from the analytic potential model described in Section 7.2.1, namely, the ratio of Eq. (7.24) to Eq. (7.22) with $\beta = \beta_s$.

Subtracting Eq. (7.63) from Eq. (7.61) yields

$$\frac{d^2 \psi}{dy^2} = \frac{q}{\varepsilon_{si}} n_i e^{q(\psi - V)/kT} - \frac{2C_{inv}(V_{gs} - V_t - V)}{\varepsilon_{si} t_{si}}. \tag{7.64}$$

This can be treated as an ordinary differential equation in the lateral or source–drain direction, with unknown functions $\psi(y)$ and $V(y)$. It is coupled to the current continuity equation, Eq. (7.62), and solved together (Hong and Taur, 2021). There are four boundary conditions: $\psi(0) = E_g/2q$ and $V(0) = 0$ at the source, and $\psi(L) = E_g/2q + V_{ds}$ and $V(L) = V_{ds}$ at the drain. I_{ds}-V_{ds} characteristics are determined in the same manner as that described in the paragraph preceding Eq. (6.78) for bulk MOSFETs.

7.2.3   Nanowire MOSFETs

While a DG MOSFET has rectangular symmetry, a nanowire MOSFET has cylindrical symmetry. Shown schematically in Figure 7.16 are the cross-sections along

---

[4] The $n = 2$ velocity saturation model can be implemented in a similar manner.

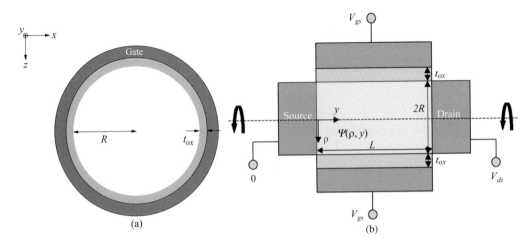

Figure 7.16 Schematic diagrams of a nanowire MOSFET: (a) cross-section, (b) a cut along the cylindrical axis (after Yu *et al.*, 2008)

different cuts of a nanowire MOSFET, also known as the surrounding gate or gate all around MOSFET. Its cylindrical symmetry lends to analytic solutions. An all-region GCA model for long-channel nanowire MOSFETs and a scale length model for short-channel nanowire MOSFETs in subthreshold have been developed.

7.2.3.1   Long-Channel Current Model

In contrast to Eq. (7.19) for DG MOSFETs, Poisson's equation in cylindrical coordinates under the GCA has the form

$$\frac{1}{\rho}\frac{d}{d\rho}\left(\rho\frac{d\psi}{d\rho}\right) = \frac{q}{\varepsilon_{si}}n(\rho) = \frac{q}{\varepsilon_{si}}n_i e^{q(\psi-V)/kT}. \tag{7.65}$$

The analytic solution satisfying the symmetry condition, $(d\psi/d\rho)|_{\rho=0} = 0$, is (Jiménez et al., 2004)

$$\psi(\rho) = V + \frac{kT}{q}\ln\left\{\frac{8\varepsilon_{si}kT(1-\beta)}{q^2 n_i R^2\left[1-(1-\beta)(\rho/R)^2\right]^2}\right\}, \tag{7.66}$$

with a constant β to be determined from the gate bias condition,

$$V_{gs} - \phi_{mi} - \psi(\rho = R) = V_{ox}. \tag{7.67}$$

To express V_{ox}, the potential drop across the oxide, in terms of the field in silicon, consider $\mathscr{E}_{ox}(\rho)$, the field in oxide in the radial direction. With no charge in the oxide, the cylindrical Laplace equation gives $\mathscr{E}_{ox}(\rho) \propto 1/\rho$. The proportional constant is determined by the continuity of displacement at the silicon–oxide interface, $\rho = R$, so that

$$\mathscr{E}_{ox}(\rho) = \frac{R}{\rho}\frac{\varepsilon_{si}}{\varepsilon_{ox}}\frac{d\psi}{d\rho}\bigg|_{\rho=R}. \tag{7.68}$$

Therefore,

$$V_{ox} = \int_R^{R+t_{ox}} \mathscr{E}_{ox}(\rho)d\rho = \frac{\varepsilon_{si}}{\varepsilon_{ox}} R \ln\left(1+\frac{t_{ox}}{R}\right)\frac{d\psi}{d\rho}\bigg|_{\rho=R}. \tag{7.69}$$

Substituting Eq. (7.66) into Eqs. (7.69) and (7.67) obtains an implicit equation for β:

$$\frac{q(V_{gs}-\phi_{mi}-V)}{2kT} - \ln\left(\frac{2}{R}\sqrt{\frac{2\varepsilon_{si}kT}{q^2 n_i}}\right) = \frac{1}{2}\ln(1-\beta) - \ln\beta + \left(\frac{2\varepsilon_{si}}{\varepsilon_{ox}}\right)\ln\left(1+\frac{t_{ox}}{R}\right)\frac{1-\beta}{\beta}. \tag{7.70}$$

The range of β is (0, 1).

The drain current is then solved by integrating the current continuity equation,

$$I_{ds} = \mu(-Q_{iL})\frac{dV}{dy}, \tag{7.71}$$

where Q_{iL} is the mobile charge density per nanowire length given by

$$-Q_{iL} = q\int_0^R n(\rho)2\pi\rho d\rho = \int_0^R 2\pi\varepsilon_{si}\frac{d}{d\rho}\left(\rho\frac{d\psi}{d\rho}\right)d\rho = 2\pi R\varepsilon_{si}\frac{d\psi}{d\rho}\bigg|_{\rho=R} = 8\pi\varepsilon_{si}\frac{kT}{q}\frac{1-\beta}{\beta}. \tag{7.72}$$

Carrying out the integration of Eq. (7.71) in a similar manner as was done in Eq. (7.23) yields the drain current,

$$I_{ds} = \mu\frac{4\pi\varepsilon_{si}}{L}\left(\frac{2kT}{q}\right)^2\left[\frac{\varepsilon_{si}}{\varepsilon_{ox}}\ln\left(1+\frac{t_{ox}}{R}\right)\left(\frac{1-2\beta}{\beta^2}\right)+\frac{1}{\beta}+\frac{\ln\beta}{2}\right]\bigg|_{\beta_d}^{\beta_s}, \tag{7.73}$$

where β_s and β_d are solutions to Eq. (7.70) for $V=0$ and V_{ds}, respectively. Note that $\beta_s < \beta_d$.

Like the analytic potential model for DG MOSFETs, Eq. (7.73) is an all-region model that contains the I_{ds}–V_{ds} characteristics in various regions of operation: subthreshold ($\beta_s \approx 1$, $\beta_d \approx 1$), linear region ($\beta_s \ll 1$, $\beta_d \ll 1$), and saturation region ($\beta_s \ll 1$, $\beta_d \approx 1$) (Jiménez et al., 2004). For nanowire MOSFETs, the current below threshold is proportional to the cross-sectional area of silicon or R^2 (volume inversion), while the current above threshold is proportional to $[\ln(1+t_{ox}/R)]^{-1}$ which, in the case of $t_{ox} \ll R$, is reduced to R/t_{ox}, proportional to the perimeter of silicon.

7.2.3.2   Scale Length of Nanowire MOSFETs

The 2-D Poisson's equation for a nanowire MOSFET in subthreshold takes the following form:

$$\frac{1}{\rho}\frac{\partial}{\partial\rho}\left(\rho\frac{\partial\psi}{\partial\rho}\right)+\frac{\partial^2\psi}{\partial y^2} = 0. \tag{7.74}$$

The boundary conditions are analogous to those for DG MOSFETs in Eqs. (7.32)–(7.36). The solutions in silicon are of the type

$$u_{SD}(\rho, y) = \sum_{n=1}^{\infty} \frac{s_n \sinh[\pi(L-y)/\lambda_n] + d_n \sinh[\pi y/\lambda_n]}{\sinh(\pi L/\lambda_n)} J_0(\pi\rho/\lambda_n), \qquad (7.75)$$

where J_0 is the zeroth-order Bessel function of the first kind that satisfies

$$\frac{1}{\rho}\frac{d}{d\rho}\left(\rho \frac{dJ_0(k\rho)}{d\rho}\right) = -k^2 J_0(k\rho). \qquad (7.76)$$

From the dielectric boundary conditions at the silicon–oxide interface, a scale length equation for nanowire MOSFETs has been derived (Oh et al., 2000). An equivalent form is given here:

$$\left(\frac{\varepsilon_{si}}{\varepsilon_{ox}}\right)\frac{J_1(\pi R/\lambda)}{J_0(\pi R/\lambda)} = \frac{J_1(\pi R/\lambda)Y_0[\pi(R+t_{ox})/\lambda] - J_0[\pi(R+t_{ox})/\lambda]Y_1(\pi R/\lambda)}{J_0(\pi R/\lambda)Y_0[\pi(R+t_{ox})/\lambda] - J_0[\pi(R+t_{ox})/\lambda]Y_0(\pi R/\lambda)}, \qquad (7.77)$$

where Y_0 and Y_1 are Bessel functions of the second kind or Neumann functions of the zeroth and first order, respectively. Figure 7.17 plots the numerical solutions of the lowest order (longest) of λ in normalized units. To gain useful insights, consider the special case of $\varepsilon_{si} = \varepsilon_{ox}$, in which Eq. (7.77) is simplified to (Yu et al, 2008)

$$J_0[\pi(R+t_{ox})/\lambda] = 0. \qquad (7.78)$$

The first zero of J_0 is $\alpha_1 = 2.405$, hence

$$\lambda = \frac{\pi}{\alpha_1}(R+t_{ox}) \approx 1.3(R+t_{ox}). \qquad (7.79)$$

The x- and y-intercepts of the straight line in Figure 7.17 for $\varepsilon_{ox}/\varepsilon_{si} = 1$ are $\alpha_1/\pi \approx 0.77$. In general, for any $\varepsilon_{ox}/\varepsilon_{si}$, λ is longer than $(\pi/\alpha_1)R$ or $(\pi/\alpha_1)t_{ox}$, whichever is longer. The minimum channel length for tolerable SCE is $L_{min} \approx 2\lambda$, the same as bulk and DG MOSFETs.

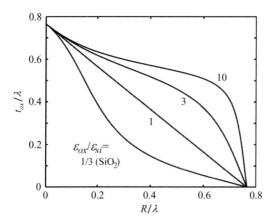

Figure 7.17 The lowest order solution to Eq. (7.77) for different values of $\varepsilon_{ox}/\varepsilon_{si}$, plotted in normalized scales (after Yu et al., 2008)

7.2.4   Scaling Limits of DG and Nanowire MOSFETs

The scaling limits of DG and nanowire MOSFETs are imposed by quantum mechanical limitations of the silicon thickness as well as the oxide thickness as they both approach atomic dimensions. The limit on the oxide thickness is the same as that discussed in Section 4.6.1 for bulk MOSFETs. The physical thickness of silicon (t_{si} or R) can be scaled below 10 nm, thinner than the gate depletion width W_{dm} in Figure 6.6 for bulk MOSFETs. The latter is limited by high body doping levels above 10^{19} cm^{-3} (Taur *et al.*, 1997). For silicon thickness much below 10 nm, however, another quantum mechanical effect needs to be considered (Frank *et al.*, 1992).

When electrons are confined within a silicon thickness t_{si} of a few nanometers, their ground-state energy can be appreciably above zero. The electron wavefunction $\varphi(x)$ is governed by Schrodinger's equation (in subthreshold),

$$-\frac{\hbar^2}{2m^*}\frac{d^2\varphi}{dx^2} = E_j\varphi, \qquad (7.80)$$

with the boundary condition $\varphi(x = \pm t_{si}/2) = 0$ (assuming infinite oxide barriers). The ground-state wavefunction is $\varphi(x) \propto \cos(\pi x/t_{si})$ which, when plugged into Eq. (7.80), yields a ground-state energy of

$$E_0 = \frac{\hbar^2}{2m^*}\left(\frac{\pi}{t_{si}}\right)^2 = \frac{h^2}{8m^* t_{si}^2}, \qquad (7.81)$$

where m^* is the effective mass of electron.

Effectively, E_0 raises the conduction band energy E_c, the lowest energy where electron states exist, and therefore the threshold voltage by E_0/q. Figure 7.18 plots

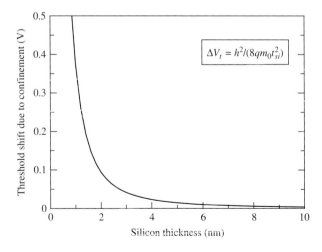

Figure 7.18 Increase of threshold voltage due to quantum confinement versus silicon film thickness for a DG MOSFET. It can be used for the nanowire case with the correspondence of $t_{si} \leftrightarrow 1.3R$

such a threshold shift versus t_{si} assuming $m^* = m_0$, the free electron mass. For lighter effective masses, the quantum effect is even stronger. The threshold voltage shift due to quantum confinement rises steeply above 0.1 V when the silicon film is thinner than 2 nm. Not only does it require gate work functions outside of those of n^+ silicon (4.1 eV) and p^+ silicon (5.2 eV) for realizing low nMOS and pMOS threshold voltages, but the sensitivity of ΔV_t to silicon film thickness also poses a serious tolerance problem. Thickness control at atomic dimensions would be required.

For nanowire MOSFETs, Schrodinger's equation takes the radial form with the electron wavefunction in the ground state given by Bessel functions $J_0(\alpha_1 \rho/R)$ where $\alpha_1 = 2.405$ is the first zero of J_0.[5] The ground-state energy is then

$$E_0 = \frac{\hbar^2}{2m^*}\left(\frac{\alpha_1}{R}\right)^2 = \frac{\hbar^2}{8m^*}\left(\frac{\alpha_1}{\pi R}\right)^2. \qquad (7.82)$$

This has the same functional dependence on R or t_{si} as Eq. (7.81) with the correspondence of $t_{si} \leftrightarrow (\pi/\alpha_1)R \approx 1.3R$. For the case of $\varepsilon_{si} = \varepsilon_{ox}$, the scale lengths are $\lambda = t_{si} + 2t_{ox}$ for DG MOSFETs and $\lambda = 1.3(R + t_{ox})$ for nanowire. With the same degree of quantum confinement, $t_{si} = 1.3R$, the silicon part of the scale length is the same for both DG and nanowire MOSFETs. But for the same oxide thickness, nanowire MOSFET has a slightly shorter scale length.

Given that the high-κ gate insulator thickness t_{ox} is limited by tunneling to $\sim$1.5–2 nm (Section 4.6.1.6) and t_{si} is limited by the threshold shift from quantum confinement to $\sim$1.5–2 nm (Figure 7.18), the minimum scale length of a DG or nanowire MOSFET would be limited to 4–6 nm, implying a channel length limit of 7–10 nm.

Exercises

7.1   Consider a symmetric double-gate nMOSFET with $t_{ox} = 2$ nm and $t_{si} = 10$ nm. The gate work functions are $q\phi_m = 4.33$ eV (half way between n^+ Si and intrinsic Si).
(a)   Plot I_{ds} versus V_{ds} for the range of $0 < V_{ds} < 2$ V, with $V_{gs} = 0, 0.5, 1, 1.5,$ and 2 V. Assume an effective mobility of 200 cm^2/V-s, $W = 1$ μm, and $L = 1$ μm.
(b)   What channel length can this device be scaled to without severe short-channel effects?

7.2 (a)   With the analytic potential model for DG MOSFETs, there is a maximum value for the center potential at $x = 0$ no matter how far V_{gs} is above the threshold. Derive this maximum center potential in terms of the silicon thickness.
(b)   Same as (a) for the nanowire MOSFET. Derive the maximum $\psi(\rho = 0)$ in terms of R.

---

[5] Isotropic effective masses are assumed here. See Yu *et al.* (2008) for the case of anisotropic effective masses.

7.3 **(a)** For V_{gs} far above the threshold, derive the surface potential in a long channel DG MOSFET in terms of V_{gs}.

   (b) Same as (a) for the nanowire MOSFET.

7.4 Consider the analytic potential model for a symmetric, long-channel DG MOSFET. Derive the expressions for the transconductance $g_m = \left.\frac{\partial I_{ds}}{\partial V_{gs}}\right|_{V_{ds}=const.}$ and the output conductance $g_d = \left.\frac{\partial I_{ds}}{\partial V_{ds}}\right|_{V_{gs}=const.}$ in terms of β_s and β_d (valid for all regions of operation).

7.5 From Eq. (7.24), obtain explicit expressions of Q_i at the source in terms of V_{gs} (no β) under the conditions of (a) subthreshold and (b) far above the threshold.

7.6 Given that $t_{si} = 10$ nm and $t_{ox} = 2.5$ nm in a symmetric DG MOSFET, derive an analytic expression for the scale length λ as an explicit function of $\varepsilon_{ox}/\varepsilon_{si}$. Check your answer for the special cases of $\varepsilon_{ox}/\varepsilon_{si} = 1$ and $\varepsilon_{ox}/\varepsilon_{si} = \infty$.

7.7 Consider a symmetric, long-channel DG MOSFET with the following parameters: $t_{si} = 10$ nm, $t_{ox} = 2$ nm, $\phi_{mi} = -0.16$ eV, $N_a = 3\times10^{15}$ cm^{-3} (p-type), and $V_{ds} = 0$.

   (a) What is the net charge density (+ or −, per gate area) in silicon if $V_{gs} = 0.2$ V (in subthreshold)?

   (b) If N_a is changed to 3×10^{18} cm^{-3}, how much does that affect the threshold voltage?

   (c) For the N_a in (b), what's the effect of dopant number fluctuations on threshold voltage (1σ uncertainty) for a device of $W = L = 30$ nm?

7.8 The inversion charge density in a DG MOSFET is often expressed as $Q_i = 2 C_{inv} \times (V_{gs} - V_t - V)$, where V is the quasi-Fermi potential. C_{inv} equals C_{ox} and C_{si} in series. Find an expression for the semiconductor capacitance C_{si} in terms of β based on the analytic potential model.

7.9 For a symmetric DG MOSFET, find an explicit expression for $V(y)$ when the device is in subthreshold, i.e., both β_s and $\beta_d \ll 1$. The source–drain bias is V_{ds}. [*Hint: Integrate the current continuity equation from 0 to y as well as from 0 to L.*]

7.10 Consider an undoped asymmetric DG MOSFET with $t_{si} = 15$ nm. The oxides are 2 nm thick under gate 1, and 3 nm thick under gate 2. The work functions are $q\phi_{m1} = 4.35$ eV for gate 1 and $q\phi_{m2} = 4.95$ eV for gate 2. If V_{gs2} is set to 0.5 V and only V_{gs1} switches, what are the subthreshold swing and the threshold voltage (when ψ_{s1} reaches $E_g/2q$) of the device?

7.11 Consider an asymmetric double-gate nMOSFET with a silicon thickness of 5 nm and an oxide thickness of 2 nm (both gates). The work function of gate 1 is the same as n$^+$ silicon. The work function of gate 2 is that of intrinsic silicon (midgap).

(a) What is the threshold voltage of channel-1 (when $\psi_{s1} = E_g/2q$) which turns on first?

(b) If, instead of switching, gate-2 is grounded to the n^+ source, what is the V_{t1} then?

(c) Following (b), what is the inverse slope of the subthreshold current in mV/decade?

(d) In (a), estimate the linear slope of the Q_i versus V_{gs} curve above the first threshold (consider ψ_{s1} pinned to $E_g/2q$).

8 CMOS Performance Factors

The performance of a CMOS VLSI chip is measured by its integration density, switching speed, and power consumption. CMOS circuits have the unique characteristic of practically zero standby power,[1] which enables higher integration levels and makes them the technology of choice for most VLSI applications. This chapter examines the various factors that determine the switching speed of basic CMOS circuit elements. It begins with a review of the basic concept of MOSFET scaling.

8.1 MOSFET Scaling

CMOS technology evolution in the past forty years has followed the path of device scaling for achieving density, speed, and power improvements. MOSFET scaling was propelled by the advancement of lithographic techniques for delineating fine lines of progressively narrower width. In Section 6.1, we discussed that reducing the source-to-drain spacing, i.e., the channel length of a MOSFET, led to short-channel effects. The rules of scaling consist of a set of simplified guidelines on how the device structure and voltage should evolve in a progressive way to deal with the ever shrinking gate length. Also projected are the factors of speed and power improvement under such scaling scenarios.

8.1.1 Constant-Field Scaling

In constant-field scaling (Dennard *et al.*, 1974), it was proposed that one can keep short-channel effects under control by scaling down the vertical dimensions (gate insulator thickness, junction depth, etc.) along with the horizontal dimensions, while also proportionally decreasing the applied voltages and increasing the substrate doping concentration (decreasing the depletion width). This is shown schematically in Figure 8.1. *The principle of constant-field scaling lies in scaling the device voltages and the device dimensions (both horizontal and vertical) by the same factor, κ (>1), so that the electric field remains unchanged*. This assures that the reliability of the scaled device is not worse than that of the original device.

---

[1] That is, until recently. See Section 4.6.1 for gate oxide tunneling and Section 6.3.1 for off current in subthreshold.

Table 8.1. Scaling MOSFET device and circuit parameters

	MOSFET Device and Circuit Parameters	**Multiplicative Factor ($\kappa > 1$)**
Scaling assumptions	Device dimensions (t_{ox}, L, W, x_j)	$1/\kappa$
	Doping concentration (N_a, N_d)	κ
	Voltage (V)	$1/\kappa$
Derived scaling behavior of device parameters	Electric field ($\mathscr{E}$)	1
	Carrier velocity (v)	1
	Depletion-layer width (W_d)	$1/\kappa$
	Capacitance ($C = \varepsilon A/t$)	$1/\kappa$
	Inversion-layer charge density (Q_i)	1
	Current, drift (I)	$1/\kappa$
	Channel resistance (R_{ch})	1
Derived scaling behavior of circuit parameters	Circuit delay time ($\tau \sim CV/I$)	$1/\kappa$
	Power dissipation per circuit ($P \sim V \times I$)	$1/\kappa^2$
	Power-delay product per circuit ($P \times \tau$)	$1/\kappa^3$
	Circuit density ($\propto 1/A$)	κ^2
	Power density (P/A)	1

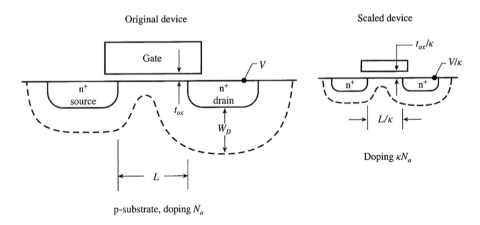

Figure 8.1 Principles of MOSFET constant-electric-field scaling ($\kappa > 1$) (after Dennard, 1986)

Table 8.1 shows the scaling rules for various device parameters and the projected circuit performance factors. The doping concentration is increased by the scaling factor κ in order to keep Poisson's equation, Eq. (6.2), invariant with respect to scaling. The maximum drain junction depletion width,

$$W_D = \sqrt{\frac{2\varepsilon_{si}(\psi_{bi} + V_{dd})}{qN_a}}, \tag{8.1}$$

from Eq. (3.15) (with $V_{app} = -V_{dd}$) scales down approximately by κ provided that the power-supply voltage V_{dd} is much greater than the built-in potential ψ_{bi}. All capacitances (including the wiring load) scale down by κ, since they are proportional to area and inversely proportional to thickness. The charge per device ($\sim C \times V$) scales down by κ^2, while the inversion-layer charge density (per unit gate area), Q_i, remains unchanged after scaling. Since the electric field at any given point is unchanged, the carrier velocity ($v = \mu_{eff} \mathscr{E}$) at any given point is also unchanged (the mobility is the same for the same vertical field). Therefore, any velocity saturation effects will be similar in the original and the scaled devices.

The drift current per MOSFET width, obtained by integrating the first term of the electron current density equation, Eq. (2.68), over the inversion layer thickness, is

$$\frac{I_{drift}}{W} = Q_i v = Q_i \mu_{eff} \mathscr{E}, \tag{8.2}$$

and is unchanged with respect to scaling. This means that the drift current scales down by κ, consistent with the behavior of both the linear and the saturation MOSFET currents in Eq. (5.23) and (5.31). A key implicit assumption is that the threshold voltage also scales down by κ. Note that the velocity saturated current, Eq. (6.49), also scales the same way, since both v_{sat} and μ_{eff} are constants, independent of scaling.

With both the voltage and the current scaled down by the same factor, it follows that the active channel resistance [e.g., Eq. (5.23)] of the scaled-down device remains unchanged. It is further assumed that the parasitic resistance is either negligible or unchanged in scaling. The circuit delay, which is proportional to RC or CV/I, then scales down by κ. **This is the most important conclusion of constant-field scaling: once the device dimensions and the power-supply voltage are scaled down, the circuit speeds up by the same factor. Moreover, power dissipation per circuit, which is proportional to VI, is reduced by κ^2.** Since the circuit density has increased by κ^2, the power density, i.e., the active power per chip area, remains unchanged in the scaled-down device. This has important technological implications in that packaging of the scaled CMOS devices does not require more elaborate heat-sinking. The power–delay product of the scaled CMOS circuit shows a dramatic improvement by a factor of κ^3 (Table 8.1).

8.1.2 Nonscaling Factors

While constant-field scaling provides a basic framework for shrinking CMOS devices to gain higher density and speed without degrading reliability and power, there are several factors that scale neither with the physical dimension nor with the operating voltage. **The primary reason for the nonscaling effects is that neither the thermal voltage kT/q nor the silicon bandgap E_g scales with the device dimension.** The former leads to subthreshold nonscaling; i.e., the threshold voltage cannot be scaled down like other parameters. The latter leads to nonscalability of the built-in potential, depletion-layer width, and short-channel effect.

In contrast to the drift current per width, Eq. (8.2), which is unchanged after applying the scaling factor κ, the diffusion current per unit MOSFET width, obtained by integrating the second term of the current density equation, Eq. (2.68),

$$\frac{I_{diff}}{W} = D_n \frac{dQ_i}{dx} = \mu_n \frac{kT}{q} \frac{dQ_i}{dx} \qquad (8.3)$$

scales up by κ, since dQ_i/dx is inversely proportional to the channel length. Therefore, the diffusion current does not scale down the same way as the drift current. This has significant implications in the nonscaling of MOSFET threshold voltage and its effect on the off current and power supply voltage, as were discussed in Sections 6.3.1 and 6.3.2.

For the nonscaling of the built-in potential, consider the maximum gate depletion width,

$$W_{dm} = \sqrt{\frac{2\varepsilon_{si}(2\psi_B)}{qN_a}}, \qquad (8.4)$$

where $2\psi_B$ is comparable to the bandgap and does not change with scaling. Consequently, scaling of N_a up by κ only reduces W_{dm} by $\sqrt{\kappa}$, which cannot contain the SCE of the scaled MOSFETs. In Section 6.1.2, we discussed that N_a needs to scale up by κ^2, which then leads to the evolution of nonuniform doping profile (Section 6.3.3) to manage the reduction of V_t as devices are scaled down (Figure 6.22).

Another nonscaling factor is related to the polysilicon gate and source–drain doping levels. In practice, they are limited by the solid solubility and cannot increase with scaling. The gate depletion effect adds a certain thickness to t_{ox}, causing the gate capacitance to scale up less than C_{ox}. It eventually led to the development of metal gate technology, as was discussed in Section 4.3.5. Likewise, because the source–drain doping level cannot be scaled up, the source–drain series resistance times width (in Ω-μm) has not been reduced in proportion to the channel resistance times width. This leads to the loss of current drive as the parasitic component becomes a more significant portion of the total resistance in the scaled device. Further discussions on the source–drain series resistance can be found in Section 8.3.1.

The most profound effects of the nonscaling factors are the higher field and power, both standby and active. The oxide field has gone up from 2 MV/cm to 10 MV/cm. Fortunately, thinner oxides can tolerate higher fields and still satisfy the reliability requirement (see Section 4.6.4). Higher power density means that the chip performance is often power limited as it reaches ~100 W/cm², the cooling capability of standard packages. Other than the density benefit derived from scaling, the industry has introduced novel process technologies to enhance device performance (Kuhn, 2012), e.g., FinFET (Section 7.2), SiGe pMOS, and strained silicon nMOS (Section 5.2.2).

8.2   Basic CMOS Circuit Elements

In a modern CMOS VLSI chip, the most important function components are CMOS static gates. In gate array circuits, CMOS static gates are used almost exclusively. In

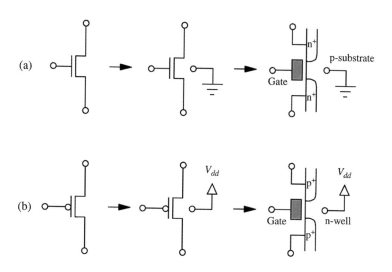

Figure 8.2 Circuit symbols and voltage terminals of (a) nMOS and (b) pMOS

microprocessors and supporting circuits of memory chips, most of the control inter-face logic is implemented using CMOS static gates. Static logic gates are the most widely used CMOS circuits because of their simplicity and noise immunity. This section describes basic static CMOS circuit elements and their switching characteristics.

Circuit symbols for nMOS and pMOS are defined in Figure 8.2. A MOSFET is a four-terminal device, although usually only three are shown. Unless specified, the body (p-substrate) terminal of an nMOS is connected to the ground (lowest voltage), while the body terminal (n-well) of a pMOS is connected to the power supply V_{dd} (highest voltage).

8.2.1 CMOS Inverters

The most basic element of digital static CMOS circuits is a CMOS inverter. A CMOS inverter is a combination of an nMOS and a pMOS, as shown in Figure 8.3 (Burns, 1964). The source terminal of the nMOS is connected to the ground, while the source of the pMOS is connected to V_{dd}. The gates of the two MOSFETs are tied together as the input node. The two drains are tied together as the output node. In such an arrangement, the complementary nature of nMOS and pMOS allows one and only one transistor to be conducting in one of the two stable states. For example, when the input voltage is high or when $V_{in} = V_{dd}$, the gate-to-source voltage of the nMOS equals V_{dd}, which turns it on. At the same time, the gate-to-source voltage of the pMOS is zero, so the pMOS is off. The output node is then pulled down to the ground potential by currents through the conducting nMOS, which is referred to as the *pull-down* transistor. On the other hand, when the input voltage is low or when $V_{in} = 0$, the nMOS is off, since its gate-to-source voltage is zero. The gate-to-source voltage of the

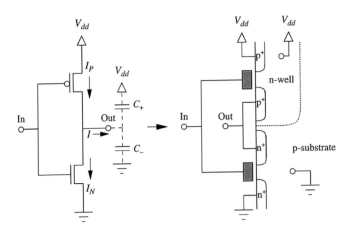

Figure 8.3 Circuit diagram and schematic cross-section of a CMOS inverter

pMOS, at the same time, is $-V_{dd}$, which turns it on (a negative gate voltage turns on a pMOS). The output node is now pulled up to V_{dd} by the conducting pMOS, which is referred to as the *pull-up* transistor. Since the output voltage is always opposite to the input voltage (V_{out} is high when V_{in} is low and vice versa), this circuit is called an inverter. Notice that *since only one of the transistors is on in the steady state, there is no static current or static power dissipation*. Power dissipation occurs only during switching transients when a charging or discharging current is flowing through the circuit.

8.2.1.1 CMOS Inverter Transfer Curve

In a CMOS inverter, both the current through the nMOS ($I_N > 0$) and the current through the pMOS ($I_P > 0$) are functions of the input voltage to the gates, V_{in}, and the output node voltage, V_{out}. A typical example is shown in Figure 8.4 where I_N and I_P are plotted versus V_{out} with V_{in} as a parameter. Note that for pMOS in Figure 8.4(b), the drain-to-source voltage is $V_{dsp} = V_{out} - V_{dd}$, and the gate-to-source voltage is $V_{gsp} = V_{in} - V_{dd}$. Both are negative or zero in normal operations. Also note that I_P enters saturation softer than I_N because holes have a more gradual velocity–field relationship than electrons (Section 6.2.1.1). The net current flowing out of the inverter is given by $I = I_P - I_N$. The output node voltage increases or decreases depending on whether $I > 0$ or $I < 0$. The directions of the currents are depicted in Figure 8.3.

In the steady state, $I = 0$, i.e., $I_P = I_N$. There are two points of operation where both I_P and I_N are zero: point A where $V_{in} = 0$ and $V_{out} = V_{dd}$, and point B where $V_{in} = V_{dd}$ and $V_{out} = 0$. For other values of V_{in} in between, the corresponding V_{out} is obtained from the intersection of two curves, $I_N(V_{in})$ and $I_P(V_{in})$, as shown in Figure 8.4(c). In this way, we can construct a V_{out} versus V_{in} curve, or a *transfer curve* for the CMOS inverter in Figure 8.5. For low values of V_{in} such as point C, V_{out} is high and the nMOS is biased in saturation while the pMOS is biased in the linear region [V_{in1} in Figure 8.4(c)]. For high V_{in} such as point D, V_{out} is low and the nMOS is in the linear

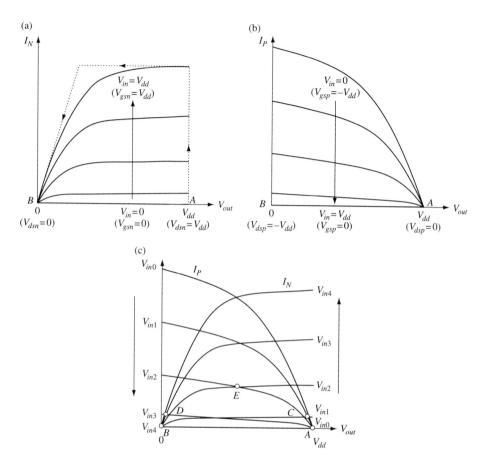

Figure 8.4 (a) nMOS current I_N and (b) pMOS current I_P in a CMOS inverter versus output node (drain) voltage for a series of input node (gate) voltages from 0 to V_{dd}. Both plots are superimposed in (c) to find the steady state points of operation (circles) given by the intersections where $I_P = I_N$ under the same V_{in} and V_{out}. The curves are labeled by the input voltages: $0 = V_{in0} < V_{in1} < V_{in2} < V_{in3} < V_{in4} = V_{dd}$, with the corresponding intercepts: A, C, E, D, B. The dotted lines in (a) depict an approximate bias point trajectory of an nMOS pull-down transition from A to B following an abrupt switching of V_{in} from 0 to V_{dd}. It is used later in Section 8.2.1.3 for discussion of the switching delay

region while the pMOS is in saturation [V_{in3} in Figure 8.4(c)]. For point E near $V_{in} = V_{dd}/2$ [V_{in2} in Figure 8.4(c)], both devices are in saturation. It is in a transition region where V_{out} changes most steeply with V_{in}.

In order for the high-to-low transition of the transfer curve to occur close to the midpoint, $V_{in} = V_{dd}/2$, it is desired for I_P and I_N to be nearly symmetrical, as illustrated in the example in Figure 8.4. This requires the threshold voltages of the nMOS and pMOS to be symmetrically matched. In addition, since the pMOS current per width, $I_{P/w} = I_P/W_p$, is inherently lower than that of nMOS, $I_{N/w} = I_N/W_n$, the device width ratio in a CMOS inverter should be

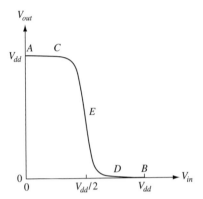

Figure 8.5 V_{out} versus V_{in} curve (transfer curve) of a CMOS inverter. Points labeled A, C, E, D, B correspond to the steady state points of operation (circles) indicated in Figure 8.4(c)

$$\frac{W_p}{W_n} = \frac{I_{N/w}}{I_{P/w}} \tag{8.5}$$

such that $I_P \approx I_N$. In the long-channel limit, $I_{N/w}/I_{P/w} \propto \mu_n/\mu_p \approx 4$ from Eq. (5.31) and Figures 5.14 and 5.16, assuming matched channel lengths and threshold voltages. For short-channel devices, however, the ratio is smaller since nMOS are more velocity saturated than pMOS. Typically, the current-per-width ratio $I_{N/w}/I_{P/w}$ is about 2–2.5 for deep-submicron CMOS technologies; therefore, **$W_p/W_n = 2$ is a good choice for CMOS inverter design**. For advanced CMOS technologies with SiGe pMOS, $I_{P/w}$ can be made close to $I_{N/w}$ to realize a symmetric $W_p = W_n$ design.

8.2.1.2   CMOS Inverter Noise Margin

Because of the nonlinear saturation characteristics of the MOSFET I_{ds}–V_{ds} curves, the V_{out}–V_{in} curve is also highly nonlinear. The maximum slope of the high-to-low transition of the V_{out}–V_{in} curve, $|dV_{out}/dV_{in}|$, referred to as the maximum voltage gain, is a measure of the g_m/g_{ds} ratio, i.e., the ratio of transconductance to output conductance, of the two transistors. From the condition $I_N(V_{gsn}, V_{dsn}) = I_P(V_{gsp}, V_{dsp})$, it can be shown that

$$\frac{dV_{out}}{dV_{in}} = -\frac{g_{mN} + g_{mP}}{g_{dsN} + g_{dsP}}, \tag{8.6}$$

where $g_{mN} \equiv \partial I_N/\partial V_{gsn}, g_{mP} \equiv -\partial I_P/\partial V_{gsp}(>0), g_{dsN} \equiv \partial I_N/\partial V_{dsn}, g_{dsP} \equiv -\partial I_P/\partial V_{dsp}(>0)$, etc.

To quantify the noise margin of a transfer curve, we consider a chain of identical inverters in cascade shown in Figure 8.6. The solid curve in Figure 8.7 represents the transfer curve of inverters #1, #3, #5, ..., i.e., V_{out1} versus V_{in1}, V_{out3} versus V_{in3}, etc. A complementary dashed curve is generated by flipping or mirror imaging the solid curve with respect to the chained line, $V_{in} = V_{out}$. It represents the inverse transfer curve of inverters #2, #4, #6, ..., i.e., V_{in2} versus V_{out2}, V_{in4} versus V_{out4}, etc. The

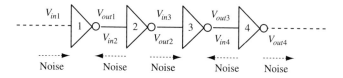

Figure 8.6 A cascade chain of identical CMOS inverters. The noise voltages at the input of each stage are for the discussion of Figure 8.8

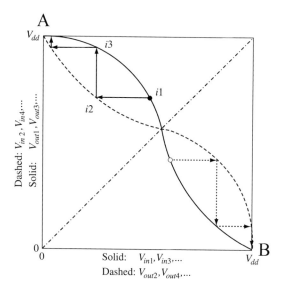

Figure 8.7 The solid transfer curve is for odd numbered inverter stages. The flipped, dashed transfer curve is for even numbered stages. The connected line segments between the curves depict the trajectory of node voltages through successive inverter stages

binary logic states are represented by point A at the upper left corner and point B at the lower right corner. Points in between are not full logic states. Using this graphical construction, we can visualize a trajectory of alternating horizontal and vertical lines between the two curves as the node voltage propagates through the inverter stages. Starting with dot $i1$ on the solid curve at coordinates (V_{in1}, V_{out1}), the next point $i2$ is on the dashed curve at coordinates (V_{out2}, V_{in2}). The line between $i1$ and $i2$ is horizontal since $V_{in2} = V_{out1}$. The next point, $i3$, is back on the solid curve with coordinates (V_{in3}, V_{out3}), and is connected to $i2$ by a vertical line as $V_{in3} = V_{out2}$, etc. In this example, the node voltage makes its transitions after each inverter stage closer and closer to the full logic state A at the upper left corner. If the starting point is below the $V_{in} = V_{out}$ intercept such as the circle in Figure 8.7, it will make transitions along the dotted line segments to the other full logic state B at the lower right corner. Such a characteristic is called "regenerative" which restores the node voltage to one of the full digital states.

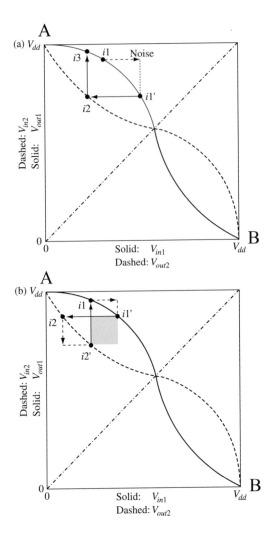

Figure 8.8 (a) Node voltage trajectory with noise added to the input to inverter #1. (b) Node voltage trajectory with positive noise at inverter #1 and negative noise at inverter #2. The shaded area represents the largest square that can be circumscribed in between the two transfer curves. The side of the square, V_{NM}, is a measure of the noise margin

To add noise to the above picture, we consider only two inverter stages with the transfer curves depicted in Figure 8.8(a). A positive noise voltage at the input to inverter #1 (Figure 8.6) shifts the starting point from $i1$ to $i1'$ on the solid curve. If there is no noise at the input to inverter #2, the output after two inverter stages will end up at point $i3$ shown. If $i3$ is to the left of $i1$, then the node voltage moves closer to the logic state A after the two inverters with noise. On the other hand, if $i3$ is to the right of $i1$, then the node voltage moves away from A and toward the other logic state B. In that case, the input voltage to the odd-numbered inverters may keep increasing through repeated cycles with noise. Finally, it will cross over the $V_{in} = V_{out}$ line and the logic state is lost (flipped from A to B). The maximum noise voltage that can be tolerated is then the one that causes $i3$ to fall back on top of $i1$.

We now add a negative going noise voltage of the same magnitude to the input of inverter #2 (Figure 8.6). Note that for this example, while a positive going noise voltage is worst at input 1, a negative going noise is worst at input 2. As shown in Figure 8.8(b), the negative noise voltage shifts the input to inverter #2 from $i2$ to $i2'$. The maximum noise magnitude that can be tolerated without eventually losing the logic state is the one that returns exactly to $i1$ after two noisy stages. Therefore, **the noise margin for a given transfer curve is measured by the size of the maximum square that can fit between itself and its complementary curve** (Hill, 1968). A different way of arriving at the same result is described in Figure 12.7 for the noise margin of SRAM cells. It is evident that for given nMOS and pMOS, a wider noise margin is achieved with the width ratio of Eq. (8.5) so that the high-to-low transition of the transfer curve happens at $V_{dd}/2$.

Since most of the noise interference in a chip environment comes from coupling of the voltage transients in the neighboring lines or devices, the noise magnitude is expected to scale with the power supply voltage (except those with other natural origins such as "soft error" due to high-energy particles). Thermal noise has too low a magnitude of concern as long as $V_{dd} \gg kT/q$. Therefore, a relevant measure of the noise margin in a CMOS circuit is the normalized V_{NM}/V_{dd}, where V_{NM} is the side of the maximum square in Figure 8.8(b). Large V_{NM}/V_{dd} (up to 0.5 in principle) is obtained with a highly skewed, symmetric transfer curve, i.e., one that has V_{out} staying high at low-to-medium V_{in}, then making an abrupt high-to-low transition at $V_{in} = V_{dd}/2$. It can be seen from the construction of the transfer curve in Figure 8.4(c) that for a given V_{dd}, V_{NM}/V_{dd} improves with a higher threshold voltage, V_t/V_{dd}. In fact, the best noise margin is achieved with subthreshold operation (Frank et al., 2001), although with poor delay performance. As V_{dd} is scaled down, V_{NM}/V_{dd} is not particularly sensitive to V_{dd} until V_{dd} becomes comparable to kT/q. In order to have the nonlinear I–V characteristics necessary for digital circuit function, a minimum V_{dd} of several kT/q, e.g., 100–200 mV, is required (Swanson and Meindl, 1972). At $V_{dd} \sim 1$ V level, the choice of power supply voltage for static CMOS logic circuits is largely based on power and performance considerations discussed in Section 8.4.3, not on noise margin.

8.2.1.3   CMOS Inverter Switching Characteristics

We now consider the basic switching characteristics of a CMOS inverter. The simplest input waveform is when the gate voltage makes an abrupt step transition from low to high or vice versa. For example, consider the inverter biased at point A in Figure 8.4(a) when V_{in} makes a step transition from 0 to V_{dd}. Before the transition, the nMOS is off and the pMOS is on. After the transition, the nMOS is on and the pMOS is off. The trajectory of V_{out} from point A to point B follows the $V_{in} = V_{dd}$ curve of the nMOS, as shown in Figure 8.4(a). If the total capacitance of the output node (including both the output capacitance of the switching inverter and the input capacitance of the next stage or stages it drives) is represented by two capacitors – one (C_-) to the ground and one

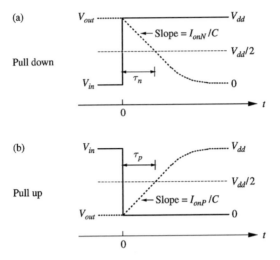

Figure 8.9 Waveforms of the output node voltage (dotted) of a CMOS inverter. (a) Pull-down transition after an abrupt rise of input voltage (solid). (b) Pull-up transition after an abrupt fall of input voltage (solid)

(C_+) to the V_{dd} rail, as illustrated in Figure 8.3 – then the pull-down switching characteristics are described by

$$C_- \frac{d(V_{out} - 0)}{dt} + C_+ \frac{d(V_{out} - V_{dd})}{dt} = -I_N(V_{in} = V_{dd}),$$

or

$$(C_- + C_+)\frac{dV_{out}}{dt} = C\frac{dV_{out}}{dt} = -I_N(V_{in} = V_{dd}), \qquad (8.7)$$

with the initial condition $V_{out}(t = 0) = V_{dd}$. Here $C = C_- + C_+$ includes both the capacitance to ground and the capacitance to V_{dd}. For simplicity, we approximate the $I_N(V_{in} = V_{dd})$ curve by two piecewise continuous lines. In the saturation region ($V_{out} > V_{dsat}$), $I_N = I_{onN}$ is a constant. In the linear region ($V_{out} < V_{dsat}$), $I_N = (I_{onN}/V_{dsat})V_{out}$, like a resistor with a resistance V_{dsat}/I_{onN}. These are shown as dotted lines in Figure 8.4(a). The solution $V_{out}(t)$ is depicted in Figure 8.9(a). Right after V_{in} switches from 0 to V_{dd}, V_{out} decreases linearly with time at a rate given by I_{onN}/C until $V_{out} = V_{dsat}$ is reached. Below that, V_{out} decreases exponentially toward zero with a time constant CV_{dsat}/I_{onN}. Similar switching characteristics are shown in Figure 8.9(b) for pMOS pull-up when V_{in} switches abruptly from V_{dd} to 0 [starting from point B in Figure 8.4(b)]. In this case, V_{out} follows the $V_{in} = 0$ curve of the pMOS, and the initial rate of increase with time is given by I_{onP}/C.

While it takes significantly longer time for the output voltage to approach zero in the pull-down case, it is conventional to define an nMOS pull-down delay τ_n as the time it takes for the output node voltage to reach $V_{dd}/2$. From Figure 8.9(a), it is clear that

$$\tau_n = \frac{CV_{dd}}{2I_{onN}} = \frac{CV_{dd}}{2W_n I_{onN/w}}, \tag{8.8}$$

where $I_{onN/w} \equiv I_{onN}/W_n$ is the nMOS on current per unit width. Similarly, the pMOS pull-up delay is

$$\tau_p = \frac{CV_{dd}}{2I_{onP}} = \frac{CV_{dd}}{2W_p I_{onP/w}}, \tag{8.9}$$

where $I_{onP/w} \equiv I_{onP}/W_p$ is the pMOS on current per unit width. *If a CMOS inverter is designed with a device width ratio given by* Eq. (8.5) *for a symmetrical transfer curve, it also follows that the pull-up and pull-down delays are equal.* The less conductive pMOS is compensated by having a width wider than that of the nMOS. The width ratio for the minimum switching delay, $\tau = (\tau_n + \tau_p)/2$, is generally different from that of Eq. (8.5) (Exercise 8.4) (Hedenstierna and Jeppson, 1987). However, the minimum is rather shallow, and the difference between the switching delay of a symmetric CMOS inverter and the minimum delay is usually no more than 5%.

8.2.1.4   Switching Energy and Power Dissipation

Switching a CMOS inverter or other logic circuit in general takes a certain amount of energy from the power supply. Let us first consider capacitor C_- between the output node and ground in Figure 8.3. During the pull-up transition of a CMOS inverter, the charge on C_- changes from zero to $\Delta Q_- = C_- V_{dd}$. This means that there is an energy of $V_{dd}\Delta Q_- = C_- V_{dd}^2$ flowing out of the power supply in the pull-up transition. Half of this energy, or $C_- V_{dd}^2/2$, is dissipated by the charging current in the pMOS resistance. The other half, another $C_- V_{dd}^2/2$, is stored in capacitor C_-. This energy stays in the capacitor until the next pull-down takes place. Then the charge on C_- drops to zero and the stored energy is dissipated by the discharging current through the nMOS resistance. Likewise, for capacitor C_+ between the output node and V_{dd} in Figure 8.3, an amount of energy $C_+ V_{dd}^2$ is supplied by the power source during the pull-down transition; half of which is dissipated by the discharging current in the nMOS resistance, while the other half $(C_+ V_{dd}^2/2)$ is stored in capacitor C_+. The stored energy is later dissipated in the pMOS resistance during the next pull-up transition.

From the above discussion, it is clear that for any capacitor C (either to the ground or to V_{dd}) to be charged or discharged, an energy of $CV_{dd}^2/2$ is dissipated irreversibly. It is a standard convention to consider a complete cycle consisting of a pair of transitions, either up–down ($0 \rightarrow V_{dd} \rightarrow 0$) or down–up ($V_{dd} \rightarrow 0 \rightarrow V_{dd}$). In that case, we say an energy of CV_{dd}^2 is dissipated per cycle. (An exception is the Miller capacitances between two switching nodes, to be discussed in Section 8.4.4.3.)

Since dc power dissipation is negligible in CMOS circuits, the only power consumption comes from switching. (Standby power dissipation of low-V_t devices was discussed in Section 6.3.1.2.) While the peak power dissipation in a CMOS inverter can reach $V_{dd}I_{onN}$ or $V_{dd}I_{onP}$, the average power dissipation depends on how often it switches. In a CMOS processor, the switching of logic gates is controlled by a clock generator of frequency f. If on the average a total equivalent capacitance C is charged and discharged within a clock cycle of period $T = 1/f$, the average power dissipation is

$$P = \frac{CV_{dd}^2}{T} = CV_{dd}^2 f. \tag{8.10}$$

Note here that each up or down transition of a capacitor within the period T contributes half of that capacitance to C. If, for example, a capacitor is switched four times (goes through the up–down cycle twice) within the clock period, its capacitance is counted twice in C. Equation (8.10) will be used in the discussion of power–delay tradeoff in Section 8.4.3.

The above simplified delay and power analysis assumes abrupt switching of V_{in}. In general, V_{in} is fed from a previous logic stage and has a finite rise or fall time associated with it. The switching trajectory from A to B or from B to A in Figure 8.4 then becomes much more complicated. Instead of staying on one constant-V_{in} curve, the bias point moves through different curves as V_{in} is ramped up or down. Furthermore, both I_N and I_P must be considered during either a pull-up or a pull-down transition, since the other transistor is not switched off completely as one transistor is being turned on. This also means that there is a *crossover*, or *short-circuit*, current that flows momentarily between the V_{dd} terminal and the ground in a switching event, which adds another power dissipation component to Eq. (8.10). One more complication is that the output node capacitances C_- and C_+ are generally voltage-dependent rather than being constant, as assumed in Eq. (8.7). More extensive numerical analysis of the general case will be given in Section 8.4.

8.2.1.5   Quasistatic Assumption

In the discussion of CMOS switching characteristics thus far, it was implicitly assumed that the device response time, i.e., the time required for charge redistribution, is fast compared with the time scale the terminal voltages change. This is called the *quasistatic* assumption. In other words, the device current responds instantaneously to an external voltage change. This assumption is valid if the input rise or fall time is long compared to the carrier transit time across the channel. In general, the carrier transit time can be expressed as

$$t_r = \int_0^L \frac{dy}{v(y)}, \tag{8.11}$$

where $v(y)$ is the carrier velocity at a point y in the channel. Current continuity requires $I = WQ_i(y)v(y)$ be a constant, independent of y. Equation (8.11) then becomes

$$t_r = \frac{W}{I} \int_0^L Q_i(y)dy = \frac{Q_I}{I}, \tag{8.12}$$

where Q_I is the total mobile charge in the device.

For a long-channel MOSFET in saturation, I is given by Eq. (5.31) and Q_I is given by Eq. (5.71). Therefore, the transit time is of the order of $L^2/\mu_{eff}V_{dd}$. For a completely velocity saturated device, the transit time approaches L/v_{sat}, which is of the order of 10 ps for 1 µm MOSFETs and 1 ps for 0.1 µm MOSFETs. These numbers are at least an order of magnitude shorter than the delay of an unloaded CMOS inverter made in the corresponding generation of technology (Taur *et al.*, 1985, 1993c). This indicates that the switching time is limited by the parasitic capacitances rather than by the time required for charge re-distribution within the transistor itself, hence validates the quasistatic approach.

8.2.2 CMOS NAND and NOR Gates

CMOS inverters described in Section 8.2.1 are used to invert a logic signal, to act as a buffer or output driver, or to form a latch (two inverters connected back to back). However, they cannot perform logic computation, since there is only one input voltage. In the static CMOS logic family, the most widely used circuits with multiple inputs are *NAND* and *NOR* gates, as shown in Figure 8.10. In a NAND gate, a number of nMOS are connected in series between the output node and the ground. The same number of pMOS are connected in parallel between V_{dd} and the output node. Each input signal is connected to the gates of a pair of nMOS and pMOS as in the case of an inverter. In this configuration, the output node is pulled to ground only if all the nMOS are turned on, i.e., only if all the input voltages are high (V_{dd}). If one of the input signals is low (zero voltage), the low-resistance path between the output node and ground is broken, but one of the pMOS is turned on, which pulls the output node to V_{dd}. On the contrary, the NOR circuit in Figure 8.10(b) consists of parallel-connected nMOS between the output node and ground, but serially connected pMOS between V_{dd} and the output node. The output voltage is high only if all the input voltages are low, i.e., all the pMOS are on and all the nMOS are off. Otherwise, the output is low.

Due to the complementary nature of nMOS and pMOS and the serial-versus-parallel connections, there is no direct low-resistance path between V_{dd} and ground except during switching. In other words, just like CMOS inverters, *there is no static current or standby power dissipation for any combination of inputs in either the CMOS NAND or NOR circuits*. The circuit output resistance is low, however, because of the conducting transistor(s).

In a CMOS technology, NAND circuits are used much more frequently than NOR. This is because it is preferable to put the transistors with the higher resistance in parallel and those with the lower resistance in series. Since pMOS have a higher resistance due to the lower hole mobility, they are rarely used in series (stacked). By connecting low-resistance nMOS in series and high-resistance pMOS in parallel, a NAND gate is more balanced in terms of the pull-up and the pull-down operations and achieves better noise immunity as well as a higher overall circuit speed.

8.2.2.1 Two-Input CMOS NAND Gate

As an example, we consider the transfer curve and the switching characteristics of a two-input NAND gate, also referred to as a two-way NAND, or NAND with a fan-in of two, shown in Figure 8.11. With the two pMOS connected in parallel between V_{dd}

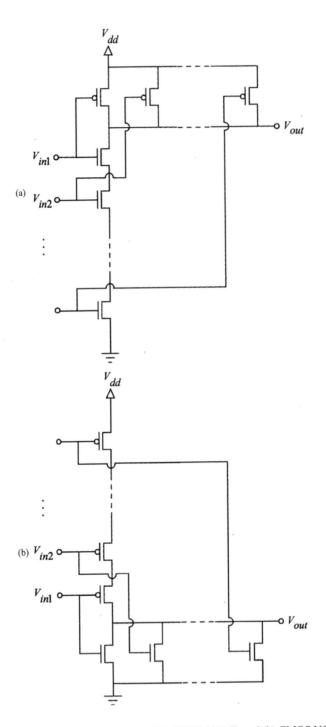

Figure 8.10 Circuit diagram of (a) CMOS NAND and (b) CMOS NOR. Multiple input signals are labeled V_{in1}, V_{in2}, $\cdots$

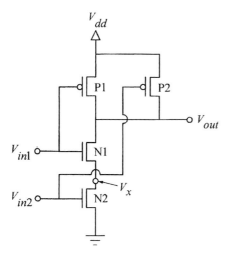

Figure 8.11 Circuit diagram of a two-input CMOS NAND. The transistors are labeled P1, P2 and N1, N2

and the output node, the pull-up operation of a two-way NAND is similar to that of an inverter. If either one of the transistors is being turned on while the other one is off, the charging current is identical to that of the pMOS pull-up in a CMOS inverter discussed in Section 8.2.1. If both transistors are pulling up, the total charging current is doubled as if the pMOS width had been increased by a factor of two. On the other hand, the two nMOS are connected in series (stacked) between the output and ground, and their switching behavior is quite different from that of the inverter. For the bottom transistor N2, its source is connected to the ground and the gate-to-source voltage is simply the input voltage V_{in2}. However, for the top transistor N1, its source is at a voltage V_x (Figure 8.11) higher than the ground. V_x plays a major role in the switching characteristics of N1, since the gate-to-source voltage that determines how far N1 is turned on is given by $V_{in1} - V_x$. Transistor N1 is also subject to the body-bias effect, as a source voltage V_x is analogous to a reverse body (substrate) bias of $V_{bs} = -V_x$ in Figure 5.19, which raises the threshold voltage of N1 as described by Eq. (5.59).

There are three possible switching scenarios, each with different characteristics. They are described below.

- *Case A. Bottom switching: Input 2 switches while input 1 stays at V_{dd}.* Initially, even though $V_{in1} = V_{dd}$, but $V_x > V_{dd} - V_t$ so that $V_{gs}(N1) = V_{dd} - V_x < V_t$ and both N1 and N2 are in subthreshold. The pull-down transition in case A when input 2 rises from 0 to V_{dd} is most similar to the nMOS pull-down in an inverter. For low input voltages $V_{in2} \leq V_{dd}/2$, transistor N2 is in saturation. Transistor N1 can be in the linear region or in saturation. In either case, N1 only acts to reduce the drain voltage of N2 with little effect on the current. The transfer curve of V_{out} versus V_{in2} in this case is similar to that of a CMOS inverter, which exhibits symmetrical characteristics if $W_p/W_n = I_n/I_p \approx 2$, as shown in Figure 8.12. For high input

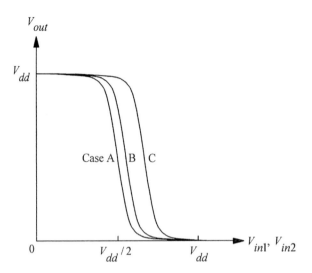

Figure 8.12 Transfer curves of a two-input CMOS NAND for different cases of switching discussed in the text. The device width ratio W_p/W_n is taken to be 2 in this illustration

voltages $V_{in2} > V_{dd}/2$, the current is somewhat degraded by the resistance of N1 as transistor N2 moves out of saturation.

- *Case B. Top switching: Input 1 switches while input 2 stays at V_{dd}.* For the pull-down transition in case B, transistor N1 is in saturation while N2 is in the linear mode during most part of the switching cycle. Transistor N2 therefore acts like a series resistance connected in the source terminal of N1. The voltage V_x between the two transistors rises slightly above the ground, depending on the current level. This degrades the pull-down current as the gate-to-source voltage of N1 is reduced to $V_{in1} - V_x$ and its threshold voltage is increased by $(m - 1)V_x$ due to the body effect. As a result, a slightly higher input voltage V_{in1} is needed to reach the high-to-low transition of the transfer curve in Figure 8.12. Even though the pull-down current in case B is slightly less than that in case A, the switching time in case B is comparable to that in case A if the output is not too heavily loaded. This is because of the additional capacitance associated with the top transistor N1 that needs to be discharged from V_{dd} to ground in case A. These factors are further discussed in detail in Section 8.4.6.
- *Case C. Both input 1 and input 2 switch simultaneously.* The worst case for pull-down in a two-input CMOS NAND is case C, in which both inputs rise from 0 to V_{dd}. It can be seen that transistor N2 is always biased in the linear region, while transistor N1 is in the linear region for small values of V_{out} and in saturation for large V_{out}. In this case, the nMOS pull-down current is reduced by approximately a factor of two from the inverter case because of the serial connection. The pull-up current, on the other hand, is twice that of the inverter case due to the parallel connection of pMOS. This moves the high-to-low transition in the transfer curve to a V_{in} significantly higher than $V_{dd}/2$, as shown in case C of Figure 8.12.

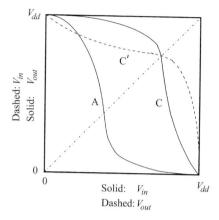

Figure 8.13 An example of worst-case noise margin of NAND circuits. Curve C' is the mirror image of C with respect to the axis $V_{in} = V_{out}$ (Liu *et al.*, 2006)

8.2.2.2 Noise Margin of NAND Circuits

Because of the spread of the transfer curves under different switching conditions, the noise margin of a CMOS NAND gate is inferior to that of a CMOS inverter. In an exaggerated case shown in Figure 8.13, curves A and C represent the extremes of all possible transfer curves. The best that can be done with the width selection (W_p/W_n) is such that A and C are symmetric on either side of $V_{dd}/2$. In the worst case scenario, one needs to consider the noise margin of a cascade chain of NAND stages with alternating switching conditions A and C. In other words, the worst case noise margin is given by the size of the smaller square that can be circumscribed between curves A and C', the flipped counterpart of C. In the example illustrated, the noise margin is severely degraded, but still positive. If the power supply voltage is too low and the number of fan-in too large, it could end up with no intersection between A and C'. That would mean eventual loss of logic state after repeated stages of worst case switching events. Higher threshold voltages are usually beneficial to the noise margin, although at the expense of switching speed. *The minimum power supply voltage for maintaining logic consistency of NAND or NOR circuits is of the order of (5–10)kT/q* (Frank *et al.*, 2001).

8.2.3 Inverter and NAND Layouts

8.2.3.1 Layout of a Single Device

Both the CMOS circuit density and the delay performance are determined by the layout ground rules of the particular technology. Figure 8.14 shows a typical layout of an isolated MOSFET and its corresponding cross-section. Only three major masking levels are shown: active region (isolation), polysilicon gate, and contact hole. To complete a CMOS process, several additional implant blockout masks are needed for doping the channel and the source–drain regions of nMOS and pMOS, respectively.

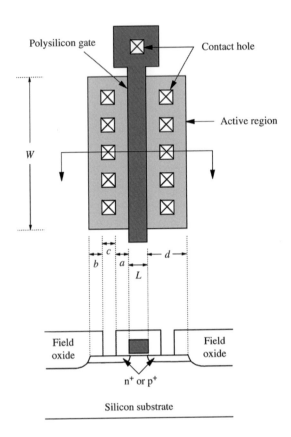

Polysilicon gate

Contact hole

Active region

W

c

b

a

d

L

Field oxide

Field oxide

n^+ or p^+

Silicon substrate

Figure 8.14 Basic layout and the corresponding cross-section of a single MOSFET, illustrating several key layout ground rules

After the device or the front-end-of-line (FEOL) process, a number of metal levels are laid down in the back-end-of-line (BEOL) process to connect the transistors into various circuits that make up the chip.

In Figure 8.14, the device length and width are indicated by L and W, respectively. The contact-hole size, represented by c, is limited by lithography. The spacings between the contact and the gate and between the contact and the edge of the active region are represented by a and b, respectively. These minimum distances are required in the ground rules to allow for alignment tolerances between the levels as well as linewidth biases and variations. Added together, a, b, and c determine the distance between the gate and the field isolation, i.e., the width of n^+ or p^+ diffusion, d. As far as CMOS delay is concerned, d should be kept as small as possible, since a larger diffusion area adds more parasitic capacitance to be charged and discharged during a switching event. In a silicided technology (Section 8.3.1.3), the diffusion area of a sufficiently wide MOSFET can be reduced by not extending the contact holes throughout the entire device width. The polysilicon contact area outside the active region is not a critical factor, as the additional capacitance introduced is negligible due to the thick field oxide (about 50 times thicker than gate oxide) underneath.

8.2.3.2   Layout of a CMOS Inverter

Figure 8.15(a) shows a simple CMOS inverter layout with $W_p/W_n \approx 2$. Four metal wires are shown, leading to V_{dd}, ground, input, and output. The pMOS receives n-well and p^+ source–drain implants with the use of two block-out masks. The nMOS receives a p-type channel implant and an n^+ source–drain implant with block-out masks of the complementary polarity. The intrinsic delay of a CMOS inverter, defined in terms of one stage driving an identical stage (fan-out = 1), is independent of the device width except for some parasitics at the ends. This is because both the current and the capacitance (gate and diffusion) are proportional to the device width in the straight-gate layout in Figure 8.15(a). *A substantial reduction in the junction contributions to the parasitic capacitance can be realized using the folded layout shown in Figure 8.15 (b).* By sandwiching the drain node between two symmetric source regions with a fork-shaped polysilicon gate, the device width for the current is effectively doubled without increasing the diffusion area. In other words, the junction capacitance per effective

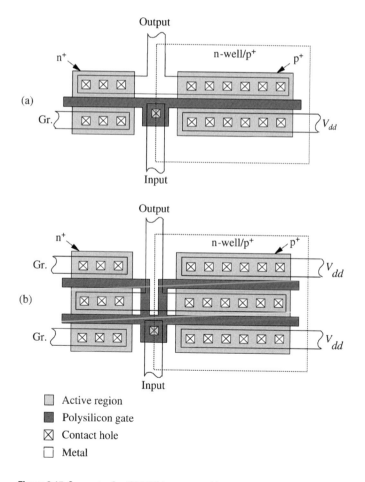

Active region
Polysilicon gate
Contact hole
Metal

Figure 8.15 Layout of a CMOS inverter with (a) straight gates and (b) folded gates for minimizing the parasitic diffusion capacitance

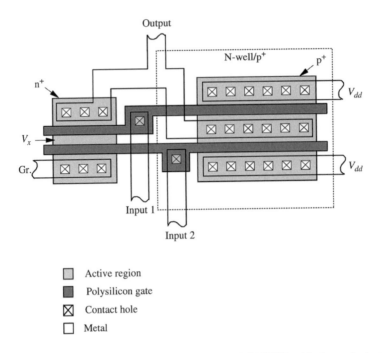

Figure 8.16 Layout example of a two-input CMOS NAND with the equivalent circuit in Figure 8.11

device width in layout (b) is about half of that in layout (a), assuming $a + b + c$ is comparable to $2a + c$ in Figure 8.14. Note that the area of the source regions is of no importance to the delay, since the source voltage is not being switched.

8.2.3.3 Layout of a Two-Input CMOS NAND

A typical layout for a two-input CMOS NAND is shown in Figure 8.16. The two parallel-connected pMOS are arranged as in the folded inverter, with the switching node sandwiched between the two input gates. This again minimizes the junction capacitance. The two nMOS are connected in series via a V_x-node between the input gates. Since no contact to the V_x-diffusion is necessary, its width can be kept as narrow as the minimum linewidth, i.e., comparable to L, c, etc., so that the capacitance associated with it is relatively small.

8.3   Parasitic Elements

There are parasitic components of resistance and capacitance in a MOSFET device structure that can adversely affect the drive current and the switching delay of a CMOS circuit. This section covers such parasitic elements as source–drain resistance, junction capacitance, overlap capacitance, gate resistance, and interconnect RC components.

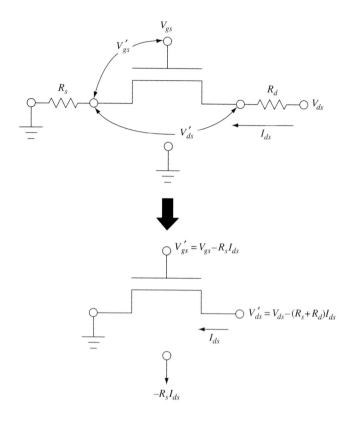

Figure 8.17 Equivalent circuit of MOSFET with source and drain series resistance. The top circuit is equivalent to the bottom circuit with redefined terminal voltages

8.3.1   Source–Drain Resistance

8.3.1.1   Effect of Source-Drain Resistance on Terminal Voltages

The effect of source–drain resistance is considered using the equivalent circuit in Figure 8.17. A source resistance R_s and a drain resistance R_d are assumed to connect an *intrinsic MOSFET* to the external terminals where voltages V_{ds} and V_{gs} are applied. The internal voltages are V'_{ds} and V'_{gs} for the intrinsic MOSFET. The current and voltages are related by the following:

$$V'_{ds} = V_{ds} - (R_s + R_d)I_{ds} \tag{8.13}$$

and

$$V'_{gs} = V_{gs} - R_s I_{ds}. \tag{8.14}$$

As shown in Figure 8.17, the actual device including the parasitic resistances is equivalent to an intrinsic MOSFET with a grounded source, with V'_{gs} and V'_{ds} at the gate and the drain terminals, and with a reverse bias $-R_s I_{ds}$ on the substrate. In a standard CMOS process, the source and drain regions are symmetrical. Therefore,

$R_s = R_d = R_{sd}/2$, where R_{sd} is the total source–drain parasitic resistance. From Eq. (5.23), the intrinsic channel resistance in the linear region is

$$R_{ch} \equiv \frac{V'_{ds}}{I_{ds}} = \frac{L/W}{\mu_{eff} C_{inv}\left(V'_{gs} - V'_t\right)}, \qquad (8.15)$$

where V'_t is the threshold voltage with the reverse bias on the substrate. R_{ch} can be expressed in terms of a channel sheet resistivity ρ_{ch} in units of $\Omega/\square$, $R_{ch} \equiv (L/W)\rho_{ch}$. For a general estimate, $\rho_{ch} \approx 1/\left[\mu_{eff}\varepsilon_{ox}\mathscr{E}_{ox}\right]$, where $\mathscr{E}_{ox}$ is the field in oxide. With $\mu_{eff} \sim 200\,\text{cm}^2/\text{V-s}$ and a maximum oxide field of 7–$10\,\text{MV/cm}$, $\rho_{ch} \sim 2000\,\Omega/\square$, insensitive to the technology generation. Source–drain series resistance has an appreciable effect on short-channel MOSFETs whose intrinsic $R_{ch} \times W = \rho_{ch} \times L$ is low. For example, if $L \sim 20\,\text{nm}$, the above figure of ρ_{ch} gives $R_{ch} \times W \sim 40\,\Omega\text{-}\mu\text{m}$, lower than the best $R_{sd} \times W$ of 100–$200\,\Omega\text{-}\mu\text{m}$ made in practice. Resistance on the source side is particularly troublesome, as it degrades the gate drive as well.

8.3.1.2 Components of Source and Drain Series Resistance

A schematic diagram of the current-flow pattern in the source or drain region of a MOSFET is shown in Figure 8.18 (Ng and Lynch, 1986). The total source or drain resistance can be divided into several parts: R_{ac} is the accumulation-layer resistance in the gate–source (or –drain) overlap region where the current mainly stays at the surface; R_{sp} is associated with current spreading from the surface layer into a uniform pattern across the depth of the source–drain; R_{sh} is the sheet resistance of the source–drain region where the current flows uniformly; and R_{co} is the contact resistance

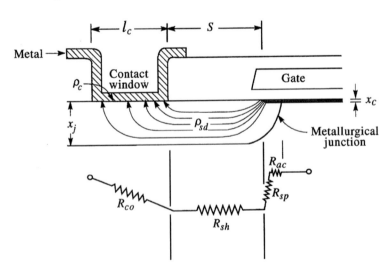

Figure 8.18 A schematic cross-section showing the pattern of current flow from a MOSFET channel through the source or drain region to the metal contact. The diagram identifies various contributions to the series resistance. The device width in the z-direction is assumed to be W (after Ng and Lynch, 1986)

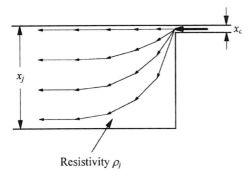

Resistivity ρ_j

Figure 8.19 Schematic diagram showing the resistance component associated with the injection region where the current spreads from a thin surface layer into a uniformly doped source or drain region (after Baccarani and Sai-Halasz, 1983)

(including the spreading resistance in silicon under the contact) in the region where the current flows into a metal line. Once the current flows into a metal line, there is very little additional resistance, since the resistivity of aluminum or copper is very low, $\rho_{Al} \approx 3 \times 10^{-6}\,\Omega$-cm and $\rho_{Cu} \approx 2 \times 10^{-6}\,\Omega$-cm. In VLSI interconnects, the aluminum thickness is typically 0.5–1.0 μm. From Eq. (2.40), the sheet resistivity is on the order of 0.05 Ω/□. This is negligible except when a long, thin wire is connected to a wide MOSFET. Figure 8.18 shows only the series resistance on one side of the device. The total source–drain series resistance per device is, of course, twice that shown in Figure 8.18, assuming that the source and drain are symmetrical. In the following paragraphs we examine the various components of the source–drain resistance.

The accumulation-layer sheet resistivity ρ_{ac} per square is lower than the channel resistivity ρ_{ch} per square. Generally speaking, current injection from the surface into the bulk takes place such that the sum of the resistances, $R_{ac} + R_{sp}$, is a minimum (Ng and Lynch, 1986). An analytical expression has been derived for R_{sp} assuming an idealized case shown in Figure 8.19 where the current spreading takes place in a uniformly doped medium with resistivity ρ_j (Baccarani and Sai-Halasz, 1983):

$$R_{sp} = \frac{2\rho_j}{\pi W}\ln\left(0.75\frac{x_j}{x_c}\right). \tag{8.16}$$

Here W is the device width, x_j and x_c are the junction depth and the inversion (or accumulation) layer thickness, respectively. For typical values of $x_j/x_c \approx 40$, we have $R_{sp} \approx 2\rho_j/W$. In practice, R_{sp} is more complex than Eq. (8.16), since current spreading usually takes place in a region where the source–drain doping falls off laterally and the local resistivity is highly nonuniform. For a heavily doped and abrupt source–drain profile, both R_{ac} and R_{sp} are low as $\rho_j \approx 10^{-3}\,\Omega$-cm for $10^{20}\,\mathrm{cm}^{-3}$ doping (Figure 2.10), hence $R_{sp} \times W \approx 20\,\Omega$-μm.

Next, we examine R_{sh} and R_{co}. In Figure 8.18, the sheet resistance of the source–drain diffusion region is simply

$$R_{sh} = \rho_{sd} \frac{S}{W}, \tag{8.17}$$

where W is the device width, S is the spacing between the gate edge and the contact edge, and ρ_{sd} is the sheet resistivity of the source–drain diffusion, typically of the order of 50–500 $\Omega/\square$. Since $\rho_{sd} \ll \rho_{ch}$ of the device, this term is usually negligible if S is kept to a minimum, limited by the overlay tolerance between the contact and the gate lithography levels. In a nonsilicided technology, $S = a$ in Figure 8.14, provided that most of the device width dimension is covered by contacts.

Based on a transmission-line model (Berger, 1972), the contact resistance can be expressed as

$$R_{co} = \frac{\sqrt{\rho_{sd}\rho_c}}{W} \coth\left(l_c\sqrt{\frac{\rho_{sd}}{\rho_c}}\right), \tag{8.18}$$

where l_c is the width of the contact window (Figure 8.18), and ρ_c is the interfacial *contact resistivity* (in Ω-cm^2) of the ohmic contact between the metal and silicon. R_{co} includes the resistance of the current crowding region in silicon underneath the contact. In a nonsilicided technology, $l_c = c$ in Figure 8.14. Equation (8.18) has two limiting cases: *short contact* and *long contact*. In the short-contact limit, $l_c \ll (\rho_c/\rho_{sd})^{1/2}$, and

$$R_{co} = \frac{\rho_c}{Wl_c} \tag{8.19}$$

is dominated by the interfacial contact resistance. The current flows more or less uniformly over the entire contact. In the long-contact limit, $l_c \gg (\rho_c/\rho_{sd})^{1/2}$, and

$$R_{co} = \frac{\sqrt{\rho_{sd}\ \rho_c}}{W}. \tag{8.20}$$

This is independent of the contact width l_c, since most of the current flows into the front edge of the contact. Once in the long-contact regime, there is no benefit increasing the contact width. The parameter $(\rho_c/\rho_{sd})^{1/2}$ is referred to as the *transfer length* in some literature.

For ohmic contacts between metal and heavily doped silicon, current conduction is dominated by tunneling or field emission (Section 3.2.3). The contact resistivity ρ_c depends exponentially on the barrier height ϕ_B and the surface doping concentration N_d [Eqs. (3.196) and (3.200)]:

$$\rho_c \propto \exp\left(\frac{4\pi\phi_B}{h}\sqrt{\frac{m^*\varepsilon_{si}}{N_d}}\right), \tag{8.21}$$

where h is Planck's constant and m^* is the electron effective mass. Depending on the doping concentration and contact metallurgy, ρ_c is typically in the range of 10^{-6}–10^{-8} Ω-cm^2.

8.3.1.3   Resistance in a Self-Aligned Silicide Technology
Both R_{sh} and R_{co} are greatly reduced in advanced CMOS technologies with self-aligned silicide process (Ting *et al.*, 1982). As shown schematically in Figure 8.20, a

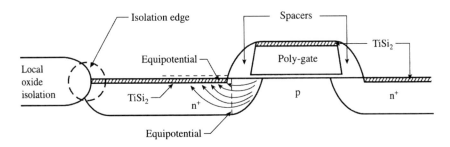

Figure 8.20 Schematic diagram of an n-channel MOSFET fabricated with self-aligned $TiSi_2$, showing the current flow pattern between the channel and the silicide (after Taur *et al.*, 1987)

highly conductive ($\approx$ 2–10 $\Omega/\square$) silicide film is formed on all the gate and source–drain surfaces separated by dielectric spacers in a self-aligned process. Since the sheet resistivity of silicide is 1–2 orders of magnitude lower than that of the source–drain, the silicide layer practically shunts all the currents, and the only significant contribution to R_{sh} is from the nonsilicided region under the spacer. This reduces the length S in Eq. (8.17) to 0.1–0.2 μm, which means that $R_{sh} \times W$ should be no more than 50 Ω-μm. At the same time, R_{co} between the source–drain and silicide is also reduced, since now the contact area is the entire diffusion. In other words, the diffusion width d in Figure 8.14 becomes the contact length l_c in Eq. (8.18). Current flow in this case is likely to be in the long-contact limit, so that Eq. (8.20) applies.

In a CMOS process, a silicide material such as $TiSi_2$ (or $CoSi_2$, NiSi) with a near-midgap work function is needed to obtain approximately equal barrier heights to n^+ and p^+ silicon. The experimentally measured ρ_c between $TiSi_2$ and n^+ or p^+ silicon is of the order of $10^{-6} - 10^{-7} \Omega$-cm^2 (Hui *et al.*, 1985). Based on Eq. (8.20), therefore, $R_{co} \times W$ for a silicided diffusion is in the range of $50 - 200 \Omega$-μm (Taur *et al.*, 1987). Contact resistance between silicide and metal is usually negligible, since the interfacial contact resistivity is of the order of 10^{-7}–$10^{-8} \Omega$-cm^2 in a properly executed process.

8.3.2 Parasitic Capacitances

A schematic diagram of the MOSFET capacitances is shown in Figure 8.21. In addition to the intrinsic capacitance C_G discussed in Section 5.4, there are also parasitic capacitances: namely, junction capacitance between the source or drain diffusion and the substrate (or n-well in the case of pMOS), and overlap capacitance between the gate and the source or drain region. These capacitances have a significant effect on CMOS delay.

8.3.2.1 Junction Capacitance

Junction or diffusion capacitance arises from the depletion charge between the source or drain and the oppositely doped substrate. As the source or drain voltage varies with time, the depletion charge increases or decreases accordingly. Note that when the

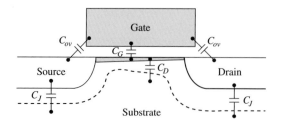

Figure 8.21 Schematic diagram of a MOSFET showing both the intrinsic capacitance C_G and the parasitic capacitances C_J, C_D, C_{ov}. The two C_Js at the source and the drain may have different values depending on the bias voltages

MOSFET is on, the channel-to-substrate depletion capacitance $C_D = WLC_d$ in Figure 8.21, where C_d is given by Eq. (4.38), can also be considered as a part of the source or drain junction capacitance. It is usually a small contribution, since the channel area of a short-channel device is generally much less than the diffusion area.

The depletion-layer width, W_{dJ}, of an abrupt p–n junction is given by Eq. (3.15). The total junction capacitance for a layout diffusion area of $W \times d$ in Figure 8.14 is therefore

$$C_J = Wd\frac{\varepsilon_{si}}{W_{dJ}} = Wd\sqrt{\frac{\varepsilon_{si}qN_a}{2(\psi_{bi} + V_J)}}, \tag{8.22}$$

where N_a is the impurity concentration of the lightly doped side, ψ_{bi} is the built-in potential, typically around 0.9 V, as seen in Figure 3.4, and V_J is the reverse bias voltage of the junction. Equation (8.22) indicates that the source or drain junction capacitance is voltage-dependent. At a higher drain voltage, the depletion layer widens and the capacitance decreases. The depletion-layer width and the capacitance at zero bias have been plotted in Figure 3.5 versus N_a. Since the junction capacitance increases with N_a, one should avoid doping the substrate (or n-well) regions under the source–drain junctions unnecessarily highly. Too low a doping concentration between the source and drain, however, would cause an excessive short-channel effect or lead to punch-through, as discussed in Section 6.1.2. For a noncontacted diffusion, d can be as small as the minimum linewidth of the lithography. The diffusion capacitance of the switching node can be reduced by a factor of two using the folded layout in Figure 8.15(b).

8.3.2.2   Overlap Capacitance

Figure 8.22 breaks the gate-to-source and gate-to-drain overlap capacitance into three components: direct overlap, outer fringe, and inner fringe. The direct overlap component is

$$C_{do} = Wl_{ov}C_{ox} = \frac{\varepsilon_{ox}Wl_{ov}}{t_{ox}}, \tag{8.23}$$

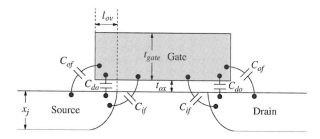

Figure 8.22 Schematic diagram showing the three components of the gate-to-diffusion overlap capacitance

where l_{ov} is the length of the source or drain region under the gate. By solving Laplace's equation analytically with proper boundary conditions, the outer and inner fringe components can be expressed as (Shrivastava and Fitzpatrick, 1982)

$$C_{of} = \frac{2\varepsilon_{ox}W}{\pi} \ln\left(1 + \frac{t_{gate}}{t_{ox}}\right), \tag{8.24}$$

and

$$C_{if} = \frac{2\varepsilon_{si}W}{\pi} \ln\left(1 + \frac{x_j}{2t_{ox}}\right), \tag{8.25}$$

where t_{gate} is the height of the gate conductor, and x_j is the depth of the source or drain junction. Equations (8.24) and (8.25) assume ideal shapes of the gate conductor and source–drain regions with square corners. For typical values of $t_{gate}/t_{ox} \approx 40$ and $x_j/t_{ox} \approx 20$, the above equations give $C_{of}/W \approx 2.3\varepsilon_{ox} \approx 0.08$ fF/μm and $C_{if}/W \approx 1.5\varepsilon_{si} \approx 0.16$ fF/μm. Even though the inner fringe component is larger due to the higher dielectric constant of silicon, it is present only when $V_{gs} < V_t$ and the region under the gate is depleted. Once $V_{gs} > V_t$, the inversion layer effectively shields the gate from electrostatic coupling to the inner edges of the source or drain junction.

From these numerical estimates, we can express the total overlap capacitance at $V_{gs} = 0$ (silicon is depleted under the gate) as

$$C_{ov}(V_{gs} = 0) = C_{do} + C_{of} + C_{if} \approx \varepsilon_{ox}W\left(\frac{l_{ov}}{t_{ox}} + 7\right). \tag{8.26}$$

Note that Eq. (8.26) assumes perfectly conducting source and drain regions. In reality, because of the lateral source–drain doping gradient at the surface, the overlap capacitance depends on the drain voltage. This is especially the case with LDD (lightly doped drain) MOSFETs.

It has been reported that a minimum length of direct overlap region of the order of $l_{ov} \approx (2\text{–}3)t_{ox}$ is needed to avoid reliability problems arising from hot-carrier injection into the ungated region (Chan *et al.*, 1987b). In other words, such a margin is required to avoid "underlap" of the gate and the source–drain. Combining this requirement with Eq. (8.26), we obtain $C_{ov}/W \approx 10\varepsilon_{ox} \approx 0.3$ fF/μm at zero gate voltage, independent of the technology generation.

8.3.3   Gate Resistance

In modern CMOS technologies, silicides are formed over polysilicon gates to lower the resistance and provide ohmic contacts to both n^+ and p^+ gates. The sheet resistivity of silicides is of the order of 2–10 $\Omega/\square$. As CMOS technologies scale down, however, the device delay improves and gate RC delays may not be negligible. Compounding the problem is a tendency for the silicide resistivity to increase in fine-line structures. This is due to either agglomeration or lack of nucleation sites to initiate the phase transformation. Gate RC delay is an ac effect not observable in dc I–V curves. It shows up as an additional delay component in ring oscillators, delay chains, and other logic circuits. In more recent generations of CMOS technology using metal gates, this problem is mitigated.

Gate RC delay can be analyzed with the distributed network in Figure 8.23 for a MOSFET device of width W and length L. The resistance per unit length R is related to the silicide sheet resistivity ρ_g ($\Omega/\square$) by

$$R = \frac{\rho_g}{L}. \tag{8.27}$$

The capacitance per unit length, C, mainly arises from the inversion charge that must be supplied (or taken away) when the voltage at a particular point along the gate increases (or decreases). To a good approximation, C is given by the gate oxide capacitance,

$$C = C_{ox}L = \frac{\varepsilon_{ox}L}{t_{ox}}. \tag{8.28}$$

For better accuracy, the overlap capacitance per unit gate width should also be included in C.

At any point x along the gate, we can write

$$V(x + dx) - V(x) = \frac{\partial V}{\partial x} dx = -I(x)R dx, \tag{8.29}$$

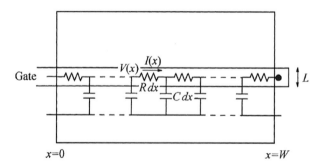

Figure 8.23 A distributed network for analysis of gate RC delay. The lower rail represents the MOSFET channel, which is connected to the source–drain. The input step voltage is applied at the left

and

$$I(x + dx) - I(x) = \frac{\partial I}{\partial x}dx = -Cdx\frac{\partial V}{\partial t}. \tag{8.30}$$

Eliminating $I(x)$ from Eqs. (8.29) and (8.30), we obtain

$$\frac{\partial^2 V}{\partial x^2} = RC\frac{\partial V}{\partial t}. \tag{8.31}$$

The differential equation that governs the RC delay of a distributed network therefore resembles the diffusion equation with a diffusion coefficient $D = 1/RC$. If a step voltage from 0 to V_{dd} is applied at $x = 0$, the boundary conditions are $V(0, t) = V_{dd}$ and $I(W, t) = 0$, which is analogous to constant-source diffusion into a finite-width medium. The numerical solution for this case is plotted in Figure 8.24 (Sakurai, 1983). For $t \ll RCW^2/8$, the solution can be approximated by a *complementary error function*,

$$V(x, t) = V_{dd}\text{erfc}\left(\frac{x}{\sqrt{4t/RC}}\right), \tag{8.32}$$

where

$$\text{erfc}(y) \equiv \frac{2}{\sqrt{\pi}}\int_y^{\infty} e^{-z^2}dz. \tag{8.33}$$

For $t \gg RCW^2/4$, the approximate solution is given by

$$V(x, t) = V_{dd}\left[1 - \frac{4}{\pi}\sin\left(\frac{\pi x}{2W}\right)\exp\left(-\frac{\pi^2 t}{4RCW^2}\right)\right]. \tag{8.34}$$

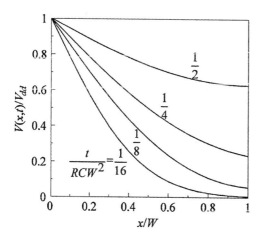

Figure 8.24 Local gate voltage versus distance along the width of the device at different time intervals after an input voltage V_{dd} is applied at $x = 0$

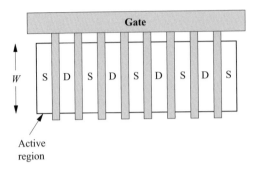

Active
region

Figure 8.25 A MOSFET layout consisting of interdigitated source and drain regions divided by multiple gate fingers. The active region is defined by the isolation mask as that shown in Figure 8.14. All the source regions are stitched together by metal wires (not shown). So are all the drain regions. By connecting them in parallel, the total drive current is equivalent to that of a MOSFET $n \times W$ wide, where n is the number of gate fingers while the gate RC delay is given by the RC delay of a single W

It can be seen from Figure 8.24 that the average value of $V(x, t)$ within $0 < x < W$ reaches $V_{dd}/2$ when $t \approx RCW^2/4$. If we take this value as the effective RC delay τ_g due to gate resistance and substitute Eqs. (8.27) and (8.28) for R and C, we obtain

$$\tau_g = \frac{\rho_g C_{ox} W^2}{4}. \tag{8.35}$$

Note that τ_g is independent of device length but is proportional to the square of the device width. In order to limit τ_g to less than 1 ps, assuming $\rho_g = 10 \ \Omega/\square$ and $t_{ox} = 1.0$ nm, the device width W must be restricted to 3.4 μm or below. *Multiple-finger gate layouts with interdigitated source and drain regions as that shown in Figure 8.25 should be used when higher-current drives are needed.* Such types of layouts also offer the benefit of reduced (by 2×) drain junction capacitance, just like the folded layout in Figure 8.15(b).

8.3.4   Interconnect R and C

Unlike the parasitic elements discussed in Sections 8.3.1–8.3.3, interconnect capacitance and resistance have very little effect on the delay of local circuits such as CMOS inverters or NAND gates described in Section 8.2. On a larger circuit macro or at the VLSI chip level, however, interconnect R and C can play a major role in the system performance, especially in standard-cell designs where wire capacitance may dominate the circuit delay. In this section, we first discuss interconnect capacitance, followed by interconnect resistance.

8.3.4.1   Interconnect Capacitance
In general, the capacitance of an interconnect line consists of three components: the parallel-plate (or area) component, the fringing-field component, and the wire-to-wire

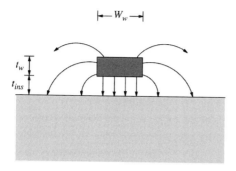

Figure 8.26 Schematic diagram showing electrostatic coupling between an isolated wire and a conducting plane. The straight field lines underneath the wire represent the parallel-plate component of the capacitance. The field lines emerging from the side and the top of the wire make up the fringing-field component of the capacitance (after Bakoglu, 1990)

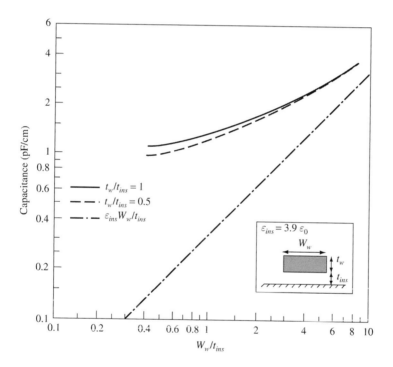

Figure 8.27 Wire capacitance per unit length as a function of width-to-gap ratio, W_w/t_{ins}, for the system in Figure 8.26. The straight chain line represents the parallel-plate component of the capacitance. The dielectric medium is assumed to be oxide with a dielectric constant of 3.9 (after Schaper and Amey, 1983)

component. Figure 8.26 shows schematically the electric field lines that constitute the parallel-plate and the fringing-field capacitance of an isolated line (Bakoglu, 1990). The total capacitance per unit length, $C_{w/l}$, calculated numerically by solving 2-D Laplace's equation, is shown in Figure 8.27 as a function of the wire width to insulator thickness ratio, W_w/t_{ins} (Schaper and Amey, 1983). Only when $W_w \gg t_{ins}$ can the total capacitance per unit length be approximated by the parallel-plate component, $\varepsilon_{ins} W_w / t_{ins}$ (the straight chain line in Figure 8.27). As $W_w/t_{ins} \to 1$, the fringing-field component becomes important and the total capacitance can be much higher than the parallel-plate component. In fact, a minimum capacitance of about 1 pF/cm (for silicon dioxide as the interlevel dielectric) is reached even if $W_w \ll t_{ins}$. This means that reducing the wire capacitance by increasing the insulator thickness becomes ineffective when the insulator thickness is comparable to the width of the wire. Decreasing the wire thickness t_w does not help much either, as is shown in Figure 8.27.

To achieve high packing density in VLSI chips, wires of minimum pitch with nearly equal line and space are often used. This increases the wiring capacitance even further due to contributions from the neighboring lines. Figure 8.28 shows the calculated total capacitance as a sum of the two components for an array of wires

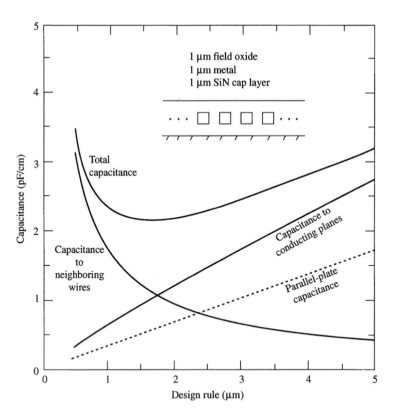

Figure 8.28 Capacitance per unit length as a function of design rules for an array of wires with equal line and space sandwiched between two conducting planes shown in the inset. The total capacitance of each wire is made up of two components: capacitance to the conducting planes, and capacitance to the neighboring wires. Both the metal and insulator thicknesses are held constant as the design rule is varied. The parallel-plate capacitance is also shown (dotted line) for reference (after Schaper and Amey, 1983)

with equal line and space sandwiched between two conducting planes. As shown in the inset, the thickness of metal lines and the thickness of insulators below (oxide) and above (nitride) them are all assumed to be 1 μm. The capacitances are calculated as a function of the metal line or space dimension. When the metal line and space are much larger than the thicknesses, the capacitance is dominated by the component (parallel-plate plus fringing) to the conducting planes above and below. When the metal pitch is much smaller than the thicknesses, however, the wire-to-wire capacitance dominates. *The total capacitance exhibits a broad minimum value of about 2 pF/cm when the metal line or space dimension is approximately equal to the insulator (and wire) thickness*. This conclusion is more general than the specific dimensions assumed. If all the line, space, metal thickness, and insulator thickness are scaled by the same factor, the result remains unchanged. The number 2 pF/cm can be understood from the capacitance per unit length between two concentric cylinders of radii a and b:

$$C_{w/l} = \frac{2\pi\varepsilon_{ins}}{\ln(b/a)}. \tag{8.36}$$

If we take $\varepsilon_{ins} = \varepsilon_{ox}$ and $b/a = 2$, the above gives $C_{w/l} \approx 2\pi\varepsilon_{ox} \approx 2\,\text{pF/cm}$. With the recently developed low-κ inter-level dielectrics, $C_{w/l}$ can be further reduced by a factor of ~2. Compared with the typical device capacitance of ~3 fF/μm in Table 8.4 from the gate oxide, junction, and overlap components, wire capacitance is at least an order of magnitude below, thus has no significant effect on local circuit delays.

8.3.4.2 Interconnect Scaling

The strategy for interconnect scaling, shown schematically in Figure 8.29 (Dennard *et al.*, 1974), is similar to that for MOSFET scaling described in Section 8.1.1. *All linear dimensions – wire length, width, thickness, spacing, and insulator thickness – are scaled down by the same factor, κ, as the device scaling factor*. Wire lengths (L_w) are reduced by κ because the linear dimension of the devices and circuits that they connect to is reduced by κ. Both the wire and the insulator thicknesses are scaled down along with the lateral dimension, for otherwise the fringe capacitance and wire-to-wire coupling (crosstalk) would increase disproportionally, as illustrated in Figure 8.28. Table 8.2 summarizes the rules for interconnect scaling. All material parameters, such as the metal resistivity ρ_w and dielectric constant ε_{ins}, are assumed to remain the same. The wire capacitance then scales down by κ, the same as the device capacitance (Table 8.1), while the wire capacitance per unit length, $C_{w/l}$, remains unchanged (approximately 2 pF/cm for silicon dioxide insulation, as already mentioned). The wire resistance, on the other hand, scales up by κ, in contrast to the device resistance, which does not change with scaling (Table 8.1). The wire resistance per unit length, $R_{w/l}$, then scales up by κ^2, as indicated in Table 8.2. It is also noted that the current density of interconnects increases with κ, which implies that the reliability issues such as electromigration may become more serious as the wire dimension is scaled down. In reality, a few material and process advances in metallurgy have taken place over the generations to keep electromigration under control in VLSI technologies.

Table 8.2. Scaling of local interconnect parameters

	Interconnect Parameters	Scaling Factor ($\kappa \geq 1$)
Scaling assumptions	Interconnect dimensions (t_w, L_w, W_w, t_{ins}, W_{sp})	$1/\kappa$
	Resistivity of conductor (ρ_w)	1
	Insulator permittivity (ε_{ins})	1
Derived wire scaling behavior	Wire capacitance per unit length ($C_{w/l}$)	1
	Wire resistance per unit length ($R_{w/l}$)	κ^2
	Wire RC delay (τ_w)	1
	Wire current density ($I/W_w t_w$)	κ

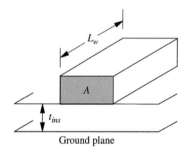

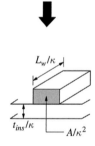

Figure 8.29 Scaling of interconnect lines and insulator thicknesses (after Dennard, 1986)

8.3.4.3   Interconnect Resistance

The interconnect *RC* delay can be analyzed using the same distributed *RC* network model introduced in Section 8.3.3. From Figure 8.24 or Eq. (8.34), the voltage at the receiving end of an interconnect line rises to $1 - e^{-1} \approx 63\%$ of the source voltage after a delay of $t = RCW^2/2$. If we take this value as the equivalent *RC* delay (τ_w) of an interconnect line and substitutes $R_{w/l}$, $C_{w/l}$, L_w for *R*, *C*, *W*, then

$$\tau_w = \frac{1}{2} R_{w/l} C_{w/l} L_w^2. \tag{8.37}$$

Using $R_{w/l} = \rho_w/W_w t_w$ and Eq. (8.36) for $C_{w/l}$ with $\ln(b/a) \approx 1$, we can express this as

$$\tau_w \approx \pi \varepsilon_{ins} \rho_w \frac{L_w^2}{W_w t_w}, \tag{8.38}$$

where W_w and t_w are the wire width and thickness, respectively. One of the key conclusions of interconnect scaling is that the wire RC delay τ_w does not change as the device dimension and intrinsic delay are scaled down. This can potentially impose a limit on VLSI performance scaling. For a quantitative assessment of the situation, take conventional aluminum metallurgy with silicon dioxide insulation so $\rho_w \approx 3 \times 10^{-6} \, \Omega - \text{cm}$ and $\varepsilon_{ins} \approx 3.5 \times 10^{-13} \, \text{F/cm}$. From Eq. (8.38),

$$\tau_w \approx \left(3 \times 10^{-18} \, \text{s}\right) \frac{L_w^2}{W_w t_w}. \tag{8.39}$$

It is easy to see that the RC delay of local wires is negligible as long as $L_w^2/W_w t_w < 3 \times 10^5$. For example, a 0.25 μm × 0.25 μm wire 100 μm long has an RC delay of 0.5 ps, which is quite negligible when compared with the ≈ 20 ps intrinsic delay of a 0.1-μm CMOS inverter (Taur et al., 1993c). The replacement of aluminum metallurgy by copper with $\rho_w \approx 2 \times 10^{-6} \, \Omega - \text{cm}$ helps even more.

In the above estimates, the wire resistivity is assumed to be the same as that of bulk metal. This is valid only if the metal linewidth is much larger than the electron mean free path in metal, ~ 30 nm. For narrower metal lines, the resistivity goes up because of surface and grain boundary scattering (Koo and Saraswat, 2011). In that case, the wire RC delay can become worse than that expected from Eq. (8.39).

8.3.4.4   *RC* Delay of Global Interconnects

Based on the above discussion, the RC delay of local wires will not seriously limit circuit speed even though it cannot be reduced through scaling. The RC delay of global wires, on the other hand, is an entirely different matter. Unlike local wires, the length of global wires, comparable to the chip dimension, does not scale down. Because the size of processor chips remains essentially constant, the RC delay of global wires scales up by κ^2 [Eq. (8.39)] if their cross-sectional area is scaled down like the local wires. For example, in a 0.25-μm CMOS technology, $L_w^2/W_w t_w \sim 10^8 - 10^9$ and $\tau_w \sim 1$ ns, severely limiting the system performance.

One of the solutions is to use repeaters to reduce the dependence of RC delay on wire length from a quadratic one to a linear one (Bakoglu, 1990). A more effective solution is to increase or not to scale the cross-sectional area of global wires. The intermetal dielectric thickness must also be increased in proportion to the wire width to keep the wire capacitance per unit length constant. Building such low-RC global wires requires additional levels of interconnects, since several levels of thin, dense local wires are still needed to make the chip wirable.

The best strategy for interconnect scaling is then to scale down the size and spacing of lower levels in step with device scaling for local wiring, and to use unscaled or even scaled-up levels on top for global wiring, as shown schematically in

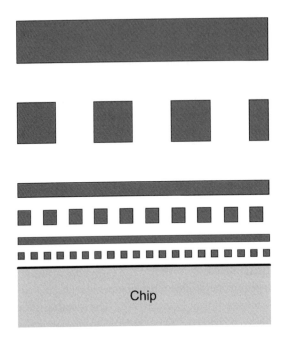

Figure 8.30 Schematic cross-section of a wiring hierarchy that addresses both the density and the global RC delay in a high-performance CMOS processor (after Sai-Halasz, 1995)

Figure 8.30 (Sai-Halasz, 1995). Unscaled wires allow the global RC delay to remain essentially unchanged, as seen from Eq. (8.39). Scaled-up (together with the insulator thickness) wires allow the global RC delay to scale down along with the device delay. Ultimately, the scaled-up global wires will approach the transmission-line limit when the inductive effect becomes more important than the resistive effect. This happens when the signal rise time is shorter than the time of flight over the length of the line. Signal propagation is then limited by the speed of electromagnetic waves, $c/(\varepsilon_{ins}/\varepsilon_0)^{1/2}$, instead of by RC delay. Here $c = 3 \times 10^{10}$ cm/s is the velocity of light in vacuum. For oxide insulators, $(\varepsilon_{ins}/\varepsilon_0)^{1/2} \approx 2$, the time of flight is approximately 70 ps/cm. Figure 8.31 shows the interconnect delay versus wire length L_w calculated from Eq. (8.39) for three different wire cross-sections. Note that the RC delays vary quadratically with L_w. Below a certain wire length, the delay is limited by the time of flight which varies linearly with L_w. For a longer global wire to reach the speed-of-light limit, a larger wire cross-section is needed. The transmission-line situation is more often encountered in packaging wires (Bakoglu, 1990).

8.4   Sensitivity of CMOS Delay to Device Parameters

This section considers the performance factors of static CMOS circuits and their sensitivity to device parameters and parasitic elements. Using 0.1 μm CMOS devices

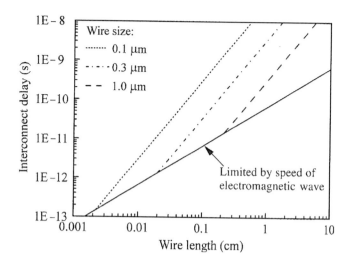

Figure 8.31 *RC* delay versus wire length for three different wire sizes (assuming square wire cross-sections). Wires become limited by electromagnetic-wave propagation when the *RC* delay equals the time of flight, $(\varepsilon_{ins}/\varepsilon_0)^{1/2}L_w/c$, over the line length L_w. An oxide insulator is assumed here

operated at 1.5 V as an example, we first define the propagation delay of an inverter chain and discuss the loading effect due to fan-out and wiring capacitances. Three performance factors – the switching resistance R_{sw}, input capacitance C_{in}, and output capacitance C_{out} – are introduced as the coefficients in a delay equation, followed by several subsections detailing their sensitivity to various device parameters. The last subsection deals with the performance factors of two-way NAND circuits.

8.4.1   Propagation Delay and Delay Equation

In this subsection, we define the propagation delay and the delay equation of a static CMOS gate. While CMOS inverters are used as an example to build the basic framework, most of the formulation and performance factors are equally applicable to other NAND and NOR circuits that perform more general logic functions.

8.4.1.1   Propagation Delay of a CMOS Inverter Chain

The basic switching characteristics of a CMOS inverter with a step input waveform have been briefly discussed in Section 8.2.1.3. In a practical logic circuit, a CMOS inverter is driven by the output of the previous stage whose waveform has a finite rise or fall time associated with it. One way to characterize the switching delay or the performance of an inverter is to construct a cascaded chain of identical inverters, as shown in Figure 8.32, and consider the propagation delay of a logic signal going through them. Load capacitors can be added to the output node of each inverter to simulate the wiring capacitance it may drive in addition to the next inverter.

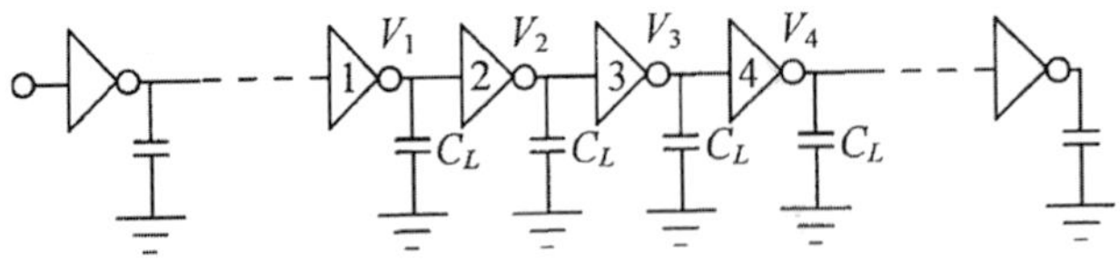

Figure 8.32 A linear chain of CMOS inverters (fan-out $= 1$). Each triangular symbol represents a CMOS inverter consisting of an nMOS and a pMOS, as shown in Figure 8.3. Power-supply connections are not shown

For a specific CMOS technology, the propagation delay is experimentally determined by constructing a *ring oscillator* with a large, odd number of CMOS inverters connected head to tail and measuring the oscillating frequency of the signal at any given point in the ring with the power-supply voltage applied. The sustained oscillation is a result of propagation of alternating logic states $(0 \rightarrow V_{dd} \rightarrow 0 \rightarrow \ldots)$ around a ring with an odd number of stages. The period of the oscillation is given by $n(\tau_n + \tau_p)$, where n is the number of stages (an odd number) and τ_n, τ_p are the inverter delays per stage for rising and falling inputs, respectively. In other words, in one period the logic signal propagates around the ring twice.

Because of the complexity of the current equations for short-channel MOSFETs and the voltage dependence of both intrinsic and extrinsic capacitances, a circuit model such as BSIM in SPICE is needed to compute the propagation delay numerically (Cai *et al.*, 2000). In order to gain insight into what the voltage and current waveforms look like during a switching event, we consider the example of a 0.1 μm CMOS inverter with the device parameters listed in Table 8.3. All lithography dimensions and contact borders, e.g., a, b, and c in Figure 8.14, are assumed to be 0.15 μm (nonfolded). The power-supply voltage is 1.5 V.

The propagation delay is evaluated by applying a step voltage signal at the input of the linear inverter chain in Figure 8.32. After a few stages, the signal waveform has become a *standardized signal*, i.e., one that has stabilized with the same rise and fall times independent of the number of stages of propagation. There are also a few stages following the ones of interest for maintaining the same capacitive loading of each stage. For any stage with input voltage V_{in} and output voltage V_{out} (see Figure 8.3),

$$C \frac{dV_{out}}{dt} = I_P(V_{in}, V_{out}) - I_N(V_{in}, V_{out}), \tag{8.40}$$

where C lumps together all the capacitances connected to the output node. (Capacitance components to a node with time-varying voltage are discussed in Section 8.4.4.3.) If $V_{in}(t)$ is known, then $V_{out}(t)$ can be solved from the above differential equation. Numerically, for given V_{in} and V_{out} at any time instant t, I_P and I_N can be evaluated, and the next V_{out} is given by

$$V_{out}(t + \Delta t) = V_{out}(t) + \frac{I_P - I_N}{C} \Delta t. \tag{8.41}$$

A $V_{out}(t)$ curve can be generated by repeating these steps.

Table 8.3. 0.1 μm CMOS parameters for circuit modeling (25 °C)

Assumed	Power supply voltage, V_{dd} (V)	1.5	
	Channel length, L (μm)	0.1	
	Lithography ground rules, a, b, c (μm)	0.15	
	Gate oxide thickness, t_{ox} (nm)	3.6	
	Linearly extrapolated threshold voltage, V_{on} (V)	±0.4	
		nMOS	**pMOS**
	Source and drain series resistance, $R_{sd} \times W$ (Ω-μm)	200	500
	Saturation velocity, v_{sat} (cm/s)	10^7	10^7
	Substrate/well doping concentration, N_a, N_d (cm^{-3})	10^{18}	10^{18}
	Gate to source or drain (per edge) overlap capacitance, C_{ov}/W (fF/μm)	0.3	0.3
	Drain-induced barrier lowering, ΔV_t between $V_{ds} = 0$ and $V_{ds} = V_{dd}$, (V)	0.05	0.11
	Body-effect coefficient, m	1.3	1.3
	Device width, W_n, W_p (μm)	1	2
Computed	Intrinsic channel capacitance per unit width, $\varepsilon_{ox}L/t_{ox}$ (fF/μm)	0.96	0.96
	Source and drain junction capacitance (@ $V_J = 0$, based on N_a, N_d), C_J/A (fF/μm^2)	2.8	2.8
	Effective mobility (@ $V_{gs} = V_{dd}/2$), μ_{eff} (cm^2/V-s)	375	85
	On current (@ $V_{gs} = V_{ds} = V_{dd}$), I_{on}/W (mA/μm)	0.56	0.25
	Off current (@ $V_{gs} = 0$ and $V_{ds} = V_{dd}$), I_{off}/W (nA/μm)	0.1	0.5

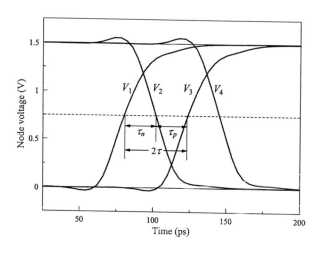

Figure 8.33 Successive voltage waveforms of the CMOS inverter chain in Figure 8.32 ($C_L = 0$). The delay is measured by their intersections with the $V_{dd}/2$ (dashed) line as shown

Figure 8.33 shows an example of the waveforms at four successive stages, V_1, V_2, V_3, V_4, for the unloaded case, $C_L = 0$. As V_1 rises, the nMOS of inverter 2 is turned on, which pulls V_2 to ground. The fall of V_2 then turns on the pMOS of inverter 3, which causes V_3 to rise, and so on. If we draw a straight line through the midpoints of all the waveforms at $V = V_{dd}/2$, we can define the pull-down propagation delay τ_n as the time interval between V_1 and V_2 along that line. Similarly, the pull-up propagation delay τ_p is the time interval between V_2 and V_3 along $V = V_{dd}/2$. The definitions here are consistent with those in Figure 8.9 for step inputs. A better-defined quantity is the CMOS propagation delay, $\tau = (\tau_n + \tau_p)/2$, which is one-half of the time delay between the parallel waveforms V_1 and V_3 or between V_2 and V_4. The time τ is also the delay measured experimentally from CMOS ring oscillators. It is equal to the oscillation period divided by twice the number of stages as stated before. In this specific example, the device width ratio is chosen to be $W_p/W_n = 2$ so that the pull-down delay equals the pull-up delay, i.e., $\tau_n = \tau_p = \tau$. In general, τ_n and τ_p may differ from each other, and the CMOS delay may be dominated by either the nMOS or the pMOS.

8.4.1.2   Bias-Point Trajectories in a Switching Event

As the logic signal arrives at the input gate of an inverter, a transient current flows in either the nMOS or the pMOS of that inverter until the output node completes its high-to-low or low-to-high transition. It is instructive to examine the bias-point trajectories through a family of I_{ds}–V_{ds} curves during a pull-down or pull-up switching event. Figure 8.34(a) plots the trajectories of nMOS (solid dots) and pMOS (open circles) currents of inverter 2 in Figure 8.32 versus the output node voltage V_2, as V_2 is pulled down from V_{dd} to ground. The points are plotted in constant 5-ps time intervals over a background of nMOS $I_{ds} - V_{ds}$ curves (I_N). The pMOS current is very low throughout the transition, indicating negligible power dissipation from the crossover current between the power-supply terminal and the ground. The output node, initially at V_{dd}, is discharged by the nMOS current, which reaches its highest value midway during switching when $V_2 \approx V_{dd}/2$ and $V_{in} \approx 0.9\,V_{dd}$. This point is also where the voltage waveform V_2 in Figure 8.33 exhibits the maximum downward slope. The peak

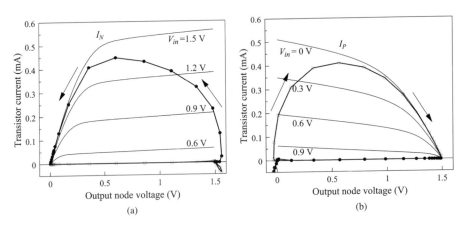

Figure 8.34 Bias-point switching trajectories of nMOS and pMOS for (a) pull-down transition of node V_2 in inverter 2, and (b) pull-up transition of node V_3 in inverter 3. Solid dots are for I_N and open circles for I_P in both (a) and (b). The bias points are plotted in equal time intervals, 5 ps for this case

current is typically 80–90% of the maximum on current at $V_{gs} = V_{ds} = V_{dd}$. The exact percentage depends on the detailed device parameters such as mobility, velocity saturation, threshold voltage, and series resistance. Likewise, Figure 8.34(b) shows the bias-point trajectories of both transistors in inverter 3 when the output node V_3 is pulled up from ground to V_{dd}. In this case, the nMOS current (solid dots) is negligible, while the pMOS current (open circles) reaches its peak value when $V_3 \approx V_{dd}/2$, as in the pull-down case. The two bias trajectories are basically similar to each other and are insensitive to loading conditions. At larger C_L, the delay time between the points in the trajectory increases, but the shape of the curve remains essentially the same.

The delay per stage, as defined in Figure 8.33, is the time duration between $V_{in} = V_{dd}/2$ and $V_{out} = V_{dd}/2$. From Figure 8.33, it is clear that when $V_{in} = V_{dd}/2$, V_{out} is just about to start switching from its prior steady-state value. During either the pull-down delay τ_n or the pull-up delay τ_p, V_{out} changes by $\approx V_{dd}/2$. In the pull-down transition, for example, I_P is negligible. Equation (8.40) can be integrated to yield

$$\tau_n = -\int_{V_{dd}}^{V_{dd}/2} \frac{C dV_{out}}{I_N} = \frac{C V_{dd}/2}{\langle I_N \rangle}, \tag{8.42}$$

where $1/\langle I_N \rangle$ is defined as the average reciprocal of the nMOS current between $V_{in} = V_{dd}/2$ and $V_{out} = V_{dd}/2$. In general, the capacitance C may have some weak dependence on V_{in} and V_{out}. We can re-define C and $\langle I_N \rangle$ to absorb that effect. Likewise,

$$\tau_p = \int_0^{V_{dd}/2} \frac{C dV_{out}}{I_P} = \frac{C V_{dd}/2}{\langle I_P \rangle} \tag{8.43}$$

for pMOS pull-up when I_N is negligible. In the step input case discussed in Section 8.2.1.3, the input rise or fall time is zero, and $\langle I_N \rangle$ or $\langle I_P \rangle$ equals the on-current I_{onN} or I_{onP}, respectively. For the propagation delay considered here, $\langle I_N \rangle$ or $\langle I_P \rangle$ is typically about 3/5 of the on-currents, as can be visually estimated from Figure 8.34 (average current between the point where $V_{gs} = V_{in} = 0.75$ V in the trajectory and the point where $V_{ds} = V_{out} = 0.75$ V). A semi-empirical expression that has been reported to work well is $\langle I_N \rangle = (1/2)\{I_N(V_{in} = V_{dd}/2, V_{out} = V_{dd}) + I_N(V_{in} = V_{dd}, V_{out} = V_{dd}/2)\}$, and vice versa for $\langle I_P \rangle$ (Na et al., 2002).

From Eqs. (8.42) and (8.43), CMOS inverter propagation delay can be written as

$$\tau = \frac{\tau_n + \tau_p}{2} = \frac{C V_{dd}}{4} \left(\frac{1}{\langle I_N \rangle} + \frac{1}{\langle I_P \rangle} \right). \tag{8.44}$$

8.4.1.3   Delay Equation: Switching Resistance, Input and Output Capacitance

The simulations plotted in Figures 8.33 and 8.34 are for the unloaded case with fan-out = 1. In general, $C_L \neq 0$, and the output of an inverter may drive more than one stage. In the latter case the fan-out is 2, 3, ..., which means that each inverter in the chain is driving 2, 3, ... stages in parallel. Each receiving stage is assumed to have the

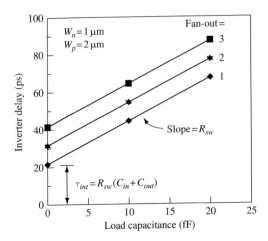

Figure 8.35 Inverter delay τ versus load capacitance C_L for fan-out of 1, 2, and 3

same widths as the sending stage. There are also situations where an inverter is driving another stage wider than its own widths. Such cases can be covered mathematically by generalizing the definition of "fan-out" to include noninteger numbers, provided that the same n- to p-width ratio is always maintained. Fan-outs greater than 3 are rarely used in CMOS logic circuits, as they lead to significantly longer delays.

Figure 8.35 plots the inverter delay τ versus the load capacitance C_L for fan-out = 1, 2, and 3, simulated with the device parameters listed in Table 8.3. Equation (8.41) suggests that the time scale or the delay should vary linearly with the capacitive loading C. This is born out in Figure 8.35 that, for each fan-out, the delay increases linearly with C_L with a constant slope independent of fan-out. The intercept with the y-axis, i.e., the delay at $C_L = 0$, in turn increases linearly with fan-out. These factors can be summarized in a general delay equation,[2]

$$\tau = R_{sw} \times (C_{out} + \text{FO} \times C_{in} + C_L), \tag{8.45}$$

where FO represents the fan-out. In this way, the *switching resistance* R_{sw} is defined as the slope of the delay-versus-load-capacitance lines in Figure 8.35, $d\tau/dC_L$. It is a direct indicator of the current drive capability of the logic gate. The *output capacitance* C_{out} represents the equivalent capacitance at the output node of the sending stage, which consists of the drain junction capacitance and the drain-to-gate capacitance including the overlap capacitance. C_{out} depends on the layout geometry. The *input capacitance* C_{in} is the equivalent capacitance presented by one-unit (FO = 1) input-gate widths of the receiving stage to the sending stage. C_{in} consists of the gate-to-source, gate-to-drain, and gate-to-substrate capacitances including both the intrinsic and the overlap components. Some of the capacitance components are subject

---

[2] M. R. Wordeman (1989). Private communication.

Table 8.4. Extracted performance factors of the 0.1-μm CMOS in Table 8.3

nMOS switching resistance, $W_n \times R_{swn}$ (Ω-μm)	2,300
pMOS switching resistance, $W_p \times R_{swp}$ (Ω-μm)	4,600
Input capacitance, $C_{in}/(W_n + W_p)$ (fF/μm)	1.4
Output capacitance, $C_{out}/(W_n + W_p)$ (fF/μm)	1.7

to the Miller effect, to be discussed in Section 8.4.4.3. The minimum unloaded delay for $C_L = 0$, or the intrinsic delay, is given by

$$\tau_{int} = R_{sw}(C_{in} + C_{out}), \qquad (8.46)$$

which is 22 ps for the 0.1-μm CMOS inverter shown in Figure 8.35.

The delay equation, Eq. (8.45), not only allows the delay to be calculated for any fan-out and loading conditions but also decouples the two important factors that govern CMOS performance: current and capacitance. The current drive capability is represented by R_{sw}, which is inversely proportional to the large-signal transconductance I_{on}/V_{dd} appropriate for digital circuits (Solomon, 1982). The switching resistance can be decomposed into R_{swn} and R_{swp} in terms of the pull-down and pull-up delays τ_n and τ_p defined in Figure 8.33, i.e., $R_{swn} \equiv d\tau_n/dC_L$ and $R_{swp} \equiv d\tau_p/dC_L$. Since $\tau = (\tau_n + \tau_p)/2$, it follows that $R_{sw} = (R_{swn} + R_{swp})/2$. From Eqs. (8.42) and (8.43),

$$R_{swn} = \frac{V_{dd}/2}{\langle I_N \rangle} \qquad (8.47)$$

and

$$R_{swp} = \frac{V_{dd}/2}{\langle I_p \rangle}, \qquad (8.48)$$

where $\langle I_N \rangle$ and $\langle I_P \rangle$ are about 3/5 of the on-currents at $V_{gs} = V_{ds} = \pm V_{dd}$, as already stated. The switching resistances extracted from the above specific example are listed in Table 8.4. For the CMOS inverters, W_p/W_n was chosen to be 2 to compensate for the difference between $I_{onN/w}$ and $I_{onP/w}$, so that $R_{swn} \approx R_{swp} \approx R_{sw}$ and $\tau_n \approx \tau_p \approx \tau$.

Both the input and the output capacitances, C_{in} and C_{out}, in Eq. (8.45) are approximately proportional to $W_n + W_p$, since both nMOS and pMOS contribute more or less equally per unit width to the node capacitance whether they are being turned on or being turned off. This assumes that all the capacitances per unit width are symmetrical between the n- and p-devices, as is the case in Table 8.3. The specific numbers for the case in Figure 8.35 are listed in Table 8.4. Note that $(C_{in} + C_{out})/(W_n + W_p)$ is about three times the intrinsic channel capacitance per unit width, 0.96 fF/μm, listed in Table 8.3.

8.4.2 Delay Sensitivity to Channel Width, Length, and Gate Oxide Thickness

The next few sections examine CMOS delay sensitivity to various device parameters, both intrinsic and parasitic, as listed in Table 8.3. This section considers the effects of physical length, width, and thickness on CMOS performance.

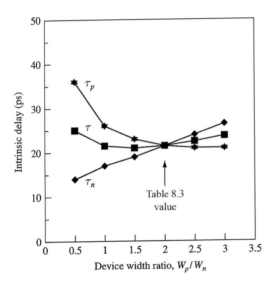

Figure 8.36 Intrinsic CMOS inverter delays τ_n, τ_p, and τ for FO $= 1$ and $C_L = 0$ versus p- to n-device width ratio

8.4.2.1   Effect of pMOS/nMOS Width Ratio

When the p- to n-device width ratio W_p/W_n is varied in a CMOS inverter, it affects the relative current drive capabilities R_{swn} and R_{swp}, and therefore τ_n and τ_p. Figure 8.36 plots the intrinsic delay of CMOS inverters as a function of the device width ratio. The rest of the device parameters are kept the same. As W_p/W_n increases, τ_p decreases but τ_n increases. At $W_p/W_n \approx 2$, the pull-up time becomes equal to the pull-down time, which gives the best noise margin, as discussed in Section 8.2.1.1. The overall delay, $\tau = (\tau_n + \tau_p)/2$, on the other hand, is relatively insensitive to the width ratio, showing a broad minimum at $W_p/W_n \approx 1.5$ (Exercise 8.4). The specific example in Section 8.4.1 uses $W_p/W_n = 2$, so that $\tau_n \approx \tau_p \approx \tau = 22$ ps, within 5% of the minimum delay at $W_p/W_n = 1.5$.

8.4.2.2   Buffer Stages for Driving a Large Load Capacitance

From the discussions in Section 8.4.1, it is clear that if W_n and W_p are scaled up by the same factor without changing the ratio W_p/W_n, the intrinsic delay remains the same. The switching resistance, $R_{sw} = d\tau/dC_L$, on the other hand, is reduced by that same factor due to the higher current drives. Therefore the delay improves when there is a capacitive load C_L. In fact, it has been argued that for high-performance purposes, one can scale up the device size until the circuit delays are mostly device-limited, i.e., approaching intrinsic delays (Sai-Halasz, 1995). This can be accomplished, if necessary, by increasing the chip size, because the capacitance due to wire loading increases only as the linear dimension of the chip (2 pF/cm in Section 8.3.4), while the effective device width can increase as the area of the chip using the corrugated (interdigitated) gate structures in Figure 8.25. Of course, delays of global interconnects, as well as chip power and cost, will go up as a result.

In practical CMOS circuits, it is a common rule not to drive a capacitive load much greater than the device's own capacitance, as that would result in delays much longer than the intrinsic delay. One solution is to insert a *buffer*, or *driver*, between the original sending stage and the load. A driver consists of one or multiple stages of CMOS inverters with progressively wider widths. To illustrate how it works, we consider an inverter with a switching resistance R_{sw}, an input capacitance C_{in}, and an output capacitance C_{out}, driving a load capacitance C_L. Without any buffer, the single-stage delay is

$$\tau = R_{sw}(C_{out} + C_L). \tag{8.49}$$

If $C_L \gg C_{in}$ and C_{out}, the delay can be improved by inserting an inverter with k (>1) times wider widths than the original inverter. Such a buffer stage would present an equivalent FO $= k$ to the sending stage but would have a much improved switching resistance, R_{sw}/k. The overall delay, including the delay of the buffer stage, would be[3]

$$
\begin{aligned}
\tau_b &= R_{sw}(C_{out} + kC_{in}) + \frac{R_{sw}}{k}(kC_{out} + C_L) \\
&= R_{sw}\left(2C_{out} + kC_{in} + \frac{C_L}{k}\right).
\end{aligned}
\tag{8.50}
$$

This has a minimum of

$$\tau_{b\,min} = R_{sw}\left(2C_{out} + 2\sqrt{C_{in}C_L}\right) \tag{8.51}$$

with the choice of $k = (C_L/C_{in})^{1/2}$ for the buffer width. Thus the buffer stage reduces the delay dependence on C_L from a linear one in Eq. (8.49) to the square root factor in Eq. (8.51). For heavy loads ($C_L \gg C_{in}$, C_{out}), τ_{bmin} can be substantially shorter than the unbuffered delay τ. To drive even heavier loads, multiple-stage buffers can be designed for best results (see Exercises 8.12, 8.13, and 8.14).

8.4.2.3 Sensitivity of Delay to Channel Length

Channel length is a key factor on CMOS performance. At shorter channel lengths, not only does the switching resistance of the driving stage decrease due to higher on-currents, the intrinsic capacitance in the receiving stage is also lower. Figure 8.37 shows the variation of inverter delay with channel length assuming the rest of the device parameters are the same (with no threshold voltage dependence on channel length). It is noted that the inverter delay improves approximately linearly with channel length at and above the 0.1-μm design point, but sub-linearly below it.

---

[3] Here we apply Eq. (8.45) as an approximation. Strictly speaking, it is not propagation delay without a few repeated stages of identical driving–receiving conditions.

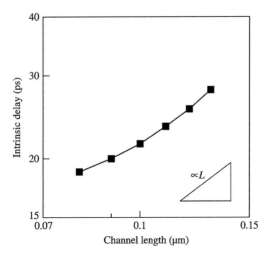

Figure 8.37 Intrinsic CMOS inverter delay versus channel length for the devices listed in Table 8.3. Both nMOS and pMOS are assumed to have the same channel length

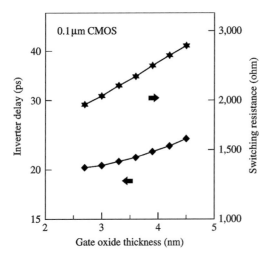

Figure 8.38 Intrinsic delay and switching resistance versus gate oxide thickness for the 0.1-µm CMOS listed in Table 8.3. Both log scales are of the same proportion for comparison

8.4.2.4   Sensitivity of Delay to Gate Oxide Thickness

Switching resistance or current drive capability can also be improved by using a thinner gate oxide. However, in contrast to shortening the channel length which helps both the resistance and the capacitance, thinner oxides lead to higher gate capacitance. It is shown in Figure 8.38 that the improvement of intrinsic delay with oxide thickness is not as much as that with channel length. Loaded delays improve more as indicated by the switching resistance curve in Figure 8.38. The R_{sw} dependence on t_{ox} is still sub-linear because of the effect of series resistance.

It should be noted that the above sensitivity study only considers t_{ox} variations at the level of circuit model, while keeping all other parameters unchanged. In other words, the interdependence between t_{ox} and V_t or L at the device level is not taken into account. From a device-design point of view, thinner oxides would allow shorter channel lengths and therefore additional performance benefits.

8.4.3 Sensitivity of Delay to Power-Supply Voltage and Threshold Voltage

This section considers the dependence of CMOS delay on power-supply voltage and threshold voltage. The effect is mainly through the switching resistance factor as the large-signal transconductance, I_{on}/V_{dd}, degrades with either higher V_t or lower V_{dd}. Both the input and output capacitances are relatively insensitive to V_{dd} and V_t. The effect of threshold voltage on the delay of 0.1-μm CMOS inverters for a given $V_{dd} = 1.5$ V was discussed in Section 6.3.1.3 and shown in Figure 6.18. In that case, the delay for the range of $V_t/V_{dd} < 0.5$ can be fitted to an empirical factor, $\propto 1/(0.6 - V_t/V_{dd})$. The dependence of inverter delay on power supply voltage for a fixed threshold voltage (Table 8.3) is shown in Figure 8.39. The delays of 2-way NAND gates exhibit a very similar V_{dd}-dependence as the inverter delay. More discussions on 2-way NAND delays can be found in Section 8.4.6.

8.4.3.1 Power and Delay Tradeoff

The delay versus supply voltage curve in Figure 8.39 is re-plotted as a power versus delay curve with V_{dd} as a parameter in Figure 8.40. Here, the active power is calculated from

$$P_{ac} = (C_{in} + C_{out})V_{dd}^2/(2\tau),$$ (8.52)

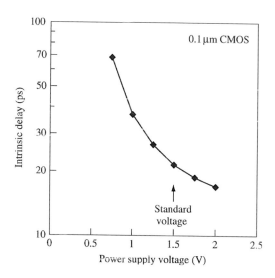

Figure 8.39 CMOS intrinsic delay versus power supply voltage for a constant threshold voltage (Table 8.3)

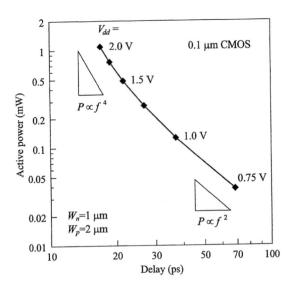

Figure 8.40 CMOS power versus delay by varying the power supply voltage while keeping the threshold voltage constant (Table 8.3)

under the assumption that the inverters are clocked at the highest frequency possible, $f = 1/(2\tau)$, where 2τ is the time it takes to complete a high-to-low-to-high switching cycle (Figure 8.33). The active power of Eq. (8.52) accounts for about 90% of the total power supplied by the source (rail-to-rail current times V_{dd}). The difference is due to the cross-over current, i.e., short-circuit power. For the devices in Table 8.3, the standby power from subthreshold leakage at room temperature is about 1 nW, negligible during the active switching transient. In Figure 8.40, lower power-delay product or switching energy is obtained at low supply voltages where $P_{ac} \propto f^2$. **For high-performance CMOS operated toward the high end of the supply voltage, premium performance comes at a steep expense of active power** ($P_{ac} \propto f^4$).

It is possible to reduce V_{dd} without a severe loss in performance if V_t is reduced as well. Of course, standby power will go up as a result. The tradeoff among performance, active power, and standby power is depicted conceptually in a V_{dd}–V_t design plane in Figure 6.23. While the standby portion of the total power stays constant with time, the active portion of the total power depends on the circuit activity factor, i.e., how often the circuit switches on average. For high-activity circuits such as clock drivers, active power dominates. Their speed can be improved by using low V_t devices (Cai et al., 2002b). The majority of circuits in a typical VLSI logic chip, however, are of the low-activity type, such as those found in static memories. High-V_t devices are needed in those circuits to limit their collective standby power.

8.4.4   Sensitivity of Delay to Parasitic Resistance and Capacitance

This section examines the sensitivity of CMOS delay to parasitic source–drain series resistance, overlap capacitance, and junction capacitance, using the 0.1-μm devices listed in Table 8.3 as an example.

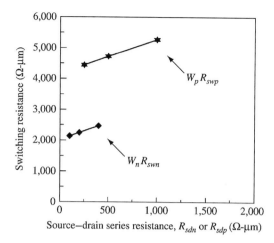

Figure 8.41 Switching resistances versus source–drain series resistance for the 0.1-μm CMOS devices listed in Table 8.3. $W_n R_{swn}$ is plotted versus R_{sdn} and $W_p R_{swp}$ is plotted versus R_{sdp}

8.4.4.1 Sensitivity of Delay to Series Resistance

The effect of source–drain series resistance on CMOS delay comes through the nMOS and pMOS currents and therefore their switching resistances. Figure 8.41 shows the sensitivity of n- and p-switching resistances to the n- and p-series resistances R_{sdn} and R_{sdp}. Since pMOS have a lower current per unit width, they can tolerate a higher series resistance for the same percentage of degradation. For the default values assumed in Table 8.3, $R_{sd} \times W = 200\ \Omega$-μm for n and 500 Ω-μm for p, both devices are degraded by about 10% in terms of their current drive capability. A simple rule of thumb for estimating the performance loss due to series resistance is to add R_{sd} to the intrinsic switching resistance, i.e., $\Delta R_{swn} \approx R_{sdn}$ and $\Delta R_{swp} \approx R_{sdp}$.

8.4.4.2 Sensitivity of Delay to Overlap Capacitance

Gate-to-drain overlap capacitance is a serious performance detractor in lightly loaded CMOS circuits. It not only enters the input capacitance but is also a component of the output capacitance, sometimes further amplified by the feedback effect. Figure 8.42 shows both the input and the output capacitances as a function of the overlap capacitance C_{ov} (per edge). The value assumed in Table 8.3 is 0.3 fF/μm, about the lowest C_{ov} that can be achieved in practice, as discussed in Section 8.3.2.2. Both the gate-to-source and the gate-to-drain capacitances contribute to the input capacitance. Only the gate-to-drain component enters the output capacitance. However, its contribution is nearly doubled from its original value due to the *Miller effect*, explained in Section 8.4.4.3. It is estimated that overall an overlap capacitance of 0.3 fF/μm accounts for about 35–40% of the intrinsic delay.

8.4.4.3 Miller Effect

The Miller effect arises when the voltages on both sides of a capacitor being charged or discharged vary with time. Figure 8.43 shows an example of three capacitors

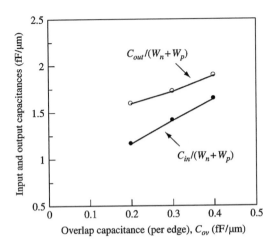

Figure 8.42 Input and output capacitances versus overlap capacitance. Both nMOS and pMOS are assumed to have the same C_{ov} per edge

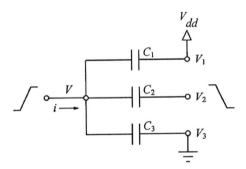

Figure 8.43 A circuit example illustrating Miller effect

connected to a node of voltage V being charged. Each capacitor is connected to a different voltage level on the other side. We can express the charging current i as

$$i = C_1 \frac{d(V - V_1)}{dt} + C_2 \frac{d(V - V_2)}{dt} + C_3 \frac{d(V - V_3)}{dt}. \tag{8.53}$$

Since both $V_1 = V_{dd}$ and $V_3 = 0$ are fixed voltages, we have

$$i = C_1 \frac{dV}{dt} + C_2 \frac{dV}{dt} - C_2 \frac{dV_2}{dt} + C_3 \frac{dV}{dt}, \tag{8.54}$$

and there is no Miller effect on C_1 and C_3. However, $dV_2/dt \neq 0$, since V_2 varies with time. If V_2 varies with time in a direction opposite to that of V, it will take more time (and charge) to charge up the node voltage V to a certain level than it otherwise would. This happens, for example, between the input gate and the output drain of a CMOS inverter, as can be seen from the waveforms in Figure 8.33. In particular, if $dV_2/dt = -dV/dt$, Eq. (8.54) becomes

$$i = (C_1 + 2C_2 + C_3)\frac{dV}{dt}. \tag{8.55}$$

In other words, the capacitor C_2 appears to have doubled its capacitance as far as charging of the node voltage V is concerned. From another angle, it takes a net flow of charge of $\Delta Q_2 = 2C_2 V_{dd}$ into the capacitor C_2 to switch it from an initial state of $V - V_2 = -V_{dd}$ to a final state of $V - V_2 = V_{dd}$.

Another effect of capacitive coupling is *feedforward*. For example, when the gate voltage rises in a CMOS inverter, the drain voltage, initially at V_{dd}, will momentarily rise to a value slightly higher than V_{dd} due to its capacitive coupling to the gate. This happens instantaneously while it takes time for the nMOS current to discharge the drain node, as can be seen from the initial overshoot of V_2 and V_4 above V_{dd} in Figure 8.33, as well as from the $I–V$ trajectory in Figure 8.34(a).

8.4.4.4 Sensitivity of Delay to Junction Capacitance

A major part of the output capacitance comes from the junction or drain-to-substrate capacitance. Figure 8.44 shows the input and output capacitances versus junction capacitance by varying the layout. In the folded layout, the junction capacitance is effectively halved as discussed in Section 8.2.3.2. This has a dramatic effect on C_{out}, but not on C_{in}. From Figure 8.44, it is estimated that the junction capacitance accounts for more than 50% of the output capacitance in the straight-gate layout and that the folded layout improves the intrinsic inverter delay (FO = 1) by about 15%.

It is instructive to break C_{in} and C_{out} for the 0.1-μm CMOS devices listed in Table 8.3 into various components: intrinsic gate capacitance, overlap capacitance, and junction capacitance. This can be done by extrapolating the simulation results in Figures 8.42 and 8.44, and the capacitance components in Figure 8.37. The results are

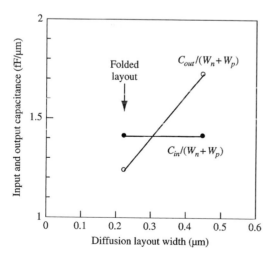

Figure 8.44 Input and output capacitances versus diffusion width d in Figure 8.14. In the straight-gate layout (default case thus far), $d = a + b + c = 0.45$ μm. In the folded-gate layout (Figure 8.15), d is effectively cut to half

Table 8.5. Components of C_{in} and C_{out}

Component	Input Capacitance (%)	Output Capacitance (%)
Intrinsic gate oxide capacitance	49	18
Overlap capacitance	51	26
Junction capacitance (nonfolded)	–	56

given in Table 8.5. Note in Table 8.4 that the values of C_{in} and C_{out} are about equal. The unloaded delay is proportional to $C_{in} + C_{out}$, in which only about a third comes from the intrinsic gate oxide capacitance.

8.4.5   Effect of Transport Parameters on CMOS Delay

Enhancement of electron and hole mobilities by strain engineering of nMOS and pMOS has been discussed in Section 5.2.2. Here we investigate the benefit of increased mobilities on CMOS delay using the same circuit model with the 0.1-μm CMOS device parameters listed in Table 8.3. The performance gain due to higher mobilities comes through the switching resistance factor and is therefore independent of fan-out and wire loading conditions. For a given set of physical dimensions and node voltages, MOSFET current is determined by three transport related parameters: mobility, saturation velocity, and series resistance. If the mobility and saturation velocity are increased by a factor $\kappa > 1$ and the series resistance decreased by $1/\kappa$, the current goes up by κ, or equivalently, the switching resistance goes down by a factor of $1/\kappa$. Empirically, we can express

$$R_{sw} \propto \mu_{eff}^{-a} v_{sat}^{-b} R_{sd}^{c}, \tag{8.56}$$

where $a + b + c = 1$, for small changes of μ_{eff}, v_{sat}, and R_{sd} with respect to their Table 8.3 values. Here, changing each parameter means changing the corresponding parameter of both nMOS and pMOS by the same factor. Figure 8.45(a) shows the simulation results of how R_{sw} varies with those parameters in a log–log scale. It is observed that $a \approx 0.61$, $b \approx 0.28$, and $c \approx 0.11$ for the 0.1-μm CMOS example considered here. ***Relatively speaking, mobility is the most influential parameter on CMOS delay.*** One of the reasons is that, based on the bias point trajectories in Figure 8.34, a significant amount of the delay time is spent in the early part of the switching event when V_{in} is around $V_{dd}/2$. The current at such low V_{gs} is not nearly as velocity saturated as the on current at $V_{gs} = V_{dd}$. The degree of velocity saturation is indicated by, e.g., the parameter z in Eq. (6.47), which is proportional to $(V_{gs} - V_t)$. The value of z at $V_{gs} = V_{dd}/2$ is only ~ 1/3 of its value at $V_{gs} = V_{dd}$, thus the mobility is a more controlling factor.

Figure 8.45(b) further breaks out the sensitivity of switching resistance to electron and hole mobilities separately. Not surprisingly, higher hole mobility is more beneficial than higher electron mobility because pMOS is not as velocity saturated

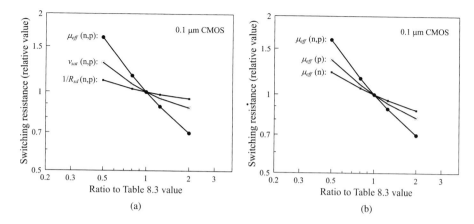

Figure 8.45 (a) Sensitivity of switching resistance to transport parameters: μ_{eff}, v_{sat}, and $1/R_{sd}$. Each curve depicts relative change of R_{sw} with respect to relative change of the specific transport parameter for both nMOS and pMOS while others are kept constant. (b) Breakdown of the mobility dependence into the electron and hole factors separately

as nMOS. Quantitatively, the mobility exponent a in Eq. (8.56) can be decomposed into $a = a_n + a_p$ such that

$$R_{sw} \propto \mu_n^{-a_n} \mu_p^{-a_p} v_{sat}^{-b} R_{sd}^{c}, \tag{8.57}$$

where $a_n = 0.24$ and $a_p = 0.37$ are found to be the case.

As channel length is further reduced toward 10 nm, R_{sd} and v_{sat} may become more dominant limiters on CMOS delay as they do not improve much with scaling. On the other hand, the mobilities will become lower because the vertical field inevitably goes up, as discussed in Sections 6.3.2 and 8.1.2.

8.4.6   Delay of Two-Way NAND Gates

So far we have been focusing on CMOS inverters, i.e., with fan-in of one, for analyzing the performance factors. Many of the basic characteristics also apply to the more general CMOS circuits. There are, however, a few additional factors associated with the multiple fan-in NAND gates in which two or more nMOS are stacked between the output node and the power-supply ground. This section discusses these factors using a two-way NAND (Figure 8.11) as an example.

8.4.6.1   Top and Bottom Switching

The simulation is set up with the layout shown in Figure 8.16 and with the same 0.1-μm CMOS parameters listed in Table 8.3, except the p- to n-device width ratio. Because the pull-down current in a NAND gate is somewhat lower than that in an inverter due to the stacking of nMOS, both the transfer curves and the up and down delays are better matched with a W_p/W_n ratio of 1.5 instead of 2. In this configuration, the two parallel pMOS are naturally folded. The nMOS are nonfolded. The width of

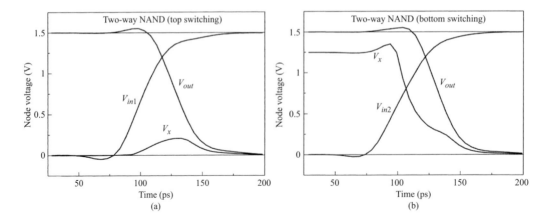

Figure 8.46 Waveforms of V_{in1} (top gate), V_{in2} (bottom gate), V_{out} (drain of the top nMOS and both pMOS), and V_x (node between the two stacked nMOS) for (a) top switching and (b) bottom switching in the pull-down event of a 2-way NAND gate. The device parameters are those listed in Table 8.3, except that $W_n/W_p = 1.0/1.5$ μm

the diffusion region (V_x-node) between the two stacked nMOS is assumed to be the minimum lithography dimension, 0.15 μm in this case. To construct a linear chain of two-way NAND gates, we must distinguish between two cases: *top switching* and *bottom switching*, as was first outlined in Section 8.2.2.1. Referring to Figure 8.11, top switching means that transistors N1 and P1 are driven by a logic transition propagated through input 1. Input 2 is tied to V_{dd} in all the stages so all the N2 are on and all P2 are off. On the other hand, in bottom switching, transistors N2 and P2 are driven by a logic signal from the output of the previous stage through input 2, while input 1 is tied to V_{dd}. These two switching modes have somewhat different delay characteristics, as discussed in Figure 8.46.

It is instructive to examine the switching waveforms of various node voltages in a two way NAND. Figure 8.46 plots the input, output, and V_x-node voltages versus time during an nMOS pull-down event. In the top-switching case in Figure 8.46(a), the V_x-node voltage starts at zero, rises momentarily to a peak about 15% of V_{dd}, then falls back to zero along with V_{out}. The rise of V_x is a result of the discharging current when the top transistor is turned on. In the bottom-switching case in Figure 8.46(b), the V_x-node voltage starts at a high value, but quite a bit lower than V_{dd}. Even though its gate is tied to V_{dd}, the top nMOS is initially biased in the subthreshold region since $V_{dd} - V_x < V_t + (m-1)V_x$ (the factor m comes from the body effect; see the discussion in Section 8.2.2.1). The exact starting value of V_x depends on a detailed matching of the subthreshold currents in the top and bottom nMOS. When the bottom nMOS is turned on, the V_x-node is pulled down to ground, followed by V_{out}.

Figure 8.47 plots the propagation delay of a two-way NAND gate (solid lines), as described in Section 8.4.1.1, versus the load capacitance C_L. The dashed line shows the delay of an inverter of the same widths for comparison. A delay equation of the same form as Eq. (8.45) also applies to two-way NAND gates, but with different values of R_{sw}, C_{in}, and C_{out}. The intrinsic delay ($C_L = 0$) of the two-way NAND is about 34% (1.34×) longer than that of the inverter, for the following reasons. First, let us consider the capacitances. The input capacitance of a two-way NAND stage is essentially the same as that of an inverter. However, a two-way NAND has a higher output capacitance. In the top-switching case, there is an additional gate-to-drain overlap capacitance C_{ov} (no Miller effect) on the pMOS side of the two-way NAND layout in Figure 8.16, compared with the inverter layout in Figure 8.15(a). In the bottom-switching case, the output capacitance is further increased by additional components on the nMOS side. These include the gate capacitance of N1, some overlap capacitance associated with the gate of N1, and a small junction capacitance of the V_x-node. In addition to the higher capacitances, the switching resistances, i.e., the slopes of the lines in Figure 8.47, of the two-way NAND are also higher than that of the inverter. This primarily stems from stacking of the two nMOS between the output node and the ground such that when one nMOS is switching, the other acts like a series resistance, which degrades the current. In terms of the switching resistances, top switching is worse than bottom switching, since the series resistance in the former is placed between the source and the ground, resulting in additional loss of gate drive. This is evident in Figure 8.47. For the intrinsic delays in this example, bottom switching is worse than top switching because the extra capacitance outweighs the slight difference in the switching resistance. Under heavy loading conditions, however, top switching is the worst case, in which the switching resistance is about 21% (1.21×) worse than that of the inverter in Figure 8.47.

The degradation of switching resistance in NAND circuits with fan-in > 1 can be roughly estimated using the following simple model. In the pull-down operation of a

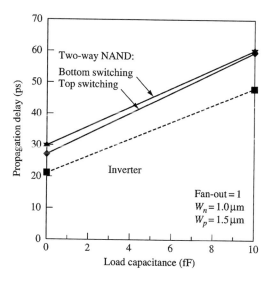

Figure 8.47 Propagation delay versus load capacitance. The two solid lines are for the top switching and the bottom switching cases of a 2-way NAND gate. The dashed line shows the delay of a CMOS inverter of the same device widths for comparison

two-way NAND, the nonswitching nMOS has its gate voltage fixed at V_{dd} and acts like a series resistor to the switching transistor. Since it operates mainly in the linear region during a switching event (see the discussion in Section 8.2.2.1), its effective resistance can be approximated by V_{dsat}/I_{onN}, where V_{dsat} and I_{onN} are the saturation voltage and current at $V_{gs} = V_{dd}$. This increases the nMOS switching resistance by roughly the same amount, i.e., $\Delta R_{swn} = V_{dsat}/I_{onN}$, based on the discussion following Figure 8.41. Using $R_{sw} = (R_{swn} + \Delta R_{swn} + R_{swp})/2$ with R_{swn} and R_{swp} for inverters given by Eqs. (8.47) and (8.48), we can express the switching resistance of a two-way NAND gate as

$$R_{sw}(\mathrm{FI} = 2) = \frac{V_{dd}}{4\langle I_N \rangle} + \frac{V_{dsat}}{2I_{onN}} + \frac{V_{dd}}{4\langle I_p \rangle}. \tag{8.58}$$

For the above 0.1-µm CMOS example, $\langle I_N \rangle \approx \langle I_P \rangle \approx (3/5)I_{onN}$, and $V_{dsat} \approx (1/3)V_{dd}$ from Figure 8.34(a). Substitution of these numbers in Eq. (8.58) yields a switching resistance about 1.2-times that of the inverter, in line with the number extracted from Figure 8.47. Equation (8.58) can be generalized to a higher number of fan-ins by inserting a multiplying factor of $(\mathrm{FI} - 1)$ in front of the V_{dsat}/I_{onN} term. Since R_{sw} degrades rapidly with the number of fan-ins, fan-ins higher than 3 are to be avoided in CMOS circuits.

8.4.6.2   Delay Sensitivity to Body Effect

The delays in Figure 8.47 are computed based on the set of device parameters in Table 8.3, in which the body-effect coefficient is $m = 1.3$. With all other device parameters being equal, the delay increases with m for two reasons. First, the device saturation current decreases with increasing m due to body effect at the drain. This can be seen from the saturation current expression, Eqs. (6.47) and (6.49), for the $n = 1$ case discussed in Section 6.2.1.2. Other values of n lead to qualitatively similar results. The dependence of the saturation current on m is stronger for less velocity saturated devices, e.g., pMOS. The fully velocity saturated current, Eq. (6.51), is independent of m. The second factor is more applicable to the stacked nMOS devices in NAND gates: when the source potential is higher than the body potential, as in transistor N1 of Figure 8.11, the threshold voltage increases because of body effect and the current decreases. The source-to-body potential of N1 is given by the voltage V_x shown in Figures 8.46(a) and (b). To a lesser degree, this effect also occurs in a CMOS inverter due to the presence of series resistances at the n- and p-source terminals.

8.5   Performance Factors of MOSFETs in RF Circuits

In Sections 8.1–8.4, we discussed the CMOS performance factors in digital circuits. In this section, we consider the performance factors of MOSFETs in RF circuits, e.g., as small signal amplifiers. For this purpose, nMOS transistors are used almost exclusively because of their superior performance over pMOS.

8.5.1   Small-Signal Equivalent Circuit

Figure 8.48 shows the small signal, two-port schematic of a MOSFET amplifier used in the common-source configuration. The dc bias circuits are not shown. The ac input goes into the gate–source port and the ac output is extracted at the drain–source port. The convention here is that the upper case symbols denote full quantities and lower case symbols small signal quantities, e.g., $v_{gs} = \delta V_{gs}$, $i_{ds} = \delta I_{ds}$, etc. All i_{gs}, i_{ds}, v_{gs}, and v_{ds} are complex numbers (phasors) with the common time dependence $e^{j\omega t}$, where ω is the angular frequency of the small signal. In the frequency domain representation, if the full time-dependent expression for the voltage across the input port is $|v_{gs}|\cos(\omega t + \alpha)$, i.e., the real part of $|v_{gs}|e^{j(\omega t + \alpha)}$, then $v_{gs} = |v_{gs}|e^{j\alpha}$. Likewise, $v_{ds} = |v_{ds}|e^{j\beta}$, etc.

The common default is to ground the body to the source. Thus the drain current is a function of V_{gs} and V_{ds}, i.e., $I_{ds}(V_{gs}, V_{ds})$. Its small-signal increment can be written as:

$$i_{ds} = \frac{\partial I_{ds}}{\partial V_{gs}} v_{gs} + \frac{\partial I_{ds}}{\partial V_{ds}} v_{ds} = g_m v_{gs} + g_{ds} v_{ds}, \tag{8.59}$$

where $g_m \equiv (\partial I_{ds}/\partial V_{gs})|_{V_{ds}}$ is the transconductance and $g_{ds} \equiv (\partial I_{ds}/\partial V_{ds})|_{V_{gs}}$ is the output conductance.

In addition to the conductive components in Eq. (8.59), there are also capacitive components from the gate to source, gate to drain, and body to drain. They give rise to displacement currents, e.g., $I_{gs}(t) = C_{gs}(dV_{gs}/dt)$ or $i_{gs} = (j\omega C_{gs})v_{gs}$, that are 90° out of phase with the sinusoidal voltage. The complete small-signal RF equivalent circuit of an intrinsic MOSFET is shown in Figure 8.49, where Eq. (8.59) is represented by a voltage-dependent current source and a conductance.

In the small-signal analysis of a two-port network in the frequency domain, an admittance matrix is often used to describe the linear relationship between the terminal currents and voltages. These relations are based on Kirchhoff's current and voltage laws applied to the equivalent circuit. For the circuit in Figure 8.49, the intrinsic admittance matrix of a MOSFET is as follows:

$$\begin{bmatrix} i_{gs} \\ i_{ds} \end{bmatrix} = \begin{bmatrix} j\omega(C_{gs} + C_{gd}) & -j\omega C_{gd} \\ g_m - j\omega C_{gd} & g_{ds} + j\omega C_{db} + j\omega C_{gd} \end{bmatrix} \begin{bmatrix} V_{gs} \\ V_{ds} \end{bmatrix}. \tag{8.60}$$

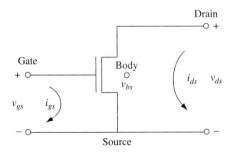

Figure 8.48 MOSFET used as a small-signal amplifier in the common-source configuration

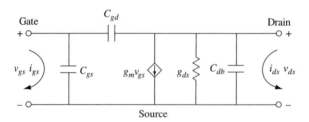

Figure 8.49 Small-signal equivalent circuit of an intrinsic MOSFET

8.5.2   Unity-Current-Gain Frequency

The intrinsic current gain β is obtained from Eq. (8.60), with the assumption that the output port is short-circuited, i.e., $v_{ds} = 0$:

$$\beta = \frac{i_{ds}}{i_{gs}}\bigg|_{v_{ds}=0} = \frac{g_m - j\omega C_{gd}}{j\omega\left(C_{gs} + C_{gd}\right)}. \tag{8.61}$$

The magnitude of the current gain is

$$|\beta| = \frac{\sqrt{g_m^2 + \left(\omega C_{gd}\right)^2}}{\omega\left(C_{gs} + C_{gd}\right)}. \tag{8.62}$$

The intrinsic unity-current-gain frequency is then obtained by setting $|\beta| = 1$ in the above equation and solving for $f_T = \omega/2\pi$:

$$f_T = \frac{g_m}{2\pi\sqrt{C_{gs}^2 + 2C_{gs}C_{gd}}}. \tag{8.63}$$

In the expressions commonly found in the literature, the square root factor in the denominator is approximated to $C_{gs} + C_{gd}$ under the assumption that $C_{gs} \gg C_{gd}$ in saturation. For the 0.1-μm nMOS example in Table 8.3, $g_m \approx 600$ mS/mm and $f_T \approx 80$ GHz.

8.5.3   Power Gain Condition of a Two-Port Network

For the power gain of a transistor, consider the general two-port network in Figure 8.50 represented by an admittance matrix:

$$\begin{bmatrix} i_1 \\ i_2 \end{bmatrix} = [Y]\begin{bmatrix} v_1 \\ v_2 \end{bmatrix} = \begin{bmatrix} Y_{11} & Y_{12} \\ Y_{21} & Y_{22} \end{bmatrix}\begin{bmatrix} v_1 \\ v_2 \end{bmatrix}. \tag{8.64}$$

Assume an output termination Y_L, then

$$i_2 + Y_L v_2 = 0, \tag{8.65}$$

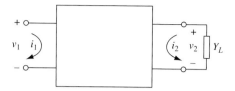

Figure 8.50 Current and voltage definitions at the input and output of a two-port network. The output is assumed to be terminated by an admittance Y_L

where $\text{Re}(Y_L) > 0$, i.e., Y_L is passive. If $v_1 = |v_1|e^{j\alpha}$, $i_1 = |i_1|e^{j\beta}$, then the power input at port 1 is the time average of $|v_1|\cos(\omega t + \alpha) \times |i_1|\cos(\omega t + \beta)$, which equals $\frac{1}{2}|v_1||i_1|\cos(\alpha - \beta)$ or $\frac{1}{2}\text{Re}(v_1 i_1^*)$. Likewise, the power output delivered at port 2 is $\frac{1}{2}\text{Re}(-v_2 i_2^*)$. By combining Eqs. (8.64) and (8.65) to express v_1, v_2, i_2 in terms of i_1, we obtain the power gain,

$$G \equiv \frac{\text{Re}(-v_2 i_2^*)}{\text{Re}(v_1 i_1^*)} = \frac{|Y_{21}|^2 \text{Re}(Y_L)}{|Y_{22} + Y_L|^2 \text{Re}(Y_{11}) - \text{Re}[Y_{12}^* Y_{21}^* (Y_{22} + Y_L)]}. \qquad (8.66)$$

Next, $Y_L (\equiv x + jy, x > 0)$ is varied to maximize G. Let $Y_{11} \equiv G_{11} + jB_{11}$ and $Y_{22} \equiv G_{22} + jB_{22}$, the above becomes

$$G = \frac{|Y_{21}|^2 x}{\left[(G_{22} + x)^2 + (B_{22} + y)^2\right]G_{11} - (G_{22} + x)\text{Re}(Y_{12}Y_{21}) - (B_{22} + y)\text{Im}(Y_{12}Y_{21})}. \qquad (8.67)$$

From basic algebra, an optimum set of $x > 0$ and y can be chosen such that G reaches its maximum value, called the maximum available gain,

$$G_{\max} = \frac{|Y_{21}|^2}{2G_{11}G_{22} - \text{Re}(Y_{12}Y_{21}) + \sqrt{[2G_{11}G_{22} - \text{Re}(Y_{12}Y_{21})]^2 - |Y_{12}Y_{21}|^2}}. \qquad (8.68)$$

The power gain criterion is $G_{\max} > 1$, which, after rearranging the terms and squaring to remove the square root, can be shown to yield $|Y_{21}/Y_{12}| > 1$ and

$$\left|Y_{12} + Y_{21}^*\right|^2 > 4G_{11}G_{22} \qquad (8.69)$$

8.5.4 Unity Power-Gain Frequency

For an intrinsic MOSFET, the power gain condition, Eq. (8.69), is always met since $\text{Re}(Y_{11}) = G_{11} = 0$ for the matrix of Eq. (8.60). Adding parasitic resistances R_g, R_s, R_d to the intrinsic MOSFET of Figure 8.49 gives the equivalent circuit of an extrinsic

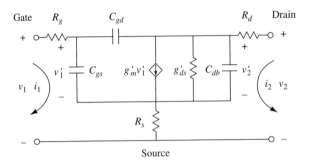

Figure 8.51 Small-signal equivalent circuit of an extrinsic MOSFET

MOSFET transistor in Figure 8.51. The convention here is that the primed parameters refer to the intrinsic device, while the unprimed parameters are for the extrinsic device, i.e., those measured externally. Also, v_1', i_1 represent v_{gs}', i_{gs}, and v_2', i_2 represent v_{ds}', i_{ds}, respectively.

The intrinsic admittance matrix of a MOSFET, Eq. (8.60), is expressed in the new notation as:

$$\begin{bmatrix} i_1 \\ i_2 \end{bmatrix} = [\tilde{Y}'] \begin{bmatrix} v_1' \\ v_2' \end{bmatrix} = \begin{bmatrix} Y_{11}' & Y_{12}' \\ Y_{21}' & Y_{22}' \end{bmatrix} \begin{bmatrix} v_1' \\ v_2' \end{bmatrix} = \begin{bmatrix} j\omega(C_{gs}+C_{gd}) & -j\omega C_{gd} \\ g_m' - j\omega C_{gd} & g_{ds}' + j\omega C_{db} + j\omega C_{gd} \end{bmatrix} \begin{bmatrix} v_1' \\ v_2' \end{bmatrix}.$$
(8.70)

The extrinsic small-signal voltages, v_1 ($=v_{gs}$) and v_2 ($=v_{ds}$), are related to the intrinsic voltages v_1', v_2' by the following matrix equation:

$$\begin{bmatrix} v_1 \\ v_2 \end{bmatrix} = \begin{bmatrix} v_1' \\ v_2' \end{bmatrix} + \begin{bmatrix} R_g + R_s & R_s \\ R_s & R_d + R_s \end{bmatrix} \begin{bmatrix} i_1 \\ i_2 \end{bmatrix}$$

$$= \left\{ [\tilde{Y}']^{-1} + \begin{bmatrix} R_g + R_s & R_s \\ R_s & R_d + R_s \end{bmatrix} \right\} \begin{bmatrix} i_1 \\ i_2 \end{bmatrix},$$
(8.71)

where the inverse of Eq. (8.70) has been used in the second step. The extrinsic admittance matrix, $[\tilde{Y}]$, can be obtained from Eq. (8.71):

$$\begin{bmatrix} i_1 \\ i_2 \end{bmatrix} = [\tilde{Y}] \begin{bmatrix} v_1 \\ v_2 \end{bmatrix} = \begin{bmatrix} Y_{11} & Y_{12} \\ Y_{21} & Y_{22} \end{bmatrix} \begin{bmatrix} v_1 \\ v_2 \end{bmatrix}$$

$$= \left\{ 1 + [\tilde{Y}'] \begin{bmatrix} R_g + R_s & R_s \\ R_s & R_d + R_s \end{bmatrix} \right\}^{-1} [\tilde{Y}'] \begin{bmatrix} v_1 \\ v_2 \end{bmatrix}.$$
(8.72)

If the parasitic resistances are not excessively large, $[\tilde{Y}]$ can be expressed to the first order of R_s, R_d, R_g as

$$[\tilde{Y}] = \begin{bmatrix} Y_{11} & Y_{12} \\ Y_{21} & Y_{22} \end{bmatrix} = [\tilde{Y}'] - [\tilde{Y}'] \begin{bmatrix} R_g + R_s & R_s \\ R_s & R_d + R_s \end{bmatrix} [\tilde{Y}']$$

$$= \begin{bmatrix} Y'_{11} - Y'^2_{11}R_g - Y'_{12}Y'_{21}R_d - (Y'_{11} + Y'_{12})(Y'_{11} + Y'_{21})R_s \\ Y'_{21} - Y'_{11}Y'_{21}R_g - Y'_{21}Y'_{22}R_d - (Y'_{21} + Y'_{11})(Y'_{21} + Y'_{22})R_s \end{bmatrix}$$

$$\begin{matrix} Y'_{12} - Y'_{11}Y'_{12}R_g - Y'_{12}Y'_{22}R_d - (Y'_{12} + Y'_{11})(Y'_{12} + Y'_{22})R_s \\ Y'_{22} - Y'^2_{22}R_d - Y'_{12}Y'_{21}R_g - (Y'_{22} + Y'_{12})(Y'_{22} + Y'_{21})R_s \end{matrix} \Bigg]$$

$$(8.73)$$

where Y'_{11}, Y'_{12}, Y'_{21}, Y'_{22} are given by Eq. (8.70).

The unity power gain frequency or the maximum oscillation frequency, f_{max}, can be solved by applying the power gain condition, Eq. (8.69), with an equal sign to the extrinsic matrix $[\tilde{Y}]$ of Eq. (8.73). By keeping only the first order terms of R_s, R_d, R_g, we obtain

$$\omega^2_{max} = \frac{g'^2_m/4}{(C_{gs} + C_{gd})^2 g'_{ds}R_g + C_{gd}(C_{gs} + C_{gd})g'_m R_g + C_{gd}\left[(C_{gd} + C_{db})g'_m + C_{gd}g'_{ds}\right]R_d + C_{gs}(C_{gs}g'_{ds} - C_{db}g'_m)R_s}.$$

$$(8.74)$$

Alternatively, it can be expressed in terms of $\omega_T = 2\pi f_T \approx g_m'/(C_{gs} + C_{gd})$ [approximation to Eq. (8.63)] as

$$\omega_{max} = \frac{\omega_T/2}{\sqrt{g'_{ds}R_g + \omega_T C_{gd}R_g + \dfrac{C_{gd} + C_{db}}{C_{gs} + C_{gd}}\omega_T C_{gd}R_d - \dfrac{C_{db}}{C_{gs} + C_{gd}}\omega_T C_{gs}R_s + \dfrac{C^2_{gd}g'_{ds}R_d + C^2_{gs}g'_{ds}R_s}{(C_{gs} + C_{gd})^2}}}.$$

$$(8.75)$$

Note that f_{max} is simply $\omega_{max}/2\pi$. An approximate expression for f_{max} commonly found in the literature is obtained by keeping only the second term in the square root:

$$f_{max} = \sqrt{\frac{f_T}{8\pi R_g C_{gd}}}.$$

$$(8.76)$$

The f_T and f_{max} figures of a modern nMOSFET with sub-50-nm channel lengths can be in the range of 200 GHz, rivaling those of modern bipolar transistors. However, as an RF amplifier, the voltage gain of a MOSFET is inferior to that of a bipolar transistor because of the lower transconductance and the output characteristics. Using Eq. (8.60) with an open circuit at the output, i.e., $i_{ds} = 0$, we can find the voltage gain at low frequencies as

$$\left|\frac{v_{ds}}{v_{gs}}\right|(\omega \to 0) = \frac{g_m}{g_{ds}}.$$

$$(8.77)$$

This has the same form as the maximum slope of an inverter transfer curve, $|dV_{out}/dV_{in}|$, discussed in Section 8.2.1.2. High g_m and f_T figures are obtained with short-channel devices which also have high g_{ds} due to the drain-induced-barrier-lowering effect. In the 0.1-μm nMOS example in Figure 8.34(a), $g_m/g_{ds} \approx 17$, significantly lower than the typical voltage gain of bipolar transistors discussed in Section 11.6.1.

Exercises

8.1 Apply constant-field scaling rules to the long-channel currents [Eq. (5.23) for the linear region and Eq. (5.31) for the saturation region], and show that they behave as indicated in Table 8.1.

8.2 Apply constant-field scaling rules to the subthreshold current, Eq. (5.39), and show that instead of decreasing with scaling ($1/\kappa$), it actually increases with scaling (note that $V_{gs} < V_t$ in subthreshold). What if the temperature is also scaled down by the same factor ($T \rightarrow T/\kappa$)?

8.3 Apply constant-field scaling rules to the saturation current from the $n = 1$ velocity saturation model [Eq. (6.49)] and the fully saturation-velocity limited current [Eq. (6.51)], and show that they behave as indicated in Table 8.1.

8.4 Consider the CMOS switching delay, $\tau = (\tau_n + \tau_p)/2$, where τ_n and τ_p are given by Eqs. (8.8) and (8.9). If the inverter is driving another stage with the same n- to p-width ratio and if both the n- and p-devices have the same capacitance per unit width, the load capacitance C is proportional to $W_n + W_p$. Show that the minimum delay τ occurs for a width ratio of $W_p/W_n = (I_{onN/w}/I_{onP/w})^{1/2}$, which is different from Eq. (8.5) for best noise margin where $\tau_n = \tau_p$.

8.5 For an RC circuit with a capacitor C connected in series with a resistor R and a switchable voltage source, solve for the waveform of the voltage across the capacitor, $V(t)$, when the voltage source is abruptly switched from 0 to V_{dd} with the initial condition $V(t = 0) = 0$. Show that when the steady state condition is established, an energy of $CV_{dd}^2/2$ has been dissipated in the resistor R and the same amount of energy is stored in C. Since the energy dissipated and the energy stored are independent of R, the same results hold even if $R = 0$. What happens if the voltage source is now switched off from V_{dd} to 0 with the initial condition $V(t = 0) = V_{dd}$?

8.6 The carrier transit time is defined as $\tau_{tr} \equiv Q/I$, where Q is the total inversion charge and I is the total conduction current of the device. For a MOSFET device biased in the linear region (low drain voltage), use Eq. (5.23) and the inversion charge expression above Eq. (5.70) to derive an expression for τ_{tr}. Similarly, use Eq. (5.31) and the expression above Eq. (5.72) to derive τ_{tr} for a long-channel MOSFET biased in saturation.

8.7 Use Eq. (6.49) and the inversion-charge expression in Exercise 6.8 to find the carrier transit time τ_{tr} for a short-channel MOSFET biased in saturation. What is the limiting value of τ_{tr} when the device becomes fully velocity-saturated as $L \rightarrow 0$?

8.8 In the top equivalent circuit of Figure 8.17, the source–drain current can be considered either as a function of the internal voltages: $I_{ds}\left(V'_{gs}, V'_{ds}\right)$ or as a function of the external voltages: $I_{ds}\left(V_{gs}, V_{ds}\right)$. The internal voltages are related to the external voltages by Eqs. (8.13) and (8.14). Show that the transconductance of the intrinsic MOSFET can be expressed as

$$g'_m \equiv \left(\frac{\partial I_{ds}}{\partial V'_{gs}}\right)_{V'_{ds}} = \frac{g_m}{1 - g_m R_s - g_{ds}(R_s + R_d)},$$

where

$$g_m \equiv \left(\frac{\partial I_{ds}}{\partial V_{gs}}\right)_{V_{ds}}$$

is the extrinsic transconductance, and

$$g_{ds} \equiv \left(\frac{\partial I_{ds}}{\partial V_{ds}}\right)_{V_{gs}}$$

is the extrinsic output conductance.

8.9 A similar distributed network to the one in Figure 8.23 can be used to formulate the *transmission-line model* of contact resistance in a planar geometry (Berger, 1972). Here we consider the current flow from a thin resistive film (diffusion with a sheet resistivity ρ_{sd}) into a ground plane (metal) with an interfacial contact resistivity ρ_c between them (Figure 8.18). Thus, in Figure 8.23, $R\,dx$ corresponds to $(\rho_{sd}/W)dx$, and $C\,dx$ is replaced by a shunt conductance $G\,dx$, which corresponds to $(W/\rho_c)dx$. Show that both the current and voltage along the current flow direction satisfy the following differential equation:

$$\frac{d^2 f}{dx^2} = RGf = \frac{\rho_{sd}}{\rho_c} f,$$

where $f(x) = V(x)$ or $I(x)$ defined in Figure 8.23.

8.10 Following the transmission-line model in Exercise 8.9, with the boundary condition $I(x = l_c) = 0$ where $x = 0$ is the leading edge and $x = l_c$ is the far end of the contact window (Figure 8.18), solve for $V(x)$ and $I(x)$ within a multiplying factor and show that the total contact resistance, $R_{co} = V(x = 0)/I(x = 0)$, is given by Eq. (8.18).

8.11 The insertion of a buffer stage (Section 8.4.2.2) between the inverter and the load is beneficial only if the load capacitance is higher than a certain value. Find, in terms of C_{in} and C_{out}, the minimum load capacitance C_L above which the single-stage buffered delay given by Eq. (8.51) is shorter than the unbuffered delay given by Eq. (8.49).

8.12 Generalize Eq. (8.50) for one-stage buffered delay to n stages: if the width ratios of the successive buffer stages are $k_1, k_2, k_3, \ldots, k_n$ (all >1), show that the n-stage buffered delay is

$$\tau_b(n) = R_{sw}\left[(n+1)C_{out} + (k_1 + k_2 + \cdots + k_n)C_{in} + \frac{C_L}{k_1 k_2 \cdots k_n}\right].$$

8.13 Following the previous exercise, show that for a given n, the n-stage buffered delay is a minimum,

$$\tau_{b\,min}(n) = R_{sw}\left[(n+1)C_{out} + (n+1)C_{in}\left(\frac{C_L}{C_{in}}\right)^{1/(n+1)}\right],$$

when $k_1 = k_2 = \cdots = k_n = (C_L/C_{in})^{1/(n+1)}$. Here $\tau_{bmin}(n)$, as expected, is reduced to Eq. (8.51) if $n = 1$.

8.14 If one plots the minimum n-stage buffered delay from the previous exercise versus n, it will first decrease and then increase with n. In other words, depending on the ratios of C_L/C_{in} and C_{out}/C_{in}, there is an optimum number of buffer stages for which the overall delay is the shortest. Show that this optimum n is given by the closest integer to

$$n = \frac{\ln(C_L/C_{in})}{\ln k} - 1,$$

where k is a solution of

$$k(\ln k\text{-}1) = \frac{C_{out}}{C_{in}}.$$

For typical C_{out}/C_{in} ratios not too different from unity, k is in the range of 3–5. Note that k also gives the optimum width ratio between the successive buffer stages, i.e., $k_1 = k_2 = \cdots = k_n = k$. Also show that the minimum buffered delay is given by

$$\tau_{b\,min} \approx kR_{sw}C_{in}\ln(C_L/C_{in}),$$

which only increases logarithmically with load capacitance.

8.15 Consider a chain of CMOS inverters with power supply V_{dd}. The propagation delay between the waveforms can be expressed by Eq. (8.45) with FO $= 1$. What is the power dissipation while the signal is propagating down the chain? If the device widths are increased or decreased by a factor of k (>1 or <1) to kW_n, kW_p while C_L remains constant, how would the delay and power vary with k?

9 Bipolar Devices

Structurally, an n–p–n bipolar transistor is formed with two p–n diodes having a common p-region which is the base of the transistor. Similarly, a p–n–p bipolar transistor is two p–n diodes having a common n-region. *Two types of bipolar transistors are covered in this book:* the traditional *vertical bipolar transistors* shown schematically in Figure 9.1(a) and the more recently developed *symmetric lateral bipolar transistors on SOI* (Sturm *et al.*, 1987; Cai *et al.*, 2011) shown schematically in Figure 9.1(b). In this chapter, we discuss the basic properties and operation of a bipolar transistor. We shall *focus on vertical n–p–n transistors* which are by far the most commonly used transistors as of this wxsriting. Any significant difference between a vertical transistor and a lateral transistor will be pointed out wherever it is important to do so.

The *steady-state characteristics of a bipolar transistor is determined by the doping profiles of the device regions (intrinsic-device regions) along the current path from emitter to collector*. For a vertical bipolar transistor, these intrinsic-device regions are shown schematically in Figure 9.2(a). The remainder regions adjacent to or surrounding the intrinsic regions, shown in Figure 9.1(a), are parasitic components needed for electrical isolation purposes and for forming the collector reach-through region which brings the collector current from below the base region to the collector contact located on the surface.

Figure 9.2(b) shows the bias condition for an n–p–n transistor in normal operation. The base–emitter diode is forward biased with a voltage V_{BE}, and the collector–base diode is reverse biased with a voltage V_{CB}. The corresponding energy-band diagram is shown schematically in Figure 9.2(c). The forward-biased base–emitter diode causes electrons to flow from the emitter into the base and holes to flow from the base into the emitter. Those electrons not recombined in the base layer arrive at the collector and give rise to a collector current. The holes injected into the emitter contribute to a base current.

Also illustrated in Figure 9.2(c) are the coordinates which we shall follow in describing the flow of electrons and holes in an n–p–n transistor. Thus, electrons flow in the x-direction, i.e., $J_n(x)$ is negative, and holes flow in the –x direction, i.e., $J_p(x)$ is also negative. The physical junction of the emitter–base diode is assumed to be located at "$x = 0$." However, to accommodate the finite thickness of the

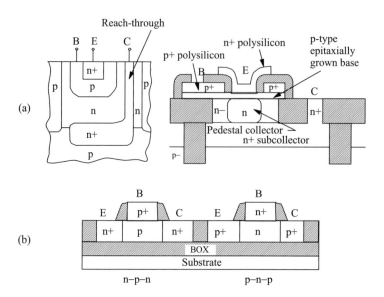

Figure 9.1 Schematics of the bipolar transistors covered in this book. (a) Vertical n−p−n bipolar transistors. The simplest, and lowest cost, is the junction-isolated device structure (left). The most modern is the self-aligned polysilicon-emitter structure with deep-trench isolation (right). (b) Symmetric lateral bipolar transistors on SOI. n−p−n and p−n−p transistors can be integrated in a manner analogous to SOI CMOS

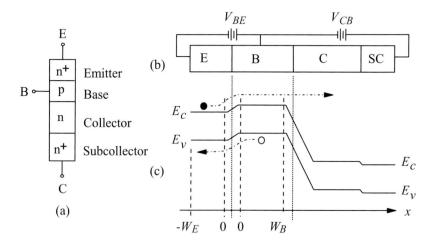

Figure 9.2 Schematics describing a vertical n−p−n bipolar transistor. (a) Vertical stack of the intrinsic-device regions, from top to bottom: emitter, base, collector, and subcollector. (b) Schematic illustrating the applied voltages in normal operation, and (c) schematic illustrating the corresponding energy-band diagram, carrier flows, and locations of the boundaries of the quasineutral emitter and base regions

space-charge layer of the emitter–base diode, the mathematical origin ($x = 0$) for the quasineutral emitter region is shifted to the left of the physical junction, as illustrated in Figure 9.2(c). Similarly, the mathematical origin ($x = 0$) for the quasineutral base region is shifted to the right of the physical junction. The emitter contact is located at $x = -W_E$, and the quasineutral base region ends at $x = W_B$. It should be noted that, due to the finite thickness of a junction space-charge layer, the widths of the quasi-neutral p- and n-regions of a diode are always smaller than their corresponding physical widths. Unfortunately, in the literature as well as here, the same symbol is often used to denote both the physical width and the quasineutral width. For example, W_B is used to denote the base width. Sometimes W_B refers to the physical base width, and sometimes it refers to the quasineutral base width. The important point to remember is that all *carrier-transport equations for p–n diodes and bipolar transistors refer to the quasineutral widths*.

A typical state-of-the-art vertical n−p−n structure is shown on the right-hand side of Figure 9.1(a). It uses oxide-filled deep trenches for electrical isolation of adjacent devices, p−n junction isolation between subcollector and wafer substrate, polysilicon for contacting the base, polysilicon emitter self-aligned to the polysilicon base-contact layer, and *in-situ* doped base region grown epitaxially. It also shows a *pedestal collector* which has a higher doping concentration in the intrinsic collector directly underneath the intrinsic base than its surrounding (extrinsic-collector) regions (Yu, 1971). A pedestal collector can be achieved quite simply by ion implantation when the emitter opening is defined in the device fabrication process. The most advanced vertical bipolar transistors employ epitaxially grown *in-situ* doped SiGe alloy for its base. Detailed discussion of SiGe-base vertical bipolar transistors is covered in Section 10.4. One way to reduce the resistance associated with the polysilicon base-contact layer is to form a silicide layer on the polysilicon layer practically everywhere (Chiu *et al.*, 1987; Iinuma *et al.*, 1995), or to form a sidewall silicide layer on the vertical edges of the polysilicon layer (Shiba *et al.*, 1991). For the polysilicon emitter, the preferred method is to use *in-situ* phosphorus-doped, instead of arsenic-doped, polysilicon and rapid thermal annealing, instead of furnace annealing, to form ultra-shallow emitters with low series resistance (Crabbé *et al.*, 1992; Nanba *et al.*, 1993; Shiba *et al.*, 1996). Since there is no oxide isolation between the subcollector and the wafer substrate, a process for integrating vertical p−n−p and n−p−n on the same chip is very complex and expensive. As a result, integration of vertical n−p−n and p−n−p is not common.

Figure 9.3(a) illustrates the typical doping profiles of the intrinsic-device regions of a vertical n–p–n transistor having a diffused, or implanted and then diffused, emitter, such as the one illustrated schematically on the left-hand side of Figure 9.1(a). The emitter junction depth x_{jE} is typically $0.2\,\mu m$ or larger (Ning and Isaac, 1980). The base junction depth is x_{jB}, and the physical base width is equal to $x_{jB} - x_{jE}$. Figure 9.3(b) illustrates the typical doping profile of a vertical n–p–n transistor with a polysilicon emitter and an implanted base. The polysilicon layer is typically about 0.2 μm or less, with an n^+ diffusion into the single-crystal region of about 30 nm or less (Nakamura and Nishizawa, 1995). That is, $x_{jE} < 30$ nm. The base widths of modern

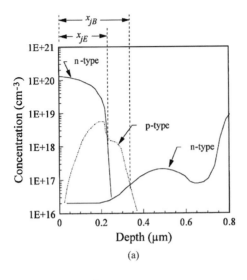

(a)

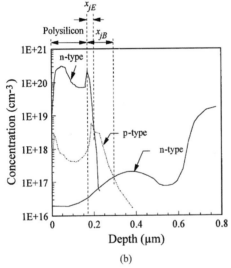

(b)

Figure 9.3 Vertical doping profiles of typical n–p–n transistors: (a) with implanted and/or diffused emitter, and (b) with polysilicon emitter

vertical bipolar transistors are typically $< 0.1\,\mu\text{m}$. The base of a polysilicon-emitter transistor can be made reproducibly much thinner than that of a diffused-emitter transistor.

It can be seen from Figure 9.1(b) that symmetrical lateral bipolar transistors on SOI have minimal parasitic regions. Electron current flows laterally from emitter to collector. Electrical isolation between adjacent devices is by STI (shallow-trench isolation) which is standard in advanced CMOS technology. Electrical isolation between device and wafer substrate is by the BOX (buried oxide). With BOX isolation, integration of n−p−n and p−n−p can be readily accomplished, just like SOI CMOS (Cai *et al.*, 2011). Figures 9.2(a), (b), and (c) can be adapted for a lateral n−p−n transistor simply by omitting the subcollector region and noting that the collector is an n$^+$-region just like the emitter.

In this book, we assume the lateral transistors to have a base doping concentration that is uniform everywhere within the base. Such a base doping profile is preferred and readily achievable. The schematics in Figure 9.1(b) show vertically straight emitter/base and collector/base boundaries, implying the emitter-to-collector spacing W_{E-C} to be the same near the n$^+$ or p$^+$ extrinsic-base regions as near the BOX. Devices with such vertically straight and abrupt emitter/base and collector/base boundaries can be obtained using epitaxially grown emitter/collector regions (Hashemi *et al.*, 2017, 2018). However, most reported lateral transistors as of this writing are fabricated using ion implantation to form emitter/collector regions. Ion implantation for forming emitter/collector results in W_{E-C} variation. For devices with small W_{E-C} by design, the variation in W_{E-C} could show up as variation in the measured collector currents.

9.1   Basic Operation of a Bipolar Transistor

As illustrated in Figure 9.2(a), a bipolar transistor physically consists of two p−n diodes connected back to back. The basic operation of a bipolar transistor, therefore, can be described by the operation of two back-to-back diodes. In normal operation, the base−emitter diode is forward biased and the base−collector diode is reverse biased, as illustrated in Figure 9.2(b). As the electrons injected from the emitter into the base reach the collector, they give rise to a collector current. The holes injected from the base into the emitter contributes to a base current. [The other components of the measured base current are due to recombination in the base layer and within the base−emitter space-charge region. They will be discussed later in Section 9.2.3.] One basic objective in bipolar transistor design is to achieve a collector current significantly larger than the base current.

To first order, the behavior of a bipolar transistor is determined by the characteristics of the forward-biased emitter–base diode, since the collector usually acts only as a sink for the carriers injected from the emitter into the base. *The base width of a typical modern bipolar transistor is much smaller than the base-region diffusion length. The carriers injected from the emitter reach the collector with negligible recombination in the base region.*

9.1.1   Modifying the Simple Diode Theory for Describing Bipolar Transistors

In order to extend the diode theory discussed in Section 3.1 to describe the behavior of a bipolar transistor quantitatively, three important effects ignored in it must be included. These are the effects of finite electric field in a quasineutral region, heavy doping, and nonuniform energy bandgap. These effects are discussed in Sections 9.1.1.1–9.1.1.3.

9.1.1.1   Electric Field in a Quasineutral Region with a Uniform Energy Bandgap

In Section 3.1.5 the current–voltage characteristics of a p–n diode due to diffusion currents were derived for the case of zero electric field in the p- and n-type quasineutral regions. The zero-field approximation is valid only where the majority-carrier current is zero and the majority-carrier concentration is uniform. For vertical bipolar transistors, as shown in Figures 9.3(a) and (b), the doping profiles are rather nonuniform. A nonuniform doping profile means nonuniform majority-carrier concentration. Furthermore, at large base–emitter forward biases, to maintain quasineutrality, the high concentration of injected minority carriers could cause significant nonuniformity in the majority-carrier concentration as well. Therefore, the effect of nonuniform majority-carrier concentration in a quasineutral region cannot be ignored in determining the current–voltage characteristics of a bipolar transistor.

For a p-type region, Eq. (2.80) gives

$$\phi_p = \psi_i + \frac{kT}{q} \ln\!\left(\frac{p_p}{n_i}\right), \tag{9.1}$$

where ϕ_p is the hole quasi-Fermi potential and ψ_i is the intrinsic potential. (Note that p_p is equal to N_a only for the case of low electron injection.) The electric field is given by Eq. (2.55), namely

$$
\begin{aligned}
\mathscr{E} &\equiv -\frac{d\psi_i}{dx} = \frac{kT}{q}\frac{1}{p_p}\frac{dp_p}{dx} - \frac{d\phi_p}{dx} \\
&= \frac{kT}{q}\frac{1}{p_p}\frac{dp_p}{dx} + \frac{J_p}{qp_p\mu_p},
\end{aligned}
\tag{9.2}
$$

where we have used Eq. (2.78), which relates $d\phi_p/dx$ to J_p. In writing Eq. (9.2), the intrinsic-carrier concentration n_i, which is a function of bandgap energy [see Eq. (2.13)], is independent of x because we are considering a region with uniform energy bandgap. The dependence of energy bandgap on x will be discussed in Section 9.1.1.2 in connection with heavy-doping effects.

Let us apply Eq. (9.2) to the intrinsic-base region of an n–p–n transistor with a typical current gain of 100. At a typical but high collector current density of 1 mA/μm^2, the base current density is 0.01 mA/μm^2, i.e., $J_p = 0.01$ mA/μm^2 in the base layer. The base doping concentration is typically on the order of 10^{18} cm^{-3} (see Figure 9.3). The corresponding hole mobility is about 150 cm^2/V-s (see Figure 2.9). That is, $p_p \approx 10^{18}$ cm^{-3} and $\mu_p \approx 150$ cm^2/V-s, and $J_p/qp_p\mu_p \approx 40$ V/cm, which is a negligibly small electric field in normal device operation. Therefore, for a p-type region, Eq. (9.2) gives

$$\mathscr{E}(\text{p-region}) \approx \frac{kT}{q}\frac{1}{p_p}\frac{dp_p}{dx} \quad \text{(for electrons in p-region).} \tag{9.3}$$

Similarly, for an n-type region,

$$\mathscr{E}(\text{n-region}) \approx -\frac{kT}{q}\frac{1}{n_n}\frac{dn_n}{dx} \quad \text{(for holes in n-region).} \tag{9.4}$$

Equations (9.3) and (9.4) show that *the electric field is negligible in a region of uniform majority-carrier concentration*. To include the effect of finite electric field, the current-density equations, Eqs. (2.68) and (2.69), which include both the drift and the diffusion components, should be used. These are repeated here for convenience:

$$J_n(x) = qn\mu_n\mathscr{E} + qD_n\frac{dn}{dx},\tag{9.5}$$

and

$$J_p(x) = qp\mu_p\mathscr{E} - qD_p\frac{dp}{dx}.\tag{9.6}$$

It should be noted that if Eq. (9.4) is substituted into Eq. (9.5), the right-hand side of Eq. (9.5) is equal to zero. Similarly, if Eq. (9.3) is substituted into Eq. (9.6), the right-hand side of Eq. (9.6) is equal to zero. What this means is that the approximations for the electric fields represented by Eqs. (9.3) *and* (9.4) *are good approximations only for describing minority-carrier currents*. They are applicable for describing carrier transport in a diode or in a bipolar transistor.

- *Built-in electric field in a nonuniformly doped base region.* Consider the electron current in the p-type base of a forward-biased emitter–base diode. Let $N_B(x)$ be the doping concentration in the base, and, for simplicity, all the dopants are assumed to be ionized. Quasineutrality requires that

$$p_p(x) = N_B(x) + n_p(x).\tag{9.7}$$

Therefore,

$$\frac{dp_p}{dx} = \frac{dN_B}{dx} + \frac{dn_p}{dx}.\tag{9.8}$$

The *built-in electric field* $\mathscr{E}_0$ is defined as the electric field from the nonuniform base dopant distribution alone, ignoring any effect of injected minority carriers. It can be obtained by substituting N_B for p_p in Eq. (9.3), namely

$$\mathscr{E}_0 \equiv \mathscr{E}(n_p = 0) = \frac{kT}{q}\frac{1}{N_B}\frac{dN_B}{dx}.\tag{9.9}$$

Substituting Eq. (9.3) into Eq. (9.5), and using Eqs. (9.8) and (9.9) and the Einstein relationship, we have, for electron current in a nonuniformly doped p-type base region,

$$J_n(x) = qn_p\mu_n\mathscr{E}_0\frac{N_B}{n_p + N_B} + qD_n\left(\frac{2n_p + N_B}{n_p + N_B}\right)\frac{dn_p}{dx}.\tag{9.10}$$

Equation (9.10) suggests that the *effective electric field* $\mathscr{E}_{eff}$ in the p-type base can be written as

$$\mathscr{E}_{eff} = \mathscr{E}_0\frac{N_B}{n_p + N_B}.\tag{9.11}$$

Equations (9.10) and (9.11) are valid for all levels of electron injection from the emitter, i.e., for all values of n_p.

- *Electric field and current density in the low-injection limit.* At low levels of electron injection from the emitter, i.e., for $n_p \ll N_B$, $\mathscr{E}_{eff}$ reduces to $\mathscr{E}_0$ and Eq. (9.10) reduces to

$$J_n(x) \approx q n_p \mu_n \mathscr{E}_0 + q D_n \frac{dn_p}{dx}, \tag{9.12}$$

which simply says that the electron current flowing in the p-type base consists of a drift component due to the built-in field from the nonuniform base dopant distribution, and a diffusion component from the electron concentration gradient in the base.

- *Electric field and current density in the high-injection limit.* When the electron injection level is very high, i.e., when $n_p \gg N_B$, $\mathscr{E}_{eff}$ becomes very small. The built-in electric field is screened out by the large concentration of injected minority carriers. Therefore, the electron current component associated with the built-in field becomes negligible, and the electron current density approaches

$$J_n(x)|_{n_p \gg N_B} \approx q 2 D_n \frac{dn_p}{dx}. \tag{9.13}$$

That is, at the high-injection limit, the minority-carrier current behaves as if it were purely a diffusion current, but with a diffusion coefficient twice its low-injection value. This is known as the *Webster effect* (Webster, 1954).

9.1.1.2   Heavy-Doping Effect

The effective ionization energy for impurities in a heavily doped semiconductor decreases with increase in doping concentration, resulting in an apparent decrease of the energy bandgap with an increase in doping concentration (see Section 2.2.3). For modeling purposes, it is convenient to define an *effective intrinsic-carrier concentration, n_{ie},* and **lump all the heavy-doping effects into a parameter called apparent bandgap narrowing, ΔE_g,** given by the equation

$$p_0(\Delta E_g) n_0(\Delta E_g) \equiv n_{ie}^2 = n_i^2 \exp(\Delta E_g / kT). \tag{9.14}$$

The heavy-doping effect increases the effective intrinsic carrier concentration. **To include the heavy-doping effect, n_i should be replaced by n_{ie}.** Thus, including the heavy-doping effect, the product pn in Eq. (2.81) becomes

$$pn = n_{ie}^2 \exp[q(\phi_p - \phi_n)/kT]. \tag{9.15}$$

It is difficult to determine ΔE_g experimentally and there is considerable scattering in the reported data in the literature (del Alamo *et al.*, 1985a). Careful analyses of the reported data suggest the following *empirical* expressions for the apparent bandgap-narrowing parameter:

$$\Delta E_g(N_d) = 18.7 \ln\left(\frac{N_d}{7 \times 10^{17}}\right) \text{meV} \tag{9.16}$$

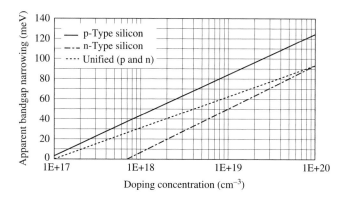

Figure 9.4 Apparent bandgap narrowing as given by the empirical expressions in Eqs. (9.16)–(9.18)

for $N_d \geq 7 \times 10^{17}$ cm^{-3}, and zero for lower doping levels, for n-type silicon (del Alamo *et al.*, 1985b), and

$$\Delta E_g(N_a) = 9\left(F + \sqrt{F^2 + 0.5}\right)\text{meV}, \qquad (9.17)$$

where $F = \ln\left(N_a/10^{17}\right)$, for $N_a > 10^{17}$ cm^{-3}, and zero for lower doping levels, for p-type silicon (Slotboom and de Graaff, 1976; Swirhun *et al.*, 1986). More recently, using a new model that treats both the majority-carrier and minority-carrier mobilities in a unified manner, Klaassen (Klaassen, 1990; Klaassen *et al.*, 1992) showed that the heavy-doping effect in both n-type silicon and p-type silicon can be described well by a unified apparent bandgap-narrowing parameter. If N represents N_d in n-type silicon and N_a in p-type silicon, then the Klaassen unified apparent bandgap-narrowing parameter is given by

$$\Delta E_g(N) = 6.92\left\{ \ln\left(\frac{N}{1.3 \times 10^{17}}\right) + \sqrt{\left[\ln\left(\frac{N}{1.3 \times 10^{17}}\right)\right]^2 + 0.5} \right\}\text{meV}. \quad (9.18)$$

Figure 9.4 is a plot of ΔE_g as a function of doping concentration, as given by Eqs. (9.16)–(9.18).

9.1.1.3 Electric Field in a Quasineutral Region with Nonuniform Energy Bandgap

Aside from heavy-doping effect, the energy bandgap can also be modified by incorporating a large amount of germanium into silicon. The Ge-containing silicon bandgap becomes narrower (People, 1986). If both heavy-doping effect and the effect of germanium are included in the parameter ΔE_g in Eq. (9.14), then the product pn given by Eq. (9.15) can be used to describe transport in heavily doped SiGe alloys.

When the energy bandgap is nonuniform, the electric field is no longer simply given by Eqs. (9.3) and (9.4), which include only the effect of nonuniform dopant distribution. When the effect of nonuniform energy bandgap is included, the electric fields are given by (van Overstraeten *et al.*, 1973)

$$\mathscr{E}(\text{p-region}) = \frac{kT}{q}\left(\frac{1}{p_p}\frac{dp_p}{dx} - \frac{1}{n_{ie}^2}\frac{dn_{ie}^2}{dx}\right) \quad \text{(for electrons in p-region)}, \tag{9.19}$$

and

$$\mathscr{E}(\text{n-region}) = -\frac{kT}{q}\left(\frac{1}{n_n}\frac{dn_n}{dx} - \frac{1}{n_{ie}^2}\frac{dn_{ie}^2}{dx}\right) \quad \text{(for holes in n-region)}. \tag{9.20}$$

Derivation of Eq. (9.20) is left as an exercise for the reader (see Exercise 9.1).

9.2   Ideal Current–Voltage Characteristics

The ideal diffusion-current characteristics of a p–n diode were derived in Section 3.1.5. The diode current has an $\exp(qV_{app}/kT)$ dependence on voltage. In practice, the measured current–voltage characteristics of a bipolar transistor appear ideal only over a certain range of V_{BE}. At small V_{BE}, the measured base current is larger than the ideal base current. At large V_{BE}, both the measured base and measured collector currents are significantly smaller than the corresponding ideal currents. In this section, the ideal current–voltage characteristics are discussed. Deviations from the ideal characteristics are discussed in Section 9.3.

Referring to Figure 9.2(a), we see that the base terminal contact is located at the side of the intrinsic-base region. When the base–emitter diode is forward biased, holes from the base terminal first flow horizontally (parallel to the emitter–base junction) into the intrinsic-base region before turning upward to enter the emitter. The *IR* drop associated with this horizontal hole-current flow in the intrinsic base causes the base–emitter junction voltage V_{BE}' to vary across the junction, with V_{BE}' largest nearest the base contact, and smallest furthest away from the base contact. This is known as *emitter current-crowding effect*. When emitter current crowding is significant, the base and collector current densities are not just a function of x [Figure 9.2(c)], but also a function of distance from the base contact. Fortunately, as shown in Section 9.2.1, *emitter current crowding is negligible in modern bipolar devices* because of their narrow emitter stripe widths. Therefore, we shall ignore emitter current-crowding effect and assume V_{BE}' to be uniform across the emitter–base junction. The implication is that both the *base and collector currents in a modern bipolar transistor can be described by simple one-dimensional transport equations* (the direction along the current flow from emitter to collector).

Let us consider the electrons injected from the emitter into the base region of an n–p–n transistor. It is convenient to reformulate the electron current density in terms of carrier concentrations (Moll and Ross, 1956). To this end, following the coordinate system illustrated in Figure 9.2(c), we start with the electron current density given by Eq. (2.77), namely

$$J_n(x) = -qn_p\mu_n \frac{d\phi_n}{dx}. \tag{9.21}$$

As we shall show later, the hole current density in the p-type base is small, being smaller than the electron current density by a factor equal to the current gain of the transistor. Also, as indicated in Figure 9.3, the base region has a reasonably high doping concentration, typically greater than 10^{18} cm^{-3} for a modern bipolar transistor. Therefore, the IR drop along the electron-current flow path (x-direction) in the p-type base is negligible, suggesting the hole quasi-Fermi potential ϕ_p is approximately constant in the x-direction (see Section 3.1.3.1). That is, we have

$$\frac{d\phi_p}{dx} \approx 0 \tag{9.22}$$

in the p-type base region. Combining Eqs. (9.21) and (9.22), we obtain

$$J_n(x) \approx qn_p\mu_n \frac{d(\phi_p - \phi_n)}{dx} \quad \text{(in p-region)}. \tag{9.23}$$

Now, Eq. (9.15) gives

$$\phi_p - \phi_n = \frac{kT}{q} \ln\left(\frac{p_p n_p}{n_{ie}^2}\right). \tag{9.24}$$

Substituting Eq. (9.24) into Eq. (9.23) and rearranging the terms, we have

$$J_n(x) = qD_n \frac{n_{ie}^2}{p_p} \frac{d}{dx}\left(\frac{n_p p_p}{n_{ie}^2}\right) \quad \text{(in p-region)} \tag{9.25}$$

for the electron current in the base.

The hole current density due to holes injected from the p-type base into the n-type emitter can be derived in a similar manner. The result is

$$J_p(x) = -qD_p \frac{n_{ie}^2}{n_n} \frac{d}{dx}\left(\frac{n_n p_n}{n_{ie}^2}\right) \quad \text{(in n-region)}. \tag{9.26}$$

9.2.1 Intrinsic-Base Resistance and Emitter Current Crowding

Consider the base–emitter diode part of a vertical bipolar transistor, the cross-section of which is shown schematically in Figure 9.5. The base current I_B enters the intrinsic base at the base contact and then spreads out, turns upward, and enters the emitter. Thus, the effective intrinsic-base resistance r_{bi} as seen by the base current depends on how the base current spreads out inside the intrinsic-base layer. One commonly used method for evaluating r_{bi} is to consider the power dissipation P in the intrinsic base (Hauser, 1968), and define r_{bi} by

$$P = I_B^2 r_{bi}. \tag{9.27}$$

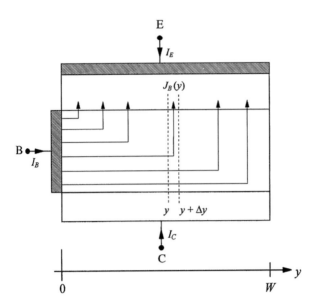

Figure 9.5 Schematic of the intrinsic part of a bipolar transistor illustrating the flow of base current. The transistor has an emitter stripe of width W and a base contact on only one side

9.2.1.1   Low-Current Intrinsic-Base Resistance

Equation (9.27) can be evaluated readily for low-current situations where there is negligible current crowding and the base-current density J_B entering the emitter can be assumed to be uniform. That this is a good assumption for modern bipolar transistors will be shown in Section 9.2.1.3.

Let us assume the emitter stripe has a width W and a length L, and the base is contacted on one side of the emitter only, as shown in Figure 9.5. Consider a slice of the intrinsic base between points y and $y + \Delta y$. The resistance of this slice as seen by the base current is

$$\Delta R = \frac{R_{Sbi}}{L} \Delta y, \tag{9.28}$$

where R_{Sbi} is the intrinsic-base sheet resistivity [see Eq. (10.5) in Chapter 10]. The base current passing through this slice is

$$i_B(y) = J_B L(W - y), \tag{9.29}$$

and the power dissipated in this slice is

$$\Delta P = i_B^2(y) \Delta R = \frac{R_{Sbi}}{L} i_B^2(y) \Delta y. \tag{9.30}$$

R_{Sbi} is a function of collector current density, which in turn is a function of the base-emitter junction voltage V'_{BE}. The assumption of negligible current crowding implies that there is negligible lateral voltage drop along the intrinsic base in the y-direction, and R_{Sbi} is independent of y. Therefore, the total power dissipated by the base current in the intrinsic base is

$$P = \frac{R_{Sbi}}{L} \int_0^W i_B^2(y)\,dy$$

$$= R_{Sbi}J_B^2L \int_0^W (W-y)^2\,dy \qquad (9.31)$$

$$= \frac{1}{3}\left(\frac{W}{L}\right)R_{Sbi}I_B^2.$$

Comparison of Eq. (9.27) with Eq. (9.31) gives

$$r_{bi} = \frac{1}{3}\left(\frac{W}{L}\right)R_{Sbi} \quad \text{(base contact on one side only).} \qquad (9.32)$$

9.2.1.2   Other Emitter Geometries

The calculations in Section 9.2.1.1 can be applied to other emitter geometries and/or base contact schemes. For the same emitter-stripe geometry, but with base contact on both sides of the emitter, the base resistance is reduced by a factor of 4, namely

$$r_{bi} = \frac{1}{12}\left(\frac{W}{L}\right)R_{Sbi} \quad \text{(base contact on two sides).} \qquad (9.33)$$

The base resistance for a square emitter with base contact on all four sides is

$$r_{bi} = \frac{1}{32}R_{Sbi} \quad \text{(four-sided base contact).} \qquad (9.34)$$

For a round emitter, it is

$$r_{bi} = \frac{1}{8\pi}R_{Sbi} \quad \text{(round emitter).} \qquad (9.35)$$

9.2.1.3   Estimation of Emitter Current-Crowding Effect

The assumption of uniform base current density used in Sections 9.2.1.1 and 9.2.1.2 is valid when the lateral IR drop in the intrinsic-base layer, between $x = 0$ and $x = W$ in Figure 9.5, caused by the lateral flow of the base current, is small compared to kT/q. If the lateral IR drop is not negligible, then the base current density becomes a function of y, with

$$J_B(y) = J_B(0)\exp\left(\frac{qV'_{BE}(y)}{kT} - \frac{qV'_{BE}(0)}{kT}\right). \qquad (9.36)$$

Since $V'_{BE}(y) \le V'_{BE}(0)$, we have $J_B(y) \le J_B(0)$. That is, the emitter current density is larger at the emitter edge than toward the middle of the emitter, i.e., there is current crowding at the emitter edge. The general expressions for $J_B(y)$ and $V'_{BE}(y)$ have been derived by Hauser (1964, 1968). The derivation is rather involved, and the interested reader is referred to the references for details. Here we simply give an upper-bound estimate of the emitter current-crowding effect.

The base current passing through the intrinsic-base slice at point y is

$$i_B(y) = L \int_y^W J_B(y')dy'. \tag{9.37}$$

The *IR* drop across an intrinsic-base slice between y and $y + \Delta y$ is

$$\Delta V'_{BE}(y) = -i_B(y)\Delta R(y)$$
$$= -\Delta y R_{Sbi}(J_C(y)) \int_y^W J_B(y')dy', \tag{9.38}$$

where Eq. (9.28) has been used for ΔR. In Eq. (9.38), R_{Sbi} is denoted explicitly as a function of J_C, which in turn is a function of y. The gradient of $V'_{BE}(y)$ along the intrinsic-base layer is therefore

$$\frac{dV'_{BE}(y)}{dy} = -R_{Sbi}(J_C(y)) \int_y^W J_B(y')dy'$$
$$= -R_{Sbi}(J_C(y))J_B(0) \int_y^W \exp\left(\frac{qV'_{BE}(y')}{kT} - \frac{qV'_{BE}(0)}{kT}\right)dy', \tag{9.39}$$

where we have used Eq. (9.36) for $J_B(y)$. An upper bound of the magnitude of this voltage gradient can be obtained by replacing the exponential inside the integral, which is smaller than unity because $V'_{BE}(y) \leq V'_{BE}(0)$, by unity, and replacing R_{Sbi} by its low-current value. [Equation (10.5) shows that R_{Sbi} decreases as collector current increases.] That is,

$$\left|\frac{dV'_{BE}(y)}{dy}\right| \leq R_{Sbi}(J_C \to 0)J_B(0) \int_y^W dy'$$
$$= R_{Sbi}(J_C \to 0)J_B(0)(W - y). \tag{9.40}$$

Integrating Eq. (9.40) gives an upper-bound estimate of the maximum *IR* drop along the intrinsic-base layer due to the lateral flow of base current. This estimate is

$$V'_{BE}(0) - V'_{BE}(W) < R_{Sbi}(J_C \to 0)J_B(0)\frac{W^2}{2}. \tag{9.41}$$

Consider a typical modern vertical bipolar transistor. To avoid speed degradation at high currents, the collector current density is typically under $1\,\text{mA}/\mu\text{m}^2$. For a typical current gain of 100, J_B is therefore typically under $0.01\,\text{mA}/\mu\text{m}^2$. The low-current value of R_{Sbi} is typically $10\,\text{k}\Omega/\square$. If the transistor has an emitter-stripe width of 0.5 μm, with base contact on both sides of the stripe, then $W = 0.25$ μm. For this typical transistor, Eq. (9.41) suggests that the maximum lateral IR drop in the intrinsic base is less than 3 mV, much smaller than kT/q, which is about 26 mV at room temperature. For a symmetric lateral bipolar transistor on SOI, the reported J_B can be high, around 0.8 mA/μm² for a device with 60 nm silicon thickness and 200 nm emitter length (Cai *et al.*, 2014). Assuming the same base sheet resistivity as the vertical bipolar transistor, the maximum lateral IR drop in the intrinsic base from Eq. (9.41) is about 15 mV. Therefore, the ***emitter current-crowding effect is negligible for modern bipolar transistors***. Of course, for transistors of earlier generations, where the emitter stripes were much wider, emitter current crowding could be significant. Our estimates are consistent with the results obtained from a more exact calculation of the internal base voltage distribution (Chiu *et al.*, 1992).

9.2.2   Collector Current

With emitter current-crowding effect being negligible, the collector current I_C is simply $J_C \times A_E$, where J_C is the collector current density and A_E is the base–emitter junction area. Referring to the current flows in the n–p–n transistor depicted in Figure 9.2(c), the quasineutral base region extends from $x = 0$ to $x = W_B$. With negligible recombination in the thin base, J_n in the base is independent of x. Therefore, Eq. (9.25) can be integrated to give

$$J_n \int_0^{W_B} \frac{p_p(x)}{qD_{nB}(x)n_{ieB}^2(x)} dx = \left.\frac{n_p p_p}{n_{ieB}^2}\right|_{x=W_B} - \left.\frac{n_p p_p}{n_{ieB}^2}\right|_{x=0}, \tag{9.42}$$

where we have added a subscript B to denote the parameters in the base region. The first term on the right-hand side is negligible compared to the second term, so Eq. (9.42) can be rewritten as

$$J_n \int_0^{W_B} \frac{p_p(x)}{qD_{nB}(x)n_{ieB}^2(x)} dx = -\frac{n_p(0)p_p(0)}{n_{ieB}^2(0)}. \tag{9.43}$$

Notice that J_n is negative, due to the fact that electrons flowing in the x-direction give rise to a negative current. In the equivalent-circuit model of a bipolar transistor [see Section 9.6], the emitter current I_E, base current I_B, and collector current I_C are all currents flowing ***into*** the transistor, with $I_E + I_B + I_C = 0$ as required by Kirchhoff's current law. The measured collector current density J_C for an n–p–n transistor is equal to $-J_n$ and is positive.

From Eqs. (3.55) and (9.15), we have

$$p_p(0)n_p(0) = n_{ieB}^2(0)\exp(qV_{BE}/kT), \tag{9.44}$$

where V_{BE} is the base–emitter junction voltage. [Again, for simplicity of writing, we are using V_{BE} to mean the junction voltage V'_{BE}.] Therefore, J_C for an n–p–n transistor is

$$J_C = \frac{q\exp(qV_{BE}/kT)}{\int_0^{W_B} \left[p_p(x)/D_{nB}(x)n_{ieB}^2(x)\right]dx}. \tag{9.45}$$

Equation (9.45) *is valid for all minority-carrier injection levels*, since no assumption of minority-carrier injection level has been made in its derivation. That is, I_C for an n–p–n bipolar transistor operating in *forward-active mode* (i.e., base–emitter diode forward biased and base–collector diode zero biased or reverse biased) can be written in the general form

$$I_C = A_E J_{C0}(V_{BE})\exp(qV_{BE}/kT), \tag{9.46}$$

with

$$J_{C0}(V_{BE}) = \frac{q}{\int_0^{W_B} \left[p_p(x)/D_{nB}(x)n_{ieB}^2(x)\right]dx}, \tag{9.47}$$

where J_{C0} has implicit and complex dependence on V_{BE}. To evaluate the integral in Eq. (9.47), we need to use the fact that $p_p = (p_{p0} + \Delta p_p)$ and $\Delta n_p (= \Delta p_p)$ given by Eq. (3.60) to determine $p_p(x)$.

It should be noted that *the collector current is a function of the base-region parameters only, and is independent of the properties of the emitter*. All the effects in the base region, such as bandgap narrowing, bandgap nonuniformity, and dopant distribution, are contained in the integral in Eq. (9.47).

The integral in Eq. (9.47) can be evaluated for a modern bipolar transistor having thin and uniformly doped base, where D_{nB} and n_{ieB} are independent of x. In this case, from Eq. (3.38) with $L_{nB}/W_B \gg 1$, the minority-electron density in the thin base has the form

$$\Delta n_p(x) = \Delta n_p(0)\left(1 - \frac{x}{W_B}\right). \tag{9.48}$$

and Eq. (3.60) gives

$$\Delta n_p(0) = \frac{N_B}{2}\left(\sqrt{1 + \frac{4n_{ieB}^2 \exp(qV_{BE}/kT)}{N_B^2}} - 1\right), \tag{9.49}$$

where N_B is the uniform base doping concentration. Using $p_p(x) = p_{p0} + \Delta p_p(x)$, we have

$$p_p(x) = N_B + \frac{N_B}{2}\left(1 - \frac{x}{W_B}\right)\left(\sqrt{1 + \frac{4n_{ieB}^2 \exp(qV_{BE}/kT)}{N_B^2}} - 1\right), \tag{9.50}$$

$$\int_0^{W_B} p_p(x)dx = N_B W_B + \frac{N_B W_B}{4}\left(\sqrt{1 + \frac{4n_{ieB}^2 \exp(qV_{BE}/kT)}{N_B^2}} - 1\right), \tag{9.51}$$

and

$$J_{C0}(V_{BE}) = \frac{qD_{nB}n_{ieB}^2}{N_B W_B}\left[1 + \frac{1}{4}\left(\sqrt{1 + \frac{4n_{ieB}^2\exp(qV_{BE}/kT)}{N_B^2}} - 1\right)\right]^{-1}, \quad (9.52)$$

for *uniformly doped thin base*.

9.2.2.1 Collector-Current Equation for a Symmetric Lateral Bipolar Transistor

The assumptions of thin and uniformly doped base fits exactly the description of the symmetric lateral bipolar transistors shown in Figure 9.1(b). Therefore, Eqs. (9.46) and (9.52) give the collector current for a symmetric lateral n−p−n transistor on SOI as

$$I_C(V_{BE}) = \frac{A_E q D_{nB}n_{ieB}^2\exp(qV_{BE}/kT)}{N_B W_B}\left[1 + \frac{1}{4}\left(\sqrt{1 + \frac{4n_{ieB}^2\exp(qV_{BE}/kT)}{N_B^2}} - 1\right)\right]^{-1}.$$

$$(9.53)$$

At V_{BE} so large that the second term inside the square-root becomes much larger than unity, Eq. (9.53) reduces to

$$I_C(V_{BE}) = \frac{A_E 2qD_{nB}n_{ieB}\exp(qV_{BE}/2kT)}{W_B} \quad \text{(large } V_{BE} \text{ limit)}. \quad (9.54)$$

Equation (9.53) describes very well the measured collector current of symmetric lateral transistors for V_{BE}' up to about 1.1 V, corresponding to a base−emitter terminal voltage of 1.5 V (Cai *et al.*, 2014).

9.2.2.2 Collector-Current Equation for a Vertical Bipolar Transistor

As illustrated in Figure 9.3, the collector doping concentration N_C of a vertical bipolar transistor is typically about 10× smaller than the peak doping concentration of the base. Also, as shown in Section 9.3.3, the density of minority carriers entering the base−collector junction space-charge region should not be higher than N_C if speed degradation with further current increase is to be avoided. That is, *a vertical bipolar transistor is normally operated at low injection of minority carriers in the base*, with $\Delta n_p \ll N_B$ for an n−p−n. Closed-form solutions can be obtained for the collector current if we limit ourselves to low injection and make the appropriate approximations first.

When low-injection approximation applies, $\Delta p_p \ll p_{p0}$, we have $p_p(x) \approx p_{p0}(x)$ and Eq. (9.47) becomes

$$J_{C0}(V_{BE}) = \frac{q}{\displaystyle\int_0^{W_B}\left[p_{p0}(x)/D_{nB}(x)n_{ieB}^2(x)\right]dx}. \quad (9.55)$$

The collector current is

$$I_C(V_{BE}) = \frac{A_E q\exp(qV_{BE}/kT)}{\displaystyle\int_0^{W_B}\left[p_{p0}(x)/D_{nB}(x)n_{ieB}^2(x)\right]dx} \quad \text{(low injection)}. \quad (9.56)$$

For a transistor with uniform base doping concentration N_B, Eq. (9.56) reduces to

$$I_C(V_{BE}) = \frac{A_E q D_{nB} n_{ieB}^2 \exp(qV_{BE}/kT)}{N_B W_B} \quad \text{(low injection, uniform doping).} \quad (9.57)$$

Equation (9.57) can also be obtained from Eq. (9.53), as expected, by noting that the second term inside the square-root is small compared to unity at low injection. [In the literature, the *base Gummel number* is often defined as the total integrated base dose, which is the product $N_B W_B$ for a uniformly doped base (Gummel, 1961).]

9.2.3   Base Current

There are ***three components in the base current*** of a bipolar transistor. For an n–p–n transistor, they are: (i) the holes injected from the base into the emitter, (ii) the holes that recombine in the quasineutral base region with the minority electrons injected from the emitter, and (iii) the holes that recombine with electrons within the base–emitter space-charge region.

The base–emitter space-charge-region current has been covered in Section 3.1.6. It is recognizable by its $\exp(qV'_{BE}/2kT)$ dependence. For silicon bipolar transistors, this space-charge-region current component is usually quite small by the time the fabrication process is well developed, noticeable only at small V_{BE} on an expanded Gummel plot.

The current due to holes recombining in the quasineutral base with injected electrons has the same V'_{BE} dependence as the collector current. For a silicon bipolar transistor, this recombination component is negligible compared with component (i) due to holes injected into the emitter (Yau *et al.*, 2016a). However, for bipolar transistor made of direct-gap materials where minority-carrier lifetimes are short compared with silicon, this may not be the case. It is left as an exercise (Exercise 9.9) for the reader to derive the equation for this recombination base-current component and verify that it is indeed negligible for a silicon bipolar transistor.

Therefore, in considering the base current of a silicon bipolar transistor, we shall ***assume the base current to be due only to minority carriers injected from the base into the emitter***. Referring to Figure 9.2(c), the emitter quasineutral region extends from $x = 0$ to $x = -W_E$. Since the emitter is usually so heavily doped that its electron concentration is not affected by the injected hole-current level, it is a good approximation to assume $n_n \approx n_{n0} = N_E$, where N_E is the emitter doping concentration. That is, ***low-injection approximation is always valid for base current***. The hole current density in the emitter, i.e., Eq. (9.26), can be rewritten as

$$J_p(x) = -qD_{pE} \frac{n_{ieE}^2}{n_{n0}} \frac{d}{dx} \left(\frac{n_{n0} p_n}{n_{ieE}^2} \right), \quad (9.58)$$

where we have added a subscript E to indicate the parameters of the emitter region. $|J_p(0)|$ is equal to the base current density. It should be noted that ***the base current is a function of the emitter-region parameters only and is independent of the properties*** *of the base region*. Thus, the base current changes as the emitter structure and design are changed. In this subsection, we shall use Eq. (9.58) to derive the base current in terms of the more familiar emitter parameters.

9.2.3.1 Shallow or Transparent Emitter

An emitter is considered *shallow* or *transparent* when its width W_E is small compared to its minority-carrier diffusion length. For a shallow emitter, there is negligible recombination in the emitter region except at the emitter contact, and the minority-carrier current density in the emitter is independent of x.

For an n–p–n transistor, the hole current density at the emitter contact is usually written in terms of the surface recombination velocity for holes, S_p, defined by

$$J_p(x = -W_E) \equiv -q(p_n - p_{n0})_{x=-W_E} S_p. \tag{9.59}$$

Notice that J_p is negative due to holes flowing in the $-x$ direction. I_B of an n–p–n transistor is positive. Since J_p is independent of x for a transparent emitter, Eq. (9.58) can be rearranged and integrated to give

$$J_p \int_{-W_E}^{0} \frac{n_{n0}}{qD_{pE}n_{ieE}^2} dx = - \left.\frac{n_{n0}p_n}{n_{ieE}^2}\right|_{x=0} + \left.\frac{n_{n0}p_n}{n_{ieE}^2}\right|_{x=-W_E}. \tag{9.60}$$

At $x = 0$, the hole concentration can be inferred from Eq. (3.61), namely

$$p_n(0) = p_{n0}(0) \exp(qV_{BE}/kT). \tag{9.61}$$

Again, for simplicity, we are using V_{BE} to mean junction voltage. Substituting Eqs. (9.59) and (9.61) into Eq. (9.60), and using the relation $n_{ieE}^2 = p_{n0}n_{n0}$, we have

$$J_p \int_{-W_E}^{0} \frac{n_{n0}}{qD_{pE}n_{ieE}^2} dx = - \exp(qV_{BE}/kT) + 1 - \frac{n_{n0}(-W_E)}{n_{ieE}^2(-W_E)qS_p} J_p$$

$$\approx - \exp(qV_{BE}/kT) - \frac{n_{n0}(-W_E)}{n_{ieE}^2(-W_E)qS_p} J_p, \tag{9.62}$$

which gives

$$J_p = \frac{-q \exp(qV_{BE}/kT)}{\displaystyle\int_{-W_E}^{0} \frac{n_{n0}}{D_{pE}n_{ieE}^2} dx + \frac{n_{n0}(-W_E)}{n_{ieE}^2(-W_E)S_p}}. \tag{9.63}$$

Equation (9.63) is valid for a transparent emitter of arbitrary doping profile and arbitrary surface recombination velocity at the emitter contact (Shibib *et al.*, 1979). The base current is therefore

$$I_B = A_E|J_p| = \frac{qA_E \exp(qV_{BE}/kT)}{\displaystyle\int_{-W_E}^{0} \frac{N_E}{D_{pE}n_{ieE}^2} dx + \frac{N_E(-W_E)}{n_{ieE}^2(-W_E)S_p}}. \tag{9.64}$$

The base current is often written in the form

$$I_B = A_E J_{B0} \exp(qV_{BE}/kT)$$
$$= A_E \frac{qn_i^2}{G_E} \exp(qV_{BE}/kT), \tag{9.65}$$

where J_{B0} is the *saturated base current density*, and G_E is the *emitter Gummel number* (de Graaff *et al.*, 1977). For a shallow or transparent emitter, Eq. (9.64) gives

$$J_{B0} = \frac{q}{\displaystyle\int_{-W_E}^{0} \frac{N_E}{D_{pE}n_{ieE}^2}\,dx + \frac{N_E(-W_E)}{n_{ieE}^2(-W_E)S_p}} \tag{9.66}$$

and

$$G_E = \int_{-W_E}^{0} \frac{n_i^2}{n_{ieE}^2}\frac{N_E}{D_{pE}}\,dx + \frac{n_i^2 N_E(-W_E)}{n_{ieE}^2(-W_E)S_p}. \tag{9.67}$$

This base current, due to minority carriers injected from base region into the emitter region, is often referred as the *ideal base current* because of its $\exp(qV'_{BE}/kT)$ dependence. ***Observed large deviations of the measured base current from ideal behavior is usually caused by the space-charge-region current component being out of control due to processing issues,*** as clearly demonstrated in a paper by Yau *et al.* (2016a).

- *Transparent emitter with uniform doping concentration and uniform energy bandgap.* Let us consider an n–p–n bipolar transistor with an emitter doping profile, as indicated in Figure 9.3(a). The emitter doping profile is not really uniform or boxlike. Even if we assume the most heavily doped region to be uniform, there is still a transition region where the emitter doping concentration drops from about 10^{20} to about 10^{18} cm^{-3} at the base–emitter junction. This transition region plays an important role in determining the base–emitter junction capacitance and the base–emitter junction breakdown voltage. However, as far as the base current is concerned, the effect of this transition region is relatively small (Roulston, 1990). This is due to the fact that the hole diffusion length in this relatively lightly doped transition region is very large compared to the thickness of the region. As a result, the transition region is almost completely transparent to the holes entering the emitter. Therefore, at least for purposes of modeling the base current, it is common to ignore this transition region and simply assume the emitter region to be uniformly doped and boxlike. Besides, such an approximation makes modeling the emitter region relatively simple. For a uniformly doped transparent emitter with uniform energy bandgap, Eq. (9.66) reduces to

$$J_{B0} = \frac{qD_{pE}n_{ieE}^2}{N_E W_E(1 + D_{pE}/W_E S_p)}, \tag{9.68}$$

and Eq. (9.67) reduces to

$$G_E = N_E \left(\frac{n_i^2}{n_{ieE}^2}\right)\left(\frac{W_E}{D_{pE}} + \frac{1}{S_p}\right). \tag{9.69}$$

For an ohmic emitter contact, S_p is infinite, and Eq. (9.68) becomes proportional to $1/W_E$. The base current increases rapidly as the emitter width, or depth, is reduced.

- *Polysilicon emitter.* The simplest model for describing a polysilicon emitter is to treat the polysilicon–silicon interface located at $x = -W_E$ as a contact with finite surface recombination velocity. In this case, Eq. (9.66) or Eq. (9.68) can be used, depending on whether the single-crystal emitter region is uniformly doped or not. Under certain conditions, a model for the polysilicon emitter can be developed which allows the surface recombination to be evaluated in terms of the properties of the polysilicon layer (Exercise 9.3). In practice, the surface recombination velocity is often used as a fitting parameter to the measured base current. The detailed physics of transport in a polysilicon emitter is very complex and is dependent on the polysilicon-emitter fabrication process. Therefore, the surface recombination velocity obtained by fitting to the measured base current is also dependent on the polysilicon-emitter fabrication process. The reader is referred to the vast published literature on polysilicon-emitter physics and technology (Ashburn, 1988; Kapoor and Roulston, 1989).

9.2.3.2 Deep Emitter with Uniform Doping Concentration and Uniform Energy Bandgap

An emitter is *deep*, or *nontransparent*, when its width is large compared to its minority-carrier diffusion length. For a deep emitter, most or all of the injected minority carriers recombine before they reach the emitter contact. For an n–p–n transistor, minority-hole current density given by Eq. (9.58) becomes rather simple if the emitter is assumed to be uniformly doped, has a uniform energy bandgap, and has an ohmic contact at $x = -W_E$. With these assumptions, Eq. (9.58) reduces to

$$J_p(x) = -qD_p \frac{dp_n}{dx}, \tag{9.70}$$

which is simply the hole diffusion current density in the uniformly doped n-side of a diode. The base current, therefore, can be obtained from the hole equivalent of Eq. (3.66), namely

$$I_B = \frac{qA_E D_{pE} n_{ieE}^2 \exp(qV_{BE}/kT)}{N_E L_{pE} \tanh(W_E/L_{pE})}, \tag{9.71}$$

where N_E is the emitter doping concentration, and L_{pE} is the hole diffusion length in the emitter. The corresponding base saturation current density and emitter Gummel number are

$$J_{B0} = \frac{qD_{pE} n_{ieE}^2}{N_E L_{pE} \tanh(W_E/L_{pE})}, \tag{9.72}$$

and

$$G_E = \left(\frac{n_i^2}{n_{ieE}^2}\right)\frac{N_E L_{pE} \tanh(W_E/L_{pE})}{D_{pE}}. \tag{9.73}$$

9.2.4   Current Gains

The static *common-emitter* current gain β_0 is defined by

$$\beta_0 = \frac{I_C}{I_B} = \frac{J_{C0}}{J_{B0}}. \tag{9.74}$$

As discussed in Section 9.2.3, due to the high emitter doping concentration, J_{B0} is derived using low-injection approximation. Also, base-emitter space-charge-region currents were ignored. As a result, J_{B0} is independent of V_{BE}. However, as shown in Eq. (9.47) and the associated discussion, J_{C0} is a function of V_{BE}. J_{C0} is independent of V_{BE} only where low-injection remains valid. Figure 9.6 is a plot of β_0 as a function of V_{BE}, normalized to its value at $V_{BE} = 0$, calculated using Eq. (9.52), which is valid for all injection levels, for J_{C0}. It shows that β_0 is relatively independent of V_{BE} until high-injection effect in the base sets in. As expected, the higher the base doping concentration, the higher is the V_{BE} value when transition to high injection starts. Note that since J_{B0} is independent of V_{BE}, Figure 9.6 *also represents the falloff of J_{C0} at large V_{BE} caused by high-injection effect in the base*.

In designing a bipolar transistor, one of the key objectives is to have adequate current gain in the flat region of Figure 9.6. This is the region of low injection in the base. Therefore, we focus on this region and *assume low-injection approximation for the collector current* in discussing current gain in the remainder of this section. In the flat region in Figure 9.6, both I_C and I_B vary as $\exp(qV_{BE}/kT)$.

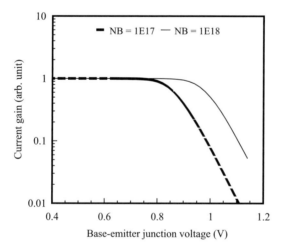

Figure 9.6 Calculated dependence of β_0 as a function of its base–emitter junction voltage. The falloff in β_0 at large voltage is due to a high-injection effect in the base. *Base current due to base-emitter space-charge-region currents are not included*. Inclusion of a space-charge region component would show up as the current gain rising from a very low level, at V_{BE} values to the left of the plot and varying as $\exp(qV_{BE}/2kT)$, to the flat region in the plot

From Eqs. (9.55) and (9.66),

$$\beta_0 = \frac{\displaystyle\int_{-W_E}^{0} \frac{N_E}{D_{pE}n_{ieE}^2} dx + \frac{N_E(-W_E)}{n_{ieE}^2(-W_E)S_p}}{\displaystyle\int_{0}^{W_B} \left(p_{p0}/D_{nB}n_{ieB}^2\right) dx} \qquad \text{(transparent emitter)}, \qquad (9.75)$$

and from Eqs. (9.55) and (9.72),

$$\beta_0 = \frac{N_E L_{pE} \tanh\left(W_E/L_{pE}\right)}{D_{pE}n_{ieE}^2 \displaystyle\int_{0}^{W_B} \left(p_{p0}/D_{nB}n_{ieB}^2\right) dx} \qquad \text{(deep emitter)}. \qquad (9.76)$$

The static *common-base current gain* α_0 is defined by

$$\alpha_0 \equiv \frac{I_C}{-I_E}, \qquad (9.77)$$

where I_E is the emitter current. Here we have defined I_E as the current flowing *into* the emitter, so that $-I_E$ is positive for an n–p–n transistor. Since $I_E + I_B + I_C = 0$, we have

$$\alpha_0 = \frac{\beta_0}{1 + \beta_0}, \qquad (9.78)$$

and

$$\beta_0 = \frac{\alpha_0}{1 - \alpha_0}. \qquad (9.79)$$

In principle, either α_0 or β_0 can be used to describe the current gain of a bipolar transistor. In practice, β_0 is often used in discussing the device characteristics, device design, and device physics. Throughout this book, we shall use β_0. (However, we shall use α_0 when we consider breakdown voltages in Section 9.7.)

β_0 is often quoted as a figure of merit for a bipolar transistor. However, it should be noted that, being the ratio of two currents, β_0 changes as either one of the currents changes. To really understand the device design and the device fabrication process, both I_C and I_B, not just β_0, should be examined.

For digital logic circuits, the circuit speed is insensitive to the current gain of the transistors (Ning *et al.*, 1981). However, for many analog circuits, a high current gain is desirable. Most transistors are designed with a current-gain target of about 100. For a given bipolar transistor fabrication process, the current gain can be increased or decreased readily by changing the base Gummel number, or the base parameters. Design considerations for the base region will be covered in Section 10.2.

For the special case of a uniformly doped emitter with $W_E/L_{pE} \gg 1$, and a uniformly doped base with concentration N_B, Eq. (9.76) reduces to

$$\beta_0 = \frac{n_{ieB}^2}{n_{ieE}^2} \frac{D_{nB}}{D_{pE}} \frac{N_E L_{pE}}{N_B W_B} \qquad \text{(uniformly doped emitter and base)}. \qquad (9.80)$$

It is instructive to estimate the magnitude of the current gain given by Eq. (9.80). If we assume $N_E = 1 \times 10^{20}\,\text{cm}^{-3}$, $N_B = 1 \times 10^{18}\,\text{cm}^{-3}$, and $W_B = 0.1\,\mu\text{m}$ for a typical deep-emitter thin-base vertical n–p–n transistor, then Figure 9.4 gives $(n_{ieB}/n_{ieE})^2 = \exp[(\Delta E_{gB} - \Delta E_{gE})/kT] \approx 0.19$ at room temperature, Figure 3.14(a) gives $D_{nB}/D_{pE} = \mu_{nB}/\mu_{pE} \approx 2.6$, $N_E/N_B = 100$, and Figure 3.14(c) gives $L_{pE}/W_B \approx 4.6$. Substituting these values into Eq. (9.80) gives $\beta_0 = 230$.

9.2.5 Ideal I_C–V_{CE} Characteristics

Figure 9.7 illustrates the ideal I_C versus V_{CE} characteristics of an n–p–n transistor, with I_B as a parameter. Each base current corresponds to a given V_{BE} value. An $I_C - V_{CE}$ plot is also referred to as the *output characteristics* of a bipolar transistor.

For $V_{CE} > V_{BE}$, the base–collector diode is reverse biased and the transistor is said to be in its normal forward-active mode of operation. All the electrons injected from the emitter into the base are collected by the collector and there is no electron injection from the collector into the base. Thus, for an ideal transistor, I_C is constant, independent of V_{CE}. The current gain, I_C/I_B, is also constant (the flat part of Figure 9.6), and the constant-I_B curves are spaced equally apart by an amount determined by the base-current step, as indicated in Figure 9.7. [However, as indicated in Figure 9.12, I_C of a real transistor, instead of being constant when $V_{CE} > V_{BE}$, keeps rising with V_{CE} for a given I_B in this region.]

For $V_{CE} < V_{BE}$, the base–collector diode is forward biased and the transistor is said to be in *saturation*. In this case, I_C is the difference of the electron current injected from the emitter into the base and the electron current injected from the collector into the base. As a result, for a given constant-I_B curve, current gain is smaller for

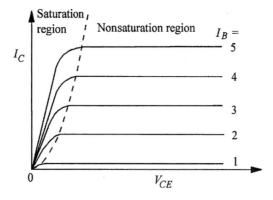

Figure 9.7 Schematic illustration of the ideal I_C versus V_{CE} characteristics of an n–p–n transistor. The dashed line marks the locus of $V_{CE} = V_{BE}$. The region to the left of the dashed curve is the saturation region. The region to the right of the curve is the nonsaturation region. It is also referred to as the forward-active region

$V_{CE} < V_{BE}$ than for $V_{CE} > V_{BE}$, decreasing as V_{CE} decreases (V_{BC} increases as V_{CE} decreases at fixed V_{BE} or fixed I_B). It can be shown (see Exercise 12.1) that the current gain becomes less than unity as V_{CE} approaches zero.

Let us consider a linear I_C versus I_B plot for a transistor connected with a resistor between its collector and a power supply (e.g., in the inverter in Figure 11.11). As we increase I_B by increasing V_{BE}, I_C increases linearly with I_B as $I_B\beta_0$, where β_0 is the current gain in the nonsaturation region, while V_{CE} decreases due to the *IR* drop across the collector resistor. At sufficiently large I_B, where $V_{CE} < V_{BE}$, I_C appears to saturate on the I_C versus I_B plot due to current gain dropping rapidly with decrease in V_{CE}. This explains the labels, saturation, and nonsaturation, for the two region in Figure 9.7.

Digital bipolar circuits are generally divided into two categories: saturated circuits and nonsaturated circuits. A circuit is said to be saturated if, during circuit operation, the bipolar transistor stays in the saturation region or switches between the saturation and nonsaturation regions. Examples of saturated circuits are the basic bipolar inverter (shown in Figure 11.11) and TTL (transistor-transistor logic) circuits. A circuit is said to be nonsaturated if, during circuit operation, the bipolar transistor stays in the nonsaturated region only. The best known nonsaturated circuit is the ECL (emitter-coupled logic) circuit (discussed in Section 11.2).

9.3   Measured Characteristics of Typical n–p–n Transistors

Measured current–voltage characteristics of typical bipolar devices are not ideal. That is, Gummel plots of measured I_C and I_B may deviate from 60 mV/decade at room temperature, and measured output characteristics may not be constant for $V_{CE} > V_{BE}$. The degree of deviation from ideal characteristics depends on the device structure, device design, device fabrication process, and the bias condition of the transistor. In this section, we examine the measured characteristics of typical n–p–n transistors and discuss the physical mechanisms governing them.

Figure 9.8(a) shows the Gummel plots for a typical vertical n–p–n transistor. The theoretical ideal base and collector currents, discussed in Section 9.2, are indicated by the dashed lines. Figure 9.8(a) shows that the measured I_C is ideal except at large V_{BE}, while the measured I_B is ideal except at small and at large V_{BE}. Figure 9.8(b) shows the Gummel plots for a typical symmetric lateral n–p–n transistor on SOI. They show qualitatively the same characteristics as the vertical transistor in (a).

Figure 9.9 is a schematic of the typical measured current gain, I_C/I_B, as a function of collector current. For the voltage range where both the base and the collector currents are approximately ideal, the current gain is approximately constant, corresponding to the flat part of Figure 9.6. At low currents, the current gain is less than its ideal value because the base current has a non-zero base–emitter space-charge-region current component, as indicated in the measured I_B in Figure 9.8. The roll off at high collector currents is due to a combination of high-injection effect in the base, corresponding to the roll off in Figure 9.6, and widening of the quasineutral base at high currents, which will be discussed in Section 9.3.3.

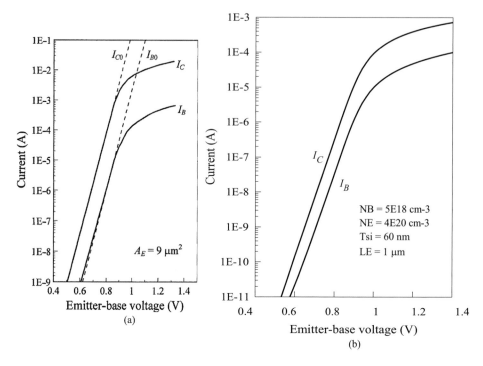

Figure 9.8 (a) Gummel plots for a typical vertical n–p–n bipolar transistor. The dashed lines represent the theoretical ideal base and collector currents (after Ning and Tang, 1984). (b) Gummel plots for a typical symmetric lateral n–p–n bipolar transistor on SOI. Notice that the measured base currents in both (a) and (b) are only slightly larger than the ideal base currents, indicating low base–emitter space-charge-region current from good quality base–emitter diodes

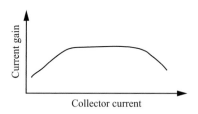

Figure 9.9 Schematic illustration of the current gain I_C/I_B as a function of collector current for a typical bipolar transistor

9.3.1   Effect of Emitter and Base Series Resistances

Figure 9.10 shows schematically the physical origins of the parasitic resistances in a typical vertical n–p–n transistor fabricated using p–n junction isolation. These resistances were ignored in Section 9.2 in the description of the ideal current–voltage characteristics. As currents flow through these parasitic resistors, voltage drops are developed, offsetting the externally applied voltages.

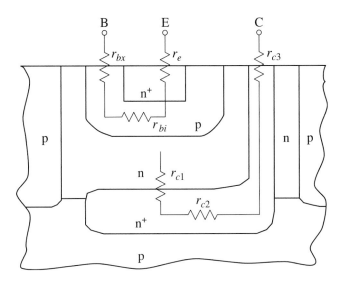

Figure 9.10 Schematic illustrating the parasitic resistances in a typical vertical n–p–n transistor

In most vertical bipolar circuits, particularly those designed for high-speed applications, the base–collector junction is designed to remain reverse biased at all times, even at high currents. From a device-structure point of view, this is accomplished by employing a heavily doped subcollector layer (to reduce r_{c2}) and a heavily doped reach-through (to reduce r_{c3}) to bring the collector contact to the surface. With the base–collector junction reverse biased, to first order, the collector resistance components shown in Figure 9.10 have no effect on the current flows in the base–emitter diode, and hence no effect on the collector current. (The effect of base–collector voltage on collector current is discussed in Section 9.3.2.)

The emitter series resistance r_e is determined primarily by the emitter contact resistance, since the resistance associated with the thin n^+ emitter region is small. The base resistance r_b can be separated into two components: the intrinsic-base resistance r_{bi}, which is determined by the design of the intrinsic-base region, and the extrinsic-base resistance r_{bx}, which includes all other resistances associated with the base terminal.

The base–emitter voltage drop due to the flow of emitter and base currents is

$$\Delta V_{BE} = -I_E r_e + I_B r_b$$
$$= I_C r_e + I_B (r_e + r_b), \tag{9.81}$$

where we have used the fact that $I_E + I_B + I_C = 0$, and $-I_E$ is a positive quantity. The relation between the base-emitter terminal voltage V_{BE} and the junction voltage V'_{BE} is

$$V'_{BE} = V_{BE} - \Delta V_{BE}. \tag{9.82}$$

To include the effect of the emitter and base series resistances, all bipolar current equations should be modified by replacing V_{BE} by V'_{BE}. This results in both the measured I_C and I_B, when plotted as a function of V_{BE}, being significantly smaller than the ideal currents at large V_{BE}. This is illustrated in Figure 9.8.

As can be seen from Eq. (9.47), J_{C0} is a function of the majority-carrier concentration in the base and the base width. Therefore, *the measured collector current is a function of ΔV_{BE} as well as a function of the base majority-carrier concentration and the base width, which in turn depend on V_{BE}*.

On the other hand, as can be seen from Eqs. (9.66) and (9.72), J_{B0} is independent of V_{BE}. Therefore, at high currents, *deviation of the base current from $\exp(qV_{BE}/kT)$ behavior is due to ΔV_{BE} alone* (Ning and Tang, 1984). The relation between the ideal base current I_{B0} and the measured base current I_B is therefore

$$I_{B0} = I_B \exp(q\Delta V_{BE}/kT). \tag{9.83}$$

Using Eq. (9.81), Eq. (9.83) can be rearranged to give

$$\frac{kT}{qI_C} \ln\left(\frac{I_{B0}}{I_B}\right) = \left(r_e + \frac{r_{bi}}{\beta_0}\right) + \frac{r_e + r_{bx}}{\beta_0}, \tag{9.84}$$

where $\beta_0 = I_C/I_B$ is the measured static common-emitter current gain. For a typical bipolar transistor, r_e and r_{bx} are constants, independent of V_{BE}. Also, it is shown in Section 10.2.1 that J_C is proportional to the intrinsic-base sheet resistivity. Therefore, r_{bi} is proportional to β_0, and the ratio r_{bi}/β_0 is constant, independent of current. Thus, if $(kT/qI_C) \ln(I_{B0}/I_B)$ is plotted as a function of $1/\beta_0$, the slope gives $r_e + r_{bx}$ and the intercept gives $r_e + r_{bi}/\beta_0$. This is illustrated in Figure 9.11.

The ratio r_{bi}/β_0 can be obtained at low currents where the individual values of r_{bi} and β_0 are relatively independent of current (see Figure 9.9). β_0 is directly measurable from I_C and I_B, and r_{bi} can be calculated as described in Section 9.2.1, or measured using specially designed test structures (Weng et al., 1992). Once the ratio r_{bi}/β_0 is determined, r_e and r_{bx} can be extracted from the intercept and the slope of the plot in

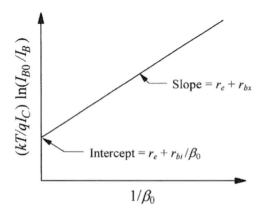

Figure 9.11 Schematic illustrating the determination of the emitter and base series resistances from Eq. (9.84)

Figure 9.11. Thus, all three resistance components can be obtained. This method was demonstrated to work well for vertical bipolar transistors (Ning and Tang, 1984). However, there are also reports of the method not yielding satisfactory results (Riccó *et al.*, 1984; Yu *et al.*, 1984; Dubois *et al.*, 1994).

A commonly used method for measuring r_e of a transistor is by measuring its open-circuit V_{CE} as a function of base current, i.e., from the relation $V_{CE}(I_C = 0) = (kT/q)\ln(1/\alpha_R) + r_e I_B$ (Ebers and Moll, 1954; Filensky and Beneking, 1981). The slope in a $V_{CE}(I_C = 0)$ versus I_B plot gives r_e. The reader is referred to Exercise 9.5 for a derivation of the method.

9.3.2 Effect of Base–Collector Voltage on Collector Current

In modern high-speed transistors where the base width is very small, the measured collector current, and hence the measured current gain, increases as the base–collector reverse-bias voltage is increased. This is due to two effects, or a combination of them. The first effect is the dependence of the quasineutral base width on V_{BC}. The second effect is avalanche multiplication in the base–collector junction. We shall discuss these two effects individually.

9.3.2.1 Modulation of Quasineutral Base Width by V_{BC}

As the reverse bias across the base–collector junction is increased, W_{dBC} increases, and hence the quasineutral base width W_B decreases. This in turn causes the collector current to increase, as can be seen from Eq. (9.47). Thus, instead of as illustrated in Figure 9.7, where I_C is independent of V_{CE} for $V_{CE} > V_{BE}$, the measured I_C of a typical bipolar transistor increases with V_{CE}, as illustrated in Figure 9.12.

- *Early voltage.* For circuit modeling purposes, I_C in the nonsaturation region is often assumed to depend linearly on V_{CE}. The collector voltage at which the linearly extrapolated I_C reaches zero is denoted by $-V_A$. As we shall show later, it is a good

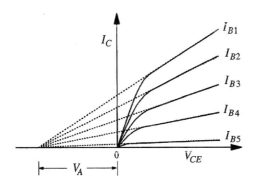

Figure 9.12 Schematic illustrating the approximately linear dependence of I_C on V_{CE}. The linearly extrapolated I_C intersects the V_{CE}-axis at $-V_A$

and useful approximation to assume V_A to be independent of I_B, i.e., independent of V_{BE}. This is illustrated in Figure 9.12. V_A is called the *Early voltage* (Early, 1952). It is defined by

$$V_A + V_{CE} \equiv I_C \left(\frac{\partial I_C}{\partial V_{CE}} \right)^{-1}. \tag{9.85}$$

In practice, except for transistors that tend to punch through (to be discussed in the next bullet point, entitled emitter–collector punch-through), V_A is much larger than the operation range of V_{CE}. Therefore, *the V_{CE} term on the left-hand side of* Eq. (9.85) *is often dropped as an approximation.*

The equations for I_C, i.e., Eqs. (9.46) and (9.47), can be written as

$$I_C = A_E J_{C0} \exp(qV_{BE}/kT) = \frac{qA_E \exp(qV_{BE}/kT)}{F(W_B)}, \tag{9.86}$$

where, for convenience, a function F has been introduced (Kroemer, 1985) which is defined by

$$F(W_B) \equiv \frac{q}{J_{C0}} = \int_0^{W_B} \frac{p_p(x)}{D_{nB}(x)n_{ieB}^2(x)} dx. \tag{9.87}$$

The majority-carrier hole charge per unit area in the base is

$$Q_{pB} = q \int_0^{W_B} p_p(x)dx. \tag{9.88}$$

Since V_{BE} is fixed for a given I_B, and $V_{CE} = V_{CB} + V_{BE}$, we have $dV_{CE} = dV_{CB}$. Therefore, Eq. (9.85) can be rewritten as

$$V_A + V_{CE} = I_C \left(\frac{-I_C}{F} \frac{\partial F}{\partial V_{CE}} \right)^{-1}$$

$$= \left(\frac{-1}{F} \frac{\partial F}{\partial W_B} \frac{\partial W_B}{\partial Q_{pB}} \frac{\partial Q_{pB}}{\partial V_{CE}} \right)^{-1} \tag{9.89}$$

$$= \left(\frac{-1}{F} \frac{\partial F}{\partial W_B} \frac{\partial W_B}{\partial Q_{pB}} \frac{\partial Q_{pB}}{\partial V_{CB}} \right)^{-1}.$$

From Eq. (3.13), and the fact that $V_{CB} = -V_{BC}$, we can write

$$-\frac{\partial Q_{pB}}{\partial V_{CB}} = \frac{\partial Q_{pB}}{\partial V_{BC}} = C_{dBC}, \tag{9.90}$$

where C_{dBC} is the base–collector junction depletion-layer capacitance per unit area. The other two derivatives in Eq. (9.89) can be evaluated directly, namely

$$\frac{\partial F}{\partial W_B} = \frac{p_p(W_B)}{D_{nB}(W_B)n_{ieB}^2(W_B)} \tag{9.91}$$

and

$$\frac{\partial W_B}{\partial Q_{pB}} = \left(\frac{\partial Q_{pB}}{\partial W_B}\right)^{-1} = \frac{1}{qp_p(W_B)}. \tag{9.92}$$

Therefore, Eq. (9.89) gives

$$V_A + V_{CE} = \frac{qD_{nB}(W_B)n_{ieB}^2(W_B)}{C_{dBC}} \int_0^{W_B} \frac{p_p(x)}{D_{nB}(x)n_{ieB}^2(x)} dx. \tag{9.93}$$

For a uniformly doped base, Eq. (9.93) reduces to

$$V_A + V_{CE} = \frac{Q_{pB}}{C_{dBC}} \quad \text{(uniformly doped base).} \tag{9.94}$$

At low injection (i.e., low I_B), the base majority-carrier concentration is approximately the same as its equilibrium value, i.e., $p_p \approx p_{p0} = N_B$, Eq. (9.93) gives

$$V_A + V_{CE} = \frac{qD_{nB}(W_B)n_{ieB}^2(W_B)}{C_{dBC}} \int_0^{W_B} \frac{N_B(x)}{D_{nB}(x)n_{ieB}^2(x)} dx \quad \text{(low } I_B\text{).} \tag{9.95}$$

Equation (9.95) is independent of V_{BE}, so that the slope of the curves in Figure 9.12 intercept the V_{CE}-axis at the same value, namely V_A, as illustrated.

It is instructive to estimate the magnitude of Eq. (9.95) for a uniformly doped base. In this case, Eq. (9.94) gives $V_A \approx qW_B N_B/C_{dBC}$. For a base of $W_B = 0.1$ μm and $N_B = 10^{18}$ cm^{-3}, we have $qW_B N_B \approx 1.6 \times 10^{-6}$ C/cm^2. For a collector with $N_C = 2 \times 10^{16}$ cm^{-3}, then, from Figure 3.5, $C_{dBC} \approx 4 \times 10^{-8}$ F/cm^2. Therefore, $V_A \approx 40$ V. In practice, V_A can vary a lot as the transistor design is "optimized." The optimization of transistor designs, and how V_A may vary with the design point, will be discussed in Chapter 10.

As can be seen in Eq. (9.93), V_A is a function of W_B, which, as discussed earlier, is a function of V_{CE}. Therefore, strictly speaking, V_A is a function of V_{CE} at which the slope is used for extrapolating to $I_C = 0$. In other words, strictly speaking, I_C does not increase linearly with V_{CE}. However, the linear dependence is a good approximation and is a useful approximation for circuit analyses and modeling purposes.

The Early voltage is a figure of merit for devices used in analog circuits. The larger the Early voltage, the more independent is I_C on V_{CE}. Another device figure of merit that has been discussed in the literature is the $\beta_0 V_A$ product (Prinz and Sturm, 1991). The reader is referred to this publication for the discussion.

It is evident from the base–collector diode doping profiles in Figure 9.3 that C_{dBC} is significantly larger for a lateral transistor than for a vertical transistor. *Therefore, V_A for a lateral transistor could be significantly smaller than that of a vertical transistor.*

- *Emitter–collector punch-through.* As shown in Eq. (9.94), V_A is proportional to Q_{pB}. As V_{CE} is increased, W_B is reduced, and hence Q_{pB} is reduced. For a device with a small Q_{pB} or small V_A to start with, it does not take much increase in V_{CE}

before Q_{pB} approaches zero, or before the collector punches through to the emitter. At collector–emitter punch-through, the collector current becomes excessively large, being limited only by the emitter and collector series resistances. The collector current at or close to punch-through is no longer controlled adequately by the base voltage for proper device operation. ***Punch-through must be avoided*** under normal device operation, by designing the device to have a sufficiently large majority-carrier base charge.

9.3.2.2   Base–Collector Junction Avalanche

If the electric field in the space-charge region of the reverse-biased base-collector junction is sufficiently large, avalanche multiplication can become appreciable (see Section 3.3.1). For a vertical n–p–n transistor, the base–collector junction avalanche process is illustrated in Figure 9.13(a) (Lu and Chen, 1989). As the electrons injected from the emitter reach the base–collector space-charge region, they can generate electron–hole pairs. The secondary electrons flow toward the collector terminal, adding to the measured collector current, while the secondary holes flow toward the base terminal, subtracting from the measured base current.

The behavior of the measured base current in Figure 9.13(b) can be explained as follows. The measured base current is positive as usual at small V_{BE}, where the secondary hole current is not large enough to completely offset the usual base current. As the electron injection from the emitter increases with increased V_{BE}, the secondary hole current increases and may reach a point at which the measured base current turns negative. At sufficiently large V_{BE}, as will be discussed in Section 9.3.3, significant widening of the base–collector space-charge layer can occur in a vertical transistor, causing the electric field in the base–collector junction to be reduced. As a result, avalanche multiplication is reduced and the measured base current returns to positive.

The magnitude of base–collector junction avalanche depends on the maximum electric field in the base–collector junction. To minimize base–collector junction avalanche, techniques for reducing the maximum electric field in a p–n junction, such as retrograding the collector doping profile or sandwiching a lightly doped layer between the base and the collector, can be used (Tang and Lu, 1989). The concept is similar to that of a p–i–n diode discussed in Section 3.1.2.3.

9.3.3   Collector-Current Falloff

There are two physical mechanisms that can cause the denominator in Eq. (9.47) for J_{C0} to increase and hence J_{C0} to decrease, as I_C is increased. As J_{C0} falls off, β_0 falls off with it since J_{B0} is independent of I_C. The two mechanisms are high-injection effect in the base, which causes p_p to increase with V_{BE}, and widening of the quasineutral base width W_B as I_C increases. The combination of increase in p_p and widening of W_B causes the sheet resistivity of the quasineutral base to decrease. This is known as the *base-conductivity modulation effect*.

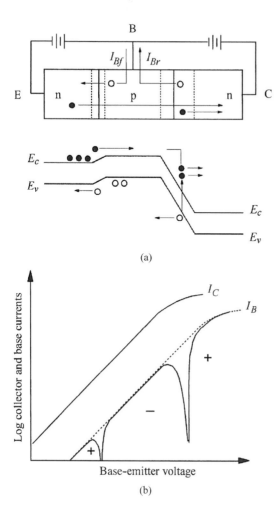

(a)

(b)

Figure 9.13 (a) Schematics of a vertical n–p–n transistor operated in the forward-active mode with electron-hole pair generation in the base–collector space-charge region. The secondary hole current, I_{Br}, subtracts from the usual forward base current, I_{Bf}. The current measured at the base terminal is $I_B = I_{Bf} - I_{Br}$. (b) Typical Gummel plot of a vertical n–p–n transistor where significant avalanche multiplication occurs in the base–collector space-charge region

The effect of high injection in the base region is built into the expression for J_{C0}, Eq. (9.47), provided we use the expression for p_p that is valid for all-injection levels, such as Eq. (9.50). The resultant β_0 falloff due to high injection in the base alone is illustrated in Figure 9.6. In Sections 9.3.3.1 and 9.3.3.2, we discuss base widening and its effect on β_0 fall off.

9.3.3.1 Base Widening and Kirk Effect

Consider the base–collector junction of a vertical n–p–n transistor. For simplicity, let us assume the base region to have a uniform doping concentration N_B, and the

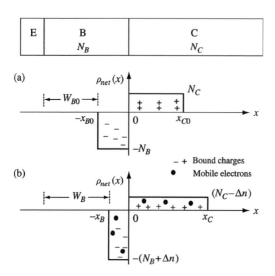

Figure 9.14 Schematics illustrating the charge distribution in the base–collector junction of a vertical n–p–n transistor. (a) Base–emitter diode not forward biased, and (b) base–emitter diode forward biased

collector region to have a uniform doping concentration N_C. When the transistor is turned off, the charge distribution in the base–collector junction is as shown schematically in Figure 9.14(a), where x_{B0} and x_{C0} are the widths of the depletion regions on the base side and on the collector side, respectively. The relationship between these widths is given by Eq. (3.8), namely,

$$x_{B0}N_B = x_{C0}N_C. \tag{9.96}$$

The maximum potential drop across the base–collector junction, ψ_{mBC}, is given by Eq. (3.9), which can be rewritten as

$$\psi_{mBC} = \frac{q}{2\varepsilon_{si}}\left(N_B x_{B0}^2 + N_C x_{C0}^2\right). \tag{9.97}$$

When the n–p–n transistor is turned on, electrons are injected into the base and flow in the base–collector space-charge region. These mobile electrons add to the space-charge. As long as this additional mobile-electron concentration is small compared with the ionized doping concentrations, depletion approximation can be used to estimate its effect. For simplicity, let us assume these mobile electrons traverse the base–collector–junction space-charge region at a saturated velocity v_{sat}. The mobile-electron concentration Δn in the space-charge region is given by the relation

$$J_C = qv_{sat}\Delta n, \tag{9.98}$$

where J_C is the collector current density. The space-charge concentration on the base side is increased from N_B to $N_B + \Delta n$, and the space-charge concentration on the collector side is decreased from N_C to $N_C - \Delta n$. As a result, the width of the depletion region on the base side is decreased to x_B, and the width of the depletion region on the collector side is increased to x_C, such that

$$x_B(N_B + \Delta n) = x_C(N_C - \Delta n). \tag{9.99}$$

This is illustrated schematically in Figure 9.14(b). The width of the quasineutral base layer is widened by an amount equal to $x_{B0} - x_B$.

An estimation of the amount of base widening can be made quantitatively if the base–emitter junction is assumed to be forward biased while V_{BC} remains unchanged, e.g., in a common-base operation of the transistor (Ghandhi, 1968). In this case, Eq. (9.97) is replaced by

$$\psi_{mBC} = \frac{q}{2\varepsilon_{si}} \left[(N_B + \Delta n)x_B^2 + (N_C - \Delta n)x_C^2 \right]. \tag{9.100}$$

Combining Eqs. (9.97) and (9.100), and assuming $\Delta n/N_C \ll 1$, we have

$$x_C = x_{C0}\sqrt{\frac{1 + (\Delta n/N_B)}{1 - (\Delta n/N_C)}} \approx \frac{x_{C0}}{\sqrt{1 - (\Delta n/N_C)}} \tag{9.101}$$

where we have used the fact that N_B is typically much larger than N_C for a vertical transistor, so that $\Delta n/N_B$ can be dropped in Eq. (9.101). Similarly

$$x_B = x_{B0}\sqrt{\frac{1 - (\Delta n/N_C)}{1 + (\Delta n/N_B)}} \approx x_{B0}\sqrt{1 - (\Delta n/N_C)}. \tag{9.102}$$

For transistors having $W_B > 100$ nm by design, this change in base width probably has no perceptible effect on the measured I_C. However, for modern vertical bipolar transistors having $W_B < 30$ nm by design, the change in base width at low current could cause a noticeable decrease in J_{C0}, resulting in the measured I_C increasing with V_{BE} somewhat more slowly than $\exp(qV_{BE}/kT)$, and the Gummel plot of the measured I_C to show a slope somewhat larger than 60 mV/decade at room temperature.

Since $N_B \gg N_C$ for a typical bipolar transistor, it can be inferred from Eq. (9.99) that as Δn approaches N_C, x_C becomes very large. That is, as Δn increases, the base–collector space-charge layer widens, with the space-charge layer boundary on the collector side moving deep into the collector, often all the way to the collector–subcollector boundary. This is known as *base-widening* or *Kirk effect* (Kirk, 1962).

In the case of a symmetric lateral transistor, with $N_C \gg N_B$, the base–collector diode behaves as a one-sided junction with a space-charge layer confined to the base side only. That is, **Kirk effect is absent in a symmetric lateral bipolar transistor on SOI.** Any widening of the quasineutral base width due to carrier injection from the emitter is limited to reduction of the base–collector space-charge layer width. The same analysis can be followed simply by setting $x_C = x_{C0} = 0$. The result is

$$x_B = x_{B0}\sqrt{N_B/(N_B + \Delta n)} \quad \text{(symmetric lateral transistor)}. \tag{9.103}$$

Comparing Eqs. (9.102) and (9.103), we see that base widening is determined by the base doping concentration in a symmetric lateral transistor and by the collector doping concentration in a vertical transistor. For a vertical transistor and a lateral transistor both having the same N_B, ***base widening effect starts in the vertical transistor at a much lower collector current density than in the lateral transistor.***

9.3.3.2   Base Widening in a Vertical Bipolar Transistor at High Currents

In a vertical bipolar transistor, as Δn approach N_C, the excess electrons in the n-type collector can induce a substantial electric field in the collector, according to Eq. (9.4), and the concept of a well-defined base–collector space-charge region is no longer valid. Also, in order to maintain quasineutrality, the excess electrons induce an excess of holes in the n-type collector. The region of the collector with excess holes becomes an extension of the p-type base. In other words, the base widens deep into the collector, until it reaches the subcollector where the excess electron concentration is small compared with the n^+-subcollector doping concentration. As the base widens toward the subcollector, J_{C0} decreases accordingly. Also, the high-field region, originally located at the physical base–collector junction, is relocated to near the collector–subcollector intersection (Poon *et al.*, 1969). The numerical simulation results shown in Figure 9.15 illustrate clearly the effects of base widening at high currents in a vertical bipolar transistor. They show that the relocation of the high-field region is accompanied by a buildup of excess electrons and holes in the collector region.

It is instructive to estimate the collector current density at which substantial base widening occurs. The saturated velocity v_{sat} for electrons in silicon is about 1×10^7 cm/s. At low collector currents, the maximum electron concentration in the n-type collector region is equal to the collector doping concentration N_C. The maximum electron current density that can be supported by an electron concentration of N_C is $J_{max} = qv_{sat}N_C$. When the current density due to injected electrons approaches J_{max}, the electron concentration has to increase to a value larger than N_C in order to support the electron current flow, i.e., there is a density of excess electrons caused by the high electron-current density. As the excess electrons build up, there is a buildup of excess holes in order to maintain quasineutrality, and a relocation of the high-field region. The results shown in Figure 9.15 suggest that significant base widening starts at a collector current density of approximately $0.3 J_{max}$. This value is consistent with the reported peak cutoff-frequency data for modern vertical bipolar devices (Crabbé *et al.*, 1993a). Thus, ***to avoid significant base widening, a vertical bipolar transistor should not be operated at collector current densities approaching J_{max}.*** For a relatively high N_C of 2×10^{17} cm^{-3}, J_{max} is about 3.2 mA/μm^2. To avoid significant base widening, J_C should be less than about 1 mA/μm^2.

9.3.4   Excess Base Current Associated with Extrinsic-Base–Emitter Junction

Figure 9.16 illustrates schematically the cross-section of the emitter–base diode of a vertical bipolar transistor. The base region directly underneath the emitter is referred to as the *intrinsic base*, and the remaining parts of the base are collectively referred to as the *extrinsic base*. The entire emitter–base diode can be considered as two diodes

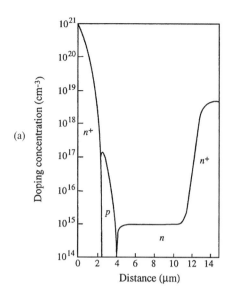

(a)

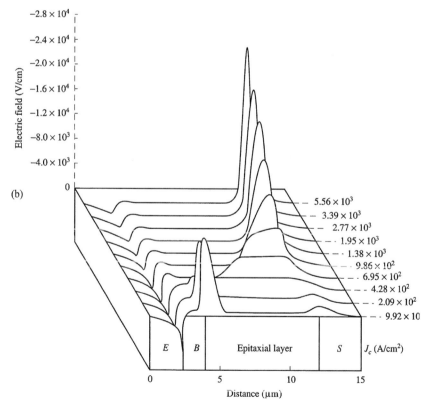

(b)

Figure 9.15 Numerical simulation results showing the effects of base widening in a vertical n–p–n transistor at high collector-current densities: (a) the doping profiles of the device simulated, (b) relocation of the high-field region from the physical base–collector junction to the collector–subcollector intersection, (c) buildup of excess holes in the collector, and (d) buildup of excess electrons in the collector (after Poon *et al.*, 1969)

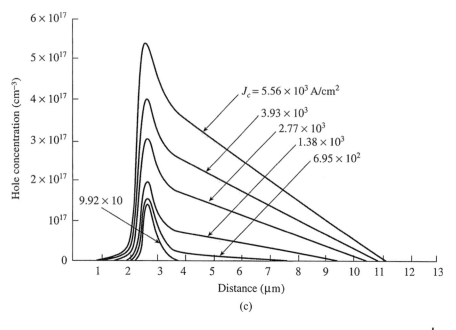

(c)

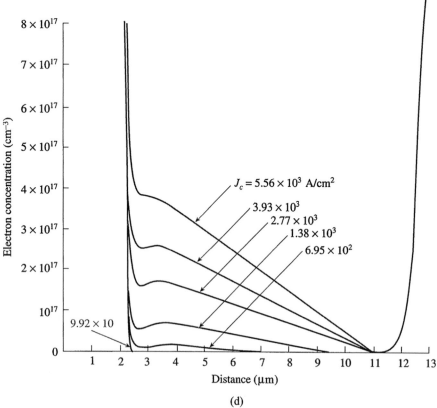

(d)

Figure 9.15 (cont.)

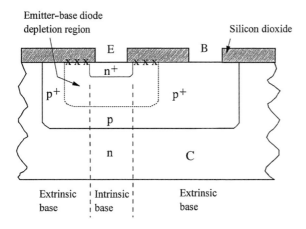

Figure 9.16 Schematic illustrating the cross-section of the emitter–base diode of a vertical n–p–n transistor. The extrinsic base is usually much more heavily doped than the intrinsic base. The presence of surface states, indicated by x x x, can cause excessive base current, as discussed in the text

connected in parallel, one formed by the emitter and the intrinsic base, and the other by the emitter and the extrinsic base. The intrinsic-base–emitter diode has been the subject of our discussion so far. The function of the extrinsic base is to provide electrical connection to the intrinsic base from the silicon surface.

To minimize parasitic resistance and to minimize electron injection from the emitter into the extrinsic-base region, the extrinsic base is usually doped much more heavily than the intrinsic base. As a result, the collector-current component due to electrons injected from the emitter into the extrinsic base and reaching the collector is small compared to the collector-current component due to electrons traversing the intrinsic base. This can be inferred readily from Eq. (9.45).

Nonetheless, the extrinsic-base–emitter diode may contribute appreciably to the measured base current. This extrinsic-base current has three components, namely (a) the current associated with the injection of holes from the extrinsic base into the emitter, (b) additional space-charge-region current associated with extrinsic-base–emitter diode, and (c) addition current due to tunneling in the extrinsic-base–emitter diode.

The current associated with the injection of holes from the extrinsic base into the emitter has the same dependence on V_{BE} as the intrinsic-base current discussed in Section 9.2.3. Therefore, this current simply adds to the intrinsic-base current, and will not show up as deviation of the measured base current from its ideal $\exp(qV_{BE}/kT)$ behavior. This additional base current is proportional to the emitter junction depth x_{jE}. It can be inferred from Figure 9.3 that this additional base current is much smaller in a polysilicon-emitter transistor than in a transistor having a diffused emitter.

The extrinsic-base–emitter space-charge-region current simply adds to any intrinsic-base–emitter space-charge-region current, both having the same $\exp(qV_{BE}/2kT)$ behavior. In well-processed devices, both space-charge-region currents should be small, noticeable only on expanded Gummel plots (see Figure 9.8).

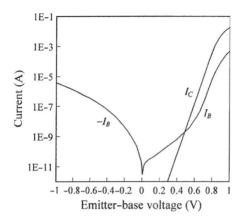

Figure 9.17 Typical current–voltage characteristics of a vertical n–p–n transistor which has excessive tunneling current in the extrinsic-base–emitter diode (after Li *et al.*, 1988)

The tunneling current in the extrinsic-base–emitter junction, on the other hand, can be very large if extrinsic base and emitter overlap at locations where both the extrinsic-base and emitter doping concentrations are heavier than intended. In addition, if the extrinsic-base–emitter space-charge region include silicon/oxide interface states, such as depicted in Figure 9.16, the interface states can assist in the tunneling process and thus enhance the tunneling current significantly. Figure 9.17 illustrates the current–voltage characteristics of a vertical bipolar transistor which has excessive base current due to tunneling (Li *et al.*, 1988). In forward bias, the base current is much larger than expected from any space-charge-region current, which has an $\exp(V_{BE}/2kT)$ dependence. In reverse bias, the excessive base current increases very rapidly with voltage, a sign that the current is due to tunneling. Fortunately, excessive base current due to tunneling can be suppressed easily by optimizing the extrinsic-base–emitter diode doping profile and device fabrication process.

9.4   Base Transit Time

Consider an n$^+$–p diode with a p-region base width W_B small compared to electron diffusion length. This diode represents the emitter–base diode of an n–p–n transistor. When a forward bias is applied to it, electrons are injected into the base, and practically all the electrons travel across the base region without recombining. The total *excess minority-carrier charge* (electrons) per unit area in the p-type base region is

$$Q_B = -q \int_0^{W_B} \left(n_p - n_{p0} \right) dx. \tag{9.104}$$

Using Eq. (9.48) in Eq. (9.104), we obtain

$$Q_B(\text{narrow base}) = -q\left(n_p - n_{p0}\right)_{x=0}\left(\frac{W_B}{2}\right)$$

$$= J_n(x=0)t_B,$$

(9.105)

where the *base-transit time* t_B is defined by

$$t_B \equiv \frac{Q_B\,(\text{narrow base})}{J_n(x=0)}$$

$$= \frac{W_B^2}{2D_n}.$$

(9.106)

In writing Eq. (9.106), we have used $J_n = qD_n(dn/dx)$ and Eq. (9.48) for the electron current. The base transit time is equal to the time needed for a steady-state electron current entering the p-type base to fill up the base with minority electrons. It is also the average time for the minority electrons to traverse the base layer. It is a key indicator of the maximum frequency response or maximum intrinsic speed of a bipolar transistor. Base-widening effects, discussed in Sections 9.3.3.1 and 9.3.3.2, cause t_B to increase, and hence the speed of a transistor to degrade.

9.5   Diffusion Capacitance in an Emitter–Base Diode

In a forward-biased diode, in addition to the depletion-layer capacitance, C_d given by Eq. (3.13), associated with the majority carriers on both sides of the space-charge layer responding to a changing applied voltage, there is also a capacitance component associated with the minority carriers on both sides of the space-charge layer responding to the changing applied voltage. This minority-carrier capacitance is called *diffusion capacitance* C_D. In the case of a bipolar transistor, the diffusion capacitance associated with the emitter–base diode being forward biased is denoted by C_{DE}.

The response of minority carriers to a changing applied voltage is through diffusion and recombination processes. These processes do not occur instantaneously, but on a time scale related to the minority-carrier lifetime or transit time. As a consequence, *when a small signal is applied to a forward-biased diode, the changes in the minority-carrier densities at different locations in the diode have different phases and cannot be lumped together and treated as a single entity*. In this section, the diffusion capacitance is derived from a small-signal analysis of the current through a diode starting from the differential equations governing the transport of minority carriers (Shockley, 1949; Lindmayer and Wrigley, 1965; Pritchard, 1967). The diffusion capacitance can also be obtained from a transmission-line analysis of a diode equivalent circuit (Bulucea, 1968).

9.5.1   Small-Signal Current in a Forward-Biased Diode

Consider an n^+–p emitter–base diode biased with a voltage $v_{BE}(t) = V_{BE} + v_{be}(t)$, where $v_{be}(t)$ is the small-signal voltage in series with the steady-state voltage V_{BE}.

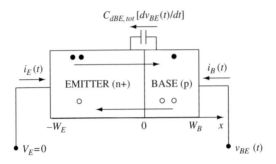

Figure 9.18 Schematic of an n$^+$–p diode showing the current components when a time-dependent voltage $v_{BE}(t)$ is applied. Here we use subscript E to denote quantities related to the emitter side and subscript B to denote quantities related to the base side

For simplicity, we assume parasitic resistances are negligible so that $v_{BE}(t)$ represents the junction voltage. The current flows, including the displacement current, are shown schematically in Figure 9.18, where $i_E(t)$ is the emitter terminal current, $i_B(t)$ is the base terminal current, and $C_{dBE,tot}$ is the depletion-layer capacitance. Kirchhoff's law requires that

$$i_E(t) + i_B(t) = 0. \tag{9.107}$$

From Eqs. (2.100), (2.93), and (2.97), the continuity equation governing the minority electrons in the p-type base is

$$A_{diode} \frac{\partial n_p(x,t)}{\partial t} = \frac{1}{q} \frac{\partial i_n(x,t)}{\partial x} - A_{diode} \frac{n_p(x,t) - n_{p0}}{\tau_{nB}}, \tag{9.108}$$

where τ_{nB} is the electron lifetime in the base and A_{diode} is the cross-sectional area of the diode. If we assume the base to be uniformly doped, so that at low electron injection currents the hole density is uniform in the p-region, then the electron current in the base does not have a drift component, only a diffusion component [see Eqs. (9.3) and (9.5), and related discussion], so that we have

$$\frac{\partial i_n(x,t)}{\partial x} = A_{diode} q D_{nB} \frac{\partial^2 n_p(x,t)}{\partial x^2}, \tag{9.109}$$

and Eq. (9.108) becomes

$$\frac{\partial n_p(x,t)}{\partial t} = D_{nB} \frac{\partial^2 n_p(x,t)}{\partial x^2} - \frac{n_p(x,t) - n_{p0}}{\tau_{nB}}. \tag{9.110}$$

If we assume the small signals to have a time dependence $e^{j\omega t}$, i.e., $v_{be}(t) = v_{be}\, e^{j\omega t}$, and the electron concentration has a form

$$n_p(x,t) = n_p(x) + \Delta n_p(x) e^{j\omega t}, \tag{9.111}$$

then Eq. (9.110) gives

$$\Delta n_p(x) j\omega e^{j\omega t} = D_{nB}\frac{d^2 n_p(x)}{dx^2} - \frac{n_p(x) - n_{p0}}{\tau_{nB}} + \left(D_{nB}\frac{d^2 \Delta n_p(x)}{dx^2} - \frac{\Delta n_p(x)}{\tau_{nB}}\right)e^{j\omega t}.$$

(9.112)

For Eq. (9.112) to be valid at all times, it must be valid for the time-independent part and the time-dependent part separately, i.e.,

$$D_{nB}\frac{d^2 n_p(x)}{dx^2} - \frac{n_p(x) - n_{p0}}{\tau_{nB}} = 0,$$

(9.113)

and

$$\Delta n_p(x) j\omega = D_{nB}\frac{d^2 \Delta n_p(x)}{dx^2} - \frac{\Delta n_p(x)}{\tau_{nB}}.$$

(9.114)

Equation (9.113) is simply the equation governing the quasisteady-state minority-electron distribution in a forward-biased diode, which has been solved in Section 3.1.3.2 [see Eq. (3.36)]. Equation (9.114) can be rearranged to give

$$\frac{d^2 \Delta n_p(x)}{dx^2} - \frac{\Delta n_p(x)}{L'^2_{nB}} = 0,$$

(9.115)

where

$$L'_{nB} \equiv \sqrt{\frac{\tau_{nB} D_{nB}}{1 + j\omega\tau_{nB}}} = \frac{L_{nB}}{\sqrt{1 + j\omega\tau_{nB}}}.$$

(9.116)

Equation (9.115) has the same form as Eq. (9.113), and hence has the same form of solution. It is left as an exercise (Exercise 9.8) for the reader to show that the solution to Eq. (9.115) is

$$\Delta n_p(x) = \frac{n_{p0} q v_{be}}{kT} \exp\left(\frac{q V_{BE}}{kT}\right) \frac{\sinh\left[(W_B - x)/L'_{nB}\right]}{\sinh\left(W_B/L'_{nB}\right)}.$$

(9.117)

Using Eq. (9.111), the electron current entering the p-type base, by diffusion, is

$$i_n(0, t) = A_{diode} q D_{nB}\left(\frac{\partial n_p(x, t)}{\partial x}\right)_{x=0}$$

$$= A_{diode} q D_{nB}\left(\frac{d n_p(x)}{dx}\right)_{x=0} + A_{diode} q D_{nB} e^{j\omega t}\left(\frac{d \Delta n_p(x)}{dx}\right)_{x=0}$$

(9.118)

$$= -I_n + \Delta i_n(0, t),$$

where

$$-I_n = A_{diode} q D_{nB}\left(\frac{d n_p(x)}{dx}\right)_{x=0} = -A_{diode}\frac{q D_{nB} n_{p0} \exp(q V_{BE}/kT)}{L_{nB}\tanh(W_B/L_{nB})}$$

(9.119)

is the steady-state electron current, and

$$\Delta i_n(0,t) = A_{diode} q D_{nB} e^{j\omega t} \left(\frac{d\Delta n_p(x)}{dx} \right)_{x=0}$$

$$= -A_{diode} \frac{q D_{nB} n_{p0} \exp(q V_{BE}/kT)}{L'_{nB} \tanh(W_B/L'_{nB})} \frac{q v_{be}(t)}{kT} \qquad (9.120)$$

$$= -\left(\frac{L_{nB}}{L'_{nB}} \right) \frac{\tanh(W_B/L_{nB})}{\tanh(W_B/L'_{nB})} \frac{q I_n}{kT} v_{be}(t)$$

is the small-signal electron current. (Note that i_n is negative because electrons flowing in the x direction give rise to a negative current. I_n is the magnitude of the steady-state electron current, a positive quantity.)

The small-signal hole current entering the emitter can be derived in a similar manner. It is

$$\Delta i_p(0,t) = -\left(\frac{L_{pE}}{L'_{pE}} \right) \frac{\tanh(W_E/L_{pE})}{\tanh(W_E/L'_{pE})} \frac{q I_p}{kT} v_{be}(t), \qquad (9.121)$$

where

$$L_{pE} = \sqrt{\tau_{pE} D_{pE}} \qquad (9.122)$$

and

$$L'_{pE} \equiv \sqrt{\frac{\tau_{pE} D_{pE}}{1 + j\omega \tau_{pE}}} = \frac{L_{pE}}{\sqrt{1 + j\omega \tau_{pE}}} \qquad (9.123)$$

are the corresponding parameters for holes in the n^+ emitter. I_p is the magnitude of the steady-state hole current flowing from the base into the emitter. (Note that i_p is negative, since holes flowing in the $-x$ direction give a negative current flow.)

Since the base is assumed to be narrow, i.e., $W_B/L_{nB} \ll 1$, we have $\tanh(W_B/L_{nB}) \approx W_B/L_{nB}$, and Eq. (9.120) gives

$$\Delta i_n(0,t) \approx -\left(\frac{W_B}{L'_{nB}} \right) \frac{1}{\tanh(W_B/L'_{nB})} \frac{q I_n}{kT} v_{be}(t) \quad \text{(narrow base).} \qquad (9.124)$$

As the emitter is assumed to be wide, we have $W_E/L_{pE} \gg 1$ and $\tanh(W_E/L_{pE}) \approx 1$. It also implies $(W_E/L'_{pE}) = (W_E/L_{pE})\sqrt{1 + j\omega \tau_{pE}} \gg 1$. Therefore, $\tanh(W_E/L'_{pE}) \approx 1$, and Eq. (9.121) gives

$$\Delta i_p(0,t) \approx -\left(\frac{L_{pE}}{L'_{pE}} \right) \frac{q I_p}{kT} v_{be}(t) = -\sqrt{1 + j\omega \tau_{pE}} \frac{q I_p}{kT} v_{be}(t) \quad \text{(wide emitter).}$$

$$\qquad (9.125)$$

From the current components shown in Figure 9.18, the base current is

$$i_B(t) = -i_n(0,t) - i_p(0,t) + C_{dBE,tot} \frac{dv_{BE}(t)}{dt}$$
$$= I_n + I_p - \Delta i_n(0,t) - \Delta i_p(0,t) + j\omega C_{dBE,tot} v_{be}(t) \qquad (9.126)$$
$$= I_B + i_b(t),$$

where

$$I_B = I_n + I_p \qquad (9.127)$$

is the steady-state base terminal current and

$$i_b(t) = -\Delta i_n(0,t) - \Delta i_p(0,t) + j\omega C_{dBE,tot} v_{be}(t) \qquad (9.128)$$

is the small-signal base terminal current.

For charging and discharging a wide-emitter region, the time constant is the lifetime τ_E for the emitter. For charging and discharging a narrow-base region, the time constant is the base transit time t_B, given in Eq. (9.106). The lifetime in the emitter is typically smaller than 10^{-9} s [see Figure 3.14(b)] while the base transit time of a modern bipolar transistor is typically smaller than 10^{-11} s. Therefore, a low-frequency assumption for the wide emitter, i.e., $\omega \tau_{pE} < 1$, also means low-frequency for the narrow base, i.e., $\omega t_B < 1$. With t_B smaller than 10^{-11} s for a modern bipolar transistor, the condition of $\omega t_B > 1$ is rarely reached. That is, in considering the diffusion capacitance of a modern bipolar transistor, we need to consider the cases of $\omega \tau_{pE} < 1$ and $\omega \tau_{pE} > 1$. In both cases, we have $\omega t_B < 1$.

9.5.2 Low-Frequency [$\omega \tau_{pE} < 1$ and $\omega t_B < 1$] Diffusion Capacitance

In the low-frequency approximation, Eq. (9.124) can be expanded to keep only the lowest terms in W_B / L_{nB} and ωt_B, i.e.,

$$\Delta i_n(0,t) \approx -\left[1 + \frac{W_B^2}{3L_{nB}^2}(1 + j\omega\tau_{nB}) + \cdots\right] \frac{qI_n}{kT} v_{be}(t) \quad \text{(narrow base)}$$
$$\simeq -\left(1 + j\frac{2\omega t_B}{3}\right) \frac{qI_n}{kT} v_{be}(t), \qquad (9.129)$$

where we have used Eq. (9.106) for the base transit time t_B. Similarly, Eq. (9.125) gives

$$\Delta i_p(0,t) \approx -\left(1 + j\frac{\omega\tau_{pE}}{2}\right) \frac{qI_p}{kT} v_{be}(t). \qquad (9.130)$$

Substituting Eqs. (9.129) and (9.130) into Eq. (9.128), we have

$$i_b(t) = \frac{q}{kT}(I_n + I_p)v_{be}(t) + j\omega\left(\frac{2qI_n t_B}{3kT} + \frac{qI_p\tau_{pE}}{2kT}\right)v_{be}(t) + j\omega C_{dBE,tot}v_{be}(t)$$
$$= \frac{1}{r_d}v_{be}(t) + j\omega(C_{Dn} + C_{Dp} + C_{dBE,tot})v_{be}(t), \qquad (9.131)$$

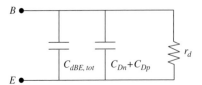

Figure 9.19 Small-signal equivalent circuit for a forward-biased diode. Parasitic resistances are ignored

where

$$\frac{1}{r_d} = \frac{q}{kT}(I_n + I_p) = \frac{qI_B}{kT} \tag{9.132}$$

is the ac conductance of the diode,

$$C_{Dn} = \frac{2qI_n t_B}{3kT} \qquad \text{(narrow base)} \tag{9.133}$$

is the diffusion capacitance due to the minority electrons in the narrow base, and

$$C_{Dp} = \frac{qI_p \tau_{pE}}{2kT} \qquad \text{(wide emitter)} \tag{9.134}$$

is the diffusion capacitance due to the minority holes in the wide emitter.

The small-signal equivalent circuit for the forward-biased emitter–base diode can be obtained from Eq. (9.131). It is shown in Figure 9.19 in the form of a lumped-component model. It should be noted that in the low-frequency approximation, the diffusion capacitances are independent of frequency.

9.5.3 Diffusion Capacitance at High Frequencies [$\omega \tau_{pE} > 1$]

In this case, Eq. (9.125) gives

$$\Delta i_p(0, t) \approx -\sqrt{j\omega \tau_{pE}} \frac{qI_p}{kT} v_{be}(t)$$

$$= -\left(\sqrt{\frac{\omega \tau_{pE}}{2}} + j\omega \sqrt{\frac{\tau_{pE}}{2\omega}}\right) \frac{qI_p}{kT} v_{be}(t), \tag{9.135}$$

which implies that the diffusion capacitance due to the minority holes in the wide emitter is

$$C_{Dp} = \frac{qI_p \tau_{pE}}{kT\sqrt{2\omega \tau_{pE}}} \qquad \text{(wide emitter)}. \tag{9.136}$$

Comparison of Eqs. (9.134) and (9.136) suggests that the emitter component of the diffusion capacitance is approximately independent of frequency until about $\omega \tau_{pE} \sim 2$, and then decreases as $1/\sqrt{\omega \tau_{pE}}$ at higher frequencies. As discussed earlier,

Eq. (9.133) is valid for the base component of the diffusion capacitance in this high-frequency régime.

9.6 Bipolar Device Models for Circuit Analyses

The merits of a bipolar device should be discussed in the context of the circuit in which it is used. For circuit applications, the device electrical characteristics are first transformed into equivalent circuit parameters. The merits of a device are then interpreted from the behavior of the circuit or from the characteristics of the equivalent-circuit parameters. In this section, the equivalent-circuit models needed for the discussion of bipolar device design, which will be covered in Chapter 10, and device optimization, which will be covered in Chapter 11, are developed. The models suitable for steady-state analyses, where capacitances have no effect and can be ignored, will be developed first, followed by the models suitable for small-signal analyses, where parasitic resistances and capacitances need to be included.

9.6.1 Basic Steady-State Model

The Ebers–Moll model (Ebers and Moll, 1954) for an n–p–n transistor is shown in Figure 9.20. It describes an n–p–n transistor as two diodes in series, arranged in the common-base mode. When a voltage V_{BE} is applied to the base–emitter diode, a forward current I_F flows in the base–emitter diode. This current causes a current $\alpha_F I_F$ to flow in the collector, where α_F is the common-base current gain in the forward direction. Similarly, when a voltage V_{BC} is applied across the base–collector diode, a reverse current I_R flows in the base–collector diode, causing a current $\alpha_R I_R$ to flow in the emitter, where α_R is the common-base current gain in the reverse direction. These currents are indicated in Figure 9.20. They are related by

$$I_E = \alpha_R I_R - I_F, \tag{9.137}$$

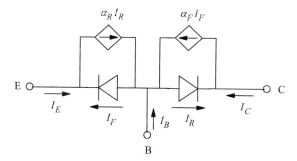

Figure 9.20 Common-base equivalent-circuit representation of the basic steady-state Ebers–Moll model of an n–p–n transistor

$$I_C = \alpha_F I_F - I_R, \tag{9.138}$$

and

$$I_B = (1 - \alpha_F)I_F + (1 - \alpha_R)I_R. \tag{9.139}$$

The emitter, base, and collector currents are related by $I_E + I_B + I_C = 0$.
From Eq. (3.66) we can write I_F and I_R in the form

$$I_F = I_{F0}[\exp(qV_{BE}/kT) - 1] \tag{9.140}$$

and

$$I_R = I_{R0}[\exp(qV_{BC}/kT) - 1]. \tag{9.141}$$

Therefore, Eqs. (9.137) and (9.138) can be rewritten as

$$I_E = -I_{F0}[\exp(qV_{BE}/kT) - 1] + \alpha_R I_{R0}[\exp(qV_{BC}/kT) - 1] \tag{9.142}$$

and

$$I_C = \alpha_F I_{F0}[\exp(qV_{BE}/kT) - 1] - I_{R0}[\exp(qV_{BC}/kT) - 1]. \tag{9.143}$$

Reciprocity characteristics of the emitter and collector terminals require the off-diagonal coefficients of the equations for I_E and I_C to be equal (Gray *et al.*, 1964; Muller and Kamins, 1977), i.e.,

$$\alpha_R I_{R0} = \alpha_F I_{F0}. \tag{9.144}$$

Alternatively, the reciprocity relationship in Eq. (9.144) can be shown as follows. We note that $\alpha_F I_{F0}$ is the saturated collector current in the forward-active mode, and $\alpha_R I_{R0}$ is the saturated collector current in the reverse active mode. For simplicity, we consider a hypothetical transistor having a unit cross-sectional area for both the base–emitter junction and the base–collector junction. In this case, $\alpha_F I_{F0}$ is given by Eq. (9.47) with the integral in the denominator being from $x = 0$ to $x = W_B$, and $\alpha_R I_{R0}$ is also given by Eq. (9.47) but with the integral in the denominator being from $x = W_B$ to $x = 0$. Since the value of an integral of a function is independent of its direction of integration, we have $\alpha_F I_{F0} = \alpha_R I_{R0}$. It should be noted that no assumption has been made about the doping and bandgap-narrowing parameters in the base, suggesting that the reciprocity relationship applies to Si-base as well as SiGe-base transistors. (See Section 10.4 for discussion of SiGe-base transistors.) Also, it has been shown through experiments and simulation studies (Rieh *et al.*, 2005) that the saturated collector currents in forward and reverse modes are approximately the same in typical vertical bipolar transistors, even though the base–collector junction area is much larger than its base–emitter junction area, as indicated in Figure 9.1(a). In the case of a symmetric lateral bipolar transistor, emitter–collector reciprocity is inherent because of its emitter/collector symmetry.

The common-emitter form of the Ebers–Moll model is often more desirable for circuit analyses. To accomplish this, let us define

$$I_{SF} \equiv \alpha_F I_F, \tag{9.145}$$

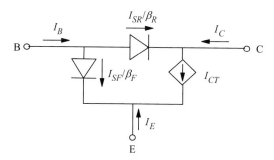

Figure 9.21 Common-emitter equivalent-circuit representation of the steady-state Ebers–Moll model of an n–p–n transistor

$$I_{SR} \equiv \alpha_R I_R, \tag{9.146}$$

$$I_{CT} \equiv I_{SF} - I_{SR}, \tag{9.147}$$

$$\beta_F \equiv \frac{\alpha_F}{1 - \alpha_F}, \tag{9.148}$$

$$\beta_R \equiv \frac{\alpha_R}{1 - \alpha_R}. \tag{9.149}$$

Comparison with Eq. (9.79) shows that β_F and β_R are the common-emitter current gains in the forward and the reverse directions, respectively. Substituting Eqs. (9.144)–(9.149) into Eqs. (9.137)–(9.139) gives

$$I_E = -I_{CT} - \frac{I_{SF}}{\beta_F}, \tag{9.150}$$

$$I_C = I_{CT} - \frac{I_{SR}}{\beta_R}, \tag{9.151}$$

$$I_B = \frac{I_{SF}}{\beta_F} + \frac{I_{SR}}{\beta_R}. \tag{9.152}$$

The equivalent-circuit model for these currents is shown in Figure 9.21.

9.6.2 Basic ac Model

To model the ac behavior of a bipolar transistor, the internal capacitances and resistances of the transistor should be included. In general, the parasitic resistances can be made rather small by using large device areas and device layout techniques, as well as fabrication process techniques. However, the internal capacitances usually can be reduced only by reducing the associated device areas. As a result, the basic ac behavior of a transistor is determined more by its capacitances than by its resistances, especially in low-current operation. We shall first neglect the resistances and consider only the capacitances. The effect due to resistances will be added back later.

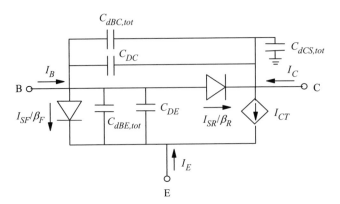

Figure 9.22 Common-emitter equivalent-circuit representation of the ac Ebers–Moll model of an n−p−n transistor. Internal capacitances are included

As discussed in Section 9.5, there are two components in the capacitance of a p−n diode. The depletion-layer capacitance C_d due to the majority carriers and the diffusion capacitance C_D associated with the minority carriers. Let $C_{dBE,tot}$ and $C_{dBC,tot}$ be the depletion-layer capacitances of the base–emitter and base–collector diodes, respectively, and C_{DE} and C_{DC} be the diffusion capacitance associated with forward-biasing the emitter–base and collector–base diodes, respectively. When these capacitances are included, the steady-state common-emitter equivalent-circuit (Figure 9.21) becomes as shown in Figure 9.22. In Figure 9.22, the depletion-layer capacitance of the collector-substrate diode, $C_{dCS,tot}$, is also included for completeness.

9.6.2.1   Model for a Transistor Operating in Forward-Active Mode

For simplicity, we shall consider only transistors biased in the forward-active mode of operation, i.e., with the base–emitter diodes forward biased and the base–collector diodes reverse biased. (Vertical bipolar transistors biased in the reverse-active mode, i.e., with the base–collector diodes forward biased, cannot be switched fast because of the large diffusion capacitance associated with the forward-biased base–collector diodes. As a result, vertical bipolar transistors in high-speed circuits are operated only in forward-active mode. Symmetric lateral bipolar transistors, on the other hand, can switch equally fast in both forward-active and reverse-active modes. Some circuits taking advantage of this unique property are discussed in Chapter 11.) In this case, I_{SR} can be neglected compared to I_{SF}, and $C_{DC} = 0$. The model in Figure 9.22 then simplifies to that shown in Figure 9.23.

If the internal resistances are now included in the equivalent circuit of Figure 9.23, the resultant equivalent circuit is shown in Figure 9.24. Here, for purposes of discussion in Chapters 10 and 11, the base resistance is shown as two parts, an intrinsic-base part r_{bi} and an extrinsic-base part r_{bx}. The depletion-layer capacitance of the base–collection diode is also separated into an intrinsic part $C_{dBCi,tot}$ and an extrinsic part

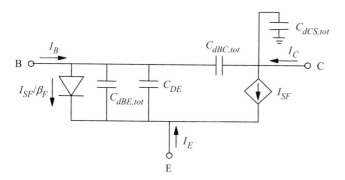

Figure 9.23 Equivalent-circuit representation of the ac Ebers–Moll model of an n–p–n transistor biased in the forward-active mode of operation. Internal capacitances are included

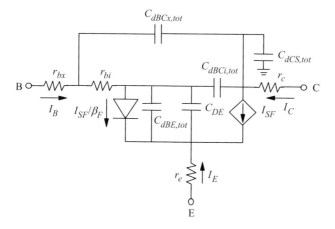

Figure 9.24 Equivalent-circuit representation of the ac Ebers–Moll model of an n–p–n transistor biased in the forward-active mode of operation. Internal resistances and capacitance are included

$C_{dBCx,tot}$. (It can be inferred from the device structure shown in Figure 9.1(b) that a symmetric lateral bipolar transistor has no $C_{dBCx,tot}$ component.)

9.6.2.2 Small-Signal Equivalent-Circuit Model

Consider a small-signal voltage applied to the input base terminal of a common-emitter equivalent circuit shown in Figure 9.23 or 9.24. It will cause small variations in currents and voltages. A small-signal equivalent-circuit model provides relationships among these current and voltage variations. We first develop the small-signal equivalent-circuit model for an intrinsic device, ignoring the transistor resistances, and then the model for an extrinsic device with these resistances included.

- *Small-signal model when resistances are negligible.* The Ebers–Moll model for an intrinsic transistor (ignoring resistances) is shown in Figure 9.23. Let us denote the steady-state base–emitter voltage by V'_{BE} and the collector–emitter voltage by V'_{CE}. The corresponding small-signal voltages are denoted as v'_{be} and v'_{ce}. Here, the convention is such that the primed parameters refer to an intrinsic device, while the unprimed parameters are for an extrinsic device. The corresponding small-signal base and collector currents are i_b and i_c, respectively. The intrinsic transconductance g'_m relates i_c to v'_{be}, i.e.,

$$g'_m = \frac{i_c}{v'_{be}} = \left.\frac{\partial I_C}{\partial V'_{BE}}\right|_{V'_{CE}} = \frac{qI_C}{kT}, \tag{9.153}$$

where we have used the fact that I_C is proportional to $\exp(qV'_{BE}/kT)$. The intrinsic input resistance r'_π relates v'_{be} to i_b, i.e.,

$$r'_\pi = \frac{v'_{be}}{i_b} = \frac{1}{\left.\frac{\partial I_B}{\partial V'_{BE}}\right|_{V'_{CE}}} = \frac{kT}{qI_B} = \frac{\beta_0}{g'_m}, \tag{9.154}$$

where we have used the fact that I_B is proportional to $\exp(qV'_{BE}/kT)$ and that $\beta_0 = I_C/I_B$. The intrinsic output resistance r'_0 relates i_c to v'_{ce}, i.e.,

$$r'_0 = \frac{v'_{ce}}{i_c} = \frac{1}{\left.\frac{\partial I_C}{\partial V'_{CE}}\right|_{V'_{BE}}} = \frac{V_A + V_{CE}}{I_C}, \tag{9.155}$$

where we have used Eq. (9.85) for the Early voltage V_A. The capacitances are designated by

$$C_\mu = C_{dBC,tot}, \tag{9.156}$$

and

$$C_\pi = C_{dBE,tot} + C_{DE}. \tag{9.157}$$

The resulting small-signal equivalent circuit is shown in Figure 9.25. This is the well-known small-signal hybrid-π model (Gray *et al.*, 1964).

- *Small-signal model including resistances.* When resistances are included, the device terminal voltages V_C, V_E, and V_B are no longer the same as the internal junction voltages V'_C, V'_E, and V'_B, respectively, owing to the *IR* drops in the

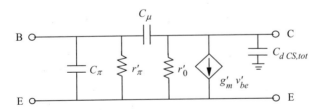

Figure 9.25 Small-signal hybrid-π model of a bipolar transistor when parasitic resistances are neglected

resistors r_e, r_c, and r_b. For simplicity, we have lumped r_{bi} and r_{bx} into r_b. The terminal voltages and the internal junction voltages are related by

$$V_C = V'_C + I_C r_c, \tag{9.158}$$

$$V_B = V'_B + I_B r_b, \tag{9.159}$$

and

$$V_E = V'_E + I_E r_e = V'_E - (I_C + I_B) r_e, \tag{9.160}$$

where we have used the fact that $I_E + I_C + I_B = 0$. Therefore,

$$V_{BE} = V'_{BE} + I_C r_e + I_B (r_b + r_e) \tag{9.161}$$

and

$$V_{CE} = V'_{CE} + I_C r_c + (I_C + I_B) r_e \approx V'_{CE} + I_C (r_e + r_c), \tag{9.162}$$

where we have neglected the IR drop due to I_B compared with that due to I_C. The extrinsic transconductance g_m relates i_c to v_{be}, i.e.,

$$
\begin{aligned}
g_m &= \frac{i_c}{v_{be}} = \frac{i_c}{v'_{be} + i_c r_e + i_b (r_b + r_e)} \\
&= \left(\frac{1}{g'_m} + r_e + \frac{r_b + r_e}{\beta_0} \right)^{-1} \approx \frac{g'_m}{1 + g'_m r_e},
\end{aligned} \tag{9.163}
$$

where we have neglected $(r_b + r_e)/\beta_0$ compared to r_e and used Eq. (9.153) for g'_m. Similarly, the extrinsic input resistance r_π relates v_{be} to i_b, i.e.,

$$
\begin{aligned}
r_\pi &= \frac{v_{be}}{i_b} = \frac{v'_{be} + i_c r_e + i_b (r_b + r_e)}{i_b} \\
&= r'_\pi (1 + g'_m r_e) + (r_b + r_e),
\end{aligned} \tag{9.164}
$$

where we have used Eq. (9.154) for r'_π. The extrinsic output resistance r_0 relates i_c to v_{ce}, i.e.,

$$
\begin{aligned}
r_0 &= \frac{v_{ce}}{i_c} = \frac{v'_{ce} + i_c (r_e + r_c)}{i_c} \\
&= r'_0 + (r_e + r_c),
\end{aligned} \tag{9.165}
$$

where we have used Eq. (9.155) for r'_0.

The device capacitance components are still the same as before, with C_μ given by Eq. (9.156) and C_π given by Eq. (9.157). It should be noted that C_μ is determined by V'_{BC}, and not V_{BC}. Similarly, C_π is determined by V'_{BE}, and not by V_{BE}. The equivalent circuit can be deduced from Eqs. (9.163)–(9.165), and is shown in Figure 9.26.

9.6.2.3 Emitter Diffusion Capacitance and Forward Transit Time

The emitter diffusion capacitance C_{DE}, a component of C_π in Eq. (9.157), is due to the minority carriers in a transistor that can respond to an ac signal when the transistor is

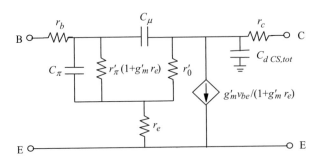

Figure 9.26 Small-signal hybrid-π model of a bipolar transistor including parasitic resistances

operated in forward-active mode. Minority carriers are present in the emitter region, the base region, as well as in the space-charge regions of the base–emitter and base–collector diodes. The total minority-carrier charge can therefore be written as the sum of these individual charges:

$$Q_{DE} = \left|Q_{E,tot,ac}\right| + \left|Q_{B,tot,ac}\right| + \left|Q_{BE,tot,ac}\right| + \left|Q_{BC,tot,ac}\right|, \tag{9.166}$$

where $Q_{E,tot,ac}$, $Q_{B,tot,ac}$, $Q_{BE,tot,ac}$, and $Q_{BC,tot,ac}$ represent the minority-carrier charge in the emitter, the base, the base–emitter space-charge region, and the base–collector space-charge region, respectively, that can contribute to the diffusion capacitance.

A note about the symbols used to denote minority-charge quantities here and elsewhere in the book is needed. As an illustration, let us consider the minority charge in the base region. In Eq. (9.104), we use Q_B to denote the minority charge per unit area in the base region of a diode. We shall use $Q_{B,tot}$ to denote the *total minority charge in the base region*. In the case of a simple diode with cross-sectional area A_{diode}, $Q_{B,tot}$ is simply $A_{diode}Q_B$. However, in general, $Q_{B,tot}$ cannot be written simply as $A_{diode}Q_B$. $Q_{B,tot}$ can be determined accurately only by using two-dimensional or three-dimensional numerical simulation. Nonetheless, it is often mathematically convenient to assume such a simple relationship, especially for explaining the basic physics governing device operation. Therefore, we shall use $A_{diode}Q_B$ to mean $Q_{B,tot}$ in many cases. It should be remembered that it is $Q_{B,tot}$ that should be used in quantitative device modeling. Similar comments apply to the other minority-charge quantities Q_E, Q_{BE}, and Q_{BC}.

The subscript 'ac' on the right-hand side of Eq. (9.166) is to distinguish these quantities from their corresponding steady-state values which can be larger. $Q_{BE,ac}$ and $Q_{BC,ac}$ are usually assumed to be the same as the corresponding steady-state quantities. This is a good assumption because of the high field in the space-charge regions. In a high-field space-charge region, the electrons travel at their saturated velocity v_{sat} which is about 1×10^7 cm/s. For a space-charge layer width of 0.1 μm, the average transit time for the electrons is on the order of 10^{-12} s. This time is short compared to $1/f$, where f is the frequency of a typical signal. That is, electrons in a space-charge region should be able to respond to an ac signal.

Notice that Q_{DE} is the sum of the *absolute values* of the individual minority-carrier charge components, and not the summation of the net charge components. As an illustration of this important distinction, consider a hypothetical transistor having a perfectly symmetrical emitter–base diode, with the n-region doping concentration equal to the p-region doping concentration (not a good transistor design, but a transistor nonetheless). In this case, we have $Q_{B,ac} = -Q_{E,ac}$. The contributions of $Q_{B,ac}$ and $Q_{E,ac}$ to Q_{DE} are $|Q_{B,ac}|$ and $|Q_{E,ac}|$, and not $Q_{B,ac} + Q_{E,ac}$, which is zero.

If the intrinsic base–emitter forward-bias voltage is $v'_{BE}(t)$ then the emitter diffusion capacitance is

$$C_{DE} \equiv \frac{\partial Q_{DE}}{\partial v'_{BE}}. \tag{9.167}$$

For modeling purposes, it is convenient to consider the collector current $i_C(t)$ being the charging current and rewrite Eq. (9.166) in the form

$$Q_{DE} \equiv \tau_F i_C(t), \tag{9.168}$$

where τ_F is referred to as the *forward transit time*. If we write $v'_{BE}(t) = V'_{BE} + v'_{be}(t)$, where V'_{BE} is the steady-state bias and $v'_{be}(t)$ is the small signal, then $i_C \propto \exp(qv'_{BE}/kT)$ at low injection and

$$i_C(t) \approx I_C \left[1 + \frac{qv'_{be}(t)}{kT} \right], \tag{9.169}$$

where I_C is the steady-state collector current. Equations (9.167)–(9.169) give

$$C_{DE} = \tau_F \frac{\partial i_C}{\partial v'_{BE}} = \tau_F \frac{qI_C}{kT} = \tau_F g'_m \quad \text{(low injection)}, \tag{9.170}$$

where g'_m is the intrinsic transconductance given by Eq. (9.153).

Once V'_{BE} exceeds the low-injection limit, base-widening cannot be ignored. In the case of a vertical bipolar transistor, Q_{DE} increases at a rate much faster than $\exp(qv'_{BE}/kT)$ due to minority-charge storage in the collector region, and τ_F *cannot be considered as a constant* (see Sections 9.3.3.1 and 9.3.3.2).

Comparing Eqs. (9.166) and (9.168), we see that τ_F has contributions from $Q_{E,tot,ac}$, $Q_{B,tot,ac}$, $Q_{BE,tot,ac}$, and $Q_{BC,tot,ac}$. To help distinguish the various contributions, τ_F is often written as the sum of these components, namely,

$$\tau_F = \tau_E + \tau_B + \tau_{BE} + \tau_{BC}. \tag{9.171}$$

In Eq. (9.171), τ_E is the *emitter delay time*, representing the contribution from $Q_{E,tot,ac}$, τ_B is the *base delay time*, representing the contribution from $Q_{B,tot,ac}$, τ_{BE} is the *base–emitter space-charge region delay time*, representing the contribution from $Q_{BE,tot,ac}$, and τ_{BC} is the *base–collector space-charge region delay time*, representing the contribution from $Q_{BC,tot,ac}$ (Ashburn, 1988). The diffusion capacitance associated with a narrow-base region and a wide-emitter region can be inferred from Eqs. (9.133) and (9.134), respectively, which in turn can be used in Eq. (9.171) for τ_B and τ_E. The space-charge region delay time is equal to the average transit time for the corresponding space-charge region. This time is $W_d/2v_{sat}$, where W_d is the space-charge layer width and v_{sat} is the saturated electron velocity (Meyer and Muller, 1987). Considerations of these delay-time components in the design of a bipolar transistor will be covered in Chapter 11.

9.7   Breakdown Voltages

The breakdown voltages of a bipolar transistor are often characterized by applying a reverse bias across two of the three device terminals, with the third device terminal left open-circuit. These breakdown voltages are denoted by

BV_{EBO} = emitter–base breakdown voltage with the collector open-circuit,
BV_{CBO} = collector–base breakdown voltage with the emitter open-circuit, and
BV_{CEO} = collector–emitter breakdown voltage with the base open-circuit.

Since bipolar transistors are usually operated with base–emitter junctions zero-biased or forward biased, their BV_{EBO} values are not important as long as they do not adversely affect the other device parameters. On the other hand, BV_{CBO} and BV_{CEO} must be adequately large for the intended circuit application. BV_{CBO} and BV_{CEO} are often determined, respectively, from the measured common-base and common-emitter current–voltage characteristics. The measurement setups for an n–p–n transistor, and the corresponding I–V characteristics, are illustrated schematically in Figure 9.27.

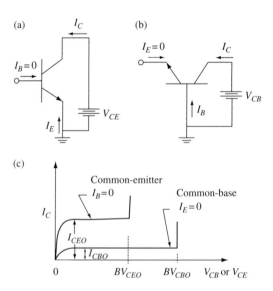

Figure 9.27 Circuit schematics for measuring (a) BV_{CEO} and (b) BV_{CBO} of an n–p–n transistor. (c) Common-emitter $I_C - V_{CE}$ characteristics at $I_B = 0$, and common-base $I_C - V_{CB}$ characteristics at $I_E = 0$

9.7.1 Common-Base Current Gain in the Presence of Base–Collector Junction Avalanche

Consider an n–p–n transistor biased in the forward-active mode, as illustrated in Figure 9.28(a). The corresponding energy-band diagram and the electron and hole current flows inside the transistor are illustrated in Figure 9.28(b), where the locations of the base–emitter junction and the base–collector space-charge layer, where avalanche multiplication takes place, are also indicated. The emitter current I_E is equal to the sum of the hole current entering the emitter from the base and the electron current leaving the emitter toward the base, i.e.,

$$I_E = A_E\left[J_n(0) + J_p(0)\right],\tag{9.172}$$

where A_E is the emitter area. It should be noted that I_E, defined as the current entering the emitter terminal, is a negative quantity for an n–p–n transistor. Both J_n and J_p are negative.

As the electrons traverse the base layer, some of them can recombine within the base layer. Only those electrons reaching $x = W_B$ contribute to the collector current. In the presence of avalanche multiplication in the reverse-biased base–collector junction, the electron current exiting the base–collector space-charge layer is a factor of M larger than that entering the space-charge layer, where M is the avalanche multiplication factor (see Section 3.3.1). That is,

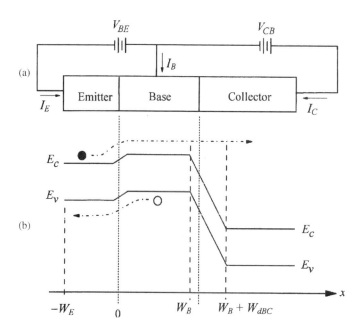

Figure 9.28 (a) Schematic illustrating the voltages and currents in an n–p–n transistor biased in the forward-active mode. (b) The corresponding energy-band diagram and illustration of the electron and hole flows inside the transistor. Also indicated are the locations of the base–emitter junction and the base–collector space-charge layer

$$J_n(W_B + W_{dBC}) = MJ_n(W_B). \tag{9.173}$$

The collector current I_C is equal to the electron current exiting the base–collector space-charge layer, i.e.,

$$I_C = -A_E J_n(W_B + W_{dBC}). \tag{9.174}$$

The minus sign in Eq. (9.174) is due to the fact that, as defined, I_C is a current entering the collector. I_C is a positive quantity for an n–p–n transistor.

Using Eqs. (9.172) to (9.174), we can rewrite the static common-base current gain α_0 [cf. Eq. (9.77)] as

$$
\begin{aligned}
\alpha_0 &\equiv \frac{I_C}{-I_E} \\
&= \frac{J_n(W_B + W_{dBC})}{[J_n(0) + J_p(0)]} \\
&= \frac{J_n(0)}{[J_n(0) + J_p(0)]} \frac{J_n(W_B)}{J_n(0)} \frac{J_n(W_B + W_{dBC})}{J_n(W_B)} \\
&= \gamma \alpha_T M,
\end{aligned}
\tag{9.175}
$$

where the *emitter injection efficiency* γ is defined by

$$\gamma \equiv \frac{J_n(0)}{J_n(0) + J_p(0)}, \tag{9.176}$$

and the *base transport factor* α_T is defined by

$$\alpha_T \equiv \frac{J_n(W_B)}{J_n(0)}. \tag{9.177}$$

When base–collector junction avalanche effect is negligible, we have $M \approx 1$, and the common-base current gain is

$$\alpha_0 = \gamma \alpha_T \quad (\text{when } M = 1). \tag{9.178}$$

If we further assume that recombination in the thin base is negligible, then the common-base current gain is simply

$$\alpha_0 = \gamma = \frac{J_n(0)}{J_n(0) + J_p(0)} \quad (\text{when } M = 1 \text{ and } \alpha_T = 1). \tag{9.179}$$

[*Note*: Throughout this chapter, by equating the collector current to the electron current entering the intrinsic base, and by equating the base current to the hole current entering the emitter, we have implicitly made the assumptions that $M = 1$ and $\alpha_T = 1$.]

9.7.2 Saturation Currents in a Transistor

Referring to Section 9.6.1, if we define I_{EBO} and I_{CBO} by (Ebers and Moll, 1954)

$$I_{EBO} \equiv I_{F0}(1 - \alpha_R \alpha_F) \tag{9.180}$$

and

$$I_{CBO} \equiv I_{R0}(1 - \alpha_R \alpha_F), \tag{9.181}$$

then Eqs. (9.142) and (9.143) give

$$I_E = -I_{EBO}[\exp(qV_{BE}/kT) - 1] - \alpha_R I_C \tag{9.182}$$

and

$$I_C = -I_{CBO}[\exp(qV_{BC}/kT) - 1] - \alpha_F I_E. \tag{9.183}$$

The physical meaning of I_{EBO} and I_{CBO} is apparent from these equations. I_{EBO} is the saturation current of the emitter–base diode when the collector is open-circuit, i.e., it is the emitter current when the emitter–base diode is reverse biased and $I_C = 0$. This is the current one measures in measuring BV_{EBO}. Similarly, I_{CBO} is the saturation current of the collector–base diode when the emitter is open-circuit, i.e., it is the collector current when the base–collector diode is reverse biased and $I_E = 0$. This is the current one measures in measuring BV_{CBO}. I_{CBO} is indicated in Figure 9.27(c).

Let us apply Eq. (9.183) to the BV_{CEO} measurement setup shown in Figure 9.27(a). We note that when V_{CE} is near BV_{CEO}, the collector–base diode is reverse biased. Also, at $I_B = 0$, $I_C = -I_E$. Therefore, Eq. (9.183) gives, for the common-emitter configuration with the base–collector junction reverse biased and $I_B = 0$,

$$I_C = \frac{I_{CBO}}{1 - \alpha_F}. \tag{9.184}$$

This is the saturation current in the common-emitter configuration. We shall denote this current by I_{CEO}, i.e.,

$$I_{CEO} = \frac{I_{CBO}}{1 - \alpha_0}, \tag{9.185}$$

where we have used the fact that $\alpha_F = \alpha_0$. This current is also indicated in Figure 9.27(c). It is clear from Eq. (9.185) that I_{CEO} is significantly larger than I_{CBO}, since α_0 is usually less than but close to unity. This is indicated in Figure 9.27(c).

9.7.3 Relation between BV_{CEO} and BV_{CBO}

The breakdown voltages are usually determined experimentally, rather than calculated from some model. The avalanche multiplication factor M in a reverse-biased diode is often expressed in terms of its breakdown voltage BV using the *empirical* formula (Miller, 1955)

$$M(V) = \frac{1}{1 - (V/BV)^m}, \tag{9.186}$$

where V is the reverse-bias voltage and m is a number between 3 and 6 depending on the material and its resistivity. Thus, for the reverse-biased collector–base diode, we have

$$M(V_{CB}) = \frac{1}{1 - (V_{CB}/BV_{CBO})^m}. \tag{9.187}$$

Equation (9.185) implies that I_{CEO} becomes infinite when $\alpha_0 = 1$. From Eq. (9.175), this means that when the collector voltage reaches BV_{CEO},

$$\gamma \alpha_T M(V_{CB}) = \gamma \alpha_T M(BV_{CEO}) = 1. \tag{9.188}$$

Equations (9.187) and (9.188) give

$$\frac{BV_{CEO}}{BV_{CBO}} = (1 - \gamma \alpha_T)^{1/m}. \tag{9.189}$$

Since $1 - \gamma \alpha_T < 1$, Eq. (9.189) indicates that BV_{CEO} can be substantially smaller than BV_{CBO}. This is indicated in Figure 9.27(c).

Another way of comparing these breakdown voltages is to note that it takes M approaching infinity to cause collector–base breakdown, while it takes M only slightly larger than unity to cause collector–emitter breakdown (see Exercise 9.7). From Eq. (9.178), $\gamma \alpha_T = \alpha_0(M = 1) = \beta_0/(1 + \beta_0)$, where we have used Eq. (9.78). Thus, Eq. (9.189) can also be written as

$$\frac{BV_{CEO}}{BV_{CBO}} = \left(\frac{1}{1 + \beta_0}\right)^{1/m} \approx \left(\frac{1}{\beta_0}\right)^{1/m}. \tag{9.190}$$

Equation (9.190) shows that **there is a tradeoff between collector–emitter breakdown voltage and current gain** of a transistor.

It should be noted that the relationship between BV_{CEO} and BV_{CBO} in Eq. (9.190) is valid only when collector–base junction breakdown is governed by the intrinsic-base–collector diode, and not the extrinsic-base–collector diode. This may or may not be the case in a typical transistor, depending on the device structure and the fabrication process employed. (The BV_{CBO} of a modern vertical bipolar transistor, with its extrinsic base formed independently of the intrinsic base and its collector optimized for minimal capacitance, is usually determined by the intrinsic-base–collector diode rather than the extrinsic-base–collector diode. The design and characteristics of modern vertical bipolar transistors are covered in Chapter 10.) Figure 9.29 is a plot

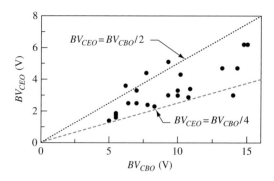

Figure 9.29 Reported BV_{CEO} versus BV_{CBO} data for recently published vertical n–p–n transistors

of BV_{CEO} versus BV_{CBO} based on data reported in recent literature for vertical n–p–n transistors. It shows BV_{CEO} is typically a factor of 2 to 4 smaller than BV_{CBO}.

9.7.4 Breakdown Voltages of Symmetric Lateral Bipolar Transistors on SOI

Consider the base–collector diode of an n–p–n transistor. For simplicity, let us treat the base–collector diode as a one-sided junction since one side of the diode is much more heavily doped than the other side. In the case of a vertical transistor, Figures 9.3 (a) and (b) indicate that the collector is the lightly doped side of the base–collector diode. In the case of a lateral transistor, Figure 9.1(b) indicates that the base is the lightly doped side of the base–collector diode. Therefore, for a vertical transistor and a lateral transistor having the same uniform base doping concentration and having the same reverse bias voltage across the base–collector diode, the maximum electric field in the base–collector space-charge region is smaller for the vertical transistor, where the maximum field is determined by its collector doping concentration, than for the lateral transistor where the maximum field is determined by its base doping concentration. Therefore, for the same uniform base doping profile, *the intrinsic-base–collector diode breakdown voltage of a lateral transistor could be appreciably smaller than that of a vertical transistor*.

The measured BV_{CEO} of a lateral transistor is a strong function of the details of the device fabrication process. Carefully tailoring the process to avoid overlap between the heavily doped extrinsic-base region with the heavily doped emitter/collector region can minimize unintended high-field at the space-charge region corners. Reported data suggest that $BV_{CEO} > 1.5$ V can be obtained readily (Cai *et al.*, 2014).

Exercises

9.1 The electric field as seen by holes in an n-type semiconductor is given by Eq. (9.20), i.e:

$$\mathscr{E}\,(\text{n-region}) = -\frac{kT}{q}\left(\frac{1}{n_n}\frac{dn_n}{dx} - \frac{1}{n_{ie}^2}\frac{dn_{ie}^2}{dx}\right).$$

Derive this equation, stating clearly the approximations made in the derivation.

9.2 The hole current density in the n-side of a p–n diode is given by Eq. (9.26), i.e.,

$$J_p(x) = -qD_p\frac{n_{ie}^2}{n_n}\frac{d}{dx}\left(\frac{n_n p_n}{n_{ie}^2}\right).$$

Derive this equation, stating clearly the approximations made in the derivation.

9.3 For a polysilicon emitter with the emitter–base junction located at $x = 0$ and the silicon–polysilicon interface located at $x = -W_E$, the emitter Gummel number is given by Eq. (9.69), namely

$$G_E = N_E \left(\frac{n_i^2}{n_{ieE}^2} \right) \left(\frac{W_E}{D_{pE}} + \frac{1}{S_p} \right).$$

One model (Ning and Isaac, 1980) for relating S_p to the properties of the polysilicon layer is to assume that there is no interfacial oxide, so that the transport of holes through the interface is simply determined by the properties of the polysilicon layer. Let W_{E1} be the thickness of the polysilicon layer, and D_{pE1} and L_{pE1} be the hole diffusion coefficient and hole diffusion length, respectively, in the polysilicon. Assume an ohmic metal–polysilicon contact.

(a)   Let $\Delta p_n(-W_E)$ be the excess hole concentration at the polysilicon–silicon interface, and let x' denote the distance from the polysilicon–silicon interface, i.e., $x' = -(x + W_E)$. Show that the excess hole distribution in the polysilicon layer, $\Delta p_n(x')$, is given by

$$\Delta p_n(x') = \Delta p_n(-W_E) \frac{\sinh\left[(W_{E1} - x')/L_{pE1}\right]}{\sinh\left(W_{E1}/L_{pE1}\right)}.$$

(b)   The relationship between S_p and the hole current density entering the polysilicon layer is given by Eq. (9.59). Show that

$$S_p = \frac{D_{pE1}}{L_{pE1} \tanh\left(W_{E1}/L_{pE1}\right)}.$$

9.4   Consider an n–p–n transistor with negligible parasitic resistances (which will be included in Exercise 9.5). Equations (9.182) and (9.183) give

$$I_E = -I_{EBO}\left[\exp\left(qV'_{BE}/kT\right) - 1\right] - \alpha_R I_C$$

and

$$I_C = -I_{CBO}\left[\exp\left(qV'_{BC}/kT\right) - 1\right] - \alpha_F I_E,$$

where V'_{BE} and V'_{BC} are the internal base–emitter and base–collector junction bias voltages. If the transistor is operated in saturation, i.e., both V'_{BE} and V'_{BC} are positive, show that the internal collector–emitter voltage, $V'_{CE} = V'_C - V'_E$, is related to the currents by

$$V'_{CE} = \frac{kT}{q} \ln\left[\frac{\alpha_F(I_{EBO} - I_E - \alpha_R I_C)}{\alpha_R(I_{CBO} - I_C - \alpha_F I_E)}\right].$$

[*Hint*: Use Eqs. (9.144), (9.180), and (9.181) to show that $I_{CBO}/I_{EBO} = \alpha_F/\alpha_R$.]

9.5   From the expression for V'_{CE} in Exercise 9.4, show that if the emitter and collector series resistances r_e and r_c are included, and if the saturation currents I_{EBO} and I_{CBO} are negligible, the voltage drop across the collector and emitter terminals, $V_{CE} = V_C - V_E$, is given by

$$V_{CE} = \frac{kT}{q} \ln\left[\frac{I_B + I_C(1 - \alpha_R)}{\alpha_R[I_B - I_C(1 - \alpha_F)/\alpha_F]}\right] + r_e(I_B + I_C) + r_c I_C$$

and that for open-circuit collector

$$V_{CE}(I_C = 0) = \frac{kT}{q} \ln\left(\frac{1}{\alpha_R}\right) + r_e I_B.$$

[The emitter resistance r_e is often determined from a plot of the saturation open-collector voltage, $V_{CE}(I_C = 0)$, as a function of I_B (Ebers and Moll, 1954; Filensky and Beneking, 1981). The collector resistance r_c can be determined in a similar way by interchanging the emitter and collector connections.]

9.6 For an n–p–n transistor, the base transport factor α_T is given in Eq. (9.177), i.e.,

$$\alpha_T \equiv \frac{J_n(x = W_B)}{J_n(x = 0)},$$

where the intrinsic-base layer is located between $x = 0$ and $x = W_B$. For a uniformly doped base, the excess-electron distribution is given by Eq. (3.38), namely

$$n_p - n_{p0} = n_{p0}\left[\exp(qV_{BE}/kT) - 1\right]\frac{\sinh[(W_B - x)/L_{nB}]}{\sinh(W_B/L_{nB})}.$$

If the electron current in the base is due to diffusion current only, show that

$$\alpha_T = \left(\cosh\frac{W_B}{L_{nB}}\right)^{-1}.$$

Use Figure 3.14(c) to estimate α_T for a uniformly doped base with $N_B = 1 \times 10^{18}\text{cm}^{-3}$ and $W_B = 100$ nm, and show that our assumption of negligible recombination in the intrinsic base is justified.

9.7 If M is the avalanche multiplication factor for the base–collector junction, and β_0 is the common-emitter current gain at negligible base–collector junction avalanche, show that the collector–emitter breakdown occurs when

$$M - 1 = \frac{1}{\beta_0}.$$

[*Hint*: Use Eqs. (9.178) and (9.188).] (It is interesting to note that since β_0 is typically about 100, collector–emitter breakdown occurs when M is only slightly larger than unity. That is, it does not take much base–collector junction avalanche to cause collector–emitter breakdown.)

9.8 Show that the solution to the equation

$$\frac{d^2 \Delta n_p(x)}{dx^2} - \frac{\Delta n_p(x)}{L_{nB}'^2} = 0,$$

where

$$L_{nB}' \equiv \sqrt{\frac{\tau_{nB} D_{nB}}{1 + j\omega\tau_{nB}}} = \frac{L_{nB}}{\sqrt{1 + j\omega\tau_{nB}}}$$

is

$$\Delta n_p(x) = \frac{n_{p0} q v_{be}}{kT} \exp\left(\frac{qV_{BE}}{kT}\right) \frac{\sinh\left[(W_B - x)/L'_{nB}\right]}{\sinh\left(W_B/L'_{nB}\right)}.$$

9.9 There are three components in the base current of a bipolar transistor. For an n–p–n transistor, one of the components is $J_{B,rec}$ due to holes in quasineutral p-type base recombining with the electrons injected from the emitter into the base region. Show that

$$J_{B,rec} = \frac{qW_B n_{ieB}^2}{2\tau_{nB}N_B} \exp(qV_{BE}/kT),$$

where τ_{nB} is the lifetime in the base region. Also show that $J_{B,rec}$ is small compared to

$$J_B = \frac{qD_{pE}n_{ieE}^2 \exp(qV_{BE}/kT)}{N_E L_{pE} \tanh\left(W_E/L_{pE}\right)},$$

which is the base current due to holes injected into the emitter, i.e., Eq. (9.71). Assume $W_E = 100$ nm, $N_E = 1 \times 10^{20}$ cm^{-3}, $N_B = 1 \times 10^{18}$ cm^{-3}, and $W_B = 50$ nm.

9.10 Consider an n$^+$–p diode, with the n$^+$ emitter side being wide compared with its hole diffusion length and the p base side being narrow compared with its electron diffusion length. The diffusion capacitance due to electron storage in the base is C_{Dn}, given by Eq. (9.133), and that due to hole storage in the emitter is C_{Dp}, given by Eq. (9.134). Assume the emitter to have a doping concentration of 10^{-20} cm^{-3} and the base to have a width of 100 nm and a doping concentration of 10^{17} cm^{-3}.

(a) If the heavy-doping effect is ignored, show that capacitance ratio C_{Dn}/C_{Dp} is

$$\frac{C_{Dn}}{C_{Dp}} = \frac{2}{3}\frac{N_E}{N_B}\frac{W_B}{L_{pE}} \quad \text{(heavy-doping effect ignored)}.$$

Evaluate this ratio for the n$^+$–p diode.

(b) Show that when the heavy-doping effect is included, the capacitance ratio becomes

$$\frac{C_{Dn}}{C_{Dp}} = \frac{2}{3}\left(\frac{n_{ieB}^2}{n_{ieE}^2}\right)\left(\frac{N_E}{N_B}\right)\frac{W_B}{L_{pE}} \quad \text{(heavy-doping effect included)},$$

where the subscript B denotes quantities in the base and the subscript E denotes quantities in the emitter. Evaluate this ratio for the n$^+$–p diode.

(This exercise demonstrates that heavy-doping effects cannot be ignored in any quantitative modeling of the switching speed of a diode.)

9.11 The collector current as a function of V_{BE} valid for low injection, $I_C(V_{BE})_{low\text{-}injection}$, is given by Eq. (9.57). The corresponding equation valid for all injection levels, $I_C(V_{BE})_{all\text{-}injection}$, is given by Eq. (9.53). Plot the ratio $I_C(V_{BE})_{low\text{-}injection}/I_C(V_{BE})_{all\text{-}injection}$ as a function of V_{BE} from $V_{BE} = 0.5$ V to $V_{BE} = 1.0$ V, for $N_B = 1 \times 10^{17}$ cm^{-3}, 1×10^{18} cm^{-3}, and 1×10^{19} cm^{-3}. (This exercise gives insight into the inaccuracy of using the low-injection approximation to modeling collector current at large V_{BE}.)

10   Bipolar Device Design

Bipolar device design can be considered in two parts. The first part deals with designing bipolar transistors in general, independent of their intended application. In this case, the goal is to reduce as much as possible, consistent with the state-of-the-art fabrication technology, all the internal resistance and capacitance components of the transistor. The second part deals with designing a bipolar transistor for a specific circuit application. In this case, the optimal device design point depends on the application. The design of a bipolar transistor in general is covered in this chapter. The optimization of a transistor for specific applications is discussed in Chapter 11.

We first consider the design of a silicon vertical bipolar transistor, where the emitter, base, and collector regions are all silicon. The design a vertical SiGe-base transistor is covered in Section 10.4. The design of a symmetric lateral bipolar transistor on SOI is discussed in Section 10.5.

10.1   Design of the Emitter of a Vertical Bipolar Transistor

The emitter parameters affect only the base current, and have no effect on the collector current (cf. Section 9.2.3). In theory, a device designer can vary the emitter design to vary the base current. In practice, this is rarely done, for two reasons. First, for digital-circuit applications, as long as current gain is greater than about 20, the performance of a bipolar transistor is rather insensitive to its base current (Ning *et al.*, 1981). For many analog-circuit applications, once the current gain is adequate, the reproducibility of the base current is more important than its magnitude. Therefore, there is really no particular reason to tune the base current of a bipolar device by tuning the emitter design, once a low and reproducible base current is obtained. Second, the emitter is formed toward the end of the device fabrication process. Any change to the emitter process to tune the base current could affect the other device parameters. As a result, once a bipolar technology is ready for manufacturing, its emitter fabrication process is usually fixed. All that a device designer can do to alter the device and circuit characteristics in this bipolar technology is to change the base and the collector designs, which can be accomplished independently of the emitter process and hence has no effect on the base current.

The objective in designing the emitter of a bipolar transistor is then to achieve a low but reproducible base current while at the same time minimizing the emitter series resistance. As illustrated in Figure 9.3, a vertical bipolar transistor has either a diffused, or an implanted-and-diffused, emitter or a polysilicon emitter.

10.1.1   Diffused or Implanted-and-Diffused Emitter

A diffused or implanted-and-diffused emitter is formed by pre-doping a surface region of the silicon above the intrinsic base and then thermally diffusing the dopant to a desired depth. As shown in Eq. (9.71), I_B is inversely proportional to N_E. Therefore, to minimize both I_B and the emitter series resistance r_e, a diffused emitter is usually doped as heavily as possible. For n–p–n transistors, arsenic, instead of phosphorus, is usually used as the dopant, because arsenic gives a more abrupt doping profile than phosphorus. A more abrupt emitter doping profile leads to a shallower emitter junction, which is needed for achieving a thin intrinsic base. Also, a shallower emitter has a smaller vertical junction area and associated capacitance. A diffused emitter typically has a peak doping concentration of about 2×10^{20} cm^{-3}, as indicated in Figure 9.3(a).

A diffused emitter is contacted either directly by a metal, or by a metal via a metal silicide layer. As discussed in Section 3.2.3, the contact resistivity is a function of the metal or metal silicide used, as well as a function of N_E at the contact. Recent advances in contact technology suggest contact resistivity approaching 10^{-9} Ω-cm^2 is possible (Zhang *et al.*, 2013). However, most reported emitter contact resistivity is about 10^{-7} Ω-cm^2.

Using the resistivity values of silicon shown in Figure 2.10, the specific series resistivity of a 0.5-μm-deep silicon region, with an averaged doping concentration of 1×10^{20} cm^{-3}, is about 4×10^{-8} Ω-cm^2. Therefore, r_e of a diffused emitter is usually dominated by its metal–silicon contact resistance; the series resistance of the doped-silicon region itself is small in comparison.

It can be inferred from Figure 9.3(a) that the intrinsic-base width W_B is related to the emitter junction depth x_{jE} and the base junction depth x_{jB} by

$$W_B = x_{jB} - x_{jE} \quad \text{(vertical transistor)}. \tag{10.1}$$

One of the objectives in the design of the intrinsic base is to minimize its width. For W_B to be well controlled, reproducible, and thin, x_{jE} should be as small as possible. If x_{jE} is larger than W_B, then W_B is given by the difference of two large numbers of comparable magnitude and hence could have large fluctuation.

Commonly used metal silicides are formed by depositing a layer of the appropriate metal on the silicon surface and then reacting the metal with the underlying silicon to form silicide. The emitter width W_E, i.e., the thickness of the heavily doped emitter layer, is therefore reduced when metal silicide is used for emitter contact, because silicon in the emitter is consumed in the metal silicide formation process. If the starting W_E is smaller than the minority-carrier diffusion length to begin with, the silicide formation process could cause I_B to increase and have large variation (Ning and Isaac, 1980). Referring to Figure 3.14(c), we see that x_{jE}, or the starting W_E, should be larger than 0.3 μm. ***Diffused emitters are therefore not suitable for base widths of less than 100 nm.***

10.1.2   Polysilicon Emitter

Practically all modern vertical bipolar transistors employ a polysilicon emitter. In this case, the emitter is formed by doping a polysilicon layer heavily and then activating the doped polysilicon layer just sufficiently to obtain reproducible I_B and low r_e. x_{jE}, measured from the silicon–polysilicon interface, can be as small as 25 nm (Warnock, 1995). With polysilicon-emitter technology, reproducible base widths of 50 nm or less can be obtained. Polysilicon-emitter process recipes are usually considered proprietary. There is a vast amount of literature on the physics of polysilicon-emitter devices (Ashburn, 1988; Kapoor and Roulston, 1989). Interested readers are referred to these publications.

The base current of a polysilicon-emitter transistor is given by Eqs. (9.65) to (9.67), with a surface recombination velocity, S_p, dependent on the details of the polysilicon-emitter fabrication process. In practice, S_p is usually used as a fitting parameter to the measured I_B. In general, I_B of a polysilicon-emitter transistor is sufficiently low so that current gains in excess of 100 are readily achievable.

The series resistance of a polysilicon emitter includes the polysilicon–silicon contact resistance, resistance of the polysilicon layer, and resistance of the metal–polysilicon contact. The specific resistivity of a metal–polysilicon contact is about the same as that of a metal–silicon contact. The polysilicon–silicon contact resistance is a strong function of the polysilicon-emitter fabrication process and can vary by large amounts (Chor *et al.*, 1985). The published data (Iinuma *et al.*, 1995; Kondo *et al.*, 1995; Uchino *et al.*, 1995; Shiba *et al.*, 1996) suggest that a total emitter specific resistivity, which includes contributions from both the polysilicon–silicon interface and the metal–polysilicon contact, of $7-50\ \Omega\text{-}\mu m^2$ should be obtainable (cf. Exercise 10.7).

The small junction depth of a polysilicon emitter implies a relatively small perimeter, or vertical, extrinsic-base–emitter junction area. The total emitter–base junction capacitance of a polysilicon emitter is therefore much smaller than that of a diffused emitter. For a 0.3-μm emitter stripe, the total emitter–base junction capacitance of a polysilicon emitter can be less than $1/3$ of that of a diffused emitter.

It should be pointed out that the junction of a polysilicon emitter is so shallow that the commonly used secondary-ion mass spectroscopy (SIMS) technique for measuring dopant concentration profiles often indicates an emitter junction deeper than it really is. The real emitter junction depth can be inferred from the p-type base SIMS profile, which shows a dip where the n-type and p-type doping concentrations are equal (Hu and Schmidt, 1968). This is illustrated schematically in Figure 10.1. These dips are also evident in Figures 9.3(a) and (b).

10.2   Design of the Base Region of a Vertical Bipolar Transistor

The design of the base of a vertical bipolar transistor can be complex, because of the tradeoffs between the desired device characteristics and the complexity of the

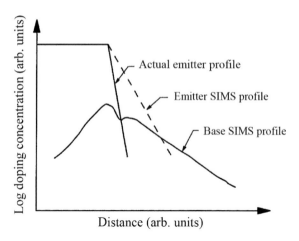

Figure 10.1 Schematic illustrating the measured SIMS doping profiles of the emitter and base of a modern vertical n–p–n transistor. The measured emitter SIMS profile is usually less abrupt than the real one

fabrication process for realizing the design. In this section, the relationship between the physical and electrical parameters of the base is derived, and the design tradeoffs are discussed. Optimization of the base design for circuit applications will be covered in Chapter 11.

Referring to Figure 9.16, the base region consists of two parts: the intrinsic base directly underneath the emitter and the extrinsic base connecting the intrinsic base to the base terminal. As a first-order but good approximation, the intrinsic base is what determines the collector current characteristics, and hence the intrinsic performance of a transistor. The extrinsic base is an integral part of a bipolar transistor. It is a parasitic component in that it does not contribute appreciably to the collector current, at least for properly designed transistors.

In general, designing the extrinsic base is very simple: the extrinsic-base area and its associated capacitance and resistance should be as small as possible. How this is accomplished depends on the fabrication process used. A major focus in bipolar-technology research and development has been to minimize the parasitic resistance and capacitance associated with the extrinsic base. The interested reader is referred to the vast literature on the subject (Nakamura and Nishizawa, 1995; Warnock, 1995; Asbeck and Nakamura, 2001; and the references therein).

Any adverse effect of the extrinsic base on the breakdown voltages of the base–emitter and base–collector diodes should be minimized. This is accomplished by having the dopant distribution of the extrinsic base not extending into the intrinsic base. Extrinsic-base encroachment on the intrinsic base will lead to a smaller collector current as well as degraded steady-state and ac characteristics (Li *et al.*, 1987; Lu *et al.*, 1987).

For an optimally designed vertical bipolar transistor, extrinsic-base encroachment is usually negligible. Therefore, only the design of the intrinsic base will be discussed further in this section. We first consider the design of a Si-base in this section. The design of a SiGe-base is covered in Section 10.4.

10.2.1 Base Sheet Resistivity and Collector Current Density

As shown in Figure 9.8, the measured I_C of a typical bipolar transistor is ideal, i.e., varying as $\exp(qV_{BE}/kT)$, for V_{BE} less than about 0.8 V. For this ideal region, J_{C0} for an n−p−n transistor is given by Eq. (9.47), which is repeated here for convenience:

$$J_{C0} = \frac{q}{\int_0^{W_B} \left[p_p(x)/D_{nB}(x)n_{ieB}^2(x) \right] dx}. \tag{10.2}$$

The effective intrinsic-carrier concentration n_{ieB} is given by Eq. (9.14). It can be used to account for heavy-doping effect as well as any bandgap-engineering effect. We shall first consider the case where n_{ieB} is used to account for heavy-doping effect. The case of using n_{ieB} to account for base-bandgap engineering will be covered in Section 10.4.

For device design purposes, it is often convenient to assume that both D_{nB} and n_{ieB} are slowly varying functions of x and hence can be approximated by some average values. That is, Eq. (10.2) is often written as

$$J_{C0} \approx \frac{q\bar{D}_{nB}\bar{n}_{ieB}^2}{\int_0^{W_B} p_p(x)dx} \qquad \text{(silicon base).} \tag{10.3}$$

At low injection, $p_p(x) = N_B(x)$, and Eq. (10.3) can be further simplified to

$$J_{C0} \approx \frac{q\bar{D}_{nB}\bar{n}_{ieB}^2}{\int_0^{W_B} N_B(x)dx} \qquad \text{(low current).} \tag{10.4}$$

The integral in the denominator of Eq. (10.4) is the total integrated base dose, also referred to as the base Gummel number. Using ion-implantation techniques for doping the intrinsic base, the base Gummel number, and hence J_{C0}, can be controlled quite precisely and reproducibly.

The sheet resistivity of the intrinsic base, R_{Sbi}, is

$$R_{Sbi} = \left(q \int_0^{W_B} p_p(x)\mu_p(x)dx \right)^{-1}. \tag{10.5}$$

Again, for device design purposes, it is convenient to assume an average mobility and rewrite Eq. (10.5) as

$$R_{Sbi} \approx \left(q\bar{\mu}_p \int_0^{W_B} p_p(x)dx \right)^{-1}. \tag{10.6}$$

Substituting Eq. (10.6) into Eq. (10.3), we obtain

$$J_{C0} \approx q^2 \bar{D}_{nB}\bar{\mu}_p\bar{n}_{ieB}^2 R_{Sbi} \qquad \text{(silicon base).} \tag{10.7}$$

That is, **the collector current of a silicon-base transistor is approximately proportional to its intrinsic-base sheet resistivity**. This direct correlation is valid for R_{Sbi} between 500 and $20 \times 10^3 \ \Omega/\square$, which is the range of interest in most bipolar device designs (Tang, 1980).

10.2.2   Ion-Implanted versus Epitaxially Grown Intrinsic Base

Many modern vertical bipolar transistors employ ion implantation, followed by thermal annealing, to form the intrinsic base. In this case, the intrinsic-base doping profile is approximately a *Gaussian* distribution, often with an exponentially decreasing tail. The collector doping concentration of a modern vertical bipolar transistor is relatively high, often in excess of $1 \times 10^{17} \ \text{cm}^{-3}$. This concentration is usually high compared with the tail of the base dopant distribution. As a result, the lightly doped tail is often clipped off by the collector dopant and has little effect on the collector current (cf. Figure 9.3). Therefore, for simplicity of discussion and analysis, we shall ignore this tail of the base dopant distribution.

If the Gaussian base dopant distribution peaks at the emitter–base junction located at $x = 0$, then the base doping concentration can be described by

$$N_B(x) = N_{B\max} \, \exp\left(-\frac{x^2}{2\sigma^2}\right), \tag{10.8}$$

where σ and $N_{B\max}$ are the standard deviation and peak concentration, respectively, of the distribution. For most vertical transistors with implanted base, the peak doping concentration in the base is approximately 10–100-times that in the collector. Here, for purposes of discussion, we assume $N_{B\max}/N_C = 100$. This implies a base width of

$$W_B \approx 3\sigma \tag{10.9}$$

for the Gaussian base dopant distribution.

The most advanced vertical transistors, for example, the one depicted on the right side of Figure 9.1(b), have an intrinsic base formed by epitaxial growth of a thin, *in situ* doped, silicon layer on top of the collector. In this case, the base dopant distribution is typically uniform, or *boxlike*.

Figure 10.2 illustrates a box profile of $N_B = 1 \times 10^{18} \ \text{cm}^{-3}$, and a Gaussian profile of the same base Gummel number and the same W_B. It shows that the peak concentration of the Gaussian profile is more than twice that of the boxlike profile. Therefore, **C_{dBE} of a Gaussian base doping profile is larger than that of a boxlike doping profile.** However, this does not imply that a box-profile base is necessarily preferred, for there are many other factors or parameters, such as base transit time and ease of fabrication, that must also be considered. Before we proceed further, we need to discuss the electric field in the quasineutral intrinsic-base region, since the transport of minority carriers in the base depends on the electric field in it.

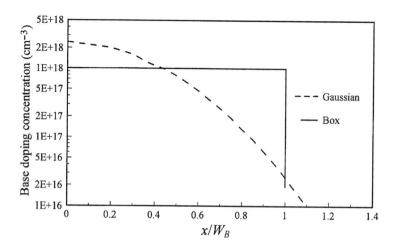

Figure 10.2 Schematic illustration of a boxlike doping profile and a Gaussian doping profile for the same W_B and the same base Gummel number. The peak doping concentration of the Gaussian profile is approximately 2.4 times that of the boxlike profile

10.2.2.1 Electric Field in the Quasineutral Intrinsic Base

The electric field as seen by electrons in the base of an n–p–n transistor is given by Eq. (9.19), namely

$$\mathscr{E}(\text{p-base}) = \frac{kT}{q}\left(\frac{1}{p_p}\frac{dp_p}{dx} - \frac{1}{n_{ieB}^2}\frac{dn_{ieB}^2}{dx}\right). \tag{10.10}$$

Using Eq. (9.14), Eq. (10.10) can also be written as

$$\mathscr{E}(\text{p-base}) = \frac{kT}{q}\frac{1}{p_p}\frac{dp_p}{dx} - \frac{1}{q}\frac{d\Delta E_{gB}}{dx}, \tag{10.11}$$

where ΔE_{gB} is the apparent bandgap-narrowing parameter due to heavy-doping effect.

At low injection currents, $p_p \approx N_B$, Eq. (10.11) gives the built-in electric field in the base caused by the base doping distribution and heavy-doping effect. As N_B increases, ΔE_{gB} increases, and hence dN_B/dx and $d\Delta E_{gB}/dx$ have the same sign. Equation (10.11) shows that *the electric field due to heavy-doping effect always tends to offset the electric field due to dopant distribution*.

When transistors are designed with base widths much larger than 100 nm, the peak N_B is usually about 10^{17} cm^{-3} or smaller. For such low concentrations, the effect of heavy doping is negligible, as shown in Figure 9.4. In this case, a graded base doping profile can result in a substantial electric field in the base, which enhances the drift component of the minority-carrier current traversing the base layer. These are so-called *drift transistors*. They have a higher cutoff frequency than transistors with a more uniform base doping profile (Ghandhi, 1968; Sze, 1981).

Modern vertical bipolar transistors, however, have peak N_B larger than 10^{18} cm^{-3}, as indicated in Figure 9.3. For these devices, the electric field due to the $d\Delta E_{gB}/dx$

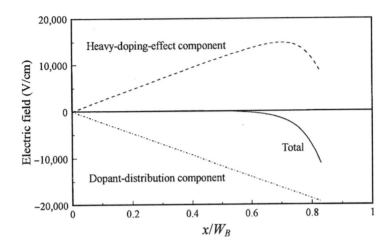

Figure 10.3 Electric fields in the quasineutral intrinsic-base region with a Gaussian doping profile. The total electric field is the sum of the dopant-distribution and the bandgap-narrowing components. The Gaussian-profile parameters are $\sigma = W_B/3$, $W_B = 100$ nm, and $N_{B\max} = 2.4 \times 10^{18}$ cm^{-3}

term must be included, which tends to cancel the electric field due to the dN_B/dx term. As a result, the net electric field in the quasineutral intrinsic base of a modern vertical bipolar transistor could be relatively small. In other words, *the drift-transistor concept is less important in thin-base vertical transistors than in wide-base vertical transistors*. This is demonstrated in Figure 10.3 for a Gaussian-profile base design.

For the Gaussian base profile given by Eq. (10.8), the electric field at low injection is

$$\mathscr{E}(\text{Gaussian-base}) = \left(-\frac{kT}{q}\frac{x}{\sigma^2}\right) + \left(-\frac{1}{q}\frac{d\Delta E_{gB}}{dx}\right). \tag{10.12}$$

The electric field components as well as the total electric field are shown in Figure 10.3. The bandgap-narrowing parameter given by Eq. (9.17) is used, and the base width is assumed to be 100 nm. Figure 10.3 shows that, for this specific Gaussian base doping profile, the effect of heavy doping almost completely offsets the effect of nonuniform dopant distribution, except for the region near the base–collector junction, where the base doping concentration is relatively small and hence the effect of heavy doping is negligible. This lightly doped region near the base–collector junction is most likely depleted in normal device operation and hence does not form part of the quasineutral base. Therefore, the net electric field in the entire quasineutral base region is negligible.

For non-Gaussian base doping profiles, the cancellation of the electric field components may not be as complete as suggested in Figure 10.3. Nonetheless, the cancellation is substantial. The reader is referred to the literature (Suzuki, 1991) for more examples of similar calculations.

10.2.3 General Expression for Base Transit Time

The base transit time of a bipolar transistor is given by Eq. (9.106), namely

$$t_B \equiv |Q_B|/|J_C|, \tag{10.13}$$

where J_C is the collector current density and Q_B is the excess minority-carrier charge per unit area stored in the base region. For an n–p–n transistor, Q_B is given by

$$Q_B = -q \int_0^{W_B} \left[n_p(x) - n_{p0}(x) \right] dx. \tag{10.14}$$

From Eq. (9.25), the collector current density in the base is

$$J_C(x) = \frac{q D_{nB}(x) n_{ieB}^2(x)}{p_p(x)} \frac{d}{dx} \left(\frac{n_p(x) p_p(x)}{n_{ieB}^2(x)} \right). \tag{10.15}$$

Since recombination in a thin base is negligible, J_C is independent of x, and Eq. (10.15) can be rearranged and then integrated to give

$$J_C \int_x^{W_B} \frac{p_p(x')}{q D_{nB}(x') n_{ieB}^2(x')} dx' = \int_x^{W_B} d \left(\frac{p_p(x') n_p(x')}{n_{ieB}^2(x')} \right)$$

$$= -\frac{p_p(x) n_p(x)}{n_{ieB}^2(x)} + \frac{p_p(W_B) n_p(W_B)}{n_{ieB}^2(W_B)}$$

$$= -\frac{p_p(x) n_p(x)}{n_{ieB}^2(x)} + \frac{p_{p0}(x) n_{p0}(x)}{n_{ieB}^2(x)}$$

$$\approx -\frac{p_p(x)}{n_{ieB}^2(x)} \left[n_p(x) - n_{p0}(x) \right], \tag{10.16}$$

where we have used the boundary condition that the density of excess electrons at W_B is zero, i.e., $n_p(W_B) = n_{p0}(W_B)$, $p_p(W_B) = p_{p0}(W_B)$, and the identity $p_{p0}(x) n_{p0}(x) = n_{ieB}^2(x)$ [cf. Eq. (9.14)], and made the approximation of $p_p(x) \approx p_{p0}(x)$ in the third line to get to the last line. Equation (10.16) can be rearranged to give

$$n_p(x) - n_{p0}(x) \approx -\frac{J_C n_{ieB}^2(x)}{q \, p_p(x)} \int_x^{W_B} \frac{p_p(x')}{D_{nB}(x') n_{ieB}^2(x')} dx'. \tag{10.17}$$

Substituting Eqs. (10.14) and (10.17) into Eq. (10.13), we obtain

$$t_B = \int_0^{W_B} \frac{n_{ieB}^2(x)}{p_p(x)} \int_x^{W_B} \frac{p_p(x')}{D_{nB}(x') n_{ieB}^2(x')} dx' dx, \tag{10.18}$$

which is the **general expression for t_B of an n–p–n transistor** where n_{ieB} accounts for heavy-doping effect and any effect due to bandgap engineering (Kroemer, 1985).

At low injection, $p_p(x) \approx N_B(x)$ and Eq. (10.18) reduces to (Suzuki, 1991)

$$t_B \approx \int_0^{W_B} \frac{n_{ieB}^2(x)}{N_B(x)} \int_x^{W_B} \frac{N_B(x')}{D_{nB}(x')n_{ieB}^2(x')} dx' dx \qquad \text{(low injection)}. \qquad (10.19)$$

In the *uniform-bandgap approximation*, n_{ieB} is independent of x and Eq. (10.19) reduces further to (Moll and Ross, 1956)

$$t_B \approx \int_0^{W_B} \frac{1}{N_B(x)} \int_x^{W_B} \frac{N_B(x')}{D_{nB}(x')} dx' dx \qquad \text{(uniform } E_g). \qquad (10.20)$$

For a boxlike profile, both N_B and D_{nB} are constant, and Eq. (10.20) reduces further to

$$t_B \approx \frac{W_B^2}{2D_{nB}} \qquad \text{(uniform } E_g \text{ and } N_B), \qquad (10.21)$$

which is the same as Eq. (9.106) for the transit time for a uniformly doped thin-base diode.

10.3   Design of the Vertical Bipolar Transistor Collector Region

The cross-section of the physical structure of a modern vertical n–p–n bipolar transistor is illustrated schematically in Figure 10.4. The collector includes all the n-type regions underneath and surrounding the p-type base. It can be subdivided into four parts. The part (shaded in Figure 10.4) directly underneath the emitter and intrinsic base is the active region of the collector. This region is usually referred to simply as the *collector*. It is the region referred to in all the transport and current equations. The horizontal heavily doped (n$^+$) region underneath the collector is the *subcollector*, and the heavily doped (n$^+$) vertical region connecting the subcollector to

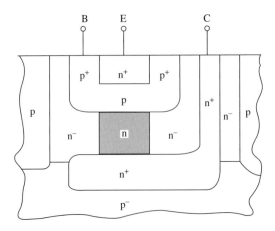

Figure 10.4 Schematic cross-section of a modern vertical n–p–n bipolar transistor, illustrating the various doped regions. This transistor employs p–n junction isolation

the collector terminal on the silicon surface is the *reach-through*. The remaining n-type regions make up the parasitic collector, which is usually relatively lightly doped (n^-) in order to minimize the total base–collector junction capacitance.

To first order, both the subcollector and the reach-through are there only to reduce the series resistance between the collector terminal and the active collector. However, as will be discussed in Sections 10.3.1 and 10.3.2, the proximity of the subcollector to the intrinsic base, that is, the thickness of the active collector layer, has a strong effect on the base–collector breakdown voltage and on the collector current characteristics at high current densities.

Just like the extrinsic base, the parasitic collector is an unavoidable part of a vertical bipolar transistor structure. In general, designing the parasitic collector is very simple: the parasitic-collector area and its associated capacitance should be as small as possible. As can be seen from Figure 10.4, the parasitic collector and the extrinsic base form a p–n diode. For a given extrinsic-base area, the parasitic-collector doping concentration should be as low as possible, in order to achieve the smallest capacitance. Interested readers are referred to the vast literature on the research and development of vertical bipolar technology, which describes methods for reducing the extrinsic-base area and reducing the parasitic-collector doping concentration (Nakamura and Nishizawa, 1995; Warnock, 1995).

The collector (we shall use the terms collector and active collector interchangeably whenever there is no confusion) and the subcollector are the only regions that affect the collector current of a vertical bipolar transistor. These are the only regions that will be discussed further in this section.

For bipolar transistors operated in the forward-active mode, i.e., with the base–collector junction reverse-biased at all times, the collector acts simply as a sink for the carriers injected from the emitter and traversing the base layer. As discussed in Section 9.3.3, how the collector and subcollector affect the collector current depends on whether or not the collector current density is large enough to cause significant base widening. Thus, the design of the collector of a vertical transistor will be discussed in two parts, one at low injection when base widening is negligible, and the other at high injection when base widening is significant.

10.3.1   Collector Design for Low-Injection Operation

As discussed in Section 9.3.3.2, to maintain negligible base widening, the collector current density J_C in a vertical bipolar transistor should be small compared to the maximum current density, $J_{\max}$, that can be supported by the collector doping concentration N_C. That is, J_C should satisfy the condition that

$$|J_C| \ll J_{\max} = q v_{sat} N_C \qquad \text{(negligible base widening)}, \qquad (10.22)$$

where v_{sat} is the saturated velocity for electrons in silicon, which is about $1 \times 10^7 \text{cm/s}$ (cf. Figure 2.11). To increase J_C without increasing the base-widening effect, N_C must be increased proportionately, which could cause degradation in certain device characteristics. Therefore, **_tradeoffs have to be made in the design of the collector_**. These design tradeoffs are discussed here.

- J_C and V_A tradeoff. The Early voltage V_A of a bipolar transistor is inversely proportional to C_{dBC} [cf. Eq. (9.94)]. As N_C is increased to allow a larger J_C, C_{dBC} is increased and V_A is decreased. Therefore, there is a tradeoff between the maximum operational collector-current design target of a vertical bipolar transistor and its Early voltage.
- Design tradeoff driven by base-collector junction avalanche effect. As discussed in Section 9.7, base–collector junction avalanche occurs when the electric field in the junction space-charge region becomes too large. Excessive base–collector junction avalanche can cause the collector current in a transistor to increase in an uncontrolled manner, potentially affecting circuit functionality. There are several ways to reduce avalanche multiplication in the base–collector junction. The most straightforward way is to reduce N_C, but that will proportionately reduce the allowed J_C. Alternatively, the base and/or the collector doping profiles, at or near the base–collector junction, can be designed to reduce the electric field in the junction.

The implanted base doping profile discussed in Section 10.2.2 has a lower electric field in the base–collector junction than the boxlike base doping profile. The collector doping profile can also be retrograded (i.e., graded with its concentration increasing with distance into the silicon) to reduce the electric field in the base–collector junction (Lee et al., 1996). A retrograded collector doping profile can be achieved readily by high-energy ion implantation. The vertical transistor doping profiles illustrated in Figures 9.3(a) and (b) show collectors with retrograded doping profiles. Qualitatively, grading the base doping profile, and/or retrograding the collector doping profile, is similar to sandwiching an i-layer between the base and collector doped regions. Introducing a thin i-layer between the p- and n-regions of a diode is quite effective in reducing the electric field in the junction, as discussed in Section 3.1.2.3.

Reducing base–collector junction avalanche, either by reducing the collector doping concentration or by grading the base doping profile and/or retrograding the collector doping profile, reduces C_{dBC} as well. This should help to improve the device and circuit performance (Lee et al., 1996). However, as can be seen from Eqs. (9.101) and (9.102), these techniques also lead to more base widening, or to base widening occurring at a lower J_C. Thus, reducing base–collector junction avalanche could lead to reduced current-density capability, and hence the maximum speed, of a vertical bipolar transistor (Lu and Chen, 1989). The tradeoff between base–collector junction avalanche effect and device and circuit speed will be discussed further in Chapter 11.

10.3.2   Collector Design for High-Injection Operation

In general, base-widening should be avoided in the operation of a vertical bipolar transistor. However, in the design of a bipolar circuit, there may be circumstances where operating with some degree of base-widening may result in an overall higher circuit speed, for example, when the transistor is driving a large load capacitance. As will be shown in Chapter 11, when a vertical bipolar transistor is operated with significant base widening, it is its emitter diffusion capacitance C_{DE}, that limits its speed. To minimize emitter diffusion capacitance, the total excess minority carriers stored in the collector should be minimized. To accomplish this goal, the total collector volume available for minority-carrier storage should also be minimized. That is, the thickness of the collector layer should be minimized, by reducing the thickness of the epitaxial layer grown after the subcollector region is formed, and the area of the extrinsic-base−collector junction area should be minimized, by adopting the self-aligned double-polysilicon deep-trench-isolated device structure indicated on the right side of Figure 9.1(a).

However, thinning the collector layer could lead to an increase in C_{dBC} if the collector-layer thickness is comparable to W_{dBC}. Also, a thin-collector design could result in reduced base−collector junction breakdown voltage. The net result is that, when operated at low injection, a thin-collector transistor may run slower than a thick-collector one. However, when operated at high injection, where base-widening is appreciable, the thin-collector transistor has smaller C_{DE} and may actually run faster than a thick-collector transistor (Tang *et al.*, 1983).

Designing the collector of a vertical bipolar transistor is a complex tradeoff process. The important point to remember is that *base-widening occurs readily in a vertical bipolar transistor; optimizing the tradeoff in the collector design is key to realizing the maximum performance of vertical bipolar circuits*.

10.4   SiGe-Base Vertical Bipolar Transistors

The energy bandgap of Ge ($\approx 0.66\,\text{eV}$) is significantly smaller than that of Si ($\approx 1.12\,\text{eV}$). When Ge is incorporated into Si, the Si energy bandgap becomes smaller (People, 1986; Van de Walle and Martin, 1986). By incorporating Ge into the base region of a Si bipolar transistor, the energy bandgap of the base region, and hence the accompanied device characteristics, can be modified (Iyer *et al.*, 1987).

The emitter of a typical SiGe-base vertical bipolar transistor is the same as that of a regular Si-base vertical bipolar transistor. In both transistors, it is a polysilicon emitter. As for the Ge distribution in the base, several variations of a graded Ge profile have been studied. The most commonly used profile is that of a triangular or linearly graded Ge distribution. This profile assumes a Ge distribution that starts at zero at the emitter end of the quasineutral base and increases at a constant rate across the base layer. It leads to a simple graded base bandgap that decreases linearly from the emitter end to the collector end.

In the fabrication of a typical SiGe-base vertical bipolar transistor, Ge is incorporated into a starting base layer prior to the polysilicon-emitter formation step. Depending on the details of the base and emitter formation steps, Ge may or may not end up in the single-crystalline region of the emitter. Once Ge ends up in the single-crystalline portion of the emitter, the Ge distribution at the emitter end of the quasineutral base will depend on the emitter junction depth x_{jE}. Therefore, a trapezoidal Ge profile, with a low but finite Ge concentration at the emitter end and a higher Ge concentration at the collector end, gives a more general description of a typical SiGe-base vertical transistor. A SiGe-base vertical transistor having a trapezoidal Ge distribution in its base can be modeled with close-form solutions. A triangular Ge profile and a constant-Ge profile can be treated as special cases of a trapezoidal Ge profile.

In Section 10.4.1, the properties of a polysilicon-emitter SiGe-base vertical transistor having a linearly graded base bandgap are discussed and compared to those of a polysilicon-emitter Si-base vertical transistor. For readers who desire only a first-order explanation of the difference between a SiGe-base transistor and a Si-base transistor, this simple description should be adequate.

The presence of Ge in the emitter may change the properties of the emitter region, which in turn may change the base current characteristics. The effect on base current due to the presence of Ge in the emitter is considered in Section 10.4.2.

The properties of a SiGe-base bipolar transistor having a trapezoidal Ge distribution in the base are discussed in greater depth in subsequent sections. The models developed there can be used for optimizing the Ge distribution beyond the simple triangular distribution, for improved device characteristics. The expressions for J_{C0}, V_A, and t_B are derived in Section 10.4.3 for a trapezoidal Ge distribution, and in Section 10.4.4 for a constant Ge distribution. The optimization of a Ge profile is discussed in Section 10.4.5. There are also subtle but interesting effects in a SiGe-base vertical transistor that are either absent or relatively unimportant in a Si-base vertical transistor. They are discussed in Sections 10.4.6 and 10.4.7. Section 10.4.8 is devoted to a discussion of the heterojunction nature of a SiGe-base vertical bipolar transistor. Finally, the device concept of a SiGe-base vertical bipolar transistor built on thin SOI is discussed in Section 10.4.9.

10.4.1   SiGe-Base Vertical Transistors Having Linearly Graded Base Bandgap

The top part of Figure 10.5(a) shows schematically the energy-band diagrams of an n^+-Si region and a physically separate p-SiGe region having a linearly graded Ge distribution. The bottom part of Figure 10.5(a) shows schematically the energy-band diagram when the n^+-Si region and the p-SiGe region are brought together to form the emitter−base diode of a SiGe-base vertical bipolar transistor. At thermal equilibrium, the Fermi level is flat across the diode. Figure 10.5(b) shows schematically the energy-band diagram of a vertical n−p−n bipolar transistor operating in the forward-active mode. The solid lines are for a transistor having Si for emitter, base, and collector. The dotted line is for a transistor having Si emitter, SiGe base, and Si collector.

A SiGe-base vertical transistor typically has the same polysilicon emitter as a Si-base vertical transistor. Therefore, *the base current of a SiGe-base transistor should be the same as that of a Si-base transistor*. This is indeed the case for most reported

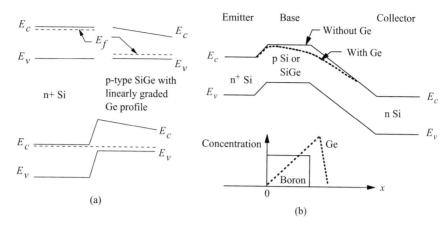

Figure 10.5 (a) Schematics of the energy-band diagrams of an n⁺-Si region and a p-SiGe region physically apart (top) and when the two regions are brought together to form a diode (bottom). (b) Schematic illustrations of the energy bands for a SiGe-base (dotted) and a Si-base (solid) vertical n–p–n transistor. Both transistors are assumed to have the same base doping profile

SiGe-base transistors. (Even when Ge ends up in the single-crystalline emitter region, the effect on base current is still small, as will be explained later in Section 10.4.2.)

Since base current is not affected by the presence of Ge in the base, we need to consider only the effect of Ge in the base on J_{C0}. The parameter n_{ieB} in Eq. (10.2) can be extended to include bandgap narrowing caused by the presence of Ge (Kroemer, 1985), i.e.,

$$n_{ieB}^2(\text{SiGe}, x) = n_{ieB}^2(\text{Si}, x)\gamma(x)\exp\left[\frac{\Delta E_{gB,SiGe}(x)}{kT}\right], \qquad (10.23)$$

where $n_{ieB}(\text{Si}, x)$ is the effective intrinsic-carrier concentration without Ge, $\Delta E_{gB,SiGe}(x)$ is the local bandgap narrowing in the base due to the presence of Ge, and the parameter

$$\gamma(x) \equiv \frac{(N_cN_v)_{SiGe}}{(N_cN_v)_{Si}} \qquad (10.24)$$

is introduced to account for any change in the density of states caused by the presence of Ge (Harame *et al.*, 1995a, 1995b). Effects due to heavy doping are contained in the parameter $n_{ieB}(\text{Si}, x)$. In addition to reducing the bandgap, the incorporation of Ge into Si also lifts the degeneracy of the valence-band and conduction-band edges (People, 1985). The result is a reduction in the densities of states N_c and N_v. That is, $\gamma(x) < 1$ except where the Ge concentration is zero. For the Ge distribution being considered here, the Ge-induced bandgap narrowing is zero at the emitter–base junction and increases linearly to ΔE_{gmax} at the base–collector junction, i.e.,

$$\Delta E_{gB,SiGe}(x) = \frac{x}{W_B}\Delta E_{gmax}. \qquad (10.25)$$

10.4.1.1   Collector Current

J_{C0} of a SiGe-base transistor can be obtained simply by substituting $n_{ieB}(SiGe, x)$ for $n_{ieB}(Si, x)$ and $D_{nB}(SiGe, x)$ for $D_{nB}(Si, x)$ in Eq. (10.2), i.e.,

$$J_{C0}(SiGe) = \frac{q}{\int_0^{W_B} \left[p_p(x) / D_{nB}(SiGe, x) n_{ieB}^2(SiGe, x) \right] dx}. \tag{10.26}$$

For a boxlike base doping profile and at low injection, $p_p(x) \approx N_B$ and is independent of x. $D_{nB}(Si, x)$ and $n_{ieB}(Si, x)$ are also independent of x. Therefore, Eqs. (10.23) and (10.26) give the ratio of J_{C0} with Ge to that without Ge as

$$\frac{J_{C0}(SiGe)}{J_{C0}(Si)} = \frac{W_B}{\int_0^{W_B} \frac{1}{\gamma(x)\eta(x)} \exp\left[-\Delta E_{gB, SiGe}(x)/kT \right] dx}$$

$$\approx \frac{\bar{\gamma}\bar{\eta} W_B}{\int_0^{W_B} \exp\left[-\Delta E_{gB, SiGe}(x)/kT \right] dx}, \tag{10.27}$$

where the ratio parameter

$$\eta(x) \equiv \frac{D_{nB}(SiGe, x)}{D_{nB}(Si, x)} \tag{10.28}$$

accounts for the effect of Ge on electron mobility in the base. $D_{nB}(SiGe)$ is proportional to $\mu_{nB}(SiGe)$, which is found to be a function of both N_B and Ge concentration (Kay and Tang, 1991). Therefore, $\eta(x)$ is a function of N_B and Ge concentration as well. In writing the last part of Eq. (10.27), we have made an assumption that $\gamma(x)$ and $\eta(x)$ inside the integral can be replaced by some average values $\bar{\gamma}$ and $\bar{\eta}$. It should be noted that Eq. (10.27) *is valid for arbitrary dependence of $\Delta E_{gB, SiGe}(x)$ on x.*

For the simple linearly graded bandgap given by Eq. (10.25), Eq. (10.27) can be integrated to give

$$\frac{J_{C0}(SiGe)}{J_{C0}(Si)} = \frac{\bar{\gamma}\bar{\eta} \Delta E_{gmax}/kT}{1 - \exp(-\Delta E_{gmax}/kT)} \quad \text{(triangular Ge profile).} \tag{10.29}$$

The value of $\gamma(x)$ varies between 1, where there is no Ge, to about 0.4, where the Ge concentration is 20% (Prinz et al., 1989). For a typical SiGe base where the N_B is in excess of 1×10^{18} cm^{-3}, $\eta(x)$ is about unity (Kay and Tang, 1991; Manku and Nathan, 1992). Therefore, the product $\bar{\gamma}\bar{\eta}$ is not far from unity. In the literature, the corrections to density of states and to electron mobility are often ignored in discussing the effect of Ge on J_{C0}, which is equivalent to setting $\bar{\gamma}\bar{\eta}$ to unity in Eq. (10.29). Equation (10.29) is the well-known result for a simple triangular Ge distribution (Harame et al., 1995a, 1995b). The current gain ratio is the same as the collector current ratio, namely

$$\frac{\beta_0(SiGe)}{\beta_0(Si)} = \frac{J_{C0}(SiGe)}{J_{C0}(Si)} = \frac{\bar{\gamma}\bar{\eta} \Delta E_{gmax}/kT}{1 - \exp(-\Delta E_{gmax}/kT)} \quad \text{(triangular Ge profile).} \tag{10.30}$$

Readily achievable values of ΔE_{gmax} are in the range of 100–150 meV, which means a SiGe-base transistor typically has a current gain that is 4–6-times that of a Si-base transistor having the same base dopant distribution. The enhanced current gain of a SiGe-base transistor can be used to tradeoff for a smaller intrinsic-base sheet resistivity [cf. Eq. (10.5)].

10.4.1.2 Early Voltage

The effect of base-bandgap grading on V_A can be obtained from Eq. (9.95) in a similar manner. The result is

$$V_A(\text{SiGe}) \approx \frac{qD_{nB}(\text{SiGe}, W_B)n_{ieB}^2(\text{SiGe}, W_B)}{C_{dBC}} \int_0^{W_B} \frac{N_B(x)}{D_{nB}(\text{SiGe}, x)n_{ieB}^2(\text{SiGe}, x)} dx.$$

(10.31)

Using Eq. (9.95) for $V_A(\text{Si})$ and substituting Eq. (10.23) into Eq. (10.31), we obtain the ratio of V_A with Ge to that without Ge as

$$\frac{V_A(\text{SiGe})}{V_A(\text{Si})}$$

$$= \frac{D_{nB}(\text{SiGe}, W_B)n_{ieB}^2(\text{SiGe}, W_B)}{W_B} \int_0^{W_B} \frac{dx}{D_{nB}(\text{SiGe}, x)n_{ieB}^2(\text{SiGe}, x)}$$

$$\approx \frac{D_{nB}(\text{SiGe}, W_B)}{\bar{D}_{nB}(\text{SiGe})} \frac{\gamma(W_B)}{\bar{\gamma}} \frac{\exp(\Delta E_{gmax}/kT)}{W_B} \int_0^{W_B} \exp\left[-\Delta E_{gB,\,SiGe}(x)/kT\right] dx$$

$$\approx \frac{\exp(\Delta E_{gmax}/kT)}{W_B} \int_0^{W_B} \exp\left[-\Delta E_{gB,\,SiGe}(x)/kT\right] dx,$$

(10.32)

where, in writing the last part of the equation, we have made a further assumption that D_{nB} and γ are independent of x. It should be noted that Eq. (10.32) *is valid for arbitrary dependence of $\Delta E_{gB,SiGe}(x)$ on x.*

For the simple linearly graded base bandgap described by Eq. (10.25), Eq. (10.32) can be integrated to give

$$\frac{V_A(\text{SiGe})}{V_A(\text{Si})} = \frac{kT}{\Delta E_{gmax}} \left[\exp(\Delta E_{gmax}/kT) - 1\right] \text{ (triangular Ge profile).}$$

(10.33)

Equation (10.33) is the well-known result for a simple triangular Ge distribution (Harame *et al.*, 1995a). For a typical value of $\Delta E_{gmax} = 100$ meV, V_A is increased by a factor of 12 at room temperature.

10.4.1.3 Base Transit Time

The graded base bandgap introduces an electric field which drives electrons across the p-type base layer. For a total bandgap narrowing of 100 meV across a base layer of 100 nm, a SiGe-base transistor has a built-in electric field of 10^4 V/cm due to the presence of Ge alone. This field is in addition to the electric fields due to base dopant distribution and heavy-doping effect discussed in Section 10.2.2.1. The base transit time at low injection can be derived in a manner similar to J_{C0} and V_A. The result is

$$
\begin{aligned}
t_B(\text{SiGe}) \\
\approx \int_0^{W_B} \frac{n_{ieB}^2(\text{SiGe}, x)}{N_B(x)} \int_x^{W_B} \frac{N_B(x')}{D_{nB}(\text{SiGe}, x')n_{ieB}^2(\text{SiGe}, x')} dx' dx \\
= \int_0^{W_B} \frac{\gamma(x)n_{ieB}^2(\text{Si}, x)}{N_B(x)} \exp\left[\Delta E_{gB,SiGe}(x)/kT\right] \\
\times \int_x^{W_B} \frac{N_B(x')}{\gamma(x')\eta(x')D_{nB}(\text{Si}, x')n_{ieB}^2(\text{Si}, x')} \exp\left[-\Delta E_{gB,SiGe}(x')/kT\right] dx' dx.
\end{aligned}
$$

$$(10.34)$$

Again, for a boxlike base dopant distribution, Eq. (10.34) simplifies to

$$
\begin{aligned}
t_B(\text{SiGe}) \approx \frac{1}{\bar{\eta}D_{nB}(\text{Si})} \int_0^{W_B} \exp\left[\Delta E_{gB,SiGe}(x)/kT\right] \\
\times \int_x^{W_B} \exp\left[-\Delta E_{gB,SiGe}(x')/kT\right] dx' dx.
\end{aligned}
$$

$$(10.35)$$

The base transit time for a Si-base transistor is $t_B(\text{Si}) = W_B^2/2D_n(\text{Si})$, given in Eq. (10.21). Therefore, the ratio of t_B with Ge to that without Ge is

$$
\begin{aligned}
\frac{t_B(\text{SiGe})}{t_B(\text{Si})} = \frac{2}{\bar{\eta}W_B^2} \int_0^{W_B} \exp\left[\Delta E_{gB,SiGe}(x)/kT\right] \\
\times \int_x^{W_B} \exp\left[-\Delta E_{gB,SiGe}(x')/kT\right] dx' dx.
\end{aligned}
$$

$$(10.36)$$

Equation (10.36) *is valid for arbitrary dependence of $\Delta E_{gB,SiGe}(x)$ on x.*

For the simple linearly graded base bandgap described by Eq. (10.25), Eq. (10.36) can be integrated to give

$$
\frac{t_B(\text{SiGe})}{t_B(\text{Si})} = \frac{2kT}{\bar{\eta}\Delta E_{gmax}} \left\{ 1 - \frac{kT\left[1 - \exp(-\Delta E_{gmax}/kT)\right]}{\Delta E_{gmax}} \right\} \quad \text{(triangular Ge profile)}.
$$

$$(10.37)$$

Again, in the literature, the diffusion coefficient correction factor $\bar{\eta}$ is often set to unity and dropped. For $\Delta E_{gmax} = 100$ meV, the base transit time of a SiGe-base transistor is about 2.5-times smaller than that of a Si-base transistor. Equation (10.37) is the well-known result for a simple triangular Ge distribution (Harame *et al.*, 1995a, 1995b).

10.4.1.4 SiGe-Base Vertical Bipolar as a RF and Analog Transistor

Some of the desirable attributes of a bipolar transistor for RF and analog applications are: small transit times, small base resistance, and large output resistance or Early

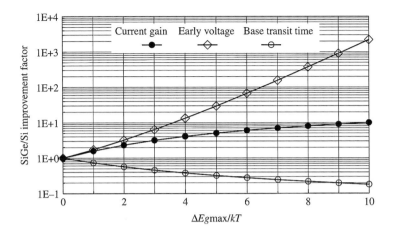

Figure 10.6 Relative improvement factors for current gain, Early voltage, and base transit time of a SiGe-base bipolar transistor having a linearly graded base bandgap over a Si-base bipolar transistor, plotted as a function of $\Delta E_{g\text{max}}/kT$. $\bar{\gamma}$ and $\bar{\eta}$ are set to unity

voltage. Figure 10.6 is a plot of the improvement factors for current gain [Eq. (10.30)], Early voltage [Eq. (10.33)], and base transit time [Eq. (10.37)], of a SiGe-base bipolar transistor over a Si-base bipolar transistor having the same base width and base dopant distribution, plotted as a function of $\Delta E_{g\text{max}}/kT$. It shows that a SiGe-base vertical bipolar transistor having a linearly graded base bandgap is a superior RF and analog transistor.

10.4.2   Base Current When Ge Is Present in the Emitter

In fabricating a SiGe-base vertical transistor, typically a p-type base layer containing the desired Ge distribution is first formed before an n$^+$ emitter polysilicon layer is formed on top. ***There is usually a Ge-free cap in the base layer to avoid exposing Ge to any oxidizing ambient*** in the polysilicon-emitter formation process. During the emitter formation process, n-type dopant from the polysilicon layer diffuses into the underlying base layer, forming a thin single-crystalline n$^+$ emitter region of depth x_{jE} [see Section 10.1.2 and Figure 9.3(b)]. The final value of x_{jE} is a function of emitter annealing condition (temperature and time), emitter dopant species (arsenic or phosphorus), emitter stripe width, and whether or not metal silicide is formed on top of the emitter-polysilicon layer (Kondo *et al.*, 2001). Depending on the thickness of the starting Ge-free cap, the final emitter–base junction may or may not extend into the Ge-containing region of the base, and hence Ge may or may not be present in the single-crystalline emitter region. Since any change in the emitter parameters can affect the base current, we want to consider what happens to the base current of a polysilicon-emitter SiGe-base bipolar transistor when Ge from the starting base layer ends up in the single-crystalline emitter region.

In the literature, there are reports of intentionally introducing Ge into the single-crystalline emitter region (Huizing *et al.*, 2001) as well as intentionally adding Ge to the emitter-polysilicon layer (Kunz *et al.*, 2002, 2003; Martinet *et al.*, 2002). Often the stated objectives are to reduce current gain of a SiGe-base transistor. The merits of these and other approaches for reducing current gain of a SiGe-base transistor will also be discussed.

10.4.2.1 Ge-Induced Bandgap Narrowing in the Emitter

The Ge distribution in the emitter and base regions of a polysilicon-emitter SiGe-base vertical bipolar transistor having a boxlike base dopant distribution is illustrated in Figure 10.7 for the case of a trapezoidal Ge distribution. The Ge distribution causes a bandgap narrowing of ΔE_{g0} near the emitter–base junction and a peak bandgap narrowing of ΔE_{gmax} at the base–collector junction. For the case illustrated in Figure 10.7, the emitter junction depth x_{jE} is larger than the starting Ge-free cap thickness, W_{cap}, resulting in a bandgap narrowing of $\Delta E_{g0be} > \Delta E_{g0}$ at the emitter–base junction. The Ge present within the single-crystalline emitter region causes a narrowing of the bandgap in the region. In Section 10.4.2.2, we examine the base current when there is Ge in the single-crystalline emitter region.

10.4.2.2 Base Current When There Is Ge in the Single-Crystalline Emitter Region

Consistent with the convention used in Section 9.2.3, Figure 10.8 shows the coordinates for modeling the current flows in the emitter region of the emitter–base diode of

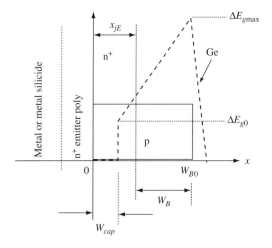

Figure 10.7 Schematic illustrating the emitter and base regions of a polysilicon-emitter SiGe-base vertical bipolar transistor having a trapezoidal Ge distribution. The starting base layer thickness is W_{B0}, including a Ge-free cap layer of thickness W_{cap}. The quasineutral base width is W_B after polysilicon-emitter drive in. The base width is $W_B = W_{B0} - x_{jE}$. The emitter–base space-charge region thickness is assumed to be zero, for simplicity of illustration. With $x_{jE} > W_{cap}$, there is no residual Ge-free region in the final quasineutral base layer, and there is Ge in the single-crystalline n$^+$ emitter region

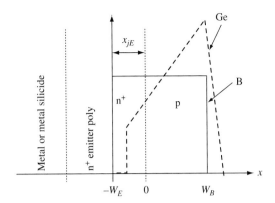

Figure 10.8 Coordinates for modeling the current flow in the emitter of the polysilicon-emitter SiGe-base bipolar transistor of Figure 10.7

Figure 10.7. The p–n junction is assumed to be located at the origin "0." The emitter is contacted by a polysilicon layer, with the polysilicon–silicon interface located at $x = -W_E$, i.e., $W_E = x_{jE}$.

Following Eqs. (9.65) and (9.67), J_{B0} for a SiGe-base bipolar transistor can be written as

$$J_{B0}(\text{SiGe}) = \frac{qn_i^2}{G_E(\text{SiGe})}, \qquad (10.38)$$

with the emitter Gummel number as

$$G_E(\text{SiGe}) = \int_{-W_E}^{0} \frac{n_i^2}{n_{ieE}^2(\text{SiGe}, x)} \frac{N_E(x)}{D_{pE}(\text{SiGe}, x)} dx + \frac{n_i^2}{n_{ieE}^2(\text{SiGe}, -W_E)} \frac{N_E(-W_E)}{S_p(\text{SiGe})}, \qquad (10.39)$$

where $S_p(\text{SiGe})$ is the surface recombination velocity for holes at the polysilicon–silicon interface, and $N_E(x)$, $D_{pE}(\text{SiGe}, x)$, and $n_{ieE}(\text{SiGe}, x)$ are the doping concentration, hole diffusion coefficient, and effective intrinsic-carrier concentration, respectively, in the single-crystalline emitter region. $S_p(\text{SiGe})$ depends on the transport of holes through the polysilicon–silicon interface and inside the polysilicon layer. For example, it is shown in Exercise 9.3 and in the literature (Ning and Isaac, 1980) that for a Si-base n–p–n bipolar transistor, $S_p(\text{Si})$ depends only on the transport of holes inside the polysilicon layer when there is no appreciable hole barrier at the polysilicon–silicon interface. In this simple case, $S_p(\text{Si})$ is given by

$$S_p(\text{Si}) = \frac{D_{pE,poly}}{L_{pE,poly} \tanh\left(W_{E,poly}/L_{pE,poly}\right)}, \qquad (10.40)$$

where $D_{pE,poly}$ and $L_{pE,poly}$ are the hole diffusion coefficient and hole diffusion length, respectively, in the emitter polysilicon, and $W_{E,poly}$ is the thickness of the emitter polysilicon layer.

Regardless of the details of the physical model for S_p, the operation of a polysilicon-emitter transistor is based on the experimentally confirmed fact that the hole current is determined primarily by the surface recombination velocity of holes at the polysilicon–silicon interface and is relatively insensitive to the transport of holes within the thin single-crystalline emitter region. That is, the operation of a polysilicon-emitter transistor is based on the assumption that G_E is determined primarily by the term containing S_p in Eq. (10.39). In other words, for a polysilicon-emitter SiGe-base bipolar transistor,

$$G_E(\text{SiGe}) \approx \frac{n_i^2 N_E(-W_E)}{n_{ieE}^2(\text{SiGe}, -W_E)S_p(\text{SiGe})}. \tag{10.41}$$

Following the procedure used in Section 10.4.1 for modeling the SiGe base region, we can write the emitter parameter $n_{ieE}(\text{SiGe}, x)$ in the form

$$n_{ieE}^2(\text{SiGe}, x) = n_{ieE}^2(\text{Si}, x)\gamma_E(x)\exp\left[\Delta E_{gE,\,SiGe}(x)/kT\right], \tag{10.42}$$

where $n_{ieE}(\text{Si}, x)$ is the effective intrinsic-carrier concentration without Ge and $\Delta E_{gE,\,SiGe}(x)$ is the local bandgap narrowing due to the presence of Ge. Also, the parameter

$$\gamma_E(x) \equiv \frac{(N_c N_v)_{SiGe}}{(N_c N_v)_{Si}} \tag{10.43}$$

is to account for any change in the densities of states in the emitter due to the presence of Ge. Effects due to heavy doping are contained in the parameter $n_{ieE}(\text{Si}, x)$. From Eqs. (10.38), (10.41), and (10.42) we can write the ratio of J_{B0} with Ge to that without Ge as (Ning, 2003)

$$\frac{J_{B0}(\text{SiGe})}{J_{B0}(\text{Si})} \approx \frac{S_p(\text{SiGe})}{S_p(\text{Si})}\gamma_E(-W_E)\exp\left[\Delta E_{gE,\,SiGe}(-W_E)/kT\right]. \tag{10.44}$$

Since the emitter polysilicon is in contact with the Ge-free cap of the starting base layer, the Ge concentration at the polysilicon–silicon interface is zero. Therefore, $\Delta E_{gE,\,SiGe}(-W_E) = 0$ and $\gamma_E(-W_E) = 1$. Furthermore, we expect $S_p(\text{Si}) \approx S_p(\text{SiGe})$ in this case. Equation (10.44) then suggests that, for a typical polysilicon-emitter SiGe-base vertical bipolar transistor, having a Ge-free cap in the starting base layer, the base current should be insensitive to the Ge distribution in the base layer, even when Ge ends up inside the single-crystalline region of the emitter. This explains why the measured base current of a polysilicon-emitter SiGe-base transistor and that of a polysilicon-emitter Si-base control are approximately the same (Prinz and Sturm, 1990; Harame *et al.*, 1995a, 1995b; Oda *et al.*, 1997).

10.4.2.3 Non-Transparent "Polysilicon Emitter"

In an attempt to reduce the current gain of a SiGe-base transistor, designers sometimes intentionally introduce a thin Ge-containing layer within the single-crystalline emitter region (Huizing *et al.*, 2001). In this case, the thin Ge-containing layer creates a local potential well for holes, causing a significant increase in Auger recombination of electrons and holes within the single-crystalline emitter region, resulting in a significant increase in base current. For such a transistor, even though a polysilicon layer is used to form a "polysilicon emitter," the single-crystalline part of the emitter is not transparent because of the large recombination in it. As a result, the conventional transparent-emitter model described in Section 9.2.3 for a polysilicon emitter does not apply.

Thus far, there is no reported data or theory suggesting that adding a high-recombination region within the single-crystalline emitter region, or using any similar techniques for reducing current gain, lead to a transistor of better performance. As a result, such non-transparent "polysilicon-emitter" devices will not be discussed any further.

10.4.2.4   Polycrystalline SiGe Emitter

In some studies (Kunz *et al.*, 2002, 2003; Martinet *et al.*, 2002), polycrystalline SiGe (polySiGe) instead of polysilicon is used to form the emitter in an attempt to reduce current gain. The energy bandgap of a polySiGe layer is smaller than that of a polysilicon layer. The reduced bandgap increases the injection of holes from the base into the n^+-polySiGe emitter region. In addition, the value of S_p for a polySiGe emitter could be quite different from that for a polysilicon emitter.

However, it is important to recognize that current gain can be changed by changing J_{C0}, J_{B0}, or both. Also, it is important to note that, compared to a polysilicon-emitter Si-base bipolar transistor, the larger current gain in a polysilicon-emitter SiGe-base bipolar transistor is due entirely to an increase in J_{C0}, and not to any significant change in J_{B0}. It will be shown in Section 10.4.5.1 that it is possible to reduce J_{C0}, and hence β_0, of a SiGe-base transistor without affecting its performance advantage over a Si-base transistor, by optimizing the Ge profile in the base.

Another effective approach to reduce J_{C0} and β_0 of a transistor is to reduce its intrinsic-base sheet resistivity [cf. Eq. (10.5)]. Reducing base resistance leads to improved device and circuit performance. Therefore, if a smaller current gain is desired, a device designer should consider reducing the intrinsic-base sheet resistivity of the transistor. This can be accomplished easily by increasing N_B.

There are no theories or experimental results to suggest that replacing a polysilicon emitter with a polySiGe emitter will lead to improved device speed. Therefore, we will not consider the polySiGe emitter any further.

10.4.3   Transistors Having a Trapezoidal Ge Distribution in the Base

Various Ge profiles have been analyzed and/or tested out experimentally by various groups (cf. Cressler *et al.*, 1993a; Harame *et al.*, 1995a, 1995b; Washio *et al.*, 2002). Here we focus on a trapezoidal Ge profile for two cases of emitter junction depths. The case of $x_{jE} > W_{cap}$ is depicted in Figure 10.7, where there is no residual Ge-free region in the final base layer. The case of $x_{jE} < W_{cap}$ is depicted in Figure 10.9, where the final base layer would contain a residual Ge-free cap of thickness

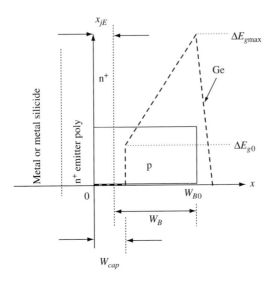

Figure 10.9 Schematic illustrating the emitter and base regions of a polysilicon-emitter SiGe-base vertical bipolar transistor having the same base dopant and Ge profiles as in Figure 10.7, but with $x_{jE} < W_{cap}$

$W_{cap} - x_{jE}$. Here we want to extend the equations developed in the previous sections to examine the ratios of J_{C0}, V_A, and t_B for the trapezoidal Ge profiles shown in Figures 10.7 and 10.9.

10.4.3.1   Collector Current for a Trapezoidal Ge Distribution

The J_{C0} ratio in Eq. (10.27) can easily be adapted to include the effect of emitter depth by noting that the quasineutral base starts at $x = x_{jE}$ and ends at $x = W_{B0}$. From Eq. (10.27), we have

$$\frac{J_{C0}(\text{SiGe}, x_{jE})}{J_{C0}(\text{Si}, x_{jE})} \approx \frac{\overline{\gamma\eta}(W_{B0} - x_{jE})}{\int_{x_{jE}}^{W_{B0}} \exp\left[-\Delta E_{gB,\text{SiGe}}(x)/kT\right] dx}. \tag{10.45}$$

- *Case of Ge in the emitter* (i.e., $x_{jE} > W_{cap}$). This is the situation depicted in Figure 10.7. The base bandgap narrowing of ΔE_{g0be} at the emitter end of the base is given by

$$\Delta E_{g0be}(x_{jE}) = \Delta E_{g0} + \left(\frac{x_{jE} - W_{cap}}{W_{B0} - W_{cap}}\right)(\Delta E_{gmax} - \Delta E_{g0}). \tag{10.46}$$

The base bandgap narrowing as a function of position in the base is given by

$$\Delta E_{gB,\text{SiGe}}(x) = \Delta E_{g0be}(x_{jE}) + \left(\frac{x - x_{jE}}{W_{B0} - x_{jE}}\right)\left[\Delta E_{gmax} - \Delta E_{g0be}(x_{jE})\right]. \tag{10.47}$$

Substituting Eqs. (10.46) and (10.47) into Eq. (10.45), we obtain

$$\frac{J_{C0}\left(SiGe, x_{jE}\right)}{J_{C0}\left(Si, x_{jE}\right)}\bigg|_{x_{jE} > W_{cap}}$$

$$= \frac{\overline{\gamma\eta}\left(W_{B0} - x_{jE}\right)}{\exp\left[\frac{-\Delta E_{g0be}\left(x_{jE}\right)}{kT}\right]\int_{x_{jE}}^{W_{B0}} \exp\left\{\frac{\left[\Delta E_{gmax} - \Delta E_{g0be}\left(x_{jE}\right)\right]\left(x_{jE} - x\right)}{\left(W_{B0} - x_{jE}\right)kT}\right\}dx}$$

$$= \overline{\gamma\eta}\left[\frac{\Delta E_{gmax} - \Delta E_{g0be}\left(x_{jE}\right)}{kT}\right]\exp\left[\frac{\Delta E_{g0be}\left(x_{jE}\right)}{kT}\right]$$

$$\times \left\{1 - \exp\left[\frac{\Delta E_{g0be}\left(x_{jE}\right) - \Delta E_{gmax}}{kT}\right]\right\}^{-1}. \tag{10.48}$$

- Case of no Ge in the emitter (i.e., $x_{jE} < W_{cap}$). When $x_{jE} < W_{cap}$, there is a residual Ge-free layer of thickness $W_{cap} - x_{jE}$ in the base. This is the situation depicted in Figure 10.9. The base bandgap-narrowing parameter is given by

$$\Delta E_{gB,SiGe}(x) = \Delta E_{g0} + \left(\frac{x - W_{cap}}{W_{B0} - W_{cap}}\right)\left(\Delta E_{gmax} - \Delta E_{g0}\right) \qquad x > W_{cap} \tag{10.49}$$
$$= 0 \qquad x < W_{cap}.$$

Substituting Eq. (10.49) into Eq. (10.45) we obtain

$$\frac{J_{C0}\left(SiGe, x_{jE}\right)}{J_{C0}\left(Si, x_{jE}\right)}\bigg|_{x_{jE} < W_{cap}} = \frac{\overline{\gamma\eta}\left(W_{B0} - x_{jE}\right)}{\left(W_{cap} - x_{jE}\right) + \int_{W_{cap}}^{W_{B0}} \exp\left[-\Delta E_{gB,SiGe}(x)/kT\right]dx}$$

$$= \frac{\overline{\gamma\eta}}{\left[\frac{W_{cap} - x_{jE}}{W_{B0} - x_{jE}}\right] + \left[\frac{W_{B0} - W_{cap}}{W_{B0} - x_{jE}}\right]\left[\frac{kT\exp\left(-\Delta E_{g0}/kT\right)}{\Delta E_{gmax} - \Delta E_{g0}}\right]\left[1 - \exp\left(\frac{\Delta E_{g0} - \Delta E_{gmax}}{kT}\right)\right]}. \tag{10.50}$$

When $x_{jE} = W_{cap}$, Eq. (10.50) has the same form as Eq. (10.48), as it should. Also, when $x_{jE} = W_{cap}$ and $\Delta E_{g0} = 0$, Eq. (10.50) reduces to Eq. (10.29), as it should.

10.4.3.2 Early Voltage for a Trapezoidal Ge Distribution

The same procedures can be followed to obtain the equations for the V_A ratio. The result is

$$\frac{V_A\left(SiGe, x_{jE}\right)}{V_A\left(Si, x_{jE}\right)} \approx \frac{\exp\left[\Delta E_{gB,SiGe}\left(W_{B0}\right)/kT\right]}{W_{B0} - x_{jE}}\int_{x_{jE}}^{W_{B0}} \exp\left[-\Delta E_{gB,SiGe}(x)/kT\right]dx. \tag{10.51}$$

- Case of Ge in the emitter (i.e., $x_{jE} > W_{cap}$). For the case of Ge in the emitter, substituting Eqs. (10.46) and (10.47) into Eq. (10.51), we obtain

$$\frac{V_A\left(SiGe, x_{jE}\right)}{V_A\left(Si, x_{jE}\right)}\bigg|_{x_{jE} > W_{cap}} = \left[\frac{kT}{\Delta E_{gmax} - \Delta E_{g0be}\left(x_{jE}\right)}\right]$$

$$\times \left\{\exp\left[\frac{\Delta E_{gmax} - \Delta E_{g0be}\left(x_{jE}\right)}{kT}\right] - 1\right\}. \tag{10.52}$$

The V_A ratio in this case depends on the bandgap energy difference

$[\Delta E_{gmax} - \Delta E_{g0be}(x_{jE})]$ across the base layer. Equation (10.52) has the same form as Eq. (10.33), where $\Delta E_{g0be}(x_{jE}) = 0$.

- *Case of no Ge in the emitter* (i.e., $x_{jE} < W_{cap}$). For the case with no Ge in the emitter, there is a residual Ge-free layer of thickness $W_{cap} - x_{jE}$ in the base. Substituting Eq. (10.49) into Eq. (10.51), we obtain

$$
\left. \frac{V_A(\text{SiGe}, x_{jE})}{V_A(\text{Si}, x_{jE})} \right|_{x_{jE} < W_{cap}} = \left(\frac{W_{cap} - x_{jE}}{W_{B0} - x_{jE}} \right) \exp\left(\frac{\Delta E_{gmax}}{kT} \right)
$$

$$
+ \left(\frac{W_{B0} - W_{cap}}{W_{B0} - x_{jE}} \right) \left(\frac{kT}{\Delta E_{gmax} - \Delta E_{g0}} \right) \tag{10.53}
$$

$$
\times \left\{ \exp\left[\frac{\Delta E_{gmax} - \Delta E_{g0}}{kT} \right] - 1 \right\}.
$$

When $x_{jE} = W_{cap}$, Eqs. (10.52) and (10.53) have the same form, as they should. Also, when $x_{jE} = W_{cap}$ and $\Delta E_{g0} = 0$, Eq. (10.53) reduces to Eq. (10.33), as it should.

10.4.3.3   Base Transit Time for a Trapezoidal Ge Distribution

The base transit time ratio can be derived in the same manner. The result is

$$
\frac{t_B(\text{SiGe}, x_{jE})}{t_B(\text{Si}, x_{jE})} = \frac{2}{\bar{\eta}(W_{B0} - x_{jE})^2} \int_{x_{jE}}^{W_{B0}} \exp\left[\Delta E_{gB, SiGe}(x)/kT \right]
$$

$$
\times \int_{x}^{W_{B0}} \exp\left[-\Delta E_{gB, SiGe}(x')/kT \right] dx' dx. \tag{10.54}
$$

- *Case of Ge in the emitter* (i.e., $x_{jE} > W_{cap}$). For the case of Ge in the emitter, substituting Eqs. (10.46) and (10.47) into Eq. (10.54), we obtain

$$
\left. \frac{t_B(\text{SiGe}, x_{jE})}{t_B(\text{Si}, x_{jE})} \right|_{x_{jE} > W_{cap}}
$$

$$
= \frac{2}{\bar{\eta}} \left[\frac{kT}{\Delta E_{gmax} - \Delta E_{g0be}(x_{jE})} \right] - \frac{2}{\bar{\eta}} \left[\frac{kT}{\Delta E_{gmax} - \Delta E_{g0be}(x_{jE})} \right]^2 \tag{10.55}
$$

$$
\times \left\{ 1 - \exp\left[\frac{\Delta E_{g0be}(x_{jE}) - \Delta E_{gmax}}{kT} \right] \right\}.
$$

Just like the V_A ratio in Eq. (10.52), the transit time ratio depends on the bandgap energy difference $[\Delta E_{gmax} - \Delta E_{g0be}(x_{jE})]$. Equation (10.55) reduces to Eq. (10.37) when $\Delta E_{g0be}(x_{jE}) = 0$, as expected.

- *Case of no Ge in the emitter* (i.e., $x_{jE} < W_{cap}$). For the case of no Ge in the emitter, there is a residual Ge-free layer of thickness $W_{cap} - x_{jE}$ in the base. Substituting Eq. (10.49) into Eq. (10.54), we obtain

$$\frac{t_B\left(SiGe, x_{jE}\right)}{t_B\left(Si, x_{jE}\right)}\Bigg|_{x_{jE}<W_{cap}} = \frac{1}{\eta}\frac{\left(W_{cap}-x_{jE}\right)^2}{\left(W_{B0}-x_{jE}\right)^2} + \frac{2}{\eta}\frac{\left(W_{B0}-W_{cap}\right)^2}{\left(W_{B0}-x_{jE}\right)^2}\left(\frac{kT}{\Delta E_{gmax}-\Delta E_{g0}}\right)$$

$$+\frac{2}{\eta}\left(\frac{W_{cap}-x_{jE}}{W_{B0}-x_{jE}}\right)\left(\frac{W_{B0}-W_{cap}}{W_{B0}-x_{jE}}\right)\left(\frac{kT}{\Delta E_{gmax}-\Delta E_{g0}}\right)$$

$$\times\left\{1-\exp\left[\frac{\Delta E_{g0}-\Delta E_{gmax}}{kT}\right]\right\}\exp\left(\frac{-\Delta E_{g0}}{kT}\right)$$

$$-\frac{2}{\eta}\frac{\left(W_{B0}-W_{cap}\right)^2}{\left(W_{B0}-x_{jE}\right)^2}\left(\frac{kT}{\Delta E_{gmax}-\Delta E_{g0}}\right)^2\left\{1-\exp\left[\frac{\Delta E_{g0}-\Delta E_{gmax}}{kT}\right]\right\}.$$

$$(10.56)$$

Equation (10.56) depends on the energy difference $\left(\Delta E_{gmax}-\Delta E_{g0}\right)$ as well as on ΔE_{g0}. This should be contrasted with the case of Ge in the emitter given by Eq. (10.55). The added dependence on ΔE_{g0} can be used to tailor the Ge profile to further improve t_B. This will be discussed in Section 10.4.5.2.

10.4.4   Transistors Having a Constant Ge Distribution in the Base

SiGe-base vertical transistors having a spatially constant Ge distribution in the quasineutral base, corresponding to a spatially constant base bandgap, are also used quite widely. A spatially constant Ge distribution is equivalent to setting $\Delta E_{g0} = \Delta E_{gmax}$ in Figure 10.7.

A cursory examination of Eq. (10.54) suggests that there should be no improvement in base transit time other than indirectly through the factor $\bar{\eta}$. Yet, in the literature, there are ample experimental data showing SiGe-base vertical transistors having supposedly constant Ge distribution in the base to be superior to Si-base vertical transistors in base transit time (cf. Hobart *et al.*, 1995; Schüppen and Dietrich, 1995; Schüppen *et al.*, 1996; Deixler *et al.*, 2001). In this section, we extend the models developed for trapezoidal Ge distribution to examine the properties of a constant-Ge SiGe-base vertical transistor more closely.

- *Case of Ge in the emitter* (i.e., $x_{jE} > W_{cap}$). The emitter and base regions are as illustrated in Figure 10.10. From Eq. (10.48), we have

$$\frac{J_{C0}\left(SiGe, x_{jE}\right)}{J_{C0}\left(Si, x_{jE}\right)}\Bigg|_{x_{jE}>W_{cap}} = \bar{\gamma}\bar{\eta}\exp\left(\Delta E_{g0}/kT\right) \quad \text{(constant bandgap).} \qquad (10.57)$$

From Eq. (10.52), we have

$$\frac{V_A\left(SiGe, x_{jE}\right)}{V_A\left(Si, x_{jE}\right)}\Bigg|_{x_{jE}>W_{cap}} = 1 \quad \text{(constant bandgap),} \qquad (10.58)$$

and from Eq. (10.55), we have

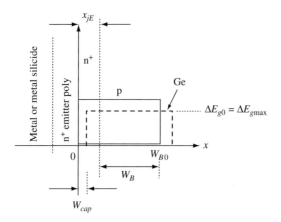

Figure 10.10 Schematic illustrating the emitter and base regions of a polysilicon-emitter SiGe-base vertical bipolar transistor having a constant-Ge distribution in the base, with $x_{jE} > W_{cap}$

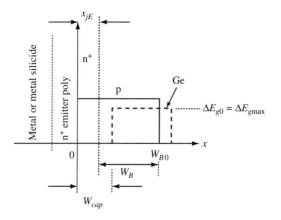

Figure 10.11 Schematic illustrating the emitter and base regions of a polysilicon-emitter SiGe-base vertical bipolar transistor having a constant Ge distribution in the base, with $x_{jE} < W_{cap}$

$$\left. \frac{t_B\left(\text{SiGe}, x_{jE}\right)}{t_B\left(\text{Si}, x_{jE}\right)} \right|_{x_{jE} > W_{cap}} = \frac{1}{\eta} \quad \text{(constant bandgap).} \tag{10.59}$$

- *Case of no Ge in the emitter* (i.e., $x_{jE} < W_{cap}$). When x_{jE} is smaller than the starting Ge-free cap thickness, the emitter and base regions are as illustrated in Figure 10.11. The energy bandgap is no longer spatially constant across the entire quasineutral base. Instead, *the base region has a high-low energy bandgap*, with the bandgap larger at the emitter end of the base where there is no Ge. The corresponding ratios of J_{C0}, V_A, and t_B can be obtained from Eqs. (10.50), (10.53), and (10.56), respectively. They are

$$\left.\frac{J_{C0}\left(SiGe, x_{jE}\right)}{J_{C0}\left(Si, x_{jE}\right)}\right|_{x_{jE}<W_{cap}} = \overline{\gamma\eta}\left\{\left(\frac{W_{cap}-x_{jE}}{W_{B0}-x_{jE}}\right)+\left(\frac{W_{B0}-W_{cap}}{W_{B0}-x_{jE}}\right)\exp\left(-\Delta E_{g0}/kT\right)\right\}^{-1},$$

(10.60)

$$\left.\frac{V_A\left(SiGe, x_{jE}\right)}{V_A\left(Si, x_{jE}\right)}\right|_{x_{jE}<W_{cap}} = \left(\frac{W_{cap}-x_{jE}}{W_{B0}-x_{jE}}\right)\exp\left(\frac{\Delta E_{g0}}{kT}\right)+\left(\frac{W_{B0}-W_{cap}}{W_{B0}-x_{jE}}\right),$$

(10.61)

and

$$\left.\frac{t_B\left(SiGe, x_{jE}\right)}{t_B\left(Si, x_{jE}\right)}\right|_{x_{jE}<W_{cap}} = \frac{1}{\eta}\left\{\left(\frac{W_{cap}-x_{jE}}{W_{B0}-x_{jE}}\right)^2+\left(\frac{W_{B0}-W_{cap}}{W_{B0}-x_{jE}}\right)^2\right.$$
$$\left.+2\left(\frac{W_{cap}-x_{jE}}{W_{B0}-x_{jE}}\right)\left(\frac{W_{B0}-W_{cap}}{W_{B0}-x_{jE}}\right)\exp\left(-\Delta E_{g0}/kT\right)\right\}.$$

(10.62)

Note that these ratios are functions of x_{jE}. $W_{B0}-x_{jE}$ is the base width and $\left(W_{cap}-x_{jE}\right)/\left(W_{B0}-x_{jE}\right)$ is the fraction of the quasineutral base with no Ge at all. Figure 10.12 is a plot of Eq. (10.62) as a function of $\left(W_{cap}-x_{jE}\right)/\left(W_{B0}-x_{jE}\right)$ for several values of $\Delta E_{g0}/kT$. It clearly shows that as long as there is a residual Ge-free region in the emitter end of the quasineutral base, the base region has a high–low bandgap and there is improvement in base transit time over a Si-base transistor. In fact, for a given value of $\Delta E_{g0}/kT$, which is equal to $\Delta E_{gmax}/kT$, the maximum improvement factor, which occurs at $\left(W_{cap}-x_{jE}\right)/\left(W_{B0}-x_{jE}\right)\sim0.5$, is quite comparable to that for a linearly graded Ge distribution (cf. Figure 10.6). It is left as

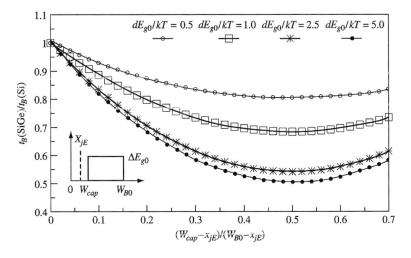

Figure 10.12 The base transit time ratio, Eq. (10.62), as a function of $\left(W_{cap}-x_{jE}\right)/\left(W_{B0}-x_{jE}\right)$ with $\Delta E_{g0}/kT$ as a parameter

an exercise (Exercise 10.9) to the reader to show that there can also be appreciable improvement in current gain and Early voltage over a Si-base transistor.

It is of interest to note that as long as the Ge concentration is sufficiently large so that $\exp(-\Delta E_{g0}/kT)$ is negligible compared to $(W_{cap} - x_{jE})/(W_{B0} - x_{jE})$, these ratios approach the values of

$$\left.\frac{J_{C0}(\text{SiGe}, x_{jE})}{J_{C0}(\text{Si}, x_{jE})}\right|_{x_{jE} < W_{cap}} \rightarrow \overline{\gamma\eta}\left(\frac{W_{B0} - x_{jE}}{W_{cap} - x_{jE}}\right), \tag{10.63}$$

$$\left.\frac{V_A(\text{SiGe}, x_{jE})}{V_A(\text{Si}, x_{jE})}\right|_{x_{jE} < W_{cap}} \rightarrow \left(\frac{W_{cap} - x_{jE}}{W_{B0} - x_{jE}}\right)\exp\left(\frac{\Delta E_{g0}}{kT}\right), \tag{10.64}$$

and

$$\left.\frac{t_B(\text{SiGe}, x_{jE})}{t_B(\text{Si}, x_{jE})}\right|_{x_{jE} < W_{cap}} \rightarrow \frac{1}{\eta}\left\{\left(\frac{W_{cap} - x_{jE}}{W_{B0} - x_{jE}}\right)^2 + \left(\frac{W_{B0} - W_{cap}}{W_{B0} - x_{jE}}\right)^2\right\}. \tag{10.65}$$

That is, both the J_{C0} ratio and the t_B ratio become independent of ΔE_{g0} for $\Delta E_{g0}/kT \gg 1$, while the V_A ratio increases exponentially with $\Delta E_{g0}/kT$. That the t_B ratio becomes less sensitive to ΔE_{g0} as $\Delta E_{g0}/kT$ increases is also evident in Figure 10.12.

So far, we have assumed the Ge distribution drops abruptly to zero at $x = W_{cap}$. In practice this never happens. (See Washio *et al.*, 2002, for an example of a realistic Ge profile that is designed to ramp up and down abruptly.) Instead of ramping down at an infinite rate to zero at the emitter end, a Ge distribution can be ramped down only at some finite rate. A model for a SiGe-base transistor having a base-region Ge distribution that ramps up from zero concentration at the emitter end to some constant concentration some distance toward the collector end is developed in Exercise 10.10. Whether the Ge distribution ramps up at some finite rate, as described in Exercise 10.10, or abruptly, as illustrated in Figure 10.11, *as long as there is some region of zero or relatively low Ge concentration at the emitter end of the quasi-neutral base, an otherwise constant-Ge SiGe-base transistor behaves like a SiGe-base transistor having a graded Ge distribution.* The transistor has larger current gain, larger Early voltage, and smaller base transit time compared to a Si-base transistor having the same polysilicon emitter, base width, and base dopant distribution. This explains why reported "constant-Ge" SiGe-base transistors usually show higher speed, higher current gain, and larger Early voltage than corresponding Si-base control transistors.

10.4.5   Some Optimal Ge Profiles

In this section, we apply the results obtained in Sections 10.4.3 and 10.4.4 to discuss tailoring the Ge profile for optimal or improved device characteristics. We first consider it from a current-gain perspective, and then from a base-transit-time perspective.

10.4.5.1 Ge Profile from Current Gain Perspective

SiGe-base transistor designers often find current gains too large and/or varying too much among transistors. The larger current gain in a SiGe-base transistor is caused by an increase in collector current, not by a decrease in base current. Therefore, any effort for reducing the current gain of a SiGe-base transistor should be focused on reducing its collector current.

- *Reducing current gain without degrading base transit time.* The most effective way to reduce collector current is to reduce the bandgap narrowing at the emitter end of the quasineutral base. To see this, let us consider the case depicted in Figure 10.7. The corresponding t_B ratio is given by Eq. (10.55), which shows that t_B is a function of the energy difference $\left[\Delta E_{g\max} - \Delta E_{g0be}(x_{jE})\right]$ across the base layer. For this transistor the J_{C0} ratio is given by Eq. (10.48), which shows that for a given energy difference $\left[\Delta E_{g\max} - \Delta E_{g0be}(x_{jE})\right]$, the J_{C0} ratio increases exponentially with increase in $\Delta E_{g0be}(x_{jE})$. Therefore, if we reduce $\Delta E_{g0be}(x_{jE})$ but keep $\left[\Delta E_{g\max} - \Delta E_{g0be}(x_{jE})\right]$ constant, we can reduce collector current, and hence current gain, without affecting base transit time.
- *Minimizing sensitivity of current gain to emitter-depth variation.* Designers often want to have a current gain that is not sensitive to the emitter formation process. This can be accomplished by using the Ge distribution illustrated in Figure 10.13 where ***the boundary of the quasineutral base on the emitter end is confined to a***

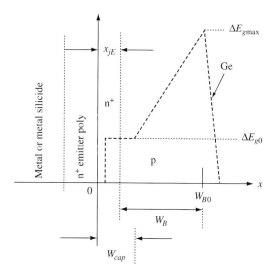

Figure 10.13 Schematic illustrating the emitter and base regions of a polysilicon-emitter SiGe-base vertical bipolar transistor having a Ge distribution that reduces the sensitivity of current gain to emitter-depth variation. For simplicity, the base dopant distribution is not shown

constant-Ge region (Oda *et al.*, 1997; Ansley *et al.*, 1998; Niu *et al.*, 2003). The model discussed in Section 10.4.3.1 for the case of no Ge in the emitter can be readily extended to this case. Instead of a Ge-free cap, we have a constant-Ge region. The base bandgap-narrowing parameter is given by

$$\Delta E_{gB,SiGe}(x) = \Delta E_{g0} + \left(\frac{x - W_{cap}}{W_{B0} - W_{cap}}\right)(\Delta E_{gmax} - \Delta E_{g0}) \quad x > W_{cap} \quad (10.66)$$
$$= \Delta E_{g0} \quad\quad\quad\quad\quad\quad\quad\quad\quad\quad\quad\quad\quad\quad\quad x < W_{cap}.$$

It is left as an exercise (Exercise 10.11) for the reader to show that the J_{C0} ratio in this case is

$$\frac{J_{C0}(SiGe, x_{jE})}{J_{C0}(Si, x_{jE})}$$
$$= \frac{\overline{\gamma\eta}\exp(\Delta E_{g0}/kT)}{\left(\dfrac{W_{cap} - x_{jE}}{W_{B0} - x_{jE}}\right) + \left(\dfrac{W_{B0} - W_{cap}}{W_{B0} - x_{jE}}\right)\left(\dfrac{kT}{\Delta E_{gmax} - \Delta E_{g0}}\right)\left[1 - \exp\left(\dfrac{\Delta E_{g0} - \Delta E_{gmax}}{kT}\right)\right]}.$$
$$(10.67)$$

Equation (10.67) is plotted in Figure 10.14 as a function of emitter-depth variation, using $\overline{\gamma\eta} = 1$. Some insights into the dependence of collector current on Ge distribution can be inferred from the plots. First, a high Ge concentration at or near the

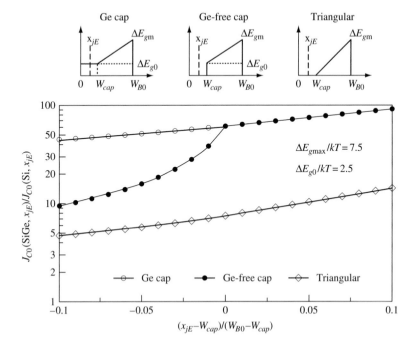

Figure 10.14 Relative improvement factors in collector current as a function of $(x_{jE} - W_{cap})/(W_{B0} - W_{cap})$ for three Ge profiles. $\overline{\gamma\eta} = 1$ assumed

emitter–base junction leads to a large J_{C0} ratio. Second, a Ge distribution that ramps down steeply or abruptly toward the emitter–base junction, as depicted by the "Ge-free cap" case in the figure, causes the J_{C0} ratio to be sensitive to emitter-depth variation. This is probably the main reason why large current-gain variations are often observed in SiGe-base vertical bipolar transistors.

10.4.5.2 Ge Profile from Base Transit Time Perspective

Here we examine the dependence of base transit time on Ge distribution in greater detail, using the models developed for the trapezoidal Ge distribution shown in Figure 10.7.

- *Dependence on ΔE_{g0}.* Let us consider the case of $x_{jE} = W_{cap}$. In this case, the t_B ratio is given by Eq. (10.55) by setting $\Delta E_{g0be}(x_{jE}) = \Delta E_{g0}$. Figure 10.15 is a plot of the ratio $t_B(\Delta E_{g0}/kT)/t_B(\Delta E_{g0}/kT = 0)$ as a function of $\Delta E_{g0}/kT$ for $\Delta E_{gmax}/kT = 7.5$. It shows that simply increasing ΔE_{g0} at constant ΔE_{gmax} increases base transit time. Therefore, from a base transit time point of view, the simple triangular Ge profile (i.e., with $\Delta E_{g0} = 0$) is preferred.

- *Case of Ge not ramped down exactly as desired.* In practice, it is difficult to achieve a truly linearly graded bandgap across the quasineutral base layer. If the Ge concentration is ramped down at a rate more slowly than intended, it will result in a finite, instead of zero, Ge concentration at the emitter end of the quasineutral base, i.e., it will result in $\Delta E_{g0be}(x_{jE}) > 0$. When that happens, the base transit time is degraded, or not improved over a Si-base transistor by as much as intended, as demonstrated in Figure 10.15.

  Now consider the case when the Ge concentration is ramped down at a rate faster than intended. Let us assume that the target design is to have a simple triangular Ge distribution with $x_{jE} = W_{cap}$, $\Delta E_{g0} = 0$, and some desired valued of ΔE_{gmax}. If the

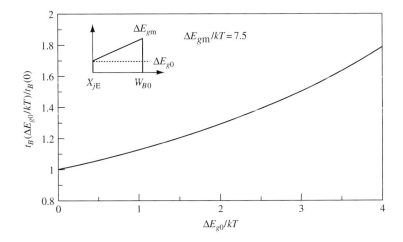

Figure 10.15 The t_B ratio for a trapezoidal Ge distribution as a function of $\Delta E_{g0}/kT$ for $\Delta E_{gmax}/kT = 7.5$, relative to the base transit time at $\Delta E_{g0}/kT = 0$. The base width is kept constant

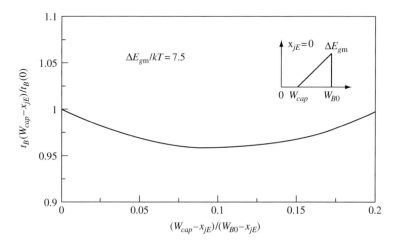

Figure 10.16 Relative change of base transit time at fixed quasineutral base width as a function of thickness of the Ge-free layer at the emitter end of the base

Ge concentration is ramped down at a rate faster than intended during growth of the base layer, there will be a finite region of the quasineutral base at the emitter end with no Ge at all. This is the case of "no Ge in emitter" described in Section 10.4.3.3. The t_B ratio can be obtained from Eq. (10.56) by setting $\Delta E_{g0} = 0$. The quasineutral base width is $W_{B0} - x_{jE}$ and the thickness of portion of the quasineutral base having no Ge is $W_{cap} - x_{jE}$. Figure 10.16 is a plot of the relative change of the base transit time as a function of $\left(W_{cap} - x_{jE}\right) / \left(W_{B0} - x_{jE}\right)$, using $\Delta E_{gmax}/kT = 7.5$. $t_B\left(W_{cap} - x_{jE}\right)$ is the base transit time when there is a Ge-free layer of thickness $W_{cap} - x_{jE}$ at the emitter end of the quasineutral base. $t_B(0)$ is the base transit time for the intended Ge distribution where the thickness of the Ge-free layer is zero. The shape of the curve is caused by the balance of the various terms in Eq. (10.56). Figure 10.16 suggests that for a given ΔE_{gmax} and base width, as the thickness of the Ge-free layer at the emitter end of the base increases from zero to some finite value, the base transit time goes through a minimum at a Ge-free layer thickness of about 10% of the base width. This result, together with the dependence on ΔE_{g0} discussed in the previous bullet point, suggests that it is preferred to ramp down the Ge distribution more rapidly than intended instead of more slowly than intended. For a given quasineutral base width and ΔE_{gmax}, *it is preferred to have a thin Ge-free layer at the emitter end of the base than to have $\Delta E_{g0be}(x_{jE}) > 0$.*

10.4.5.3   Current Gain and Base Transit Time Tradeoff

The Ge distribution illustrated in Figure 10.13 can be represented as the sum of a constant-Ge distribution and a graded-Ge distribution. That is, instead of Eq. (10.66), the base bandgap-narrowing parameter can be written as

$$\Delta E_{gB, SiGe}(x) = \Delta E_{g0} + \Delta E'_{gB, SiGe}(x), \tag{10.68}$$

where

$$\Delta E'_{gB,SiGe}(x) = \left(\frac{x - W_{cap}}{W_{B0} - W_{cap}}\right)\left(\Delta E_{gmax} - \Delta E_{g0}\right) \qquad x > W_{cap}$$

$$= 0 \qquad x < W_{cap}.$$

(10.69)

Substituting Eq. (10.68) into Eq. (10.36), we see that the part containing ΔE_{g0} drops out, and only the part containing $\Delta E'_{gB,SiGe}(x)$ remains in the integrals. That is, the constant-Ge part has no effect on t_B. Only $\Delta E'_{gB,SiGe}(x)$ contributes to t_B enhancement. For a given ΔE_{gmax}, t_B is smallest when $\Delta E_{g0} = 0$. The example illustrated in Figure 10.16 thus corresponds to an optimal Ge distribution from a t_B perspective.

However, the constant-Ge part causes the J_{C0} to increase in proportion to $\exp(\Delta E_{g0}/kT)$, as can be inferred from Eq. (10.27). Therefore, for a given value of ΔE_{gmax}, there is a tradeoff between base transit time and current gain. Current gain can be increased readily by increasing ΔE_{g0}. But it will lead to an increase in base transit time.

10.4.6   Base-Width Modulation by V_{BE}

The space-charge layer width W_{dBE} of an emitter–base diode is a function of V_{BE}. That means W_B is modulated by V_{BE}. In Section 9.3.2.1, the modulation of W_B by V_{BC} was discussed. The degree of W_B modulation by V_{BC} is indicated by the Early voltage. As a result, W_B modulation by V_{BE} is sometimes referred to as *reverse Early effect* (Crabbé *et al.*, 1993b; Salmon *et al.*, 1997; Deixler *et al.*, 2001). (This should not be confused with the Early voltage of a transistor operated in the reverse-active mode, which will be discussed in Section 10.4.7.)

Since $N_E \gg N_B$, the emitter–base diode can be treated as a one-sided diode, with W_{dBE} residing in the base side only. From Eq. (3.15), this width is

$$W_{dBE}(V_{BE}) = \sqrt{\frac{2\varepsilon_{si}(\psi_{bi} - V_{BE})}{qN_B}}.$$

(10.70)

The base-side boundary of the emitter–base space-charge region is also the emitter-side boundary of the quasineutral base layer. As a bipolar transistor is turned on by increasing V_{BE} from zero to some positive value, W_{dBE} is reduced by an amount $W_{dBE}(0) - W_{dBE}(V_{BE})$, causing the W_B to increase by the same amount. Such widening of the base at the emitter end has negligible effect on a Si-base bipolar transistor. However, in a SiGe-base bipolar transistor, the effect can be readily observable because it is amplified by the base bandgap profile near the emitter end. It explains why base widening at the emitter end has rarely been discussed in the literature until SiGe-base bipolar transistors are studied in detail (Crabbé *et al.*, 1993b; Cressler *et al.*, 1993a, 1993b; Paasschens *et al.*, 2001).

In all of the models developed and discussed so far, we have implicitly ignored W_{dBE} altogether by assuming that the quasineutral base extends from $x = x_{jE}$ at the emitter end to $x = W_{B0}$ at the collector end. This assumption is valid as long as effects caused by variation of W_{dBE} are negligible. Ignoring W_{dBE} greatly simplifies the schematics illustrating the emitter–base diode and depicting the Ge distribution within the quasineutral base layer.

When variation of W_{dBE} cannot be ignored, as is the case being considered here, the equations derived in Sections 10.4.3–10.4.5 are still valid provided that we treat the quasineutral base as extending from $x = x_{jE}$ at the emitter end to $x = W_{B0}$ at the collector end. That is, **the parameter x_{jE} is now used to denote the sum of the emitter depth in the single-crystalline region and W_{dBE}.** When W_{dBE} varies with V_{BE}, the parameter x_{jE} varies with V_{BE} by the same amount. Equation (10.70) can be used to calculate the variation of x_{jE} as a function of V_{BE}. Widening of the quasineutral base at the emitter end can then be treated as "emitter-depth variation."

10.4.6.1 J_{C0} Roll Off at Low Currents

It can readily be inferred from Eq. (10.26) that a certain degree of J_{C0} roll off caused by base widening at the emitter end is inherent in the operation of a bipolar transistor. For a Si-base transistor, this effect is quite small and usually ignored. As an illustration, consider a Si-base bipolar transistor having a boxlike base dopant distribution with an average concentration of $N_B = 5 \times 10^{18} \, \text{cm}^{-3}$. The change in W_B between $V_{BE} = 0.3$ V and $V_{BE} = 0.7$ V, estimated from Eq. (10.70), is about 4 nm. For a transistor having $W_B = 70$ nm at $V_{BE} = 0.3$, this change is about 6%. That is J_{C0} at $V_{BE} = 0.7$ V is reduced by only about 6% compared to J_{C0} at $V_{BE} = 0.3$ V. If the transistor has a larger N_B, or if the base dopant distribution has a higher concentration at the emitter end, which is typically the case for an implanted base, the amount of roll off in J_{C0} is even smaller.

However, in the case of a SiGe-base transistor, as V_{BE} is increased, the base bandgap at the emitter end of the base may change. This is demonstrated in Figure 10.17, where W_B at two values of V_{BE} are illustrated for two SiGe-base transistors.

A roll off in J_{C0} should lead to a roll off in current gain. Figure 10.18 is a plot of measured current gain in a SiGe-base vertical bipolar transistor (Crabbé et al., 1993b). The initial rise in current gain at very low currents is caused by the base−emitter space-charge-region current component of the base current being finite, instead of zero (see Figure 9.8 and related discussion).

Next, let us focus on the data at 300 K in Figure 10.18. The rapid rolloff in current gain at current densities greater than about 2 mA/μm^2 is caused by Kirk effect, which is base widening at the collector end of a vertical bipolar transistor (see Figure 9.9 and related discussion). The slow current gain roll off at current densities less than 2 mA/μm^2 is caused by base-width modulation by V_{BE}, which is base widening at the emitter end.

Figure 10.18 also shows that the current gain roll off due to V_{BE}-induced base widening (from J_C of about 10^{-2} mA/μm to about 1 mA/μm) increases rapidly as temperature decreases. This is to be expected because the values of $\Delta E_{g\text{max}}$ and ΔE_{g0} are fixed for a given SiGe-base transistor, but the values of $\Delta E_{g\text{max}}/kT$ and $\Delta E_{g0}/kT$

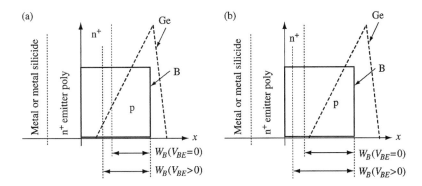

Figure 10.17 Schematics illustrating base widening at the emitter end for two SiGe-base transistors. (a) For this transistor, the Ge concentration ramps down to zero at a point beyond the emitter end of the quasineutral base layer when the transistor is biased at $V_{BE} = 0$. As V_{BE} is increased, $\Delta E_{g0be}(V_{BE})$ decreases. (b) For this transistor, the Ge concentration ramps down to zero before reaching the emitter end of the quasineutral base layer when the transistor is biased at $V_{BE} = 0$. As V_{BE} is increased, $\Delta E_{g0be}(V_{BE})$ remains zero

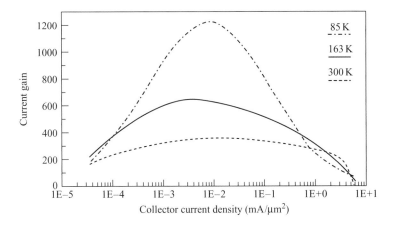

Figure 10.18 Measured current gain in a typical SiGe-base transistor as a function of collector current density (after Crabbé *et al.*, 1993b)

increase as temperature decreases, causing the "emitter-depth variation" effect to increase.

10.4.6.2 V_{BE} As a Reference

Circuit designers often use V_{BE} as a reference. V_{BE}-referenced circuits are based on the assumption that, for a given emitter area, the collector current I_C of a bipolar transistor is determined by V_{BE}. Circuit designers often refer to "V_{BE}" of a transistor as the V_{BE} value needed to achieve a target I_C value. For a Si-base bipolar transistor, the relationship between I_C and V_{BE} is simply $I_C = A_E J_{C0} \exp(q V_{BE}/kT)$, and J_{C0} is independent of V_{BE} for V_{BE} in the low-injection limit. However, J_{C0} of a SiGe-base vertical bipolar transistor is a strong function of its Ge profile in the base near the emitter end. It has been shown that a small change in the Ge profile shape can cause an appreciable change in "V_{BE}" (Salmon *et al.*, 1997; Deixler *et al.*, 2001). Therefore, special attention should be paid in designing V_{BE}-referenced circuits using SiGe-base vertical bipolar transistors. If needed, the discussion in Section 10.4.5.1 can be followed in designing the Ge profile for minimum V_{BE} sensitivity.

10.4.7 Reverse-Mode *I–V* Characteristics

Bipolar transistors are normally operated in the forward-active mode. Occasionally, a transistor is operated in the reverse-active mode either unintentionally or by design. For instance, when a transistor goes into saturation in a circuit, its base–collector junction becomes forward biased and its collector current is the difference between a forward component and a reverse component. In this section we compare the reverse-mode currents and the forward-mode currents.

The normal (or forward) and reverse modes of operation of a SiGe-base vertical bipolar transistor having a triangular Ge distribution in the base are depicted schematically in Figure 10.19. As discussed in Section 9.6.1, the reciprocity relationship between emitter and collector implies that the magnitude of the collector current in normal mode is equal to the magnitude of the collector current in reverse mode. That is, we have *in theory*

$$I_C(\text{forward}) = I_C(\text{reverse}). \tag{10.71}$$

Figure 10.20 shows the measured forward and reverse Gummel characteristics for two SiGe-base vertical bipolar transistors. The transistor in Figure 10.20(a) has $I_C(\text{reverse})$ about the same as $I_C(\text{forward})$, but its $I_B(\text{reverse})$ is significantly larger than its $I_B(\text{forward})$. The transistor in Figure 10.20(b) has both I_C and I_B appreciably larger

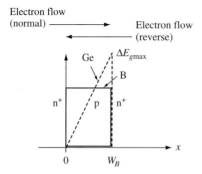

Figure 10.19 Schematic illustrating a SiGe-base bipolar transistor operated in the forward and reverse modes. In the forward mode, the left n$^+$ region is the emitter and the right n$^+$ region is the collector. In the reverse mode, the right n$^+$ region is the emitter and the left n$^+$ region is the collector

in reverse mode than in forward mode. The physical mechanisms governing these subtle differences are discussed next.

Let us first examine the base currents. I_B of a vertical transistor in forward and reverse modes are quite different because the emitter parameters in forward mode are quite different from those in reverse mode. In forward mode, the transistor has a polysilicon emitter. In reverse mode, the transistor has a complex n-type region (the pedestal collector region and the heavily doped subcollector region) as an emitter. [See Figure 9.1(a)]. Also, the base–emitter diode area in reverse mode (intrinsic-base area plus extrinsic-base area) is much larger than the base–emitter diode area in forward mode (intrinsic-base area only). The net is that I_B should be larger in reverse mode than in forward mode.

Regarding the collector currents, the two collector currents for the transistor in Figure 10.20(a) are about the same for V_{BE} less than about 0.8 V where the currents are reasonably ideal. This is consistent with Eq. (10.71). At V_{BE} values larger than about 0.8 V, the two collector currents are significantly below their ideal values because of a combination of series resistance effect and Kirk effect. The emitter series resistance in forward mode is smaller than that in reverse mode. Thus, on the basis of emitter resistance alone, we expect I_C(forward) to be larger than I_C(reverse) in the non-ideal region. On the other hand, the Kirk effect is significantly smaller in reverse mode because of the absence of a lightly doped region in the "collector." Thus, on the basis of Kirk effect alone, we expect I_C(forward) to be smaller than I_C(reverse). The fact that the measured I_C(reverse) drops below I_C(forward) at high V_{BE} suggests the drop is due entirely to the large "emitter" series resistance in reverse mode.

Even in the voltage range where series resistance and Kirk effects are negligible, there are two other effects that can cause the measured collector currents not to be equal, as predicted by Eq. (10.71). These are the effect of base-width modulation due to V_{BE} (cf. Section 10.4.6) and base-width modulation effect due to V_{BC} (cf. Section 9.3.2.1). We shall examine these two effects separately.

As explained in Section 10.4.6, base-width modulation due to V_{BE} causes the collector current to differ from its ideal value. In forward mode, W_{dBE} resides primarily in the base side because N_B is much smaller than N_E. As V_{BE} is increased, W_{dBE} is reduced, causing W_B to widen toward the emitter by the same amount. In the case of a SiGe-base vertical transistor, the resulting reduction in J_{C0}(forward) is amplified by the graded Ge profile (cf. Section 10.4.6.1).

However, in reverse mode, with the collector functioning as emitter, "N_E" is much smaller than N_B. As a result, "W_{dBE}" resides primarily in the "emitter" side instead of in the base side. As "V_{BE}" is increased in reverse mode, the change in "W_{dBE}" is absorbed in the "emitter" side instead of in the base side. That is, there should be negligible "base widening at the emitter end" when a vertical transistor is operated in reverse mode. That is, on the basis of base-width modulation alone, J_{C0}(reverse) and J_{C0}(forward) may be different for a SiGe-base vertical bipolar transistor.

The effect of base-width modulation by V_{BC} in a transistor is characterized by its Early voltage (cf. Section 9.3.2.1). It can be inferred readily from Figure 9.12 that, for the same I_B or same V_{BE}, the smaller the Early voltage the larger the collector current.

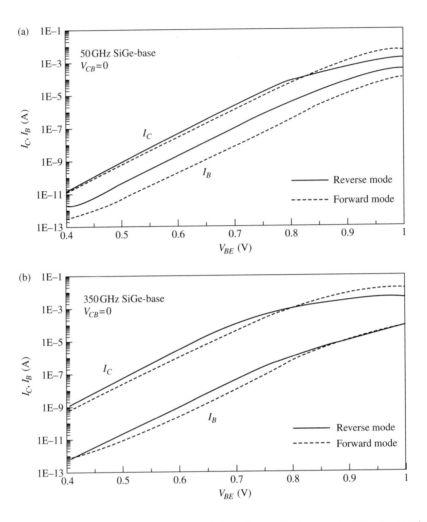

Figure 10.20 Measured Gummel characteristics of two SiGe-base vertical bipolar transistors in normal forward-active mode and in reverse-active mode. (a) A transistor having $f_T = 50$ GHz, $V_A(\text{forward}) = 55.7$ V, and $V_A(\text{reverse}) = 1.34$ V. (b) A transistor having $f_T = 350$ GHz, $V_A(\text{forward}) = 22.2$ V, and $V_A(\text{reverse}) = 0.16$ V (after Rieh *et al.*, 2005)

Also, it can be shown that, for a SiGe-base transistor having a graded base bandgap, the Early voltage in reverse mode is much smaller than that in forward mode (cf. Exercise 10.15). The smaller Early voltage in reverse mode definitely contributes to the observed $I_C(\text{reverse})$ being larger than $I_C(\text{forward})$ in a typical SiGe-base transistor.

Returning to Figure 10.20, it is apparent that a SiGe-base transistor having a higher f_T has greater asymmetry between $I_C(\text{forward})$ and $I_C(\text{reverse})$ than a transistor having a lower f_T. This observation is consistent with the two base-width-modulation effects, namely the base-width modulation by V_{BE} and the base-width modulation by V_{BC}, in a vertical SiGe-base bipolar transistor. The important point is that the subtle difference in the forward and reverse *I–V* characteristics of a SiGe-base vertical transistor should be included in modeling bipolar circuits which involve operation in reverse and/or saturation modes.

10.4.8   Heterojunction Nature of a SiGe-Base Vertical Bipolar Transistor

In a wide-gap-emitter *heterojunction bipolar transistor* (HBT), the small base current, and hence the large current gain, is due to the large energy barrier for base current injection at the emitter–base junction (Kroemer, 1957). A wide-gap-emitter HBT, even without base-bandgap grading, usually has a large Early voltage as well because the small base current allows the intrinsic-base layer to be doped heavily and still maintain sufficient current gain. A heavily doped intrinsic base implies a large base Gummel number, which in turn leads to large Early voltage [cf. Eq. (9.94)]. Both adequate current gain and large Early voltage can be obtained simultaneously in a wide-gap-emitter HBT without base-bandgap grading. (Of course, base bandgap grading leads to even larger Early voltage.)

In the literature, most SiGe-base vertical transistors have, by design, either a graded base bandgap or a constant but narrowed base bandgap. Both the graded-bandgap and the constant-bandgap SiGe-base transistors are often referred to as HBTs as well. In this section, we want to examine the heterojunction nature of these two types of SiGe-base transistors, and contrast their properties with a wide-gap-emitter HBT.

10.4.8.1   Heterojunction Nature of a Graded-Bandgap SiGe-Base Vertical Transistor

Consider a SiGe-base vertical transistor having a simple linearly graded base band-gap. Compared to a Si-base transistor, the linearly graded base bandgap leads to a larger J_{C0}, a larger V_A, and a smaller t_B. There is no change in I_B. Equation (10.71) implies that the two SiGe-base bipolar transistors illustrated in Figure 10.21 should have the same basic current–voltage characteristics (ignoring Early voltage effect). Transistor (a) is a usual SiGe-base transistor. Transistor (b) is the same transistor as (a) in every respect except that its Ge distribution is retrograded. That is, for transistor (b), ΔE_{gmax} is located at the emitter end of the base layer. The energy-band diagrams corresponding to the emitter–base diodes of these two transistors are illustrated in Figure 10.22.

To first order, the two transistors have the same collector current. However, once we include second order effects, such as base-width modulation caused by V_{BE} and V_{BC}, the two transistors in Figure 10.21 have readily distinguishable characteristics, as discussed in Section 10.4.7. In addition, transistor (b) has a retarding field in its base region which gives it a larger base transit time than transistor (a) (cf. Exercise 10.15). In other words, the characteristics of a SiGe-base vertical transistor having a nonuniform base bandgap are determined more by the direction of the base bandgap grading and the amount of bandgap difference across the base, and less by the electron and hole injection barriers at the emitter–base junction.

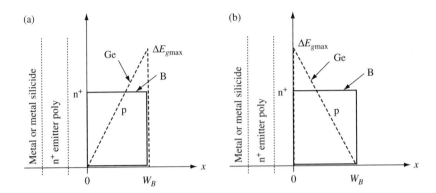

Figure 10.21 Schematics illustrating the emitter–base diodes of two SiGe-base bipolar transistors. Transistor (a) is a usual SiGe-base bipolar transistor having a graded Ge distribution, with ΔE_{gmax} at the collector end of its quasineutral base. Transistor (b) is the same transistor except that the Ge distribution is retrograded, with ΔE_{gmax} at the emitter end of its quasineutral base

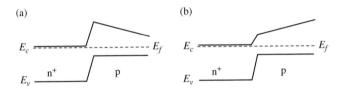

Figure 10.22 Energy-band diagrams corresponding to the emitter–base diodes illustrated in Figure 10.21, at $V_{BE} = 0$

10.4.8.2 Heterojunction Nature of a Constant-Bandgap SiGe-Base Vertical Transistor

To clearly distinguish a constant-Ge SiGe-base transistor from a graded-Ge SiGe-base transistor, we assume the Ge distribution in the constant-Ge transistor to ramp down abruptly near the emitter end, such as those illustrated in Figures 10.10 and 10.11, instead of at some finite rate.

- *Case of Ge in the emitter* (i.e., $x_{jE} > W_{cap}$). The Ge distribution and the dopant distributions in the emitter and base regions are illustrated schematically in Figure 10.10. The corresponding energy-band diagram is illustrated in Figure 10.23. The emitter bandgap is large compared to the base bandgap. Indeed, one can think of this transistor as a wide-gap-emitter HBT. As discussed in Section 10.4.4, compared with a Si-base transistor, this transistor has larger current gain, consistent with a wide-gap-emitter transistor, but the larger current gain is due to an increase in collector current, not a reduction in base current.
- *Case of no Ge in the emitter* (i.e., $x_{jE} < W_{cap}$). In this case, the Ge distribution and the dopant distributions in the emitter and base regions are as illustrated schematically in Figure 10.11. The corresponding energy-band diagram is

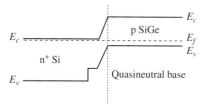

Figure 10.23 Energy-band diagram corresponding to the emitter–base diodes illustrated in Figure 10.10, at $V_{BE} = 0$

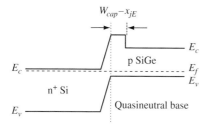

Figure 10.24 Energy-band diagram corresponding to the emitter–base diodes illustrated in Figure 10.11, at $V_{BE} = 0$

illustrated in Figure 10.24. If we focus at the region near the emitter–base junction, the transistor appears not to be a wide-gap-emitter device at all because the emitter and the emitter end of the quasineutral base have the same energy bandgap. Compared to a Si-base transistor, this transistor has the same base current. However, as discussed in Section 10.4.4 and shown in Exercise 10.10, this transistor not only has higher β_0, due to its larger I_C, but also larger V_A and smaller t_B. The high–low energy gap in the base makes this transistor behave more like a graded-base-bandgap transistor than a constant-base-bandgap transistor.

The net of all these is that ***it is best to think of a SiGe-base vertical bipolar transistor as a graded-base-bandgap transistor instead of an HBT***. By focusing on the base-bandgap grading characteristics instead of the emitter–base junction parameters, we focus on the device properties that depend on electron injection into and transport across the base. It is the electron injection and transport across the base region that makes a SiGe-base vertical transistor superior to a Si-base vertical transistor.

10.4.9   SiGe-Base Vertical Bipolar Transistor on Thin SOI

Designers often prefer BiCMOS technology (bipolar and CMOS integrated on the same chip) for mixed-signal applications where the CMOS is used primarily for the digital logic functions and the bipolar is used primarily for the analog and RF functions. From a mixed-signal system perspective, SOI is attractive because an SOI

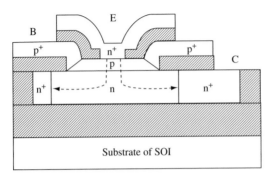

Figure 10.25 Schematic cross-section of a thin-silicon SOI SiGe-base vertical n−p−n transistor. The dotted arrows indicate the path of electrons from the emitter to the collector reach-through (after Cai *et al.*, 2003)

substrate provides good electrical isolation between the digital and the RF and analog components, particularly when a thick insulator or a high-resistivity substrate is used (see, e.g., Washio *et al.*, 2000) or when substrate engineering is applied (see, e.g., Burghartz *et al.*, 2002).

It is possible to build a SiGe-base vertical bipolar transistor on SOI using silicon thickness compatible with high-speed SOI CMOS (Cai *et al.*, 2002a, 2003; Cai and Ning, 2006). In fact, with thin SOI, it is possible to integrate SiGe-base vertical n−p−n, SiGe-base vertical p−n−p, and SOI CMOS, enabling designers to choose among bipolar circuits, complementary bipolar circuits, CMOS circuits, and BiCMOS circuits for design optimization (Duvernay *et al.*, 2007). Such a thin-silicon SOI SiGe-base vertical transistor, shown schematically in Figure 10.25, requires no subcollector layer. The transistor has the same emitter and base structure as a regular SiGe-base vertical transistor. Only the collector region behaves differently. The collector reach-through is only as deep as the silicon thickness, like the source/drain diffusion of SOI CMOS.

If a large positive voltage is applied to the substrate of an n−p−n transistor, an electron accumulation layer can be formed in the collector at the BOX–silicon interface. The electron accumulation layer reduces the collector series resistance and has little effect on other device parameters. The reduced collector resistance leads to improved f_T, which in turn leads to improved f_{max}. The improvement in peak f_T can be quite large. As for f_{max}, the improvement may not be as large because of the increase in C_{dBC} due to the formation of the electron accumulation layer. This is illustrated in Figure 10.26 for an experimental thin-silicon SOI SiGe-base vertical bipolar transistor (Cai *et al.*, 2003).

10.5　Design of Symmetric Lateral Bipolar Transistors on SOI

The schematics of the device structure and its associated parasitic resistances for a symmetric lateral n−p−n bipolar transistor on SOI is shown in Figure 10.27. The

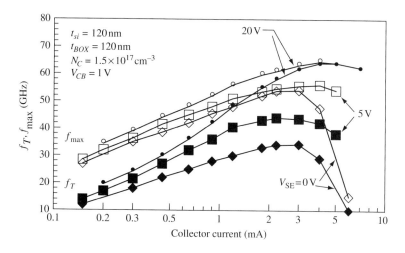

Figure 10.26 Measured f_T and f_{max} as a function of collector current for a thin-silicon SOI SiGe-base vertical n−p−n transistor for three values of substrate−emitter bias voltage. The transistor has an emitter area of $0.16 \times 16 \ \mu m^2$ (after Cai *et al.*, 2003)

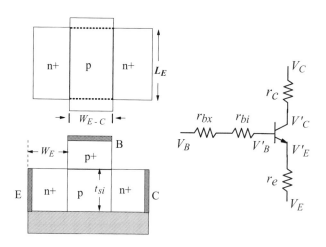

Figure 10.27 Schematic of the device-structure top view (top left), cross-sectional view (bottom left), and device junction and terminal voltages and parasitic resistances (right) of a symmetric lateral n−p−n transistor on SOI. For modeling purposes, the emitter contact is assumed to be located at a distance W_E from the emitter-base junction (after Ning and Cai, 2013)

equation for J_{C0} when the transistor is operated in forward-active mode is given by Eq. (9.52). The same equation applies to J_{C0}(reverse) as well when the transistor is operated in reverse-active mode. This should be contrasted with the analytic equations for a vertical bipolar transistor which are valid only when the transistor is operated in forward-active mode. In this section, we develop the analytic equations for describing the I−V characteristics and various device parameters for a symmetric lateral n−p−n transistor in operation.

10.5.1   Relationship Governing Emitter-to-Collector Spacing and Base Width

Figure 10.28 shows the quasineutral regions, the space-charge regions, as well as the energy-band diagram of a symmetric lateral n−p−n transistor at thermal equilibrium. With emitter/collector symmetry, the emitter-to-collector spacing W_{E-C} is fixed for a given lateral transistor, and the quasineutral base width W_B is governed by the relationship (Ning and Cai, 2013)

$$W_{E-C} = W_B \left(V'_{BE}, V'_{BC} \right) + W_{dBE} \left(V'_{BE} \right) + W_{dBC} \left(V'_{BC} \right). \tag{10.72}$$

When a lateral transistor in operation is biased with a collector−emitter terminal voltage V_{CE}, W_{dBE} and W_{dBC} are largest, and hence W_B is smallest, when the device is turned off, i.e., when $V'_{BE} = V_{BE} = 0$ and $V'_{BC} = V_{CE}$. The value of this minimum W_B depends on N_B and V_{CE}. Equation (10.72) *sets several design guidelines for a symmetric lateral bipolar transistor*, which are discussed next.

- $W_B \left(V'_{BE} = 0, V'_{BC} = V_{CE} \right)$ *must remain finite in the worst case*: The nominal value of W_{E-C} is determined by the lithography and patterning techniques available for device fabrication, and by the process used to form the emitter/collector regions. Fabrication process variation means there is variation in W_{E-C} among the transistors on a chip. Since V_A of a transistor decreases as W_B decreases [cf. Eq. (9.94)], the combination of W_{E-C} and N_B must be sufficiently large so that the worst-case $W_B \left(V'_{BE} = 0, V'_{BC} = V_{CE} \right)$ meets the minimum V_A required for the target circuit application.

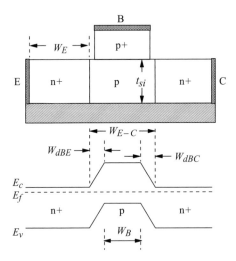

Figure 10.28 Schematics of a symmetric lateral n−p−n bipolar transistor on SOI, showing the device cross-section and energy-band diagram at thermal equilibrium ($V_{BE} = V_{BC} = 0$). The relationship between the dimensions of the physical regions and quasineutral regions are as indicated. W_E is the emitter "junction depth" for modeling purposes (after Cai *et al.*, 2011)

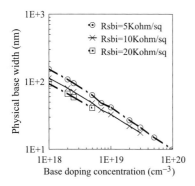

Figure 10.29 Physical base width W_{E-C} for a lateral bipolar transistor. W_{E-C} is calculated at $V_{BE} = V_{BC} = 0$. The R_{Sbi} values are for zero bias across B–E and B–C diodes (after Ning and Cai, 2013)

- *Meeting R_{Sbi} design target*: Vertical bipolar transistors are usually designed to have an intrinsic-base sheet resistivity in the range of 1–20 KΩ/□. For a lateral n−p−n transistor at thermal equilibrium, Eq. (10.5) gives

$$R_{Sbi}\left(V'_{BE} = 0, V'_{BC} = 0\right) = \frac{1}{qN_B\mu_{pB}W_B\left(V'_{BE} = 0, V'_{BC} = 0\right)}, \qquad (10.73)$$

where μ_{pB} is the hole mobility in the p-type base layer. The combination of N_B and W_B in Eq. (10.73) must satisfy Eq. (10.72). Figure 10.29 is a plot of W_{E-C} as a function of N_B, calculated from Eqs. (10.72) and (10.73) for several commonly used values of R_{Sbi}. It shows that for $W_{E-C} = 100$ nm, R_{Sbi} in the range of 5-20 KΩ/□ can be achieved for N_B of $1 - 2 \times 10^{18}$ cm^{-3}.

- *Consideration of W_{E-C} patterning technology*: The fabrication of a lateral bipolar transistor is quite similar to and compatible with SOI CMOS. The extrinsic-base p$^+$ region is patterned in a manner similar to the gate stack of a CMOS transistor. For a device with a target of $W_B = 10$ nm at zero bias, $N_B = 1 \times 10^{19}$ cm^{-3} and $R_{Sbi} = 9$ KΩ/□, we have $W_{E-C} = 34$ nm. For a device with a target of $W_B = 10$ nm at zero bias, $N_B = 5 \times 10^{18}$ cm^{-3} and $R_{Sbi} = 15$ KΩ/□, we have $W_{E-C} = 44$ nm. The message is that *lateral bipolar transistors with $W_B = 10$ nm can readily be fabricated using CMOS patterning technology of the 45-nm and 32-nm nodes.*

10.5.2   Analytic Model for Collector and Base Currents

The collector current of a symmetric lateral transistor operating in forward-active mode is given by Eq. (9.53), which is restated here:

$$I_C\left(V'_{BE}, V'_{BC}\right) = I_{C0}\left(V'_{BE}, V'_{BC}\right) \exp\left(qV'_{BE}/kT\right), \qquad (10.74)$$

with

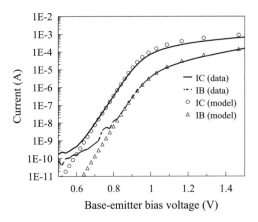

Figure 10.30 Analytical model fits to measured I_B and I_C of a symmetric lateral n−p−n bipolar transistor on SOI. The excess measured currents at low V_{BE} is due to process-related defects in this experimental device. The device has $t_{si} = 60$ nm, $N_B = 2.5 \times 10^{18}$ cm^{-3}, $L_E = 200$ nm, and $N_E = 4 \times 10^{20}$ cm^{-3}. $r_e = 267\,\Omega$ determined from open-collector V_{CE}-vs-I_B measurement. From comparing measured and calculated currents, the other device parameters are $W_{E-C} = 57.3$ nm, $W_B = 10.3$ nm, $W_E = 48.5$ nm, and $r_{bx} = 325$ (after Cai *et al.*, 2014)

$$I_{C0}(V'_{BE}, V'_{BC}) = \frac{qA_E D_{nB} n_{ieB}^2}{N_B W_B(V'_{BE}, V'_{BC})} \left[1 + \frac{1}{4} \left(\sqrt{1 + \frac{4n_{ieB}^2 \exp(qV'_{BE}/kT)}{N_B^2}} - 1 \right) \right]^{-1}.$$

(10.75)

The base current can be inferred from Eq. (9.71) for a uniformly doped ***shallow emitter*** (see discussion of Figure 10.30), which is restated here:

$$I_B(V'_{BE}) = I_{B0} \exp(qV'_{BE}/kT),$$

(10.76)

with

$$I_{B0} = \frac{qA_E D_{pE} n_{ieE}^2}{N_E W_E} \qquad \text{(uniformly doped shallow emitter).}$$

(10.77)

Notice that in Eq. (10.75), the base width W_B is explicitly indicated as a function of the junction voltages, to be consistent with the condition set by Eq. (10.72). Equations (10.74) ***and*** (10.75) ***are valid for all injection levels***. Therefore, it is possible to develop analytic models for calculating the base and collector currents of a symmetric lateral transistor for qV'_{BE} values close to the silicon bandgap energy (Cai *et al.*, 2014). This model is outlined next.

For modeling a transistor operating beyond the low-injection limit, the mobile carriers in a space-charge region must be included in determining the space-charge region width. For a symmetric lateral n−p−n transistor, the base−emitter and base−collector diodes can be treated as one-sided diodes, with the corresponding space-charge layer widths given by

$$W_{dBE}\left(V'_{BE}\right) = \sqrt{2\varepsilon_{si}\left(\psi_{bi} - V'_{BE}\right)/q(N_B + \Delta n)}, \qquad (10.78)$$

and

$$W_{dBC}\left(V'_{BC}\right) = \sqrt{2\varepsilon_{si}\left(\psi_{bi} - V'_{BC}\right)/q(N_B + \Delta n)}, \qquad (10.79)$$

where Δn is given by

$$J_C = qv_{sat}\Delta n. \qquad (10.80)$$

In Eq. (10.80), J_C is the collector current density and v_{sat} is the electron saturated velocity.

The intrinsic-base resistance r_{bi} can be calculated from Eq. (9.32) and the device parameters. The result is

$$r_{bi} = (t_{si}/3L_E)\rho_B/W_B\left(V'_{BE}, V'_{BC}\right), \qquad (10.81)$$

where L_E is the emitter stripe length indicated in Figure 10.27, ρ_B is the base-region resistivity, and t_{si} is the silicon layer thickness. The total base resistance r_b is

$$r_b = r_{bx} + r_{bi}, \qquad (10.82)$$

where r_{bx} is the extrinsic-base resistance. As discussed in Section 9.3.1 and illustrated in Exercise 9.5, the emitter resistance r_e can be determined from a plot of the measured open-collector voltage versus base current.

With r_e (which is equal to r_c) measured, Eqs. (10.72)–(10.82) can be used to calculate the collector and base currents of a symmetric lateral transistor. The transistor base width W_B can be determined from fitting the measured and calculated Gummel-plot collector currents in the ideal region (the region of V_{BE} where I_C increases at 60 mV/decade at room temperature). Once W_B is determined, W_{E-C} can be calculated from Eq. (10.72). The emitter width W_E can be determined from comparing the calculated and measured Gummel-plot base currents in the ideal region. The extrinsic-base resistance r_{bx} is determined from fitting the calculated and measured Gummel-pot currents at V_{BE} values beyond the ideal region.

Figure 10.30 shows the Gummel plots comparing the measured currents with the currents calculated using the analytic model described above. The only fitting parameters are W_E and r_{bx}. Figure 10.30 demonstrates that the analytic model explains the measured currents well up to $V_{BE} = 1.5\,\text{V}$, corresponding to V'_{BE} of almost 1.1 V.

Notice that the emitter-width parameter W_E from fitting the measured base current is 48.5 nm. For the transistor in Figure 10.31 (see Section 10.5.3), W_E is 45.5 nm. These values are smaller than the hole diffusion length in the emitter [cf. Figure 3.14 (c)]. It suggests that *the emitter behaves like a uniformly doped shallow emitter*, and hence Eq. (10.77) for I_{B0}.

10.5.3   Analytic Ebers-Moll Model Equations

The Ebers-Moll model in Section 9.6.1 is suitable for modeling the output characteristics which include the saturation region where both base−emitter and

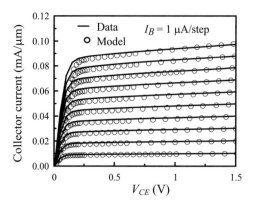

Figure 10.31 Comparison of calculated and measured output characteristics of a symmetric lateral n–p–n transistor on SOI. The device has $W_{E-C} = 98$ nm, $t_{si} = 60$ nm, $L_E = 1\,\mu$m, $N_E = 4 \times 10^{20}$ cm^{-3}, and $N_B = 5 \times 10^{18}$ cm^{-3}. The parameters from fitting to the measured Gummel plots are $W_E = 45.5$ nm, $r_e = r_c = 450\,\Omega$, and $r_{bx} = 400\,\Omega$ (after Ning and Cai, 2013)

base–collector junctions are forward biased. The Ebers-Moll equations for I_B and I_C are given by Eqs. (9.139) and (9.143), respectively. For a bipolar transistor with emitter/collector symmetry, these equations reduce to

$$I_C = I_{C0}\left(e^{qV'_{BE}/kT} - 1\right) - (I_{C0} + I_{B0})\left(e^{qV'_{BC}/kT} - 1\right), \qquad (10.83)$$

and

$$I_B = I_{B0}\left(e^{qV'_{BE}/kT} - 1\right) + I_{F0}\left(e^{qV'_{BC}/kT} - 1\right), \qquad (10.84)$$

where the collector saturated current I_{C0} is given by Eq. (10.75), and the base saturated current I_{B0} is given by Eq. (10.77).

The depletion-layer widths W_{dBE} and W_{dBC}, which are needed to calculate the base width W_B from Eq. (10.72) are given by Eqs. (10.78) and (10.79). In the saturation region, mobile electrons are injected into the space-charge regions from both the emitter and the collector. To properly reflect the mobile-charge density in the space-charge regions when the transistor is operated in the saturation region, we should add the mobile charges coming from both the emitter and the collector. Therefore, instead of Eq. (10.80), we should use

$$\Delta n = (J_{CF} + J_{CR})/qv_{sat} \quad \text{(allowing for operation in saturation region)} \qquad (10.85)$$

for the density of mobile electrons in Eqs. (10.78) and (10.79) to calculate W_{dBE} and W_{dBC}. In Eq. (10.85),

$$J_{CF} = J_{C0}\left(e^{qV'_{BE}/kT} - 1\right) \qquad (10.86)$$

is the collector current density injected in the forward direction, i.e., from the emitter, and

$$J_{CR} = J_{C0}\left(e^{qV'_{BC}/kT} - 1\right) \qquad (10.87)$$

is the collector current density injected in the reverse direction, i.e., from the collector.

Figure 10.31 compares the measured and calculated output characteristics of a symmetric lateral n−p−n transistor on SOI. The calculated characteristics compare well with the measured data, even in the saturation region where the base−collector diode is forward biased.

10.5.4   Early Voltage and Emitter−Collector Spacing

From Eq. (9.94), the Early voltage for a symmetric lateral transistor is

$$V_A + V_{CE} = q N_B W_B \left(V'_{BE}, V'_{BC} \right) / C_{dBC} \left(V'_{BC} \right). \tag{10.88}$$

Equation (10.88) suggests that an estimation of the emitter−collector spacing W_{E-C} may be obtained from measurements of V_A, using the following procedure.

Consider the typical measured Gummel currents, e.g., those shown in Figure 10.30. There is a range of V_{BE} where both I_C and I_B are ideal, varying with V_{BE} at 60 mV/decade at room temperature. This is the range where the effect of parasitic resistance is negligible, and the device terminal voltages can be considered the same as the junction voltages. As a specific example, let us take $I_B = 0.1\ \mu A$ from the device in Figure 10.30, corresponding to $V_{BE} = 0.87$ V. The corresponding $I_C = 4\ \mu A$ and $I_C \times r_e = 1$ mV. That is, the error in using the terminal voltages V_{BE} and V_{BC} for V'_{BE} and V'_{BC} is about 1 mV.

The $I_C - V_{CE}$ curves, e.g., those in Figure 10.31, are plotted at fixed I_B steps. From the Gummel plots, e.g., those in Figure 10.30, select an I_B value with its corresponding V_{BE} within the ideal region. For this (I_B, V_{BE}) pair, V_{BC} is simply $(V_{BE} - V_{CE})$. With V_{BC} thus determined, C_{dBC} can be calculated from Eq. (3.26), namely

$$C_{dBC} \left(V'_{BC} \right) = \varepsilon_{si} / W_{dBC} \left(V'_{BC} \right), \tag{10.89}$$

where W_{dBC} is given by Eq. (10.79). Therefore, using the measured V_A from an $I_C - V_{CE}$ plot, and knowing N_B, the corresponding value of W_B can be determined from Eq. (10.88). W_{E-C} of the transistor can then be estimated from Eq. (10.72).

The estimated emitter-to-collector spacing of a symmetric lateral transistor may be used to shed light on the device fabrication process. This property is especially useful at the technology development stage when the fabrication steps are not yet controlled and/or reproducible.

10.5.5   Analytic Model for the Transit Times

A figure-of-merit that indicates the maximum speed of a transistor is its forward transit time τ_F, discussed in Section 9.6.2.3. The four components of τ_F are given in Eq. (9.171), namely

$$\tau_F = \tau_E + \tau_B + \tau_{BE} + \tau_{BC}. \tag{10.90}$$

As discussed in Section 10.5.2, the measured I_B of typical symmetric lateral bipolar transistors are consistent with transistors having uniformly doped shallow or transparent emitters. The low-injection current gain can be inferred from Eq. (9.80), which is for a deep or non-transparent emitter, by replacing L_{pE} in the numerator by W_E. Therefore, by inferring from Eq. (9.133) for a narrow base, and adapting Eq. (9.80) for a shallow emitter, we can write

$$\tau_E(\text{shallow emitter}) = \frac{2I_B}{3I_C}\frac{W_E^2}{2D_{pE}} = \frac{W_E W_B\left(V'_{BE}, V'_{BC}\right)N_B n_{ieE}^2}{3D_{nB}N_E n_{ieB}^2}$$

(10.91)

for the emitter delay time, and

$$\tau_B = \frac{2t_B}{3} = \frac{W_B^2\left(V'_{BE}, V'_{BC}\right)}{3D_{nB}}$$

(10.92)

for the base delay time. From the discussion in Section 9.6.2.3, we can write

$$\tau_{BE} = t_{BE} = \frac{W_{dBE}\left(V'_{BE}\right)}{2v_{sat}}$$

(10.93)

as the base-emitter space-charge region delay time, and

$$\tau_{BC} = t_{BC} = \frac{W_{dBC}\left(V'_{BC}\right)}{2v_{sat}}$$

(10.94)

as the base-collector space-charge region delay time. The device parameters W_E and W_B, as explained in Section 10.5.2, can be determined from comparison of the calculated and measured Gummel-plot currents in the ideal-V_{BE} region. Thus, all four components of τ_F can be calculated. The calculated τ_F in turn can be used to calculate the f_T and $f_{\max}$ of a lateral transistor (to be shown and discussed in Chapter 11).

10.5.6   On the Fabrication of Thin-Base Symmetric Lateral Transistors

As discussed in Section 10.5.1, symmetric lateral transistors with $W_B = 10$ nm can be fabricated using CMOS technology of the 45-nm and 32-nm nodes. Therefore, the challenge of fabricating a transistor with $W_B = 10$ nm is not in lithography and patterning, but in the emitter/collector formation process for producing a 10-nm base that is uniform.

   As of this writing, most reported experimental symmetric lateral bipolar transistors on SOI were fabricated using high-dose ion implantation to form the emitter/collector regions. While the results show excellent device characteristics, such as high drive current and low emitter/collector series resistance, they also reveal the need to avoid high-dose ion implantation for emitter/collector formation if the goal is to achieve controlled and uniform W_B of nanometer dimensions. High-dose emitter/collector implantation has large lateral implantation straggle and requires high-temperature (typically 1,000 °C or higher) post-implant anneal, resulting in large variation in W_B device-to-device and large non-uniformity of W_B within a device. This is illustrated in Figure 10.32.

   One method for fabricating a lateral transistor having uniform base width is to pattern the extrinsic-base, followed by reactive-ion etching the silicon layer outside

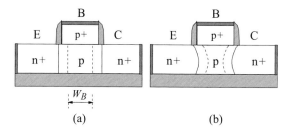

Figure 10.32 Schematics illustrating the structure of a symmetric lateral n−p−n transistor on SOI. The dashed lines denote the boundaries of the quasineutral base region. (a) Illustration of a uniform base width necessary for thin-base transistors. (b) Illustration of a typical base-width profile using high-dose ion implantation for emitter/collector formation

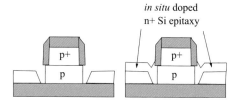

Figure 10.33 Schematics illustrating the notch-assisted RIE and epitaxy growth scheme for forming emitter/collector. Left: After notch-assisted RIE. Right: After epitaxial growth of emitter/collector (after Hashemi *et al.*, 2017, 2018)

the extrinsic-base region all the way down to the BOX, and then use *in situ* doped lateral silicon epitaxy to form the emitter/collector regions, using the exposed vertical intrinsic-base surfaces as seed for lateral epitaxial growth. By using a novel notch-assisted reactive-ion-etching and epitaxy scheme, illustrated in Figure 10.33, the laterally grown emitter/collector regions are automatically connected to their extension regions for metal contact and/or for electrical probing (Hashemi *et al.*, 2017, 2018). The resultant transistor has a uniform base width similar to that in Figure 10.32(a).

10.5.7   SiGe-on-Insulator Symmetric lateral n–p–n Transistors

In the low-injection approximation, the dependence of the collector current on V_{BE} has the form

$$I_C \propto \exp\left[(qV_{BE} - E_{gB})/kT\right], \tag{10.95}$$

where E_{gB} is the base-region bandgap energy. As long as $(qV_{BE} - E_{gB})$ is the same, I_C remains about the same when a transistor is built using a semiconductor having a different bandgap. Therefore, a path for reducing V_{BE} without reducing I_C is to use a narrow-gap semiconductor for the transistor base region. Lower V_{BE} for the same I_C implies a circuit implemented in narrow-base-bandgap bipolar transistors can operate

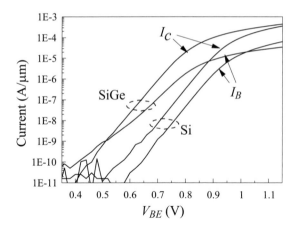

Figure 10.34 Gummel plots of a symmetric lateral SiGe n−p−n transistor and a symmetric lateral Si n−p−n control transistor, both having an emitter/collector formed by ion implantation, with $N_E = 4 \times 10^{20}$ cm^{-3}. The SiGe transistor achieves the same I_C as the Si transistor at about 130 mV lower V_{BE} (after Yau *et al.*, 2015)

at higher speed and/or reduced power dissipation compared to the same circuit implemented in Si bipolar transistors.

SiGe has a smaller energy bandgap than Si (cf. Section 10.4). Symmetric lateral SiGe n−p−n bipolar transistors, having SiGe emitter/collector and uniform SiGe base, have been demonstrated using SiGe-OI (SiGe-on-Insulator) as starting wafer substrate. For the transistors shown in Figure 10.34, the SiGe transistor reaches the same current as a control Si transistor at 130 mV lower V_{BE}, suggesting the SiGe has an effective energy bandgap about 130 meV smaller than Si (Yau *et al.*, 2015).

Ge is probably the candidate semiconductor with the smallest bandgap energy for building symmetric lateral bipolar transistors. An attempt to demonstrate lateral n−p−n bipolar transistors using a Ge-on-insulator as starting wafer substrate and ion-implantation for forming emitter/collector yielded devices with poor device characteristics (Yau *et al.*, 2016c). Interested readers are referred to the reference publication for discussion of the fabrication issues encountered.

10.5.8   Symmetric Si-Emitter/Collector SiGe-Base Lateral HBT

Conceptually, if the *in situ* doped lateral silicon epitaxy process illustrated in Figure 10.33 were used to form the emitter/collector of a SiGe-base transistor, the result would be a symmetric Si-emitter/collector uniform SiGe-base lateral HBT on insulator. The energy-band diagram of such an n−p−n transistor would be as illustrated in Figure 10.35.

As of this writing, there is no reported demonstration of such a transistor. However, there is a "partial-heterojunction" transistor, where the SiGe layer is etched only partially toward the BOX before the emitter/collector is formed by *in situ* doped Si epitaxial growth (Yau *et al.*, 2015). The resultant transistors do not show much

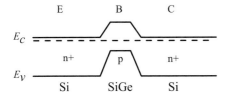

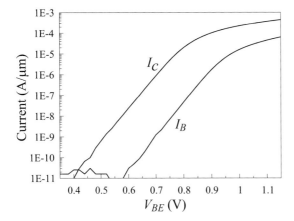

Figure 10.35 Schematic illustrating the energy-band diagram of Si-emitter/collector SiGe-base symmetric lateral n−p−n HBT (after Ning, 2013)

Figure 10.36 Expected collector and base currents of a hypothetical symmetric Si-emitter/collector SiGe-base lateral n−p−n HBT. The SiGe-base parameters are assumed to be the same as those of the SiGe lateral transistor in Figure 10.34

improvement in current gain, consistent with the fact that they are not true HBT, but a homojunction transistor and a HBT in parallel. However, the resultant transistors do show excellent base−emitter and base−collector diode characteristics, suggesting that symmetric Si-emitter/collector SiGe-base heterojunction lateral transistors on the insulator should be a viable technology. The expected Gummel currents of a true Si-emitter/collector SiGe-base lateral HBT would be as illustrated in Figure 10.36, where the collector current is taken from the SiGe transistor in Figure 10.34 while the base current is taken from the control Si transistor. The transistor would have a current gain of about 1,000.

Exercises

10.1 This exercise is designed to show the sensitivity of the current gain of a polysilicon-emitter transistor to the emitter-polysilicon thickness. For the polysilicon-emitter model described in Exercise 9.3, the emitter Gummel number is a function of the single-crystalline emitter-region thickness W_E and the polysilicon thickness W_{E1}, i.e.,

$$G_E(W_E, W_{E1}) = N_E \left(\frac{n_i^2}{n_{ieE}^2}\right)\left(\frac{W_E}{D_{pE}} + \frac{L_{pE1}\tanh(W_{E1}/L_{pE1})}{D_{pE1}}\right).$$

It is reasonable to assume the lifetimes in heavily doped silicon and heavily doped polysilicon to be the same for the same doping concentration, since both are determined by Auger recombination. It is also reasonable to assume the mobility in polysilicon to be smaller than that in silicon, since there is additional grain-boundary scattering in polysilicon.

(a)   A typical nonpolysilicon emitter has $N_E = 10^{20}\,\mathrm{cm}^{-3}$, $W_E = 300$ nm, and $W_{E1} = 0$. Estimate G_E (300 nm, 0) using the hole mobility and lifetime values in Figures 3.14(a) and (b).

(b)   A typical polysilicon emitter has $N_E = 10^{20}\,\mathrm{cm}^{-3}$ and $W_E = 30$ nm. Let us assume the hole mobility in polysilicon to be $1/3$ that in silicon, and assume the hole lifetimes in silicon and polysilicon to be the same. Estimate G_E (30 nm, W_{E1}) for $W_{E1} = 50, 100, 200,$ and 300 nm.

(c)   Graph the ratio G_E (30 nm, W_{E1})/G_E (300 nm, 0) as a function of W_{E1}.

10.2   The intrinsic-base sheet resistivity is given by Eq. (10.5), namely

$$R_{Sbi} = \left(q \int_0^{W_B} p_p(x)\mu_p(x)dx\right)^{-1}.$$

Most n–p–n transistors have a low-current R_{Sbi} value of about $10^4\ \Omega/\square$. Assuming a boxlike base doping profile, graph N_B as a function of W_B for W_B between 50 and 300 nm for $R_{Sbi} = 10^4\,\Omega/\square$. This graph illustrates how the base doping concentration varies with the intrinsic-base width in scaling in many practical device designs.

10.3   The base transit time for a box profile is given by

$$t_B \approx \frac{W_B^2}{2D_{nB}}.$$

For a fixed intrinsic-base sheet resistivity of $10^4\,\Omega/\square$, calculate and plot t_B as a function of W_B for W_B between 50 and 300 nm. Use the mobility and lifetime values in Figures 3.14(a) and (b) to estimate D_{nB} (see Exercise 10.2 for the relation between W_B and N_B).

10.4   This exercise illustrates the tradeoff between collector current density and Early voltage in an optimized bipolar device design. The Early voltage for a boxlike base doping profile is given by $V_A \approx qN_BW_B/C_{dBC}$ [cf. Eq. (9.94)]. To maintain negligible base widening in scaling, we assume the collector current density is maintained at $J_C = 0.3\,qv_{sat}N_C$, where $v_{sat} = 10^7$ cm/s is the electron saturated velocity. Thus, as N_C is increased to increase J_C, C_{dBC} is increased and V_A is decreased.

(a)   Use the one-sided junction approximation for the base–collector diode, and assume $V_{CB} = 2$ V (for purposes of calculating C_{dBC}). Plot C_{dBC} as a function of J_C for J_C between 0.1 and 5 mA/μm^2.

(b) For a base design with $qN_BW_B = 1.6 \times 10^{-6} C/cm^2$ (e.g., $N_B = 10^{18}$ cm^{-3} and $W_B = 100$ nm), estimate and plot V_A as a function of J_C for J_C between 0.1 and 5 mA/μm^2.

10.5 Consider an n–p–n transistor with a wide base of $W_B = 500$ nm. Suppose the base doping concentration is linearly graded, i.e., N_B has the form $N_B(x) = A - \alpha x$, with $N_B(0) = 2 \times 10^{17} cm^{-3}$ and $N_B(W_B) = 2 \times 10^{16} cm^{-3}$. For such light doping concentrations, the effect of heavy doping is negligible. Plot the built-in electric field due to the dopant distribution as a function of distance between $x = 0$ and $x = W_B$.

10.6 Consider an n–p–n transistor with a linearly graded base doping profile (cf. Exercise 10.5). The doping concentration at the emitter–base junction is $N_B(0) = 5 \times 10^{18} cm^{-3}$. The doping concentration at the base–collector junction is $N_B(W_B) = 5 \times 10^{17} cm^{-3}$, and $W_B = 100$ nm. For such high doping concentrations, the heavy-doping effect cannot be ignored. Plot the electric fields due to the dopant distribution and due to heavy-doping effect, as well as the total electric field, as a function of distance from $x = 0$ to $x = W_B$. [Use the bandgap-narrowing parameter in Eq. (9.17).]

10.7 The emitter series resistance r_e of a polysilicon-emitter n–p–n transistor, with negligible polysilicon–silicon interface oxide, has three components, namely, the resistance due to the single-crystal emitter region, the resistance due to the polysilicon layer, and the resistance due to the metal–polysilicon contact. Consider a polysilicon emitter with $N_E = 10^{20} cm^{-3}$, a single-crystal region of width $W_E = 30$ nm, a poly-silicon layer of thickness $W_{E1} = 200$ nm, and a metal–polysilicon contact resistivity of $2 \times 10^{-7} \Omega\text{-}cm^2$. Assume that, for the same doping concentration, the resistivity of polysilicon is 3-times that of single-crystal silicon. Calculate the series resistance components, as well as the total series resistance r_e, for an emitter 1 μm^2 in area. (Use the resistivity for n-type silicon shown in Figure 2.10.)

10.8 In the literature, the heavy-doping effect in the emitter is well recognized, but in the base it is often ignored. The saturated collector current density for an n–p–n transistor at low injection is [cf. Eq. (9.55)]

$$J_{C0} = \frac{q}{\displaystyle\int_0^{W_B} \frac{p_{p0}(x)}{D_{nB}(x)n_{ieB}^2(x)} dx}.$$

Assume a uniformly doped base with $N_B = 10^{18} cm^{-3}$ and $W_B = 100$ nm. Also assume $p_{p0} = N_B$. Estimate J_{C0} for the following two cases: (a) heavy-doping effect in the base is neglected, i.e., $n_{ieB} = n_i$, and (b) heavy-doping effect in the base is included. (This exercise demonstrates that heavy doping in the intrinsic base of modern bipolar transistors cannot be ignored.)

10.9 Plot the collector current ratio given by Eq. (10.60) and the Early voltage ratio given by Eq. (10.61) as a function of the ratio $(W_{cap} - x_{jE})/(W_{B0} - x_{jE})$ from $(W_{cap} - x_{jE})/(W_{B0} - x_{jE}) = 0$ to $(W_{cap} - x_{jE})/(W_{B0} - x_{jE}) = 1$, for $\Delta E_{g0}/kT = 0.5, 1, 2.5,$ and 5.

10.10 In practice, the Ge concentration in a constant-Ge SiGe-base transistor does not ramp down abruptly at the emitter end of the base. Instead it is ramped down at some finite rate. If the emitter depth is not deep enough to extend beyond the ramp part, the Ge ramp has to be included in modeling an otherwise constant-Ge SiGe-base transistor. Figure Ex. 10.10 illustrates the Ge distribution, with the emitter–base junction right at the foot of the Ge ramp. The quasineutral base width is W_B, which includes the Ge ramp region of width W_{B1}. Show that, compared to a Si-base transistor having the same polysilicon emitter, base width and base dopant distribution,

$$\frac{J_{C0}(\text{SiGe})}{J_{C0}(\text{Si})} = \overline{\gamma\eta}\left\{ \frac{kT}{\Delta E_{g0}}\frac{W_{B1}}{W_B}\left[1 - \exp(-\Delta E_{g0}/kT)\right]\right.$$

$$\left. + \left(1 - \frac{W_{B1}}{W_B}\right)\exp(-\Delta E_{g0}/kT)\right\}^{-1},$$

$$\frac{V_A(\text{SiGe})}{V_A(\text{Si})} = \frac{kT}{\Delta E_{g0}}\frac{W_{B1}}{W_B}\left[\exp(\Delta E_{g0}/kT) - 1\right] + \left(1 - \frac{W_{B1}}{W_B}\right),$$

and

$$\frac{t_B(\text{SiGe})}{t_B(\text{Si})} = \frac{1}{\eta}\left\{\left(1 - \frac{W_{B1}}{W_B}\right)^2 + \frac{2kT}{\Delta E_{g0}}\left(\frac{W_{B1}}{W_B}\right)^2\right\}$$

$$+ \frac{1}{\eta}\left\{\frac{2kT}{\Delta E_{g0}}\frac{W_{B1}}{W_B}\left(1 - \frac{W_{B1}}{W_B} - \frac{kT}{\Delta E_{g0}}\frac{W_{B1}}{W_B}\right)\left[1 - \exp(-\Delta E_{g0}/kT)\right]\right\}.$$

Note that the base transit time ratio reduces to $1/\overline{\eta}$ when W_{B1} is reduced to zero, as expected. Our transit time ratio is different from that in Eq. (4) of Cressler *et al.* (1993b) which does not reduce to the expected value as W_{B1} is reduced to zero.

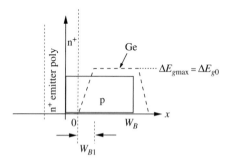

Figure Ex. 10.10

10.11 Show that the collector current ratio for the SiGe-base bipolar transistor illustrated in Figure 10.13 is

$$\frac{J_{C0}\left(\mathrm{SiGe},x_{jE}\right)}{J_{C0}\left(\mathrm{Si},x_{jE}\right)}$$

$$=\frac{\overline{\gamma\eta}\exp\left(\Delta E_{g0}/kT\right)}{\left(\dfrac{W_{cap}-x_{jE}}{W_{B0}-x_{jE}}\right)+\left(\dfrac{W_{B0}-W_{cap}}{W_{B0}-x_{jE}}\right)\left(\dfrac{kT}{\Delta E_{gmax}-\Delta E_{g0}}\right)\left[1-\exp\left(\dfrac{\Delta E_{g0}-\Delta E_{gmax}}{kT}\right)\right]}.$$

10.12 Show that the base transit time ratio for the Ge distribution illustrated in Figure 10.13 is

$$\left.\frac{t_B\left(\mathrm{SiGe},x_{jE}\right)}{t_B\left(\mathrm{Si},x_{jE}\right)}\right|_{x_{jE}<W_{cap}}$$

$$=\frac{1}{\eta}\frac{\left(W_{cap}-x_{jE}\right)^2}{\left(W_{B0}-x_{jE}\right)^2}+\frac{2}{\eta}\frac{\left(W_{cap}-x_{jE}\right)}{\left(W_{B0}-x_{jE}\right)}\frac{\left(W_{B0}-W_{cap}\right)}{\left(W_{B0}-x_{jE}\right)}\left(\frac{kT}{\Delta E_{gmax}-\Delta E_{g0}}\right)$$

$$\times\left[1-\exp\left(\frac{\Delta E_{g0}-\Delta E_{gmax}}{kT}\right)\right]$$

$$+\frac{2}{\eta}\frac{\left(W_{B0}-W_{cap}\right)^2}{\left(W_{B0}-x_{jE}\right)^2}\left(\frac{kT}{\Delta E_{gmax}-\Delta E_{g0}}\right)$$

$$\times\left\{1-\left(\frac{kT}{\Delta E_{gmax}-\Delta E_{g0}}\right)\left[1-\exp\left(\frac{\Delta E_{g0}-\Delta E_{gmax}}{kT}\right)\right]\right\}.$$

Note that the transit time is a function of the energy difference $\left(\Delta E_{gmax}-\Delta E_{g0}\right)$ only, and does not depend on ΔE_{gmax} or ΔE_{g0} individually. Changing ΔE_{gmax} and ΔE_{g0} together such that the difference remains the same has no effect on the base transit time.

10.13 It is instructive to use the equations obtained in Exercise 10.10 to study a "constant-Ge" SiGe-base transistor having a finite Ge-ramp region that starts at the emitter end of its quasineutral base, i.e., Figure Ex. 10.10. Plot the ratios for collector current, Early voltage, and base transit time as a function of W_{B1}/W_B from $W_{B1}/W_B=0$ to $W_{B1}/W_B=1$, for $\Delta E_{g0}/kT=2.5,5$, and 10.

10.14 Referring to Figure 10.19, show that

$$\frac{J_{C0}(\mathrm{SiGe,reverse})}{J_{C0}(\mathrm{Si,reverse})}=\frac{\overline{\gamma\eta}\Delta E_{gmax}/kT}{1-\exp\left(-\Delta E_{gmax}/kT\right)}$$

by carrying out the integration in Eq. (10.27). Comparison of this result with Eq. (10.29) shows that $J_{C0}(\text{forward})=J_{C0}(\text{reverse})$ for a SiGe-base bipolar transistor. This result is expected from the reciprocity relationship between the emitter and collector of an ideal bipolar transistor.

10.15 Referring to Figure 10.21, show that
(a)

$$\frac{V_A(\text{transistor b})}{V_A(\text{transistor a})}=\exp\left(\frac{-\Delta E_{gmax}}{kT}\right),$$

and

(b)

$$
\frac{t_B(\text{transistor b})}{t_B(\text{transistor a})} = \frac{\dfrac{kT}{\Delta E_{g\text{max}}} \exp\left(\dfrac{\Delta E_{g\text{max}}}{kT}\right)\left[1 - \exp\left(\dfrac{-\Delta E_{g\text{max}}}{kT}\right)\right] - 1}{1 - \dfrac{kT}{\Delta E_{g\text{max}}}\left[1 - \exp\left(\dfrac{-\Delta E_{g\text{max}}}{kT}\right)\right]}.
$$

11 Bipolar Performance Factors

In Chapter 10, the design of the individual regions and parameters of a bipolar transistor was discussed. It was noted that, during device operation, an individual device region is not isolated from and independent of the other device regions. Optimization of one device parameter often adversely affects the other device parameters. Thus, *optimization of the design of a bipolar transistor is a tradeoff process*. This design tradeoff should be done at the circuit and/or chip level, for the optimum design of a transistor is a function of its application and environment. In this chapter, we first discuss some figures of merit for evaluating a bipolar transistor for typical analog and digital circuit applications, and then discuss the tradeoffs in the design of a bipolar transistor for specific applications.

When we consider the performance of a circuit, the wires connecting the transistors and other elements of the circuit must be included. The signal propagation delays associated with the interconnect wires have been discussed in Section 8.3.4 in connection with CMOS circuits. The reader is referred to it for details. In this chapter, the wire capacitance which acts as a load on a bipolar circuit is included when we consider the performance and optimization of bipolar transistors and circuits.

We first discuss several commonly used figures of merit of bipolar transistors. This is followed by a discussion of performance factors in designing with vertical bipolar transistors. Significant differences between designing with a vertical bipolar transistor and a symmetric lateral bipolar transistor on SOI will be pointed out.

From the contrast between vertical bipolar transistors and symmetric lateral bipolar transistors on SOI, it should become apparent that the unique characteristics of symmetric lateral bipolar transistors suggest to us to rethink bipolar circuits and applications. Some circuit and application opportunities enabled by the advent of symmetric lateral bipolar transistors on SOI are discussed in Section 11.8.

11.1 Figures of Merit of a Bipolar Transistor

It is desirable to consider the merit of a transistor in terms of some simple, and preferably readily measurable, parameters. However, it is important to note that the relevance of a particular figure of merit depends on the application. Some of the commonly used figures of merit are discussed here.

11.1.1   Cutoff Frequency

For small-signal applications, the *cutoff frequency, or transition frequency, or unity-current-gain frequency,* f_T, is probably the most often used figure-of-merit. It is defined as the transition frequency at which the common-emitter, short-circuit load, small-signal current gain drops to unity. It is a measure of the maximum useful frequency of a transistor when it is used as an amplifier.

The cutoff frequency can be readily obtained from an analysis of a two-port network representation of a bipolar transistor (Exercise 11.6). Here we use a physically intuitive approach to obtain commonly used approximations for the cutoff frequency.

For simplicity, we shall first neglect the internal parasitic resistances and use the equivalent circuit shown in Figure 9.25 to determine the intrinsic cutoff frequency f_T'. Here, the convention is such that the primed parameters refer to an intrinsic device, while the unprimed parameters are for an extrinsic device. With the output shorted, r_0 and $C_{dCS,tot}$ have no influence, and the resulting equivalent circuit is shown in Figure 11.1.

From this equivalent circuit, the small-signal collector and base currents can be written as

$$i_c = g_m' v_{be}' - j\omega C_\mu v_{be}' \tag{11.1}$$

and

$$i_b = \left(\frac{1}{r_\pi'} + j\omega C_\pi + j\omega C_\mu\right) v_{be}', \tag{11.2}$$

where g_m' is the intrinsic transconductance, r_π' is the intrinsic input resistance, $C_\mu = C_{dBC,tot}$ is the base-collector junction depletion-layer capacitance, and $C_\pi = C_{dBE,tot} + C_{DE}$ is the sum of the base–emitter junction depletion-layer capacitance and the emitter diffusion capacitance, v_{be}' is the applied small-signal input voltage, and i_b and i_c are the small-signal base and collector currents due to v_{be}'. (See Section 9.6.2.2 for a derivation of the small-signal equivalent-circuit model.) The small-signal frequency-dependent common-emitter current gain is

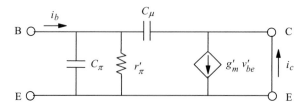

Figure 11.1 Small-signal equivalent circuit for determining the cutoff frequency of a bipolar transistor. Parasitic resistances are not included

$$\beta'(\omega) = \frac{i_c}{i_b} = \frac{g'_m - j\omega C_\mu}{(1/r'_\pi) + j\omega(C_\pi + C_\mu)}. \tag{11.3}$$

In the low-frequency limit, Eq. (11.3) gives

$$\beta'(\omega = 0) = g'_m r'_\pi, \tag{11.4}$$

which, according to Eq. (9.154) is simply the static common-emitter current gain β_0.

The cutoff frequency f'_T of the intrinsic transistor is given by setting $\beta'(\omega = \omega'_T) = 1$. That is, from Eq. (11.3), we have

$$1 = \sqrt{\frac{g'^2_m + \omega'^2_T C^2_\mu}{\dfrac{g'^2_m}{\beta^2_0} + \omega'^2_T (C_\pi + C_\mu)^2}}, \tag{11.5}$$

which can be rearranged to give

$$\omega'_T = 2\pi f'_T = \sqrt{\frac{g'^2_m - \dfrac{g'^2_m}{\beta^2_0}}{(C_\pi + C_\mu)^2 - C^2_\mu}} \approx \frac{g'_m}{\sqrt{(C_\pi + C_\mu)^2 - C^2_\mu}}. \tag{11.6}$$

For a typical modern bipolar transistor which has a relatively small $C_\mu (= C_{dBC,tot})$, we have $C_\pi (= C_{dBE,tot} + C_{DE}) \gg C_\mu$, and Eq. (11.6) reduces to the commonly used approximation of

$$\omega'_T = 2\pi f'_T \approx \frac{g'_m}{C_\pi + C_\mu} \tag{11.7}$$

or

$$\frac{1}{2\pi f'_T} = \tau_F + \frac{kT}{qI_C}(C_{dBE,tot} + C_{dBC,tot}) \qquad \text{(resistances not included)}, \tag{11.8}$$

where $\tau_F (= C_{DE}/g'_m)$ is the forward transit time given by Eq. (9.170), and I_C is the collector current.

When parasitic resistances are included, the equivalent circuit shown in Figure 9.26 should be used. In this case, a similar analysis can be followed to obtain the extrinsic cutoff frequency f_T. A commonly used approximation is given by (see Exercise 11.1)

$$\frac{1}{2\pi f_T} = \tau_F + \frac{kT}{qI_C}(C_{dBE,tot} + C_{dBC,tot}) + C_{dBC,tot}(r_e + r_c), \tag{11.9}$$

where r_c is the collector series resistance and r_e is the emitter series resistance.

Equations (11.8) and (11.9) can also be obtained from a small-signal analysis of a two-port network, using an admittance matrix representation, of a bipolar transistor in a procedure similar to that discussed in Section 8.5 for a MOSFET. The details are left as an exercise to the reader [see Exercise 11.6].

Equation (11.9) is often used to determine the value of τ_F. This is done by plotting the measured values of $1/f_T$ as a function of $1/I_C$, as illustrated in Figure 11.2 for a

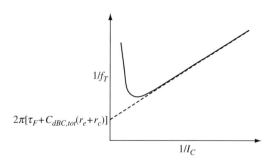

Figure 11.2 Schematic illustration of a plot of measured $1/f_T$ versus $1/I_C$ for a vertical bipolar transistor

vertical bipolar transistor. At low currents, $1/f_T$ varies linearly with $1/I_C$. Equation (11.9) suggests that extrapolation of the linear portion of $1/f_T$ to $(1/I_C) = 0$ gives $2\pi \left[\tau_F + C_{dBC,tot}(r_e + r_c)\right]$. At large currents, the measured f_T decreases rapidly as current is increased, causing $1/f_T$ to increase rapidly. This is also illustrated in Figure 11.2. This rapid drop in f_T is due to Kirk effect in a vertical transistor (see Section 9.3.3).

11.1.2   Maximum Oscillation Frequency

The cutoff frequency is a good indicator of the intrinsic speed of a transistor at low currents. However, as a figure of merit, it does not include the effect of base resistance, which is very important in determining the transient response of a bipolar transistor in circuit operation. Consequently, other figures of merit have been proposed and discussed in the literature (Taylor and Simmons, 1986; Hurkx, 1994, 1996). One that is relatively simple and commonly used is the *maximum oscillation frequency*, f_{max}, which is the frequency at which the unilateral power gain becomes unity. A commonly used approximation is given by (Pritchard, 1955; Thornton *et al.*, 1966; Roulston, 1990)

$$f_{max} = \left(\frac{f_T'}{8\pi r_b C_{dBC,tot}}\right)^{1/2}, \tag{11.10}$$

where r_b is the base resistance. Equation (11.10) can be obtained using two-port network analysis of an extrinsic transistor (in a manner similar to those in Sections 8.5.3 and 8.5.4 for a MOSFET). The procedure for arriving at Eq. (11.10) is quite tedious, and is left as an exercise (Exercise 11.7) to the interested reader.

The important point is that both f_T and f_{max} should be considered only as qualitative indicators of the frequency-response capability of a transistor. There are many other elements that can impact the performance of a transistor, and the magnitude of the impact depends on the circuit application and on the operation point of the transistor, which will be discussed in Sections 11.3–11.7.

11.1.3   Logic Gate Delay

For large-signal logic-circuit applications, neither f_T nor f_{max} is really a good indicator of device performance (Taylor and Simmons, 1986). For a logic circuit, the gate delay or circuit switching delay itself is often used as a figure of merit for the transistors in the circuit. (For digital circuits, the terms "circuit" and "gate" are used interchangeably.) The simplest way of measuring the switching delay of a logic circuit is by constructing a ring oscillator, as described in Section 8.4.1.1, and measuring the oscillation frequency.

The achievable delay is a function of the logic circuit and its design point. That is, *a transistor optimized for one circuit may not be optimum for another circuit, and a transistor optimized for one design point of a particular circuit may not be optimum for another design point.* In Section 11.2, the optimization of a bipolar transistor for different circuit design points are discussed. We focus our discussion on vertical bipolar transistors. Any significant differences between a vertical bipolar transistor and a symmetric lateral bipolar transistor will be pointed out as we come across them.

11.2   ECL Circuit and Delay Components

For high-performance logic applications, the most commonly used bipolar circuit is the *emitter-coupled logic* or *ECL* circuit. An ECL gate with fan-in of 1 and fan-out of 1 is shown in Figure 11.3. Both the inverted output V_{out} and the noninverted output $\overline{V}_{out}$ are shown. In this circuit configuration, the voltage V_S and the resistor R_S together set the *switch current* I_S of the ECL gate. This current is constant, i.e., it does not change when the circuit switches. The two resistors R_L are the load resistors of the gate. The capacitor C_L represents the total external load capacitance connected to the output. The two resistors R_E together with transistors Q_3 and Q_5 form the two emitter followers. (In an emitter follower, the emitter voltage follows the base voltage. Thus,

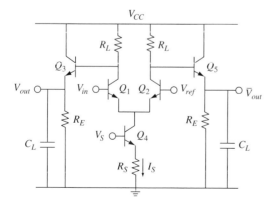

Figure 11.3 Schematic of an ECL gate with fan-in = fan-out = 1 and a load capacitance C_L. Both the inverting and the noninverting outputs are shown

if the base voltage of Q_3 goes up, the emitter voltage of Q_3 goes up by the same amount.) The input voltage V_{in} and the output voltages V_{out} and $\bar{V}_{out}$ swing above and below a fixed reference voltage V_{ref}, usually approximately symmetrically, by one-half of the logic swing ΔV.

When V_{in} is high, i.e., when $V_{in} = V_{ref} + \Delta V/2$, Q_1 is turned on much harder than Q_2, causing I_S to flow mainly through Q_1 and its load resistor. The IR drop across this load resistor in turn lowers the base voltage of Q_3. The output voltage V_{out} follows the base voltage of Q_3 to low. At the same time, with negligible current flowing through Q_2 and its load resistor, the base voltage of Q_5 is pulled up to high. The output voltage $\bar{V}_{out}$ follows the base voltage of Q_5 to high. Thus, V_{out} is inverting, while $\bar{V}_{out}$ is noninverting. A similar analysis shows that when V_{in} is switched to low, V_{out} is switched to high and $\bar{V}_{out}$ is switched to low.

When the gate switches, the switch current I_S is steered, or switched, from one load resistor to the other. (For its current-switching characteristics, an ECL gate is sometimes called a *current-switch emitter-follower circuit*.) The logic swing ΔV is equal to the IR drop in one of the load resistors, i.e., $\Delta V = I_S R_L$.

Instead of using a fixed voltage as the reference for the input signal, the inversion of the input signal can be used as the reference. That is, in Figure 11.3, V_{ref} can be replaced by $\bar{V}_{in}$. For instance, if V_{in} changes from 0 to say $+200$ mV, then $\bar{V}_{in}$ changes from 0 to -200 mV. An ECL circuit having an inverted input signal as the reference voltage is called a *differential-ECL* or *differential-current-switch circuit*. With V_{in} and $\bar{V}_{in}$ moving in opposite directions, V_{in} in a differential-ECL circuit needs to swing only 200 mV to result in the same change in transistor current as a 400-mV swing in a regular ECL circuit. That is, a differential-ECL circuit can have a logic swing that is one half that of a regular ECL circuit. Compared to a regular ECL circuit, the relatively small signal swing of a differential-ECL leads to superior speed and lower power dissipation (Eichelberger and Bello, 1991). However, the wire connection for V_{in} and that for $\bar{V}_{in}$ must be routed together on a chip. As a result, it takes more wiring channels and/or wiring levels to wire up a chip using differential ECL circuits than a chip using ECL circuits.

The switching delay of an ECL gate can be expressed as a linear combination of all the time constants of the circuit, with each time constant weighed by a factor that is determined by the detailed arrangement of the circuit (Tang and Solomon, 1979; Chor et al., 1988). For the ECL gate depicted in Figure 11.3, the switching delay can therefore be written as

$$T_{delay} = \sum_i K_i R_i C_i + \sum_j K_j \tau_j, \tag{11.11}$$

where the first sum is over all the resistances and capacitances of the transistors in the circuit, and the second sum includes the forward and reverse transit times of the transistors. For designs using vertical transistors, which is the case for all reported ECL circuits as of this writing, the vertical transistors are all biased in the forward-active mode. Therefore, only the forward transit times need to be included in Eq. (11.11).

For simplicity of discussion, it is often assumed that all the transistors in the circuit are the same. In this case, Eq. (11.11) is reduced to (Chor *et al.*, 1988)

$$
\begin{aligned}
T_{delay} = \ & K_1 \tau_F + r_{bi}(K_2 C_{dBCi,tot} + K_3 C_{dBCx,tot} + K_4 C_{dBE,tot}) \\
& + r_{bx}(K_6 C_{dBCi,tot} + K_7 C_{dBCx,tot} + K_8 C_{dBE,tot}) \\
& + R_L(K_{10} C_{dBCi,tot} + K_{11} C_{dBCx,tot} + K_{12} C_{dBE,tot} + K_{13} C_{dCS,tot} + K_{14} C_L) \\
& + r_c(K_{15} C_{dBCi,tot} + K_{16} C_{dBCx,tot} + K_{18} C_{dCS,tot}) \\
& + r_e(K_{19} C_{dBCi,tot} + K_{20} C_{dBCx,tot} + K_{21} C_{dBE,tot} + K_{23} C_{dCS,tot} + K_{24} C_L) \\
& + C_{DE}(K_5 r_{bi} + K_9 r_{bx} + K_{17} r_c + K_{22} r_e),
\end{aligned}
$$

$$(11.12)$$

where the same numbering system for the *K*-factors as in Chor *et al.* is followed. The internal resistances and capacitances of the transistors are illustrated in Figure 9.24. The circuit resistances and capacitances are shown in Figure 11.3. It should be noted that transistor Q_4 and resistor R_S, functioning only to set the switch current, are not involved in the switching of the circuit and hence do not enter into Eq. (11.12).

In practice, the performance of a bipolar logic gate is often characterized as a function of the operating current of its transistors. Since the power dissipation of a logic gate is proportional to the total current passing through the transistors, the performance of a logic gate can also be characterized as a function of its power dissipation. That is, in principle, the delay-versus-current and the delay-versus-power dissipation characteristics contain the same information, and either one can be used to describe the behavior of the transistors in the circuit. The circuit delay-versus-current or delay-versus-power dissipation characteristics are usually obtained by varying the resistor values in the circuit, keeping the transistor geometries and parameters fixed. In so doing, the collector current density becomes proportional to the collector current. In the published literature, sometimes the circuit delay is plotted as a function of the collector current density, and sometimes as a function of the collector current. In any event, the delay-versus-current or delay-versus-power characteristics reflects the performance of a *fixed* transistor design as a function of its collector current density.

The relative magnitudes of the delay components represented in Eq. (11.12) have been evaluated for vertical bipolar transistors (Tang and Solomon, 1979; Chor *et al.*, 1988). The results are illustrated schematically in Figure 11.4. Each delay component depends on a key device or circuit parameter. Analysis of these delay components provides a guide to optimizing the device design.

11.2.1 Transit-Time Delay Component

The first term in Eq. (11.12) is the delay due to the forward transit time τ_F. As discussed in Section 9.6.2.3, τ_F is constant at low injection. However, once I_C exceeds the low-injection regime, the total stored minority charge Q_{DE} in a vertical transistor increases rapidly as I_C is further increased. In Figure 11.4, this is indicated by the τ_F component increasing rapidly with I_C. High-speed logic circuits are usually designed to have delays limited by the transit-time component (Tang and Solomon, 1979; Chor *et al.*, 1988).

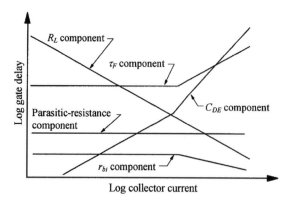

Figure 11.4 Schematic illustration of the relative magnitudes of the ECL gate delay components implemented in vertical bipolar transistors, and their dependence on collector current. For an ECL gate implemented in symmetric lateral bipolar transistors, the curves would look similar except the τ_F and C_{DE} components would not increase as rapidly at high currents

11.2.2   Intrinsic-Base-Resistance Delay Component

The second term in Eq. (11.12) is due to the RC time constants associated with the intrinsic-base resistance r_{bi}. In a bipolar transistor, r_{bi} is independent of collector current at low injection. However, when I_C is beyond the low-injection regime, base widening or Kirk effect in a vertical transistor causes r_{bi} to decrease appreciably with a further increase in I_C. This is illustrated in Figure 11.4. The r_{bi} delay component is usually quite small (Tang and Solomon, 1979; Chor *et al.*, 1988). In the case of symmetric lateral bipolar transistors, the r_{bi} delay component should remain relatively constant even at high collector currents due to the absence of Kirk effect.

11.2.3   Parasitic-Resistance Delay Components

The third, fifth, and sixth terms in Eq. (11.12) are due to the RC time constants associated with the extrinsic-base resistance r_{bx}, the collector resistance r_c, and the emitter resistance r_e, respectively. Since these parasitic resistors, to first order, are all independent of I_C, these delay components are independent of I_C, as illustrated in Figure 11.4. The parasitic-resistance delay components are also quite small (Tang and Solomon, 1979; Chor *et al.*, 1988).

11.2.4   Load-Resistance Delay Component

The fourth term in Eq. (11.12) is due to all the RC time constants associated with the load resistors R_L. For circuits with a large load capacitance C_L, the R_L delay component is often dominated by the $R_L C_L$ term. For this reason, the R_L delay component is also referred to as the load-capacitance delay component.

Referring to Figure 11.3, it can be seen that the logic swing ΔV, the switch current I_S, and the load resistor R_L are interrelated by

$$R_L = \frac{\Delta V}{I_S}. \qquad (11.13)$$

Since ΔV is fixed, the R_L delay component is inversely proportional to I_S. This is illustrated in Figure 11.4. Most ECL circuits are designed to operate at large currents in order to minimize the R_L delay component.

11.2.5 Diffusion-Capacitance Delay Component

The last term in Eq. (11.12) is associated with the emitter diffusion capacitance C_{DE}. As shown in Eq. (9.168), the stored minority charge can be written as $Q_{DE} = \tau_F I_C = \tau_F I_S$. For modeling purposes, C_{DE} is often approximated by (stored minority-carrier charge)/(average change in input voltage when the gate changes state) $= 2Q_{DE}/\Delta V$ (Tang and Solomon, 1979; Chor *et al.*, 1988), i.e.,

$$C_{DE} \approx \frac{2\tau_F I_S}{\Delta V}. \qquad (11.14)$$

Thus, the C_{DE} delay component is proportional to I_C at low injection. Once I_C is beyond the low-injection regime in a vertical transistor, Q_{DE} increases rapidly with a further increase in I_C, causing the C_{DE} delay component to increase rapidly. This is illustrated in Figure 11.4.

11.3 Speed-versus-Current Characteristics of Bipolar Transistors

Most publications show the frequency-response, represented by f_T and/or f_{max}, of a bipolar transistor as a function of I_C or J_C. Often the gate delay of a high-performance logic circuit, e.g., an ECL gate or a CML (current-mode logic) gate, is also shown as a function of I_C or J_C of the transistors employed to implement the circuit. In this section, we discuss typical speed-versus-current characteristics of a bipolar transistor, and contrast the characteristics of a vertical transistor with those of a symmetric lateral transistor.

11.3.1 f_T and f_{max} as a Function of Collector Current

Figure 11.5 shows the measured f_T and f_{max} of a state-of-the-art SiGe-base vertical bipolar transistor as a function I_C. Both f_T and f_{max} increase with I_C at low currents, as expected from Eqs. (11.9) and (11.10). At high currents, f_T and f_{max} both peak at about 4 mA. Beyond the peak, both f_T and f_{max} fall off very rapidly as I_C is further increased. This rapid fall off with current is due to Kirk effect in a vertical bipolar transistor (see Section 9.3.3). Thus, the ***maximum attainable f_T and f_{max} of a vertical bipolar transistor are limited by Kirk effect.***

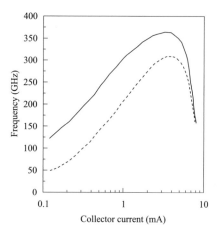

Figure 11.5 Measured f_T (dashed) and f_{max} (solid) of a high-performance SiGe-base vertical bipolar transistor. The transistor emitter has a mask dimension of $0.1 \times 2.0 \ \mu m^2$ (after Pekarik *et al.*, 2014)

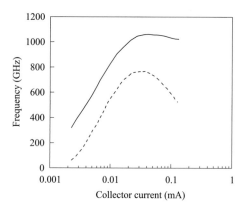

Figure 11.6 Simulated f_T (dashed) and f_{max} (solid) for a symmetric lateral Si-emitter/collector, constant-SiGe base, lateral n−p−n transistor on SOI. The transistor has emitter−collector spacing $W_{E-C} = 32$ nm, $t_{si} = 20$ nm, and emitter length $L_E = 100$ nm, corresponding to an emitter area $A_E = 0.02 \times 0.1 \ \mu m^2$ (after Raman *et al.*, 2015)

Kirk effect is absent in a symmetric lateral bipolar transistor because $N_C \gg N_B$. Base widening in a symmetric lateral transistor is limited to within the emitter−base and collector−base space-charge regions (see discussion in Section 9.3.3.1). As a result, we expect the f_T and f_{max} of a symmetric lateral transistor not to fall off as rapidly at currents beyond the peaks. This has been confirmed for measured f_T (Cai *et al.*, 2013). Figure 11.6 shows the f_T and f_{max} characteristics for a symmetric lateral transistor obtained by simulation (Raman *et al.*, 2015). Both f_T and f_{max} reach peak values at about 0.035 mA, corresponding to $J_C \sim 17 \ \mu A/\mu m^2$. Comparing with the vertical transistor in Figure 11.5, two distinguishing features of a lateral transistor are apparent. First, the peak f_T and f_{max} of a symmetric lateral transistor can be quite high, even without a graded Ge distribution in the base. Second, both f_T and f_{max} beyond the peaks fall off relatively slowly with further increase in current.

11.3.2 Logic Gate Delay as a Function of Collector Current

The solid line in Figure 11.7 shows schematically the delay-versus-current plot for a vertical-bipolar ECL gate. The delay is simply the sum of all the components illustrated in Figure 11.4. Qualitatively, *all vertical bipolar logic circuits exhibit similar characteristics*. The gate delay decreases more-or-less linearly with increase in current at low currents, where the delay is dominated by the R_L component. At some intermediate current level, the delay saturates and reaches some minimum determined by the transit-time component. At higher currents, the delay increases with further increase in current due to Kirk effect, the same physical mechanism causing f_T and f_{max} to fall off rapidly beyond their peaks in Figure 11.5.

As of this writing, there is no report of measured bipolar logic circuits implemented in symmetric lateral transistors on SOI. However, based on the fact that the key difference between a vertical bipolar transistor and a symmetric lateral transistor is the absence of Kirk effect in a symmetric lateral transistor, the expected delay-versus-current behavior for an ECL gate implemented in lateral transistors is as illustrated by the dotted-line curve in Figure 11.7. After reaching a minimum, the circuit delay increases only slowly with a further increase in current, due to the high-injection effect in the base and limited base widening due to mobile charge in the space-charge regions (see discussion in Sections 9.3.3.1 and 10.5.2). The smaller delay is due to the fact that, for the same emitter area, a lateral transistor has smaller parasitic capacitance than a vertical transistor. The physics responsible for the difference in gate delay behavior between a vertical transistor and a lateral transistor in Figure 11.7 is the same as those responsible for the difference in f_T and f_{max} in Figures 11.5 and 11.6.

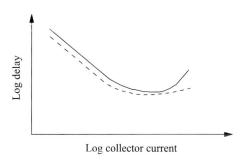

Figure 11.7 Schematic illustrations of log-delay-versus-log-current characteristics of typical bipolar logic circuits. The solid curve is for a circuit implemented in vertical bipolar transistors, and the dotted curve is for the same circuit implemented in symmetric lateral bipolar transistors

11.4   Vertical-Transistor Optimization from Data Analyses

Since there is no report of measured symmetric lateral bipolar circuit characteristics as of this writing, we shall only consider optimization of vertical bipolar devices in this section, based on published device data.

- *Collector thickness effect.* Figure 11.8 shows the measured gate delays for a non-threshold logic (NTL) inverter chain designed using two different vertical bipolar transistors, one with a collector thickness (i.e., the distance between the intrinsic base and the subcollector) of 270 nm, and one with a collector thickness of 670 nm (Tang *et al.*, 1983). The thin-collector device has a larger C_{dBC} than the thick-collector device. As a result, at low J_C, where delay is dominated by capacitance, the thin-collector device leads to a larger gate delay. At sufficiently large J_C, the gate delay is dominated by C_{DE}. When that happens, the thin-collector device, with less collector volume than the thick-collector device for minority-carrier charge storage, leads to a smaller C_{DE} and hence faster circuits. The data in Figure 11.8 shows clearly that the optimized design of a bipolar transistor depends on the intended operation point of the transistor in a circuit.
- *SiGe-base versus Si-base.* Figure 11.9 shows the ECL gate delays versus current for a Si-base vertical transistor and a SiGe-base vertical transistor of the same design rule and approximately the same base dopant distribution (Harame *et al.*, 1995a). As expected, the SiGe-base transistor is faster than the Si-base transistor, with the speed advantage larger at high currents than at low currents. However, the maximum speed of both transistors is limited by Kirk effect, as evidenced by the increase in delay at large currents. Optimizing the collector design to minimize Kirk effect should improve the maximum performance of both transistors.
- *Emitter length effect.* In general, the emitter stripe width is designed to be as narrow as possible, consistent with the available emitter patterning technique. A small emitter stripe width leads to a small intrinsic-base resistance (see Section 9.2.1). For

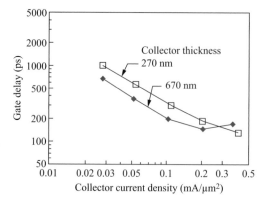

Figure 11.8 Typical switching delay of a vertical bipolar circuit as a function of collector current density, with collector thickness as a parameter (after Tang *et al.*, 1983)

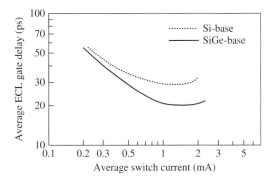

Figure 11.9 Comparison of measured ECL ring oscillator gate delays made of a Si-base vertical transistor with those made of a SiGe-base vertical transistor as a function of switch current. Each ECL gate has FI = FO = 1, and an external wire load of 5 fF. All the transistors have the same mask emitter area ($0.5 \times 12.5\ \mu m^2$). The SiGe-base devices have a triangular Ge distribution in the quasineutral base (after Harame *et al.*, 1995a)

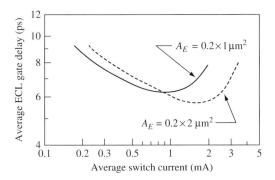

Figure 11.10 Typical delay-versus-current characteristics of an ECL ring oscillator showing the effect of emitter size, implemented in SiGe-base vertical transistors. The two transistors have the same device parameters other than emitter length. For a given current, the short-emitter device has a current density twice that of the long-emitter device (after Washio *et al.*, 2002)

a given emitter stripe width, the emitter area can be varied by varying the emitter length. Figure 11.10 shows the measured ECL gate delay versus switch current for two SiGe-base transistors, one having twice the emitter length of the other (Washio *et al.*, 2002).

In Figure 11.10, the delay is plotted as a function of current. That means, for a given current, the short-emitter device is operating at twice the current density as the long-emitter device. At low currents where the R_L delay component dominates the circuit delay, the short-emitter device leads to a faster ECL gate because of its smaller junction area and associated capacitance.

At large currents, the delay is no longer dominated by the R_L component, but by the τ_F and C_{DE} components. It is evident from the discussion on base widening and

Kirk effect in Section 9.3.3.1 that whether base widening starts to become appreciable is determined by J_C. Since for a given current on the plots in Figure 10.10, J_C for the short-emitter device is twice that of the long-emitter device, the detrimental effect due to base widening, which causes C_{DE} to increase rapidly with further increase in current, should hit the short-emitter device at $2\times$ lower current than the long-emitter device. This is indeed the case in Figure 10.10.

The net is that, in designing a vertical bipolar transistor to avoid a significant base-widening effect in circuit operation, *J_C of the transistor should be much less than $J_{max} = qv_{sat}N_C$ (see Section 9.3.3.2). The emitter length should be adjusted accordingly to meet the target maximum I_C needed for operation of the circuit.*

11.5   Bipolar Device Scaling for Logic Circuits

Since the optimal design of a bipolar transistor depend on its circuit application, scaling of a bipolar transistor should be discussed in the context of its circuit application as well. The basic concept in logic-circuit device scaling is to reduce the dominant resistance and capacitance components in a coordinated manner so that all the dominant delay components are reduced proportionally as the horizontal dimensions of the transistor are scaled down. In this way, if a transistor is optimized for a given circuit design point before, it remains optimized after scaling. While this basic concept applies equally to scaling a vertical transistor and a symmetric lateral transistor, the details of how the device structure is modified in scaling for a vertical transistor and for a lateral transistor are quite different. In this section, we first discuss the scaling of a vertical transistor for an ECL circuit, and then discuss why and how scaling of a symmetric lateral logic transistor are different.

11.5.1   Vertical-Transistor Scaling for ECL

The goal in scaling the vertical transistors in a high-performance ECL circuit is to reduce the dominant resistance and capacitance components in a coordinated manner so that the circuit speed improves in proportion to the scaling factor. This is accomplished by requiring the capacitance ratio $C_{DE}/C_{dBC,tot}$ and the resistance ratio r_b/R_L to be constant in scaling. Here $C_{dBC,tot} = C_{dBCi,tot} + C_{dBCx,tot}$ and $r_b = r_{bi} + r_{bx}$ (Solomon and Tang, 1979).

There are several additional constraints in bipolar scaling. First of all, because of the exponential dependence of current on voltage, the turn-on voltage of a diode is insensitive to the diode area (for a given target current, V_{BE} changes by about 20 mV for $2\times$ change in diode area). To first order, one can assume the diode turn-on voltage to be constant in scaling. As a result, the voltages in a bipolar circuit, including the logic swing ΔV, remain constant in scaling. If the voltages are already optimally small to begin with, they remain unchanged in scaling (increased by 10–20 mV if needed). Secondly, N_C in a vertical bipolar transistor should be varied in proportion to J_C in order to maintain the same degree of Kirk effect in scaling. Thirdly, to avoid emitter–

collector punch-through as W_B is reduced in scaling, N_B must be increased. The base is depleted on the emitter side as well as on the collector side, but the depletion on the emitter side is usually more severe than on the collector side because $N_E \gg N_B \gg N_C$ for a vertical transistor. To avoid excessive base-region depletion near the emitter–base junction, W_{dBE} should remain the same fraction of W_B as W_B is reduced. From the dependence of W_{dBE} on N_B [see Eq. (3.15)], we see that this requirement is met if N_B is increased such that $N_B \propto W_B^{-2}$.

As shown in Eq. (11.13), for a given logic swing, R_L is inversely proportional to the switch current I_S. The requirement of r_b/R_L being constant means that r_b should be varied inversely proportional to I_S as well, which would greatly complicate the device layout and design, and would also greatly narrow the device design window. It is much more practical to drop this resistance-ratio requirement and only keep the capacitance-ratio requirement in scaling. This approximation is quite reasonable, since, as discussed in Section 11.2.1, the r_b-component of the gate delay is relatively small to start with. Furthermore, the base resistance can be reduced readily, if desired, by modifying the physical layout of the transistor.

The switch current is often kept constant in scaling in order to achieve circuit delay reduction in proportion to the scaling factor. In this case, to maintain a constant ΔV in scaling, R_L is also kept constant. Therefore, the ratio r_b/R_L is constant if r_b is kept constant. For a given emitter geometry, r_b is constant if the intrinsic-base sheet resistivity R_{Sbi} is constant. In the case of a Si-base transistor, R_{Sbi} is indeed often maintained at around 10 K $\Omega/\square$, partly to maintain a current gain of about 100 and partly to maintain a sufficiently large Early voltage. In the case of a SiGe-base transistor, R_{Sbi} is usually smaller, typically in the 3–5 K $\Omega/\square$ range. Thus, without requiring special effort, r_b/R_L is more-or-less constant in the scaling of vertical bipolar transistors for high-speed logic circuits.

The scaling constraints, together with the requirement on the capacitance ratio in scaling, are summarized in Table 11.1. The resulting scaling rules for the device design and circuit delay are summarized in Table 11.2. These rules are for the case where the ECL gate delay is reduced by the scaling factor (Solomon and Tang, 1979).

11.5.2 Symmetric-Lateral-Transistor Scaling for Logic Circuits

The concept of scaling is driven by reduction of horizontal dimensions. The idea is to increase the areal densities of devices and circuits in scaling. It is apparent from the

Table 11.1. Constraints and requirements in ECL scaling for vertical bipolar transistors

Parameter	Constraint or Requirement
Voltage	$V, \Delta V = \text{constant}$
Capacitance	$C_{DE}/C_{dBC,tot} = \text{constant}$
Base doping concentration	$N_B \propto W_B^{-2}$
Collector doping concentration	$N_C \propto J_C$

Table 11.2. Scaling rules for ECL circuits

Parameter	Scaling Rule[*]
Feature size or emitter-stripe width	$1/\kappa$
Base width W_B	$1/\kappa^{0.8}$
Collector current density J_C	κ^2
Circuit delay	$1/\kappa$

*Scaling factor $\kappa > 1$.

device schematics in Figures 9.1(b) and 10.27 that the active area (emitter area) of a symmetric lateral transistor does not take up areal space. The emitter area lies in a vertical plane, defined by the emitter length L_E and the silicon thickness t_{si}. In reducing the areal size of the device, the only boundary condition that must be met is Eq. 10.72, namely $W_{E-C} = W_B(V'_{BE}, V'_{BC}) + W_{dBE}(V'_{BE}) + W_{dBC}(V'_{BC})$. As shown in Figure 10.29, W_{E-C} can be scaled down to about 20 nm by designing with $N_B \approx 1 \times 10^{19} \text{cm}^{-3}$.

Of the four scaling constraints and requirements in Table 11.1 for a vertical bipolar transistor, only the ones related to voltage and N_B apply to a symmetric lateral bipolar transistor. The results in Figure 10.29 are for designs that meet the N_B requirement. The capacitance constraint does not apply since Kirk effect is absent in a symmetric lateral transistor where C_{DE} is always minimal and not really a device design parameter. Also, the N_C requirement does not apply either because $N_C \gg N_B$ for a symmetric lateral transistor. The net is that *a symmetric lateral bipolar transistor has fewer scaling constraints than a vertical bipolar transistor* for logic circuits.

11.5.3   Power-Dissipation Issues with Resister-Load Bipolar Logic Circuits

The scaling constraint of constant voltage (Table 11.1) and the scaling rule of J_C varying as κ^2 (Table 11.2) imply that in order for an ECL circuit to achieve speed improvement by the scaling factor κ in areal dimension scaling, the circuit power dissipation remains unchanged in scaling. SiGe-base vertical-bipolar current-switch circuits can achieve delays of 3 ps at about 10 mW per circuit (Böck *et al.*, 2004). Such large power dissipation implies vertical bipolar logic circuits are not suitable for systems requiring large circuit counts. The last generation of high-speed vertical-bipolar ECL developed for high-end computer systems was reported in 1992 (Brown *et al.*, 1992). Since then all high-end computing systems have been built using scaled CMOS.

There are two reasons why the power dissipation of a high-speed vertical-bipolar ECL circuit, or a high-speed vertical-bipolar resistor-load circuit in general, is so large. First, it is apparent from the layout schematic in Figure 9.1(a) that the total area of a vertical bipolar transistor is much larger than its active region represented by the emitter area. As a result, a vertical bipolar transistor has a large device area and associated device capacitance, and large capacitance associated with the wires connecting the devices and circuits on chip. A vertical transistor must be operated at correspondingly large current in order to achieve the intended high speed. Second, to avoid operation of a vertical transistor in the saturation region, which would cause unacceptable speed degradation, designers tend to increase the power supply voltage. The reason for this will be made clear next. The combination of large current and large power supply voltage results in large power dissipation.

There are two reasons why resistor-load logic circuits using symmetric lateral bipolar transistors, instead of vertical bipolar transistors, can achieve the same high speed at much lower power dissipation. First, it is apparent from comparing the lateral transistor schematic in Figure 9.1(b) with the vertical transistor schematic in Figure 9.1(a) that, for the same emitter area, a lateral transistor has much smaller device area and associated device capacitance than a vertical transistor. The implication is that a lateral transistor can be operated at a much smaller collector current to achieve the same circuit speed as a vertical transistor. Second, without Kirk effect, a lateral transistor can be operated in deep saturation without speed degradation. This allows the same logic circuit to be powered with a smaller supply voltage, as explained next.

Figure 11.11 shows a basic resistor-load bipolar inverter, which is the "switch" in current-switch circuits, e.g., in the current-switch emitter-follower circuit in Figure 11.3. Let us assume the inverter input swings between ground and ΔV. When $V_{in} = 0$, the transistor is off and $V_{out} = V_{cc}$. When $V_{in} = \Delta V$, the transistor is turned on and a current I_C flows in the resistor, resulting in $V_{out} = V_{cc} - I_C R_L$. In a properly designed current switch, V_{out} swing is equal to V_{in} swing, i.e., $I_C R_L = \Delta V$. To avoid the transistor going into saturation, we need to avoid $V_C < V_B$ during circuit operation. For the inverter in Figure 11.11, the minimum required value of V_{cc} without the transistor entering the saturation region is $2\Delta V$. Designers typically use V_{cc} larger than the "required minimum" to ensure no transistor on the chip enters the saturation region in the worst-case situation. This leads to larger power dissipation for all the circuits on the chip.

Since the collector–base diode and the emitter–base diode switches equally fast in a symmetric lateral bipolar transistor, a symmetric lateral transistor can be operated in full saturation, i.e., with $V_{BC} = V_{BE}$, without speed degradation. Therefore, a lateral-

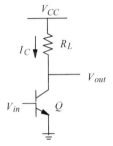

Figure 11.11 Circuit schematic of a bipolar inverter with resistor load

transistor current-switch circuit can be operated with a smaller power supply voltage than an equivalent vertical-transistor circuit. For the inverter shown in Figure 11.11, the minimum V_{cc} is ΔV for a lateral-transistor circuit. [This is to be compared with minimum V_{cc} of $2\Delta V$ for a vertical-transistor circuit, as discussed in the previous paragraph.] It is estimated that the power-delay product of a lateral-transistor current-switch circuit could be 100-times lower than a vertical-transistor current-switch circuit (LeRoy et al., 2015; Raman et al., 2015).

11.6   Vertical-Transistor Design Optimization for RF and Analog Circuits

In general, the techniques used to optimize a bipolar transistor for digital circuits, such as minimizing the base resistance and collector capacitance, are also applicable to optimizing a bipolar transistor for RF and analog circuits. The current-density limit in vertical transistors due to Kirk effect apply for digital as well as RF and analog circuits. However, there are some device design differences between digital circuits and RF or analog circuits.

For a digital circuit, the overall circuit speed and power dissipation are the most important factors governing the device design. For RF and analog circuits, perhaps the most important device parameters or figures of merit are f_T, f_{max}, and V_A. In this section, we first consider a bipolar transistor as an amplifier for examining the merit of large V_A. We then discuss the design tradeoffs among f_T, f_{max}, and V_A. The design tradeoffs and optimization for symmetric lateral bipolar transistors are covered in Section 11.7.

11.6.1   The Single-Transistor Amplifier

Insights into the figures of merit of an analog transistor can be obtained by examining the single-transistor amplifier. The circuit configuration of a bipolar transistor biased to operate as an amplifier is shown in Figure 11.12(a), where R_L is the load resistor. For simplicity, let us consider the low-frequency case. In this case, the small-signal equivalent circuit in Figure 9.25 can be adapted to give the small-signal equivalent circuit shown in Figure 11.12(b) for the amplifier. If no load is attached to the output, the output voltage is

$$v_o = -i_o \left(r_0' \| R_L \right) = -g_m' \left(r_0' \| R_L \right) v_i, \tag{11.15}$$

where $\left(r_0' \| R_L \right)$ denotes the resistance of the two resistors r_0' and R_L in parallel. The minus sign is for the fact that a positive v_i leads to a negative v_o. Thus, the open-circuit or unloaded voltage gain is

$$a_v = \frac{v_o}{v_i} = -g_m' \left(r_0' \| R_L \right). \tag{11.16}$$

The maximum low-frequency open-circuit voltage gain is given by letting R_L be very large, i.e.,

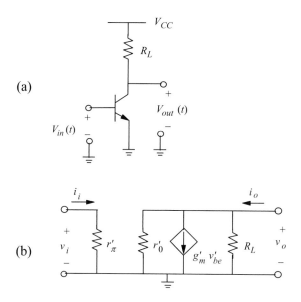

Figure 11.12 (a) Circuit configuration of a single-transistor bipolar amplifier. (b) The small-signal low-frequency equivalent circuit

$$\lim_{R_L \to \infty} a_v = -g'_m r'_0. \tag{11.17}$$

For a bipolar transistor, substituting Eqs. (9.153) and (9.155) into Eq. (11.17) gives

$$\lim_{R_L \to \infty} a_v = -\frac{qV_A}{kT}, \tag{11.18}$$

where V_A is the Early voltage. In Eq. (11.18), we have ignored the V_{CE} term in Eq. (9.155) which is small compared to V_A. That is, the intrinsic voltage amplification capability of a bipolar transistor is proportional to its Early voltage, provided that the transistor has sufficiently large breakdown voltages to handle the amplified voltage. For a Si-base vertical bipolar transistor, V_A is typically about 40 V (see Section 9.3.2.1), implying an intrinsic voltage gain of about 1,600. This is significantly larger than the intrinsic voltage gain of a MOSFET which is about 17 for a 0.1 μm nMOSFET (see Section 8.5.4).

11.6.2 Maximizing f_T of a Vertical Transistor

The cutoff frequency f_T is given by Eq. (11.8) or (11.9). To maximize f_T, the capacitances $C_{dBE,tot}$ and $C_{dBC,tot}$ and the forward transit time τ_F should all be minimized. The simplest way to minimize the capacitances is to use advanced device structures that have small parasitic capacitance, the same as for digital circuits. The most important component of τ_F is t_B, which can be reduced effectively by reducing

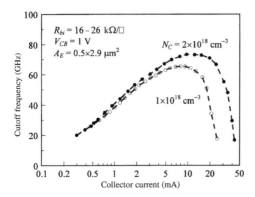

Figure 11.13 Typical measured f_T of a vertical bipolar transistor as a function of I_C, with N_C as a parameter (after Crabbé *et al.*, 1993a)

W_B. However, reducing W_B alone will lead to a larger intrinsic-base resistance and a smaller Early voltage.

In the low-injection regime, τ_F is a function of the intrinsic-device doping profiles. It is independent of the horizontal device dimensions and geometry. That is, for a given vertical device doping profile, a large-emitter device has the same τ_F as a small-emitter device, if both devices are operated at the same J_C. Equation (11.8) shows that the maximum intrinsic f_T is determined by τ_F. However, to reach the maximum f_T value determined by τ_F, the transistor must be operated at sufficiently high current such that the term $kT(C_{dBE,tot} + C_{dBC,tot})/qI_C$ is small compared to τ_F. Unfortunately, base widening becomes important at large J_C in a vertical bipolar transistor. Once base widening becomes appreciable, τ_F is no longer a constant. Instead, it increases rapidly with further increase in J_C. When base widening happens, instead of decreasing with increasing current, f_T actually decreases rapidly with further increase in current, as demonstrated by the measured f_T in Figure 11.5.

Figure 11.13 shows the measured f_T as a function of I_C for two vertical transistors of different N_C (Crabbé *et al.*, 1993a). It clearly shows that the f_T rolloff characteristics shift toward higher current in proportion to the increased N_C, as expected from base-widening effects. It also shows that a simple way to increase the peak f_T of a vertical transistor is to increase N_C, provided that the device breakdown voltages remain acceptable.

f_T is degraded by the parasitic emitter and collector resistances r_e and r_c, respectively [see Eq. (11.9)]. Thus, to maximize f_T, it is important that r_e and r_c are kept small.

11.6.3   Minimizing r_{bi} of a Vertical Transistor

As shown in Section 9.2.1, the intrinsic-base resistance r_{bi} for an emitter stripe of width W and length L, is proportional to $(W/L)R_{Sbi}$, where R_{Sbi} is the sheet resistivity of the intrinsic-base layer. Thus, r_{bi} can be reduced by making the emitter stripe as narrow as possible and using long emitter stripes. Furthermore, contacting the intrinsic base on both sides of an emitter stripe, instead of simply on one side, reduces r_{bi} by a factor of four. In practice, all high-performance vertical bipolar transistors have base contacts on both sides of their emitter stripes. With a polysilicon emitter, current gain is usually not an issue. As a result, the intrinsic-base layer can be doped rather heavily to reduce R_{Sbi}, which in turn reduces r_{bi}.

11.6.4 Maximizing $f_{\max}$ of a Vertical Transistor

The maximum oscillation frequency $f_{\max}$ is given by Eq. (11.10). It is a function of f_T, r_b, and $C_{dBC,tot}$. Therefore, designs that increase f_T, at the expense of increasing r_b, could result in a decrease in $f_{\max}$. In fact, if a slightly reduced f_T allows a significantly reduced r_b, the net result could be a larger $f_{\max}$.

Reducing $C_{dBC,tot}$ increases $f_{\max}$. However, if $C_{dBC,tot}$ is reduced by reducing the collector doping concentration, base widening will set in at a lower J_C, which in turn could reduce the maximum f_T of the transistor (see Figure 11.13). Thus, maximizing $f_{\max}$ is a complex tradeoff process. This point will be discussed further in Section 11.6.5.

11.6.5 Maximizing V_A of a Vertical Transistor

The Early voltage V_A for a uniformly doped intrinsic base is given by the ratio Q_{pB}/C_{dBCi} in Eq. (9.94). An obvious way to increase V_A is to increase Q_{pB}. If this is done at a fixed W_B by increasing N_B, then it has relatively little effect on τ_F, and it will reduce r_{bi}. The net result is relatively little effect on f_T, but a higher $f_{\max}$. However, it will also result in reduced current gain, which is not an issue for polysilicon-emitter transistors. On the other hand, if Q_{pB} is increased by increasing W_B, it will decrease f_T, and could adversely affect $f_{\max}$ as well.

V_A can also be increased by reducing C_{dBCi}. However, reducing C_{dBCi} by reducing N_C and/or increasing the collector layer thickness renders the transistor more susceptible to Kirk effect, resulting in reduced maximum values of f_T and $f_{\max}$.

11.6.6 Examples of Vertical-Transistor RF and Analog Design Tradeoffs

A boxlike intrinsic-base doping profile can be obtained readily by using an epitaxially grown intrinsic-base layer doped *in situ* with boron during growth. By incorporating carbon into the base layer, diffusion of boron in the base layer can be suppressed during subsequent process steps (Stolk *et al.*, 1995; Lanzerotti *et al.*, 1996). The results from a tradeoff study for a boxlike base doping profile (Yoshino *et al.*, 1995) are shown in Figures 11.14 and 11.15. These results clearly show that f_T can be traded off for a larger $f_{\max}$ and/or a larger V_A.

Base widening limits the maximum J_C without performance degradation, and BV_{CEO} limits how heavily the collector can be doped to minimize base widening. Figure 11.16 is a plot of BV_{CEO} versus f_T for some vertical n–p–n transistors reported in the literature. Data for Si-base and SiGe-base transistors, as well as some InP-based HBTs, are

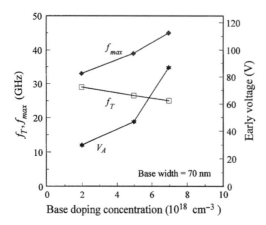

Figure 11.14 Experimental results showing the dependence of f_T, f_{max}, and V_A on base doping concentration N_B of a vertical bipolar transistor. The base has a boxlike doping profile, formed by epitaxial growth of the intrinsic-base layer (after Yoshino *et al.*, 1995)

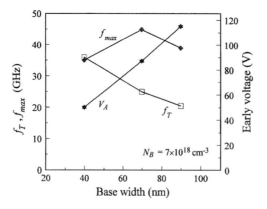

Figure 11.15 Experimental results showing the dependence of f_T, f_{max}, and V_A on intrinsic-base layer thickness of a vertical bipolar transistor. The intrinsic base is formed by epitaxial growth of silicon (after Yoshino *et al.*, 1995)

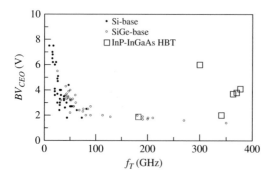

Figure 11.16 BV_{CEO} versus f_T for some vertical n–p–n transistors reported in the literature

included. It shows that there is definitely a tradeoff between BV_{CEO} and f_T. As expected, for the same BV_{CEO} value, the SiGe-base transistors can reach much higher f_T.

11.7 Symmetric-Lateral-Transistor Design Tradeoffs and Optimization for RF and Analog Circuits

As discussed in Section 10.5, unlike vertical transistors, most of the ac and steady-state parameters of a symmetric lateral transistor can be reduced to closed-form expressions in terms of device dimensions and doping concentrations. These parameters in turn allow many figures of merit for RF, analog, and logic circuits to be calculated. Here we examine the *calculated characteristics and design tradeoffs* of symmetric lateral transistors for RF and analog applications.

To calculate f_T and f_{max}, we need the device parameters $C_{dBC,tot}$, $C_{dBE,tot}$, τ_F, r_e, and r_c [see Eq. (11.9)]. Referring to the device schematics in Figure 10.27, the total-capacitance parameters are given by

$$C_{dBC,tot} = A_E C_{dBC}(V'_{BC}) + L_E C_{BC,fringe}, \tag{11.19}$$

and

$$C_{dBE,tot} = A_E C_{dBE}(V'_{BE}) + L_E C_{BE,fringe}, \tag{11.20}$$

where $C_{dBC}(V'_{BC})$ and $C_{dBE}(V'_{BE})$ are the base–collector and base–emitter junction capacitance per unit area, respectively, and $C_{BC,fringe}$ and $C_{BE,fringe}$ are the base–collector and base–emitter fringing capacitance per unit emitter length, respectively. For CMOS, the fringing capacitance is typically 0.08 fF/μm. The same value should apply to symmetric lateral bipolar transistors. The emitter area is $A_E = L_E \times t_{si}$. The forward transit time τ_F in Eqs. (11.9) and (11.10) are given by Eqs. (10.90)–(10.94). The intrinsic-base resistance r_{bi} is given by Eq. (10.81). Typical values for r_e, r_c, r_{bx}, and W_E can be taken from the measured device currents, such as those shown in Figures 10.30 and 10.31.

At high collector currents where mobile charge density in the base–emitter and base–collector space-charge regions cannot be ignored, the space-charge-region widths are given by Eqs. (10.78) and (10.79), using Δn given by Eq. (10.80) for a device operating in the linear region, or Δn given by Eqs. (10.85)–(10.87) for a device operating in and out of the saturation region. In this case, the calculation of W_{dBC} and W_{dBE} as a function of V_{BE} and V_{BC} is more involved because of the interdependence among C_{dBC}, C_{dBE}, V_{BE} and V_{BC} (see Section 10.5.1).

11.7.1 Calculated Low-Injection f_T and f_{max} of Symmetric Lateral n–p–n

However, if we limit the calculation to V_{BE} values satisfying the *low-injection criteria* of

$$(n_{ieB}^2/N_B) \exp(qV'_{BE}/kT) < N_B, \tag{11.21}$$

then $\Delta n < N_B$, and W_{dBC} and W_{dBE} may be assumed to be independent of Δn. In this case, Eqs. (10.78) and (10.79) reduce to the familiar approximate forms of

$$W_{dBE}\left(V'_{BE}\right) = \sqrt{2\varepsilon_{si}\left(\psi_{bi} - V'_{BE}\right)/qN_B},$$
(11.22)

and

$$W_{dBC}\left(V'_{BC}\right) = \sqrt{2\varepsilon_{si}\left(\psi_{bi} - V'_{BC}\right)/qN_B}.$$
(11.23)

The junction voltage V'_{BE} satisfying Eq. (11.21) is a function of N_B. For $N_B = 2.5 \times 10^{18}\,\mathrm{cm}^{-3}$, the corresponding voltage and current density are $V'_{BE} = 0.96\,\mathrm{V}$ and $J_C = 5.74\,\mathrm{mA/\mu m^2}$. For $N_B = 1.0 \times 10^{19}\,\mathrm{cm}^{-3}$, the values are $V'_{BE} = 1.01\,\mathrm{V}$ and $J_C = 15.7\,\mathrm{mA/\mu m^2}$. That is, for symmetric lateral transistors with high N_B, the low-injection approximation is valid for describing operation at fairly high J_C. The examples shown in Figures 11.17 and 11.18 are calculated for V'_{BE} and J_C ranges where the low-injection approximation is valid (Ning and Cai, 2013).

Figure 11.17 shows the f_T and f_{max} as a function of collector current density, calculated as discussed in the previous paragraph, for sample symmetric lateral n–p–n transistors. Figure 11.17(a) shows that for a given combination of W_{E-C} and t_{si}, f_T and f_{max} decrease as N_B is increased. Increasing N_B at fixed W_{E-C} increases W_B and junction capacitance, but reduces r_{bi}. The increase in W_B and junction capacitance explains the decrease in f_T. The reduction in r_{bi} by itself should increase f_{max}. The fact that f_{max} is reduced in Figure 11.17(a) is due to the fact that the benefit of reduced r_{bi} is not enough to offset the degradation caused by the increase in W_B and junction capacitance.

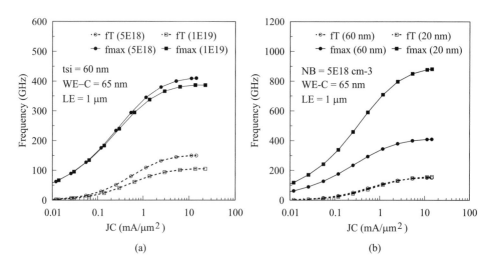

Figure 11.17 (a) Calculated f_T and f_{max} for $t_{si} = 60\,\mathrm{nm}$ and two N_B values. (b) Calculated f_T and f_{max} for $N_B = 5 \times 10^{18}\,\mathrm{cm}^{-3}$ and two values of t_{si}. $V_{BC} = -1\,\mathrm{V}$, and the resistances assumed are $r_{bx} = 20\,\Omega$ and $r_e = r_c = 100\,\Omega$ (after Ning and Cai, 2013)

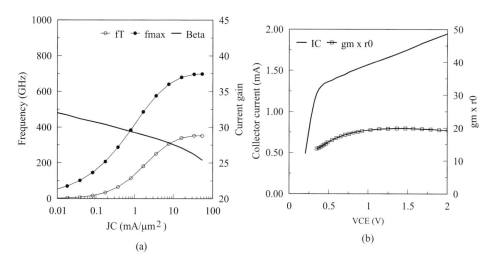

Figure 11.18 (a) Calculated f_T, f_{max}, and current gain for a lateral bipolar device with $L_E = 1$ μm, $N_B = 1 \times 10^{19}$ cm^{-3}, $W_{E-C} = 38$ nm, and $t_{si} = 40$ nm. The target is a device with peak f_T of 350 GHz and peak f_{max} of 700 GHz. (b) Calculated output current of the same device at $I_B = 80$ μA and the corresponding self gain $g_m \times r_0$ (after Ning and Cai, 2013)

Figure 11.17(b) compares two designs of different t_{si}. It shows that reducing t_{si} alone increases f_{max} due to r_{bi} reduction. Reduction of t_{si} alone has little effect on f_T because there is no change in W_B, τ_F, C_{dBE}, or C_{dBC}. At $t_{si} = 20$ nm, the reduction in r_{bi} enables peak f_{max} to approach 900 GHz.

The design with $t_{si} = 20$ nm in Figure 11.17(b) has peak $f_{max} \sim 900$ GHz and peak $f_T \sim 300$ GHz, i.e., peak f_{max} about 3× peak f_T. For a better balance between f_T and f_{max}, W_{E-C} can be reduced to increase f_T and t_{si} can be increased to reduce f_{max}. The reduction in V_A and self gain (the product $g_m \times r_0$) arising from reducing W_{E-C} can be minimized by increasing N_B, if desired. Figure 11.18 shows one such design example where the target is for a device with peak f_T of 350 GHz and peak f_{max} of 700 GHz. Figure 11.18(a) shows the calculated f_T, f_{max}, and current gain of the device. Figure 11.18(b) shows the calculated output characteristics at $I_B = 80$ μA and the self gain of the same device. The values shown in Figure 11.18 are significantly higher than those reported for 32-nm SOI CMOS (Lee et al., 2012), or those reported for 22-nm FinFET CMOS (Sell et al., 2017).

The current gain in Figure 11.18(a) may be a bit low for some designs. If Si-emitter/collector SiGe-base HBT structures were used (see discussion in Section 10.5.8), current gain would be larger than 100.

11.7.2 Fin-Structure Symmetric Lateral Transistors for $f_{max} > 1$ THz

CMOS technology has evolved to the so-called FinFET device structure where inversion channels are formed on opposite sides of a vertical silicon fin (see Section 7.2). The CMOS-FinFET-on-SOI device structure can be adapted to symmetric lateral

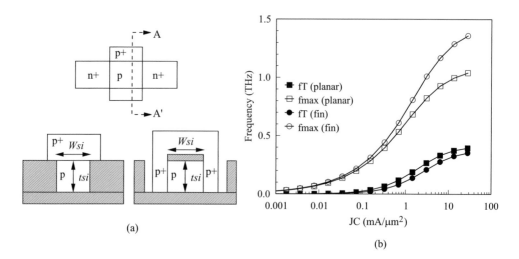

(a)

(b)

Figure 11.19 (a) Side-view schematics along AA′ of a planar-structure (bottom left) and a fin-structure (bottom right) SOI symmetric lateral bipolar transistor. For the planar-structure device, the p^+ extrinsic base contacts the intrinsic base only on the top surface. For the fin-structure, the contact between the p^+ extrinsic base and the p-type base can be on two sides, as illustrated here, or on three sides. The top view (top) is the same for both planar-structure and fin-structure devices. (b) Calculated f_T and f_{max} as a function of collector current density for a planar-structure device and for a fin-structure device with extrinsic base contacting the two vertical surfaces of the intrinsic base (after Ning, 2016)

bipolar transistors on SOI as well. In a fin-structure lateral-bipolar device, the intrinsic base can be contacted on two or three sides, leading to reduction of base resistance by more than 50% (see Section 9.2.1.2). However, for the same emitter area, a fin-structure device has larger base–emitter and base–collector fringing capacitance than a planar-structure device because of the longer emitter–extrinsic-base perimeter. The net result of adopting a fin-structure in place of a planar-structure is slightly reduced f_T due to the increase in fringing capacitance, but greatly increased f_{max} due to the reduced base resistance. The simulation results comparing a fin-structure device with base contact on two vertical sides and a planar device are shown in Figure 11.19. Both devices have $t_{si} = 20$ nm and $A_E = 20 \times 20$ nm^2 (i.e., fin-width $W_{si} = 20$ nm), $N_B = 5 \times 10^{18}$ cm^{-3} and $W_B = 10$ nm, corresponding to $R_{Sbi} = 15$ KΩ/□. This shows f_{max} exceeding 1 THz can be achieved in a fin-structure device (Ning, 2016).

11.7.3   Noise Reduction with Substrate Bias

Low-frequency noise is an important issue in transistors designed for analog applications. In the case of symmetric lateral bipolar transistors on SOI, it is important to minimize any residual defects associated with the emitter/collector formation process and residual defects at the semiconductor-oxide interfaces. First, it is preferred to use *in situ* doping, instead of high-dose ion implantation, to dope the extrinsic-base layer. This will avoid unintentionally forming any insufficiently doped extrinsic-base region adjacent to the intrinsic base (Hashemi *et al.*, 2018; Yau *et al.*, 2018). Second, proper post-metallization anneal in hydrogen-containing environments is important to minimize interface-state defects (Yau *et al.*, 2016a). Third, application of a substrate bias ($V_x < 0$ for an n–p–n and $V_x > 0$ for a p–n–p) can cause the base region adjacent to the BOX to be accumulated (Yau *et al.*, 2016b). A substrate bias causing accumulation of the base region adjacent to the BOX reduces low-frequency noise caused by surface defects at the Si/BOX interface (Hu *et al.*, 2018).

11.8   Unique Opportunities from Symmetric Lateral Bipolar Transistors

The merits of a symmetric lateral bipolar transistor compared to a vertical bipolar transistor in terms of conventional resistor-load current-switch logic circuits have been discussed in Section 11.5, and in terms of RF and analog circuit parameters in Section 11.7. The combination of small device capacitance and absence of Kirk effect leads to significantly lower power dissipation and higher maximum performance for symmetric lateral transistors.

The high drive-current capability and the relatively simple process of integrating n–p–n and p–n–p devices on the same chip offer interesting, and potentially exciting, opportunities from lateral bipolar transistors that are simply not possible from vertical bipolar transistors. They suggest a ***need to rethink bipolar circuits and applications***. In this section, some of these opportunities are discussed.

11.8.1   Symmetric Lateral Bipolar Transistor as a High-Drive-Current Device

As CMOS is scaled down in areal dimensions, the maximum drive-current-per-gate-width has not improved much, within about $3\times$, over many generations. To first order, the drive current of a MOSFET is proportional to the product of its inversion-layer carrier density and the mobility of those charge carriers. For example, for the 1.8-volt CMOS generation, reported drive currents for n-channel MOSFET were about 0.8 mA/μm (Su *et al.*, 1996). At the 22-nm node, the highest reported drive current for a planar n-channel MOSFET is about 2 mA/μm (Freeman *et al.*, 2015). This drive current is small compared to the measured drive current of 4.8 mA/μm shown in Figure 10.30 for a symmetric lateral n–p–n transistor having $t_{si} = 60$ nm.

The reasons why symmetric lateral bipolar transistors on SOI have a much higher drive current than CMOS are as follows:

- Unlike CMOS, there is no gate-insulator breakdown concern for a bipolar transistor. As a result, a bipolar transistor can be over-driven until qV'_{BE} close to the base-region bandgap energy. The terminal V_{BE} needed to achieve this depends on the parasitic emitter and base resistances. For the transistor in Figure 10.30, the maximum-current point corresponds to V'_{BE} of about 1.1 V (Cai *et al.*, 2014).
- The emitter area of a lateral transistor is $A_E = L_E \times t_{si}$. That is, the drive current is proportional to t_{si}. The transistor in Figure 10.30 has $t_{si} = 60$ nm. For the same

device design with $t_{si} = 100$ nm, the drive current at $V_{BE} = 1.5$ V would be about 8 mA/µm. For $t_{si} = 300$ nm, the drive current would be about 24 mA/µm. Using the *in situ* doped lateral silicon epitaxy scheme discussed in Section 10.5.6 to form the emitter/collector regions, it should be possible to fabricate symmetric lateral bipolar transistors on rather thick SOI.

As of this writing, the need for a CMOS-compatible transistor having significantly higher drive current than CMOS is most apparent in the quest for a "universal" random-access memory. A universal RAM is meant to have operational speed comparable to SRAM, density comparable to DRAM, and is nonvolatile. Two of the universal RAM technologies in focus in the semiconductor industry are Spin-Transfer-Torque Magnetoresistive RAM (STT-MRAM) and Phase Change Memory (PCM). A memory cell consists of one access transistor and one memory element (a magnetic tunnel junction in the case of STT-MRAM and a phase-change resistor in the case of PCM). The memory element is usually "hidden" above the access transistor, and the memory cell area is determined primarily by the area of the access transistor.

In the case of STT-MRAM, the access transistor must be able to deliver a programming current large enough to satisfy the thermal stability factor of the cell (Jin *et al.*, 2014). In the case of PCM, the access transistor must be able to deliver the power needed to cause the resistor memory element to reach melting temperature during resetting (Wong *et al.*, 2010). When implemented in a CMOS platform, which is the common practice in the industry, the width of the access transistors are large for both SST-MRAM and PCM, resulting in large memory cells. For example, the SST-MRAM implemented in a 22-nm FinFET CMOS logic platform has a cell layout of 216-nm × 225-nm pitch grid or cell area of 0.0486 µm^2 (Golonzka *et al.*, 2018), which is about 55% of a six-transistor SRAM cell size of 0.087 µm^2 (Sell *et al.*, 2017). The PCM implemented in a 28-nm FDSOI CMOS logic platform has a cell area of 0.036 µm^2 (Arnaud *et al.*, 2018), which is about 30% the size of a six-transistor SRAM cell of 0.120 µm^2 (Planes *et al.*, 2012). While these large cell areas, implemented in logic platforms, may not represent cell areas in stand-alone large-capacity high-density universal memories, it is well-recognized that minimum-size FETs do not have the drive current needed for universal memories. Therefore, the high-drive-current property should make **symmetric lateral bipolar transistors a prime candidate technology platform for low-cost large-capacity high-density universal memories.**

To be truly useful to system designers, a high-drive-current device must be a desirable logic device as well as a dense access transistor for universal memory cells. That is, the high-drive-current device must also form a competitive logic technology platform for microprocessors and microcontrollers. In Sections 11.5.2 and 11.5.3, it is shown that as far as resistor-load current-switch logic circuits, such as ECL and CML, are concerned, symmetric lateral bipolar is far superior to vertical bipolar in terms of power, performance, and cost. With the integration of p−n−p and n−p−n, as illustrated in Figure 9.1(b), symmetric lateral bipolar transistors on SOI offer additional interesting logic technology platforms previously unimaginable with vertical bipolar transistors. In Sections 11.8.2–11.8.4, we discuss some of the opportunities that have been reported (Ning and Cai, 2013, 2015; Ning, 2016).

11.8.2 Revisit Integrated Injection Logic Circuits and SRAM

Between the mid and late 1970s, there was much excitement in the VLSI industry about the prospect of Integrated Injection Logic (I2L) (Hart and Slob, 1972), aka Merged Transistor Logic (MTL) (Wiedmann and Berger, 1972). I2L is by far the densest circuit. It uses minimum-size devices, and requires one p−n−p per gate for current injection and one n−p−n per fan-out. Thus, an inverter with FO = 3 requires only four transistors.

One reason for the high circuit density of I2L circuits is the fact that, when implemented in vertical bipolar technology, the bottom n-region of the vertical n−p−n transistor functions as the emitter and the top n-region functions as the collector. The n−p−n transistors in an I2L circuit are connected in a common-emitter (i.e., common bottom n-regions) configuration, as indicated in Figure 11.20 (a), which enables many I2L gates to share one emitter (bottom n-region) terminal contact. For vertical n−p−n transistors, the bottom-n-region-as-emitter configuration means the transistors are operated in reverse-active mode, resulting in huge emitter diffusion capacitance and severe speed limitation. The most advanced vertical-bipolar I2L has minimum delays > 200 ps (Tang *et al.*, 1979; Aufinger *et al.*, 2011). As a result of this speed limitation, and the rapid progress in CMOS scaling in the early 1980s, vertical-bipolar I2L ceased to be of interest to system designers.

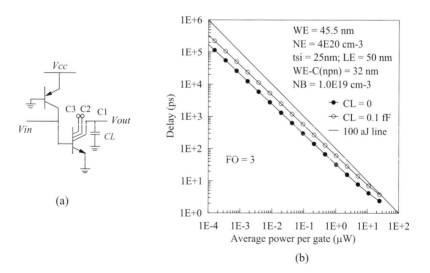

Figure 11.20 (a) Schematic of an I2L gate with three fan-outs. (b) Calculated delay versus upper bound estimate of average standby power dissipation for an I2L gate with FO = 3 implemented in symmetric lateral bipolar transistors on SOI. The n−p−n device parameters are as indicated (after Ning and Cai, 2013)

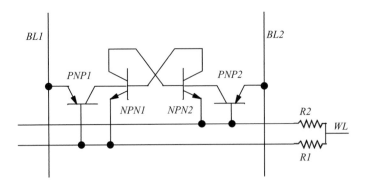

Figure 11.21 Circuit schematic of an I2L SRAM cell (after Wiedmann and Tang, 1981)

Symmetric lateral bipolar transistors have equal switching speed in forward-active and reverse-active modes, suggesting lateral-bipolar I2L should be able to attain speed comparable to resistor-load current-switch circuits. The circuit schematic for an I2L gate with FO = 3 is shown in Figure 11.20(a). As explained in Section 10.5, the simplicity of the symmetric lateral device structure allows the currents, resistances, and capacitances to be expressed in closed-form one-dimensional equations, which in turn allow inverter delays and power dissipations to be calculated relatively easily. This was done for the I2L inverter in Figure 11.20(a) (Ning and Cai, 2013) using the terminal-oriented model of Berger and Wiedmann (1974). The calculated gate delay as a function of average power dissipation is shown in Figure 11.20(b). It shows a power-delay product of less than 100 aJ for a load of $C_L = 0.1$ fF.

There are two reasons for the very small power-delay product: (i) the small-size transistors and associated capacitance, and (ii) the small power supply voltage, with $V_{cc} = V_{BE}$ of the p−n−p injector transistor. With further reduction of device dimensions, it should be able to achieve power-delay products of less than 10 aJ.

The circuit schematic of an I2L SRAM cell is shown schematically in Figure 11.21. The memory cell requires only four transistors (two n−p−n plus two p−n−p) (Wiedmann and Tang, 1981; Wiedmann et al., 1981). In comparison, a standard CMOS SRAM cell requires six transistors. SRAM cell areas are best compared in terms of the half-pitch F of the wires employed to connect the cell in an array. A sample layout of an I2L SRAM cell in symmetric lateral bipolar transistors shows an area of $4F \times 9F = 36\ F^2$ (Ning, 2016). For comparison, a 14-nm FinFET CMOS SRAM cell has an area of 73.8 F^2 (Jan et al., 2015), and a 22-nm SOI CMOS SRAM cell has an area of 90 F^2 (Freeman et al., 2015). Thus, compared to CMOS SRAM cells, I2L SRAM cells are much denser.

11.8.3   Complementary Bipolar Logic Circuits

The concept of CMOS-like complementary bipolar (CBipolar) circuits has been around for a long time (Berger and Wiedmann, 1976). The basic building block is a

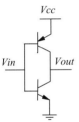

Figure 11.22 Circuit schematic of a complementary bipolar inverter

CBipolar inverter shown schematically in Figure 11.22. It operates with a voltage V_{cc} equal to the V_{BE} for a transistor. In operation, the n−p−n transistor is in full saturation when $V_{in} = V_{cc}$, and the p−n−p transistor is in full saturation when $V_{in} =$ ground. As explained earlier, vertical bipolar transistors are simply not suitable for operation in full-saturation mode, and hence there has been no report of development of CBipolar circuits all these years. Symmetric lateral bipolar transistors on SOI are ideal for CBipolar because there is no speed penalty operating in full-saturation mode.

It is apparent from Figure 9.1(b) that CBipolar circuits layout in a manner similar to CMOS circuits. In CMOS designs, the most commonly used circuit is the NAND gate. A simulation study confirmed that CBipolar NAND gates function properly when implemented in symmetric lateral bipolar transistors (Ning and Cai, 2015).

For the same functional circuit using devices of the same drive current, a CBipolar circuit should layout more densely than a CMOS circuit for two reasons: (i) the drive current density (on current per unit MOSFET gate width) of a lateral bipolar transistor is higher than that of a MOSFET, as discussed in Section 11.8.1; (ii) the metal contact to the base of a lateral transistor can be located directly on top of the intrinsic base while the metal contact to the gate of a MOSFET is located away from the thin gate insulator. In other words, CBipolar circuits are expected to have significant density advantage over CMOS circuits.

Figure 11.23 shows the calculated delay and standby power dissipation of CBipolar inverters as a function of its power supply voltage V_{cc}, for an all-silicon case and for a case of Si-emitter/collector SiGe-base HBT described in Section 10.5.8. The log-delay versus V_{cc} characteristics for the CBipolar in Figure 11.23 are similar to the log-delay versus log-power characteristics for I2L in Figure 11.20(b). In both cases, the minimum delay is about 2 ps. The similarity in delay characteristics are not unexpected since in both circuits the delay is governed by the switching of one symmetric lateral transistor.

The standby power dissipation of CBipolar, being governed by the base current of the n−p−n and p−n−p transistors, is the same for the all-silicon case and the SiGe-base case since both transistors have the same silicon emitter/collector. The standby power dissipation, shown in Figure 11.23, is large compared to CMOS. Thus, ***CBipolar is not a technology to replace CMOS which has negligible standby power dissipation.***

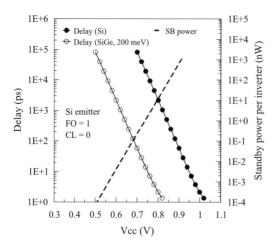

Figure 11.23 Comparison of calculated inverter delay and standby power dissipation for all-Si CBipolar with projections for Si-emitter/collector SiGe-base HBT CBipolar. All devices are assumed to have the same Si emitter region, $t_{si} = 60$ nm, $L_E = 100$ nm, $N_B = 1 \times 19$ cm^{-3}, and $W_B = 10$ nm. The SiGe-base delay is projected by shifting the Si plot to the left by 200 mV, equivalent to assuming the SiGe bandgap to be 200 meV smaller than that of Si (after Ning and Cai, 2015)

Perhaps Si-emitter/collector SiGe-base CBipolar, operated at V_{cc} around 0.5 V for low standby power dissipation, could be an important logic technology platform for applications where the primary objective is not to replace CMOS, but to offer higher density (due to the higher drive current of lateral bipolar transistors) and lower cost (due to the simple fabrication process) than CMOS. Also, it takes advantage of the existing CMOS design tools and design methodology. One candidate application could be the high-density universal memory mentioned in Section 11.8.1.

11.8.4   Performance-On-Demand Designs with I2L or CBipolar Circuits

The need for high-speed systems, or high-speed subsystems within an overall ultra-low-power system, has not gone away. At the device level, it is challenging to develop a CMOS transistor that satisfies the requirements of both high-speed and ultra-low-power systems. CMOS designers typically offer devices with several threshold-voltage options to enable chip designers some room for power-performance tradeoff and optimization. As an example, the published 14-nm FinFET (Jan *et al.*, 2015) offers a high-speed transistor with about 3× higher on current at about 7,000× larger off current than an ultra-low-power transistor. Another common practice in CMOS system design is to enable "turbo mode" operation. In turbo mode, most of the computing cores in the system are powered down to enable just one or two cores to run at higher speed.

I2L and CBipolar circuits do not employ resistors as loads. The speed of a given I2L or CBipolar circuit can be dialed up or down simply by adjusting V_{cc}. Adjusting V_{cc} by 60 mV could change the circuit speed by a factor of 10, without changing the transistor size or design. The simulations shown in Figure 11.20(b) show that the same physical I2L gate can be operated with delays from larger than 100 ns to smaller than 2 ps. Similar conclusions apply to the CBipolar simulations shown in Figure 11.23.

The performance-on-demand characteristics of I2L and CBipolar represent an intriguingly interesting opportunity for system designers. For example, one could imagine a processor chip containing many cores, with most cores running at base speed in ultra-low-power mode, some cores running at 10–100× base speed, and a couple of cores running at 100–1,000× base speed as accelerators. The transistors and circuit layouts for all these variable-speed cores could be identical, thus greatly reducing the design cost and chip area, and hence chip and system cost. Of course, care should be exercised to ensure there are no reliability issues, such as electromigration in the wire contacts and connections. The performance-on-demand of a vertical-bipolar I2L 1024 divider circuit was demonstrated (Aufinger et al., 2011). The operating frequency of the divider was changed from 10 MHz to 1.2 GHz simply by changing the p−n−p injector current. The highest operating frequency (1.2 GHz) was limited by the fact that the I2L divider was built using vertical transistors. More possibilities of taking advantage of such performance-on-demand properties of I2L and CBipolar circuits are discussed elsewhere (Ning, 2016).

Exercises

11.1 The small-signal equivalent circuit for a bipolar transistor, including its internal series resistances, is shown in Figure 9.26. Ignore the collector-substrate capacitance $C_{dCS,tot}$, which is quite small in modern bipolar transistors. Show that the cutoff frequency is given by Eq. (11.9), i.e.,

$$\frac{1}{2\pi f_T} = \tau_F + \frac{kT}{qI_C}(C_{dBE,tot} + C_{dBC,tot}) + C_{dBC,tot}(r_e + r_c).$$

State the assumptions used in the derivation.

11.2 Consider a vertical n–p–n bipolar transistor, with an emitter area of $A_E = 1 \times 2\ \mu m^2$ and a base–collector junction area of $A_{BC} = 10\ \mu m^2$, a deep emitter with $N_E = 10^{20}\ cm^{-3}$, a boxlike intrinsic base with $N_B = 1 \times 10^{18}\ cm^{-3}$ and $W_B = 100$ nm, and a uniformly doped collector of $N_C = 5 \times 10^{16}\ cm^{-3}$. Assume one-sided junction approximation for all the junctions, and that the transistor is biased with $V_{BE} = 0.8$ V (for purposes of t_{BE} and $C_{dBE,tot}$ calculations) and $V_{CB} = 2$ V (for purposes of t_{BC} and $C_{dBC,tot}$ calculations).

(a) Derive the equations for τ_E in terms of τ_{pE} and β_0, and τ_B in terms of t_B, from Eqs. (9.133) and (9.134). Estimate the low-injection values of the delay or transit times τ_E, τ_B, t_{BE}, t_{BC}, and τ_F, assuming $\beta_0 = 100$.

(b) Estimate $C_{dBE,tot}$ and $C_{dBC,tot}$.

(c) The maximum cutoff frequency can be estimated if we assume the transistor is operated at its maximum current density without significant base widening, i.e., at $J_C = 0.3qv_{sat}N_C$. (Note that the maximum J_C increases with N_C.) Estimate the maximum obtainable cutoff frequency from

$$\frac{1}{2\pi f_T} = \tau_F + \frac{kT}{qI_C}(C_{dBE,tot} + C_{dBC,tot}).$$

(d) The effect of the emitter and collector series resistances on the cutoff frequency can be estimated from

$$\frac{1}{2\pi f_T} = \tau_F + \frac{kT}{qI_C}(C_{dBE,tot} + C_{dBC,tot}) + C_{dBC,tot}(r_e + r_c).$$

Assume $(r_e + r_c) = 50\,\Omega$. Estimate the maximum cutoff frequency.

11.3 This exercise is designed to illustrate the advantage of the pedestal-collector design for a vertical bipolar transistor. Consider the n–p–n transistor of Exercise 11.2. Let $N_C(\text{int}) = 5 \times 10^{16} \text{cm}^{-3}$ be the collector doping concentration directly underneath the emitter and intrinsic base, and $N_C(\text{ext}) = 5 \times 10^{15}\text{ cm}^{-3}$ be the doping concentration of the extrinsic part of the collector. $C_{dBC,tot}$ of the pedestal-collector design is smaller than that of the non-pedestal-collector design. Repeat Exercise 11.2 for this pedestal-collector transistor.

11.4 This exercise is designed to illustrate the sensitivity of the maximum cutoff frequency to the pedestal-collector doping concentration in a vertical bipolar transistor. Consider the pedestal-collector transistor of Exercise 11.3. The maximum collector current density without significant base widening can be increased by increasing N_C (int). Repeat Exercise 11.3 for the case of $N_C(\text{int}) = 1 \times 10^{17} \text{ cm}^{-3}$.

11.5 The incorporation of Ge into the intrinsic base of a vertical transistor reduces the base transit time and the emitter delay time. For a linearly graded Ge profile, the transit time is reduced by a factor [see Eq. (10.37)]

$$\frac{t_B(\text{SiGe})}{t_B(\text{Si})} = \frac{2kT}{\Delta E_{g\,\text{max}}}\left\{1 - \frac{kT}{\Delta E_{g\,\text{max}}}[1 - \exp(-\Delta E_{g\,\text{max}}/kT)]\right\},$$

where $\Delta E_{g,\,\text{max}}$ is the maximum bandgap narrowing due to Ge. Since the emitter delay time τ_E is inversely proportional to current gain, τ_E is therefore reduced by a factor

$$\frac{\tau_E(\text{SiGe})}{\tau_E(\text{Si})} = \frac{\beta_0(\text{Si})}{\beta_0(\text{SiGe})}.$$

The current gain ratio is given by Eq. (10.30), namely

$$\frac{\beta_0(\text{SiGe})}{\beta_0(\text{Si})} = \frac{\Delta E_{g,\,\text{max}}/kT}{1 - \exp(-\Delta E_{g,\,\text{max}}/kT)}.$$

Repeat Exercise 11.4 for a SiGe-base transistor with $\Delta E_{g,\,\text{max}} = 100\text{ meV}$. (Assume the same V_{BE} of 0.8 V.) (Comparison of the results from Exercises 11.4 and 11.5 illustrates the effect of SiGe-base technology on f_T.)

11.6 From the small-signal equivalent circuit for an intrinsic transistor (biased in forward-active mode) shown in Figure 9.25, set up a two-point network representation of an intrinsic bipolar transistor (dropping the collector-substrate capacitor, which is small compared to $C_{dBC,tot}$).

(a) From the admittance matrix representation

$$\begin{bmatrix} i_1 \\ i_2 \end{bmatrix} = \begin{bmatrix} Y'_{11} & Y'_{12} \\ Y'_{21} & Y'_{22} \end{bmatrix} \begin{bmatrix} v'_1 \\ v'_2 \end{bmatrix},$$

where the primes denote parameters for an intrinsic transistor (parasitic resistances neglected), write the equations for the elements $Y'_{11}, Y'_{12}, Y'_{21}$, and Y'_{22}.

(b) When resistances r_e, r_c, and r_b are included, the admittance matrix representation is

$$\begin{bmatrix} i_1 \\ i_2 \end{bmatrix} = \begin{bmatrix} Y_{11} & Y_{12} \\ Y_{21} & Y_{22} \end{bmatrix} \begin{bmatrix} v_1 \\ v_2 \end{bmatrix}.$$

Write the equations for Y_{11}, Y_{12}, Y_{21}, and Y_{22} in terms of $Y'_{11}, Y'_{12}, Y'_{21}$, and Y'_{22} and the resistances.

(c) From (a), derive Eq. (11.8) for an intrinsic transistor. From (b), derive Eq. (11.9) for a transistor including resistances. State the approximations made.

11.7 From Eq. (8.69) and the discussion in Section 8.5.4, $f_{\max}$ is given by the equation

$$4\text{Re}(Y_{11})\text{Re}(Y_{22}) = \left| Y_{12} + Y^*_{21} \right|^2.$$

Derive Eq. (11.10), stating the approximations made.

11.8 Consider a symmetric lateral n−p−n transistor on SOI having the following physical parameters: $W_E = 45$ nm, $W_{E-C} = 40$ nm, $N_E = 4 \times 10^{20}$ cm^{-3}, $N_B = 5 \times 10^{18}$cm^{-3}, $L_E = 2$ μm, and $t_{si} = 100$ nm (i.e., $A_E = 0.1 \times 2$ μm^2).

(a) The transit times are given by Eqs. (10.90)–(10.94), namely,

$$\tau_F = \tau_E + \tau_B + \tau_{BE} + \tau_{BC},$$

$$\tau_E(\text{shallow emitter}) = \frac{2I_B}{3I_C}\frac{W_E^2}{2D_{pE}} = \frac{W_E W_B \left(V'_{BE}, V'_{BC}\right) N_B n^2_{ieE}}{3D_{nB} N_E n^2_{ieB}},$$

$$\tau_B = \frac{2t_B}{3} = \frac{W_B^2\left(V'_{BE}, V'_{BC}\right)}{3D_{nB}},$$

$$\tau_{BE} = t_{BE} = \frac{W_{dBE}\left(V'_{BE}\right)}{2v_{sat}},$$

and

$$\tau_{BC} = t_{BC} = \frac{W_{dBC}\left(V'_{BC}\right)}{2v_{sat}}.$$

Calculate W_B, W_{dBE}, W_{dBC}, τ_E, τ_B, τ_{BE}, τ_{BC}, and τ_F, for $V'_{BE} = 0.88, 0.90, 0.92,$ $0.94, 0.96,$ and 0.98 V, assuming $V'_{BC} = -10$ V and low-injection approximation (i.e., ignoring mobile electrons in the junction space-charge regions).

(b) The collector current can be calculated from Eqs. (10.74) and (10.75), namely

$$I_C(V'_{BE}, V'_{BC}) = I_{C0}(V'_{BE}, V'_{BC}) \exp(qV'_{BE}/kT),$$

with

$$I_{C0}(V'_{BE}, V'_{BC}) = \frac{qA_E D_{nB} n_{ieB}^2}{N_B W_B(V'_{BE}, V'_{BC})} \left[1 + \frac{1}{4}\left(\sqrt{1 + \frac{4n_{ieB}^2 \exp(qV'_{BE}/kT)}{N_B^2}} - 1\right)\right]^{-1}.$$

Calculate I_C at the V'_{BE} and V'_{BC} conditions in (a).

(c) From the results of (a) and (b), calculate the corresponding values of f_T and f'_T, assuming $(r_e + r_c) = 200$ Ω. Compare the results with the measured f_T of a typical high-performance SiGe-base vertical bipolar transistor shown in Figure 11.5.

[Fair comparison of the intrinsic merits of transistors should be made at the same J_C. Since both the SiGe-base transistor in Figure 11.5 and our SOI lateral transistor have the same emitter area, comparison can be made at the same collector current.]

(d) The comparison in (c) should show that the SOI lateral transistor has significantly higher f_T than the SiGe-base transistor at low currents, but only comparable or smaller f_T at large currents. What is the physics for this observation (i.e., f_T (lateral) appreciably larger than f_T(SiGe) at low currents, but only comparable at large currents)?

12   Memory Devices

The previous chapters have considered the operation of CMOS and bipolar devices mainly in the context of logic circuits. This chapter addresses another basic functional block in modern VLSI chips – memory. A predominant majority of the VLSI devices produced today are in various forms of *random-access memory* (RAM).

Viewed from the operation standpoint, a RAM functional unit is usually organized into an *array* of memory cells (or *bits*) together with its supporting circuits for selecting, writing, and reading the memory cells. In an array, the bits on the same row are selected by a *word* signal. A schematic block diagram of a RAM unit is shown in Figure 12.1. The array consists of W words with B bits each, for a total memory capacity of $W \times B$ bits. A random bit in the array can be accessed through signals applied to its *wordline* and *bitline*.

Depending on the retention of information in the cells of a memory array, random-access memories can be classified into three categories: *static random-access memory* (SRAM), *dynamic random-access memory* (DRAM), and *nonvolatile random-access memory* (NVRAM). NVRAM is often referred to as nonvolatile memory for short. SRAMs have fast access times. They retain data as long as they are connected to the power supply. Practically every VLSI chip contains a certain amount of SRAM which is usually built using basically the same devices as in the logic circuits. DRAMs have relatively slow access times. They require periodic refresh in order to prevent loss of data. On a per-bit basis, DRAMs have a much lower cost than SRAMs because a DRAM cell is typically only about one tenth the size of an SRAM cell. For systems that require much more SRAM than can be contained on the logic chip, stand-alone SRAM chips are often used to meet the need. However, in order to reduce system cost and size, designers often use stand-alone DRAM chips instead of stand-alone SRAM chips. In that case, some form of memory-hierarchy architecture is usually employed to minimize the impact of the relatively slow DRAM on the system performance. Both SRAMs and DRAMs are volatile in that data are lost once the power supply to the chip is disconnected.

For systems that must retain all or some of their data at all times, nonvolatile memories are often used in place of or in addition to DRAMs. Nonvolatile memories can be classified into three categories: read-only or non-programmable, programmable once, and erasable and programmable. By far the most versatile nonvolatile memory technology is the erasable and programmable type, particularly the *electrically erasable and programmable read-only memory* (EEPROM).

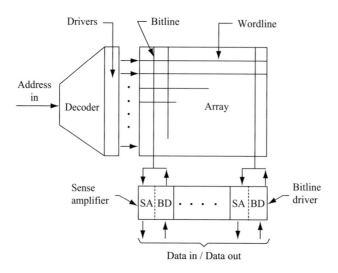

Figure 12.1 Block diagram of a random-access memory functional unit. In this schematic, a bitline represents either a single line or a pair of lines (after Terman, 1971)

Semiconductor memory development reached a critical high point when bipolar SRAMs were used as the main memory in the IBM System 370 Model 145 mainframe computer first shipped in 1971 (Pugh *et al.*, 1981). However, as explained in Section 11.5.3, the standby power of bipolar circuits is high, making bipolar devices unsuitable for memory applications that require a very large number of bits on a single chip. CMOS devices, with their low standby power characteristics, are uniquely suitable for building large SRAM arrays. As a result, CMOS is now used to build SRAM functions whenever CMOS devices are available on the chip. Relatively small arrays of bipolar SRAMs are still used in applications built with bipolar-only technology.

The papers by Terman (1971) and Sah (1988) give a detailed account of the early efforts on exploring various kinds of dynamic semiconductor memory cells. The one-transistor, one-capacitor memory cell (Dennard, 1968) is by far the densest dynamic memory cell. It has been subsequently adopted universally as the standard DRAM cell. In the case of NVRAM and EEPROM, the early development of many of the device concepts has been summarized in the literature (Sah, 1988; Hu, 1991). They remain areas of very active research.

One emerging memory device topic that has been pursued very actively by the semiconductor industry, for well over ten years now, is *universal memory*. Universal memory refers to a computer data storage device which has the cost of DRAM, the speed of SRAM, is non-volatile, and has infinite durability. Such a memory device, if it ever becomes commercially viable, is expected to have a far-reaching impact on the computer, communication, and semiconductor industry. At the memory cell level, a universal memory cell typically consists of an *access transistor*, which can be a MOSFET or a bipolar transistor, and a memory element specific of the memory technology. For example, the memory element is a magnetic tunnel junction in the case of MRAM (magnetoresistive random-access memory), a chalcogenide glass filament in the case of PCM (phase-change memory), and a layer of resistive dielectric solid-state material in the case of RRAM or ReRAM (resistive random-access memory). R&D efforts in universal memory focus mainly on these memory elements. It is beyond the scope of this book to cover universal memory, which would require in-depth discussion of the physical mechanisms, materials science, and engineering of these evolving memory elements. Interested readers are referred to the vast literature on these subjects (e.g., Tang and Lee, 2010, Wong *et al.*, 2010, 2012, Tang and Pai, 2020, and the references therein). Nonetheless, it is pointed out in Section 11.8.1 that symmetric lateral bipolar transistors on SOI could be a preferred access-transistor technology platform for low-cost large-capacity high-density universal memories.

In this chapter, we discuss the basic operational principles, device design, and scaling issues of the CMOS SRAM cell, the one-transistor, one-capacitor DRAM cell, and several commonly used EEPROM devices. Only the most commonly used bipolar SRAM cell will be discussed.

12.1   Static Random-Access Memory

In principle, any device or arrangement of devices that can be programmed into two distinct, electrically stable states can be used as a storage element of an SRAM cell. Depending on the storage element, one or more access transistors are connected to the storage element to make an SRAM cell. In this section, we first discuss the basic operation of CMOS SRAM cells and the design and scaling issues of the constituent devices. After that, the basic operation of the commonly used bipolar SRAM cell is given.

12.1.1   CMOS SRAM Cell

In CMOS VLSI designs, the most commonly used SRAM storage element is the *bistable latch* consisting of two cross-coupled CMOS inverters shown in Figure 12.2. It can be built using a standard CMOS logic fabrication process. Inverter 1 consists of nMOSFET $Q1$ and pMOSFET $Q3$, while inverter 2 consists of nMOSFET $Q2$ and pMOSFET $Q4$. The two stable states can be readily recognized by plotting the transfer curves (Section 8.2.1.1) of the two inverters back to back, as illustrated in Figure 12.3, often referred to as the "butterfly" plot of a pair of cross-coupled inverters.

In Figure 12.2, one of the inverters has its input at high and output at low, while the other inverter has its input at low and output at high. The first inverter, with its output at low, keeps the second inverter in the state described above, and vice versa. Thus, a CMOS SRAM storage element has two stable states: one at the intersection A of the two inverter transfer curves in Figure 12.3 with $V_{out1} = V_{in2} = V_{dd}$, and the other at the intersection B with $V_{out2} = V_{in1} = V_{dd}$. The two stable states can be interpreted as logical "0" and "1." Here we designate logical "1" as $V_{out1} = 0$ and $V_{out2} = V_{dd}$, i.e., point B, and logical "0" as $V_{out1} = V_{dd}$ and $V_{out2} = 0$, i.e., point A. A bistable latch

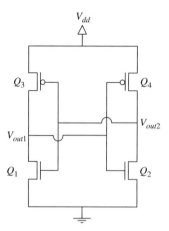

Figure 12.2 Circuit schematic of two cross-coupled CMOS inverters. The output of inverter 1 ($Q1$ and $Q3$) is the input to inverter 2 ($Q2$ and $Q4$), i.e., $V_{in2} = V_{out1}$, and vice versa

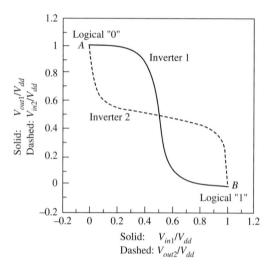

Figure 12.3 Butterfly plot for two cross-coupled CMOS inverters. The transfer curve of inverter 1 (solid) is plotted as V_{out1} versus V_{in1}, and inverter 2 (dashed) as V_{in2} versus V_{out2}. Identical and symmetric inverters are assumed. In a CMOS SRAM cell, nMOSFETs $Q1$, $Q2$ are usually made stronger than pMOSFETs $Q3$, $Q4$ and the high-to-low transition of the transfer curves is not symmetric (see Figure 12.7(c))

will remain in one of its two stable states until it is forced by an external signal to flip to the other stable state.

The most commonly used SRAM cell is a six-transistor cell consisting of two cross-coupled CMOS inverters and two access transistors. The circuit schematic for a CMOS SRAM cell is shown in Figure 12.4. The cross-coupled inverters are connected

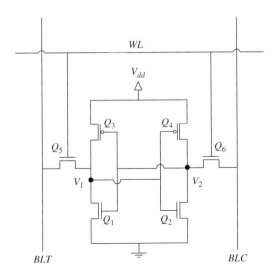

Figure 12.4 Circuit configuration of a CMOS SRAM cell. In the text, we assume the cell is storing a "1" when V_2 = high (V_1 = low) and a "0" when V_1 = high (V_2 = low)

to two bitlines, *BLT* (bitline true) and *BLC* (bitline complement), through n-channel access transistors $Q5$ and $Q6$. The access transistors are controlled by the wordline (WL) voltage. In the standby mode, *WL* is kept low ($V_{WL} = 0$ V), thus turning off the access transistors and isolating the bitlines from the cross-coupled inverter pair.

12.1.1.1   Basic Operation of a CMOS SRAM Cell

Here we give a description of the basic read and write operations of a CMOS SRAM cell. The reader is referred to the literature for details on the circuits involved in operating an SRAM array or chip (see e.g., Itoh, 2001).

- *Read operation.* In a read operation, the wordline is kept high ($V_{WL} = V_{dd}$), thus turning on the access transistors and allowing the logical state of the cell, represented by the values of V_1 and V_2, to be sensed via the bitlines. The signal voltages involved in the read operation and their timing are shown in Figure 12.5. Prior to the wordline being selected (V_{WL} raised from 0 to V_{dd}), both bitlines are precharged to V_{dd} via low-impedance loads Z. Let us first consider reading a logical "0." In this case, prior to the wordline being selected, we have V_{BLT}, V_{BLC}, and node V_1 all at V_{dd}, and node V_2 at 0. After the wordline is selected, $Q5$ and $Q6$ are on. Charge flows from *BLC* to V_2 through the conducting $Q6$, causing V_2 to rise and V_{BLC} to drop. The bitline voltage difference, $V_{BLT} - V_{BLC} > 0$, is read by the sense amplifier connected to the bitlines. In reading a logical "1," the bitline voltage difference is such that $V_{BLT} - V_{BLC} < 0$. At the end of the read operation, the wordline is turned off, thus isolating the cell from the bitlines and allowing V_1 and V_2 to return to their standby values before the read cycle. It should be noted that voltages V_1 and V_2 deviate from their standby values during the read operation. This

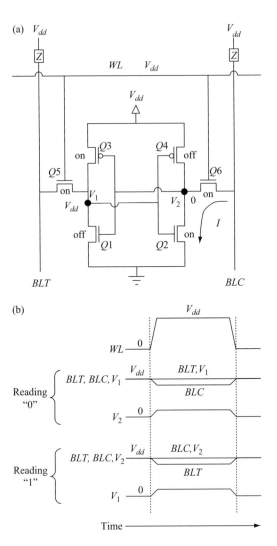

Figure 12.5 Read operation of a CMOS SRAM cell. (a) Voltage and current in reading a "0". (b) Node waveforms in reading a "0" and reading a "1"

makes the SRAM cell less stable while being read. Proper design needs to be exercised to ensure that a memory bit is not flipped by the read operation itself. The read instability issue will be discussed later.

- *Write operation.* In a write operation, appropriate write voltages are applied to the bitlines to force the cell into the intended logical state. As an example, let us consider writing a cell from logical "0" to logical "1." The signal voltages involved and their timing are illustrated in Figure 12.6. The bitlines are precharged to V_{dd} prior to the wordline being selected. A "write enable" signal is given at time t_{WE} when a voltage $V = V_{dd}$ is applied to BLC and a voltage $V = 0$ is applied to BLT. The voltage on BLT forces V_1 to 0, while the voltage on BTC forces V_2 to V_{dd}, thus

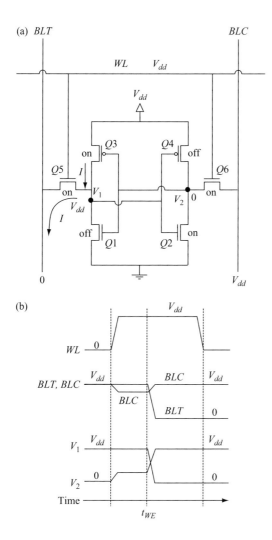

Figure 12.6 Write operation of a CMOS SRAM cell. (a) Voltage and current in writing a "1" to a cell originally storing a "0," i.e., flipping (V_1, V_2) from $(V_{dd}, 0)$ to $(0, V_{dd})$. The "on," "off" labels refer to the transistor states before flipping. (b) Node waveforms in writing a "1" to a cell storing a "0"

writing a logical "1" to the cell. At the end of the write operation, the wordline is turned off, thus leaving the isolated cell in a logical "1" state.

12.1.1.2 Device Sizing for a CMOS SRAM Cell

From density consideration, it is desirable to have all the transistors in a CMOS SRAM cell as small as possible. However, for stability in read and instability in write operations, the devices must have the correct on conductance or "strength" relative to one another. As a result, not all transistors can be of minimum size.

Let us consider the current path during a read "0" operation illustrated in Figure 12.5(a). We want the current through transistors $Q6$ and $Q2$ to cause V_{BLC} to drop but without causing V_2 to rise too much to affect the stability of the cell. That is, we want the on resistance of $Q2$ to be small compared to that of $Q6$, i.e., we want $R(Q2) < R(Q6)$, or $Q2$ to be stronger (wider) than $Q6$. By symmetry, $Q1$ needs to be stronger than $Q5$, or $R(Q1) < R(Q5)$.

The relative device strength for the write operation is considered in Figure 12.6(a), where a "1" is written to a cell originally storing a "0." Here we want to pull V_1 readily from high to low and flip the cell. Hence we want $Q5$ to be strong compared with $Q3$, or $R(Q5) < R(Q3)$. Similarly, we want $Q6$ to be stronger than $Q4$, or $R(Q6) < R(Q4)$.

For CMOS using unstrained silicon, it is straightforward to make $Q3$ and $Q4$ the weakest transistors in a CMOS SRAM cell because they are pMOSFETs, which have about half the mobility as nMOSFETs (see Figures 5.14 and 5.16). For the same device channel length and threshold voltage magnitude, a pMOSFET has about half of the current per width of an nMOSFET. Thus, by using minimum-size pMOSFETs for $Q3$ and $Q4$ and minimum-size nMOSFETs for $Q5$ and $Q6$, we meet the device sizing requirement for write operation. To meet the device sizing requirement for read operation, $Q1$ and $Q2$ must be larger than minimum size. Designers typically make $Q1$ and $Q2$ about twice the width of $Q5$ and $Q6$ (see e.g., Seevinck *et al.*, 1987).

For CMOS using strained silicon, the mobility of the pMOSFETs and the mobility of the nMOSFETs could both be increased and by different amounts, as discussed in Section 5.2.2. Care needs to be exercised in designing a CMOS SRAM cell to ensure $Q3$ and $Q4$ to be the weakest transistors in the cell.

12.1.1.3   Static Noise Margin of a CMOS SRAM Cell

A memory cell should maintain its logical state when the memory array or chip is in use in a system environment. The stability of a memory cell is often characterized by its *static noise margin* (SNM), i.e., by the magnitude of noise voltage needed to cause the memory bit to flip to the other logical state. As discussed earlier in connection with Figure 12.4, when a memory cell is in standby, it is isolated from the bitlines. However, when the cell is accessed during a read or write operation, it is coupled to the bitlines through the access transistors. We shall discuss the noise margins of a CMOS SRAM cell when it is in the standby mode, when during a read operation, and when during a write operation.

- *SNM in standby mode.* In the standby mode, the access transistors are turned off. Consider the cell in a logical "0" state, i.e., at the stable point A with $V_2 = 0$ and $V_1 = V_{dd}$. Let us assume there is a noise voltage of magnitude V_n that tends to flip the cell, i.e., the noise adds a bias of $+V_n$ to the input of inverter 1 and a bias of $-V_n$ to the input of inverter 2, as illustrated schematically in Figures 12.7(a) and 12.7(b). The corresponding transfer curves are illustrated in Figure 12.7(c). Because read and write operations require that the bottom nMOSFETs be stronger than the top pMOSFETs, the transfer curve is not symmetric, with its high-to-low transition at $V_{in} < V_{dd}/2$. The effect of the noise voltages is to shift the transfer curve of inverter

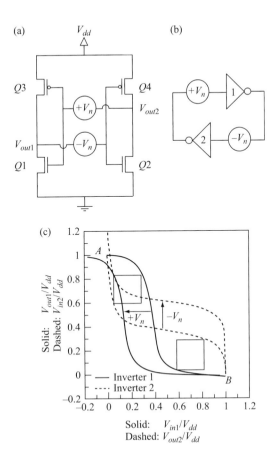

Figure 12.7 A CMOS SRAM cell having a noise voltage that tends to flip the cell from "0" to "1." (a) Schematic showing the transistor connections. (b) Schematic showing the circuit configuration. (c) Effect of the noise on the transfer curves. Note that the high-to-low transitions of the transfer curves before noise occur at $V_{in}/V_{dd} < 0.5$, as discussed in the text

1 horizontally to the left and the transfer curve of inverter 2 vertically upward. The SNM is the minimum noise voltage that shifts the two transfer curves until they no longer intersect at point A, i.e., the only intersection is at point B. The cell has flipped from "0" (stable point A) to "1" (stable point B). *Graphically, the static noise margin is measured by the side of the maximum square that can be nested between the two transfer curves*, as indicated in Figure 12.7(c) (Lohstroh *et al.*, 1983), just like that of a cascade chain of inverters discussed in Section 8.2.1.2. The static noise margin for flipping a cell from "1" to "0" can be derived in a similar manner by switching the noise voltage polarities. In that case, the transfer curve of inverter 1 is shifted horizontally to the right while that of inverter 2 is shifted vertically downward until they only intersect at point A.

- *SNM during read access.* As shown in Figure 12.5, when a CMOS SRAM cell is being read, the node that is low is pulled up by the bitline while the node that is high

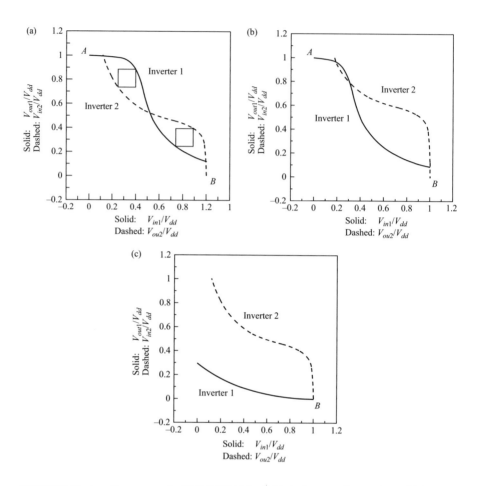

Figure 12.8 The transfer curves of a CMOS SRAM cell (a) during a read operation with identical inverters, (b) during a read operation with mismatched inverters, and (c) during a write operation

remains high. When the cell is reading a "0" (V_1 = high and V_2 = low), $V_{out1} = V_1$ remains high (= V_{dd}), but $V_{out2} = V_2$ is pulled up by the bitline to a value > 0. Similarly, when the cell is reading a "1" (V_1 = low and V_2 = high), $V_{out2} = V_2$ remains high (= V_{dd}), but $V_{out1} = V_1$ is pulled up by the bitline to a value > 0. The resulting butterfly curves during a read operation are illustrated in Figure 12.8(a). Because of the added connection of the output node through the turned-on access transistor to the bitline at V_{dd}, the high-to-low transition of each inverter becomes less steep and the output voltage at input high (V_{dd}) does not go all the way to zero. *__In practice, a CMOS SRAM cell is more vulnerable to noise disturbance during the read operation because of the smaller noise margin__*.

For nominal bits with nearly symmetric inverters and transfer curves, the normalized SNM, V_{NM}/V_{dd}, does not change significantly with V_{dd} until V_{dd} is only a few kT/q, as discussed in Section 8.2.1.2. However, for the worst-case bits with severely mismatched devices such that one of the noise margins is barely above zero (Figure 12.8(b)), reduction of V_{dd} even slightly could push those bits over the edge, i.e., one of their SNM goes to zero and they fail to function properly. This is further discussed in Section 12.1.1.4.

- *SNM during write access.* In the write operation shown in Figure 12.6, voltages $(0, V_{dd})$ are applied to (BLT, BLC) with the access transistors turned on to flip the SRAM cell from a "0" state to a "1" state, i.e., flipping (V_1, V_2) from $(V_{dd}, 0)$ to $(0, V_{dd})$. This can be illustrated in terms of the static transfer curves plotted in Figure 12.8(c). Since $Q5$ is stronger than $Q3$, i.e., $R(Q5) < R(Q3)$, the connection to $BLT = 0$ causes $V_{out1}(V_{in1} = 0)$ to fall far enough below V_{dd} that the noise margin for state "0" (intersection A) completely vanishes and the only allowed state for the cross-coupled latch is state "1" (intersection B).

12.1.1.4   Scaling Issues of CMOS SRAM Cells

The scaling properties of CMOS devices and technology have been discussed in Sections 6.3 and 8.1, with the device parameter trends in scaling shown in Figure 6.22. In general, the overdrive ratio V_{dd}/V_t has decreased as both V_{dd} and V_t are scaled down with the gate length. This means that the I–V characteristics and hence the transfer curves become more sensitive to V_t variations that do not scale with V_{dd}. Since the SNM of a CMOS SRAM cell depends critically on device matching and since a typical SRAM array has a very large number of cells with a wide statistical distribution of device parameters, these factors pose specific scaling issues for CMOS SRAMs, as discussed next.

- *Threshold voltage variation due to short-channel effect.* Because the built-in potential ψ_{bi} is nearly a constant for silicon, the short-channel threshold-voltage rolloff, Eq. (6.35), does not scale with V_{dd}. Also, any process variation of t_{ox} and doping concentration can add to the V_t variation. They can cause device mismatch and reduction of the SNM of SRAM cells. Gate length mismatch due to lithography can be minimized by laying out the transistors within a cell in a symmetrical manner. Other process variations usually track well within the close proximity of a cell.

- *Threshold voltage variation due to statistical dopant fluctuation.* Threshold voltage variation caused by statistical fluctuation of the number of dopant atoms in a small MOSFET has been covered in Section 6.3.4. For CMOS SRAM devices with a minimum width of about 100-nm, the standard deviation σ_{Von} of V_t fluctuation is of the order of 30 mV (Wong and Taur, 1993; Frank *et al.*, 1999; Bhavnagarwala *et al.*, 2001). Threshold voltage variation due to dopant number fluctuation is completely random and hence cannot be minimized by placing transistors close to one another or by layout. In fact, V_t fluctuation is usually determined experimentally by using matched pairs of adjacent transistors and measuring the V_t difference between the two transistors in a pair (Mizuno *et al.*, 1994; Tuinhout *et al.*, 1996, 1997). Threshold voltage fluctuation can cause one or more of the transistor pairs in a CMOS SRAM cell to become mismatched in V_t, which in the worst case can cause one of the SNMs of the cell to vanish.

- *Threshold voltage variation due to high-field effects.* As discussed in Section 6.4, the characteristics of a CMOS device can change due to high-field effects. In general, high-field degradation results in an increase of V_t magnitude. Thus, even if a device pair is well matched as fabricated, their characteristics may become mismatched appreciably after burn-in stress or during operation. For advanced CMOS generations, threshold voltage instability in pMOSFETs due to negative-bias-temperature instability (NBTI) is of the most concern. NBTI can cause relatively large V_t shifts as well as V_t mismatch in pMOSFETs (Rauch III, 2002).

For a given CMOS SRAM array, there is a minimum supply voltage V_{min} below which at least one bit in the array ceases to function. The failed bit usually has severely mismatched devices and inverters. For example, one inverter (#1 in Figure 12.8(b)) may have its high-to-low transition shifted to $V_{in} < V_{dd}/2$ because the nMOSFET, with its V_t at the low end of the distribution, is much stronger than the pMOSFET whose V_t (magnitude) happens to be on the high side. The other inverter (2 in Figure 12.8(b)) in the cross-coupled cell could be on the opposite side of the spectrum, i.e., its low-V_t (magnitude) pMOSFET is much stronger than its high-V_t nMOSFET such that the high-to-low transition is at $V_{in} > V_{dd}/2$. Reduction of V_{dd} has a more pronounced effect percentage-wise on the current drive of the high-V_t device whose overdrive ratio V_{dd}/V_t is already low, than on the current drive of the low-V_t device. In other words, reduction of V_{dd} makes the weak device even weaker, thus further worsens the mismatch until it becomes so severe that one of the SNMs goes to zero.

For a large array of the order of 10^9 devices with a Gaussian distribution, on average two of the devices will have V_t deviated more than $6\sigma_{Von}$ from the nominal value. This is of the order of 0.2 V due to dopant number fluctuations alone. With a scaled down $V_{dd} \sim 1$ V and $V_t \sim 0.3$ V in the sub-100-nm CMOS generations, it is difficult to design the transistors in an SRAM cell to guard against such a large percentage of V_t variation (Bhavnagarwala *et al.*, 2001). Tradeoffs in device size (larger width) and V_t (may impact either standby power or read speed) are often necessary to ensure functionality of the SRAM array in the intended voltage range.

12.1.2 Other Bistable MOSFET SRAM Cells

As explained in Section 12.1.1, the storage element in a CMOS SRAM cell is the bistable latch consisting of two cross-coupled CMOS inverters. Other bistable latches can be employed as storage elements to make other types of SRAM cells. Figure 12.9 shows schematically three of the common types that are in use in applications built using older generations of technologies. The access transistors, not shown, are connected in the same manner as in a full CMOS SRAM cell. In Figure 12.9(a), the depletion-mode nMOSFETs are normally on devices with a negative V_t, in contrast to the normally off enhancement-mode nMOSFETs with a positive V_t. In Figure 12.9(c), the thin-film transistors (TFT) are low-mobility MOSFETs made in polysilicon film. A TFT-load cell offers density advantage since the p-channel TFTs can be stacked on top of the regular nMOSFETs in the cell. The noise margin for these storage elements

Table 12.1. Comparison of the characteristics of SRAM cells

	Full CMOS	**Depletion Load**	**Resistor Load**	**TFT Load**
Standby current	Low	High	Medium	Low
Cell stability	High	High	Low	Medium
Cell density	Low	Medium	High	High

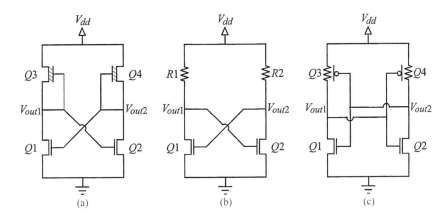

Figure 12.9 Schematics of three other bistable storage elements that can be used to form SRAM cells. (a) A depletion-load cell, where $Q3$ and $Q4$ are depletion-mode nMOSFETs. (b) A resistor-load cell, where $R1$ and $R2$ are high-resistance resistors. (c) A TFT-load cell, where $Q3$ and $Q4$ are p-channel thin-film transistors

can be analyzed in a similar manner, and the standby current is determined by the off current of the load devices ($Q3$ and $Q4$ in the depletion-load cell and the TFT-load cell, and $R1$ and $R2$ in the resistor-load cell). Table 12.1 gives a comparison of these SRAM cells with the full CMOS SRAM cell (Itoh, 2001). The only cell that has a noise margin comparable to the full CMOS SRAM cell is the depletion-load cell. However, its standby current is much too large for modern VLSI applications where the number of SRAM cells on a chip can be larger than 20 MB. The resistor-load cell inherently has an inferior noise margin compared to the full CMOS cell (Seevinck et al., 1987). From a process complexity point of view, the full CMOS SRAM cell is free in that it is fabricated using a CMOS logic process without modification. With best noise margin and standby power characteristics, the full CMOS cell is the SRAM cell of choice in high performance scaled technologies.

12.1.3 Bipolar SRAM Cell

The storage element in a conventional bipolar SRAM is a bistable latch shown in Figure 12.10(a). Each bipolar inverter consists of one bipolar transistor and a load resistor. In normal operation, one of the transistors in the latch is on while the other transistor is off. Let us assume transistor $Q1$ to be on. There is a relatively large current

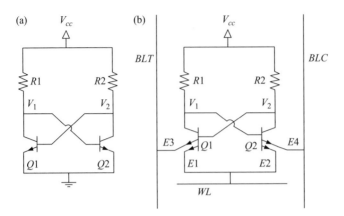

Figure 12.10 (a) A bipolar latch. (b) An emitter-coupled bipolar SRAM cell

flow through resistor $R1$. The IR drop in $R1$ means that the base voltage (which is V_1) of transistor $Q2$ is low, keeping $Q2$ in the off state. Similarly, if $Q2$ is on, the IR drop in $R2$ keeps the base voltage (which is V_2) of $Q1$ low and hence keeps $Q1$ in the off state.

To form an SRAM cell, each bipolar transistor in the latch is coupled to a bitline through an additional emitter. This emitter-coupled bipolar SRAM cell is illustrated in Figure 12.10(b). Instead of having just one emitter, each of the bipolar transistors in the latch now has two emitters, one for forming the basic bistable latch, and one for coupling to a bitline. The emitters of $Q1$ are labeled $E1$ and $E3$, and the emitters of $Q2$ are labeled $E2$ and $E4$. $E1$ and $E2$ are used to form the bistable latch, while $E3$ and $E4$ couple the true bitline BLT and the complement bitline BLC, respectively, to the latch.

The operation of an emitter-coupled bipolar SRAM cell is based on the fact that *the current flow in a multi-emitter transistor is carried primarily by the emitter that has the largest base–emitter forward bias voltage* V_{BE}. Consider the transistor $Q1$ in Figure 12.10(b). Let us assume both emitters $E1$ and $E3$ to have the same area. Let V_{BE1} and V_{BE3} be the base–emitter forward bias voltage of $E1$ and $E3$, respectively. The collector current is carried by the two emitters. The portion of collector current in $E1$ is proportional to $\exp(qV_{BE1}/kT)$, and the portion in $E3$ is proportional to $\exp(qV_{BE3}/kT)$. At room temperature, the collector current changes by $10\times$ for 60 mV change in V_{BE}. Therefore, if $V_{BE1} - V_{BE3} > 60$ mV, the collector current can be assumed to be carried entirely by emitter $E1$. Similarly, if $V_{BE3} - V_{BE1} > 60$ mV, the collector current can be assumed to be carried entirely by emitter $E3$.

There are other means of coupling a bipolar latch to bitlines to form an SRAM cell. However, the emitter-coupled cell is the simplest because it can be built using just bipolar transistors and resistors, the same as used for building fast bipolar logic circuits. Furthermore, bipolar SRAM is now used only in niche applications where process simplicity is more important than power dissipation and/or density. As a result, even though the emitter-coupled cell is not as low power as some other cells (Lynes and Hodges, 1970), it is the most commonly used bipolar SRAM cell. Here we discuss the behavior of a bipolar transistor in the operation of an emitter-coupled cell. The reader is referred to the literature for discussion on other bipolar memory cells (Wiedmann and Berger, 1971; Farber and Schlig, 1972; Hodges, 1972; Nokubo *et al.*, 1983). In addition, it was pointed out in Section 11.8.2 that the emerging technology of symmetric lateral bipolar transistors on SOI may lead to rethinking of bipolar for SRAM. Only time will tell how this novel lateral-bipolar SRAM technology will be used in applications.

12.1.3.1 Bipolar Transistor as an Ideal On–Off Switch

In considering bipolar circuits, designers find it convenient to think of a bipolar transistor as an ideal switch which is off when V_{BE} is less than V_{on} and on when V_{BE} is larger than V_{on}. This approach works well because the collector current of a bipolar transistor increases exponentially with V_{BE}, with the collector current changing $10\times$ for every 60 mV change in V_{BE} at room temperature. Once the desired on current has been established, the transistor current increases or decreases by large amounts for just a small change in V_{BE}, a property expected of an ideal switch. The exact value of V_{on} is determined by the target on current for the circuit application. For a modern silicon-base bipolar transistor, V_{on} is typically about 0.9 V, corresponding to a collector current density of about $1\,\text{mA}/\mu\text{m}^2$ (see Figure 9.8). [In the literature, V_{on} is often taken as 0.8 V or smaller (Meyer *et al.*, 1968). The smaller value for V_{on} in those older publications is primarily caused by the larger emitter areas, and hence smaller collector current densities, of the bipolar devices used at the time. For instance, if the emitter area is $10\,\mu\text{m}^2$, and the target on current is 1 mA, then the collector current density is only $0.1\,\text{mA}/\mu\text{m}^2$, and hence V_{on} can be taken as 0.84 V instead of 0.9 V.]

12.1.3.2 Operation of a Resistor-Load Bipolar Inverter

Let us consider one of the inverters used to build the bistable latch in Figure 12.10(a). This inverter by itself is shown in Figure 12.11(a). It consists of an n–p–n bipolar transistor Q and a collector resistor R_C, connected between power supply V_{cc} and

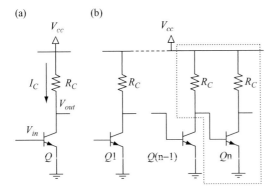

Figure 12.11 (a) An isolated bipolar inverter. (b) A chain of bipolar inverters

ground. This inverter is the basic building block for forming a *direct-coupled transistor logic* (DCTL) circuit (Meyer *et al.*, 1968). For simplicity, we assume the parasitic resistances of the transistor to be negligible. We will return to discuss the effect of parasitic resistances later. It is apparent from Figure 12.11(a) that the output voltage V_{out} is always lower than V_{cc} by $I_C R_C$. When the input voltage V_{in} is much smaller than V_{on}, I_C is negligibly small and V_{out} approaches V_{cc}. As V_{in} is increased, I_C increases exponentially with V_{in}, and V_{out} decreases in proportion to I_C. Above a certain V_{in} value, we have V_{out} lower than V_{in}. For proper operation in a logic circuit, the bias condition must be such that the logic swing, which is equal to $V_{in} - V_{out}$ when the transistor is on, has an adequate noise margin for the intended application.

Consider an inverter chain shown schematically in Figure 12.11(b). If we assume transistor Qn to be on, then transistor $Q(n - 1)$ is off. In this case, the current flowing through the load resistor of inverter $n - 1$ is not the collector current of transistor $Q(n - 1)$, but the base current of transistor Qn. In other words, the load resistor of inverter $n - 1$ determines the bias current for the base node of inverter n. The quiescent voltage of an inverter in the chain is determined by the circuit configuration enclosed within the dotted line in Figure 12.11(b). We consider the voltages of this circuit configuration next.

From Figure 12.11(b), the biasing scheme for a bipolar inverter in an inverter chain is shown in Figure 12.12. The emitter is held at ground potential, i.e., $V_E = 0$. I_B is the base current and I_C is the collector current. We have

$$V_{BE} \equiv V_B - V_E = V_{cc} - I_B R_C, \tag{12.1}$$

and

$$V_{CE} \equiv V_C - V_E = V_{cc} - I_C R_C. \tag{12.2}$$

For a given combination of V_{cc} and R_C, Eqs. (12.1) and (12.2) relate the transistor currents I_B and I_C to the transistor terminal voltages V_{BE} and V_{CE}. These relationships

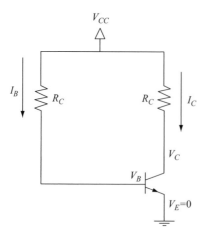

Figure 12.12 Biasing of a basic bipolar inverter in an inverter chain. There is no external resistor at the emitter terminal

are rather complex because the currents themselves are functions of the terminal voltages. The difference between V_B and V_C gives the logic swing ΔV of the inverter circuits in the chain, i.e.,

$$\Delta V \equiv V_B - V_C = V_{BE} - V_{CE} = (I_C - I_B)R_C. \tag{12.3}$$

For proper circuit operation, $\Delta V > 0$ or $V_{BE} > V_{CE}$. That is, a bipolar transistor in an inverter circuit is operated in the saturation region. It can be shown (see Exercise 12.1) that I_C remains much larger than I_B even in deep saturation. Therefore, the logic swing is, to first order, given by the $I_C R_C$ drop. In the rest of this subsection, we examine the characteristics of a transistor operated in saturation and the dependence of the inverter terminal voltages on load resistance and power supply voltage.

- *Current–voltage characteristics in the saturation region for a vertical transistor.*
  The current equations derived in Chapters 9–11 are for a transistor operated in the nonsaturation region. Here we modify them for the saturation region so that they are applicable to a bipolar inverter circuit. With V_B larger than V_C, the collector–base diode is forward biased, causing an electron current to flow from the collector to the emitter and a hole current to flow from the base into the collector. The complete collector current in the Ebers–Moll model, i.e., Eq. (9.143), can be rearranged to give

$$
\begin{aligned}
I_C &= \alpha_F I_{F0}\left(e^{qV_{BE}/kT} - 1\right) - I_{R0}\left(e^{qV_{BC}/kT} - 1\right) \\
&= I_{COF}\left(e^{qV_{BE}/kT} - 1\right) - (I_{COR} + I_{B0R})\left(e^{qV_{BC}/kT} - 1\right) \\
&= I_{COF}\left(e^{qV_{BE}/kT} - 1\right) - (I_{COF} + I_{B0R})\left(e^{qV_{BC}/kT} - 1\right).
\end{aligned}
\tag{12.4}
$$

- In Eq. (12.4), $I_{COF} = \alpha_F I_{F0}$ is the saturated collector current in the forward active mode, I_{COR} is the saturated collector current in the reverse active mode, and I_{B0R} is the saturated base current in the reverse active mode. Notice that the base current in the forward-biased emitter–base diode does not contribute to the collector current. Also, in writing the third line of Eq. (12.4), we have used the reciprocity characteristics of the emitter and collector for electron injection into the base, which give $I_{COR} = I_{COF}$ (see Section 9.6.1). For a properly designed inverter, we have V_{BE} and V_{BC} much larger than kT/q, so that Eq. (12.4) can be simplified to

$$
\begin{aligned}
I_C &\approx I_{COF}e^{qV_{BE}/kT} - I_{COF}e^{qV_{BC}/kT} - I_{B0R}e^{qV_{BC}/kT} \\
&= I_{B0F}e^{qV_{BE}/kT}\left[\beta_0\left(1 - e^{-qV_{CE}/kT}\right) - \frac{I_{B0R}}{I_{B0F}}e^{-qV_{CE}/kT}\right],
\end{aligned}
\tag{12.5}
$$

where I_{B0F} is the saturated base current in the forward mode, and $\beta_0 = I_{COF}/I_{B0F}$ is the common-emitter current gain in the nonsaturation region. In a similar manner, we can write the base current approximately as

$$
\begin{aligned}
I_B &\approx I_{B0F}e^{qV_{BE}/kT} + I_{B0R}e^{qV_{BC}/kT} \\
&= I_{B0F}e^{qV_{BE}/kT}\left(1 + \frac{I_{B0R}}{I_{B0F}}e^{-qV_{CE}/kT}\right),
\end{aligned}
\tag{12.6}
$$

where the first term is for hole injection into the emitter and the second term is for hole injection into the collector. Notice that I_C is the difference of the forward and reverse electron currents, while I_B is the sum of the forward and reverse hole currents. Also, in general, we have $I_{BOR} > I_{BOF}$ for a vertical bipolar transistor because the forward base current sees a polysilicon emitter while the reverse base current sees a regular n^+ silicon region. In addition, the junction area for base current injection is significantly larger for the reverse mode than for the forward mode, as evident from the device cross-section in Figure 9.1(a). For some transistors, such as the one in Fig. 10.20(a), I_{BOR} can be ten times as large as I_{BOF} (Rieh *et al.*, 2005).

- V_{CE} *for a bipolar inverter.* Equations (12.1) and (12.6) can be combined to give

$$V_{cc} - V_{BE} = R_C I_{BOF} e^{qV_{BE}/kT} \left(1 + \frac{I_{BOR}}{I_{BOF}} e^{-qV_{CE}/kT} \right), \tag{12.7}$$

which can be rearranged to give

$$V_{CE} = -\frac{kT}{q} \ln \left\{ \frac{[\beta_0(V_{cc} - V_{BE})/R_C I_{COF} e^{qV_{BE}/kT}] - 1}{I_{BOR}/I_{BOF}} \right\}. \tag{12.8}$$

Similarly, Eqs. (12.2) and (12.5) can be combined to give

$$V_{cc} - V_{CE} = R_C I_{BOF} e^{qV_{BE}/kT} \left[\beta_0 \left(1 - e^{-qV_{CE}/kT} \right) - \frac{I_{BOR}}{I_{BOF}} e^{-qV_{CE}/kT} \right], \tag{12.9}$$

which can be rearranged to give

$$V_{BE} = \frac{kT}{q} \ln \left\{ \frac{V_{cc} - V_{CE}}{R_C [I_{COF}(1 - e^{-qV_{CE}/kT}) - (I_{BOR}/I_{BOF})(I_{COF}/\beta_0)e^{-qV_{CE}/kT}]} \right\}. \tag{12.10}$$

Equation (12.8) gives V_{CE} as a function of V_{BE}, while Eq. (12.10) gives V_{BE} as a function of V_{CE}. They can be used to determine V_{BE} and V_{CE} in terms of V_{cc}, R_C, and the transistor parameters I_{COF}, I_{BOR}/I_{BOF}, and β_0. The relationship between V_{BE} and V_{CE} with load resistance and V_{cc} as parameters is illustrated in Figure 12.13.

Let us examine Figure 12.13(b). It shows that increasing the load resistance from 2 KΩ to 20 KΩ decreases V_{CE} from about 0.022 V to about 0.015 V, while decreasing V_{BE} from about 0.90 V to about 0.84 V. Equation (12.2) suggests that the collector current is reduced by about 10×, consistent with the fact that V_{BE} is reduced by about 60 mV. The logic swing, given by Eq. (12.3), is also reduced by about 60 mV, but remains more than adequate. The power dissipation of the inverter circuit is reduced by about 10×. In general, the choice of the load resistance in a bipolar inverter circuit is based primarily on speed or drive current requirement. There is an approximately linear relationship between power dissipation and drive current.

- *Effect of parasitic collector and base resistances.* The effect of parasitic resistances on the device terminal voltages is discussed in Section 9.3.1. Here we extend the discussion to a bipolar inverter. If the device has an internal collector resistance r_c that is appreciable, then R_C in Eq. (9.2) should be replaced by $R_C + r_c$. The presence of r_c causes the collector current to be reduced from $\sim V_{cc}/R_C$ to

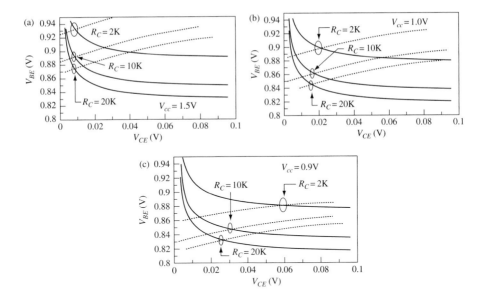

Figure 12.13 Simulated relationship between V_{CE} and V_{BE} for a bipolar transistor in an inverter chain at room temperature. The transistor parameters used in the simulation are: $\beta_0 = 100$, $I_{B0R}/I_{B0F} = 10$, and I_{C0F}, chosen such that the forward collector current I_{CF} is 1 mA at $V_{BE} = 0.9$ V. The V_{CE} curves (dashed) are given by Eq. (12.8) and the V_{BE} curves (solid) are given by Eq. (12.10). R_C is in units of Ω. (a) For $V_{cc} = 1.5$V, (b) for $V_{cc} = 1.0$ V, and (c) for $V_{cc} = 0.9$ V. For each value of R_C, the V_{BE} and V_{CE} values for the inverter are determined by the intersection (circled) of the V_{CE} and V_{BE} curves

$\sim V_{cc}/(R_C + r_c)$. The intrinsic V'_{CE} value remains very small, but the terminal V_{CE} value is now given by $V_{CE} = V'_{CE} + r_c I_C$. There should be little change in V_{BE} because even if r_c is sufficiently large to cause a $2\times$ decrease in I_C, V_{BE} is reduced by only about 18 mV. We can treat parasitic base resistance in a similar manner. If the device has an appreciable internal base resistance r_b, then R_C in Eq. (12.1) should be replaced by $R_C + r_b$. The base current is reduced somewhat, according to Eq. (12.1), but there should be little change to the intrinsic V'_{BE} value or to V_{CE}.

- *Effect of parasitic emitter resistance or an external emitter resistor.* For a given power supply voltage V_{cc}, the effective voltage across a bipolar inverter can be reduced by adding an external resistor to the emitter node. This is illustrated in Figure 12.14. An emitter resistor R_E has been added to the inverter. For a given value of V_{cc}, the IR drop in the emitter resistor has the effect of raising the emitter voltage V_E from ground to $R_E(I_C + I_B)$ above ground. As far as the inverter operation is concerned, the effective power supply is now $V_{cc} - V_E = V_{cc} - R_E(I_C + I_B)$. It can be inferred from comparison of the three V_{cc} cases in Figure 12.13 that reducing the power supply causes a larger decrease in V_{BC} than in V_{BE}, resulting in the transistor operating in a less saturated region. It is standard practice in bipolar circuit design to **add an external resistor to the emitter node to reduce saturation effect and hence improve circuit speed.** A bipolar SRAM cell

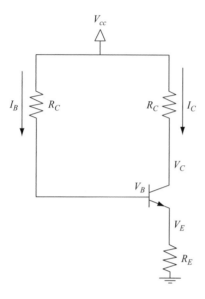

Figure 12.14 A bipolar inverter circuit having an emitter resistor

having an emitter resistor in the bistable latch is faster than one without an emitter resistor (Mayumi *et al.*, 1974).

12.1.3.3   Basic Operation of a Bipolar SRAM Cell

For simplicity, we choose a cell without an external emitter resistor, i.e., the cell in Figure 12.10(b). In standby mode, one of the transistors in the latch is on and one is off. Let us assume the on transistor has $V_{BE} = V_{on}$. To read a cell, we want to transfer the current in the on transistor from one emitter to another without upsetting the off transistor, and without upsetting the memory state of any of the other cells in the array. To write a cell, we want to force its off transistor to turn on by forcing its V_{BE} to become larger than V_{on}, without upsetting the memory state of any of the other cells in the array. There are many schemes for biasing the wordline and bitlines in standby and during read and write operations. In general, V_{cc} for the latch is typically more than 2 V above the standby wordline voltage, but the wordline standby voltage varies with design. The wordline in standby can be ground, positive, or even negative. Here, for simplicity, we assume the wordline standby voltage is ground, and examine how the bitlines and wordline should be varied relative to ground in order to operate the SRAM cell.

- *Standby mode and cell selection.* Referring to Figure 12.10(b), if $Q1$ is on, then $V_{BE1} = V_{on}$. In standby mode, the wordline is at ground while the bitlines are kept at a voltage higher than V_{on} to ensure that no significant amount of current flows to either bitline. The current in the latch flows through $E1$ to the wordline. There is no current flow in $E3$ because bitline *BLT* is at a voltage higher than V_{on}. There is no current flow in $Q2$ which is off. A cell is selected by raising its wordline voltage, reducing the voltage across the latch. As can be seen from the simulation results in Figure 12.13, both V_{BE} and V_{CE} of a transistor in an inverter are relatively insensitive to the voltage across the inverter. That is, to first order, *V_B of the on transistor and V_B of the off transistor are shifted upward by about the same amount as the wordline*. Now, consider the cells connected to a bitline. With the V_B of the transistors of the selected cell higher than V_B of the transistors of the nonselected cells, a voltage can be applied to the bitline to read or write the selected cell without upsetting the nonselected cells.

- *Read operation*. A cell is read by raising its wordline and lowering its bitlines to the point where the wordline becomes higher than the bitlines. Referring to Figure 12.10(b), if $Q1$ is on, raising the wordline voltage to above the bitlines causes $V_{BE3} > V_{BE1}$, forcing the current to transfer from $E1$ to $E3$ and a current to flow in bitline *BLT*. All the while, $Q2$ remains off and no current flows in bitline *BLC*. Thus, the state of the memory cell is read. At the end of the read operation, the wordline and bitlines are returned to their respective standby voltages.
- *Write operation*. Note that transistor $Q2$ can be turned on by forcing a current through one or both of its emitters $E2$ and $E4$. Assuming $Q1$ is on and we want to write the cell such that $Q2$ becomes on (and $Q1$ becomes off). The cell is selected by raising its wordline, but keeping it several kT/q below the bitline standby voltage. At this raised voltage, with bitlines at standby, the cells are not disturbed and no current flows in the bitlines. $Q2$ is then forced to turn on by lowering the voltage on bitline *BLC* so that V_{BE4} becomes larger than V_{BE1}, causing the flip-flop current to switch from $E1$ of $Q1$ to $E4$ of $Q2$. In this way, $Q2$ is forced to turn on while $Q1$ is forced to turn off. The write operation is complete after *BLC* is returned to standby.

12.2 Dynamic Random-Access Memory

A DRAM cell consists of one MOSFET and one capacitor. It is considerably smaller in silicon area than a CMOS SRAM cell which consists of six MOSFETs. The cell is in state "0" when there is no charge and in state "1" when there is charge in the capacitor. The charge stored in the capacitor leaks away over time if left alone. Therefore, periodic read and refresh cycles are necessary to restore the charge state ("0" or "1") in the cell. In this section, we describe the basic operation of a DRAM cell and discuss its design and scaling issues.

12.2.1 Basic DRAM Cell and Its Operation

A DRAM cell is shown schematically in Figure 12.15. The MOSFET Q is for accessing and transferring charge into and out of the capacitor C. The MOSFET is often referred to as the *access device* or the *transfer device*. It is usually an n-channel device. In a memory array arrangement, the gate electrode of Q is connected to the

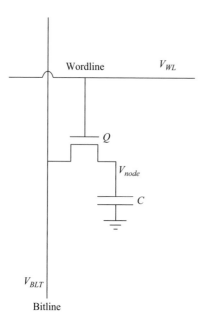

Figure 12.15 Schematic of a DRAM cell

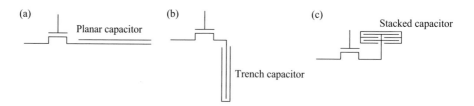

Figure 12.16 Schematics showing three DRAM cell structures: (a) planar-capacitor cell, (b) trench-capacitor cell, and (c) stacked-capacitor cell

wordline while the source and drain regions are connected to the capacitor and the bitline. In Figure 12.15, V_{node} denotes the voltage on the storage capacitor.

- *Cell structures.* The storage capacitance is determined by the minimum memory cell charge required for data retention and read operation, which will be discussed in Section 12.2.2. A typical value for the storage capacitance per cell is 30 fF. If the capacitor is a simple planar structure constructed using silicon dioxide of 10 nm thick, then the capacitor will take up an area of $A_C = Ct_{ox}/\varepsilon_{ox} \sim 9 \ \mu m^2$. Such an area is excessively large for design rules of less than 1 μm. Since the generation of 4 Mb DRAM, built using 1.2 μm minimum lithographic features, either a trench capacitor buried deep beneath the silicon surface (Lu *et al.*, 1985, 1986) or a stacked capacitor constructed above the transfer device (Koyanagi *et al.*, 1978) has been used to significantly reduce the area taken up by the storage capacitor. DRAM cells using these capacitor structures are illustrated schematically in Figure 12.16. Also, insulators with a higher dielectric constant than SiO_2 are often used to increase the capacitance per area. Most of the effort in the development of DRAM technology is devoted to finding manufacturable means for making the cell area, particularly the storage capacitor area, small.

- *Write operation.* It is straightforward to write a "0" to a DRAM cell. Referring to Figure 12.15, all one has to do is to turn on the access device by applying V_{dd} to the wordline and then set $V_{BLT} = 0$ to discharge any charge stored on C until $V_{node} = 0$. In writing a "1," one would like to charge the storage capacitor to $V_{node} = V_{dd}$. However, it is not sufficient to just set both V_{WL} and V_{BLT} to V_{dd}. This is because the transfer device Q turns off, i.e., goes into subthreshold, when V_{node} reaches $V_{dd} - V_t$. When V_{node} rises above $V_{dd} - V_t$, the gate-to-source voltage of Q becomes $V_{gs} = V_{WL} - V_{node} < V_t$, and the charging rate drops precipitously. To ensure that V_{node} reaches V_{dd} during a write operation, the wordline must be *boosted* to $V_{WL} > V_{dd} + V_t$ so that Q stays on through $V_{node} = V_{BLT} = V_{dd}$. After the storage node is fully charged or discharged, the wordline is brought back to its standby voltage, thus turning Q off and isolating the storage capacitor from the bitline. (In most designs, the standby voltage for the wordline is zero. However, in some designs, the standby voltage for the wordline can be negative. See discussion in Section 12.2.2.)

- *Read operation.* The read operation of a DRAM cell is illustrated in Figure 12.17. Figure 12.17(a) shows the bitline (*BLT*) and bitline complementary (*BLC*) connections to a cross-coupled CMOS sense amplifier that not only senses the stored bit in the cell but also writes the same bit back to the cell at the end of the read cycle. The *BLC* acts to provide a reference voltage to make up the differential signal. It can be the bitline of another array of cells not on the same wordline as the cell to be read (see folded bitline in Figure 12.18). The waveforms of the voltages involved in a read operation are shown in Figure 12.17(b). First, both *BLT* and *BLC* are precharged to $V_{dd}/2$. Then the access transistor Q is turned on. The same boost of the wordline to $V_{WL} > V_{dd} + V_t$ is applied to ensure a full V_{dd} write-back later. If a "0" bit is stored in the cell, $V_{node} = 0$, the bitline voltage V_{BLT} is discharged to below $V_{dd}/2$ and a sense signal $V_s = V_{BLT} - V_{BLC} < 0$ is developed. If a "1" bit is stored in the cell, $V_{node} = V_{dd}$, the bitline voltage V_{BLT} is charged to above $V_{dd}/2$ and a sense signal $V_s = V_{BLT} - V_{BLC} > 0$ is developed. These are shown in Figure 12.17(b). Before turn-on of the wordline, the sense amplifier is biased in a neutral state with both V_{SP} and V_{SN} at $V_{dd}/2$. After a differential signal is developed between *BLT* and *BLC*, the sense amplifier is activated by setting V_{SN} to 0 and V_{SP} to V_{dd}. This turns the sense amplifier into a cross-coupled latch that must settle into one of the two stable states, depending on the polarity of the signal V_s. If $V_s < 0$, i.e., if $V_{BLT} < V_{BLC}$, then the latch settles into the state of $V_{BLT} = 0$ and $V_{BLC} = V_{dd}$. Conversely, if $V_s > 0$, the latch ends up in the other stable state, namely, $V_{BLT} = V_{dd}$ and $V_{BLC} = 0$. This also restores the cell voltage V_{node} back to either 0 or V_{dd}, the same as its starting value. After the read cycle, the wordline is turned off and all *BLT*, *BLC*, V_{SP}, and V_{SN} are returned to their neutral voltage of $V_{dd}/2$. The cell, now isolated, is back to the originally stored "0" or "1" state.

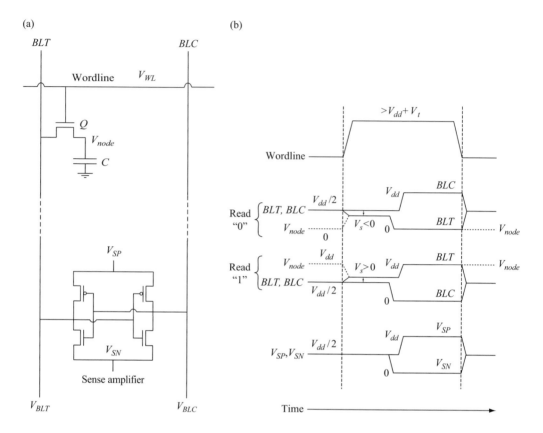

(a)

(b)

Figure 12.17 (a) Schematic connection of a DRAM cell to sense amplifier for read and write back. (b) Voltage waveforms for read and write back operation. After V_s is developed, V_{SN} is set to 0 first, which causes the lower of *BLT*, *BLC* to fall to 0. The higher of *BLT*, *BLC* is then pulled up from $V_{dd}/2$ to V_{dd} by setting V_{SP} to V_{dd}.

- *The read signal.* Let us assume that C_{cell} is the capacitance of the storage capacitor and $C_{bitline}$ is the capacitance associated with the bitline. Then the differential signal developed between *BLT* and *BLC* is given by

$$V_s = \left(V_{node} - \frac{V_{dd}}{2} \right) \frac{C_{cell}}{C_{cell} + C_{bitline}}, \qquad (12.11)$$

where $C_{cell}/(C_{cell} + C_{bitline})$ is referred to as the *transfer ratio*. To obtain a large read signal, V_{node} should be close to V_{dd} for a "1" state and close to zero for a "0" state, and $C_{cell}/C_{bitline}$ should not be too small. To minimize the chip area taken up by sense amplifiers and associated circuits, designers usually hang 256–1,024 cells on one bitline. The capacitance contribution of each unselected cell to the bitline can be estimated to be ~ 1 fF/μm (per device width) from the drain-to-gate fringe capacitance and the drain junction capacitance of an off-state MOSFET (Table 8.3). The total $C_{bitline}$ would then be of the order of 100–300 fF assuming an effective transfer device width of ~ 0.3 μm. For a typical DRAM design with a choice of

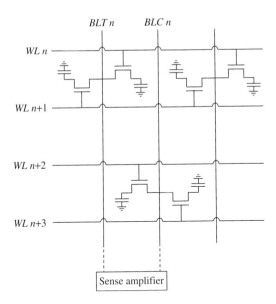

Figure 12.18 Schematic showing the cell arrangement in a folded-bitline architecture of DRAM array. Cells on adjacent bitlines are not on the same wordline

$C_{cell} = 30$ fF and a maximum $V_{node} = 1$ V, the transfer ratio is about 0.2 and the maximum V_s is about 100 mV.

- *Folded-bitline architecture.* Even though the maximum read signal may not be small, the DRAM circuits are usually designed to detect just a fraction of the signal before it is fully developed to achieve a short read time. Noise on a bitline can make it difficult to read a memory cell fast and reliably. To minimize the effect of noise, designers often employ the so-called *folded-bitline architecture*, illustrated schematically in Figure 12.18. In this case, the *BLC* is the bitline of another array of cells not connected to the same wordline as the cell to be read. Both *BLT* and *BLC* cross over the same wordline so that any voltage transient on the wordline is coupled equally to both bitlines. Thus, the signal being sensed is insensitive to noise originating from the wordlines. In a folded-bitline architecture, the minimum area taken up by one memory bit is $8F^2$, where $2F$ is the pitch of the wordlines and bitlines.

12.2.2   Device Design and Scaling Considerations for a DRAM Cell

Even though the storage capacitor may be fully charged in a write operation, leakage current can significantly degrade the signal by the time the cell is read. The design of a DRAM cell is primarily driven by the desire to achieve simultaneously the smallest cell possible consistent with the lithography ground rules and the specified read signal and data *retention time* requirements. Data retention time is the time interval between data refreshes. More frequent read and refresh cycles mean higher chip power.

A typical worst-case data retention time specified for a DRAM chip is about 100 ms. The retention time requirement sets the upper limit of the total leakage current allowed in a DRAM cell. For instance, for a capacitor storing 30 fC of charge (30 fF capacitor charged to 1 V), if we want to limit the charge loss to less than 10% between data refreshes, the maximum allowed leakage current from all sources is 30 fA. The insulator forming the storage capacitor is typically sufficiently thick so that the tunneling current through it is negligible. Also negligible is the diffusion-controlled reverse-biased junction leakage, of the order of 10^{-16} A/μm^2 at 100 °C (Section 3.1.8), as the junction area is only a fraction of 1 μm^2. The leakage current requirement mainly affects the design and scaling of the transfer device, which are discussed here.

- *Threshold voltage of the transfer device.* It was discussed in Section 6.3.1.2 that the MOSFET current at threshold is approximately 10^{-7} A for $W = L = 0.1\,\mu$m, insensitive to temperature, and that the inverse slope of subthreshold current is 100 mV/decade at 100 °C. To satisfy the 30 fA off-current requirement, the threshold voltage of the transfer device would have to be at least 0.6 V at 100 °C or 0.7 V at 25 °C. This V_t value for the transfer device cannot be reduced in scaling unless the retention time requirement is reduced. In some designs, a transfer device with a natural V_t of less than 0.7 V can be used as long as the wordline at standby is held at a voltage more than 0.7 V below the natural V_t. This can be accomplished by holding the wordline at a negative voltage in standby.
- *Gate insulator thickness of the transfer device.* The gate insulator of the transfer device must be sufficiently thick so that the gate leakage current is less than 30 fA. Under the standby condition, the gate is low and the MOSFET is off. Electrons could tunnel from the gate to the positively biased drain if charge is stored (a "1" state). If we assume a gate-to-drain tunneling area of 0.1×0.01 μm^2, then the gate leakage current should be less than 10^{-3} A/cm^2. Figure 4.41 suggests that the gate oxide should be thicker than ~2 nm for $V_{dd} = 1$ V. This means that the DRAM transfer device cannot be scaled to as short a gate length as the high-performance logic device, which can use a 1 nm thick gate oxide (see Figure 6.22).

12.3   Nonvolatile Memory

In theory, any bistable device that retains its state when the power supply is disconnected can make a nonvolatile memory cell. If the memory cell cannot be reprogrammed, e.g., a *fuse* (which changes from a conducting state to a nonconducting state in programming) or an *antifuse* (which changes from a nonconducting state to a conducting state in programming), it is called a *programmable read-only memory* (PROM). A nonvolatile memory that can be erased and reprogrammed is called an *erasable programmable read-only memory* (EPROM). In the literature, the term EPROM includes nonvolatile memories that are reprogrammable but the memory erasure is done by nonelectrical means, e.g., by exposure to ultra-violet light.

A nonvolatile memory that can be programmed and erased electrically is called an *electrically erasable and programmable read-only memory* (EEPROM). Notice that, as suggested by their full names, these nonvolatile memories are read-only memories. This means that, when used in a system, these nonvolatile memories really function as storage for data and program codes, and not as memories (like SRAM and DRAM) for running computer program codes. The reason for this read-only restriction is, as will be evident in the discussions to follow, that these nonvolatile memories do not have the write and/or endurance properties like SRAM and DRAM.

The field of nonvolatile memory technology is extremely broad and rapidly evolving since many bistable elements or devices can retain their state when disconnected from their power supplies. The technical considerations of a nonvolatile memory technology are: (i) memory speed, which includes access time, program time, and erase time; (ii) memory retention time, which measures how long a memory bit retains its state after being programmed; (iii) memory endurance, which measures how many cycles a memory bit can be programmed and erased while still functioning properly; (iv) power, which includes the power dissipation in programming, accessing, and erasing a memory bit; (v) power supply voltages, which include the voltages needed for program and erase operations; (vi) memory cell size; and (vii) scaling properties of the memory technology. The choice of a nonvolatile memory technology depends on the application requirements and the cost involved. In this section, we discuss MOSFET-based nonvolatile memory devices and the basic principles of their operation. The reader is referred to the vast literature for further reading on the circuit and chip design aspects of the technology (e.g., Hu, 1991; Itoh, 2001, and the references therein).

12.3.1   MOSFET Nonvolatile Memory Devices

Figure 12.19 illustrates the basic principles of a MOSFET nonvolatile memory device. It is shown in Section 4.5.1 that the flat-band voltage depends on the amount and distribution of charge in the gate insulator. Let us assume that by some programming means we are able to inject a charge distribution $\rho_{net}(x)$ in the gate insulator (Figure 4.27). This charge distribution causes a shift in the gate voltage needed to maintain the flatband condition, which means a shift in the device threshold voltage. From Eq. (4.79), the shift in threshold voltage is given by

$$\Delta V_t = -\frac{1}{\varepsilon_{ox}} \int_0^{t_{ox}} x\rho_{net}(x)dx = -\frac{1}{C_{ox}} \int_0^{t_{ox}} \frac{x}{t_{ox}}\rho_{net}(x)dx, \tag{12.12}$$

where $x = 0$ is the gate–oxide interface and $x = t_{ox}$ is the oxide–silicon interface. The injected charge is trapped in the gate insulator with a retention time of years, without the need of a power supply.

Figure 12.19(b) shows schematically the $I_{ds}-V_{gs}$ characteristics before and after charge injection. For electron injections in an nMOSFET, ΔV_t is positive, i.e., the threshold voltage increases from $V_{t,low}$ to $V_{t,high}$ after charge injection. Typically, ΔV_t

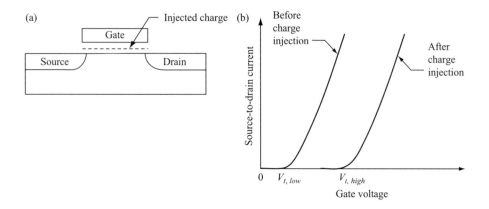

Figure 12.19 (a) Schematic diagram of a MOSFET nonvolatile memory device. (b) The MOSFET threshold voltage shifts from $V_{t,low}$ to $V_{t,high}$ after electron injection

can be a few volts. One can readily identify one of the states as logical "0" and the other logical "1." It is straightforward to read out the bit stored in the MOSFET by setting the gate voltage between $V_{t,low}$ and $V_{t,high}$. In one logical state, the MOSFET is on or conducting, and in the other, the MOSFET is off or nonconducting.

It is indicated in Figure 4.2 that the energy barrier for electron injection into SiO_2 is 3.1 eV, significantly lower than that for hole injection, 4.6 eV. As a result, MOSFET-based nonvolatile devices usually employ electrons instead of holes for memory programming and erasure. In programming, the threshold voltage is shifted in a positive direction by injecting electrons from the channel region into the gate insulator and storing part or all of the injected charge within the gate insulator. In memory erasure, the stored charge is neutralized, usually by tunneling electrons out of the gate insulator region. In the subsections here, we consider the device physics related to charge injection, charge storage, and charge erasure in a MOSFET-based nonvolatile memory.

12.3.1.1 Charge Injection

- *By hot electrons.* Electron injection from silicon into SiO_2 can be by tunneling or by hot electron injection. If an n-channel MOSFET is used, usually channel hot electron injection is employed. Typical gate current and channel current in an n-channel MOSFET are shown in Figure 12.20(a) and those in a p-channel MOSFET are shown in Figure 12.20(b) (Hsu *et al.*, 1992). Note that in both cases the gate current is an electron current.

  As discussed in Section 6.4.1, the substrate current is a direct measure of the amount of secondary carriers generated by impact ionization, which in turn is an indirect measure of the amount of primary hot carriers in the device channel region. From Figure 6.31, we see that the substrate current, hence the amount of secondary carriers, is largest at gate voltages slightly above threshold voltage. In the case of an n-channel MOSFET, this is the region where the voltage difference between the

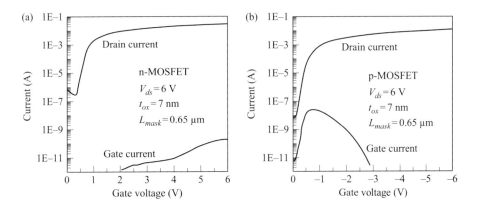

Figure 12.20 Typical drain and gate current characteristics in MOSFETs. (a) n-channel MOSFET. The gate current source is primarily the hot electrons flowing from source to drain. (b) p-channel MOSFET. The gate current is due to the injection of electrons generated via avalanche multiplication (after Hsu *et al.*, 1992)

gate and the drain, $V_{gs}-V_{ds}$, is negative [see Figure 5.2] and therefore does not favor injection of hot electrons into the gate insulator. Hot electron injection becomes more favorable as V_{gs} approaches or exceeds V_{ds}, but, as explained in Section 6.4.1, the maximum electric field in the silicon decreases in that case. This mismatch between the gate voltage for maximum electric field in the channel and the gate voltage for maximum electric field in the gate insulator for hot electron injection makes the channel hot electron induced gate current in an nMOSFET quite low and the injection process highly inefficient. For the example in Figure 12.20(a), the maximum electron current into the gate is about 10^{-10} A and the maximum I_g/I_{ds} ratio is only about 10^{-8}.

The situation is quite different when a p-channel MOSFET is used. In this case, electron injection is accomplished by injecting secondary hot electrons created by avalanche multiplication. Just like the nMOSFET in Figure 6.31, the pMOSFET substrate current, which is a measure of the amount of avalanche generated electrons, also increases with gate voltage and peaks at V_{gs} slightly above V_t (magnitude-wise). At low gate voltages where the substrate current is rising or at its peak, the voltage difference between the gate and the drain, $V_{gs}-V_{ds}$, is positive (i.e., V_{ds} more negative than V_{gs}) and therefore in favor of hot electron injection from the silicon into the gate insulator. That is, the gate voltage favorable for generating avalanche hot electrons is also favorable for injection of hot electrons into the gate insulator. The hot electron gate current in Figure 12.20(b) indeed has very similar dependence on gate voltage as the substrate current of nMOSFET in Figure 6.31. Note that at the gate voltages where the avalanche hot electron gate current rises and peaks, the drain current is low. As a result, avalanche hot electron injection from silicon into gate insulator in a p-channel MOSFET is much more efficient. For the example in Figure 12.20(b), the maximum gate current is about 10^{-8} A and the maximum I_g/I_{ds} ratio is about 10^{-4}.

Avalanche hot carrier injection can also be induced with the source of the MOSFET floating, i.e., without any channel current flowing. If a p-channel MOSFET source is left floating, the drain behaves like a p^+–n gated diode. As explained in Section 4.6.3, in a gated-diode mode, substrate carrier (electrons in a p-channel MOSFET) can be injected into the gate insulator if the gate is biased toward surface accumulation.

The injection currents shown in Figure 12.20 do not tell the whole story. The fact that there is avalanche hot electron injection in a p-channel MOSFET at low gate voltages suggests that there should also be avalanche hot hole injection at low gate voltages in an n-channel MOSFET. Indeed, avalanche hot hole injection in an n-channel MOSFET at low gate voltages can be measured using ultra-sensitive current monitors (Takeda *et al.*, 1983; Nissan-Cohen, 1986). The expected gate current dependence on gate voltage is illustrated in Figure 12.21. At low gate voltages, the electric field in the gate insulator favors the injection of avalanche hot holes. At somewhat higher gate voltages, the field in the gate insulator becomes less favorable for hot hole injection and more favorable for avalanche hot electron injection. As the gate voltage is increased further, the amount of hot carriers produced by avalanche multiplication actually decreases, as evidenced by the decrease in the substrate current in Figure 6.31, and the field in the insulator becomes even more favorable for hot electron injection. Thus, the gate current becomes more and more dominated by channel hot electron injection as the gate voltage increases. Similarly, for a p-channel memory device, we should see avalanche hot electron injection at low gate voltages and channel hot hole injection at high gate voltages.

In an n-channel nonvolatile memory device using channel hot electron injection for programming, the injection of secondary hot holes at low gate voltages can have unintended consequences on non-selected cells on the same bitline. The injected

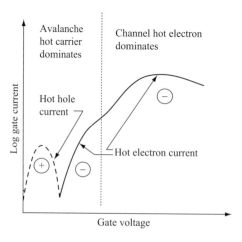

Figure 12.21 Schematic illustrating the injection of hot holes and hot electrons into the gate insulator region in an n-channel MOSFET as a function of gate voltage at large V_{ds}

positive charge shifts the V_t of non-selected devices in the negative direction, which, after repeated write cycles, may either unintentionally erase a previously programmed bit or cause device leakage when the V_t becomes too low. Care should be exercised to keep the wordline voltage of non-selected devices far enough below the threshold to avoid such "write disturbs" (Yamada *et al.*, 1996).

- *By Fowler–Nordheim tunneling.* Electron injection in nonvolatile memory devices can also be accomplished by Fowler–Nordheim tunneling, discussed in Section 4.6.1.1. As shown in Figure 4.39(b), electrons in the silicon conduction band tunnel through a triangular energy barrier into the oxide conduction band. The tunneling current density is a strong function of the electric field in the oxide, as plotted in Figure 4.40. A typical field for programming is in the range of 8–10 MV/cm, employing an oxide 10 nm thick. Too thick an oxide would require higher voltages for programming. Too thin an oxide would lead to direct tunneling and charge leakage. Typical programming time for a nonvolatile memory cell is on the order of 1 μs to 1 ms. Since there is no unintended current flow in the MOSFET during Fowler-Nordheim tunneling, *power dissipation in programming is much lower by Fowler-Nordheim tunneling than by hot electron injection.*

12.3.1.2 Charge Storage

- *In silicon nitride layer.* In theory, electron traps in silicon dioxide can be used for charge storage for nonvolatile memory applications. However, the capture efficiency, which is the product of the trap density and the capture cross-section, is just too small for silicon dioxide to be an effective electron storage medium for such an application (Ning and Yu, 1974). An oxide–nitride–oxide (ONO) composite layer that can store plenty of electrons is commonly used instead. Electrons are stored mostly in the nitride layer. If charge injection is uniform and if ΔQ is the stored charge per unit area in the thin nitride layer at an average distance t_q from the gate electrode, then the threshold voltage shift is $\Delta V_t = -t_q \Delta Q / \varepsilon_{ox}$ from Eq. (12.12). The effect on V_t in the case of nonuniform charge injection is discussed in Sections 12.3.3.5 and 12.3.3.6.

- *In floating gate.* Another commonly used technique for enhancing charge storage within the gate insulator region of a MOSFET is to embed a conductive *floating gate*, typically just a thin layer of polysilicon, in the gate insulator between the silicon and the gate electrode. Electrons are stored in the floating gate, illustrated in Figure 12.22. The stored charge spreads uniformly over the entire floating gate. When a floating gate is present, the usual gate electrode is referred to as the *control gate*, to differentiate it from the floating gate. The oxide between the floating gate and the silicon is called the tunnel oxide, typically about 10 nm thick. The oxide between the control gate and the floating gate is called the inter-poly oxide, typically about 20 nm thick. The overlap between the floating gate and the control gate can extend beyond the device region.

   The potential of the floating gate, V_{FG}, is determined by the stored charge and the capacitive coupling of the floating gate to other electrodes surrounding it and their voltages. From the capacitive equivalent circuit in Figure 12.22 (Wang, 1979),

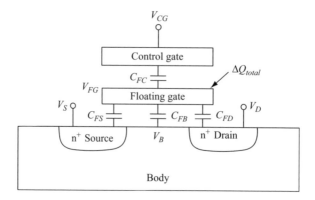

Figure 12.22 Capacitive coupling of the floating gate to other electrodes in an n-channel MOSFET nonvolatile memory device. Any depletion capacitance in the silicon body is absorbed in C_{FB}

$$\Delta Q_{total} = C_{FC}(V_{FG} - V_{CG}) + C_{FS}(V_{FG} - V_S) \\ + C_{FB}(V_{FG} - V_B) + C_{FD}(V_{FG} - V_D). \tag{12.13}$$

Therefore,

$$V_{FG} = \frac{\Delta Q_{total} + C_{FC}V_{CG} + C_{FS}V_S + C_{FB}V_B + C_{FD}V_D}{C_{FC} + C_{FS} + C_{FB} + C_{FD}}. \tag{12.14}$$

One can define coupling factors of the various electrodes to the floating gate by the ratio of their individual capacitances to the total capacitance in the denominator. The presence of ΔQ_{total} shifts the floating gate potential by $\Delta V_{FG} = \Delta Q_{total}/ (C_{FC} + C_{FS} + C_{FB} + C_{FD})$. From the control gate point of view, an additional $\Delta V_{CG} = -\Delta Q_{total}/C_{FC}$ needs to be applied to offset the effect of the stored charge and restore V_{FG} to the zero-charge value. In other words, the threshold voltage shift due to the stored charge ΔQ_{total} on the floating gate is

$$\Delta V_t = -\frac{\Delta Q_{total}}{C_{FC}}. \tag{12.15}$$

Note that a larger control-gate coupling factor, $C_{FC}/(C_{FC} + C_{FS} + C_{FB} + C_{FD})$, gives more control of the floating gate potential to the control gate and yields a higher V_{FG} and therefore a higher field in the tunnel oxide for a given applied V_{CG}. High C_{FC}, however, means a smaller threshold voltage shift for a given charge injection.

12.3.1.3   Charge Erasure

In an EPROM, if erasure is desired, it can be accomplished by exposing the device to high-energy photons, such as ultra-violet light or X-rays. The high-energy photons excite the stored electrons in the gate insulator or in the floating gate to an energy level above the oxide conduction band, thus allowing the excited electrons to flow back into

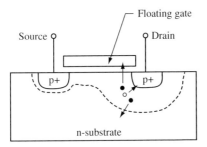

Figure 12.23 Schematic of a FAMOS device. The dotted line indicates the boundary of the depletion region

the silicon substrate. It is cumbersome to carry out such an operation for a packaged chip. In practice, EPROM is usually used in applications where the memory bits are to be programmed only once and never erased. Perhaps the simplest CMOS-compatible EPROM is the *floating-gate avalanche-injection MOS* (FAMOS) device (Frohman-Bentchkowsky, 1971). Structurally, it is simply a p-channel MOSFET with the gate floating, as shown schematically in Figure 12.23. Programming is accomplished by avalanche hot electron injection.

In EEPROM devices, erasure is done mostly by one of two ways. One way is by Fowler–Nordheim tunneling of electrons from the floating gate back to the control gate or to the source or drain region. In some cases, erasure is done by neutralizing the trapped electrons with hole injection. However, since hole injection is very inefficient, this erasure technique is not commonly used.

For erasure by tunneling back to the control gate, often the overlap between the control gate and the floating gate is engineered for enhanced electron tunneling. Similarly, for tunneling back to the source and drain regions, often the overlap between the floating gate and the source and/or drain region is engineered for enhanced electron tunneling.

For tunneling from a floating gate into the source and/or drain region, the source or drain is typically biased at a positive voltage with the control gate grounded. For an applied erasure voltage V_S to the source, the potential difference across the tunnel oxide, $V_S - V_{FG}$, is obtained from Eq. (12.14) with $V_{CG} = 0$ (ignoring the V_B, V_D terms),

$$V_S - V_{FG} = \left(1 - \frac{C_{FS}}{C_{FC} + C_{FS} + C_{FB} + C_{FD}}\right) V_S$$
$$- \frac{\Delta Q_{total}}{C_{FC} + C_{FS} + C_{FB} + C_{FD}}. \tag{12.16}$$

A small coupling factor from the source to the floating gate, $C_{FS}/(C_{FC} + C_{FS} + C_{FB} + C_{FD})$, is desired for achieving a high field in the tunnel oxide for erasure. Note that the stored negative charge adds to the field and helps the initial erasure speed. (By the same token, negative charge on the floating gate retards the field during programming and slows down the programming speed toward the end.) For an erasure voltage V_S in the range of 10–12 V, the erasure time is typically 0.1–1 s.

The common practice is to carry out the erasure operation for a large block of cells simultaneously. This is discussed in Section 12.3.2.

12.3.2   Flash Memory Arrays

So far we have discussed the basic program and erase operations of an individual bit in a nonvolatile memory array. In designing a nonvolatile memory array, it is desirable to be able to erase at once a block of memory bits or the entire array. In the case of UV-erased EPROM, the entire array is erased when it is exposed to radiation. In the case of EEPROM, this can be accomplished by connecting the bits to be erased together in an erasure process. Masuoka first proposed to provide a special erase gate to an EEPROM array such that "the contents of all memory cells are simultaneously erased by using field emission of electrons from a floating gate to an erase gate in a *flash*" (Masuoka *et al.*, 1984). Since then, flash erasure of various forms has become common in most if not all EEPROM designs. Furthermore, the terms flash, flash memory, and EEPROM are now used interchangeably.

12.3.2.1   Write, Read, and Erase Operations

The write, read, and erase operations of a typical stacked-gate flash memory array are illustrated in Figure 12.24. In a write or program operation, a large positive voltage is applied to turn on the selected wordline and a high drain voltage is applied to the selected bitline to generate hot electrons near the drain where they are injected onto the floating gate. In a read operation, the selected wordline is biased at a voltage (5 V in this example) between $V_{t,low}$ and $V_{t,high}$ depicted in Figure 12.19(b) and a positive voltage is applied to the selected bitline. The current in the bitline reflects the threshold voltage and therefore the state of charge storage in the cell. Note that all the non-selected cells on the same bitline are biased below $V_{t,low}$ to ensure that they are off regardless of their charge state. In an erase operation, all the sources are connected to a large positive voltage with all the control gates grounded. This causes the electrons stored on the floating gates to tunnel back to the source and erases a large block of cells simultaneously. Because the Fowler–Nordheim tunneling current is very sensitive to the field (Figure 4.40), a slight variation in the thickness of the tunnel oxide could lead to a large spread of the erasure speed. It is usually necessary to verify that all the bits are back to their un-programmed state after each erasure. Repetitive erasure operations are carried out if there are bits not completely erased the first time.

A common problem encountered in a flash memory array is *over erasure*. This happens when the number of electrons tunneled out of the floating gate is larger than the number of electrons injected into the floating gate during programming. It can also happen when holes tunnel or are injected into the floating gate (see Figure 12.21). Over erasure results in the threshold voltage being lower than $V_{t,low}$ in Figure 12.19(b) for an unprogrammed device, causing a higher device off current than intended. In a memory array, if many of the non-selected cells (e.g., those with zero wordline voltage in Figure 12.24(b)) on the same bitline as the selected cell are over-erased, their combined leakage current may be so large as to hinder sensing the on–off state of the selected cell.

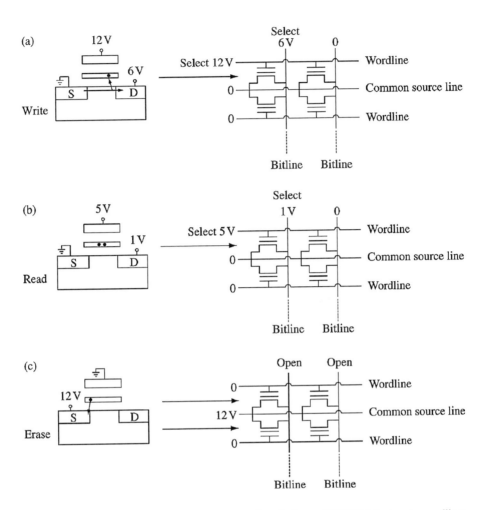

Figure 12.24 Schematics showing the connection of stacked-gate EEPROM devices to wordlines and bitlines in a memory array and their bias voltages for (a) write, (b) read, and (c) erase operations. This is a *NOR* array (after Itoh, 2001)

That is, over erasure can render parts of a memory array nonfunctional. Problems due to over erasure can be avoided by using a split-gate device (see Figures 12.28(b) and 12.29). With a sufficiently high threshold voltage on the single-gate part of the split-gate memory device, the cell off current can be controllably low, independent of the charge level in the floating gate. The disadvantage of a split-gate device is its larger area.

12.3.2.2 *NOR* and *NAND* Architecture

The cell array architecture shown in Figure 12.24 is called a *NOR* configuration, in which cells on the same bitline are connected in parallel and the unselected devices are turned off, i.e., with their wordlines biased below $V_{t,low}$ (Figure 12.19(b)). The device terminals, namely control gate, source, and drain, are all connected to voltage lines. Depending on the specific memory device design and structure, to be discussed further

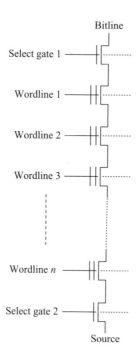

Bitline

Select gate 1

Wordline 1

Wordline 2

Wordline 3

Wordline n

Select gate 2

Source

Figure 12.25 Schematic showing the serial connection of EEPROM devices in one bitline in a *NAND* array

in Section 12.3.3, programming can be done by hot electron injection or by Fowler–Nordheim tunneling.

Another commonly employed nonvolatile memory array architecture is called the *NAND* configuration, shown in Figure 12.25. In a *NAND* configuration, memory cells on the same bitline are connected in series and the unselected devices are turned on, i.e., with their wordlines biased above $V_{t,high}$ (Masuoka *et al.*, 1987; Itoh *et al.*, 1989). In fact, $V_{t,low}$ is often made negative so unprogrammed devices are normally on nMOSFETs. For read operation, the selected wordline is biased between $V_{t,low}$ and $V_{t,high}$ which can be zero if $V_{t,low} < 0 < V_{t,high}$.

Write and erase of a *NAND* array are done by Fowler–Nordheim tunneling. For write, the selected wordline is biased at a high positive voltage, say 20V, while the unselected wordlines are biased moderately above $V_{t,high}$, say 10V. To program a cell for charge injection, its bitline is grounded so that a high field for tunneling is established in the tunnel oxide. The field in the unselected cells on the same bitline is not high enough for charge injection. If no charge injection to the selected cell is desired, the bitline is biased positively at, say, 10V, so there is no high field in any of the cells on that bitline. For block erase, a high positive voltage of 20V is applied to the bitline and the substrate (to avoid junction breakdown) with all the wordlines tied to zero voltage to tunnel the stored electrons back to silicon.

Since no contacts to the source and drain of the memory devices are needed in a *NAND* architecture, it has significant density advantage over the *NOR* configuration. However, with the memory bits connected in series, the read current in a *NAND* array is low, resulting in relatively long access times for the memory array. NAND flash is used primarily for data storage. In a typical application, data from NAND flash are usually transferred to SRAM or DRAM for program code execution by a computer processor. NOR flash typically has an access speed comparable to SRAM or DRAM, and hence can be used directly for code execution.

12.3.2.3 Endurance

The programming and erasure of an EEPROM device involve electric fields that are much higher than encountered in the normal operation of a MOSFET. As discussed in Sections 4.6 and 6.4, device degradation accompanies hot-carrier injection and electron tunneling into the gate oxide of a MOSFET. In the case of an EEPROM device, oxide degradation results in collapsing of the memory window (the threshold voltage difference between the programmed state and the erased state) as the device goes through many repetitive program and erase cycles. This is illustrated schematically in Figure 12.26. The endurance of an EEPROM device is measured in terms of the number of program and erase cycles before the memory window is reduced to the point of inadequate margin. For a device that passes electrons through the same oxide location during both programming and erasure, the endurance is the lowest, often in the range of 10^3 to 10^4 cycles. The endurance can be improved by passing electrons in programming and in erasure through different oxide locations (e.g., programming through the drain and erasure through the source). Most flash memories on the market have endurance specifications of 10^3 to 10^6 cycles.

It is also important to characterize the data retention of an EEPROM device after many program and erase cycles. As discussed in Section 4.6, when defects build up in an oxide layer, the oxide layer becomes leaky and its data retention characteristics degrade.

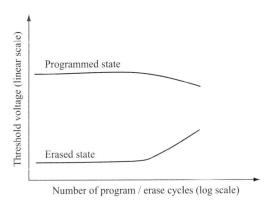

Figure 12.26 Schematic illustrating the collapse of the memory window as a function of the number of program and erase cycles

12.3.2.4 Modern NAND Flash Technologies

One of the most amazing developments in VLSI technologies over the past 10+ years is the wide adoption of *Solid State Drive* (SSD), as a new data storage medium or as a replacement of magnetic *Hard Disk Drive* (HDD), for data storage in all sorts of electronic systems, ranging from consumer systems, like mobile phones and personal computers, to enterprise computing and communication systems, like servers in data centers. The backbone of these SSD products is the floating-gate EEPROM device connected in the NAND architecture (Figure 12.25). Many innovations in the fabrication process for NAND flash chips and in circuit and system design have contributed to the dominance of SSD for data storage. These NAND flash innovations result in a steady but impressive improvement of chip-level NAND flash memory bit density and chip-level memory capacity over time, while overcoming the endurance limitation inherent in NAND flash memory devices (see Section 12.3.2.3). The concepts behind three of these innovations are discussed briefly here.

- *Multi-bit per cell.* Referring to Figure 12.19, the two trapped-charge states of the MOSFET, one with a distribution of V_t centered around $V_{t,low}$ and one with a distribution of V_t centered around $V_{t,high}$, are used to represent logic "0" and logic "1," respectively, of a memory bit. Thus, one memory bit takes up one MOSFET device. It is a *single level cell* (SLC). The operation margin between "0" and "1" is represented by the smallest separation between the tail of the V_t distribution for "0" and that for "1." This operation margin is a function of the process technology and the memory chip and system design, and is smaller than the voltage difference $\left(V_{t,high} - V_{t,low}\right)$.

  Employing advanced memory controllers and using sophisticated charge-level sensing techniques, NAND flash designers have been able to divide the total trapped charge, represented by the voltage difference $\left(V_{t,high} - V_{t,low}\right)$ to represent more than one memory bit. For example, a MOSFET device designed to represent two bits, referred to as a *multi-level cell* (MLC), would have four logic states: "0,0," "0,1," "1,0," and "1,1." Again, each logic state would be represented by a narrow distribution of V_t centered around a corresponding V_t. Thus, four narrow V_t distributions, each centered around $V_{t,00}$, $V_{t,01}$, $V_{t,10}$, and $V_{t,11}$ would represent the logic states "0,0," "0,1," "1,0," and "1,1," respectively, of an MLC.

  If the same floating-gate MOSFET is employed in a SLC and in an MLC, it is apparent that the number of stored electrons representing one bit is smaller for an MLC design than for a SLC design. Also, the operation margin for an MLC design is also smaller than that for a SLC design. As a result, MLC NAND flash are typically slower and more susceptible to programing error than SLC NAND flash. Thus, MLC NAND flash and SLC NAND flash do not necessarily address the same application market.

  As of this writing, NAND flash products based on three bits per cell, referred to as *tri-level cell* (TLC), have been in production for several years already. And products based on four bits per cell, referred to as *quad-level cell* (QLC), are just coming onto the market. Such multi-bit per cell technologies is one factor driving up the memory bit density at the chip level, and hence driving down the cost per bit for NAND flash memory.

- *3D NAND*. Besides relying on the development of ever finer patterning techniques, the semiconductor industry has long been pursuing three-dimension (3D) integration as a means for increasing chip-level device and circuit density, for system performance improvement and/or for cost reduction. Thus far, NAND flash process developers are the only ones that have mastered the art of chip-level 3D integration in volume manufacturing.

  A typical 3D NAND fabrication process (Fukuzumi *et al.*, 2007; Ishiduki *et al.*, 2009; Kim *et al.*, 2009) starts with a regular silicon wafer which contains the first layer of memory cells together with control logic circuits. Multiple layers of memory cells are then added on top of each other and on top of the base-layer arrays, along with interconnections between the layers. Some manufacturers combine 3D NAND with multi-bit-per-cell designs, resulting in additional chip-level bit density. The greatly increased chip-level bit density enables 3D NAND to achieve much shorter interconnect, and hence much higher speed than conventional 2D NAND.

  The first generation SSD based on 24-layer 3D NAND MLC technology was introduced around 2013. As of this writing, SSD based on 3D NAND with 100+ layers are available on the market.

- *Wear leveling.* A special feature of flash memories is that a memory chip is organized into erase blocks. As discussed in Section 12.3.2.3 , the typical endurance of a NAND flash bit is only 1,000–10,000 cycles. When a flash memory system is in use, an erase block that sees a high concentration of write cycles would tend to have bits that fail to function properly sooner than blocks that see fewer write cycles. Flash memory controllers are often designed with a *wear leveling* function to work around this limitation by arranging data so that write and erase are distributed evenly across the entire flash memory chip and/or system. In this way, no single erase block fails much sooner, due to the high concentration of write cycles, than other erase blocks. The memory controller keeps track of the wear of each block and the movement of data across segments of the memory chip and/or system. With wear leveling, SSD can achieve effective endurance of 100,000 cycles, adequate for many industrial or enterprise applications.

12.3.3   Devices for a *NOR* Array

As discussed in Section 12.3.2.2 and shown in Figure 12.24, the memory device for a NOR flash array is operated with voltages connected to its control gate, source, and drain terminals. Programming can be done by hot electron injection or by Fowler-Nordheim tunneling. Many NOR flash devices have been developed over the years. Since NOR flash can be read fast enough for program code execution, it is often embedded, i.e., integrated with CPU and SRAM on the same chip. The selection of one device over another is governed by many factors, such as read and write speed, power dissipation in programming, compatibility with the CMOS logic process, including the ability to scale with the CMOS logic platform, and cost. Here we focus only on several devices that can be used to highlight the basic physics and operation principles of nonvolatile devices suitable for a *NOR* array.

12.3.3.1 Floating-Gate Devices with Enhanced Tunneling

To reduce the voltage needed for erasure in floating-gate MOSFETs, designers sometimes use an insulator that is easier for electrons to tunnel through between the device channel and the floating gate, or between the floating gate and the control gate. For example, a silicon-rich oxide can be used to enhance tunneling (Hsu *et al.*, 1992). Another technique to enhance tunneling from a floating gate is to use a thinner oxide in an extended overlap area between the floating gate and the source diffusion (Johnson *et al.*, 1980; Lai *et al.*, 1991). Figure 12.27 illustrates one such device concept, with programming by channel hot electron injection near the drain region and erasure by electron tunneling from the floating gate to the source region.

12.3.3.2 Devices Using Source-Side Injection

As shown in Section 12.3.1.1, channel hot electron injection in nMOSFET is a very inefficient process, with only a tiny fraction of the channel current actually injected into the gate insulator. The injection efficiency can be greatly enhanced by using the device structures shown in Figure 12.28. The structure in Figure 12.28(a) has a sidewall floating gate at the source end (Wu *et al.*, 1986). The device is in effect

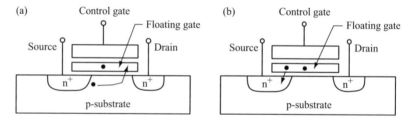

Figure 12.27 Schematic diagrams of a floating-gate nonvolatile memory device, with (a) programming by channel hot electron injection, and (b) erasure by electron tunneling from floating gate to source

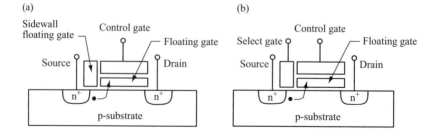

Figure 12.28 Schematics of floating-gate devices using source-side hot electron injection. (a) A stacked-gate device having a second floating gate at the source end. (b) A stacked-gate device having a second gate (select gate) at the source end

two MOSFETs in series. The left (source side) device has a floating gate but no control gate, and the right (drain side) device is a stacked-gate device. The sidewall floating gate is weakly coupled to the control gate. A positive control gate voltage causes a relatively strong surface inversion in the stacked-gate device and a relatively weak surface inversion in the floating-gate device. Being strongly inverted, the surface channel of the stacked-gate device behaves like an extended drain for the floating-gate device. When a large drain voltage is applied, the peak electric field in the silicon is located in the stacked-gate channel region but close to the sidewall floating-gate. Therefore, as the sidewall floating-gate device is turned on through coupling to the control gate, channel hot electrons are injected into the floating gate of the stacked-gate device at the source end. In the device structure shown in Figure 12.28(b), instead of left floating, the sidewall gate is contacted to form a second control gate, called the select gate in the figure (Naruke *et al.*, 1989). The sidewall device can be turned on and off independently of the stacked-gate device, giving an additional degree of freedom in the operation of the memory device.

For a given drain voltage applied to a source-side injection device, the maximum electric field in the silicon channel increases with the control-gate voltage. Also, as the control-gate voltage increases, the electric field extracting the hot electrons from the channel to the floating gate increases. This is in contrast to the drain-side injection case discussed in Section 12.3.1.1 where the maximum channel field for hot carrier production decreases with gate voltage (see Figure 6.31). Compared to a simple stacked-gate device with hot electron injection on the drain side, source-side injection devices have much better hot electron injection efficiency (Wu *et al.*, 1986; Naruke *et al.*, 1989).

12.3.3.3 Split- and Stacked-Gate Devices

If the select gate and the control gate in the device in Figure 12.28(b) are electrically tied together, e.g., by forming them with the same polysilicon layer, then we have a *split-gate* EEPROM device, shown schematically in Figure 12.29. The device consists of a regular MOSFET (device 1) and a stacked-gate MOSFET (device 2) in series with a common gate electrode. Several variations of the split-gate EEPROM device have

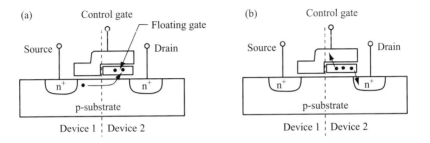

Figure 12.29 Schematic diagrams of a stacked-gate nonvolatile memory device having a split control gate. (a) Programming by channel hot electron injection. (b) Erasure can be accomplished by tunneling electrons from the floating gate to the drain region or to the control gate

been reported, each having a different overlap coupling between the control gate and floating gate and between the floating gate and source–drain diffusions. In the version closest to that illustrated in Figure 12.29 (Samachisa *et al.*, 1987), the channel region of device 1 behaves like a source extension for device 2, and channel hot electrons can be injected into the floating gate near the drain region. Erasure can be done by tunneling the electrons from the floating gate to the drain region. In another version, the floating gate is purposely made to couple strongly to the drain diffusion so that the maximum electric field in the silicon is located close to the source end of the floating gate, thus achieving source-side channel hot electron injection (Kianian *et al.*, 1994). Furthermore, by intentionally shaping the floating gate edges to enhance the electric field locally where the control gate overlaps the floating-gate edge, erasure by electron tunneling from the floating gate to the control gate can be accomplished.

12.3.3.4   MNOS Device Using Tunneling Injection

The simplest EEPROM device based on charge storage in the gate insulator region is one where programming and erasure are accomplished by tunneling electrons and/or holes into and out of a nitride layer. In the literature, such a device is often referred to as an MNOS (metal–nitride–oxide–semiconductor) device. Depending on whether it is a p-channel MOSFET or an n-channel MOSFET, electrons or holes can be injected from the device channel into the nitride layer for programming. In erasure, the stored charge can be driven to tunnel from the nitride layer either to the silicon or to the gate electrode. Since the charge is injected into the gate insulator region uniformly over the device channel, the effect on the surface potential is uniform over the entire channel region. Details of the transport of electrons and holes in MNOS devices can be found in the literature (e.g., Suzuki *et al.*, 1989, and the references therein).

12.3.3.5   MNOS Device Using Channel Hot Electron Injection

If channel hot electrons are used to program an MNOS device, then the stored charge is localized near the drain. (Channel hot electron effect is discussed in Section 6.4.1.) This is illustrated in Figure 12.30(a). The effect of stored charge on the surface potential in silicon is also localized. This makes the source and drain regions of the MOSFET asymmetric. Its threshold voltage then depends on the bias condition for the measurement. In the normal mode and with a high drain voltage, i.e., the same bias condition as that for programming, there is very little effect of the stored charge on the threshold voltage. This is because a MOSFET reaches threshold when the surface potential on the *source* side reaches $2\psi_B$ [Eq. (5.24)]. However, if the MOSFET is measured in the reversed source–drain mode, i.e., with the stored charge on the source side, the stored charge induces a retarding potential for electron flow from the source and there is a pronounced positive shift of the threshold voltage. These are schematically illustrated in Figure 12.30(b) (Abbas and Dockerty, 1975; Ning *et al.*, 1977b). Thus, the memory effect is readily recognizable when the device is read in the reverse mode. The memory can be erased either by injecting hot holes to neutralize the stored electrons or by tunneling the stored electrons back to the device channel region or to the drain region (Chan *et al.*, 1987c).

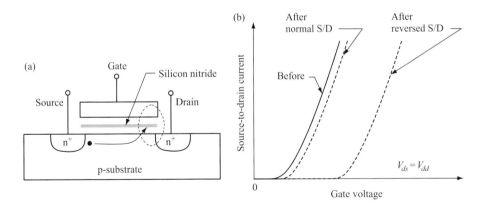

Figure 12.30 MNOS memory device using channel hot electron for programming. (a) Schematic of device cross-section showing electrons being stored near the drain end after channel hot electron injection. (b) Schematic showing the current–voltage characteristics at large V_{ds} for three situations, namely, prior to programming, after programming with the device operated in the normal mode, and after programming with the device operated in the source–drain-reversed mode

12.3.3.6   MNOS Device Storing Two Bits per Cell

The source and the drain are structurally identical in an unprogrammed MOSFET. Therefore, we can also inject electrons into the nitride layer near the source end by operating the MNOS device with the source and drain reversed. Furthermore, we can inject electrons into both the right (drain) end and the left (source) end of the nitride layer, thus creating two localized electron distributions, one at each end of the device channel. This is illustrated in Figure 12.31(a). As long as the two charge distributions do not overlap much, the two localized charge distributions can be considered as distinctly separate and hence can be used to represent two bits of memory information. In this way, two memory bits can be realized in one MNOS memory device (Eitan *et al.*, 2000). The expected current–voltage characteristics measured at large V_{ds} are illustrated schematically in Figure 12.31(b). The key point is, when measurement is made at large V_{ds}, the measured device threshold voltage is determined primarily by the amount of charge stored in the gate insulator near the diffusion region used as the source for the measurement. At large V_{ds}, the charge stored near the diffusion region used as the drain for the measurement has relatively little influence on the measured device threshold voltage.

12.3.3.7   Devices with Other Charge Storage Material

Other charge storage media besides silicon nitride have been explored. Among them, the most interesting one is to embed a thin region of nanocrystals of silicon within the gate insulator (Tiwari *et al.*, 1996). Charge is stored in the nanocrystals. The nanocrystals can enhance the tunneling of electrons into and out of the gate insulator. They can be used to implement the concept of two bits per device as well (Kim *et al.*, 2003). No nonvolatile memory product employing nanocrystals for charge storage has yet been developed.

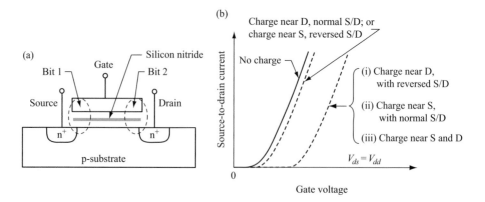

Figure 12.31 MNOS memory device using channel hot electron injection for programming. (a) Schematic of device cross-section showing electrons being stored at both the drain and the source ends. (b) Schematic showing the expected current-voltage characteristics measured at large V_{ds}

Exercise

12.1   Consider a bipolar transistor biased to operate in the saturation region. Ignoring parasitic resistances, the collector current is given by Eq. (12.5) and the base current is given by Eq. (12.6), namely,

$$I_C = I_{B0F}e^{qV_{BE}/kT}\left[\beta_0\left(1 - e^{-qV_{CE}/kT}\right) - \frac{I_{B0R}}{I_{B0F}}e^{-qV_{CE}/kT}\right],$$

and

$$I_B = I_{B0F}e^{qV_{BE}/kT}\left(1 + \frac{I_{B0R}}{I_{B0F}}e^{-qV_{CE}/kT}\right).$$

The current gain is

$$\beta \equiv \frac{I_C}{I_B} = \frac{\beta_0\left(1 - e^{-qV_{CE}/kT}\right) - \frac{I_{B0R}}{I_{B0F}}e^{-qV_{CE}/kT}}{1 + \frac{I_{B0R}}{I_{B0F}}e^{-qV_{CE}/kT}}.$$

Plot the current gain β as a function of qV_{CE}/kT from $qV_{CE}/kT = 0$ to $qV_{CE}/kT = 4$, using $\beta_0 = 100$ and $I_{B0R}/I_{B0F} = 1$. Repeat using $\beta_0 = 100$ and $I_{B0R}/I_{B0F} = 10$. This exercise shows that for practical transistors the current gain is less than 1 only for very small V_{CE} values.

References

Abbas, S. A. (1974). Substrate current – a device and process monitor, *IEEE IEDM Technical Digest*, 404–407.

Abbas, S. A. and Dockerty, R. C. (1975). Hot-carrier instability in IGFETs, *Appl. Phys. Lett.* **27**, 147–148.

Andrews, J. M. (1974). The role of the metal–semiconductor interface in silicon integrated technology, *J. Vac. Sci. Technol.* **11**, 972–984.

Andrews, J. M. and Phillips, J. C. (1975). Chemical bonding and structure of metal–semiconductor interfaces, *Phys. Rev. Lett.* **35**, 56–59.

Ansley, W. E., Cressler, J. D., and Richey, D. M. (1998). Base-profile optimization for minimum noise figure in advanced UHV/CVD SiGe HBTs, *IEEE Trans. Microwave Theory and Techniques* **46**, 653–660.

Arnaud, F., Zuliani, P., Reynard, J. P., *et al.* (2018). Truly innovative 28 nm FDSOI technology for automotive micro-controller applications embedding 16 MB phase change memory, *IEEE IEDM Technical Digest*, 18.4.1–18.4.4.

Arora, N. (1993). *MOSFET Models for VLSI Circuit Simulation*. Wien: Springer-Verlag.

Arora, N. D. and Gildenblat, G. S. (1987). A semi-empirical model of the MOSFET inversion layer mobility for low-temperature operation, *IEEE Trans. Electron Devices* **ED-34**, 89–93.

Asbeck, P. M. and Nakamura, T. (2001). Bipolar transistor technology: Past and future directions, *IEEE Trans. Electron Devices* **48**, 2455–2456.

Ashburn, P. (1988). *Design and Realization of Bipolar Transistors*. Chichester: Wiley.

Aufinger, K., Knapp, H., Boguth, S., Gerasika, O., and Lachner, R. (2011). Integrated injection logic in a high-speed SiGe bipolar technology, Proc. Bipolar/BiCMOS Circuit & Technology Meeting, *IEEE*, 87–90.

Auth, C., Allen, C., Blattner, A., *et al.* (2012). A 22 nm high performance and low power CMOS technology featuring fully-depleted tri-gate transistors, self-aligned contacts, and high density MIM capacitors, Symp. VLSI Technology Digest of Tech. Papers, *IEEE*, 131–132.

Baccarani, G. and Sai-Halasz, G. A. (1983). Spreading resistance in submicron MOSFETs, *IEEE Electron Device Lett.* **EDL-4**, 27–29.

Baccarani, G. and Wordeman, M. R. (1983). Transconductance degradation in thin-oxide MOSFETs, *IEEE Trans. Electron Devices*, **ED-30**, 1295–1304.

Bakoglu, H. B. (1990). *Circuits, Interconnections, and Packaging for VLSI*. Reading, MA: Addison-Wesley.

Balk, P., Burkhardt, P. G., and Gregor, L. V. (1965). Orientation dependence of built-in surface charge on thermally oxidized silicon, *IEEE Proc.*, **53**, 2133.

Bao, R., Hung, S., Wang, M., *et al.* (2018). Novel materials and processes in replacement metal gate for advanced CMOS technology, *IEEE IEDM Technical Digest*, 11.4.1–11.4.4.

Bardeen, J. (1947). Surface states and rectification at a metal semiconductor contact, *Phys. Rev.* **71**, 717–727.

Berger, H. H. (1972). Models for contacts to planar devices, *Solid-State Electron.* **15**, 145–158.

Berger, H. H. and Wiedmann, S. K. (1974). Terminal-oriented model for merged transistor logic (MTL), *IEEE J. Solid-State Electronics* **SC-9**, 211–217.

Berger, H. H. and Wiedmann, S. K. (1976). Complementary transistor circuit for carrying out Boolean functions, U.S. Patent 3,956,641.

Berglund, C. N. and Powell, R. J. (1971). Photoinjection into SiO_2: Electron scattering in the image force potential well, *J. Appl. Phys.* **42**, 573–579.

Berry, C., Bell, B., Jatkowski, A., *et al.* (2020). IBM z15: A 12-core 5.2GHz microprocessor, *IEEE ISSCC Digest of Technical Papers*, 54–56.

Bhavnagarwala, A. J., Tang, X., and Meindl, J. D. (2001). The impact of intrinsic device fluctuations on CMOS SRAM cell stability, *IEEE J. Solid-State Circuits* **36**, 658–665.

Blakemore, J. S. (1982). Approximations for Fermi-Dirac integrals, especially the function $F_{1/2}(\eta)$ used to describe electron density in a semiconductor, *Solid-State Electronics* **25**, 1067–1076.

Blat, C. E., Nicollian, E. H., and Poindexter, E. H. (1991). Mechanism of negative-bias-temperature instability, *J. Appl. Phys.* **69**, 1712–1720.

Böck, J., Schafer, H., Knapp, H., *et al.* (2004). 3.3 ps SiGe bipolar technology, *IEEE IEDM Technical Digest*, 255–258.

Brews, J. R. (1978). A charge sheet model of the MOSFET, *Solid-State Electron.* **21**, 345–355.

Brews, J. R. (1979). Threshold shifts due to nonuniform doping profiles in surface channel MOSFETs, *IEEE Trans. Electron Devices* **ED-26**, 1696–1710.

Brown. K. H., Grose, D. A., Lange, R. C., Ning, T. H., and Totta, P. A. (1992). Advancing the state of the art in high-performance logic and array technology, *IBM J. Res. Dev.* **36**, 821–828.

Bulucea, C. D. (1968). Diffusion capacitance of p–n junctions and transistors, *Electron. Lett.* **4**, 559–561.

Burghartz, J. N., Bartek, M., Rajaei, B., *et al.* (2002). Substrate options and add-on process modules for monolithic RF silicon technology, Proc. Bipolar/BiCMOS Circuit & Technology Meeting, *IEEE*, 17–23.

Burns, J. R. (1964). Switching response of complementary-symmetry MOS transistor logic circuits, *RCA Review* **25**, 627.

Cai, J. and Ning, T. H. (2006). SiGe HBTs on CMOS-compatible SOI. In *Silicon Heterostructure Handbook*, ed. J. D. Cressler. New York: CRC Press, 233–247.

Cai, J., Ajmera, A., Ouyang, C., *et al.* (2002a). Fully-depleted-collector polysilicon-emitter SiGe-base vertical bipolar transistor on SOI, Symp. VLSI Technology Digest of Tech. Papers, *IEEE*, 172–173.

Cai, J., Taur, Y., Huang, S.-F., Frank, D. J., Kosonocky, S., and Dennard, R. H. (2002b). Supply voltage strategies for minimizing the power of CMOS processors, Symp. VLSI Technology Digest of Tech. Papers, *IEEE*, 102–103.

Cai, J., Kumar, M., Steigerwalt, M., *et al.* (2003). Vertical SiGe-base bipolar transistors on CMOS-compatible SOI substrate, Proc. Bipolar/BiCMOS Circuit & Technology Meeting, *IEEE*, 215–218.

Cai, J., Ning, T. H., D'Emic, C., *et al.* (2011). Complementary thin-base symmetric lateral bipolar transistor on SOI, *IEEE IEDM Technical Digest*, 386–389.

Cai, J., Ning, T. H., D'Emic C., *et al.* (2013). SOI lateral bipolar transistor with drive current > 3 mA/μm, *IEEE SOI-3D-Subthreshold Microelectronics Technology Unified Conference (S3S)*, 1–2.

Cai, J., Ning, T. H., D'Emic C., *et al.* (2014). On the device design and drive-current capability of SOI lateral bipolar transistors, *IEEE J. Electron Devices Soc.* **2**, 105–113.

Cai, Y., Sato, T., Orshansky, M., Sylvester, D., and Hu, C. (2000). New paradigm of predictive MOSFET and interconnect modeling for early circuit simulation, *Proc. Custom Integrated Circuit Conference (CICC)*, 201–204.

Caughey, D. M. and Thomas, R. E. (1967). Carrier mobilities in silicon empirically related to doping and field, *Proc. IEEE* **55**, 2192.

Celler, G. K. and Cristoloveanu, S. (2003). Frontiers of silicon-on-insulator, *J. Appl. Phys.* **93**, 4955–4978.

Chan, T. Y., Chen, J., Ko, P. K., and Hu, C. (1987a). The impact of gate-induced drain leakage current on MOSFET scaling, *IEEE IEDM Technical Digest*, 718–721.

Chan, T. Y., Wu, A. T., Ko, P. K., and Hu, C. (1987b). Effects of the gate-to-drain/source overlap on MOSFET characteristics, *IEEE Electron Device Lett.* **EDL-8**, 326–328.

Chan, T. Y., Young, K. K., and Hu, C. (1987c). A true single-transistor oxide-nitride-oxide EEPROM device, *IEEE Electron Device Lett.* **EDL-8**, 93–95.

Chang, C.-P., Vuang, H.-H., Baker, M. R., *et al.* (2000). SALVO process for sub-50 nm low V_T replacement gate CMOS with KrF lithography, *IEEE IEDM Technical Digest*, 53–56.

Chang, C. Y., Fang, Y. K., and Sze, S. M. (1971). Specific contact resistance of metal–semiconductor barriers, *Solid-State Electron.* **14**, 541–550.

Chang, L. L., Stiles, P. J., and Esaki, L. (1967). Electron tunneling between a metal and a semiconductor: Characteristics of Al–Al$_2$O$_3$–SnTe and –GeTe junctions, *J. Appl. Phys.* **38**, 4440–4445.

Chang, W.-H., Davari, B., Wordeman, M. R., *et al.* (1992). A high-performance 0.25-µm CMOS technology: I – Design and characterization, *IEEE Trans. Electron Devices* **39**, 959–966.

Chen, H.-P., Yuan, Y., Yu, B., *et al.* (2012). Interface-state modeling of Al$_2$O$_3$-InGaAs MOS from depletion to inversion, *IEEE Trans. Electron Devices* **59**, 2383–2389.

Chen, H.-P., Yuan, Y., Yu, B., *et al.* (2013). Re-examination of the extraction of MOS interface state density by C-V stretchout and conductance methods, *Semiconductor Science and Technology* **28**, 085008 (5 pp).

Chen, I. C., Holland, S., and Hu, C. (1986). Oxide breakdown dependence on thickness and hole current-enhanced reliability of ultra-thin oxides, *IEEE IEDM Technical Digest*, 660–663.

Cheng, K. and Khakifirooz, A. (2015). FDSOI technology and its implications to analog and digital design. In *Digitally-Assisted Analog and Analog-Assisted Digital IC Design*, ed. X. Jiang. Cambridge: Cambridge University Press, 56–97.

Chiu, T.-Y., Chin, G. M., Lau, M. Y., *et al.* (1987). A high speed super self-aligned bipolar-CMOS technology, *IEEE IEDM Technical Digest*, 24–27.

Chiu, T.-Y., Tien, P. K., Sung, J., and Liu, T.-Y. M. (1992). A new analytical model and the impact of base charge storage on base potential distribution, emitter current crowding and base resistance, *IEEE IEDM Technical Digest*, 573–576.

Cho, H.-J., Seo, K.-I., Jeong, W. C., *et al.* (2011). Bulk planar 20 nm high-k/metal gate CMOS technology platform for low power and high performance applications, *IEEE IEDM Technical Digest*, 350–353.

Chor, E. F., Ashburn, P., and Brunnschweiler, A. (1985). Emitter resistance of arsenic- and phosphorus-doped polysilicon emitter transistors, *IEEE Electron Device Lett.* **EDL-6**, 516–518.

Chor, E. F., Brunnschweiler, A., and Ashburn, P. (1988). A propagation-delay expression and its application to the optimization of polysilicon emitter ECL processes, *IEEE J. Solid-State Circuits* **23**, 251–259.

Chung, J. E., Jeng, M.-C., Moon, J. E., Ko, P.-K., and Hu, C. (1990). Low-voltage hot-electron currents and degradation in deep-submicrometer MOSFETs, *IEEE Trans. Electron Devices* **37**, 1651–1657.

Chynoweth, A. G. (1957). Ionization rates for electrons and holes in silicon, *Phys. Rev.* **109**, 1537–1540.

Chynoweth, A. G., Feldmann, W. L., Lee, C. A., Logan, R. A., and Pearson, G. L. (1960). Internal field emission at narrow silicon and germanium p–n junctions, *Phys. Rev.* **118**, 425–434.

Coen, R. W. and Muller, R. S. (1980). Velocity of surface carriers in inversion layers on silicon, *Solid-State Electron.* **23**, 35–40.

Cowley, A. M. and Sze, S. M. (1965). Surface states and barrier height of metal–semiconductor systems, *J. Appl. Phys.* **36**, 3212–3220.

Crabbé, E. F., Comfort, J. H., Lee, W., *et al.* (1992). 73-GHz self-aligned SiGe-base bipolar transistors with phosphorus-doped polysilicon emitters, *IEEE Electron Device Lett.* **13**, 259–261.

Crabbé, E. F., Meyerson, B. S., Stork, J. M. C., and Harame, D. L. (1993a). Vertical profile optimization of very high frequency epitaxial Si- and SiGe-base bipolar transistors, *IEEE IEDM Technical Digest*, 83–86.

Crabbé, E. F., Cressler, J. D., Patton, G. L., *et al.* (1993b). Current gain rolloff in graded-base SiGe heterojunction bipolar transistors, *IEEE Electron Device Lett.* **14**, 193–195.

Cressler, J. D., Comfort, J. H., Crabbé, E. F., *et al.* (1993a). On the profile design and optimization of epitaxial Si- and SiGe-base bipolar technology for 77K applications – Part I: Transistor dc design considerations, *IEEE Trans. Electron Devices* **40**, 525–541.

Cressler, J. D., Crabbé, E. F., Comfort, J. H., Stork, J. M. C., and Sun, J. Y.-C. (1993b). On the profile design and optimization of epitaxial Si- and SiGe-base bipolar technology for 77K applications – Part II: Circuit performance issues, *IEEE Trans. Electron Devices* **40**, 542–554.

Cristoloveanu, S., Athanasiou, S., Bawedin, M., and Galy, Ph. (2017). Evidence of super-coupling effect in ultrathin silicon layers using a four-gate MOSFET, *IEEE Electron Device Lett.*, **38**, 157–159.

Crowell, C. R. (1965). The Richardson constant for thermionic emission in Schottky barrier diodes, *Solid-State Electron.* **8**, 395–399.

Crowell, C. R. (1969). Richardson constant and tunneling effective mass for thermionic and thermionic-field emission in Schottky barrier diodes, *Solid-State Electron.* **12**, 55–59.

Crowell, C. R. and Rideout, V. L. (1969). Normalized thermionic-field (T-F) emission in metal–semiconductor (Schottky) barriers, *Solid-State Electron.* **12**, 89–105.

Crowell, C. R. and Sze, S. M. (1966a). Current transport in metal-semiconductor barriers, *Solid-State Electron.* **9**, 1035–1048.

Crowell, C. R. and Sze, S. M. (1966b). Temperature dependence of avalanche multiplication in semiconductors, *Appl. Phys. Lett.* **9**, 242–244.

Crupi, F., Ciofi, C., Germano, A., *et al.* (2002). On the role of interface states in low-voltage leakage currents of metal–oxide–semiconductor structures, *Appl. Phys. Lett.* **80**, 4597–4599.

de Graaff, H. C., Slotboom, J. W., and Schmitz, A. (1977). The emitter efficiency of bipolar transistors, *Solid-State Electron.* **20**, 515–521.

Deal, B. E., Sklar, M., Grove, A. S., and Snow, E. H. (1967). Characteristics of the surface-state charge of thermally oxidized silicon, *J. Electrochem. Soc.* **114**, 266–274.

Degraeve, R., Groeseneken, G., Bellens, R., Depas, M., and Maes, H. E. (1995). A consistent model for the thickness dependence of intrinsic breakdown in ultra-thin oxide, *IEEE IEDM Technical Digest*, 863–866.

Degraeve, R., Groeseneken, G., Bellens, R., *et al.* (1998). New insights in the relation between electron trap generation and the statistical properties of oxide breakdown, *IEEE Trans. Electron Devices* **43**, 904–911.

Deixler, P., Huizing, H. G. A., Donkers, J. J. T. M., *et al.* (2001). Explorations for high performance SiGe-heterojunction bipolar transistor integration, Proc. Bipolar/BiCMOS Circuit & Technology Meeting. *IEEE*, Minneapolis, MN, 30–33.

del Alamo, J., Swirhun, S., and Swanson, R. M. (1985a). Measuring and modeling minority carrier transport in heavily doped silicon, *Solid-State Electron.* **28**, 47–54.

del Alamo, J., Swirhun, S., and Swanson, R. M. (1985b). Simultaneous measurement of hole lifetime, hole mobility and band-gap narrowing in heavily doped n-type silicon, *IEEE IEDM Technical Digest*, 290–293.

Dennard, R. H. (1968). Field-effect transistor memory, U.S. Patent 3,387,286 issued June 4.

Dennard, R. H. (1984). Evolution of the MOSFET dynamic RAM – A personal view, *IEEE Trans. Electron Devices* **ED-31**, 1549–1555.

Dennard, R. H. (1986). Scaling limits of silicon VLSI technology. In *The Physics and Fabrication of Microstructures and Microdevices*, ed. M. J. Kelly and C. Weisbuch. Berlin: Springer-Verlag, 352–369.

Dennard, R. H., Gaensslen, F. H., Yu, H. N., *et al.* (1974). Design of ion-implanted MOSFETs with very small physical dimensions, *IEEE J. Solid-State Circuits* **SC-9**, 256–268.

Depas, M., Nigam, T., and Heyns, M. H. (1996). Soft breakdown of ultrathin gate oxide layers, *IEEE Trans. Electron Devices* **43**, 1499–1504.

DiMaria, D. J. (1987). Correlation of trap creation with electron heating in silicon dioxide, *Appl. Phys. Lett.* **51**, 655–657.

DiMaria, D. J. and Cartier, E. (1995). Mechanism for stress-induced leakage currents in thin silicon dioxide films, *J. Appl. Phys.* **78**, 3883–3894.

DiMaria, D. J. and Stathis, J. H. (1997). Explanation for the oxide thickness dependence of breakdown characteristics of metal-oxide semiconductor structures, *Appl. Phys. Lett.* **70**, 2708–2710.

DiMaria, D. J., Cartier, E., and Arnold, D. (1993). Impact ionization, trap creation, degradation, and breakdown in silicon dioxide films on silicon, *J. Appl. Phys.* **73**, 3367–3384.

DiStefano, T. H. and Shatzkes, M. (1974). Impact ionization model for dielectric instability and breakdown, *Appl. Phys. Lett.* **25**, 685–687.

Dubois, E., Bricout, P.-H., and Robilliart, E. (1994). Accuracy of series resistances extraction schemes for polysilicon bipolar transistors, Proc. Bipolar/BiCMOS Circuit & Technology Meeting, *IEEE*, 148–151.

Duvernay, J., Brossard, F., Borot, G., *et al.* (2007). Development of a self-aligned pnp HBT for a complementary thin-SOI SiGeC BiCMOS technology, Proc. Bipolar/BiCMOS Circuit & Technology Meeting, *IEEE*, pp. 34–37.

Dziewior, J. and Schmid, W. (1977). Auger coefficients for highly doped and highly excited silicon, *Appl. Phys. Lett.* **31**, 346–348.

Dziewior, J. and Silber, D. (1979). Minority-carrier diffusion coefficients in highly doped silicon, *Appl. Phys. Lett.* **35**, 170–172.

Early, J. M. (1952). Effects of space-charge layer widening in junction transistors, *Proc. IRE* **40**, 1401–1406.

Ebers, J. J. and Moll, J. L. (1954). Large-signal behavior of junction transistors, *Proc. IRE* **42**, 1761–1772.

Eichelberger, E. B. and Bello, S. E. (1991). Differential current switch – High performance at low power, *IBM J. Res. Develop.* **35**, 313–320.

Eitan, B., Pavan, P., Bloom, I., *et al.* (2000). NROM: A novel localized trapping, 2-bit nonvolatile memory cell, *IEEE Electron Device Lett.* **21**, 543–545.

EMIS Datareviews Series No. 4 (1988). *Properties of Silicon*. London: INSPEC, The Institute of Electrical Engineers.

Fair, R. B. and Wivell, H. W. (1976). Zener and avalanche breakdown in As-implanted low voltage Si n–p junctions, *IEEE Trans. Electron Devices* **ED-23**, 512–518.

Farber, A. S. and Schlig, E. S. (1972). A novel high-performance bipolar monolithic memory cell, *IEEE J. Solid-State Circuits* **SC-7**, 297–298.

Filensky, W. and Beneking, H. (1981). New technique for determination of static emitter and collector series resistances of bipolar transistors, *Electron. Lett.* **17**, 503–504.

Fischetti, M. V., Laux, S. E., and Crabbé, E. (1995). Understanding hot-electron transport in silicon devices: Is there a shortcut? *J. Appl. Phys.* **78**, 1058–1087.

Forbes, L. (1977). *Gold in Silicon: Characterisation and Infra-red Detector Applications.* Available from: https://link.springer.com/content/pdf/10.1007%2FBF03215429.pdf (last accessed June 14, 2021).

Frank, D. J., Laux, S. E., and Fischetti, M. V. (1992). Monte Carlo simulation of a 30 nm dual-gate MOSFET: how short can silicon go? *IEEE IEDM Technical Digest*, 553–556.

Frank, D. J., Taur, Y. and Wong, H.-S. (1998). Generalized scale length for two-dimensional effects in MOSFETs, *IEEE Electron Device Lett.* **19**, 385–387.

Frank, D. J., Taur, Y., Ieong, M., and Wong, H.-S. P. (1999). Monte Carlo modeling of threshold variation due to dopant fluctuations, Symp. VLSI Technology Digest of Tech. Papers, *IEEE*, 171–172.

Frank, D. J., Dennard, R. H., Nowak, E., *et al.* (2001). Device scaling limits of Si MOSFETs and their application dependencies, *IEEE Proc.* **89**, 259–288.

Freeman, G., Chang, P., Engbrecht, E. R., *et al.* (2015). Performance-optimized gate-first 22-nm SOI technology with embedded DRAM, *IBM J. Res. Develop.* **59**(1), 5:1–5:14.

Frohman-Bentchkowsky, D. (1971). A fully decoded 2048-bit electrically programmable FAMOS read-only memory, *IEEE J. Solid-State Circuits* **SC-6**, 301–306.

Fukuzumi, Y., Matsuoka, Y., Kito, M., *et al.* (2007). Optimal integration and characteristics of vertical array devices for ultra-high density, bit-cost scalable flash memory, *IEEE IEDM Technical Digest*, 449–452.

Gannavaram, S., Pesovic, N., and Ozturk, C. (2000). Low temperature ($\leq 800°C$) recessed junction selective silicon-germanium source/drain technology for sub-70 nm CMOS, *IEEE IEDM Technical Digest*, 437–440.

Ghandhi, S. K. (1968). *The Theory and Practice of Microelectronics*. New York: Wiley.

Gildenblat, G., Li, X., Wu, W., *et al.* (2006). PSP: An advanced surface-potential-based MOSFET model for circuit simulation, *IEEE Trans. Electron Devices* **ED-53**, 1979–1993.

Golonzka, O., Alzate, J.-G., Arslan, U., *et al.* (2018). MRAM as embedded non-volatile memory solution for 22FFL FinFET technology, *IEEE IEDM Technical Digest*, 18.1.1–18.1.4.

Good, R. H. Jr. and Müller, E. W. (1956). Field emission. In *Handbuck der Physik*, vol. XXI, ed. S. Flugge. Berlin: Springer-Verlag, 176–231.

Grant, W. H. (1973). Electron and hole ionization rates in epitaxial silicon at high electric fields, *Solid-State Electron.* **16**, 1189–1203.

Gray, P. E., DeWitt, D., Boothroyd, A. R., and Gibbons, J. F. (1964). *Physical Electronics and Circuit Models of Transistors*, vol. 2, Semiconductor Electronics Education Committee. New York: Wiley.

Green, M. A. (1990). Intrinsic concentration, effective densities of states, and effective mass in silicon, *J. Appl. Phys.* **67**, 2944–2954.

Grove, A. S. (1967). *Physics and Technology of Semiconductor Devices*. New York: Wiley.

Grove, A. S. and Fitzgerald, D. J. (1966). Surface effects on p–n junctions – Characteristics of surface space-charge regions under non-equilibrium conditions, *Solid-State Electron.* **9**, 783–806.

Gummel, H. K. (1961). Measurement of the number of impurities in the base layer of a transistor, *Proc. IRE* **49**, 834.

Gummel, H. K. (1967). Hole-electron product of p–n junctions, *Solid-State Electron.* **10**, 209–212.

Hall, R. N. (1952). Electron–hole recombination in germanium, *Phys. Rev.* **87**, 387.

Harame, D. L., Comfort, J. H., Cressler, J. D., *et al.* (1995a). Si/SiGe epitaxial-base transistors – Part I: Materials, physics, and circuits, *IEEE Trans. Electron Devices* **42**, 455–468.

Harame, D. L., Comfort, J. H., Cressler, J. D., *et al.* (1995b). Si/SiGe epitaxial-base transistors – Part II: Process integration and analog applications, *IEEE Trans. Electron Devices* **42**, 469–482.

Harari, E. (1978). Dielectric breakdown in electrically stressed thin films of thermal SiO_2, *J. Appl. Phys.* **49**, 2478–2489.

Hart, K. and Slob, A. (1972). Integrated injection logic: A new approach to LSI, *IEEE J. Solid-State Circuits* **7**, 346–351.

Hashemi, P., Yau, J.-B., Chan, K. K., Ning, T. H., and Shahidi, G. G. (2017). First demonstration of symmetric lateral npn transistors on SOI featuring epitaxially-grown emitter/collector regions, *IEEE SOI-3D-Subthreshold Microelectronics Technology Unified Conference (S3S)*, 1–2.

Hashemi, P., Yau, J.-B., Chan, K. K., Ning, T. H., and Shahidi, G. G. (2018). Demonstration of symmetric lateral npn transistors on SOI featuring epitaxially grown emitter/collector regions, *IEEE J. Electron Devices Soc.* **6**, 537–542.

Hauser, J. R. (1964). The effects of distributed base potential on emitter current injection density and effective base resistance for stripe transistor geometries, *IEEE Trans. Electron Devices* **ED-11**, 238–242.

Hauser, J. R. (1968). Bipolar transistors. In *Fundamentals of Silicon Integrated Device Technology*, vol. II, ed. R. M. Burger and R. P. Donovan. Englewood Cliffs: Prentice-Hall.

Hedenstierna, N. and Jeppson, K. O. (1987). CMOS circuit speed and buffer optimization, *IEEE Trans. Computer-Aided Design*, **CAD-6**, 270.

Heine, V. (1965). Theory of surface states, *Phys. Rev.* **138**, A1689–A1696.

Henisch, H. K. (1984). *Semiconductor Contacts*. Oxford: Clarendon Press.

Hill, C. F. (1968). Noise margin and noise immunity in logic circuits, *Microelectronics*, 16–21.

Hisamoto, D., Lee, W.-C., Kedzierski, J., *et al.* (2000). FinFET – A self-aligned double-gate MOSFET scalable to 20 nm, *IEEE Trans. Electron Devices* **47**, 2320–2325.

Hobart, K. D., Kub, F. J., Papanicoloau, N. A., Kruppa, W., and Thompson, P. E. (1995). $Si/Si_{1-x}Ge_x$ heterojunction bipolar transistors for microwave power applications, *J. Crystal Growth* **157**, 215–220.

Hodges, D. A., ed. (1972). *Semiconductor Memories*. New York: IEEE Press.

Hong, D. C. and Taur, Y. (2021). An above threshold model for short channel DG MOSFETs, *IEEE Trans. Electron Devices* **ED-68**, 3734–3739.

Hsu, C. C.-H. and Ning, T. H. (1991). Voltage and temperature dependence of interface trap generation by hot electrons in p- and n-poly gated MOSFETs, Symp. VLSI Technology Digest of Tech. Papers, *IEEE*, 17–18.

Hsu, C. C.-H., Acovic, A., Dori, L., *et al.* (1992). A high-speed low-power p-channel flash EEPROM using silicon-rich oxide as tunneling dielectric, *Int. Conf. Solid-State Devices and Materials, Extended Abstract*, 140–142.

Hu, C., ed. (1991). *Nonvolatile Semiconductor Memories Technologies, Design, and Applications*. New York: IEEE Press.

Hu, C., Tam, S. C., Hsu, F.-C., *et al.* (1985). Hot-electron-induced MOSFET degradation – Model, monitor, and improvement, *IEEE Trans. Electron Devices* **32**, 375–385.

Hu, Q., Chen, X., Norstrom, H., *et al.* (2018). Current gain and low-frequency noise of symmetric lateral bipolar junction transistors on SOI, *Proc. European Solid-State Device Research Conf. (ESSDERC)*, 258–261.

Hu, S. M. and Schmidt, S. (1968). Interactions in sequential diffusion processes in semiconductors, *J. Appl. Phys.* **39**, 4272–4283.

Hui, J., Wong, S., and Moll, J. (1985). Specific contact resistivity of $TiSi_2$ to p^+ and n^+ junctions, *IEEE Electron Device Lett.* **EDL-6**, 479–481.

Huizing, H. G. A., Klootwijk, J. H., Aksen, E., and Slotboom, J. W. (2001). Base current tuning in SiGe HBT's by SiGe in the emitter, *IEEE IEDM Technical Digest*, 899–902.

Hurkx, G. A. M. (1994). Bipolar transistor physics. In *Bipolar and Bipolar-MOS Integration*, ed. P. A. H. Hart. Amsterdam: Elsevier, 73–175.

Hurkx, G. A. M. (1996). The relevance of fT and fmax for the speed of a bipolar CE amplifier stage, Proc. Bipolar/BiCMOS Circuit & Technology Meeting, *IEEE*, 53–56.

Iinuma, T., Itoh, N., Nakajima, H., *et al.* (1995). Sub-20 ps high-speed ECL bipolar transistor with low parasitic architecture, *IEEE Trans. Electron Devices* **42**, 399–405.

Im, H.-J., Ding, Y., and Pelz, J. P. (2001). Nanometer-scale test of the Tung model of Schottky-barrier height inhomogeneity, *Phys. Rev. B* **64**, 0753101–0753109.

Irvin, J. C. and Vanderwal, N. C. (1969). Schottky barrier devices. In *Microwave Semiconductor Devices and Their Circuit Applications*, ed. H. A. Watson. New York: McGraw-Hill, 349–369.

Ishiduki, M., Fukuzumi, Y., Katsumata, R., *et al.* (2009). Optimal device structure for pipe-shaped BiCS flash memory for ultra high density storage device with excellent performance and reliability, *IEEE IEDM Technical Digest*, 625–628.

Itoh, K. (2001). *VLSI Memory Chip Design*. Berlin: Springer.

Itoh, Y., Momodomi, M., Shirota, R., *et al.* (1989). An experimental 4Mb CMOS EEPROM with a NAND structured cell, *IEEE ISSCC Digest of Technical Papers*, 134–135.

Iyer, S. S., Patton, G. L., Delage, S. S., Tiwari, S., and Stork, J. M. C. (1987). Silicon-Germanium base heterojunction bipolar transistors by molecular beam epitaxy, *IEEE IEDM Technical Digest*, 874–876.

Jan, C.-H., Al-amoody, F., Chang, H.-Y., *et al.* (2015). A 14 nm SoC platform technology featuring 2nd generation tri-gate transistors, 70 nm gate pitch, 52 nm metal pitch, and 0.0499 μm2 SRAM cells, optimized for low power, high performance and high density SoC products, Symp. VLSI Technology Digest of Tech. Papers, *IEEE*, T12–T13.

Jeppson, K. O. and Svensson, C. M. (1977). Negative bias stress of MOS devices at high electric fields and degradation of MNOS devices, *J. Appl. Phys.* **48**, 2004–2014.

Jiménez, D., Iñiguez, B., Suñé, J., Marsal, L. F., Pallarès, J., Roig, J. and Flores, D. (2004). Continuous analytic I–V model for surrounding-gate MOSFETs, *IEEE Electron Device Lett.* **25**, 571–573.

Jin, Y., Shihab, M., and Jung, M. (2014). Area, power, and latency considerations of STT-MRAM to substitute for main memory, in *The Memory Forum*, a workshop in conjunction with ISCA 2014, June 14, 2014, Minneapolis, MN. Available from www.cs.utah.edu/theme moryforum/jin.pdf (last accessed June 14, 2021).

Joardar, K., Gullapalli, K. K., McAndrew, C. C., Burnham, M. E., and Wild, A. (1998). An improved MOSFET model for circuit simulation, *IEEE Trans. Electron Devices* **45**, 134–148.

Johnson W., Perlegos, G., Renninger, A., Kuhn, G., and Ranganath, T. (1980). A 16Kb electrically erasable nonvolatile memory, *IEEE ISSCC Digest of Technical Papers*, 152–153.

Jund, C. and Poirier, R. (1966). Carrier concentration and minority carrier lifetime measurement in semiconductor epitaxial layers by the MOS capacitance method, *Solid-State Electron.* **9**, 315–318.

Kaczer, B., Degraeve, R., Groeseneken, G., *et al.* (2000). Impact of MOSFET oxide breakdown on digital circuit operation and reliability, *IEEE IEDM Technical Digest*, 553–556.

Kahng, D. and Atalla, M. M. (1960). Silicon dioxide field surface devices, *Presented at IEEE Device Research Conference*, Pittsburgh.

Kane, E. O. (1961). Theory of tunneling, *J. Appl. Phys.* **32**, 83–91.

Kapoor, A. K. and Roulston, D. J., eds. (1989). *Polysilicon Emitter Bipolar Transistors.* New York: IEEE Press.

Kay, L. E. and Tang, T.-W. (1991). Monte Carlo calculation of strained and unstrained electron mobilities in $Si_{1-x}Ge_x$ using an improved ionized-impurity model, *J. Appl. Phys.* **70**, 1483–1488.

Kerwin, R. E., Klein, D. L., and Sarace, J. C. (1969). Method for making MIS structures, U.S. Patent 3,475,234 issued October 28.

Khakifirooz, A., Cheng, K., Reznicek, A., *et al.* (2012) Scalability of extremely thin SOI (ETSOI) MOSFETs to sub-20 nm gate length, *IEEE Electron Device Lett.*, **33**, 149–151.

Kianian, S., Levi, A., Lee, D., and Hu, Y.-W. (1994). A novel 3 volts-only, small sector erase, high density flash E2PROM, Symp. VLSI Technology Digest of Tech. Papers, *IEEE*, 71–72.

Kim, I.-G., Yanagidarira, K., and Hiramoto, T. (2003). Integration of fluorinated nano-crystal memory cells with $4.6F^2$ size by landing plug polysilicon contact and direct-tungsten bitline, *IEEE IEDM Technical Digest*, 605–608.

Kim, K. (2008). Future memory technology: challenges and opportunities, *Proc. Symp. on VLSI Technology, Systems, and Applications*, 5–9.

Kim, W., Choi, S., Sung, J., *et al.* (2009). Multi-layered vertical gate NAND flash overcoming stacking limit for terabit density storage, Symp. VLSI Technology Digest of Tech. Papers, *IEEE*, 188–189.

Kircher, C. J. (1975). Comparison of leakage currents in ion-implanted and diffused p–n junctions, *J. Appl. Phys.* **46**, 2167–2173.

Kirk Jr., C. T. (1962). A theory of transistor cutoff frequency (f_T) falloff at high current densities, *IEEE Trans. Electron Devices* **ED-9**, 164–174.

Kittel, C. (1976). *Introduction to Solid State Physics.* New York: Wiley.

Klaassen, D. B. M. (1990). A unified mobility model for device simulation, *IEEE IEDM Technical Digest*, 357–360.

Klaassen, D. B. M., Slotboom, J. W., and de Graaff, H. C. (1992). Unified apparent bandgap narrowing in n- and p-type silicon, *Solid-State Electron.* **35**, 125–129.

Ko, P. K., Muller, R. S., and Hu, C. (1981). A unified model for hot-electron currents in MOSFETs, *IEEE IEDM Technical Digest*, 600–603.

Kolhatkar, J. S. and Dutta, A. K. (2000). A new substrate current model for submicron MOSFET's, *IEEE Trans. Electron Devices* **47**, 861–863.

Kondo, M., Kobayashi, T., and Tamaki, Y. (1995). Hetero-emitter-like characteristics of phosphorus doped polysilicon emitter transistors – Part I: Band structure in the polysilicon emitter obtained from electrical measurements, *IEEE Trans. Electron Devices* **42**, 419–426.

Kondo, M., Shimamoto, H., and Washio, K. (2001). Variation in emitter diffusion depth by TiSi$_2$ formation on polysilicon emitters of Si bipolar transistors, *IEEE Trans. Electron Devices* **48**, 2108–2117.

Koo, K.-H. and Saraswat, K. C. (2011). Study of performances of low-k Cu, CNTs, and optical interconnects. In *Nanoelectronic Circuit Design*, ed. N. K. Jha and D. Chen. New York: Springer, 377–408.

Koyanagi, M., Sunami, H., Hashimoto, N., and Ashikawa, M. (1978). Novel high density stacked capacitor MOS RAM, *IEEE IEDM Technical Digest*, 348–351.

Kroemer, H. (1957). Theory of a wide-gap emitter for transistors, *Proc. IRE* **45**, 1535–1537.

Kroemer, H. (1985). Two integral relations pertaining to the electron transport through a bipolar transistor with a non-uniform energy gap in the base region, *Solid-State Electron.* **28**, 1101–1103.

Kuhn, K. J. (2012). Considerations for ultimate CMOS scaling, *IEEE Trans. Electron Devices*, **59**, 1813–1828.

Kuhn, K. J., Avci, U., Cappellani, A., *et al.* (2012). The ultimate CMOS device and beyond, *IEEE IEDM Technical Digest*, 171–174.

Kunz, V. D., de Groot, C. H., Hall, S., *et al.* (2002). Application of polycrystalline SiGe for gain control in SiGe HBTs, *Proc. European Solid-State Device Research Conf. (ESSDERC)*, 171–174.

Kunz, V. D., de Groot, C. H., Hall, S., and Ashburn, P. (2003). Polycrystalline silicon–germanium emitters for gain control, with application to SiGe HBTs, *IEEE Trans. Electron Devices* **50**, 1480–1486.

Kyung, M. (2005). Charge sheet models for MOSFET current: Examining their deviations from the Pao–Sah integral. Master thesis, Dept. Electrical and Computer Engineering, University of California, San Diego.

La Rosa, G., Guarin, F., Rauch, S., *et al.* (1997). NBTI-channel hot carrier effects in pMOSFETs in advanced CMOS technologies, *Proc. IEEE Int. Reliability Phys. Symp.*, 282–286.

Lai, S., Mielke, N., Atwood, G., *et al.* (1991). Highly reliable E2PROM cell fabricated with ETOX flash process, Symp. VLSI Technology Digest of Tech. Papers, *IEEE*, 59–60.

Lanzerotti, L. D., Sturm, J. C., Stach, E., *et al.* (1996). Suppression of boron out diffusion in SiGe HBTs by carbon incorporation, *IEEE IEDM Technical Digest*, 249–252.

Laux, S. E. and Fischetti, M. V. (1988). Monte-Carlo simulation of submicrometer Si n-MOSFET's at 77 and 300 K, *IEEE Electron Device Lett.* **EDL-9**, 467–469.

Lee, J.-H., Kang, W.-G., Lyu, J.-S., and Lee, J. D. (1996). Modeling of the critical current density of bipolar transistor with retrograde collector doping profile, *IEEE Electron Device Lett.* **17**, 109–111.

Lee, S., Johnson, J., Greene, B., *et al.* (2012). Advanced modeling and optimization of high performance 32 nm HKMG SOI CMOS for RF/analog SoC applications, Symp. VLSI Technology Digest of Tech. Papers, *IEEE*, 135–136.

Lenzlinger, M. and Snow, E. H. (1969). Fowler–Nordheim tunneling into thermally grown SiO_2, *J. Appl. Phys.* **40**, 278–283.

LeRoy, M. R., Raman, S., Chu, M., et al. (2015). High-speed reconfigurable circuits for multirate systems in SiGe HBT technology, *Proc. IEEE*, 103, 1181–1196.

LeVeque, R. J. (2007). *Finite Difference Methods for Ordinary and Partial Differential Equations*. Philadelphia: Society for Industrial and Applied Mathematics (SIAM).

Li, G. P., Chuang, C. T., Chen, T. C., and Ning, T. H. (1987). On the narrow-emitter effect of advanced shallow-profile bipolar transistors, *IEEE Trans. Electron Devices* **35**, 1942–1950.

Li, G. P., Hackbarth, E., and Chen, T.-C. (1988). Identification and implication of a perimeter tunneling current component in advanced self-aligned bipolar transistors, *IEEE Trans. Electron Devices* **ED-35**, 89–95.

Li, M. F., He, Y. D., Ma, S. G., *et al.* (1999). Role of hole fluence in gate oxide breakdown, *IEEE Electron Devices Lett.* **20**, 586–588.

Liang, X. and Taur, Y. (2004). A 2-D analytical solution for SCEs in DG MOSFETs, *IEEE Trans. Electron Devices* **ED-51**, 1385–1391.

Liboff, R. L. (2003). *Introductory Quantum Mechanics*, 4th ed. San Francisco: Addison-Wesley.

Lim, H.-K., and Fossum, J. G. (1983). Threshold voltage of thin-film silicon-on-insulator (SOI) MOSFETs, *IEEE Trans. Electron Devices* **ED-30**, 1244–1251.

Lin, B.-J. (2004). Immersion lithography and its impact on semiconductor manufacturing, in *Proc. SPIE 5377, Optical Microlithography XVII*, May 28, 2004.

Lin, H.-H. and Taur, Y. (2017). Effect of source-drain doping on subthreshold characteristics of short-channel DG MOSFETs, *IEEE Trans. Electron Devices* **ED-64**, 4856–4860.

Linder, B. P., Frank, D. J., Stathis, J. H., and Cohen, S. A. (2001). Transistor-limited constant voltage stress of gate dielectrics, Symp. VLSI Technology Digest of Tech. Papers, *IEEE*, 93–94.

Linder, B. P., Lombardo, S., Stathis, J. H., Vayshenker, A., and Frank, D. J. (2002). Voltage dependence of hard breakdown growth and the reliability implication in thin dielectrics, *IEEE Electron Device Lett.* **23**, 661–663.

Lindmayer, J. and Wrigley, C. Y. (1965). *Fundamentals of Semiconductor Devices*. New York: Van Norstrand Reinhold.

Liu, M., Cai, M., and Taur, Y. (2006). Scaling limit of CMOS supply voltage from noise margin considerations, *Proc. 2006 International Conference on Simulation of Semiconductor Process and Devices (SISPAD)*, 287–289.

Lo, S.-H., Buchanan, D. A., and Taur, Y. (1999). Modeling and characterization of quantization, polysilicon depletion, and direct tunneling effects in MOSFETs with ultrathin oxides. *IBM J. Res. Develop.* **43**, 327–337.

Lo, S.-H., Buchanan, D. A., Taur, Y., and Wang, W. (1997). Quantum-mechanical modeling of electron tunneling current from the inversion layer of ultra-thin-oxide nMOSFET's, *IEEE Electron Device Lett.* **18**, 209–211.

Lochtefeld, A. and Antoniadis, D. A. (2001). On experimental determination of carrier velocity in deeply scaled NMOS: how close to the thermal limit?, *IEEE Electron Device Lett.* **22**, 95–97.

Lohstroh, J., Seevinck, E., and de Groot, J. (1983). Worst-case static noise margin criteria for logic circuits and their mathematical equivalence, *IEEE J. Solid-State Circuits* **SC-18**, 803–807.

Lombardo, S., Stathis, J. H., and Linder, B. P. (2003). Breakdown transient in ultrathin gate oxides: transition in the degradation rate, *Phys. Rev. Lett.* **90**, 1676011–1676014.

Lu, N., Cottrell, P., Craig, W., *et al.* (1985). The SPT cell – A new substrate-plate trench cell for DRAMs, *IEEE IEDM Technical Digest*, 771–772.

Lu, N. C. C., Cottrell, P. E., Craig, W. J., *et al.* (1986). A substrate-plate trench-capacitor (SPT) memory cell for dynamic RAM's, *IEEE J. Solid-State Circuits* **SC-21**, 627–734.

Lu, P.-F., and Chen, T.-C. (1989). Collector-base junction avalanche effects in advanced double-poly self-aligned bipolar transistors, *IEEE Trans. Electron Devices* **ED-36**, 1182–1188.

Lu, P.-F., Li, G. P., and Tang, D. D. (1987). Lateral encroachment of extrinsic base dopant in submicron bipolar transistors, *IEEE Electron Device Lett.* **8**, 496–498.

Lu, W.-Y. and Taur, Y. (2006). On the scaling limit of ultrathin SOI MOSFETs, *IEEE Trans. Electron Devices* **ED-53**, 1137–1141.

Lundstrom, M. (1997). Elementary scattering theory of the Si MOSFET, *IEEE Electron Device Lett.* **18**, 361–363.

Lundstrom, M. and Jeong, C. (2013). *Near-Equilibrium Transport: Fundamentals and Applications*. Singapore: World Scientific.

Lynes, D. J. and Hodges, D. A. (1970). Memory using diode-coupled bipolar transistor cells, *IEEE J. Solid-State Circuits* **SC-5**, 186–191.

Manku, T. and Nathan, A. (1992). Electron drift mobility model for devices based on unstrained and coherently strained $Si_{1-x}Ge_x$ grown on $\langle 001 \rangle$ silicon substrate, *IEEE Trans. Electron Devices* **39**, 2082–2089.

Many, A., Goldstein, Y., and Grover, N. B. (1965). *Semiconductor Surfaces*. New York: John Wiley & Sons.

Martinet, B., Romagna, F., Kermarrec, O., *et al.* (2002). An investigation of the static and dynamic characteristics of high speed SiGe:C HBTs using a poly-SiGe emitter, Proc. Bipolar/BiCMOS Circuit & Technology Meeting, *IEEE*, 147–150.

Masuoka, F., Asano, M., Iwahashi, H., Komuro, T., and Tanaka, S. (1984). A new flash E^2 PROM cell using triple polysilicon technology, *IEEE IEDM Technical Digest*, 464–467.

Masuoka, F., Momodomi, M., Iwata, Y., and Shirota, R. (1987). New ultra high density EPROM and flash EEPROM with NAND structure cell, *IEEE IEDM Technical Digest*, 552–555.

Mayumi, H., Nokubo, J., Okada, K., and Shiba, H. (1974). A 25-ns read access bipolar 1-kbit TTL RAM, *IEEE J. Solid-State Circuits* **SC-9**, 283–284.

Mazhari, B., Cristoloveanu, S., Ioannou, D. E., and Caviglia, A. L. (1991). Properties of ultrathin wafer-bonded silicon on insulator MOSFETs, *IEEE Trans. Electron Devices* **38**, 1289–1295.

Mead, C. A. and Spitzer, W. G. (1964). Fermi level position at metal–semiconductor interfaces, *Phys. Rev.* **134**, A713–A716.

Meyer, C. S., Lynn, D. K., and Hamilton, D. J. (eds.) (1968). *Analysis and Design of Integrated Circuits*. New York: McGraw-Hill.

Meyer, R. G. and Muller, R. S. (1987). Charge-control analysis of the collector–base space-charge-region contribution to bipolar-transistor time constant τ_T, *IEEE Trans. Electron Devices* **ED-34**, 450–452.

Mii, Y., Wind, S., Taur, Y., *et al.* (1994). An ultra-low power 0.1 μm CMOS, Symp. VLSI Technology Digest of Tech. Papers, *IEEE*, pp. 9–10.

Miller, S. L. (1955). Avalanche breakdown in Germanium, *Phys. Rev.* **99**, 1234–1241.

Mizuno, T., Okamura, J., and Toriumi, A. (1994). Experimental study of threshold voltage fluctuation due to statistical variation of channel dopant number in MOSFET's, *IEEE Trans. Electron Devices* **41**, 2216–2221.

Moll, J. L. (1964). *Physics of Semiconductors.* New York: McGraw-Hill.

Moll, J. L. and Ross, I. M. (1956). The dependence of transistor parameters on base resistivity, *Proc. IRE* **44**, 72–78.

Muller, R. S. and Kamins, T. I. (1977). *Device Electronics for Integrated Circuits.* New York: Wiley.

Na, M. H., Nowak, E. J., Haensch, W., and Cai, J. (2002). The effective drive current in CMOS inverters, *IEEE IEDM Technical Digest*, 121–124.

Nakamura, T. and Nishizawa, H. (1995). Recent progress in bipolar transistor technology, *IEEE Trans. Electron Devices* **ED-42**, 390–398.

Nanba, M., Uchino, T., Kondo, M., *et al.* (1993). A 64-GHz f_T and 3.6-V BV_{CEO} Si bipolar transistor using *in situ* phosphorus-doped and large-grained polysilicon emitter contacts, *IEEE Trans. Electron Devices* **ED-40**, 1563–1565.

Naruke, K., Yamada, S., Obi, E., Taguchi, S., and Wada, M. (1989). A new flash-erase EEPROM cell with a sidewall select-gate on its source side, *IEEE IEDM Technical Digest*, 603–606.

Natori, K. (1994). Ballistic metal-oxide-semiconductor field effect transistor, *J. Appl. Phys.* **76**, 4879–4890.

Ng, K. K. and Lynch, W. T. (1986). Analysis of the gate-voltage dependent series resistance of MOSFETs, *IEEE Trans. Electron Devices* **ED-33**, 965–972.

Nguyen, T. N. (1984). Small-geometry MOS transistors: physics and modeling of surface- and buried-channel MOSFETs. Unpublished PhD thesis, Stanford University.

Nicollian, E. H. and Brews, J. R. (1982). *MOS Physics and Technology.* New York: Wiley.

Ning, T. H. (2003). Polysilicon-emitter SiGe-base bipolar transistors – What happens when Ge gets into the emitter? *IEEE Trans. Electron Devices* **50**, 1346–1352.

Ning, T. H. (2013). A perspective on future nanoelectronic devices, *Int. Symp. VLSI Technology, Systems and Application (VLSI-TSA)*, 1–2.

Ning, T. H. (2016). A perspective on SOI symmetric lateral bipolar transistors for ultra-low-power systems, *IEEE J. Electron Devices Soc.* **4**, 227-235.

Ning, T. H. and Cai, J. (2013). On the performance and scaling of symmetric lateral bipolar transistors on SOI, *IEEE J. Electron Devices Soc.* **1**, 21–27.

Ning, T. H. and Cai, J. (2015). A perspective on symmetric lateral bipolar transistors on SOI as a complementary bipolar logic technology, *IEEE J. Electron Devices Soc.* **3**, 24–36.

Ning, T. H. and Isaac, R. D. (1980). Effect of emitter contact on current gain of silicon bipolar devices, *IEEE Trans. Electron Devices* **ED-27**, 2051–2055.

Ning, T. H. and Sah, C. T. (1971). Multivalley effective-mass approximation for donor states in silicon, I. Shallow-level group-V impurities, *Phys. Rev.* **B4**, 3468–3481.

Ning, T. H. and Tang, D. D. (1984). Method for determining the emitter and base series resistances of bipolar transistors, *IEEE Trans. Electron Devices* **ED-31**, 409–412.

Ning, T. H. and Yu, H. N. (1974). Optically induced injection of hot electrons into SiO_2, *J. Appl. Phys.* **45**, 5373–5378.

Ning, T. H., Osburn, C. M., and Yu, H. N. (1976). Threshold instability in IGFET's due to emission of leakage electrons from silicon substrate into silicon dioxide, *Appl. Phys. Lett.* **29**, 198–200.

Ning, T. H., Osburn, C. M., and Yu, H. N. (1977a). Emission probability of hot electrons from silicon into silicon dioxide, *J. Appl. Phys.* **48**, 286–293.

Ning, T. H., Osburn, C. M., and Yu, H. N. (1977b). Effect of electron trapping on IGFET characteristics, *J. Electronic Materials* **6**, 65–76.

Ning, T. H., Cook, P. W., Dennard, R. H., *et al.* (1979). 1-μm MOSFET VLSI technology: part IV. hot-electron design constraints, *IEEE Trans. Electron Devices* **ED-26**, 346–353.

Ning, T. H., Isaac, R. D., Solomon, P. M., *et al.* (1981). Self-aligned bipolar transistors for high-performance and low-power-delay VLSI, *IEEE Trans. Electron Devices* **ED-28**, 1010–1013.

Nissan-Cohen, Y. (1986). A novel floating-gate method for measurement of ultra-low hole and electron gate currents in MOS transistors, *IEEE Electron Devices Lett.* **EDL-7**, 561–563.

Niu, G., Zhang, S., Cressler, J. D., *et al.* (2003). Noise modeling and SiGe profile design tradeoffs for RF applications, *IEEE Trans. Electron Devices* **7**, 2037–2044.

Noble, W. P., Voldman, S. H., and Bryant, A. (1989). The effects of gate field on the leakage characteristics of heavily doped junctions, *IEEE Trans. Electron Devices* **ED-36**, 720–726.

Nokubo, J., Tamura, T., Nakamae, M., *et al.* (1983). A 4.5-ns access time 1k × 4 bit ECL RAM, *IEEE J. Solid-State Circuits* **SC-18**, 515–520.

Oda, K., Ohue, E., Tanabe, M., *et al.* (1997). 130-GHz f_T SiGe HBT technology, *IEEE IEDM Technical Digest*, 791–794.

Ogura, S., Codella, C. F., Rovedo, N., Shepard, J. F., and Riseman, J. (1982). A half-micron MOSFET using double-implanted LDD, *IEEE IEDM Technical Digest*, 718–721.

Oh, S.-H., Monroe, D., and Hergenrother, J. M. (2000). Analytic description of short-channel effects in fully-depleted double-gate and cylindrical, surrounding-gate MOSFETs, *IEEE Electron Device Lett.* **9**, 445–447.

Ohkura, Y. (1990). Quantum effects in Si n-MOS inversion layer at high substrate concentration, *Solid-State Electron.* **33**, 1581–1585.

Okada, K. (1997). Extended time dependent dielectric breakdown model based on anomalous gate area dependence of lifetime in ultra-thin dioxides, *Jpn. J. Appl. Phys.* **36**, 1143–1147.

Paasschens, J. C. J., Kloosterman, W. J., and Havens, R. J. (2001). Modeling two SiGe HBT specific features for circuit simulation, Proc. Bipolar/BiCMOS Circuit & Technology Meeting, *IEEE*, 38–41.

Padovani, F. A. and Stratton, R. (1966). Field and thermionic–field emission in Schottky barriers, *Solid-State Electron.* **9**, 695–707.

Pao, H. C. and Sah, C. T. (1966). Effects of diffusion current on characteristics of metal-oxide (insulator)-semiconductor transistors, *Solid-State Electron.* **9**, 927–937.

Pandey, N., Lin, H.-H., Nandi, A., and Taur, Y. (2018). Modeling of short-channel effects in DG MOSFETs: Green's function method versus scale length model, *IEEE Trans. Electron Devices* **ED-65**, 3112–3119.

Pavan, P., Bez, R., Olivo, P., and Zanoni, E. (1997). Flash memory cells – An overview, *IEEE Proc.* **85**, 1248–1271.

Pekarik, J. J., Adkisson, J., Gray, P., *et al.* (2014). A 90nm SiGe BiCMOS technology for mm-wave and high-performance analog applications, Proc. Bipolar/BiCMOS Circuit & Technology Meeting, *IEEE*, 92−95.

People, R. (1985). Indirect band gap of coherently strained Ge_xSi_{1-x} bulk alloys on ⟨001⟩ silicon substrate, *Phys. Rev. B* **32**, 1405–1408.

People, R. (1986). Physics and applications of Ge_xSi_{1-x}/Si strained-layer heterostructures, *IEEE J. Quantum Electron.* **QE-22**, 1696–1710.

Planes, N., Weber, O., Barral, V., *et al.* (2012). 28 nm FDSOI technology platform for high-speed low-voltage digital applications, Symp. VLSI Technology Digest of Tech. Papers, *IEEE*, 133–134.

Poon, H. C., Gummel, H. K., and Scharfetter, D. L. (1969). High injection in epitaxial transistors, *IEEE Trans. Electron Devices* **ED-16**, 455–457.

Prinz, E. J. and Sturm, J. C. (1990). Base transport in near-ideal graded-base $Si/Si_{1-x}Ge_x/Si$ heterojunction bipolar transistors from 150 K to 370 K, *IEEE IEDM Technical Digest*, 975–978.

Prinz, E. J. and Sturm, J. C. (1991). Current gain–early voltage products in heterojunction bipolar transistors with non-uniform base bandgaps, *IEEE Electron Device Lett.* **12**, 661–663.

Prinz, E. J., Garone, P. M., Schwartz, P. V., Xiao, X., and Sturm, J. C. (1989). The effect of base-emitter spacers and strain-dependent densities of states in $Si/Si_{1-x}Ge_x/Si$ heterojunction bipolar transistors, *IEEE IEDM Technical Digest*, 639–642.

Pritchard, R. L. (1955). High-frequency power gain of junction transistors, *Proc. IRE* **43**, 1075–1085.

Pritchard, R. L. (1967). *Electrical Characteristics of Transistors.* New York: McGraw-Hill.

Pugh, E. W., Critchlow, D. L., Henle, R. A., and Russell, L. A. (1981). Solid state memory development in IBM, *IBM J. Res. Develop.* **25**, 585–602.

Ranade, P., Ghani, T., Kuhn, K., *et al.* (2005). High performance 35nm L_{gate} CMOS transistors featuring NiSi metal gate (FUSI), uniaxial strained silicon channels and 1.2nm gate oxide, *IEEE IEDM Technical Digest*, 217–220.

Rao, G. S., Gregg, T. A., Price, C. A., Rao, C. L., and Repka, S. J. (1997). IBM S/390 Parallel Enterprise Servers G3 and G4, *IBM J. Res. Develop.* **41**, 397–403.

Rauch III, S. E. (2002). The statistics of NBTI-induced V_T and β mismatch shifts in pMOSFETs, *IEEE Trans. Device and Materials Reliability* **2**, 89–93.

Raman, S., Sharma, P., Neogi, T. G., *et al.* (2015). On the performance of lateral SiGe heterojunction bipolar transistors with partially depleted base, *IEEE Trans. Electron Devices*, **62**, 2377–2383.

Razouk, R. R. and Deal, B. E. (1979). Dependence of interface state density on silicon thermal oxidation process variables, *J. Electrochem. Soc.* **126**, 1573–1581.

Ren, Z. and Taur, Y. (2020). Non-GCA modeling of near-threshold I-V characteristics of DG MOSFETs, *Solid-State Electron.* **166**, 107766.

Riccó, B., Stork, J. M. C., and Arienzo, M. (1984). Characterization of non-ohmic behavior of emitter contacts of bipolar transistors, *IEEE Electron Device Lett.* **EDL-5**, 221–223.

Rideout, V. L., Gaensslen, F. H., and LeBlanc, A. (1975). Device design consideration for ion-implanted n-channel MOSFETs, *IBM J. Res. Develop.* **19**, 50–59.

Rieh, J.-S., Cai, J., Ning, T., Stricker, A., and Freeman, G. (2005). Reverse active mode current characteristics of SiGe HBTs, *IEEE Trans. Electron Devices* **52**, 1291–1222.

Rim, K., Welser, J., Hoyt, J. L., and Gibbons, J. F. (1995). Enhanced hole mobilities in surface-channel strained-Si p-MOSFETs, *IEEE IEDM Technical Digest*, 517–520.

Rios, R. and Arora, N. D. (1994). Determination of ultra-thin gate oxide thickness for CMOS structures using quantum effects, *IEEE IEDM Technical Digest*, 613–616.

Robertson, J. and Wallace, R. M. (2015). High-k materials and metal gates for CMOS applications, *Materials Science and Engineering R*, **88**, 1–41.

Roulston, D. J. (1990). *Bipolar Semiconductor Devices.* New York: McGraw-Hill.

Sabnis, A. G. and Clemens, J. T. (1979). Characterization of the electron mobility in the inverted $\langle 100 \rangle$ Si surface, *IEEE IEDM Technical Digest*, 18–21.

Sah, C. T. (1966). The spatial variation of the quasi-Fermi potential in p–n junctions, *IEEE Trans. Electron Devices* **ED-13**, 839–846.

Sah, C. T. (1988). Evolution of the MOS transistor – From concept to VLSI, *Proc. IEEE* **76**, 1280–1326.

Sah, C. T. (1991). *Fundamentals of Solid-State Electronics*. Singapore: World Scientific.

Sah, C. T., Noyce, R. N., and Shockley, W. (1957). Carrier generation and recombination in p–n junction and p–n junction characteristics, *Proc. IRE* **45**, 1228–1243.

Sah, C. T., Ning, T. H., and Tschopp, L. L. (1972). The scattering of electrons by surface oxide charges and by lattice vibrations at the silicon–silicon dioxide interface, *Surface Sci.* **32**, 561–575.

Sai-Halasz, G. A. (1995). Performance trends in high-end processors, *IEEE Proc.* **83**, 20–36.

Sai-Halasz, G. A., Wordeman, M. R., Kern, D. P., Rishton, S., and Ganin, E. (1988). High transconductance and velocity overshoot in NMOS devices at the 0.1 μm gate-length level, *IEEE Electron Device Lett.* **EDL-9**, 464–466.

Sakurai, T. (1983). Approximation of wiring delay in MOSFET LSI, *IEEE J. Solid-State Circuits* **SC-18**, 418.

Salmon, S. L., Cressler, J. D., Jaeger, R. C., and Harame, D. L. (1997). The impact of Ge profile shape on the operation of SiGe HBT precision voltage references, Proc. Bipolar/BiCMOS Circuit & Technology Meeting, *IEEE*, 100–103.

Samachisa, G., Su, C.-S., Kao, Y.-S., *et al.* (1987). A 128K flash EEPROM using double-polysilicon technology, *IEEE J. Solid-State Circuits* **SC-22**, 676–683.

Schaper, L. W. and Amey, D. I. (1983). Improved electrical performance required for future MOS packaging, *IEEE Trans. Components, Hybrids, and Manufacturing Tech.* **CHMT-6**, 283.

Schroder, D. K. (1990). *Semiconductor Material and Device Characterization*. New York: Wiley.

Schroder, D. K. and Babcock, J. A. (2003). Negative bias temperature instability: Road to cross in deep submicron silicon semiconductor manufacturing, *J. Appl. Phys.* **96**, 1–18.

Schuegraf, K. F. and Hu, C. (1994). Metal–oxide–semiconductor field-effect-transistor substrate current during Fowler–Nordheim tunneling stress and silicon dioxide reliability, *J. Appl. Phys.* **76**, 3695–3700.

Schuegraf, K. F., King, C. C. and Hu, C. (1992). Ultra-thin dioxide leakage current and scaling limit, Symp. VLSI Technology Digest of Tech. Papers, *IEEE*, pp. 18–19.

Schüppen, A. and Dietrich, H. (1995). High speed SiGe heterobipolar transistors, *J. Crystal Growth* **157**, 207–214.

Schüppen, A., Dietrich, H., Gerlach, S., *et al.* (1996). SiGe – Technology and components for mobile communication systems, *Proc. Bipolar/BiCMOS Circuit & Technology Meeting, IEEE*, pp. 147–150.

Seevinck, E., List, F. J., and Lohstroh, J. (1987). Static-noise margin analysis of MOS SRAM cells, *IEEE J. Solid-State Circuits* **SC-22**, 748–754.

Sell, B., Bigwood, B., Cha, S., *et al.* (2017). 22FFL: A high performance and ultra low power FinFET technology for mobile and RF applications, *IEEE IEDM Technical Digest*, 29.4.1–29.4.4.

Selmi, L., Sangiorgi, E., Bez, R., and Riccò, B. (1993). Measurement of the hot hole injection probability from Si into SiO_2 in p-MOSFET, *IEEE IEDM Technical Digest*, 333–336.

Shahidi, G. G., Antoniadis, D. A., and Smith, H. I. (1989). Indium channel implants for improved MOSFET behavior at the 100 nm channel length regime, DRC Abstract, *IEEE Trans. Electron Devices* **ED-36**, 2605.

Shahidi, G. G., Ajmera, A., Assaderaghi, F., *et al.* (1999). Partially-depleted SOI technology for digital logic, *IEEE ISSCC Digest of Technical Papers*, 426–427.

Shang, H., Jain, S., Josse, E., *et al.* (2012). High performance bulk planar 20 nm CMOS technology for low power mobile applications, Symp. VLSI Technology Digest of Tech. Papers, *IEEE*, 129–130.

Shatzkes, M., Av-Ron, M., and Anderson, R. M. (1974). On the nature of conduction and switching in SiO_2, *J. Appl. Phys.* **45**, 2065–2077.

Shiba, T., Tamaki, Y., Onai, T., *et al.* (1991). SPOYEC – A sub-10-μm^2 bipolar transistor structure using fully self-aligned sidewall polycide base technology, *IEEE IEDM Technical Digest*, 455–458.

Shiba, T., Uchino, T., Ohnishi, K., and Tamaki, Y. (1996). In-situ phosphorus-doped poly-silicon emitter technology for very high-speed, small emitter bipolar transistors, *IEEE Trans. Electron Devices* **43**, 889–897.

Shibib, M. A., Lindholm, F. A., and Therez, R. (1979). Heavily doped transparent-emitter regions in junction solar cells, diodes, and transistors, *IEEE Trans. Electron Devices* **ED-26**, 959–965.

Shimizu, A., Hachimine, K., Ohki, N., *et al.* (2001). Local mechanical-stress control (LMC): A new technique for CMOS-performance enhancement, *IEEE IEDM Technical Digest*, 433–436.

Shockley, W. (1949). The theory of p–n junctions in semiconductors and p–n junction transistors, *Bell Syst. Tech. J.* **28**, 435–489.

Shockley, W. (1950). *Electrons and Holes in Semiconductors*. Princeton, NJ: D. Van Nostrand.

Shockley, W. (1952). A unipolar "field-effect" transistor, *Proc. of the IRE*, 1365–1376.

Shockley, W. (1961). Problems related to p-n junctions in silicon, *Solid-State Electron.* **2**, 35–67.

Shockley, W. and Read, W. T. (1952). Statistics of the recombination of holes and electrons, *Phys. Rev.* **87**, 835–842.

Shrivastava, R. and Fitzpatrick, K. (1982). A simple model for the overlap capacitance of a VLSI MOS device, *IEEE Trans. Electron Devices* **ED-29**, 1870–1875.

Sleva, S. and Taur, Y. (2005). The influence of source and drain junction depth on the short-channel effect in MOSFETs, *IEEE Trans. Electron Devices* **ED-52**, 2814–2816.

Slotboom, J. W. and de Graaff, H. D. (1976). Measurements of bandgap narrowing in Si bipolar transistors, *Solid-State Electron.* **19**, 857–862.

Sodini, C. G., Ekstedt, T. W., and Moll, J. L. (1982). Charge accumulation and mobility in thin dielectric MOS transistors, *Solid-State Electron.* **25**, 833–841.

Sodini, C. G., Ko, P. K., and Moll, J. L. (1984). The effect of high fields on MOS device and circuit performance, *IEEE Trans. Electron Devices* **ED-31**, 1386–1393.

Solomon, P. M. (1982). A comparison of semiconductor devices for high speed logic, *IEEE Proc.* **70**, 489.

Solomon, P. M. and Tang, D. D. (1979). Bipolar circuit scaling, *IEEE ISSCC Digest of Technical Papers*, 86–87.

Solomon, P. M., Jopling, J., Frank, D. J., *et al.* (2004). Universal tunneling behavior in technologically relevant p–n junction diodes, *J. Appl. Phys.* **95**, 5800–5812.

Stathis, J. H. (1999). Percolation models for gate oxide breakdown, *J. Appl. Phys.* **86**, 5757–5766.

Stathis, J. H. and DiMaria, D. J. (1998). Reliability projection for ultrathin oxides at low voltage, *IEEE IEDM Technical Digest*, 167–170.

Stern, F. (1972). Self-consistent results for n-type Si inversion layers, *Phys. Rev. B* **5**, 4891–4899.

Stern, F. (1974). Quantum properties of surface space-charge layers, *CRC Crit. Rev. Solid-State Sci.* **4**, 499.

Stern, F. and Howard, W. E. (1967). Properties of semiconductor surface inversion layers in the electric quantum limit, *Phys. Rev.* **163**, 816–835.

Stolk, P. A., Eaglesham, D. J., Gossmann, H.-J., and Poate, J. M. (1995). Carbon incorporation in silicon for suppressing interstitial-enhanced boron diffusion, *Appl. Phys. Lett.* **66**, 1370–1372.

Sturm, J. C., McVittie, J. P., Gibbons, J. F., and Pfeiffer, L. (1987). A lateral silicon-on-insulator bipolar transistor with a self-aligned base contact, *IEEE Electron Device Lett.* **8**, 104–106.

Su, L. T., Jacobs, J. B., Chung, J., and Antoniadis, D. A. (1994). Deep-submicrometer channel design in silicon-on-insulator (SOI) MOSFETs, *IEEE Electron Device Lett.* **15**, 183–185.

Su, L., Subbanna, S., Crabbe, E., *et al.* (1996). A high-performance 0.08 μm CMOS, Symp. VLSI Technology Digest of Tech. Papers, *IEEE*, 12–13.

Suehle, J. S. (2002). Ultrathin gate oxide reliability: physical models, statistics, and characterization, *IEEE Trans. Electron Devices* **49**, 958–971.

Sun, J. Y.-C., Taur, Y., Dennard, R. H., and Klepner, S. P. (1987). Submicrometer-channel CMOS for low-temperature operation, *IEEE Trans. Electron Devices* **ED-34**, 19–27.

Sun, Y., Thompson, S., and Nishida, T. (2007). Physics of strain effects in semiconductors and metal-oxide-semiconductor field-effect transistors, *J. Appl. Phys.* **101**, 104503.

Suñé, J. and Wu, E. Y. (2002). Statistics of successful breakdown events for ultra-thin gate oxides, *IEEE IEDM Technical Digest*, 147–150.

Suñé, J., Wu, E. Y., and Lai, W. L. (2004). Successive oxide breakdown statistics; correlation effects, reliability methodologies, and their limits, *IEEE Trans. Electron Devices* **51**, 1584–1592.

Suzuki, K. (1991). Optimum base doping profile for minimum base transit time, *IEEE Trans. Electron Devices* **38**, 2128–2133.

Suzuki, E., Miura, K., Hayasi, Y., Tsay, R.-P., and Schroder, D. K. (1989). Hole and electron current transport in metal–oxide–nitride–oxide–silicon memory structures, *IEEE Trans. Electron Devices* **36**, 1145–1149.

Swanson, R. M. and Meindl, J. D. (1972). Ion-implanted complementary MOS transistors in low-voltage circuits, *IEEE J. Solid-State Circuits* **SC-7**, 146–153.

Swirhun, S. E., Kwark, Y.-H., and Swanson, R. M. (1986). Measurement of electron lifetime, electron mobility and band-gap narrowing in heavily doped p-type silicon, *IEEE IEDM Technical Digest*, 24–27.

Sze, S. M. (1981). *Physics of Semiconductor Devices*. New York: Wiley.

Sze, S. M., Crowell, C. R., and Kahng, D. (1964). Photoelectric determination of the image force dielectric constant for hot electrons in Schottky barriers, *J. Appl. Phys.* **35**, 2534–2536.

Takagi, S., Iwase, M., and Toriumi, A. (1988). On universality of inversion-layer mobility in n- and p-channel MOSFETs, *IEEE IEDM Technical Digest*, 398–401.

Takeda, E., Suzuki, N., and Hagiwara, T. (1983). Device performance degradation due to hot-carrier injection at energies below the Si-SiO$_2$ energy barrier, *IEEE IEDM Technical Digest*, 396–399.

Takeda, E., Yang, C. Y., and Miura-Hamada, A. (1995). *Hot-Carrier Effects in MOS Devices*. San Diego: Academic Press.

Tang, D. D. (1980). Heavy doping effects in pnp bipolar transistors, *IEEE Trans. Electron Devices* **ED-27**, 563–570.

Tang, D. D. and Lee, Y.-J. (2010). *Magnetic Memory: Fundamentals and Technology*. Cambridge: Cambridge University Press.

Tang, D. D. and Lu, P.-F. (1989). A reduced-field design concept for high-performance bipolar transistors, *IEEE Electron Device Lett.* **10**, 67–69.

Tang, D. D. and Pai, C.-F. (2020), *Magnetic Memory Technology, Spin-Torque Transfer MRAM and Beyond*. Hoboken, NJ: Wiley.

Tang, D. D. and Solomon, P. M. (1979). Bipolar transistor design for optimized power-delay logic circuits, *IEEE J. Solid-State Circuits* **SC-14**, 679–684.

Tang, D. D., MacWilliams, K. P., and Solomon, P. M. (1983). Effects of collector epitaxial layer on the switching speed of high-performance bipolar transistors, *IEEE Electron Device Lett.* **EDL-4**, 17–19.

Tang, D. D., Ning, T. H., Isaac, R. D., *et al.* (1979). Subnanosecond self-aligned I2L/MTL circuits, *IEEE IEDM Technical Digest*, 201–204.

Taur, Y. (2000). An analytical solution to a double-gate MOSFET with undoped body, *IEEE Electron Device Lett.* **21**, 245–247.

Taur, Y. (2001). Analytic solutions of charge and capacitance in symmetric and asymmetric double-gate MOSFETs, *IEEE Trans. Electron Devices* **ED-48**, 2861–2869.

Taur, Y. and Lin, H.-H. (2018). Modeling of DG MOSFET I-V characteristics in the saturation region, *IEEE Trans. Electron Devices* **ED-65**, 1714–1720.

Taur, Y., Hu, G. J., Dennard, R. H., *et al.* (1985). A self-aligned 1 μm channel CMOS technology with retrograde n-well and thin epitaxy, *IEEE Trans. Electron Devices* **ED-32**, 203–209.

Taur, Y., Sun, J. Y.-C., Moy, D., *et al.* (1987). Source-drain contact resistance in CMOS with self-aligned $TiSi_2$, *IEEE Trans. Electron Devices* **ED-34**, 575–580.

Taur, Y., Hsu, C. H., Wu, B., *et al.* (1993a). Saturation transconductance of deep-submicron-channel MOSFETs, *Solid-State Electron.* **36**, 1085–1087.

Taur, Y., Cohen, S., Wind, S., *et al.* (1993b). Experimental 0.1-μm p-channel MOSFET with p^+ polysilicon gate on 35-Å gate oxide, *IEEE Electron Device Lett.* **EDL-14**, 304–306.

Taur, Y., Wind, S., Mii, Y., *et al.* (1993c). High performance 0.1 μm CMOS devices with 1.5 V power supply, *IEEE IEDM Technical Digest*, 127–130.

Taur, Y., Mii, Y.-J., Frank, D. J., *et al.* (1995). CMOS scaling into the 21th century: 0.1 μm and beyond, *IBM J. Res. Develop.* **39**, 245–260.

Taur, Y., Buchanan, D. A., Chen, W., *et al.* (1997). CMOS scaling into the nanometer regime, *IEEE Proc.* **85**, 486–504.

Taur, Y., Wann, C. H., and Frank, D. J. (1998). 25 nm CMOS design considerations, *IEEE IEDM Technical Digest*, 789–792.

Taur, Y., Liang, X., Wang, W., and Lu, H. (2004). A continuous, analytic drain-current model for double-gate MOSFETs, *IEEE Electron Device Lett.* **25**, 107–109.

Taur, Y., Chen, H.-P., Xie, Q., *et al.* (2015). A unified two-band model for oxide traps and interface states in MOS capacitors, *IEEE Trans. Electron Devices* **62**, 813–820.

Taur, Y., Choi, W., Zhang, J., and Su, M. (2019). A non-GCA MOSFET model continuous into the velocity saturation region, *IEEE Trans. Electron Devices* **66**, 1160–1166.

Taylor, G. W. and Simmons, J. G. (1986). Figure of merit for integrated bipolar transistors, *Solid-State Electron.* **29**, 941–946.

Terman, L. M. (1962). An investigation of surface states at a silicon/silicon oxide interface employing metal-oxide-silicon diodes, *Solid-State Electron.* **5**, 285–299.

Terman, L. M. (1971). MOSFET memory circuits, *Proc. IEEE* **59**, 1044–1058.

Thompson, S. E., Sun, G., Choi, Y. S., and Nishida, T. (2006). Uniaxial-process-induced strained-Si: Extending the CMOS roadmap, *IEEE Trans. Electron Devices*, **53**, 1010–1020.

Thornton, R. D., DeWitt, D., Gray, P. E., and Chenette, E. R. (1966). *Characteristics and Limitations of Transistors*, vol. 4, Semiconductor Electronics Education Committee. New York: Wiley.

Ting, C. Y., Iyer, S. S., Osburn, C. M., Hu, G. J., and Schweighart, A. M. (1982). The use of $TiSi_2$ in a self-aligned silicide technology. *Symp. VLSI Science and Technology*, 224–231.

Tiwari, S., Rana, F., Hanafi, H., *et al.* (1996). A silicon nanocrystals based memory, *Appl. Phys. Lett.* **68**, 1377–1379.

Trumbore, F. A. (1960). Solid solubilities of impurity elements in germanium and silicon, *Bell Syst. Tech. J.* **39**, 205.

Tuinhout, H., Pelgrom, M., de Vries, R. P., and Vertregt, M. (1996). Effects of metal coverage on MOSFET matching, *IEEE IEDM Technical Digest*, 735–738.

Tuinhout, H. P., Montree, A. H., Schmitz, J., and Stolk, P. A. (1997). Effects of gate depletion and boron penetration on matching of deep submicron CMOS transistors, *IEEE IEDM Technical Digest*, 631–634.

Tung, R. T. (1992). Electron transport at metal–semiconductor interfaces: General theory, *Phys. Rev. B* **45**, 13509–13523.

Uchida, K., Zednik, R., Lu, C.-H., *et al.* (2004). Experimental study of biaxial and uniaxial strain effects on carrier mobility in bulk and ultrathin-body SOI MOSFETs, *IEEE IEDM Technical Digest*, 229–232.

Uchino, T., Shiba, T., Kikuchi, T., *et al.* (1995). Very-high-speed silicon bipolar transistors with *in-situ* doped polysilicon emitter and rapid vapor-phase doping base, *IEEE Trans. Electron Devices* **42**, 406–412.

Van de Walle, C. G. and Martin, R. M. (1986). Theoretical calculations of heterojunction discontinuities in the Si/Ge system, *Phys. Rev. B* **34**, 5621–5634.

van Dort, M. J., Woerlee, P. H., and Walker, A. J. (1994). A simple model for quantization effects in heavily-doped silicon MOSFETs at inversion conditions, *Solid-State Electron.* **37**, 411–414.

van Overstraeten, R. and de Man, H. (1970). Measurement of the ionization rates in diffused silicon p–n junctions, *Solid-State Electron.* **13**, 583–608.

van Overstraeten, R. J., de Man, H. J., and Mertens, R. P. (1973). Transport equations in heavy doped silicon, *IEEE Trans. Electron Devices* **ED-20**, 290–298.

Verwey, J. F. (1972). Hole currents in thermally grown SiO_2, *J. Appl. Phys.* **43**, 2273–2277.

Wang, S. T. (1979). On the I–V characteristics of floating-gate MOS transistors, *IEEE Trans. Electron Devices* **26**, 1292–1294.

Wanlass, F. and Sah, C. T. (1963). Nanowatt logic using field-effect metal-oxide-semiconductor triodes, *IEEE ISSCC Digest of Technical Papers*, 32–33.

Warnock, J. D. (1995). Silicon bipolar device structures for digital applications: Technology trends and future directions, *IEEE Trans. Electron Devices* **42**, 377–389.

Washio, K., Ohue, E., Shimamoto, H., *et al.* (2000). A 0.2-μm 180GHz f_{max} 6.7-ps-ECL SOI/HRS self-aligned SEG SiGe HBT/CMOS technology for microwave and high-speed digital applications, *IEEE IEDM Technical Digest*, 741–744.

Washio, K., Ohue, E., Shimamoto, H., *et al.* (2002). A 0.2-μm 180-GHz-f_{max} 6.7-ps-ECL SOI-HRS self-aligned SEG SiGe HBT/CMOS technology for microwave and high-speed digital applications, *IEEE Trans. Electron Devices* **49**, 271–278.

Webster, W. M. (1954). On the variation of junction-transistor current amplification factor with emitter current, *Proc. IRE* **42**, 914–920.

Weng, J., Holz, J., and Meister, T. F. (1992). New method to determine the base resistance of bipolar transistors, *IEEE Electron Device Lett.* **13**, 158–160.

Wiedmann, S. K. and Berger, H. H. (1971). Small-size low-power bipolar memory cell, *IEEE J. Solid-State Circuits* **SC-6**, 283–288.

Wiedmann, S. K. and Berger, H. H., (1972). Merged-transistor logic (MTL) – a low-cost bipolar logic concept, *IEEE J. Solid-State Circuits* **7**, 340–346.

Wiedmann, S. K. and Tang, D. D. (1981). High-speed split-emitter I^2L/MTL memory cell, *IEEE ISSCC Digest of Technical Papers*, 158–159.

Wiedmann, S. K., Tang, D. D., and Beresford, R. (1981). High-speed split-emitter I^2L/MTL memory cell, *IEEE J. Solid-State Circuits* **SC-16**, 429–434.

Wong, C. Y., Sun, J. Y.-C., Taur, Y., *et al.* (1988). Doping of n^+ and p^+ polysilicon in a dual-gate CMOS process, *IEEE IEDM Technical Digest*, 238–241.

Wong, H.-S., and Taur, Y. (1993). Three-dimensional atomistic simulation of discrete random dopant distribution effects in sub-0.1 μm MOSFETs, *IEEE IEDM Technical Digest*, 705–708.

Wong, H.-S., Frank, D. J., Taur, Y., and Stork, J. M. C. (1994). Design and performance considerations for sub-0.1 μm double-gate SOI MOSFETs, *IEEE IEDM Technical Digest*, 747–750.

Wong, H.-S., Chan, K. K., Lee, Y., Roper, P., and Taur, Y. (1997). Fabrication of ultrathin, highly uniform thin-film SOI MOSFETs with low series resistance using pattern-constrained epitaxy, *IEEE Trans. Electron Devices* **44**, 1131–1135.

Wong, H.-S., Frank, D. J., Solomon, P. M., Wann, H. J., and Welser, J. J. (1999). Nanoscale CMOS, *IEEE Proc.* **87**, 537–570.

Wong, H.-S. P., Lee, H.-Y., Yu, S., *et al.* (2012). Metal-oxide RRAM, *IEEE Proc.* **100**, 1951–1970.

Wong, H.-S. P., Raoux, S., Kim, S.-B., *et al.* (2010). Phase change memory, *IEEE Proc.* **98**, 2201–2227.

Wordeman, M. R. (1986). Design and modeling of miniaturized MOSFETs. Unpublished PhD thesis, Columbia University.

Wu, A. T., Chan, T. Y., Ko, P. K., and Hu, C. (1986). A novel high-speed 5-volt programming EPROM structure with source-side injection, *IEEE IEDM Technical Digest*, 584–587.

Xie, Q., Lee, C.-J., Xu, J., *et al.* (2013). Comprehensive analysis of short-channel effects in ultra-thin SOI MOSFETs, *IEEE Trans. Electron Devices* **60**, 1814–1819.

Xie, Q., Xu, J., and Taur, Y. (2012). Review and critique of analytic models of MOSFET short-channel effects in subthreshold, *IEEE Trans. Electron Devices* **59**, 1569–1579.

Yamada, S., Yamane, T., Amemiya, K., and Naruke, K. (1996). A self-convergence erase for NOR flash EEPROM using avalanche hot carrier injection, *IEEE Trans. Electron Devices* **43**, 1937–1941.

Yan, R. H., Ourmazd, A., Lee, K. F., *et al.* (1991). Scaling the Si metal-oxide-semiconductor field-effect transistor into the 0.1-μm regime using vertical doping engineering, *Appl. Phys. Lett.* **59**, 3315–3317.

Yang, X. and Schroder, D. K. (2012). Some semiconductor device physics considerations and clarifications, *IEEE Trans. Electron Devices* **59**, 1993–1996.

Yau, J.-B., Cai, J., Hashemi, P., *et al.* (2018). A study of process-related electrical defects in SOI lateral bipolar transistors fabricated by ion implantation, *J. Appl. Phys.* **123**(16), Art. 161526.

Yau, J.-B., Cai, J., and Ning, T. H. (2016a). On the base current component in SOI symmetric lateral bipolar transistors, *IEEE J. Electron Devices Soc.* **4**, 116–123.

Yau, J.-B., Cai, J., and Ning, T. H. (2016b). Substrate-voltage modulation of currents in symmetric SOI lateral bipolar transistors, *IEEE Trans. Electron Devices* **63**, 1835–1839.

Yau J.-B., Cai, J., Yoon, J., *et al.* (2015). SiGe-on-insulator symmetric lateral bipolar transistors, *IEEE SOI-3D-Subthreshold Microelectronics Technology Unified Conference (S3S)*, 1–2.

Yau, J.-B., Yoon, J., Cai, J., *et al.* (2016c). Ge-on-insulator lateral bipolar transistors, Proc. Bipolar/BiCMOS Circuit & Technology Meeting, *IEEE*, 130–133.

Yoshino, C., Inou, K., Matsuda, S., *et al.* (1995). A 62.8-GHz fmax LP-CVD epitaxially grown silicon-base bipolar transistor with extremely high Early voltage of 85.7 V, Symp. VLSI Technology Digest of Tech. Papers, *IEEE*, 131–132.

Yu, A. Y. C. (1970). Electron tunneling and contact resistance of metal-silicon contact barriers, *Solid-State Electron.* **13**, 239–247.

Yu, B., Song, J., Yuan, Y., Lu, W.-Y., and Taur, Y. (2008a). A unified analytic drain current model for multiple-gate MOSFETs, *IEEE Trans. Electron Devices* **ED-55**, 2157–2163.

Yu, B., Wang, L., Yuan, Y., Asbeck, P. M., and Taur, Y. (2008b). Scaling of nanowire transistors, *IEEE Trans. Electron Devices* **ED-55**, 2846–2858.

Yu, H. N. (1971). Transistor with limited-area base-collector junction, U.S. Patent 27,045, reissued February 2.

Yu, Z., Riccó, B., and Dutton, R. W. (1984). A comprehensive analytical and numerical model of polysilicon emitter contacts in bipolar transistors, *IEEE Trans. Electron Devices* **ED-31**, 773–785.

Yuan, Y., Yu, B., Ahn, J., *et al.* (2012). A distributed bulk-oxide trap model for Al_2O_3-InGaAs MOS devices, *IEEE Trans. Electron Devices* **ED-59**, 2100–2106.

Zafar, S., Lee, B. H., Stathis, J., and Ning, T. (2004). A model for negative bias temperature instability (NBTI) in oxide and high-κ pFETs, Symp. VLSI Technology Digest of Tech. Papers, *IEEE*, 208–209.

Zhang, Z., Koswatta, S. O., Bedell, S. W., *et al.* (2013). Ultra low contact resistivities for CMOS beyond 10-nm node, *IEEE Electron Devices Lett.* **34**, 723–725.

Index